Sustainable Development Goals Series

The **Sustainable Development Goals Series** is Springer Nature's inaugural cross-imprint book series that addresses and supports the United Nations' seventeen Sustainable Development Goals. The series fosters comprehensive research focused on these global targets and endeavours to address some of society's greatest grand challenges. The SDGs are inherently multidisciplinary, and they bring people working across different fields together and working towards a common goal. In this spirit, the Sustainable Development Goals series is the first at Springer Nature to publish books under both the Springer and Palgrave Macmillan imprints, bringing the strengths of our imprints together.

The Sustainable Development Goals Series is organized into eighteen subseries: one subseries based around each of the seventeen respective Sustainable Development Goals, and an eighteenth subseries, "Connecting the Goals", which serves as a home for volumes addressing multiple goals or studying the SDGs as a whole. Each subseries is guided by an expert Subseries Advisor with years or decades of experience studying and addressing core components of their respective Goal.

The SDG Series has a remit as broad as the SDGs themselves, and contributions are welcome from scientists, academics, policymakers, and researchers working in fields related to any of the seventeen goals. If you are interested in contributing a monograph or curated volume to the series, please contact the Publishers: Zachary Romano [Springer; zachary.romano@springer.com] and Rachael Ballard [Palgrave Macmillan; rachael.ballard@palgrave.com].

Zeba Khanam • Saleha Jamal
Azizur Rahman Siddiqui
Editors

Climate Change, Resilience and Ecosystem Services

Concepts, Models and Implementation

Springer

Editors
Zeba Khanam
Jawaharlal Nehru University
New Delhi, Delhi, India

Saleha Jamal
Department of Geography
Aligarh Muslim University
Aligarh, Uttar Pradesh, India

Azizur Rahman Siddiqui
Department of Geography
University of Allahabad
Prayagraj, Uttar Pradesh, India

ISSN 2523-3084 ISSN 2523-3092 (electronic)
Sustainable Development Goals Series
ISBN 978-3-031-96976-8 ISBN 978-3-031-96977-5 (eBook)
https://doi.org/10.1007/978-3-031-96977-5

This Springer imprint is published by the registered company Springer Nature Switzerland AG
The registered company address is: Gewerbestrasse 11, 6330 Cham, Switzerland

Foreword

Climate change is essentially a terrestrial phenomenon. The Earth's surface has faced several glacial and interglacial periods. Although the interglacial periods tend to be relatively stable, several pieces of evidence show that Holocene climate variations have been larger and more frequent than previously recognized. Moreover, a series of rapid climate change events have occurred in the last 11,500 years BP, and not all sites have shown to respond synchronously or equally. The issue of whether the temperature rise of the previous 100 years crossed over the warm limit of the boundary defined by the Medieval Climate Anomaly has been a controversial topic in the science community. However, at present, the global average temperature during the last few decades was indeed warmer than any comparable period in the previous 400 years. These rapid climate change events affect the ecosystem and destabilize ecological balances and amplify vulnerabilities across the globe. That is why climate-resilient ecosystems should be introduced and preserved. By preserving ecological stability and minimizing susceptibility to degradation, these systems enhance the adaptive capacity of communities and economies. From this point of view, the present volume, *Climate Change, Resilience and Ecosystem Services: An Introduction to Interconnections* edited by Azizur Rahman Siddiqui, Saleha Jamal, and Zeba Khanam, is a timely contribution to address the present-day problems through a multidisciplinary approach.

The volume is divided into five parts. In the first part, the editors have tried to elaborate on the interconnections between climate change and ecosystem stability. The second part contains five contributions on the impact of climate change on surface- and ground-water, river delta formation, and the nexus between the increase of greenhouse gases and climate change. Because of increasing urbanization, the nature of land use and land cover has severely been affected. This has a direct impact on the climate system and ecosystems. All these issues have scientifically been addressed by nine different authors in the third part of the book. The fourth part of the book is designed to address the issue of disaster risk and management due to climate change. This part spans from cyclone and flood risk to the influence of human activities on the mangrove ecosystem. The authors have certainly tried to justify the theme of the book through their scientific contributions. Lastly, the role of geospatial technologies on climate change and sustainable development has been addressed in the fifth part by eight authors.

I hope and believe that this volume will certainly attract the attention of the wide scientific community, including geomorphologists, environmentalists, environmental activists, and planners.

Department of Geography
North Eastern Hill University,
Shillong, Meghalaya, India

Sunil Kumar De

Contents

Editorial

Climate Change, Resilience, and Ecosystem Services: An Introduction to Interconnections

Azizur Rahman Siddiqui, Saleha Jamal, and Zeba Khanam

1 Introduction: Climate Change and Ecosystem Resilience

Climate change, an escalating global phenomenon of twenty-first century, has profound implications for natural systems and societies (Burroughs, 2003). Rapid fluctuations in thermal patterns, erratic precipitation patterns, and an increasing frequency of extreme weather events are destabilizing ecological balances and amplifying vulnerabilities across the globe (Upadhyay, 2020). Amid these mounting challenges, ecosystems play a pivotal role in mitigating these impacts. The resistance of ecosystems, "their ability to withstand environmental disturbances while maintaining existential functions," forms the backbone for resilience against climatic challenges. Resistant ecosystems act as natural safeguards, absorbing shocks and mitigating risks associated with climatic-induced stressors, such as floods, droughts, and heatwaves (Climate Resilience Portal). By preserving ecological stability and minimizing susceptibility to degradation, these systems enhance the adaptive capacity of communities and economies. Enhancing ecosystem resilience is not merely a simple conservation effort; it is a strategic investment in climate resilience. It provides communities with natural instruments for mitigating unanticipated climate extremes and promoting long-term adaptation pathways. Understanding and strengthening resilience in this complex interplay between ecosystems and climate is critical to creating a more resilient and harmonious future.

2 Climate Resilience: A Holistic Framework for Ecosystem Sustainability

In the context of climate change, resilience is defined as the ability of social, economic, and ecological systems to handle and adapt to the effects of climate variability and extremes (United Nations Climate Change). This includes anticipating, planning for, absorbing, recovering from, and adapting to the impacts. Resilience is more than just returning to a previous state; it also entails modifying and adapting to changing environmental conditions. For ecosystems, resilience is defined by two important components: resistance (the ability to withstand disturbances) and recovery (the speed and degree with which equilibrium is restored following disruptions).

A. R. Siddiqui
Department of Geography, University of Allahabad, Prayagraj, Uttar Pradesh, India

S. Jamal
Department of Geography, Faculty of Science, Aligarh Muslim University, Aligarh, Uttar Pradesh, India

Z. Khanam (✉)
Jawaharlal Nehru University, New Delhi, India

Z. Khanam et al. (eds.), *Climate Change, Resilience and Ecosystem Services*, Sustainable Development Goals Series, https://doi.org/10.1007/978-3-031-96977-5_1

These two characteristics are critical for conserving ecosystem services, which include the providing, regulating, cultural, and supporting functions that support human well-being.

Improving climate resilience requires a multidisciplinary strategy that combines scientific knowledge, policy frameworks, and active community participation. Leveraging ecosystems' inherent resilience gives opportunities to reduce the negative effects of climate change while encouraging adaptive methods that benefit both natural systems and human communities. This integrated viewpoint emphasizes the significance of combining ecological sustainability and socioeconomic growth to achieve a resilient and equitable future.

3 The Vital Role of Ecosystem Services in Enhancing Climate Resilience

Ecosystem services are the multifaceted benefits that humans derive from the natural environment, ranging from the provision of food and water to the regulation of climate and natural hazards (Munang et al., 2013). As climate-related disturbances such as extreme weather events, droughts, and wildfires become more frequent and intense, the ecosystem services that people rely on will become increasingly vital (Vignola et al., 2015). Thus, these services are not only crucial for human well-being but also play a vital role in enhancing the resilience of communities to the impacts of climate change.

Ecosystem services can be categorized into four main types: provisioning, regulating, supporting, and cultural (Faith et al., 2010). Provisioning services are the tangible products obtained from ecosystems, such as food, water, timber, and medicinal resources. Resilient ecosystems ensure the sustainable availability of these resources, even in the face of climatic variability. For example, diverse agricultural systems are better equipped to withstand droughts or pests, ensuring food security during adverse conditions. Regulating services, however, manage and moderate environmental processes, providing stability in the face of climate fluctuations. Wetlands, for instance, act as natural buffers by absorbing excess rainfall, reducing flood risks, and sequestering carbon to mitigate global warming. Similarly, forests help regulate local climates and purify air and water, enhancing resilience against extreme weather events (Balvanera et al., 2016). Supporting services underpin the functioning of ecosystems and are the foundational processes, such as soil formation, nutrient cycling, and habitat provision, that sustain all other ecosystem services. These underpin the health of ecosystems, ensuring their ability to recover from disruptions and continue delivering essential benefits. For example, healthy soils support robust plant growth, which in turn stabilizes ecosystems against erosion and climate-induced stress., while cultural services encompass the non-material benefits that people derive from nature, including inspiration, recreation, and spiritual fulfillment (Boyd & Banzhaf, 2007).

These services are the myriad benefits that nature provides to humanity, forming the bedrock of environmental health and human well-being. These are essential for building climate resilience, as they can help mitigate the impacts of climate change and enhance the adaptive capacity of communities. For example, well-functioning ecosystems can regulate the climate by sequestering carbon, reducing the severity of extreme weather events, and maintaining water supply. Moreover, the provisioning of food, water, and other resources from natural systems can help communities withstand the shocks and stresses associated with climate change (Boyd & Banzhaf, 2007). In developing countries, where economies and livelihoods often rely heavily on ecosystem services, the importance of incorporating these services into climate adaptation strategies is particularly salient. An international workshop on "Adaptation to Climate Change: The Role of Ecosystem Services" held in Costa Rica in 2008 highlighted the need for policies in these regions to explicitly account for the capacity of ecosystem services to increase the resilience of society to the impacts of a changing climate (Vignola et al., 2015).

Ecosystem services are essential for building climate resilience because they provide natural means to lessen the effects of climate change and increase communities' ability to adapt. Ecosystem services act as an important buffer against climate-induced stressors by maintaining the long-term provision of resources, regulating environmental processes, supporting ecological functions, and increasing human well-being through cultural advantages. It is not just a conservation priority but also a calculated investment in sustainable development to incorporate them into climate adaption plans, especially in areas that are at risk. In order to create a future that balances ecological stability with socioeconomic advancement and guarantees a more sustainable future, it is imperative to strengthen ecosystem resilience.

4 Climate Change Initiatives: Way Forward Toward Sustainability

As the various climatic feedback mechanisms are exacerbating over spatial and temporal dimensions, different climatic events are also changing and their severity and frequency of occurrence are also changing. To combat with the situation, different updated adaptive measures are suggested:

1. Renewable Energy Transition: Many countries like Costa Rica and Norway are leading examples in transitioning to renewable energy sources, thus reducing dependence on fossil fuels, which are a major cause of climate change.
2. Carbon Pricing: Mechanisms like carbon taxes and cap-and-trade systems will help to put a price on carbon emissions, encouraging reductions.
3. Climate-Resilient Infrastructure: Various cities like Rotterdam and Singapore are investing in climate-resilient infrastructure, such as sea walls and green roofs, even smart cities in India proceed with the concept of green walls.
4. Resilience-Building Initiatives Like Community-Based Adaptation: Programs like the UNDP's Adaptation Fund support community-led adaptation initiatives, enhancing local resilience.
5. Climate-Smart Agriculture: Practices like agroforestry and conservation agriculture help farmers adapt to climate change, improving food security.
6. Urban Planning and Design: Cities like Copenhagen and Vancouver are incorporating green spaces and climate-resilient design into urban planning.
7. Ecosystem Services Conservation Like Protected Areas: Establishing and maintaining protected areas, like national parks and wildlife reserves, helps conserve biodiversity and ecosystem services.
8. Sustainable Forest Management: Practices like selective logging and reforestation help maintain forest ecosystem services, including carbon sequestration.
9. Wetland Restoration: Restoring degraded wetlands helps maintain ecosystem services, including water filtration and coastal protection.
10. Ecosystem-based adaptation like restoring mangroves and coral reefs help communities adapt to climate change while maintaining ecosystem services.
11. Climate-Smart Conservation: Programs like the Nature Conservancy's climate-resilient conservation approach will help to protect ecosystem services while supporting climate change adaptation.
12. Integrated Landscape Management: Approaches like landscape restoration and agroforestry help balance human needs with ecosystem conservation, promoting resilience and sustainable development.

These examples illustrate the importance of addressing climate change, resilience, and ecosystem services in an integrated and interconnected manner. Would you like me to elaborate on any of these initiatives or provide more examples?

5 Organization of the Book

The book, *Climate Change, Resilience, and Ecosystem Services: Concepts, Models, and Implementation,* is structured into five thematic parts, offering an in-depth exploration of climate impacts, urbanization, disaster risks, and geospatial methodologies. The chapters are systematically arranged to provide clarity and coherence across the discussed themes.

5.1 Part I: Editorial

The **editorial section**, authored by Azizur Rahman Siddiqui, Saleha Jamal, and Zeba Khanam, introduces the book's central themes. It emphasizes the critical need for interdisciplinary approaches and innovative frameworks to address the global environmental challenges tied to climate change, urbanization, and ecosystem service management. The editorial sets the tone by highlighting the interconnections among these themes and their implications for sustainable development.

5.2 Part II: Climate Change: Understanding Impacts to Adaptation and Transformation

This section, comprising Chaps. 2, 3, 4, 5 and 6, delves into the multifaceted impacts of climate change on ecosystems, human health, and socio-economic systems, while exploring adaptive strategies. Chapter 2 by Avinash Dass et al. examines the impacts of climate change on groundwater quality, contamination risks, and human health, proposing a health risk assessment framework to safeguard water resources. Chapter 3, authored by Manu Raj Sharma et al., investigates how rainfall variability affects wetland ecosystems in Madhubani, Bihar, highlighting the need for sustainable adaptation measures. Chapter 4 by Mohammad Afsar Alam et al. reviews the relationship between climate change, natural disasters, and migration in the South Pacific, illustrating the socioeconomic complexities of climate-induced displacement. In Chap. 5, Hue Thi Thu Ha et al. discuss coastal communities' adaptation to hazards in Vietnam's Red River Delta, emphasizing resilience-building strategies. Finally, Chap. 6 by Sana Rafi et al. explores the nexus between greenhouse gases and global temperature, advocating for urgent emission reduction measures to combat global warming.

5.3 Part III: Climate Change, Urbanization, and Ecosystem Services: A Complex Systems Approach

Spanning Chapters 7, 8, 9, 10, 11, 12, 13, 14 and 15, this part examines the environmental impacts of urbanization and the critical role of ecosystem services. Chapter 7 by Azizur Rahman Siddiqui et al. evaluates ecosystem service values in response to land use and land cover changes, emphasizing declining trends in forest contributions. Chapter 8, authored by Dr. Fouzia Rahman, assesses the impact of land use and land cover dynamics on land surface temperature in Araria, Bihar, underscoring the importance of sustainable land use policies. Chapter 9 by Sadaf focuses on urban growth in Ayodhya, highlighting the ecological and infrastructural challenges posed by unplanned expansion. Chapters 10 and 15 by Dr. Deepti Khubalkar and Madhurima Purkait, respectively, examine the dynamics of urban green spaces and land use changes in the Mahananda River Basin, emphasizing the importance of ecological balance and adaptive management. Chapters 11 and 13 by Rukhsar Anjum et al. and Manal Ahmad et al. discuss wetland shrinkage and human encroachment, illustrating the adverse impacts of urbanization on aquatic ecosystems. Chapters 12 and 14 by Prof. Rashid Aziz Faridi and Md Babor Ali et al. explore the integration of green spaces into sustainable urban planning frameworks in Noida and Kolkata, emphasizing their role in mitigating pollution and enhancing resilience.

5.4 Part IV: Disaster Risk Reduction and Management: From Preparedness to Recovery

This section, comprising Chaps. 16, 17, 18, 19, 20 and 21, focuses on innovative approaches to disaster risk management and resilience-building. Chapter 15 by Farhana Parvin et al. proposes a GIS-based model for micro solar power plant site selection in the Sundarban Delta, offering renewable energy solutions to enhance resilience. Chapter 17 by Mujahid Husain et al. employs machine learning and GIS to model flood susceptibility in Banda District, Uttar Pradesh, focusing on protecting agricultural land. In Chap. 18, Toushif Jaman et al. develop a Cyclone Coping Capacity Index for Bhadrak District, Odisha, using geospatial techniques to assess community resilience. Chapter 19 by Azim Jamal et al. addresses urban flood risk in Delhi through HEC-RAS modeling, offering actionable insights for flood mitigation. Chapter 20, authored by Anusha et al., evaluates the influence of natural and anthropogenic factors on mangrove ecosystems in Kerala, highlighting the need for improved conservation strategies. Chapter 21 by Sana Rafi et al. reviews India's advancements in cyclone early warning systems, positioning the country as a global leader in disaster preparedness.

5.5 Part V: Climate Change, Geospatial Technologies, and Sustainable Development: A Framework for Action

Chapters 22, 23, 24, 25, 26, 27, 28 and 29 showcase the transformative potential of geospatial technologies in addressing diverse environmental challenges. Chapter 22 by Ashis Kumar Paul et al. explores tidal flat environments in the Indian Sundarban, employing geospatial techniques to assess climate impacts. Chapter 23 by Bishnu Hazr et al. analyzes soil erosion in the Kangsabati River Basin, using remote sensing to inform conservation planning. Chapter 24 by Pritam Bar et al. investigates forest fragmentation in Uttarakhand, providing insights for sustainable forest management. Chapter 25 by Tanzeel Khatoon et al. prioritizes sub-watersheds for soil erosion risk mitigation in Lalitpur using morphometric analysis. Chapter 26 by Mohd Saqib et al. evaluates agricultural drought in Punjab and Haryana through a multi-index approach, emphasizing adaptive strategies. In Chap. 27, Bharghav Ghosh et al. examine wetland of Kolkata using geospatial strategies. Chapter 28 by Sania Farrukh et al. employs the CA-Markov model to project future land use trends in Bulandshahr, Uttar Pradesh. Finally, Chap. 29 by L. Manawadu et al. studies the spatiotemporal dynamics of urban green spaces and their role in carbon storage, showcasing examples from Colombo, Sri Lanka.

This book concludes by synthesizing its core themes, highlighting the urgency of addressing climate change and fostering resilience through interdisciplinary research, technological innovation, and sustainable practices. By offering diverse case studies and cutting-edge methodologies, this book serves as a valuable resource for understanding and mitigating the complex challenges of climate change, urbanization, and ecosystem management. It provides actionable insights for researchers, policymakers, and practitioners working toward a more sustainable and resilient future.

References

Balvanera, P., Quijas, S., Martín-López, B., Barrios, E., Dee, L., Isbell, F., et al. (2016). The links between biodiversity and ecosystem services. In *Routledge handbook of ecosystem services* (pp. 45–61). Routledge.

Boyd, J., & Banzhaf, S. (2007). What are ecosystem services? The need for standardized environmental accounting units. *Ecological Economics, 63*(2–3), 616–626.

Burroughs, W. (Ed.). (2003). *Climate: Into the 21st century*. Cambridge University Press.

Centre for Climate and Energy Solutions (Climate Resilience Portal). https://www.c2es.org/content/climate-resilience-overview/#:~:text=Climate%20resilience%20is%20the%20ability,better%20cope%20with%20these%20risks

Faith, D. P., Magallon, S., Hendry, A. P., Conti, E., Yahara, T., & Donoghue, M. J. (2010). Evosystem services: An evolutionary perspective on the links between biodiversity and human well-being. *Current Opinion in Environmental Sustainability, 2*(1–2), 66–74.

Munang, R., Thiaw, I., Alverson, K., Liu, J., & Han, Z. (2013). The role of ecosystem services in climate change adaptation and disaster risk reduction. *Current Opinion in Environmental Sustainability, 5*(1), 47–52.

United Nations Climate Change. https://unfccc.int/topics/adaptation-and-resilience/the-big-picture/introduction

Upadhyay, R. K. (2020). Markers for global climate change and its impact on social, biological and ecological systems: A review. *American Journal of Climate Change, 9*(03), 159.

Vignola, R., Harvey, C. A., Bautista-Solis, P., Avelino, J., Rapidel, B., Donatti, C., & Martinez, R. (2015). Ecosystem-based adaptation for smallholder farmers: Definitions, opportunities and constraints. *Agriculture, Ecosystems & Environment, 211*, 126–132.

Climate Change: Understanding Impacts to Adaptation and Transformation

Climate Impacts on Groundwater: Contamination Pathways and Health Risk Assessment

Avinash Dass, Nighat Parveen, and Deepak Soni

1 Introduction

Climate change, an undeniable reality of our time, profoundly alters Earth's systems (Dass et al., 2024). While extreme weather events, rising temperatures, and shifting rainfall patterns dominate the headlines, their impact on groundwater, a hidden yet essential resource, is often overlooked. Groundwater within subterranean aquifers provides a lifeline for billions worldwide, particularly in regions facing water scarcity or surface water contamination. However, the escalating effects of climate change are jeopardizing this critical resource, threatening human health and livelihoods.

Despite not being a new phenomenon, climate change is worsening groundwater contamination in various ways. Rising sea levels drive saltwater intrusion into coastal aquifers, rendering freshwater supplies brackish and unusable (Mishra, 2023; Nicholls & Cazenave, 2010). Increased precipitation intensity and flooding are mobilizing surface pollutants and transporting them into groundwater systems (Kundzewicz, 2007). On the other hand, drought concentrates existing contaminants within aquifers, leading to elevated concentrations of harmful substances (Kubicz et al., 2021). These alterations in groundwater quality have far-reaching implications for human health, as contaminated groundwater can act as a vector for various diseases and ailments. A complex interplay of factors further intensifies the degradation of groundwater quality due to climate change. Shifts in precipitation patterns can diminish groundwater replenishment, leading to heightened concentrations and persistence of contaminants (Dao et al., 2024). Elevated temperatures can accelerate the chemical reactions of pollutants within aquifers, potentially transforming them into more toxic forms (Kløve et al., 2014; Neidhardt & Shao, 2023). Additionally, extreme weather incidents, such as cyclones and hurricanes, can damage infrastructure, releasing hazardous substances into the environment and subsequently infiltrating groundwater systems (Wahl et al., 2015).

The intricate relationship between climate change and groundwater contamination necessitates a comprehensive analysis. This chapter explores the various ways in which a shifting climate is deteriorating the quality of groundwater. It focuses on natural and human- induced processes. It also assesses the influence of extreme weather incidents, particularly droughts and floods, on replenishing and contaminating groundwater resources. The chapter also explores the role of climate change in mobilizing geogenic contaminants, such as fluoride and arsenic (As), which naturally occur in cer-

A. Dass (✉) · N. Parveen
Department of Environmental Science, Central University of South Bihar, Gaya, Bihar, India
e-mail: avinashdass@cusb.ac.in

D. Soni
Department of Environmental Sciences, Savitribai Phule Pune University, Pune, Maharashtra, India

Z. Khanam et al. (eds.), *Climate Change, Resilience and Ecosystem Services*, Sustainable Development Goals Series, https://doi.org/10.1007/978-3-031-96977-5_2

tain geological formations. These contaminants can leach into groundwater under specific conditions, posing significant health risks to populations relying on these sources (Mukherjee et al., 2024; Lapworth et al., 2017). Factors like alkalinity and temperature influence the mobilization of fluoride from geogenic sources (Ali et al., 2016).

Additionally, recent advancements have been made in understanding the enrichment mechanism of geogenic substances in groundwater (Wang et al., 2023). Geogenic arsenic contamination in groundwater is a significant global concern, impacting the health of millions worldwide (Amini et al., 2008). Regions such as South and Southeast Asia are particularly vulnerable, with geogenic arsenic being a major contributor to health issues (Klump et al., 2006). This type of contamination is primarily linked to the microbial breakdown of arsenic-containing minerals and the oxidation of organic carbon within aquifers (Nghiem et al., 2019). This chapter showcases significant case studies from various regions across India to demonstrate the potential contamination pathways and their impacts. These case studies provide valuable insights into how climate change impacts groundwater resources, highlighting the need of addressing this concern. The chapter further emphasizes the importance of understanding the health risks associated with contaminated groundwater. A comprehensive health risk assessment (HRA) framework is outlined, encompassing hazard identification, exposure assessment, dose–response relationships, and risk characterization, aligning with established methodologies (Du et al., 2024; Nordberg & Fowler, 2015). This framework enables the quantification of both carcinogenic and non-carcinogenic risks linked to contaminated groundwater exposure, serving as a valuable source for decision-makers and public health officials in risk management and mitigation strategies. Given the escalating risks to groundwater quality in the era of climate change, proactive measures are imperative.

This chapter concludes by offering recommendations for sustainable water resource management, advocating for implementing advanced monitoring systems, and developing innovative treatment technologies. An integrated, comprehensive approach can ensure the protection of groundwater resources, preserving human health and guaranteeing access to clean water for future generations (Fig. 1).

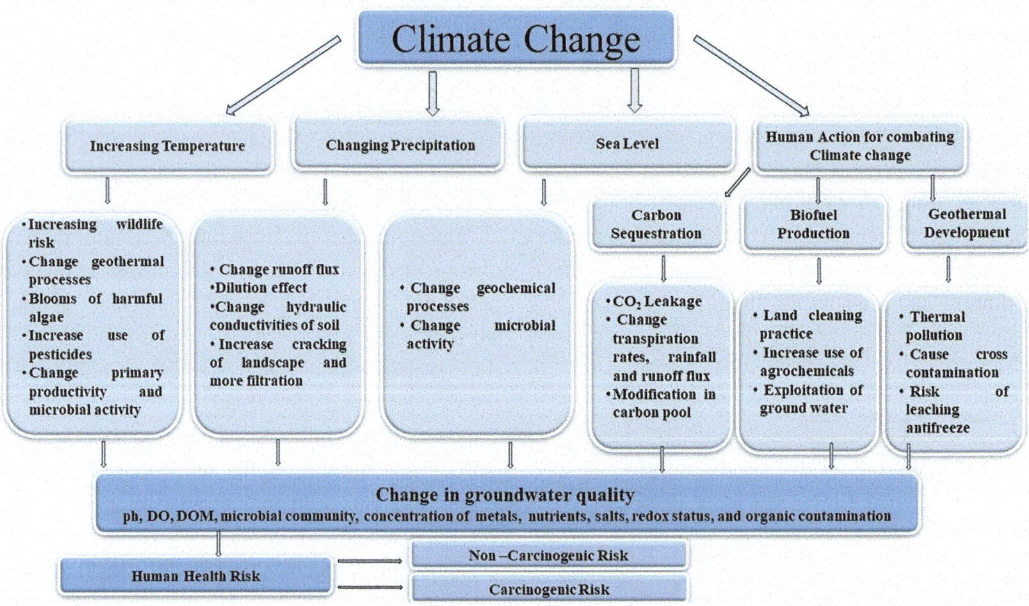

Fig. 1 Interconnected impacts of climate change and human interventions on groundwater quality and associated health risks

2 Climate Change and Its Impact on Groundwater Systems

2.1 Overview of Climate Change

Climate change pertains to sustained modifications in rainfall patterns, temperature, and other facets of the Earth's climate system, predominantly resulting from human endeavors such as industrial operations, deforestation, and fossil fuel combustion (du Plessis, 2019; du Plessis & du Plessis, 2019). Because of these activities, the atmospheric concentration of greenhouse gases (GHGs) such as carbon dioxide (CO_2), nitrous oxide (N_2O), and methane (CH_4) has increased to an unparalleled degree (Jain et al., 2020). The outcome is an acceleration of Earth's climatic warming, which has an extensive range of effects on natural systems, including groundwater supplies.

The average global temperature has increased by roughly 1.1 °C over pre-industrial levels, according to unambiguous data from the Intergovernmental Panel on Climate Change (IPCC). If GHG emissions are not drastically decreased, forecasts show that this warming trend will continue, with an increase of 1.5–2 °C predicted over the following few decades (Hoegh Guldberg et al., 2018). Global warming is not happening uniformly everywhere; some places are warming more rapidly than others. The hydrological cycle, which controls the distribution and flow of water throughout the ecosystem, is being significantly impacted by this temperature rise as well as modifications to precipitation patterns (Kundzewicz, 2008; Yang et al., 2021).

The hydrological cycle is one of the main ways that groundwater systems are impacted by climate change (Kumar, 2012). There is an increasing degree of unpredictability in precipitation patterns; certain areas are seeing more rain while others are facing protracted droughts (Maleksaeidi et al., 2017). These modifications have an immediate impact on groundwater recharge rates, which are essential for preserving groundwater levels and guaranteeing a steady supply of fresh water (Goderniaux et al., 2009; Gorelick & Zheng, 2015). Groundwater recharge can be drastically decreased in areas where precipitation is decreasing or becoming more erratic, which can result in dropping groundwater levels and more competition for available water resources (Cavé et al., 2003).

Elevated temperatures are also linked to elevated levels of evapotranspiration, which is the total amount of water lost through transpiration from plants and evaporation from water and land surfaces (Baron et al., 1998; Kadlec, 2006). The quantity of water that is available for soil infiltration and subsequent groundwater aquifer recharge may be decreased by this process (Han et al., 2017). Furthermore, warmer weather can make water shortage worse in arid and semi-arid areas by pushing people to use more water for activities like irrigation and other water-intensive projects, which puts more pressure on groundwater supplies (Shevah, 2015; Ward & Ruckstuhl, 2017).

The frequency and intensity of extreme weather events, including floods and hurricanes, are increasing due to climate change, and this has a substantial impact on groundwater systems (Bolan et al., 2023). These occurrences have the potential to contaminate groundwater supplies by allowing contaminants from surface runoff to seep in, especially in places with insufficient infrastructure to handle them. For instance, pesticides, fertilizers, and other pollutants from agricultural fields can be washed into groundwater sources by severe rains and flooding, posing serious dangers to populace health and water quality (Ritter et al., 2002; Akhtar et al., 2021).

Groundwater systems in coastal locations are particularly vulnerable to sea level increase brought on by climate change (Michael et al., 2013; Rasmussen et al., 2013). Freshwater supplies become salinized as a result of increased likelihood of saltwater interruption into coastal aquifers brought on by rising sea levels. Communities that depend on coastal aquifers for agriculture and drinking water should be especially concerned about this situation (Abedin et al., 2014; Carrard et al., 2019). It is difficult and expensive to clean up saltwater contamination in aquifers, which frequently makes the water unfit for human use (Zeidan, 2017).

Because of its effects on human activity, climate change also indirectly affects groundwater systems. There is frequently more pressure to exploit groundwater resources, which can result in over-extraction, as surface water sources become less dependable due to climate change (Sharma et al., 2021).

2.2 Impact on Groundwater Recharge and Levels

Due to changes in temperature, evapotranspiration rates, and rainfall patterns, climate change has a substantial influence on groundwater recharge and levels (Kumar, 2012). Reduced or irregular rainfall regions see a drop in groundwater recharge rates because less water percolates through the soil to refill aquifers (Bhattacharya, 2010). This may result in lower groundwater levels, which would exacerbate the scarcity of water, particularly in regions where groundwater is extensively used for agriculture and drinking (Mukherjee et al., 2021).

Elevated temperatures exacerbate these difficulties by augmenting evapotranspiration, thereby diminishing the quantity of water accessible for replenishment (Banerjee & Sharma, 2021). Higher temperatures and protracted droughts can significantly restrict groundwater recharge in arid and semi-arid regions, which can result in over-extraction and aquifer depletion (Zeydalinejad & Nassery, 2023).

Extreme weather events, like intense storms and flooding, can disrupt recharge processes, causing either rapid, unsustainable recharge or contamination from surface runoff (El-Rawy et al., 2024). On the other hand, in regions with increased rainfall, recharge may improve, but it can also elevate risks of contamination as more pollutants are washed into the groundwater system (Ritter et al., 2002).

2.3 Climate Change-Induced Contamination Pathways

Water quality issues are increasing as a result of climate change's increasing influence on groundwater contamination pathways. An important consequence is the increase in extreme weather events, such as flooding and heavy rain, which can move pollutants from urban areas, industrial sites, and agricultural regions (Bolan et al., 2023). These pollutants, which include industrial chemicals, fertilizers, and pesticides, can seep into the soil and eventually into groundwater, lowering the quality of the water (Khan et al., 2018).

Another effect of climate change is extended droughts, which concentrate already-existing pollutants in groundwater and worsen contamination. Because there is less water available for dilution, pollutants like salts and nitrates can build up to dangerous levels and endanger the health of people who depend on groundwater for irrigation and drinking (Al-Sudani, 2019).

Freshwater aquifers in coastal regions are experiencing saltwater interruption due to increasing sea levels. As a result of this process, formerly fresh groundwater is progressively becoming salinized and unfit for human consumption or agricultural usage (Ondrasek & Rengel, 2021). Furthermore, as permafrost in polar regions thaws, contained pollutants are being released (Kerényi et al., 2020). These pollutants have the potential to penetrate groundwater systems as a result of their increased mobility in the changing climate.

3 Groundwater Contamination Pathways

Groundwater, while comprising a relatively small fraction (0.8%) of the Earth's freshwater resources, is a vital and readily available source of fresh water, second only to glaciers in magnitude. Its essential role in sustaining life on Earth cannot be overemphasized. Recent observations have indicated significant and growing changes in both the accessibility and quality of groundwater resources. Consequently, there is now a necessity for systematic and thorough monitoring. Regular surveillance plays a crucial role in evaluating the quality of water resources, monitoring the existence of pollutants, and efficiently preventing contamination of groundwater sources (Fioreze & Mancuso, 2019). Groundwater is a crucial resource that is vital in supplying clean and easily accessible drinking water to over two billion people across the globe (Kundzewicz & Döll, 2009). Every year, there are 1.23 million deaths globally due to insufficient water supplies. Low-income countries bear a more significant burden of contaminated water, which is twice as severe as other nations (Adelodun et al., 2021). The increasing presence of human-made pollutants is a significant concern, driven by population growth, urbanization, industrialization, and intensified agriculture (Akhtar et al., 2021; Sharma et al., 2024). Groundwater is a serious freshwater resource in regions characterized by limited surface water availability and low precipitation, mainly in arid and semi-arid environments (Priyan, 2021). Over the last 30 years, a persistent problem of chemical pollution has been routinely recorded in research on groundwater. However, making up only 0.8% of the Earth's freshwater resources, it is crucial in providing clean, accessible drinking water to over two billion people worldwide. In addition to naturally occurring pollutants from the Earth's crust, human-induced pollutants have become a pressing concern due to the rapid growth of the global population and associated economic activities, particularly in rapidly developing nations mainly located in the Eastern Hemisphere. These warrant tailored studies addressing the challenges of natural and human-made contaminants in groundwater across different regions, including India, Turkey, Bangladesh, Ethiopia, and Nigeria. Groundwater contamination, caused by the introduction of unwanted substances from human activities and natural sources, poses unique challenges in detection and remediation. Unlike surface water contamination, groundwater pollution remains invisible, making it more challenging to address with current technological capabilities. The prolonged and costly cleanup process further underscores the urgency of addressing groundwater contamination, as purification efforts can span multiple decades. Groundwater can contain many pollutants from natural and human sources, such as chemicals, biological substances, and radioactive materials. Protecting this essential resource requires ongoing efforts to monitor, use sustainable practices, and make well-informed decisions. Contamination of groundwater resources presents a significant risk to both public health and the environment, necessitating urgent action to guarantee the accessibility of clean water for current and future populations. Groundwater pollution is a substantial obstacle to human populations. Moreover, it presents a significant chance for scholars to gain a more profound understanding of the development of subsurface aquifers. This information can guide policymakers in developing effective strategies for the sustainable management and safeguard of groundwater resources. A significant number of pollutants found in groundwater are naturally occurring and originate from the dissolution of the outermost layer of the Earth, which contains deposits of minerals (Basu et al., 2014; Pandey et al., 2016; Subba Rao et al., 2020). Various factors, including chemicals, bacteria, viruses, medications, fertilizers, and petrol, can contribute to this phenomenon. The pollutants present in groundwater generally do not possess any discernible color or odor; moreover, the detrimental effects of contaminated groundwater on human welfare are enduring and provide substantial difficulties in identification (Chakraborti et al., 2015). Remediation of contaminated groundwater can be complex and expensive because groundwater is found in underground geological layers and can take a long time to cleanse (National Research Council et al., 1999; McCarty, 2010). Typically, restoring contaminated groundwater might span multiple decades despite eliminating the cause of contamination (Tatti et al., 2019). The number of types of pollutants identified in groundwater is continually growing. Groundwater pollution can come from natural sources like salty brines and water bodies such as rivers, lakes, and oceans. This can happen when these water sources have below-average quality due to geological formations containing valuable minerals. When people disrupt the ecosystem's natural balance, certain natural sources can become major contributors to pollution. This can include the reduction of aquifers preceding to saltwater intrusion, the exploitation of mineral resources resulting in acid mine drainage, and excessive irrigation causing the leaching of hazardous chemicals (Mohanavelu et al., 2021). Inorganic nitrogen-based contaminants include ammonia nitrogen nitrate, nitrite, and nitrate that mainly originate from human activities, such as agriculture (e.g., fertilizers, manure) and domestic wastewater. Nitrate contamination of groundwater resources is a widespread concern that has been extensively reported in numerous geographical locations worldwide.

Groundwater may occasionally include high concentrations of salt, which refers to the collective quantity of water containing both inorganic and organic substances. Toxic metals and metalloids stand a substantial risk to human health and the environment. Various chemical elements are commonly found in groundwater, including metals like zinc, lead, mercury, chromium, cadmium, and metalloids like selenium and arsenic. High exposure concentrations can result in severe poisoning, although certain elements are crucial micronutrients when administered in lesser quantities (Hashim et al., 2011). For example, chromium (Cr^{+6}) raises cancer risk in human beings. Arsenic is classified as a group 1 human cancer-causing agent by both the USEPA (United States Environment Protection Agency) and the IARC (International Association for Study on Cancer), underscoring its significant health risk. The trivalent form of arsenic (As^{+3}) is particularly harmful due to its ability to disrupt cellular processes and induce apoptosis by binding to sulfhydryl groups present in vital proteins and enzymes (Abbas et al., 2018; Rebelo & Caldas, 2016). Metals are persistent environmental contaminants that can bioaccumulate in organisms, with concentrations increasing at higher trophic levels within the food web (Hashim et al., 2011). These compounds are considered to be carcinogens or disruptors of the endocrine system. According to various studies, over 200 organic pollutants have been found in groundwater, which continues to rise (Yang Lesser et al., 2018; Jurado et al., 2012). Specific organic contaminants can naturally break down over time, while others remain in the environment longer. The primary source of renewable organic contaminants is home drainage and industrial effluent. Organic materials can be spontaneously obtained from carbs, proteins, lipids, and oils. Microorganisms can transform these chemicals into durable, inorganic substances. Although they do not have any immediate detrimental impacts on living creatures, they can decrease the dissolved oxygen concentration in groundwater. Organic pollutants can manifest in several states, such as petroleum products, halogen-containing chemicals, plasticizers, pesticides, medicines, household goods, and natural estrogens (Lapworth et al., 2015). Several halogenated chemicals, such as chlorinated, brominated, and fluorinated compounds, exhibit environmental stability and can collect and concentrate in organisms. Accumulation of this substance may negatively impact organisms at higher trophic levels, involving humans (Gwenzi et al., 2018; Schulze et al., 2019). The primary sources of persistent organic pollutants include agricultural and industrial practices and compounds used in human health protection (Lapworth et al., 2015). The extended deterioration or absence of these compounds can permanently compromise the condition of groundwater intended for human consumption (Schulze et al., 2019). Groundwater may include radioactive pollutants from natural deposits of radionuclides or human actions, such as dumping medical radioisotopes, testing nuclear weapons, and inappropriate management of nuclear power plant waste (Schlüter et al., 2004). There are several routes via which radioactive chemicals might enter the human body, and one of these is by consuming drinking water. Radioactive pollutants in groundwater pose little danger to human health, as they are seldom found at detrimental levels. Biological pollutants refer to various species, such as algae, bacteria, viruses, and protozoa. Various studies found that human and animal waste contains over 400 bacteria and 100 viruses (Greening & Cannon, 2016; Kurtz & Lee, 2007). Certain microbial organisms are derived from natural sources, while others are minute organisms that live alongside natives. Consuming water that is polluted with patho-

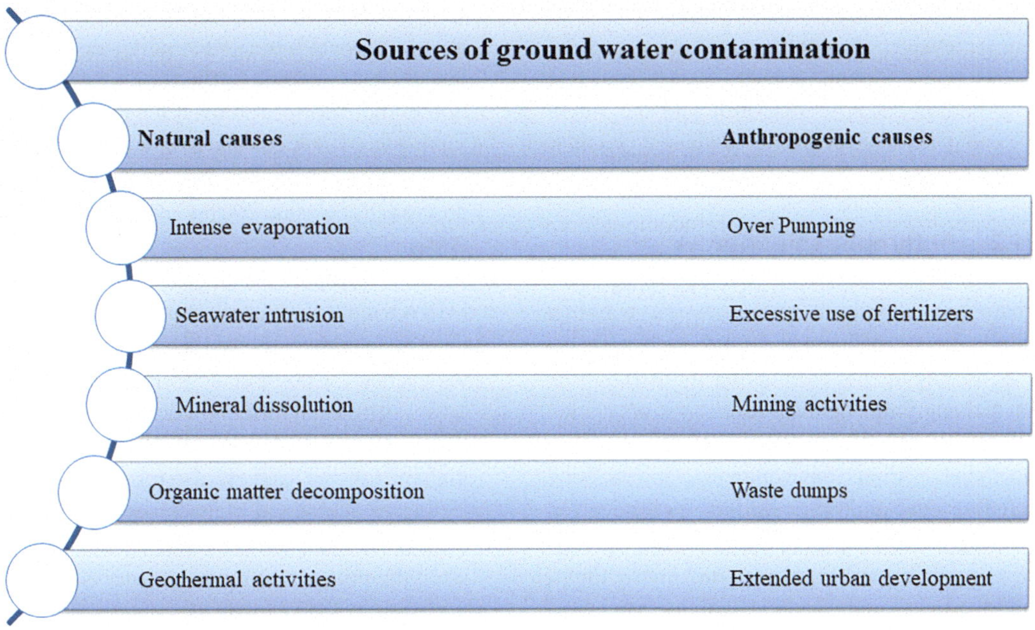

Fig. 2 Different sources of groundwater contamination

genic microbes can lead to a range of human illnesses, including severe digestive problems such as typhoid and cholera. The Centers for Disease Control and Prevention (CDC) states that the coronavirus mainly spreads through respiratory droplets when people are in close contact with water polluted with the virus, potentially threatening human safety (Bhowmick et al., 2020; Lokhandwala & Gautam, 2020) (Fig. 2).

3.1 Ways to Classify Groundwater Contamination Pathways

The origin of certain groundwater contaminants arises from either natural occurrences or human activities.

3.1.1 Natural Sources

Natural factors like recurring cycles or natural phenomena can contaminate groundwater. This group of materials encompasses the following sources: Certain types of rocks, such as gypsum and mineral salt, have a high tendency to dissolve readily. A high evaporation rate in shallow aquifers increases groundwater levels and salt deposition. In geothermal and volcanic locations, the water source can become degraded. Other factors contributing to water source degradation include rock oxidation, seawater intrusion, and the decay of radioisotopes in uranium. The bedrock is rich in minerals and undergoes chemical reactions when exposed to air and water.

(a) **Intense evaporation:** Managing Groundwater in dry or semi-arid locations is difficult because there is not enough water available, rainfall is erratic and low, evaporation rates are high, recharge is limited, and climate change has an impact (DeNicola et al., 2015; Shrestha et al., 2018). Water quality degradation often puts significant pressure on water distribution and management systems. Evaporation in arid and semi-arid areas substantially raises salt content in groundwater and soil (Re et al., 2017).

(b) **Seawater intrusion:** Besides potential mainland-based urban, industrial, or agricultural contamination, a coastal aquifer is particularly vulnerable to seawater mixing and quality degradation since one side of its boundary directly touches the sea. This issue is prevalent in many regions of the globe. Excessive pumping along coastlines is a common cause of overdraft and the subsequent formation of a seawater gradient inland.

3.1.2 Anthropogenic Sources

Anthropogenic sources are the predominant contributors to pollution, encompassing activities including waste effluent and solid waste management, managing and eliminating industrial effluent, and applying chemical fertilizers, pesticides, and pest control substances. These activities encompass various environmental interactions, including over-pumping aquifers, excessive fertilizer application, mining operations, waste disposal practices, urban expansion, microclimate modifications, and water basin management. Moreover, the improper disposal of organic and inorganic elements and the storage of sewage (infiltration) further exemplify the range and impact of human activities on the environment.

(a) **Over-pumping:** Excessive extraction may lead to a drop in the groundwater level, making it difficult for wells to obtain water. As a result, pumping water over longer distances becomes necessary, leading to higher energy consumption and increased costs. In extreme cases, the utilization of such wells may become financially unfeasible.
(b) **Excessive use of fertilizers:** Nitrogen, which exists as nitrate, is a prevalent contaminant in surface and groundwater sources. Nitrate-N can easily percolate through agricultural soils, extending beyond the area where plant roots may access it and ultimately making its way into groundwater and surface water sources. When the amounts of nitrate-N go beyond the acceptable limits, the groundwater is unsuitable for drinking.
(c) **Mining activities:** Mining operations significantly influence both the quantity and condition of water sources in their surrounding environment. Dewatering processes associated with mining can result in a lowering of the water table, potentially causing the depletion of specific springtime or a reduction in their flow. Moreover, the cessation of mining activities can have an adverse effect on groundwater quality due to the recovery process (Razowska, 2001; Gandy & Younger, 2007).
(d) **Waste dumps:** Municipal solid waste (MSW) landfills heavily pollute groundwater. The leading cause of this is the discharge of leachate, a complicated blend of pollutants with a significant chemical oxygen demand and a high concentration of ammonium nitrogen. Leachate also possesses long-lasting toxicological characteristics. Cuevas et al. (2012) and Li et al. (2008) have emphasized the potential for groundwater contamination from MSW landfills.
(e) **Extended urban development:** Urbanization is a global phenomenon that has been rapidly increasing due to economic growth and industrialization, leading to a significant shift in the distribution of the world's populace toward urban areas (Satterthwaite et al., 2010; Bettencourt et al., 2007). This trend is predictable to continue, with estimates indicating that by 2030, around 60% of humanity will reside in urban regions, and this figure is estimated to rise to 70% by 2050 (Satterthwaite et al., 2010). Before the 1980s, there was a widespread belief that urban expansion leads to decreased aquifer recharge because of the non-porous barrier formed by asphalt, concrete, and roof-topped structures across extensive land surface areas. Multiple studies have unequivocally shown that urbanization restricts the natural replenishment of aquifers through direct recharge, which occurs when rainwater is transferred through the soil zone (Sharma, 2017; Henao Casas, 2023).
(f) **Change of climatic conditions (microclimate):** Compared to surface water, less hydrological investigation has been dedicated to understanding the complex relationship between climate change and

groundwater, including the reciprocal effects of groundwater on the climate system (Kundzewicz & Döll, 2009; Green et al., 2011). Anticipated warming trends are expected to influence both the amount and timing of groundwater recharge. Specifically, changes in rainfall patterns and evapotranspiration rates due to a warming climate are projected to impact the proportion of rainfall replenishing aquifers seasonally, a critical factor in forecasting annual groundwater recharge rates (Jasechko et al., 2014). The renewable component of groundwater storage, directly influenced by climate-induced modifications in recharge, plays a pivotal role in determining groundwater availability (Crosbie et al., 2013). Thus, climate change is poised to significantly alter groundwater recharge patterns, potentially impacting global renewable groundwater resources (Döll, 2009). These changes can be observed through fluctuations in average groundwater levels across seasons and years, corresponding to shifts in rainfall distribution and snowmelt dynamics (Jyrkama & Sykes, 2007; Okkonen & Kløve, 2010).

Studies have highlighted climate change's direct and indirect impacts on water resources such as groundwater, emphasizing changes in recharge patterns and overall availability (Morgan et al., 2022). Projections indicate potential increases or decreases in groundwater recharge under different climate change scenarios, suggesting a complex interplay between climate variables and groundwater dynamics (Ngo et al., 2024). Evaluating the climatic impact on groundwater restoration can require insights into the effects of climate change on groundwater resources (Yang & Bertetti, 2023) (Fig. 3).

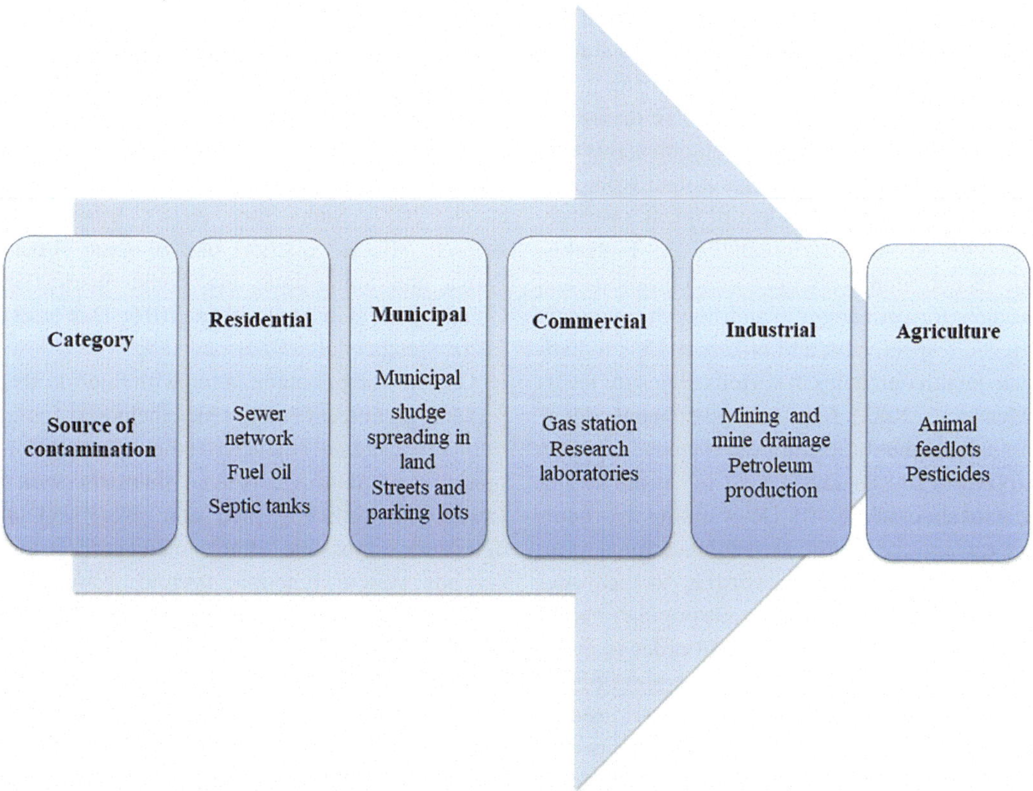

Fig. 3 Different categories and sources of groundwater contamination

4 Human Health Risk Associated with Groundwater Contamination

Assessing the potential health risks posed by groundwater contamination involves a systematic process that can be broken down into four key stages: hazard identification, dose–response assessment, exposure assessment, and risk characterization (Williams & Paustenbach, 2024). The initial step, hazard identification, focuses on identifying contaminants present in groundwater that could pose a threat to human health. This may include metals, metalloids, pesticides, radioactive materials, and other harmful substances (Valskys et al., 2022; Rajan et al., 2024). A review of toxicological data aids in understanding these impacts, such as carcinogenicity and toxicity (Harwood et al., 2007). Regulatory standards from establishments like the USEPA and the World Health Organization (WHO) guide acceptable contaminant levels (Van Dam et al., 2019). Dose–response assessment studies the association between pollutant dosages and health effects (Tsuji et al., 2019). To determine thresholds for various health outcomes, this step needs to: examine toxicological data from animal studies, human epidemiological research, and clinical trials (Hernández & Tsatsakis, 2017); determine reference concentrations and dosages for carcinogenic and non-carcinogenic effects, respectively; and utilize models to estimate health outcomes at various exposure levels (Men et al., 2023). Exposure assessments determine the amount, time, and frequency of human exposure to contaminants in groundwater (Baciocchi et al., 2010). Determining the routes of exposure (ingestion, skin contact, inhalation) and creating scenarios according to the water consumption patterns of various populations are critical (Grellier et al., 2015). In order to quantify exposure, techniques such as point-of-use measurements and biomonitoring are used to measure the amounts of contaminants present, the amount of water used, and the duration of exposure (Smith et al., 2016). To quantify health risks, risk characterization incorporates data from exposure assessment, dose–response assessment, and hazard identification (Williams & Paustenbach, 2024). The risks of both cancer and non-cancer outcomes were analyzed, with cancer risk quantified as the lifetime probability of developing cancer, and non-cancer risk evaluated by comparing estimated exposure levels with reference doses through the hazard quotient and hazard index approach (Valcke et al., 2017).

Groundwater contamination with arsenic can have serious health effects, such as skin lesions like keratosis and pigmentation changes, cancers of the skin, bladder, lung, and kidney, an increased risk of cardiovascular disease and hypertension, neurological effects like peripheral neuropathy and cognitive deficits, a relationship to type 2 diabetes, and probable congenital disabilities and developmental delays in children (Chakraborti et al., 2015; Palma-Lara et al., 2020). The most severely affected areas in India are the eastern parts of Uttar Pradesh, including Ballia and Gazipur; Assam's Brahmaputra River basin areas, like Jorhat and Dhemaji districts; West Bengal, with districts like Murshidabad, Malda, Nadia, and North 24 Parganas displaying arsenic levels frequently exceeding 300 μg/L, far above the WHO guideline of 10 μg/L (Chakraborti et al., 2017; Sinha Ray, 2024); Bihar, with severely affected districts like Bhojpur, Buxar, Patna, and Samastipur; and some districts in Chhattisgarh (Ghosh & Bose, 2016; Das et al., 2020; Shukla et al., 2020).

Groundwater contamination with fluoride also presents significant health risks, including dental fluorosis, which is characterized by tooth discoloration and mottling; skeletal fluorosis, which causes joint difficulty, pain, and severe skeletal deformities; possible effects on cognition and nervous system function; thyroid dysfunction associated with hypothyroidism because fluoride interferes with iodine absorption; and gastrointestinal suffering from excessive fluoridization (Singh & Singh, 2021; Choubisa et al., 2023). The most severely affected areas in India are the districts of Ajmer, Jaipur, Jhunjhunu, Sikar, and Nagaur in Rajasthan; the districts of Anantapur, Nalgonda, and Prakasam in Andhra Pradesh;

Telangana, particularly Nalgonda; Mehsana, Sabarkantha, and Banaskantha in Gujarat; Amritsar, Bathinda, and Ferozepur in Punjab; Mahendragarh and Rewari in Haryana; Dharmapuri, Krishnagiri, and Salem in Tamil Nadu; and Gulbarga, Bellary, and Tumkur in Karnataka. These areas also have widespread dental and skeletal fluorosis (Krishan et al., 2015; Singh et al., 2018; Yadav et al., 2019 Kashyap et al., 2022).

In India, prolonged groundwater consumption has led to widespread health issues like arsenicosis and fluorosis (Baboo et al., 2022). These conditions are directly linked to elevated levels of arsenic and fluoride in water sources, presenting significant health risks to local communities. The implications are severe, affecting these area's public health and agricultural productivity. To address these challenges, articles such as those by Ding et al. (2020) emphasize the significant need for improved management of groundwater resources.

India focused on assessing groundwater quality, revealing that while a substantial portion of the water is unfit for direct consumption, about 80% remains appropriate for irrigation purposes (Guo et al., 2021). Despite this, children are identified as a particularly vulnerable group due to higher exposure risks compared to adults, necessitating targeted interventions to mitigate health impacts (Xu et al., 2012). Shifting the focus to South India's Shanmuganadhi River Basin, Karunanidhi et al. (2024) investigated the quality of water consumption using the Water Quality Index (WQI). Their study revealed alarming results, with 52% of sampled water falling into categories deemed poor, very poor, or unsuitable for consumption. This situation poses heightened health risks, particularly impacting children who are more sensitive to the adverse effects of contaminated water sources. The articles collectively advocate for sustainable water management practices in response to these challenges. Recommendations include promoting the consumption of iodized salt to counteract fluoride- related health issues and advocating for the integrated utilization of groundwater and surface water sources in basin and mountain areas. These strategies are crucial for enhancing water quality, promoting resilient health outcomes, and ensuring the effectiveness of groundwater management practices. In their study of southwestern Punjab (India), Kumar et al. (2020) shed light on the origin, dispersion, and probable health hazards associated with increased uranium and fluoride levels in groundwater. These contaminants, originating mainly from geogenic sources, are further exacerbated by minor anthropogenic activities, leading to significant potential health dangers for people in the area. The incidence of prominent concentrations of uranium and fluoride in groundwater raises concerns about their probable detrimental impacts on human health, encompassing dental and skeletal fluorosis, kidney damage, and other systemic health issues. Kumar et al. (2020) revealed that the carcinogenic health risks posed by arsenic (As) and chromium (Cr) exceeded the thresholds established by USEPA. This finding underscores the serious need for intervention to mitigate these risks and safeguard people's health. To address these challenges, the study recommended that local governments take necessary actions, such as providing potable water facilities and conducting awareness campaigns to educate the population about the potential health risks related with polluted water sources. In addition, Subba Rao (2021) conducted an assessment of statistical non-carcinogenic risks in an arid rural region of South India. This region is characterized by high nitrate levels (30–95 mg/L) and fluoride (0.7–2.8 mg/L) in groundwater. Subba Rao's study highlighted the hazardous health risks faced by adult males, adult females, and children due to their varying sensitivities to contaminants and different body masses. Elevated levels of nitrate and fluoride in groundwater pose significant health risks, including methemoglobinemia (blue baby syndrome) from nitrate contamination and dental and skeletal fluorosis from fluoride contamination. The health hazards are especially worrying for pregnant women, infants, and children. These susceptible groups may be at a higher risk of experiencing harmful effects from exposure to these contaminants due to their unique physiological factors and still developing immune systems. Non-carcinogenic risk assessment for both ingestion and skin contact exposure in adults and children indicates significant risk levels in several states of India. The Target Hazard Quotient (THQ) values for both adults and children in Andhra Pradesh, Rajasthan, Punjab, Karnataka, Madhya Pradesh, and Chhattisgarh exceed 1, indicating a considerable risk to consumers in these states. In contrast, children in Bihar, Haryana, Maharashtra, Odisha, Tamil Nadu, Telangana, and Uttar Pradesh are at significant risk (THQ > 1). In contrast, adults in these states are within the safe range (THQ < 1) (Ali et al., 2019). The calculated pooled fluoride concentration in India's groundwater exceeds the Indian and WHO standard limits. This study revealed that fluoride concentration in remote location groundwater is 1.9 times higher compared to town areas. Additionally, it was noted that increased rainfall significantly decreases fluoride concentrations in groundwater ($p < 0.05$).

The risk assessment findings suggest that adult males, adult females, and children in most Indian states are exposed to significant non-carcinogenic risks, with children being particularly vulnerable to fluorosis due to their lower body mass (Mridha et al., 2021). Therefore, it is imperative to implement fluoride mitigation initiatives, especially in regions with limited rainfall. Providing drinking water with fluoride concentrations below the permissible limit is essential for the highly affected states. To address these challenges, comprehensive strategies are needed to safeguard access to potable and clean drinking water for all populations disturbed by groundwater contamination. This includes implementing effective water treatment technologies, promoting sustainable agricultural practices to reduce nitrate contamination, and raising awareness about the potential health risks correlated with contaminated water sources.

In summary, these studies underscore the urgent prerequisite for coordinated efforts among policymakers, researchers, and local communities to address water quality challenges effectively. By prioritizing sustainable management practices and implementing targeted interventions, it is possible to mitigate health risks associated with groundwater contamination and ensure access to potable consumption water for all populations affected by these environmental health concerns.

5 Human Health Risk Assessment Framework

The estimation of possible health effects from environmental risks, such as those resulting from groundwater contamination, is facilitated by the use of health risk assessments (HRAs) (Ritter et al., 2002). A strong background for risk assessment and mitigation of human health is necessary when climate change disrupts pathways of contamination and increases the number of contaminants present in groundwater systems (Bolan et al., 2023). In light of climate change, a thorough HRA framework has been developed expressly to combat groundwater contamination.

5.1 Hazard Identification

Hazard identification, which entails identifying the pollutants in groundwater that may be destructive to human health, is the first stage in the HRA process. Agricultural runoff, industrial discharges, wastewater leaks, and naturally occurring materials like fluoride and arsenic are just a few of the potential causes of these dangers (Karri et al., 2021). Emerging contaminants like pharmaceuticals, personal hygiene items, and microplastics may also become significant in the situation of climate change as a result of modifications to land use, waste disposal techniques, and environmental factors (Pozzebon & Seifert, 2023).

Due to changes in hydrological models carried about by climate change, the presence of these toxins is made worse by more frequent and severe contaminant movement into groundwater systems (Swain et al., 2022). Recognizing these dangers at an early stage is essential to comprehending the possible health consequences they may cause.

5.2 Exposure Assessment

Quantifying the level of human population exposure to recognized pollutants is part of the exposure assessment process. This phase necessitates knowing the routes by which toxins enter people, such as through contaminated soil contact, food irrigation, or drinking water. It also entails making estimates about the length, frequency, and intensity of exposure (Elert et al., 2011).

As a result of altered water use, a greater reliance on groundwater during droughts, and modifications to agricultural practices, exposure patterns may change in a changing climate (Kløve et al., 2014). For instance, populations may become more reliant on groundwater in areas where surface water availability is limited, increasing their exposure to any pollutants that may be present. Exposure can also be directly impacted by changes in groundwater levels and recharge rates, which can alter the concentration of contaminants (Ritter et al., 2002).

Assessing exposure in this dynamic context requires up-to-date monitoring data and predictive models that account for climate-related variables, such as changes in temperature, rainfall, and extreme weather events (Zscheischler et al., 2020).

5.3 Dose–Response Assessment

The study of dose–response assessment looks at the connection between the probability of negative health impacts and the amount of exposure to pollutants. Finding the threshold values at which a contaminant presents serious health risks requires this step. For many groundwater contaminants, established dose–response relationships exist, such as the link between high nitrate levels and methemoglobinemia (blue baby syndrome) or between arsenic exposure and cancer (Danaraj et al., 2022).

However, climate change can alter these relationships by influencing the behavior of contaminants and affecting population vulnerability. For instance, rising temperatures may enhance the toxicity of certain chemicals or promote the growth of pathogens in groundwater (Lipczynska-Kochany, 2018). Furthermore, vulnerable groups such as children, the elderly, or individuals with pre-existing health conditions may face more serious health impacts even at small exposure amounts (Makri & Stilianakis, 2008). This necessitates adjustments to dose–response models to account for these heightened risks.

5.4 Risk Characterization

Risk characterization put together the findings from hazard identification, exposure assessment, and dose–response assessment to estimate the whole health risks posed by contaminated groundwater (Williams & Paustenbach, 2024). This step involves calculating risk metrics, such as the estimated dietary intake (EDI), THQ, and cancer risk, which help quantify the potential impact on human health (Miranzadeh Mahabadi et al., 2020).

Risk characterization in the perspective of climate change needs to take into account the increased uncertainty and variability in exposure and health outcomes (Gallina et al., 2016). For example, the long-term effects of repeated exposure during droughts or the variability in contaminant concentrations due to extreme weather events can complicate risk estimates (Bolan et al., 2023). To capture the maximum range of potential risks, sensitivity analyses and probabilistic approaches are essential.

6 Conclusions

Climate change is undeniably exacerbating groundwater contamination, posing a significant threat to human health and livelihoods. This chapter has comprehensively explored the intricate relationship between climate change and groundwater quality deterioration, highlighting the various pathways through which this contamination occurs. Climate change amplifies the natural processes of saltwater intrusion, driven by rising sea levels and the mobilization of geogenic contaminants like arsenic and fluoride. These contaminants pose severe health risks, including various cancers, dental and skeletal fluorosis, and neurological disorders. Additionally, human activities like over-pumping, excessive fertilizer use, mining, and improper waste disposal further contribute to groundwater pollution, intensified by climate change-induced variations in rainfall patterns, temperature fluctuations, and extreme weather events. This chapter has presented compelling case studies from diverse regions across India, illustrating the far-reaching consequences of groundwater contamination on human health. The employed health risk assessment framework in these studies has allowed for quantifying both carcinogenic and non-carcinogenic risks linked to exposure to contaminated groundwater. The findings underscore the urgency of addressing this issue and implementing proactive measures to safeguard groundwater sources and protect human health. A multifaceted approach is imperative to mitigate the escalating risks to groundwater condition in the era of climate change. This includes:

- **Sustainable Water Resource Management:** Implementing strategies that promote water conservation, efficient irrigation practices, and responsible waste management is crucial in minimizing the anthropogenic contribution to groundwater contamination.
- **Advanced Monitoring Systems:** Developing and deploying sophisticated monitoring networks that can track changes in groundwater quality, identify contamination sources, and provide early warning signals are essential for timely intervention and remediation.
- **Innovative Treatment Technologies:** To guarantee access to safe drinking water, resources must be dedicated to research and development that creates innovative and economically viable technologies for removing groundwater contaminants.
- **Public Awareness and Education:** Raising community awareness about the impacts of climate change on groundwater quality and promoting responsible water use practices are essential for fostering a collective effort to protect this valuable resource.
- **Policy and Regulatory Frameworks:** Strengthening policy and regulatory frameworks that address groundwater contamination, promote sustainable water use, and incentivize the adoption of best practices is crucial for long-term groundwater protection.

7 Specific Policy Recommendations

- **Water Pricing:** Implementing tiered water pricing structures that incentivize efficient water use and discourage wastage.
- **Agricultural Subsidies:** Reforming agricultural subsidies to promote sustainable farming practices that reduce groundwater pollution.
- **Industrial Regulations:** Enforcing stricter environmental regulations on industries to minimize their discharge of pollutants into groundwater.
- **Land Use Planning:** Developing comprehensive land use plans that protect groundwater recharge areas and prevent contamination from hazardous activities.
- **Community-Based Groundwater Management:** Empowering local communities to participate in groundwater management decisions and implement sustainable practices.
- **International Collaboration:** Fostering international collaboration and knowledge sharing on groundwater contamination research, monitoring, and remediation strategies can accelerate progress toward sustainable solutions.

By adopting a holistic and integrated approach encompassing these recommendations, including specific policy interventions, we can safeguard groundwater resources, protect human health, and ensure the availability of clean water for current and future generations. The challenges posed by climate change-induced groundwater contamination are substantial. However, with concerted efforts and a commitment to sustainable practices, we can mitigate these risks and secure a water-secure future for all.

References

Abbas, G., Murtaza, B., Bibi, I., Shahid, M., Niazi, N. K., Khan, M. I., Amjad, M., Hussain, M., & Natasha. (2018). Arsenic uptake, toxicity, detoxification, and speciation in plants: Physiological, biochemical, and molecular aspects. *International Journal of Environmental Research and Public Health, 15*(1), 59. https://doi.org/10.3390/ijerph15010059

Abedin, M. A., Habiba, U., & Shaw, R. (2014). Community perception and adaptation to safe drinking water scarcity: Salinity, arsenic, and drought risks in coastal Bangladesh. *International Journal of Disaster Risk Science, 5*, 110–124. https://doi.org/10.1007/s13753-014-0021-6

Adelodun, B., Ajibade, F. O., Ighalo, J. O., Odey, G., Ibrahim, R. G., Kareem, K. Y., Bakare, H. O., Tiamiyu, A. O., Ajibade, T. F., Abdulkadir, T. S., Adeniran, K. A., & Choi, K. S. (2021). Assessment of socioeconomic inequality based on virus-contaminated water usage in developing countries: A review. *Environmental Research, 192*, 110309. https://doi.org/10.1016/j.envres.2020.110309

Akhtar, N., Syakir Ishak, M. I., Bhawani, S. A., & Umar, K. (2021). Various natural and anthropogenic factors responsible for water quality degradation: A review. *Water, 13*(19), 2660. https://doi.org/10.3390/w13192660

Ali, S., Thakur, S. K., Sarkar, A., & Shekhar, S. (2016). Worldwide contamination of water by fluoride. *Environmental Chemistry Letters, 14*(3), 291–315. https://doi.org/10.1007/s10311-016-0563-5

Ali, S., Fakhri, Y., Golbini, M., Thakur, S. K., Alinejad, A., Parseh, I., Shekhar, S., & Bhattacharya, P. (2019). Concentration of fluoride in groundwater of India: A systematic review, meta-analysis and risk assessment. *Groundwater for Sustainable Development, 9*, 100224. https://doi.org/10.1016/j.gsd.2019.100224

Al-Sudani, H. I. Z. (2019). A review on groundwater pollution. *International Journal of Recent Engineering Science (IJRES), 6*(5), 14–22.

Amini, M., Abbaspour, K. C., Berg, M., Winkel, L., Hug, S. J., Hoehn, E., Yang, H., & Johnson, C. A. (2008). Statistical modeling of global geogenic arsenic contamination in groundwater. *Environmental Science & Technology, 42*(10), 3669–3675. https://doi.org/10.1021/es702859e

Baboo, H., Patel, T., Faldu, R., Shah, M., & Shah, H. (2022). A comprehensive and systematic study of fluoride and arsenic contamination and its impacts in India. *Sustainable Water Resources Management, 8*(4), 122. https://doi.org/10.1007/s40899-022-00688-z

Baciocchi, R., Berardi, S., & Verginelli, I. (2010). Human health risk assessment: Models for predicting the effective exposure duration of on-site receptors exposed to contaminated groundwater. *Journal of Hazardous Materials, 181*(1), 226–233. https://doi.org/10.1016/j.jhazmat.2010.05.001

Banerjee, C., & Sharma, A. (2021). Decline in terrestrial water recharge with increasing global temperatures. *Science of the Total Environment, 764*, 142913. https://doi.org/10.1016/j.scitotenv.2020.142913

Baron, J. S., Hartman, M. D., Kittel, T. G., Band, L. E., Ojima, D. S., & Lammers, R. B. (1998). Effects of land cover, water redistribution, and temperature on ecosystem processes in the South Platte Basin. *Ecological Applications, 8*(4), 1037–1051. https://doi.org/10.1890/1051-0761(1998)008

Basu, A., Saha, D., Saha, R., Ghosh, T., & Saha, B. (2014). A review on sources, toxicity and remediation technologies for removing arsenic from drinking water. *Research on Chemical Intermediates, 40*(2), 447–485. https://doi.org/10.1007/s11164-012-1000-4

Bettencourt, L. M. A., Lobo, J., Helbing, D., Kühnert, C., & West, G. B. (2007). Growth, innovation, scaling, and the pace of life in cities. *Proceedings of the National Academy of Sciences, 104*(17), 7301–7306. https://doi.org/10.1073/pnas.0610172104

Bhattacharya, A. K. (2010). Artificial ground water recharge with a special reference to India. *International Journal of Research Review of Applied Sciences, 4*(2), 214–221.

Bhowmick, G.D., Dhar, D., Nath, D., Ghangrekar, M.M., Banerjee, R., Das, S. and Chatterjee, J. (2020). Coronavirus disease 2019 (COVID-19) outbreak: some serious consequences with urban and rural water cycle. NPJ Clean Water, *3*(1), 32. https://doi.org/10.1038/s41545-020-0079-1

Bolan, S., Padhye, L. P., Jasemizad, T., Govarthanan, M., Karmegam, N., Wijesekara, H., Amarasiri, D., Hou, D., Zhou, P., Biswal, B. K., Balasubramanian, R., & Bolan, N. (2023). Impacts of climate change on the fate of contaminants through extreme weather events. *Science of the Total Environment, 168388*. https://doi.org/10.1016/j.scitotenv.2023.168388

Carrard, N., Foster, T., & Willetts, J. (2019). Groundwater as a source of drinking water in Southeast Asia and the Pacific: A multi-country review of current reliance and resource concerns. *Water, 11*(8), 1605. https://doi.org/10.3390/w11081605

Cavé, L., Beekman, H. E., & Weaver, J. (2003). Impact of climate change on groundwater recharge estimation. In *Groundwater recharge estimation in Southern Africa* (pp. 189–197). UNESCO.

Chakraborti, D., Rahman, M. M., Mukherjee, A., Alauddin, M., Hassan, M., Dutta, R. N., Pati, S., Mukherjee, S. C., Roy, S., Quamruzzman, Q., Rahman, M., Morshed, S., Islam, T., Sorif, S., Selim, M., Islam, M. R., & Hossain, M. M. (2015). Groundwater arsenic contamination in Bangladesh—21 years of research. *Journal of Trace Elements in Medicine and Biology, 31*, 237–248. https://doi.org/10.1016/j.jtemb.2015.01.003

Chakraborti, D., Rahman, M. M., Das, B., Chatterjee, A., Das, D., Nayak, B.,... & Kumar, M. (2017). Groundwater arsenic contamination and its health effects in India. Hydrogeology Journal, 25(4), 1165–1181. https://doi.org/10.1007/s10040-017-1556-6

Choubisa, S. L., Choubisa, D., & Choubisa, A. (2023). Fluoride contamination of groundwater and its threat to health of villagers and their domestic animals and agriculture crops in rural Rajasthan, India. *Environmental Geochemistry and Health, 45*(3), 607–628. https://doi.org/10.1007/s10653-022-01267-z

Crosbie, R. S., Scanlon, B. R., Mpelasoka, F. S., Reedy, R. C., Gates, J. B., & Zhang, L. (2013). Potential climate change effects on groundwater recharge in the High Plains Aquifer, USA. *Water Resources Research, 49*(7), 3936–3951. https://doi.org/10.1002/wrcr.20292

Cuevas, J., Ruiz, A. I., de Soto, I. S., Sevilla, T., Procopio, J. R., Da Silva, P., Gismera, M. J., Regadío, M., Sánchez Jiménez, N., Rodríguez Rastrero, M., & Leguey, S. (2012). The performance of natural clay as a barrier to the diffusion of municipal solid waste landfill leachates. *Journal of Environmental Management, 95*, S175–S181. https://doi.org/10.1016/j.jenvman.2011.02.014

Danaraj, J., Ushani, U., Packiavathy, S. V., Dharmadhas, J. S., Karuppiah, T., Anandha Kumar, S., & Aooj, E. S. (2022). Climate change impacts of nitrate contamination on human health. In *Climate change impact on groundwater resources: Human health risk assessment in arid and semi-arid regions* (pp. 257–278). Springer. https://doi.org/10.1007/978-3-031-04707-7_14

Dao, P. U., Heuzard, A. G., Le, T. X. H., Zhao, J., Yin, R., Shang, C., & Fan, C. (2024). The impacts of climate change on groundwater quality: A review. *Science of the Total Environment, 912*, 169241. https://doi.org/10.1016/j.scitotenv.2023.169241

Das, A., Das, A., Mukherjee, M., Das, B., Mukherjee, S. C., Pati, S., Dutta, R. N., Quamruzzaman, Q., Saha, K. C., & Rahman, M. M. (2020). Status of groundwater arsenic contamination in the gmb plain. In *Managing water resources and hydrological systems* (2nd ed.). CRC Press.

Dass, A., Mishra, A. K., Santos, G. A. D. A., & Ranjan, R. K. (2024). Spatio-temporal variation of atmospheric CO2 and its association with anthropogenic, vegetation, and climate indices over the state of Bihar, India. *Environmental Advances, 16*, 100513. https://doi.org/10.1016/j.envadv.2024.100513

DeNicola, E., Aburizaiza, O. S., Siddique, A., Khwaja, H., & Carpenter, D. O. (2015). Climate change and water scarcity: The case of Saudi Arabia. *Annals of Global Health, 81*(3), 342–353. https://doi.org/10.1016/j.aogh.2015.08.005

Ding, G., Chen, G., Liu, Y., Li, M., & Liu, X. (2020). Occurrence and risk assessment of fluoroquinolone antibiotics in reclaimed water and receiving groundwater with different replenishment pathways. *Science of the Total Environment, 738*, 139802. https://doi.org/10.1016/j.scitotenv.2020.139802

Döll, P. (2009). Vulnerability to the impact of climate change on renewable groundwater resources: A global-scale assessment. *Environmental Research Letters, 4*(3), 035006. https://doi.org/10.1088/1748-9326/4/3/035006

du Plessis, A. (2019). Primary water quality challenges, contaminants and the world's dirtiest places. In A. du Plessis (Ed.), *Water as an inescapable risk: Current global water availability, quality and risks with a specific focus on South Africa* (pp. 79–114). Springer. https://doi.org/10.1007/978-3-030-03186-2_5

du Plessis, A., & du Plessis, A. (2019). Climate change: Current drivers, observations and impacts on the Globe's natural and human systems. In *Water as an inescapable risk: Current global water availability, quality and risks with a specific focus on South Africa* (pp. 27–53). Springer. https://doi.org/10.1007/978-3-030-03186-2_3

Du, Z., Li, H., Nie, L., Yao, Z., Zhang, X., Liu, Y., & Chen, S. (2024). High-solution emission characters and health risks of volatile organic compounds for sprayers in automobile repair industries. *Environmental Science and Pollution Research, 31*(15), 22679–22693. https://doi.org/10.1007/s11356-024-32478-9

Elert, M., Bonnard, R., Jones, C., Schoof, R. A., & Swartjes, F. A. (2011). Human exposure pathways. In *Dealing with contaminated sites: from theory towards practical application* (pp. 455–515). Springer. https://doi.org/10.1007/978-90-481-9757-6_11

El-Rawy, M., Wahba, M., & Fathi, H. (2024). Rainwater harvesting for managed aquifer recharge and flood mitigation in the MENA region. In *Managed aquifer recharge in MENA countries: Developments, applications, challenges, strategies, and sustainability* (pp. 47–72). Springer. https://doi.org/10.1007/978-3-031-58764-1_3

Fioreze, M., & Mancuso, M. A. (2019). MODFLOW and MODPATH for hydrodynamic simulation of porous media in horizontal subsurface flow constructed wetlands: A tool for design criteria. *Ecological Engineering, 130*, 45–52. https://doi.org/10.1016/j.ecoleng.2019.01.012

Gallina, V., Torresan, S., Critto, A., Sperotto, A., Glade, T., & Marcomini, A. (2016). A review of multi-risk methodologies for natural hazards: Consequences and challenges for a climate change impact assessment. *Journal of Environmental Management, 168*, 123–132. https://doi.org/10.1016/j.jenvman.2015.11.011

Gandy, C. J., & Younger, P. L. (2007). Predicting groundwater rebound in the South Yorkshire Coalfield, UK.

Mine Water and the Environment, 26(2), 70–78. https://doi.org/10.1007/s10230-007-0153-7

Ghosh, A., & Bose, N. (2016). The clean water problem—Example of arsenic contamination in the Gangetic Plain of India. In *Innovative materials and methods for water treatment*. CRC Press.

Goderniaux, P., Brouyère, S., Fowler, H. J., Blenkinsop, S., Therrien, R., Orban, P., & Dassargues, A. (2009). Large scale surface–subsurface hydrological model to assess climate change impacts on groundwater reserves. *Journal of Hydrology, 373*(1–2, 122), –138. https://doi.org/10.1016/j.jhydrol.2009.04.017

Gorelick, S. M., & Zheng, C. (2015). Global change and the groundwater management challenge. *Water Resources Research, 51*(5), 3031–3051. https://doi.org/10.1002/2014WR016825

Green, T. R., Taniguchi, M., Kooi, H., Gurdak, J. J., Allen, D. M., Hiscock, K. M., Treidel, H., & Aureli, A. (2011). Beneath the surface of global change: Impacts of climate change on groundwater. *Journal of Hydrology, 405*(3), 532–560. https://doi.org/10.1016/j.jhydrol.2011.05.002

Greening, G. E., & Cannon, J. L. (2016). Human and animal viruses in food (Including taxonomy of enteric viruses). In S. M. Goyal & J. L. Cannon (Eds.), *Viruses in foods* (pp. 5–57). Springer. https://doi.org/10.1007/978-3-319-30723-7_2

Grellier, J., Rushton, L., Briggs, D. J., & Nieuwenhuijsen, M. J. (2015). Assessing the human health impacts of exposure to disinfection by-products—A critical review of concepts and methods. *Environment International, 78*, 61–81. https://doi.org/10.1016/j.envint.2015.02.003

Guo, H., Li, M., Wang, L., Wang, Y., Zang, X., Zhao, X., Wang, H., & Zhu, J. (2021). Evaluation of groundwater suitability for irrigation and drinking purposes in an agricultural region of the North China Plain. *Water, 13*(23), 3426. https://doi.org/10.3390/w13233426

Gwenzi, W., Mangori, L., Danha, C., Chaukura, N., Dunjana, N., & Sanganyado, E. (2018). Sources, behaviour, and environmental and human health risks of high-technology rare earth elements as emerging contaminants. *Science of the Total Environment, 636*, 299–313. https://doi.org/10.1016/j.scitotenv.2018.04.235

Han, D., Currell, M. J., Cao, G., & Hall, B. (2017). Alterations to groundwater recharge due to anthropogenic landscape change. *Journal of Hydrology, 554*, 545–557.

Harwood, M., Danielewska-Nikiel, B., Borzelleca, J. F., Flamm, G. W., Williams, G. M., & Lines, T. C. (2007). A critical review of the data related to the safety of quercetin and lack of evidence of *in vivo* toxicity, including lack of genotoxic/carcinogenic properties. *Food and Chemical Toxicology, 45*(11), 2179–2205. https://doi.org/10.1016/j.fct.2007.05.015

Hashim, M. A., Mukhopadhyay, S., Sahu, J. N., & Sengupta, B. (2011). Remediation technologies for heavy metal contaminated groundwater. *Journal of Environmental Management, 92*(10), 2355–2388. https://doi.org/10.1016/j.jenvman.2011.06.009

Henao Casas, J. D. (2023). *Managed aquifer recharge as a low-regret measure for climate change adaptation*. Phd, E.T.S. de Ingeniería Agronómica, Alimentaria y de Biosistemas (UPM). https://oa.upm.es/73999/

Hernández, A. F., & Tsatsakis, A. M. (2017). Human exposure to chemical mixtures: Challenges for the integration of toxicology with epidemiology data in risk assessment. *Food and Chemical Toxicology, 103*, 188–193. https://doi.org/10.1016/j.fct.2017.03.012

Hoegh Guldberg, O., Jacob, D., Taylor, M., Bindi, M., Brown, S., Camilloni, I. A., Djalante R, Ebi KL, Engelbrecht F., & Zhou, G. (2018). *Impacts of 1.5 C global warming on natural and human systems*. ISBN: 978-92-9169-151-7.

Jain, A., Xu, X., & Hewitt, N. (2020). Global warming and climate change science. In *Atmospheric science for environmental scientists* (Vol. 140, p. 367). Springer.

Jasechko, S., Birks, S. J., Gleeson, T., Wada, Y., Fawcett, P. J., Sharp, Z. D., McDonnell, J. J., & Welker, J. M. (2014). The pronounced seasonality of global groundwater recharge. *Water Resources Research, 50*(11), 8845–8867. https://doi.org/10.1002/2014WR015809

Jurado, A., Vàzquez-Suñé, E., Carrera, J., López de Alda, M., Pujades, E., & Barceló, D. (2012). Emerging organic contaminants in groundwater in Spain: A review of sources, recent occurrence and fate in a European context. *Science of the Total Environment, 440*, 82–94. https://doi.org/10.1016/j.scitotenv.2012.08.029

Jyrkama, M. I., & Sykes, J. F. (2007). The impact of climate change on spatially varying groundwater recharge in the grand river watershed (Ontario). *Journal of Hydrology, 338*(3), 237–250. https://doi.org/10.1016/j.jhydrol.2007.02.036

Kadlec, R. H. (2006). Water temperature and evapotranspiration in surface flow wetlands in hot arid climate. *Ecological Engineering, 26*(4), 328–340. https://doi.org/10.1016/j.ecoleng.2005.12.010

Karri, R. R., Ravindran, G., & Dehghani, M. H. (2021). Wastewater—Sources, toxicity, and their consequences to human health. In *Soft computing techniques in solid waste and wastewater management* (pp. 3–33). Elsevier. https://doi.org/10.1016/B978-0-12-824463-0.00001-X

Karunanidhi, D., Aravinthasamy, P., Roy, P., Subramani, T., & Chandra Jayasena, H. (2024). Nitrate contamination in groundwater and its evaluation of non-carcinogenic health hazards from Arjunanadi River basin, South India. *Groundwater for Sustainable Development, 25*, 101153. https://doi.org/10.1016/j.gsd.2024.101153

Kashyap, P., Kaushal, J., Mahajan, P., & John, P. (2022). Critical analysis of fluoride contaminated water in the Malwa belt of Punjab and its impact on health. *AIP Conference Proceedings, 2357*, 030018. https://doi.org/10.1063/5.0080618

Kerényi, A., McIntosh, R. W., Kerényi, A., & McIntosh, R. W. (2020). Changes on earth as a result of interaction between the society and nature. In *Sustainable development in changing complex earth systems* (pp. 75–202). Springer. https://doi.org/10.1007/978-3-030-21645-0_4

Khan, M. N., Mobin, M., Abbas, Z. K., & Alamri, S. A. (2018). Fertilizers and their contaminants in soils, surface and groundwater. *Encyclopedia of the Anthropocene, 5*, 225–240.

Kløve, B., Ala-Aho, P., Bertrand, G., Gurdak, J. J., Kupfersberger, H., Kværner, J., Muotka, T., Mykrä, H., Preda, E., Rossi, P., Uvo, C. B., Velasco, E., & Pulido-Velazquez, M. (2014). Climate change impacts on groundwater and dependent ecosystems. *Journal of Hydrology, 518*, 250–266. https://doi.org/10.1016/j.jhydrol.2013.06.037

Klump, S., Kipfer, R., Cirpka, O. A., Harvey, C. F., Brennwald, M. S., Ashfaque, K. N., Badruzzaman, A. B. M., Hug, S. J., & Imboden, D. M. (2006). Groundwater dynamics and arsenic mobilization in Bangladesh assessed using noble gases and tritium. *Environmental Science & Technology, 40*(1), 243–250. https://doi.org/10.1021/es051284w

Krishan, G., Rao, R. S. M., Gupta, S., & Tiwari, P. K. (2015). Fluoride, iron and nitrate affected areas of Punjab. *Suresh Gyan Vihar University Journal of Climate Change and Water, 1*(1), 1–5.

Kubicz, J., Lochyński, P., Pawełczyk, A., & Karczewski, M. (2021). Effects of drought on environmental health risk posed by groundwater contamination. *Chemosphere, 263*, 128145. https://doi.org/10.1016/j.chemosphere.2020.128145

Kumar, C. P. (2012). Climate change and its impact on groundwater resources. *International Journal of Engineering and Science, 1*(5), 43–60. ISSN: 2278-4721.

Kumar, R., Mittal, S., Peechat, S., Sahoo, P. K., & Sahoo, S. K. (2020). Quantification of groundwater–agricultural soil quality and associated health risks in the agri-intensive Sutlej River Basin of Punjab, India. *Environmental Geochemistry and Health, 42*(12), 4245–4268. https://doi.org/10.1007/s10653-020-00636-w

Kundzewicz, Z. W. (2007). Global freshwater resources for sustainable development. *Ecohydrology & Hydrobiology, 7*(2), 125–134. https://doi.org/10.1016/S1642-3593(07)70178-7

Kundzewicz, Z. W. (2008). Climate change impacts on the hydrological cycle. *Ecohydrology & Hydrobiology, 8*(2–4), 195–203. https://doi.org/10.2478/v10104-009-0015-y

Kundzewicz, Z. W., & Döll, P. (2009). Will groundwater ease freshwater stress under climate change? *Hydrological Sciences Journal, 54*(4), 665–675. https://doi.org/10.1623/hysj.54.4.665

Kurtz, J. B., & Lee, T. W. (2007). Astroviruses: Human and animal. In G. Bock & J. Whelan (Eds.), *Novartis Foundation symposia* (1st ed., pp. 92–107). Wiley. https://doi.org/10.1002/9780470513460.ch6

Lapworth, D. J., MacDonald, A. M., Krishan, G., Rao, M. S., Gooddy, D. C., & Darling, W. G. (2015). Groundwater recharge and age-depth profiles of intensively exploited groundwater resources in Northwest India. *Geophysical Research Letters, 42*(18), 7554–7562. https://doi.org/10.1002/2015GL065798

Lapworth, D. J., Krishan, G., MacDonald, A. M., & Rao, M. S. (2017). Groundwater quality in the alluvial aquifer system of Northwest India: New evidence of the extent of anthropogenic and geogenic contamination. *Science of the Total Environment, 599–600*, 1433–1444. https://doi.org/10.1016/j.scitotenv.2017.04.223

Li, Q., Mahendra, S., Lyon, D. Y., Brunet, L., Liga, M. V., Li, D., & Alvarez, P. J. J. (2008). Antimicrobial nanomaterials for water disinfection and microbial control: Potential applications and implications. *Water Research, 42*(18), 4591–4602. https://doi.org/10.1016/j.watres.2008.08.015

Lipczynska-Kochany, E. (2018). Effect of climate change on humic substances and associated impacts on the quality of surface water and groundwater: A review. *Science of the Total Environment, 640*, 1548–1565. https://doi.org/10.1016/j.scitotenv.2018.05.376

Lokhandwala, S., & Gautam, P. (2020). Indirect impact of COVID-19 on environment: A brief study in Indian context. *Environmental Research, 188*, 109807. https://doi.org/10.1016/j.envres.2020.109807

Makri, A., & Stilianakis, N. I. (2008). Vulnerability to air pollution health effects. *International Journal of Hygiene and Environmental Health, 211*(3–4), 326–336. https://doi.org/10.1016/j.ijheh.2007.06.005

Maleksaeidi, H., Keshavarz, M., Karami, E., & Eslamian, S. (2017). Climate change and drought: Building resilience for an unpredictable future. In *Handbook of drought and water scarcity* (pp. 163–186). CRC Press. ISBN: 9781315226781.

McCarty, P. L. (2010). Groundwater contamination by chlorinated solvents: History, remediation technologies and strategies. In H. F. Stroo & C. H. Ward (Eds.), *In situ remediation of chlorinated solvent plumes* (pp. 1–28). Springer. https://doi.org/10.1007/978-1-4419-1401-9_1

Men, S.-H., Xie, X., Zhao, X., Zhou, Q., Chen, J.-Y., Jiao, C.-Y., & Yan, Z.-G. (2023). The application of reference dose prediction model to human health water quality criteria and risk assessment. *Toxics, 11*(4), 318. https://doi.org/10.3390/toxics11040318

Michael, H. A., Russoniello, C. J., & Byron, L. A. (2013). Global assessment of vulnerability to sea-level rise in topography-limited and recharge-limited coastal groundwater systems. *Water Resources Research, 49*(4), 2228–2240. https://doi.org/10.1002/wrcr.20213

Miranzadeh Mahabadi, H., Ramroudi, M., Asgharipour, M. R., Rahmani, H. R., & Afyuni, M. (2020). Assessment of heavy metals contamination and the risk of target hazard quotient in some vegetables in Isfahan. *Pollution, 6*(1), 69–78.

Mishra, R. K. (2023). Fresh water availability and its global challenge. *British Journal of Multidisciplinary and Advanced Studies, 4*(3), 1–78. https://doi.org/10.37745/bjmas.2022.0208

Mohanavelu, A., Naganna, S. R., & Al-Ansari, N. (2021). Irrigation induced salinity and sodicity hazards on soil and groundwater: An overview of its causes, impacts and mitigation strategies. *Agriculture, 11*(10), 983. https://doi.org/10.3390/agriculture11100983

Morgan, H., Hussien, H. M., Madani, A., & Nassar, T. (2022). Delineating groundwater potential zones in hyper-arid regions using the applications of remote sensing and GIS modeling in the eastern desert, Egypt. *Sustainability, 14*(24), 16942. https://doi.org/10.3390/su142416942

Mridha, D., Priyadarshni, P., Bhaskar, K., Gaurav, A., De, A., Das, A., Joardar, M., Chowdhury, N. R., & Roychowdhury, T. (2021). Fluoride exposure and its potential health risk assessment in drinking water and staple food in the population from fluoride endemic regions of Bihar, India. *Groundwater for Sustainable Development, 13*, 100558. https://doi.org/10.1016/j.gsd.2021.100558

Mukherjee, A., Scanlon, B. R., Aureli, A., Langan, S., Guo, H., & McKenzie, A. (2021). Global groundwater: From scarcity to security through sustainability and solutions. In *Global groundwater* (pp. 3–20). Elsevier. https://doi.org/10.1016/B978-0-12-818172-0.00001-3

Mukherjee, A., Coomar, P., Sarkar, S., Johannesson, K. H., Fryar, A. E., Schreiber, M. E., Ahmed, K. M., Alam, M. A., Bhattacharya, P., Bundschuh, J., Burgess, W., Chakraborty, M., Coyte, R., Farooqi, A., Guo, H., Ijumulana, J., Jeelani, G., Mondal, D., Nordstrom, D. K., et al. (2024). Arsenic and other geogenic contaminants in global groundwater. *Nature Reviews Earth & Environment, 5*(4), 312–328. https://doi.org/10.1038/s43017-024-00519-z

National Research Council, Division on Earth and Life Studies, Commission on Geosciences, Environment and Resources, & Committee on Technologies for Cleanup of Subsurface Contaminants in the DOE Weapons Complex. (1999). *Groundwater and soil cleanup: Improving management of persistent contaminants*. National Academies Press.

Neidhardt, H., & Shao, W. (2023). Impact of climate change-induced warming on groundwater temperatures and quality. *Applied Water Science, 13*(12), 235. https://doi.org/10.1007/s13201-023-02039-5

Nghiem, A. A., Stahl, M. O., Mailloux, B. J., Mai, T. T., Trang, P. T., Viet, P. H., Harvey, C. F., Van Geen, A., & Bostick, B. C. (2019). Quantifying riverine recharge impacts on redox conditions and arsenic release in groundwater aquifers along the red river, Vietnam. *Water Resources Research, 55*(8), 6712–6728. https://doi.org/10.1029/2019WR024816

Ngo, T.-M.-L., Wang, S.-J., & Chen, P.-Y. (2024). Assessment of future climate change impacts on groundwater recharge using hydrological modeling in the choushui river alluvial fan, Taiwan. *Water, 16*(3), 419. https://doi.org/10.3390/w16030419

Nicholls, R. J., & Cazenave, A. (2010). Sea-level rise and its impact on coastal zones. *Science, 328*(5985), 1517–1520. https://doi.org/10.1126/science.1185782

Nordberg, G. F., & Fowler, B. A. (2015). Risk assessment. In *Handbook on the toxicology of metals* (pp. 461–486). Elsevier. https://doi.org/10.1016/B978-0-444-59453-2.00022-6

Okkonen, J., & Kløve, B. (2010). A conceptual and statistical approach for the analysis of climate impact on ground water table fluctuation patterns in cold conditions. *Journal of Hydrology, 388*(1), 1–12. https://doi.org/10.1016/j.jhydrol.2010.02.015

Ondrasek, G., & Rengel, Z. (2021). Environmental salinization processes: Detection, implications & solutions. *Science of the Total Environment, 754*, 142432. https://doi.org/10.1016/j.scitotenv.2020.142432

Palma-Lara, I., Martínez-Castillo, M., Quintana-Pérez, J. C., Arellano-Mendoza, M. G., Tamay-Cach, F., Valenzuela-Limón, O. L., García-Montalvo, E. A., & Hernández-Zavala, A. (2020). Arsenic exposure: A public health problem leading to several cancers. *Regulatory Toxicology and Pharmacology, 110*, 104539. https://doi.org/10.1016/j.yrtph.2019.104539

Pandey, P., Shinde, V. N., Deopurkar, R. L., Kale, S. P., Patil, S. A., & Pant, D. (2016). Recent advances in the use of different substrates in microbial fuel cells toward wastewater treatment and simultaneous energy recovery. *Applied Energy, 168*, 706–723. https://doi.org/10.1016/j.apenergy.2016.01.056

Pozzebon, E. A., & Seifert, L. (2023). Emerging environmental health risks associated with the land application of biosolids: A scoping review. *Environmental Health, 22*(1), 57. https://doi.org/10.1186/s12940-023-01008-4

Priyan, K. (2021). Issues and challenges of groundwater and surface water management in semi-arid regions. In C. B. Pande & K. N. Moharir (Eds.), *Groundwater resources development and planning in the semi-arid region* (pp. 1–17). Springer. https://doi.org/10.1007/978-3-030-68124-1_1

Rajan, M., Karunanidhi, D., Jaya, J., Preethi, B., Subramani, T., & Aravinthasamy, P. (2024). A comprehensive review on human health hazards due to groundwater contamination: A global perspective. *Physics and Chemistry of the Earth, Parts A/B/C, 135*, 103637. https://doi.org/10.1016/j.pce.2024.103637

Rasmussen, P., Sonnenborg, T. O., Goncear, G., & Hinsby, K. (2013). Assessing impacts of climate change, sea level rise, and drainage canals on saltwater intrusion to coastal aquifer. *Hydrology and Earth System Sciences, 17*(1), 421–443. https://doi.org/10.5194/hess-17-421-2013

Razowska, L. (2001). Changes of groundwater chemistry caused by the flooding of iron mines (Czestochowa Region, Southern Poland). *Journal of Hydrology, 244*(1), 17–32. https://doi.org/10.1016/S0022-1694(00)00420-0

Re, V., Sacchi, E., Kammoun, S., Tringali, C., Trabelsi, R., Zouari, K., & Daniele, S. (2017). Integrated socio-hydrogeological approach to tackle nitrate contamination in groundwater resources. The case of Grombalia Basin (Tunisia). *Science of the Total Environment, 593–594*, 664–676. https://doi.org/10.1016/j.scitotenv.2017.03.151

Rebelo, F. M., & Caldas, E. D. (2016). Arsenic, lead, mercury and cadmium: Toxicity, levels in breast milk and the risks for breastfed infants. *Environmental Research, 151*, 671–688. https://doi.org/10.1016/j.envres.2016.08.027

Ritter, K. S., Sibley, P., Hall, K., Keen, P., Mattu, G., & Beth Linton, L. (2002). Sources, pathways, and relative risks of contaminants in surface water and groundwater: A perspective prepared for the Walkerton inquiry. *Journal of Toxicology and Environmental Health Part A, 65*(1), 1–142. https://doi.org/10.1080/152873902753338572

Satterthwaite, D., McGranahan, G., & Tacoli, C. (2010). Urbanization and its implications for food and farming. *Philosophical Transactions of the Royal Society B: Biological Sciences, 365*(1554), 2809–2820. https://doi.org/10.1098/rstb.2010.0136

Schlüter, M., Sauter, E. J., Andersen, C. E., Dahlgaard, H., & Dando, P. R. (2004). Spatial distribution and budget for submarine groundwater discharge in Eckernförde Bay (Western baltic sea). *Limnology and Oceanography, 49*(1), 157–167. https://doi.org/10.4319/lo.2004.49.1.0157

Schulze, S., Zahn, D., Montes, R., Rodil, R., Quintana, J. B., Knepper, T. P., Reemtsma, T., & Berger, U. (2019). Occurrence of emerging persistent and mobile organic contaminants in European water samples. *Water Research, 153*, 80–90. https://doi.org/10.1016/j.watres.2019.01.008

Sharma, S. (2017). Effects of urbanization on water resources-facts and figures. *International Journal of Scientific and Engineering Research, 8*(4), 433–459.

Sharma, R., Kumar, R., Agrawal, P. R., & Gupta, G. (2021). Groundwater extractions and climate change. In *Water conservation in the era of global climate change* (pp. 23–45). Elsevier. https://doi.org/10.1016/B978-0-12-820200-5.00016-6

Sharma, K., Rajan, S., & Nayak, S. K. (2024). Chapter 1: Water pollution: Primary sources and associated human health hazards with special emphasis on rural areas. In S. Madhav, A. L. Srivastav, S. C. Izah, & E. van Hullebusch (Eds.), *Water resources management for rural development* (pp. 3–14). Elsevier. https://doi.org/10.1016/B978-0-443-18778-0.00014-3

Shevah, Y. (2015). Water resources, water scarcity challenges, and perspectives. In *Water challenges and solutions on a global scale* (pp. 185–219). American Chemical Society Publications. eISBN: 9780841231054.

Shrestha, U. B., Sharma, K. P., Devkota, A., Siwakoti, M., & Shrestha, B. B. (2018). Potential impact of climate change on the distribution of six invasive alien plants in Nepal. *Ecological Indicators, 95*, 99–107. https://doi.org/10.1016/j.ecolind.2018.07.009

Shukla, A., Awasthi, S., Chauhan, R., & Srivastava, S. (2020). The status of arsenic contamination in India. In S. Srivastava (Ed.), *Arsenic in drinking water and food* (pp. 1–12). Springer. https://doi.org/10.1007/978-981-13-8587-2_1

Singh, A., & Singh, J. (2021). Chapter 1—Effects on human health due to fluoride. In M. Hadi Dehghani, R. Karri, & E. Lima (Eds.), *Green technologies for the defluoridation of water* (pp. 1–16). Elsevier. https://doi.org/10.1016/B978-0-323-85768-0.00006-3

Singh, G., Kumari, B., Sinam, G., Kriti Kumar, N., & Mallick, S. (2018). Fluoride distribution and contamination in the water, soil and plants continuum and its remedial technologies, an Indian perspective– A review. *Environmental Pollution, 239*, 95–108. https://doi.org/10.1016/j.envpol.2018.04.002

Sinha Ray, S. P. (2024). High arsenic groundwater contamination issues, causes, and mitigation strategies with successful case histories. In S. P. Sinha Ray & A. Acharyya (Eds.), *Ground water contamination in India: Adverse effects on habitats* (pp. 325–340). Springer Nature. https://doi.org/10.1007/978-3-031-49092-7_18

Smith, A. E., Lincoln, R. A., Paulu, C., Simones, T. L., Caldwell, K. L., Jones, R. L., & Backer, L. C. (2016). Assessing arsenic exposure in households using bottled water or point-of-use treatment systems to mitigate well water contamination. *Science of the Total Environment, 544*, 701–710. https://doi.org/10.1016/j.scitotenv.2015.11.136

Subba Rao, N. (2021). Spatial distribution of quality of groundwater and probabilistic non- carcinogenic risk from a rural dry climatic region of South India. *Environmental Geochemistry and Health, 43*(2), 971–993. https://doi.org/10.1007/s10653-020-00621-3

Subba Rao, N., Ravindra, B., & Wu, J. (2020). Geochemical and health risk evaluation of fluoride rich groundwater in Sattenapalle Region, Guntur district, Andhra Pradesh, India. *Human and Ecological Risk Assessment: An International Journal, 26*(9), 2316–2348. https://doi.org/10.1080/10807039.2020.1741338

Swain, S., Taloor, A. K., Dhal, L., Sahoo, S., & Al-Ansari, N. (2022). Impact of climate change on groundwater hydrology: A comprehensive review and current status of the Indian hydrogeology. *Applied Water Science, 12*(6), 120. https://doi.org/10.1007/s13201-022-01652-0

Tatti, F., Petrangeli Papini, M., Torretta, V., Mancini, G., Boni, M. R., & Viotti, P. (2019). Experimental and numerical evaluation of Groundwater Circulation Wells as a remediation technology for persistent, low permeability contaminant source zones. *Journal of Contaminant Hydrology, 222*, 89–100. https://doi.org/10.1016/j.jconhyd.2019.03.001

Tsuji, J. S., Chang, E. T., Gentry, P. R., Clewell, H. J., Boffetta, P., & Cohen, S. M. (2019). Dose-response for assessing the cancer risk of inorganic arsenic in drinking water: The scientific basis for use of a threshold approach. *Critical Reviews in Toxicology, 49*(1),

36–84. https://doi.org/10.1080/10408444.2019.1573804

Valcke, M., Bourgault, M.-H., Rochette, L., Normandin, L., Samuel, O., Belleville, D., Blanchet, C., & Phaneuf, D. (2017). Human health risk assessment on the consumption of fruits and vegetables containing residual pesticides: A cancer and non-cancer risk/benefit perspective. *Environment International, 108*, 63–74. https://doi.org/10.1016/j.envint.2017.07.023

Valskys, V., Hassan, H. R., Wołkowicz, S., Satkūnas, J., Kibirkštis, G., & Ignatavičius, G. (2022). A review on detection techniques, health hazards and human health risk assessment of arsenic pollution in soil and groundwater. *Minerals, 12*(10), 1326. https://doi.org/10.3390/min12101326

Van Dam, R. A., Hogan, A. C., Harford, A. J., & Humphrey, C. L. (2019). How specific is site-specific? A review and guidance for selecting and evaluating approaches for deriving local water quality benchmarks. *Integrated Environmental Assessment and Management, 15*(5), 683–702. https://doi.org/10.1002/ieam.4181

Wahl, T., Jain, S., Bender, J., Meyers, S. D., & Luther, M. E. (2015). Increasing risk of compound flooding from storm surge and rainfall for major US cities. *Nature Climate Change, 5*(12), 1093–1097. https://doi.org/10.1038/nclimate2736

Wang, Y., Yuan, S., Shi, J., Ma, T., Xie, X., Deng, Y., Du, Y., Gan, Y., Guo, Z., Dong, Y., Zheng, C., & Jiang, G. (2023). Groundwater quality and health: Making the invisible visible. *Environmental Science & Technology, 57*(13), 5125–5136. https://doi.org/10.1021/acs.est.2c08061

Ward, C., & Ruckstuhl, S. (2017). *Water scarcity, climate change and conflict in the Middle East: Securing livelihoods, building peace*. Bloomsbury Publishing.

Williams, P. R. D., & Paustenbach, D. J. (2024). Risk characterization. In D. J. Paustenbach & K. Feinberg (Eds.), *Human and ecological risk assessment* (1st ed., pp. 263–331). Wiley. https://doi.org/10.1002/9781119742975.ch6

Xu, Z., Sheffield, P. E., Hu, W., Su, H., Yu, W., Qi, X., & Tong, S. (2012). Climate change and children's health—A call for research on what works to protect children. *International Journal of Environmental Research and Public Health, 9*(9), 3298–3316. https://doi.org/10.3390/ijerph9093298

Yadav, S., Bansal, S. K., Yadav, S., & Kumar, S. (2019). Fluoride distribution in underground water of district Mahendergarh, Haryana, India. *Applied Water Science, 9*(3), 62. https://doi.org/10.1007/s13201-019-0935-7

Yang, C., & Bertetti, F. P. (2023). Climate elasticity assessment on groundwater recharge to the Edwards Balcones Fault Zone Aquifer, United States. *JAWRA Journal of the American Water Resources Association, 59*(6), 1273–1286. https://doi.org/10.1111/1752-1688.13142

Yang Lesser, L. E., Mora, A., Moreau, C., Mahlknecht, J., Hernández-Antonio, A., Ramírez, A. I., & Barrios-Piña, H. (2018). Survey of 218 organic contaminants in groundwater derived from the world's largest untreated wastewater irrigation system: Mezquital Valley, Mexico. *Chemosphere, 198*, 510–521. https://doi.org/10.1016/j.chemosphere.2018.01.154

Yang, D., Yang, Y., & Xia, J. (2021). Hydrological cycle and water resources in a changing world: A review. *Geography and Sustainability, 2*(2), 115–122. https://doi.org/10.1016/j.geosus.2021.05.003

Zeidan, B. A. (2017). Groundwater degradation and remediation in the Nile Delta Aquifer. *The Nile Delta*, 159–232. https://doi.org/10.1007/698_2016_128

Zeydalinejad, N., & Nassery, H. R. (2023). A review on the climate-induced depletion of Iran's aquifers. *Stochastic Environmental Research and Risk Assessment, 37*(2), 467–490. https://doi.org/10.1007/s00477-022-02278-z

Zscheischler, J., Martius, O., Westra, S., Bevacqua, E., Raymond, C., Horton, R. M., van den Hurk, B., AghaKouchak, A., Jézéquel, A., Mahecha, M. D., & Maraun, D. (2020). A typology of compound weather and climate events. *Nature Reviews Earth & Environment, 1*(7), 333–347. https://doi.org/10.1038/s43017-020-0060-z

Climate Change and Sustainable Adaptation: An Analysis of the Effects of Rainfall Variability on Wetland Ecosystems in Madhubani, Bihar, India

Manu Raj Sharma, Sonu Kumar Das,
Anand Shankar, and
Vishwa Bandhu Singh Chandel

1 Introduction

Climate change is a comprehensive term that encompasses any alteration in the climate over any period, whether driven by natural or human-induced factors (Mitsch et al., 2013; Salimi et al., 2021). It can impact wetlands directly through effects such as rising temperatures and variations in the frequency and intensity of rainfall, and indirectly through extreme hydroclimatic conditions such as storms, flooding, and meteorological droughts (Birnbaum et al., 2021; Moomaw et al., 2018; O"Neill et al., 2020). Recognized as a significant threat to species diversity and ecosystem integrity worldwide (Erwin, 2009), climate change can significantly alter wetland hydrology and biogeochemistry affecting functions to the point where essential primary productivity may be disrupted (Salimi et al., 2021; Sharma & Priya, 2023).

Wetlands are eco-sensitive areas that are inundated with water permanently or seasonally (Balak Yadav, 2020; Keddy, 2010; Salimi et al., 2021). They are among the most valuable natural resources in the world and are recognized as significant ecosystems in terms of biological diversity and nutrient cycling (Bullock & Acreman, 2003; Junk et al., 2006; Salimi et al., 2021). These ecosystems are regarded as the most productive human heritage in the world and offer a great variety of habitats (Chenchouni et al., 2015). Humans benefit directly from wetland ecosystems in addition to indirect gains on an economic level (Mitsch & Gosselink, 2000; Pasupalati et al., 2017). They are vital to the process of climate change because of their ability to regulate concentrations of greenhouse gases including methane, nitrous oxide, sulfur dioxide, and carbon dioxide, which in turn can escalate the effects of global warming (IPCC AR-6, 2022). Wetlands are capable of actively storing carbon as biomass or rich organic matter in the soil through the Calvin cycle of photosynthesis. The soggy conditions are characteristics of wetlands, where organic synthesis occurs more rapidly than slow decomposition. This anoxic environment leads to significant carbon accumulation, making wetlands one of the most

M. R. Sharma (✉) · S. K. Das
University Department of Geography, Advanced Research Centre, L. N. Mithila University, Darbhanga, Bihar, India

A. Shankar
India Meteorological Department, Ministry of Earth Sciences, Government of India, Patna, India

V. B. S. Chandel
Department of Geography, Panjab University, Chandigarh, Punjab, India

Z. Khanam et al. (eds.), *Climate Change, Resilience and Ecosystem Services*, Sustainable Development Goals Series, https://doi.org/10.1007/978-3-031-96977-5_3

effective natural carbon sinks (Mitsch & Gosselink, 2000; Salimi et al., 2021; Were et al., 2019).

As a great technological tool for evaluating the Earth's surface, remote sensing has made it possible to discern, characterize, and map the distribution of wetlands at a variety of spatial, temporal, and spectral resolutions. These tools are reliable, efficient, and relatively cheaper. It supports comprehensive and detailed assessments of wetlands, enhancing our ability to study and manage these critical ecosystems (Alawadi et al., 2023; Tiner et al., 2015; Rebelo et al., 2009). When considering wide geographic regions, these methods are frequently less expensive and quicker than traditional field mapping (Kaplan & Avdan, 2017). Different methods and approaches have been used in wetland monitoring such as the Normalized Difference Vegetation Index (NDVI) for vegetation (Debanshi & Pal, 2020; Rousse Jr et al., 1973), and the Normalized Difference Water Index (NDWI) for soil wetness (Bian et al., 2021; McFeeters, 1996). Multispectral sensors (MSS) acquire data with 30–60-meter resolution in the visible and near-infrared (NIR), which makes them suitable for mapping and monitoring water and vegetation (Kaplan & Avdan, 2017; Merwade, 2019; Rebelo et al., 2009). Sentinel-2 provides 10–60-meter resolution satellite imagery (Dehni & Lounis, 2012; Kaplan & Avdan, 2017; Pradeepraju et al., 2022), which contains three Vegetation Red Edge bands with a 15-meter spatial resolution, better spectral resolution in the near-infrared, and better spatial resolution than the most recent Landsat series. Many recent studies utilized a ratio index called "Modified Normalized Difference Water Index" (MNDWI), to map open water areas in the study region (Laonamsai et al., 2023; Pernollet et al., 2015; Szabó et al., 2016). Wetland monitoring has become more accurate and efficient because of the ensembling of remote sensing technologies and machine learning algorithms, which have provided better conservation and management strategies (Aslam et al., 2024; Peng et al., 2023).

Since drought monitoring is linked to regular water availability and helps prevent the degradation of these essential habitats, it is crucial for the sustainable maintenance of wetland ecosystems. To evaluate the sustainability of wetlands, a review of the conceptual framework of droughts, their trends, and the methodology of modeling techniques, such as the use of hydro-meteorological variables, drought indices, and climatic indices, is crucial (Halder et al., 2022; Jiang et al., 2019; Mishra & Singh, 2011). For long-trend detection of yearly rainfall, the most often used non-parametric methods are Sen's slope estimator and the Mann–Kendall (MK) trend test (Sharma & Priya, 2023; Tripathi et al., 2014). Although the impact of human activity and climate change on streamflow has been extensively studied, there is little research examining their effects on hydrological drought (Jiang et al., 2019; Xi et al., 2021). Several studies have characterized rainfall behavior through the calculation of indices such as the Standard Precipitation Index (SPI), Rainfall Anomaly Index (RAI), and Seasonality Index (SI) to study the interrelationship of rainfall with the sustainability of wetlands (Mehta & Yadav, 2022; Mishra & Singh, 2011). The Standardized Precipitation Index (SPI) is a commonly used metric for characterizing and attributing drought experiences across various climatic zones (Mckee et al., 1993; Sharma & Priya, 2023).

Worldwide across geographical boundaries of 172 countries, there are over 2500 Wetlands of International Importance (Ramsar Sites), encompassing approximately 257 million hectares (ha), or roughly 21% of the total wetland area (Das & Sharma, 2024). As of January 2024, there were 80 Ramsar sites in India, encompassing around 1,332,200 hectares, or 5.6% of the country's total land area (Ramsar sites in India, 2024). Bihar is a complex hydrological environment with big rivers that give rise to a large number of wetlands. About 80% of its population depends on the southwest monsoon and western disturbances to sustain an economy based on farming, aquaculture, and allied activities. Significant disruptions in the annual rainfall coupled with soaring peak temperatures and a rise in the frequency of extreme weather like droughts and floods over the past two decades have had a profound impact on Bihar's hydrological balance (Aslam et al., 2021; Singh & Sinha, 2020). These changes have adversely affected the region's wetlands, altering their hydrological cycles, reducing biodiversity, and impairing their ability to provide essential ecosystem services (Aslam et al., 2021; Mandal et al., 2023; Mitsch & Gosselink, 2000). The National Wetlands Conservation Programme (NWCP) lists 115 wetland areas in India, including the Bihar wetlands of Kusheshwarasthan of Darbhanga, Kabar Tal of Begusarai, and Baraila Wetland of Vaishali. The estimated total wetland area in Madhubani district is 5367 hectares or roughly 2.5% of the district's total land area. Here, the major wetland types comprise of oxbow lakes/cut-off meanders (159 ha), tanks/ponds (407 ha), rivers/streams (4645 ha), and naturally occurring waterlogged areas (135 ha) (Das & Sharma, 2024; "National Wetland Atlas of Bihar," 2010).

North Bihar's wetlands are enriched by rainfall (Indian summer monsoon and western disturbances) carried downstream from Nepal by the streams of the Adhwara group, Gandak, Burhi Gandak, Baghmati, and Kosi. The wetlands are undergoing rapid transformation, resulting in a significant loss of area due to factors such as reduced annual rainfall, building activities in the vicinity, and river caging caused by embankment construction. This paper examines the effects of rainfall variability on wetland ecosystems in Madhubani, Bihar, by analyzing spatiotemporal changes in wetlands and evaluating the impact of climate change using meteorological drought indicators, with a focus on sustainable adaptation strategies.

2 Study Area

Madhubani district lying in the north part of the Mithila plains in Bihar is an important seat of cultural learning endowed with plenty of natural water resources. It lies between longitudes 85°43′–86°42′ E and latitudes 25°59′–26°39′ N (Fig. 1).

It forms an international border with Nepal to its north, and shares administrative boundaries with the Indian districts of Darbhanga to the south, Supaul to the east, and Sitamarhi to the west. The topography of the area has a general slope from north to almost south. This part of Mithila plains is drained by several large rivers, including the Dhauns, Kamla, Kamla-Balan, Bhuthi-Balan, Tiljuga, and Kosi, which originate in the Nepal Himalayan range and flow down southward (Sinha & Friend 1994; Singh & Pandey 2014; Das & Sharma 2024).

These plains' topography is shaped by sediment loads carried by rivers bringing fresh alluvium from Nepal (S. K. Singh & Pandey, 2014). Madhubani's climate is classified as *sub-tropical monsoon type (Cwa)* where its planetary wind flow is influenced by its proximity to both the Himalayan mountains and the Bay of Bengal. It experiences summer temperatures reaching as high as 42 °C and winter temperatures dipping to as low as 4 °C (Sharma & Priya, 2023). It is one of Bihar's rainiest districts, with a mean annual rainfall of above 1200 mm (IMD Annual Report, 2022). Both diverse aquatic and terrestrial vegetation can be found flourishing here owing to favorable climate conditions. Water mosses (*Salvinia natans*), water chestnuts (*Trapa bispinosa*), lotus (*Nelumbo nucifera*), water hyacinth (*Eichhornia crassipes*), foxnut (*Euryale ferox*), and blue water lily (*Nymphaea nouchali*) comprise some of the major aquatic plants (Kumari, 2019; O"Hare et al., 2018), while major terrestrial plants include peepal (*Ficus religiosa*), mango (*Mangifera indica*), rosewood (*Dalbergia sissoo*), and bamboo (*Bambusa vulgaris*) (Anamika Kumari, 2019; Smita Jha, 2019).

The Madhubani district has a total area of 3501 km^2 and about 44.8 lakh people live there. The population density is 1282 persons per km^2, which is higher than the population density of Bihar. Ninety-six percent of the population lives in rural areas, and the literacy rate is 58.62% (Census, 2011; Amrita Datta, 2016), which is much lower than both the national average (74.04%) and of Bihar (61.8%).

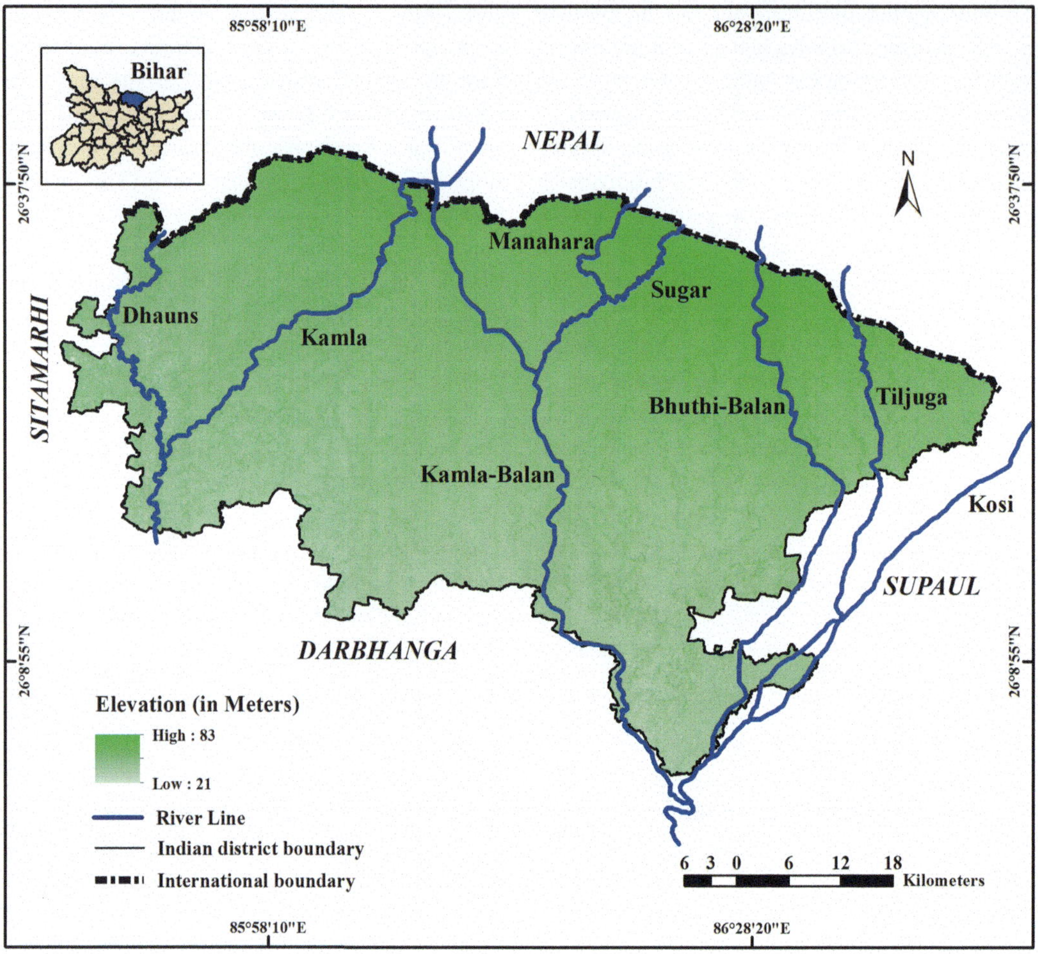

Fig. 1 Relief map of Madhubani (Bihar)

3 Objectives

The study attempts to assess the spatiotemporal dynamics of wetlands in Madhubani by analyzing changes in wetland extent and characteristics over time. By investigating the long-term trends and projections in rainfall, it aims to calculate the frequency and intensity of meteorological drought occurrences and their influence on wetland sustainability. Moreover, it also aims to understand the impact of inter-annual variability of rainfall on hydrological changes and future implications in wetland ecosystems.

4 Data Sources and Methodology

In this study, wetlands in the study area were assessed spatiotemporally using satellite-based data from the United States Geological Survey (USGS), namely the Landsat series data (Table 1). On "Top of the Atmospheric" (TOA) corrections, a technique called pre-processing has been applied for all Landsat data to remove atmospheric effects and for quality enhancement. Spatiotemporal and physical assessments of wetlands were undertaken based on drainage pattern areas and the Normalized Difference Water Index (NDWI).

The drainage patterns within the study area were mapped within ArcGIS-pro environments utilizing the ArcHydro Tool. Figure 2 depicts the various steps entailed in generating drainage patterns in Madhubani, utilizing the SRTM-DEM (Shuttle Radar Topography Mission - Digital Elevation Model). The Normalized Difference Water Index (NDWI) furnishes crucial insights into water presence and distribution within vegetation, aiding ecologists in assessing plant stress, irrigation requirements, and broader water management strategies (Laonamsai et al., 2023; Szabó et al., 2016). Computed from reflectance values acquired from remote sensing imagery, particularly in the near-infrared (NIR) and green spectrums, NDWI monitoring over time facilitates the tracking of water availability, early detection of water-related issues, and optimization of irrigation strategies. The systematic utilization of tools for analyzing hydrological and climatic evaluation of wetlands is illustrated in Table 2.

Monthly and annual rainfall data for Madhubani were acquired for the years 1901–2022 from the India Meteorological Department (IMD), Patna. For the long-term trend detection of annual rainfall, Mann–Kendall trend test (non-parametric) and Sen's slope estimator were undertaken. Several indices, including the Seasonality Index (SI), Rainfall Anomaly Index (RAI), and Standard Precipitation Index (SPI), were computed to better understand rainfall behavior and how it affects wetlands' sustainability. SPI is an index frequently used to understand meteorological drought over a variety of timescales. It has a strong relationship with soil moisture on short durations, and it may have indirect relationships with underground water and reservoir reserves over longer periods.

In this study, SPI was computed for precipitation data for periods of 3, 6, 9, and 12 months. Van-Rooy (1965) developed RAI to calculate the inter-annual variability of rainfall over a terrain. With just one input (precipitation) that can be examined on a monthly, seasonal, and annual basis, it is relatively simple to compute. The Seasonality Index (SI) is used to assess the variability of rainfall in an area within a year and helps in predicting weather-related impacts.

Table 1 Utilized Landsat series data

Sl. no.	Data	Acquisition date	Row	Path	Resolution
1	Landsat 2	12/03/1975	150	42	60 m
2	Landsat 2	20/11/1975	150	42	60 m
3	Landsat 5	28/03/2000	140	42	30 m
4	Landsat 7	14/10/2000	140	42	30 m
5	Landsat 8	25/03/2022	140	42	30 m
6	Landsat 8	20/11/2022	140	42	30 m

Source: https://earthexplorer.usgs.gov/

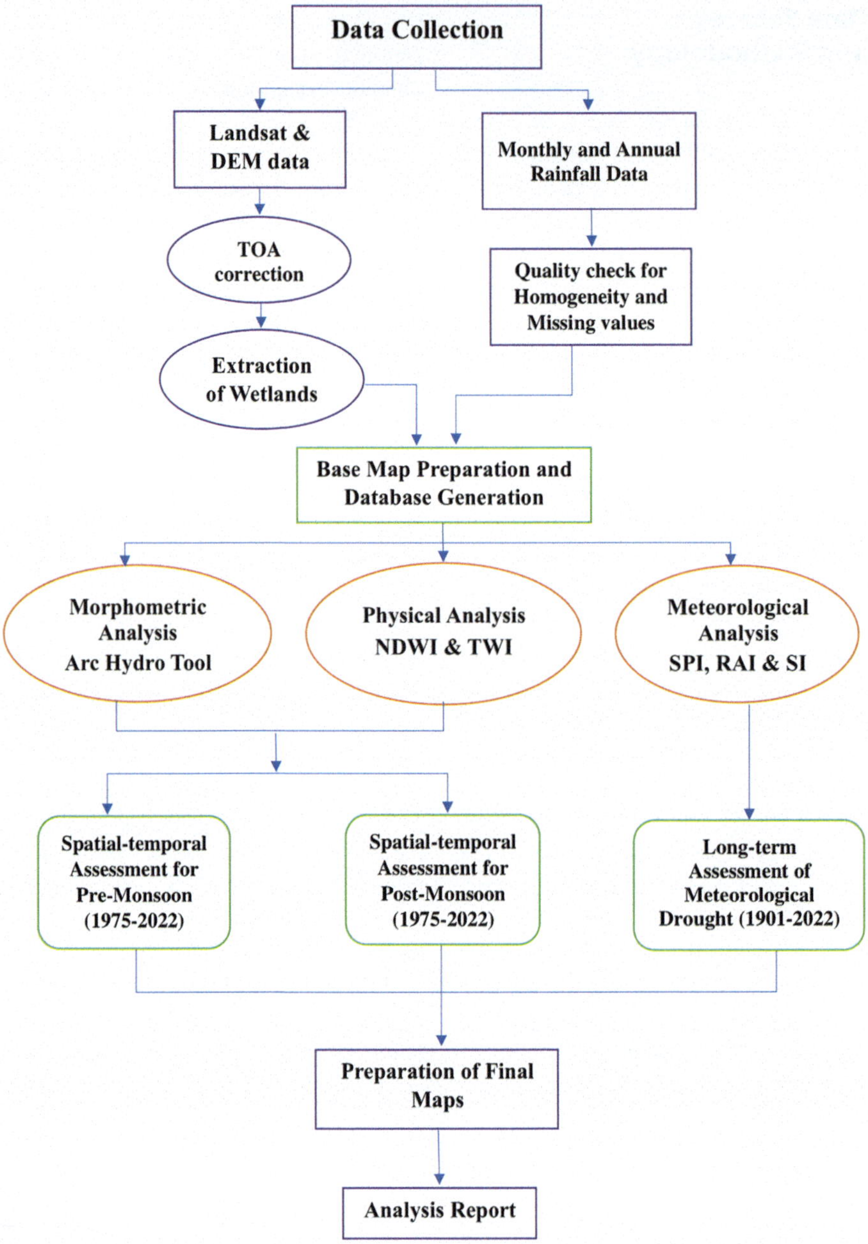

Fig. 2 Methodology for hydrometeorological and physical assessment of wetlands

Table 2 Hydrological and climatic indices

Sl no.	Indices	Equation/steps	Description	References
1	NDWI	NDWI = Green—NIR/Green + NIR The NDWI has a value between −1 and + 1; all positive values indicate moist circumstances, while dry conditions of the landscape are represented by all negative values	The main purpose of NDWI is to identify and track minute variations in the water content of water bodies	Das and Sharma (2024)
2	Autoregressive integrated moving average (ARIMA) model	$Y_t = C + \phi_1 y_{t-1} + \phi_2 y_{t-2} + \ldots + \phi_p y_{t-p} + \varepsilon t$ Where, Y_t is the data of the ARIMA model, and parameters ϕ_1, ϕ_2, and so on are ARIMA coefficients	This statistical analysis technique makes use of time series data to forecast future trends or to gain a deeper understanding of the dataset	Shumway and Stoffer (2017)
3	SPI	$SPI = (R—R^*)/\sigma_p$ Where, R stands for precipitation, R* for mean precipitation, and σp for the standard deviation of precipitation	The SPI computes the intensity of droughts over any desired interval	Elbeltagi et al. (2023)
4	RAI	Positive RAI = $3[(N—N^-)/(M^-—N^-)]$ Negative RAI = $-3[(N—N^-)/(X^-—N^-)]$ Where, M^- is the average of the 10 highest annual rainfall, X^- is the average of the 10 lowest annual rainfall, N is the current annual rainfall, and N^- is the annual average rainfall	The frequency and intensity of the wet and dry years are examined using the RAI	Elbeltagi et al. (2023)
5	SI	SI = average annual rainfall/overall average annual rainfall	The SI is generally used to study the variability of rainfall in an area within a year	Pascale et al. (2015)

5 Result and Discussion

5.1 Annual Rainfall Variability and Trend Analysis of Madhubani District

The analysis of the variability and trend of annual rainfall helps understand the climate dynamics of any region. Rainfall is an essential element of climate that directly affects the water quality, spatial distribution, and productivity of wetlands. In this study, long-term (1901–2022) annual rainfall trends and variability are analyzed to understand their effect on the natural wetlands of Madhubani. The variability of annual rainfall is calculated with the help of standard deviation and variance from the mean. Sen's slope estimator and the Mann–Kendall (MK) test are used to identify any significant trend in annual rainfall and assess the rate and direction of change.

This important non-parametric test finds a significant decreasing trend in Madhubani ($Z_s = -3.21$). The Sen's slope computed for the study period detects a decreasing rainfall trend at 2.57 mm per year ($Qs = -2.57$). Figure 3 shows a decreasing trend of annual rainfall over the period.

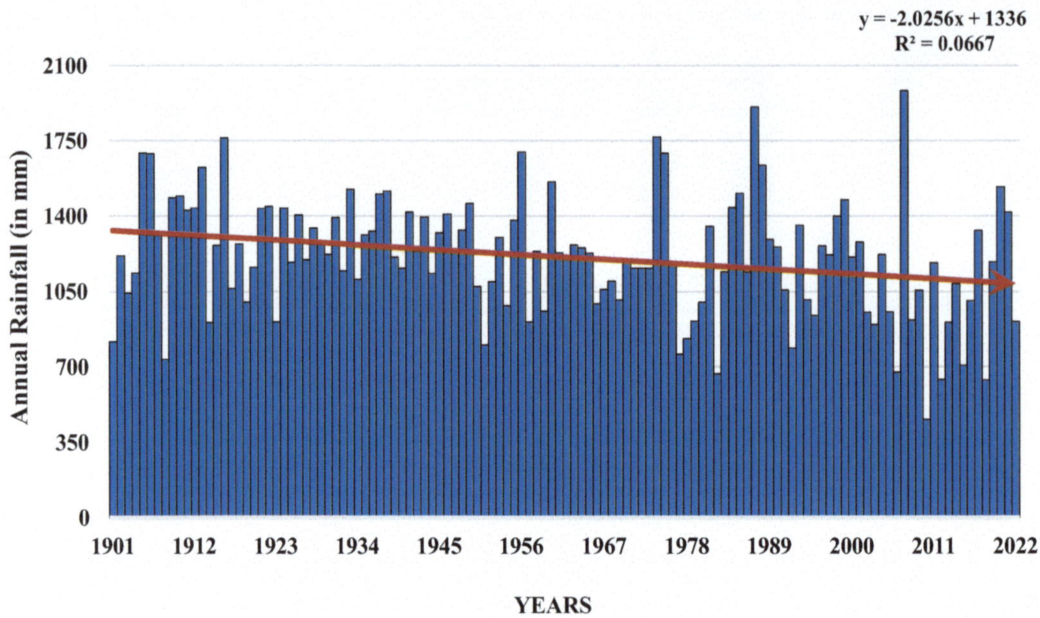

Fig. 3 Annual rainfall trend of Madhubani district (1901–2022)

The last two decades (2002–2022) have witnessed a substantial decrease in rainfall compared to all other periods indicating an occurrence of pronounced long dry spells. For time series forecasting, an Autoregressive Integrated Moving Average (ARIMA) model with five auto- regressive terms (5, 1, 0) that represented the associations between the current observations and the lag observations was also employed for the next two decades. The model finds that annual rainfall over Madhubani during 2023–2043 shall remain below its long-term average with a deficit of 15–20 mm annually.

Madhubani receives rainfall mainly from the southwest monsoon. Lying in the Terai region of the Himalayas, Madhubani receives rainfall in monsoon and winter. For the period 1901–2022, the variability of the annual rainfall of Madhubani has been observed to be highly fluctuating and decreasing significantly, while only monsoonal rainfall has shown a sharp increase among all seasons (Fig. 4). Under conditions of a significant decreasing trend of rainfall and intensive agriculture, water levels in wetlands and underground water have been severely strained. This may lead to unforeseen conditions of water crisis and drought, especially during summer. The overall rainfall variability during this period remained about 103%. High rainfall variability (above 120%) adversely affects the sustainability of land-water surfaces such as wetlands. During the study period high annual rainfall variability has been recorded in Madhubani, especially from the period 1954 to 1974 and 2002 to 2020. The increase in variability during these periods indicates that annual rainfall has deviated periodically from its mean value and this in turn has affected the process of annual recharge of wetlands. High rainfall variability means lesser reliability, which in turn affects the net primary productivity of wetlands.

5.2 Droughts and Dry Spells in Madhubani District (1901–2022)

A period is considered to be dry or drought-like when an area receives less than the long-term average rainfall. Important drought indices like the Seasonality Index (SI), Rainfall Anomaly Index (RAI), and Standard Precipitation Index (SPI) were used in the present study to analyze precipitation patterns and changes in wetlands.

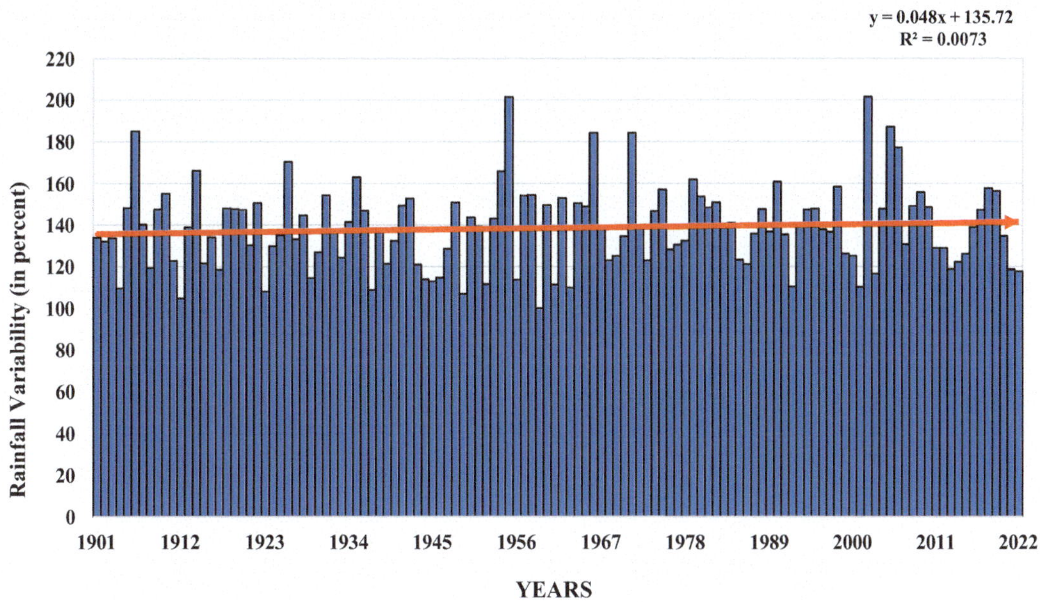

Fig. 4 Annual rainfall variability of Madhubani district (1901–2022). (Source: IMD Patna, 2022)

5.2.1 Analysis of Standard Precipitation Index in Madhubani (1901–2022)

Drought assessments with the help of SPI are done for a variety of periods, ranging from short-term (3 months) to long-term (12 months or more). Positive SPI values indicate increased precipitation, ensuring water availability in wetlands and thus fostering aquatic ecosystems and biodiversity. Conversely, extended periods of negative SPI values signify drought conditions, triggering reduced water levels in wetlands, thereby threatening habitat destruction, loss of biodiversity, and weakened ecosystem functions. For Madhubani, a heatmap has been generated to visually depict drought conditions using the Standardized Precipitation Index (SPI) calculated over different periods (3, 6, 9, and 12 months) for the period spanning 1901–2022. This analysis is done based on SPI3 (calculated using 3 months of rainfall data), SPI6 (using 6 months of data), SPI9 (using 9 months of data), and SPI12 (using 12 months of data).

In Fig. 5, the high value of SPI represents rainfall surplus or moist conditions and is represented by yellow-green colors, while the low values of SPI indicate rainfall insufficiency and drought-like conditions, which are represented by violet-blue colors. Between the period 1901 and the mid-1940s, SPI values displayed large fluctuations, indicating varying levels of moisture availability, where numerous short periods are marked by prolonged dry spells while others by increased precipitation levels. From the mid-1940s to the early 1960s, a relatively stable phase was recorded, with SPI values remaining around zero or with insignificant shifts, reflecting a more balanced precipitation distribution. However, from the mid-1960s to the late 1970s, Madhubani was characterized by an increased affinity toward aridity, with SPI values consistently remaining below zero throughout the period, indicating prolonged dry spells and the sudden onset of drought conditions. Subsequently, from the late 1970s to the early 1990s, SPI values remained fluctuating, indicating an alternating occurrence of dry and wet conditions, though some years were prominently marked with continuous dry periods. Moving toward the end of the twentieth century, from the late 1990s to the early 2000s, an evident shift toward wetter conditions occurred, with SPI values showing positive readings of above-normal precipitation. Beginning from the mid-2000s, SPI values exhibit an escalated variability, with a greater number of years witnessing extreme dryness (SPI values < zero) and others depicting extreme wetness (SPI values > zero), indicating a trend toward more intense precipitation behavior.

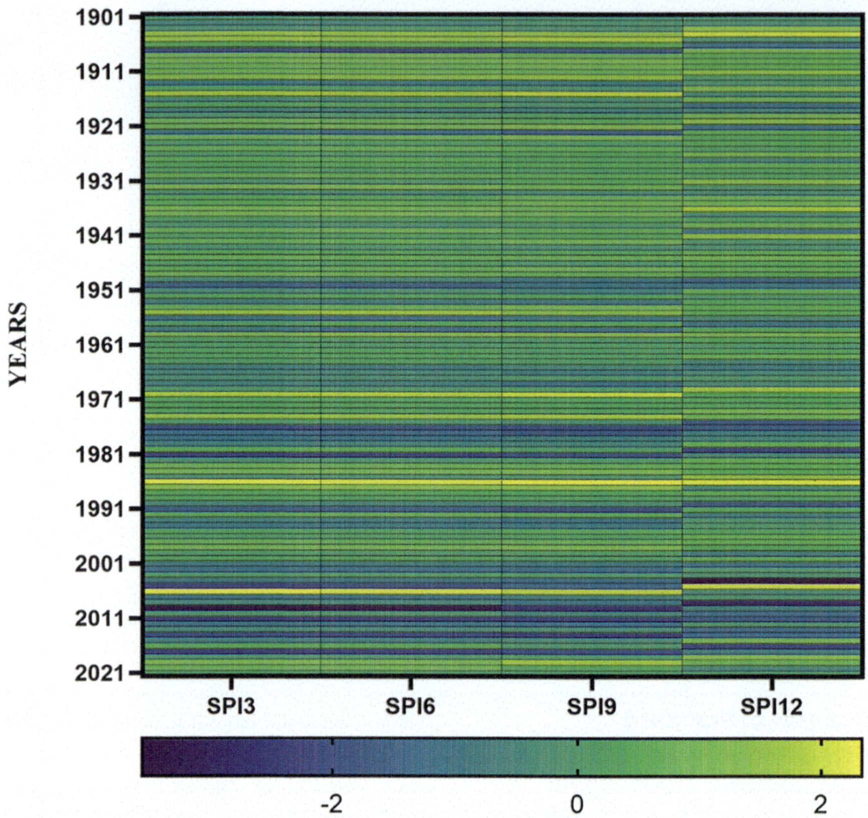

Fig. 5 Heat map of SPI in Madhubani (1901–2022). (Source: IMD Patna, 2022)

5.2.2 Rainfall Anomaly Index (RAI) of Madhubani (1901–2022)

The Rainfall Anomaly Index (RAI) is another important index that is used to evaluate and track variations in rainfall patterns and helps in addressing cyclic fluctuations in rainfall with alternating dry and wet spells that impact agriculture, hydrology, forest resources, ecosystems, and allied sectors. The RAI is considered an efficient tool, especially with data records on small timescales, and a large number of missing records or zero probability data (Raziei, 2021). Positive RAI values indicate wet periods with sufficient rainy years, ensuring water availability for wetlands recharge and high productivity, while negative RAI values indicate dry periods, indicating low receipt of rainfall with reduced water availability for aquatic ecosystems. In this study, heavy rainfall events and dry periods have been studied as interdependent forces, each playing a complementary role in the ecosystem.

In Fig. 6, the positive values of RAI represent extremely humid conditions or rainy years (represented by blue color), while the negative values of RAI indicate very dry conditions (represented by red color). Between the period 1920 and 1950, the values of RAI remained mostly positive and ranged between −1 and +2 representing the humid climate and facilitating good water availability for wetlands ecosystem. The fluctuation in values of RAI tends to be higher (−3 to +4) during the period 1950 to 2000 indicating alternating periods of very humid and very dry spells. The significant variability in annual rainfall in this region harmed the primary productivity of local biodiversity. A prolonged dry spell persisted from 1975 to 1982, severely impacting the availability of water in wetland ecosystems. Similarly, from 2000 to 2022, the consistently low RAI values indicate an intense and prolonged dry period, adversely affecting aquatic plants and animals due to habitat loss, degraded water quality, and increased vulnerability to predation. As a result of

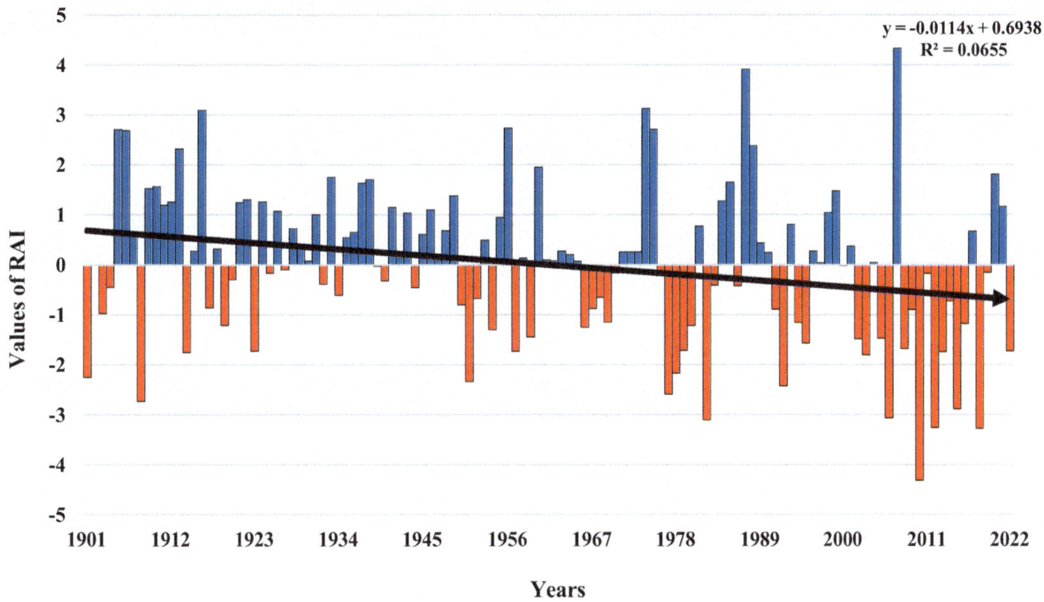

Fig. 6 Rainfall anomaly index in Madhubani (1901–2022). (Source: IMD Patna, 2022)

more open spaces, this decreasing trend in rainfall directly impacts wetland-dependent economies like Makhana, Singhara, and fish.

The RAI highlights periods of significant fluctuations in rainfall patterns over the past 122 years in Madhubani. Notably, the most substantial variations have occurred within the last two decades, indicating a prolonged dry spell that has adversely affected water availability for wetlands recharge. These fluctuations in RAI reflect the dynamic nature of the region's climate, with higher values correlating with increased humidity and lower values indicating intensified drought conditions. As illustrated in Fig. 7, Madhubani experienced mild to severe drought-like conditions for about 56 years out of the total 122 years. This underscores the vulnerability of the region to climate variability and highlights the need for adaptive strategies to mitigate the impacts of such fluctuations on water resources and ecosystems.

5.2.3 Seasonality Index of Madhubani (1901–2022)

The Seasonality Index (SI) is another important tool like SPI and RAI for precipitation behavior assessment. It is generally used to study the variability of rainfall in an area within a year. A seasonality index value higher than 1 indicates above-normal rainfall for that particular year, while a value below 1 indicates below-normal rainfall. For the period 1901–2022, the SI index shows a trend of alternating periods of above- and below-average rainfall over time. For 53 years, SI values remained between 0.4 and 1.0 indicating seasonal rainfall with a marked dry season. This pattern suggested fluctuations in wetland water levels, potentially causing habitat modifications or loss for species dependent on wetland ecosystems. For about 48 years, SI remained between 1.0 and 1.19 signaling the concentrated temporal distribution of rainfall in less than 3 months and high dependence on monsoons. This high dependence on monsoonal rainfall can lead to significant challenges for wetland sustainability, as water availability becomes unpredictable and may not align with the ecological needs of wetland ecosystems. And about 20 years exhibited extreme seasonality with values exceeding 1.2, with rainfall occurring in 1–2 months.

In Table 3, the comparison of RAI, SPI, and SI indicates that these indices, though all assessing moisture conditions, may have varying sensitivities and thresholds for classifying wet and dry periods. RAI appears to show a higher frequency of years that are significantly wetter or

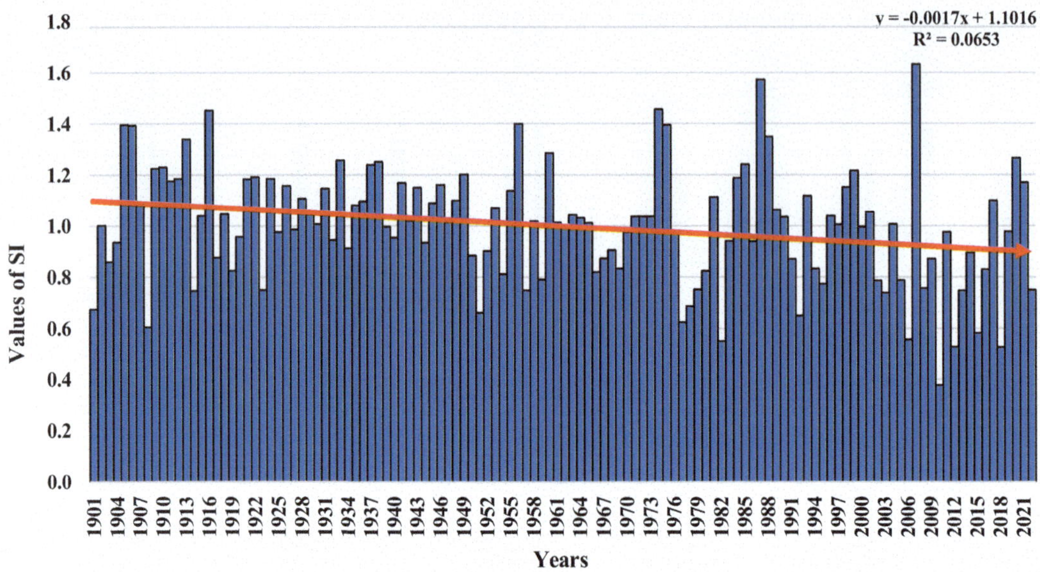

Fig. 7 Seasonality index (SI) in Madhubani (1901–2022). (Source: IMD Patna, 2022)

Table 3 Temporal distribution of RAI, SPI, and SI in Madhubani (1901–2022)

RAI		SPI		SI	
Classes	No. of years	Classes	No. of years	Classes	No. of years
Extremely humid (>4)	1	Extremely wet (<2)	0	Extreme seasonality, with almost all the rainfall occurring in 1–2 months (≥1.2)	20
Very humid (2–4)	9	Very wet (1.5–1.99)	1	Most rainfall occurs in less than 3 months (1.00–1.19)	48
Humid (0–2)	56	Moderate wet (1.0–1.49)	7	Marked seasonal with a long dry season (0.80–0.99)	30
Dry (−2 to 0)	44	Near normal (−0.99 to 0.99)	10(6)	Seasonal (0.60–0.79)	18
Very dry (−2 to −4)	11	Moderate dry (−1.0 to −1.49)	7	Rather seasonal with a short dry season (0.40–0.59)	5
Extremely dry (< −4)	1	Extremely dry (> −2)	1	Rainfall spread throughout the year but with a definite wetter season (0.0–0.39)	1

drier than normal, whereas SPI tends to classify most years as “near normal.” This means that SPI may be less sensitive to variations that RAI categorizes as “Humid” or “Dry.” However, this fluctuating nature of SPI values over a long period, characterized by episodic fluctuations and extended periods of either dryness or wetness, poses significant challenges to the sustainability and conservation of wetlands. There is no single year in which there has been rainfall that is uniformly distributed across all seasons in a year except the year 2010, which recorded rainfall with “a definite wet season.” For the last two decades (2002–2022), SI values largely remained between 0.8 and 1.0 indicating prolonged dry seasons. This emergent trend highlights the increasing pressure on wetland resources due to evolving rainfall patterns, which can lead to habitat degradation and loss of biodiversity.

5.3 Hydrological Profile of Wetlands in Madhubani Districts (1975–2022)

In this section, the hydrological characteristics of the study area are analyzed and discussed. The Normalized Difference Wetness Index (NDWI) serves as an important tool for the analysis of hydrological changes in wetlands. This section employs NDWI to elucidate the hydrological dynamics of wetlands spanning from 1975 to 2022 (Fig. 8).

During the pre-monsoon period of 1975, the lower portions of catchment areas A2 and B2

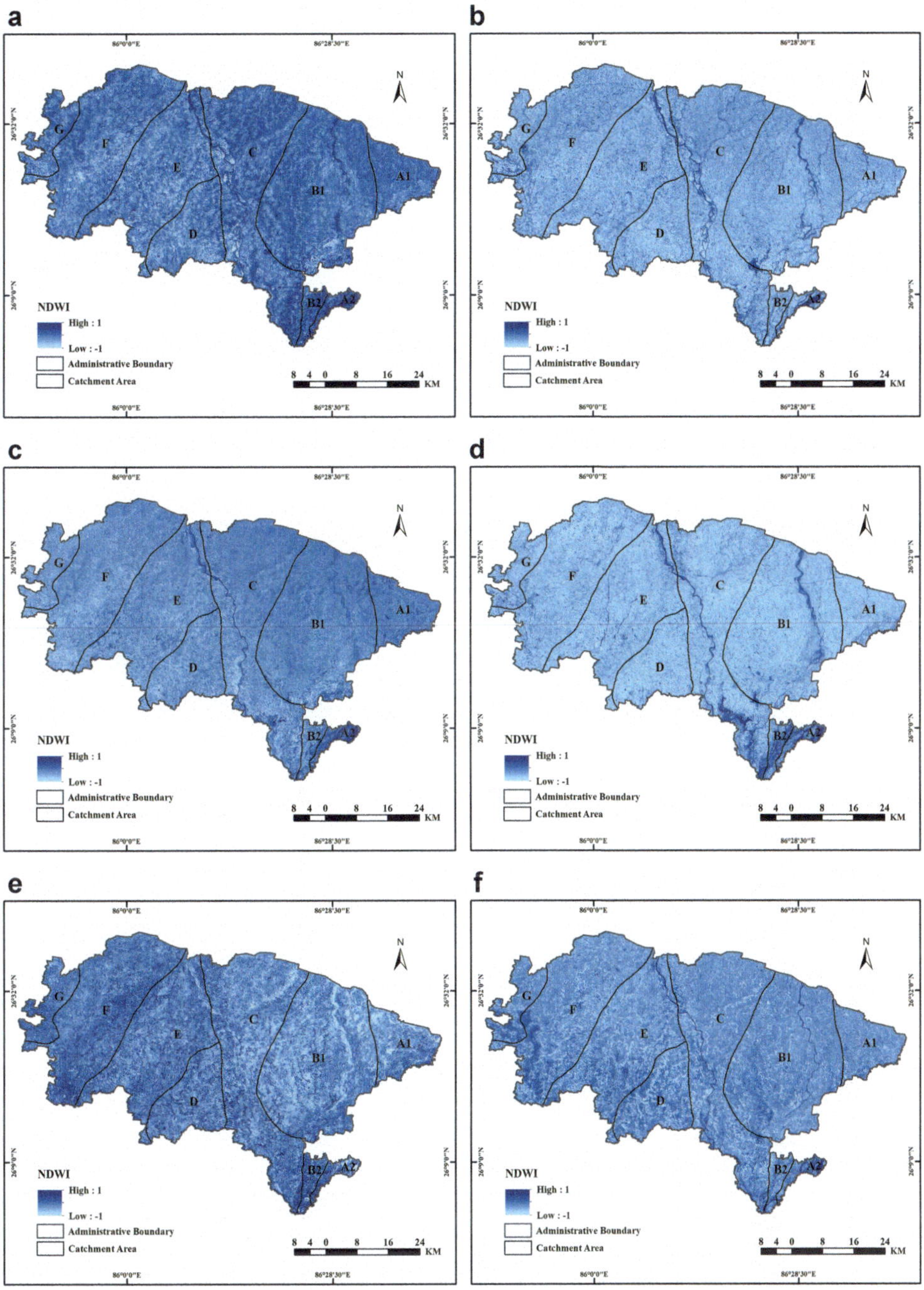

Fig. 8 Spatial distribution of wetness in Madhubani (1975–2022)

were noticeably moist, while central parts of Madhubani exhibited comparatively lower wetness levels (Fig. 8a). Following the monsoon season in the same year, the lower portions of catchment area C and the southwest of catchment area F experienced significant moisture (Fig. 8b). In contrast to decreasing wetness levels in the northeastern portions of catchment areas B1 and C, enhanced wetness was visible in the lower reaches of catchment areas B and C during the pre-monsoon period of 2000. Elevated wetness levels were also observed in southwestern parts of Madhubani (Fig. 8c). Conversely, during post-monsoon (2000), maximum wetness tended to be concentrated in the lower part of catchment areas B1 and C, while the southern regions of Madhubani and catchment areas A2 and B2 exhibited lower wetness (Fig. 8d). In the pre-monsoon phase of 2022, increased wetness was prominent in catchment areas A2 and B2, as well as the lower sections of F, with lower wetness levels observed in catchment areas B1 and A1 (Fig. 8e). Finally, post-monsoon conditions in 2022 revealed the highest wetness in transition zones of catchment areas F and G, while the lowest wetness was evident in catchment areas A1–A2 and B1–B2 (Fig. 8f).

The analysis of hydrological changes in Madhubani's wetlands, spanning from 1975 to 2022, reveals dynamic patterns of wetness distribution during pre-monsoon and post-monsoon periods. Wetlands provide vital habitats for diverse flora and fauna, and sudden changes in moisture levels can disrupt these ecosystems, leading to shifts in species composition, altered nutrient cycling, and potential habitat loss. Also, the observed variations in wetness patterns may indicate broader climate change impacts on the region. Local communities (particularly "*Nishads*") rely on these wetlands for their livelihoods, including pisciculture, aquaculture, and horticulture. These lowered levels of wetness in the area can directly impact their livelihoods, affecting income, food security, and social practices. Understanding these changes will serve as guidelines for building resilience and sustainability in these communities.

6 Future Implications of Rainfall Variability and Wetland Sustainability

The future implications of rainfall variability on the sustainability of wetlands are multidimensional. Under scenarios of projected global warming (IPCC Sixth Assessment Report), more extreme and unpredictable rainfall is expected. The short, intense, and significant rate of decreasing rainfall in North Bihar may alter the hydrology of sensitive ecosystems of wetlands. Increased rainfall variability with unexpected prolonged periods of droughts (especially during the last two decades) followed by short-duration intense rain will affect floral and faunal biodiversity as well as geochemical cycling. For understanding future scenarios of precipitation behavior in Madhubani, an Autoregression Integrated Moving Average (ARIMA) model was used based on annual rainfall data of the last 122 years. Using defined time series data, ARIMA is a powerful statistical model that helps forecast near-future trends and project the behavior of precipitation. In Fig. 9, historical rainfall data is represented with a blue line while future rainfall estimation is done with a red line. The figure shows that for the period 1901–2022, the mean annual rainfall remained about 1214 mm. Periodic fluctuations in a cyclic manner are observed in data series that define the stationarity of annual rainfall records. No multicollinearity in the data series is visible, which ensures that despite high rainfall variability, models do not suffer from over-estimation. The model effectively detects the temporal dependence structure of the annual rainfall data and finds that for the period 2023–2043, the decreasing trend of rainfall will continue with high annual variability and is predicted to remain around 1200 mm (below long-term normal).

With multi-indicator analysis of precipitation behavior with indices like SPI, RAI, SI, and trend detection and estimation tools like the Mann–Kendall Trend Test and ARIMA Prediction model, it has become clear that during the last century in Madhubani district, annual rainfall has

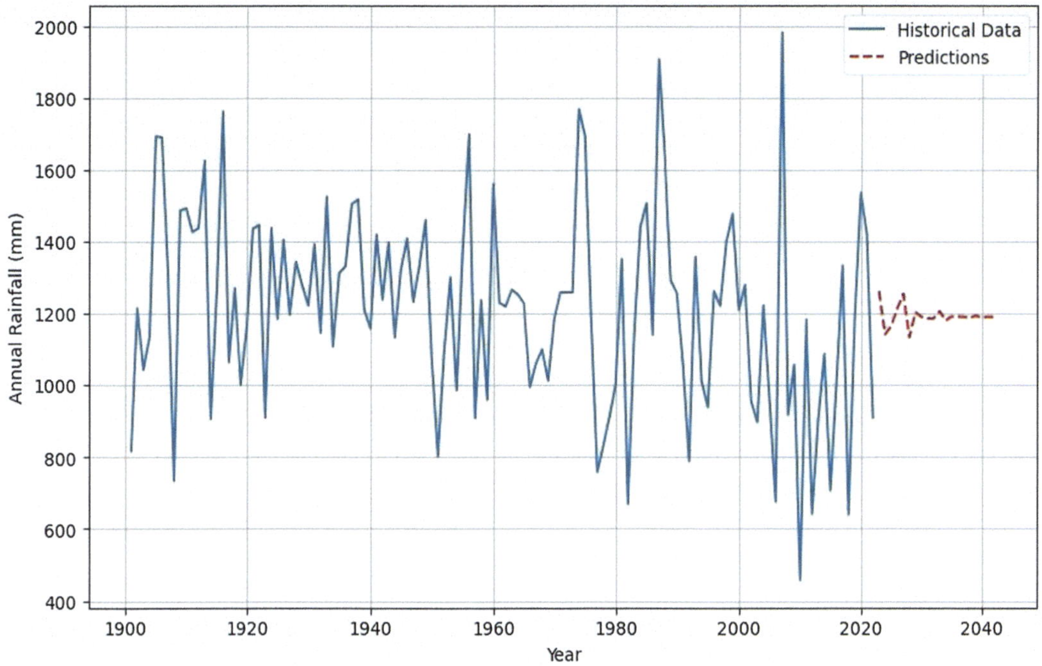

Fig. 9 ARIMA predictions for annual rainfall in Madhubani (2023–2043). (Source: IMD Patna, 2022)

decreased with highly variable seasonal rainfall. As a result, dry spells leading toward drought-like conditions have increased. The decreasing trend of rainfall and inter-annual seasonal variations of rainfall have raised alarms regarding the degradation of wetlands. This in turn has adversely affected the natural ecosystems, agricultural production, fisheries, horticulture, aquaculture, and allied economic activities.

Madhubani lies at the heart of the agricultural zone of North Bihar, where farmers largely depend on monsoon rainfall for paddy cultivation. With recently witnessed dry spells occurring over the last two decades, high dependence on groundwater for agricultural activities has increased. This unchecked tapping of groundwater has aggravated the problem of water crisis during summers and may have negative implications for water availability in the near future. Owing to its large-scale cultivation of "*Makhana*" (foxnut, a species of water lily), this district stands different in the Mithila-Makhana region (a "Geographical Indicator" tag given by the Government of India in the year 2022). This district has been long known for its association with wetlands-based aquaculture of fruits like Makhana, and Singhara, and local fishes like Rohu, Hilsa, Salmon, and Barracuda.

The Bihar State Action Plan on Climate Change, 2015, has stressed sustainable utilization of resources, the study of habitat potential, the creation of green shelterbelts around wetlands, scientific water, treatment of catchment areas, water quality assessment, fisheries management through Panchayati Raj Institutions (PRIs), and community awareness. It has also allocated funds under the 12th Five-Year Plan, but it failed to address the direct impact of climate change on the fragile and very sensitive ecological system of the wetlands of Bihar. Although climate-resilient agriculture has gained attention in the past decade, sustainable development of horticulture practices around wetlands recharged by rivers Kamla, Kamla-Balan, Tiljuga, and Kosi needs an independent strategy for the conservation of indigenous crop varieties viz. makhana. Effective community-based programs focusing on people's participation (particularly fishermen communities like *Sahanis*) must be promoted in the protection and conservation of wetlands. Thus, the significant trend of decreasing rainfall and increased variability in wetlands of North Bihar, emphasizes the urgent need for holistic conservation strategies and community involvement to ensure their sustainability and safeguard local livelihoods.

7 Conclusion

The detailed analysis presented in this study highlights the critical importance of understanding annual rainfall trends and variability for comprehending climatic dynamism and its impacts on wetland ecosystems. Annual rainfall patterns in Madhubani reveal a significant decreasing trend over the past century, accompanied by increased variability, particularly in the past two decades. This declining trend threatens the sustainability of wetlands, which are important for the maintenance of biodiversity, agricultural activities, and local communities.

The trend detection test of Mann–Kendall and similar indices for characterizing the behavior of precipitation such as SPI, RAI, and SI shed light on the changing hydrological dynamics of the region. The study of NDWI further emphasizes the importance of understanding moisture dynamics for ecosystem sustainability. The results find changes in wetness patterns and increasing frequency of drought-like conditions, influenced by both precipitation and river flow, may aggravate associated risks to wetland ecosystems and hydrological resources. A future projection-based ARIMA model suggests that this trend of decreasing rainfall in the region is likely to continue, warranting adaptive strategies to mitigate its long-term impacts.

Way forward, the implications of rainfall variability on wetland sustainability are multifaceted, requiring holistic conservation strategies and community involvement. Action-oriented initiatives such as the "Bihar State Action Plan on Climate Change" provide a framework for addressing these challenges, but more focused efforts are required to ensure the resilience of wetland ecosystems. In conclusion, the findings of this study underpin the urgent need for immediate measures to conserve and manage wetlands in the face of evolving climate patterns. Through proactive efforts encircling lawmakers, local communities, and the scientific realm, it must be treated urgently to preserve these vital ecosystems and secure the well-being of coming generations.

References

Alawadi, W. A., Raheem, Z. A. H. A., & Yaseen, D. A. (2023). Use of remote sensing techniques to assess water storage variations and flood-related inflows for the Hawizeh wetland. *Environmental Monitoring and Assessment, 195*(10), 1246. https://doi.org/10.1007/s10661-023-11838-x

Amrita Datta. (2016). Migration, remittances and changing sources of income in rural Bihar (1999–2011): Some findings from a longitudinal study. *Economic and Political Weekly, 15*(31), 85–93.

Anamika Kumari. (2019). Study of some aquatic plants of Pandaul block of Madhubani District, Bihar with special reference to their ethnobotany. *Journal of Emerging Technologies and Innovative Research (JETIR), 6*(6), 1–4.

Aslam, A., Parthasarathy, P., & Ranjan, R. K. (2021). Ecological and societal importance of wetlands. In *Wetlands conservation* (pp. 55–86). Wiley. https://doi.org/10.1002/9781119692621.ch4

Aslam, R. W., Shu, H., Naz, I., Quddoos, A., Yaseen, A., Gulshad, K., & Alarifi, S. S. (2024). Machine learning-based wetland vulnerability assessment in the Sindh Province Ramsar site using remote sensing data. *Remote Sensing, 16*(5), 928. https://doi.org/10.3390/rs16050928

Balak Yadav, S. (2020). Wetland environment problems and management of Bihar. *International Journal of Geography, 2*(2), 81–83. http://www.geojournal.net/

Bian, L., Melesse, A. M., Leon, A. S., Verma, V., & Yin, Z. (2021). A deterministic topographic wetland index based on lidar-derived DEM for delineating open-water wetlands. *Water (Switzerland), 13*(18). https://doi.org/10.3390/w13182487

Birnbaum, C., Waryszak, P., & Farrer, E. C. (2021). Direct and indirect effects of climate change in coastal wetlands: Will climate change influence wetlands by affecting plant invasion? *Wetlands, 41*(5), 59. https://doi.org/10.1007/s13157-021-01456-z

Bullock, A., & Acreman, M. (2003). The role of wetlands in the hydrological cycle. *Hydrology and Earth System Sciences, 7*(3), 358–389. https://doi.org/10.5194/hess-7-358-2003

Census 2011. (2011). *District hand book census 2011 in Bihar.*

Chenchouni, H., Menasria, T., Neffar, S., Chafaa, S., Bradai, L., Chaibi, R., Mekahlia, M. N., Bendjoudi, D., & Si Bachir, A. (2015). Spatiotemporal diversity, structure and trophic guilds of insect assemblages in a semi-arid Sabkha ecosystem. *PeerJ, 3*, e860. https://doi.org/10.7717/peerj.860

Das, S. K., & Sharma, M. R. (2024). Mapping long-term transformation of wetlands and annual rainfall variability in Madhubani District (1975–2022). *Current World Environment, 19*(1), 251–268. https://doi.org/10.12944/CWE.19.1.22

Dehni, A., & Lounis, M. (2012). Remote sensing techniques for salt affected soil mapping: Application to the Oran region of Algeria. *Procedia Engineering, 33*, 188–198. https://doi.org/10.1016/j.proeng.2012.01.1193

Debanshi, S., & Pal, S. (2020). Wetland delineation simulation and prediction in deltaic landscape. Ecological Indicators, 108, 105–757. https://doi.org/10.1016/j.ecolind.2019.105757

Elbeltagi, A., Kumar, M., Kushwaha, N. L., Pande, C. B., Ditthakit, P., Vishwakarma, D. K., & Subeesh, A. (2023). Drought indicator analysis and forecasting using data driven models: Case study in Jaisalmer, India. *Stochastic Environmental Research and Risk Assessment, 37*(1), 113–131. https://doi.org/10.1007/s00477-022-02277-0

Erwin, K. L. (2009). Wetlands and global climate change: The role of wetland restoration in a changing world. *Wetlands Ecology and Management, 17*(1), 71–84. https://doi.org/10.1007/s11273-008-9119-1

Halder, S., Roy, M. B., & Roy, P. K. (2022). Modelling drought vulnerability tracts under changed climate scenario using fuzzy DEMATEL and GIS techniques. *Theoretical and Applied Climatology, 150*(1–2), 425–452. https://doi.org/10.1007/s00704-022-04165-7

Jiang, S., Wang, M., Ren, L., Xu, C., Yuan, F., Liu, Y., Lu, Y., & Shen, H. (2019). A framework for quantifying the impacts of climate change and human activities on hydrological drought in a semiarid basin of Northern China. *Hydrological Processes, 33*(7), 1075–1088. https://doi.org/10.1002/hyp.13386

Junk, W. J., Brown, M., Campbell, I. C., Finlayson, M., Gopal, B., Ramberg, L., & Warner, B. G. (2006). The comparative biodiversity of seven globally important wetlands: a synthesis. Aquatic Sciences, *68*(3), 400–414. https://doi.org/10.1007/s00027-006-0856-z

Kaplan, G., & Avdan, U. (2017). Mapping an monitoring wetlands using Sentinel-2 satellite imagery. *ISPRS Annals of the Photogrammetry, Remote Sensing and Spatial Information Sciences, 4*(4W4), 271–277. https://doi.org/10.5194/isprs-annals-IV-4-W4-271-2017

Keddy, P. A. (2010). *Wetland ecology*. Cambridge University Press. https://doi.org/10.1017/CBO9780511778179

Kumari, A. (2019). Issue 6 www.jetir.org (ISSN-2349-5162). *JETIR1906E99 Journal of Emerging Technologies and Innovative Research* (Vol. 6). JETIR. www.jetir.org979

Laonamsai, J., Julphunthong, P., Saprathet, T., Kimmany, B., Ganchanasuragit, T., Chomcheawchan, P., & Tomun, N. (2023). Utilizing NDWI, MNDWI, SAVI, WRI, and AWEI for estimating erosion and deposition in Ping River in Thailand. *Hydrology, 10*(3), 70. https://doi.org/10.3390/hydrology10030070

Mandal, S., Bandyopadhyay, A., & Bhadra, A. (2023). Dynamics and future prediction of LULC on Pare River basin of Arunachal Pradesh using machine learning techniques. *Environmental Monitoring and Assessment, 195*(6). https://doi.org/10.1007/s10661-023-11280-z

McFeeters, S. K. (1996). The use of the normalized difference water index (NDWI) in the delineation of open water features. *International Journal of Remote Sensing, 17*(7), 1425–1432. https://doi.org/10.1080/01431169608948714

Mckee, T. B., Doesken, N. J., & Kleist, J. (1993). The relationship of drought of frequency and duration to time scales. In *Eighth conference on applied climatology.*

Mehta, D., & Yadav, S. M. (2022). Temporal analysis of rainfall and drought characteristics over Jalore District of S-W Rajasthan. *Water Practice and Technology, 17*(1), 254–267. https://doi.org/10.2166/wpt.2021.114

Merwade, V. (2019). *Terrain processing using ArcHydro/GeoHMS.*

Mishra, A. K., & Singh, V. P. (2011). Drought modeling—A review. *Journal of Hydrology, 403*(1–2), 157–175. https://doi.org/10.1016/j.jhydrol.2011.03.049

Mitsch, W. J., & Gosselink, J. G. (2000). The value of wetlands: Importance of scale and landscape setting. *Ecological Economics, 35*(1), 25–33. https://doi.org/10.1016/S0921-8009(00)00165-8

Mitsch, W. J., Bernal, B., Nahlik, A. M., Mander, Ü., Zhang, L., Anderson, C. J., Jørgensen, S. E., & Brix, H. (2013). Wetlands, carbon, and climate change. *Landscape Ecology, 28*(4), 583–597. https://doi.org/10.1007/s10980-012-9758-8

Moomaw, W. R., Chmura, G. L., Davies, G. T., Finlayson, C. M., Middleton, B. A., Natali, S. M., Perry, J. E., Roulet, N., & Sutton-Grier, A. E. (2018). Wetlands in a changing climate: Science, policy and management. *Wetlands, 38*(2), 183–205. https://doi.org/10.1007/s13157-018-1023-8

O"Hare, M. T., Aguiar, F. C., Asaeda, T., Bakker, E. S., Chambers, P. A., Clayton, J. S., Elger, A., Ferreira, T. M., Gross, E. M., Gunn, I. D. M., Gurnell, A. M., Hellsten, S., Hofstra, D. E., Li, W., Mohr, S., Puijalon, S., Szoszkiewicz, K., Willby, N. J., & Wood, K. A. (2018). Plants in aquatic ecosystems: Current trends and future directions. *Hydrobiologia, 812*(1), 1–11. https://doi.org/10.1007/s10750-017-3190-7

O"Neill, B. C., Carter, T. R., Ebi, K., Harrison, P. A., Kemp-Benedict, E., Kok, K., Kriegler, E., Preston, B. L., Riahi, K., Sillmann, J., van Ruijven, B. J., van

Vuuren, D., Carlisle, D., Conde, C., Fuglestvedt, J., Green, C., Hasegawa, T., Leininger, J., Monteith, S., & Pichs- Madruga, R. (2020). Achievements and needs for the climate change scenario framework. *Nature Climate Change, 10*(12), 1074–1084. https://doi.org/10.1038/s41558-020-00952-0

Pascale, S., Lucarini, V., Feng, X., Porporato, A., & Hasson, S. ul. (2015). Analysis of rainfall seasonality from observations and climate models. Climate Dynamics, 44(11–12), 3281–3301. https://doi.org/10.1007/s00382-014-2278-2.

Pasupalati, N., Nath, M., Sharan, A., Narayanan, P., Bhatta, R., Ramachandran, R., & Ramachandran, P. (2017). Economic valuation of wetland ecosystem goods and services. In *Wetland science* (pp. 259–284). Springer India. https://doi.org/10.1007/978-81-322-3715-0_14

Peng, K., Jiang, W., Hou, P., Wu, Z., Ling, Z., Wang, X., Niu, Z., & Mao, D. (2023). Continental-scale wetland mapping: A novel algorithm for detailed wetland types classification based on time series Sentinel-1/2 images. *Ecological Indicators, 148*, 110113. https://doi.org/10.1016/j.ecolind.2023.110113

Pernollet, C. A., Guelmami, A., Green, A. J., Curcó Masip, A., Dies, B., Bogliani, G., Tesio, F., Brogi, A., Gauthier-Clerc, M., & Guillemain, M. (2015). A comparison of wintering duck numbers among European rice production areas with contrasting flooding regimes. *Biological Conservation, 186*, 214–224. https://doi.org/10.1016/j.biocon.2015.03.019

Pradeepraju, N., Nagaraju, D., & Sudeep, S. R. (2022). Suitable site selections for artificial recharge structure in Bandalli watershed. Chamaraja Nagar District, Karnataka, India using remote sensing, and GIS techniques. *Current World Environment, 17*(3), 727–742. https://doi.org/10.12944/cwe.17.3.20

Raziei, T. (2021). Revisiting the rainfall anomaly index to serve as a simplified standardized precipitation index. *Journal of Hydrology, 602*, 126761. https://doi.org/10.1016/j.jhydrol.2021.126761

Rebelo, L.-M., Finlayson, C. M., & Nagabhatla, N. (2009). Remote sensing and GIS for wetland inventory, mapping and change analysis. *Journal of Environmental Management, 90*(7), 2144–2153. https://doi.org/10.1016/j.jenvman.2007.06.027

Rousse, J. W., Jr., Haas, R. H., Deering, D. W., Schell, J. A., & Harlan, J. C. (1973). Monitoring the vernal advancement and retrogradation (green wave effect) of natural vegetation. *NASA-CR, 132982*.

Salimi, S., Almuktar, S., & Scholz, M. (2021). Impact of climate change on wetland ecosystems: A critical review of experimental wetlands. *Journal of Environmental Management, 286*. https://doi.org/10.1016/j.jenvman.2021.112160

Sharma, M. R., & Priya, S. (2023). Long-term assessment of precipitation behaviour in Bihar (1901-2021): Patterns, Trends and Observed Variability. *Current World Environment, 18*. https://doi.org/10.12944/CWE.18.2.19

Shumway, R. H., & Stoffer, D. S. (2017). ARIMA models. In *Time series analysis and its applications* (Springer texts in statistics) (pp. 75–163). Springer. https://doi.org/10.1007/978-3-319-52452-8_3

Singh, S. K., & Pandey, A. C. (2014). Geomorphology and the controls of geohydrology on waterlogging in Gangetic Plains, North Bihar, India. *Environmental Earth Sciences, 71*(4), 1561–1579. https://doi.org/10.1007/s12665-013-2562-1

Singh, M., & Sinha, R. (2020). Distribution, diversity, and geomorphic evolution of floodplain wetlands and wetland complexes in the Ganga plains of North Bihar, India. *Geomorphology, 351*, 106960. https://doi.org/10.1016/j.geomorph.2019.106960

Sinha, R., & Friend, P. F. (1994). River systems and their sediment flux, indo-Gangetic plains, northern Bihar, India. *Sedimentology, 41*(4), 825–845. https://doi.org/10.1111/j.1365-3091.1994.tb01426.x

Smita Jha, K. (2019). Studies of dicotyledonous Flora of Madhubani District with special reference to some important medicinal plants. *International Journal of Advanced Research in Science, Communication and Technology (IJARSCT), 7*(1) www.ijarsct.co.in

Szabó, S., Gácsi, Z., & Balázs, B. (2016). Specific features of NDVI, NDWI and MNDWI as reflected in land cover categories. *Landscape & Environment, 10*(3–4), 194–202. https://doi.org/10.21120/le/10/3-4/13

Tiner, R. W., Lang, M. W., & Klemas, V. V. (2015). *Remote sensing of wetlands: Applications and advances*. CRC Press.

Tripathi, P., Kumar, A., Singh, K. K., & Singh, A. (2014). Trend of rainfall in different sectors of Uttar Pradesh under present scenario of climate change. *52 publications 143 citations free profile*. https://www.researchgate.net/publication/300080324

Were, D., Kansiime, F., Fetahi, T., Cooper, A., & Jjuuko, C. (2019). Carbon sequestration by wetlands: A critical review of enhancement measures for climate change mitigation. *Earth Systems and Environment, 3*(2), 327–340. https://doi.org/10.1007/s41748-019-00094-0

Xi, Y., Peng, S., Ciais, P., & Chen, Y. (2021). Future impacts of climate change on inland Ramsar wetlands. *Nature Climate Change, 11*(1), 45–51. https://doi.org/10.1038/s41558-020-00942-2

Van-rooy, M. P. (1965). A Rainfall Anomaly Index (RAI), Independent of the Time and Space. Notos, *14*, 43–48.

Climate Change, Natural Disasters, and Migration Problems in the South Pacific Region: A Systematic Review

Mohammad Afsar Alam
and Azizur Rahman Siddiqui

1 Introduction

There has been a lot of debate and interest in the link between climate-related catastrophes, migration, and shocks in recent years. According to one theory, more people are forced to leave their homes as a result of natural catastrophes brought on by climate change. On the other hand, others argue that the impact of climate change on migration might be minimal at best. The effective management of natural disaster consequences requires evidence-based policy recommendations, which can only be made if we know whether climate change and natural disasters cause more migration. This knowledge is critical for understanding the various pathways between climatic shocks and migration (IZA World of Labour, November 2023).

If they are unable to adapt to the changing climate or endure the effects of a natural catastrophe, are compelled to relocate as a result of the shock, or can pay the expenditures associated with migrating, people will migrate. As a result, public and private mechanisms, such as migration and remittances, must be incorporated in a comprehensive approach to disaster management. Reducing the human impact of climate change requires enhanced international cooperation, in addition to supporting households in building resilience through measures such as investing in vulnerable areas' infrastructure, providing social protection, and efficiently and quickly distributing aid (IZA World of Labour, November 2023).

Recent work from the Pacific Climate Change Science Programme (PCCSP) includes an article discussing how global warming is influencing migratory patterns in the region (ABM & CSIRO, 2011). According to the Agent-Based Model, a moderate climate change scenario Representative Concentration Pathway (RCP 6) and population growth will cause Kiribati and Tuvalu migration to surge by 2055. There will be more migrants in nations like Australia, Fiji, and New Zealand if migration keeps up its current rate of growth. Because of their already-scarce resources, nations like Fiji find this a major reason for alarm. An increase in the level of pressure might have far-reaching consequences for both society and the economy.

Because of climate change and natural disasters, 80% of the migratory population stays within the same country or area, while 20% moves outside (Books, 2014). Ekstrom (2010) found that urbanization, brain drain, and rural–urban drift are all directly impacted by climate change migration. Cities have seen the negative

M. A. Alam (✉)
Department of Geography, Jamia Millia Islamia, New Delhi, India
e-mail: malam18@jmi.ac.in

A. R. Siddiqui
Department of Geography, University of Allahabad, Prayagraj, UP, India

Z. Khanam et al. (eds.), *Climate Change, Resilience and Ecosystem Services*, Sustainable Development Goals Series, https://doi.org/10.1007/978-3-031-96977-5_4

effects of recent migration. From 24% in 1970 to 42% in 2007, and projected to reach 50% in the next decade, with a high coastal concentration, the urban population in Asia has grown substantially (United Nations 2018). Like many other regions, the Pacific nations are very susceptible due to their large volcanic islands or low-lying atolls (Books, 2014).

There is mounting evidence that climate change is already having an impact on migratory patterns and will have an even greater impact in the future. According to Bronen (2010), "in recent decades, there has been an enormous relocation around Asia and Pacific, which is a direct result of Climate Change and the debacle it has brought, due to which there is an incredible change in the portability, monetary and social changes." This study aims to lay the groundwork for a discussion about how climate change is impacting evacuations in the Pacific and Fiji. A new finding implies that climate change will significantly impact future travel plans. Migration patterns, whether ecological or monetary in origin, localized or global in extent, long-term or short-lived, will characterize the immediate response to environmental change (Cloud, 2002).

Environmental change influences migration because people want success, prosperity, happiness, and a better future. It is easy to lose sight of the interconnected nature of migration and climate change when thinking about how to improve policy and practice on a national and international scale. If plans and guidelines can be put in place regarding migration in light of climate change, "humanitarian emergencies will be minimized, clashes avoided, and nations can profit" (Clouds, 2014).

Relocation is impacted by climate change because people in the area may have experienced disasters like floods, intense tropical cyclones, dry spells, and other hazards that had a direct impact on their plants and animals. About 80% of these climate change impacts are domestic or within the same country, whereas 20% are global in scope (Books, 2014).

Climate change has prompted migration for many different causes—at first, due to the widespread destruction of crops and plants. Because of the impact of climate change on these industries, many Asian and Pacific countries rely on ranching and farming, and people in these professions are constantly on the move in search of better growing conditions and other opportunities. Areas in Fiji are prone to flooding, especially Ba, Rakiraki, Labasa, and Vanua Levu. As a result of environmental change, people in these areas are fleeing to safer countries. As migration continues, policies and regulations are needed to control population expansion and preserve resources. Furthermore, people choose to relocate away from areas where environmental change has an impact on the resources available to them. People are forced to leave their homes in various parts of Asia because of environmental issues (Bacolod, 2014).

2 Study Objectives

The following objectives are intended to be achieved by the study.

- To establish a connection between migration, climate change, and catastrophes caused by climate change.
- To learn about the South Pacific region's potential repercussions from climate change, natural disasters, and the resulting migration.
- Determine whether natural catastrophes brought on by climate change are driving up migration rates.
- For the purpose of effectively managing the aftermath of natural catastrophes, we must develop policy suggestions supported by evidence.

3 Materials and Methods

This descriptive research draws on secondary source data on migration, climate change, and natural disasters. Public records offices, academic journals, published books, and unpublished digital versions obtained via the web and software comprise the sources of this material.

Publications from various dates on the web have also been reviewed. However, the Pacific SPC and Regional Infrastructure Facility are the primary sources of data used in this analysis. Additionally, research projects and research articles are included to enhance the study.

4 Analysis

In order to make the study more accessible and understandable for academics and researchers in the field, all of the materials that were gathered from different sources have been carefully examined, processed, organized, and tabulated logically and systematically under the appropriate row and column headings. Whenever they are useful, the research also includes graphical representations in the form of bar and pie charts.

5 Limitations of the Study

The precise nature of the link between climate change, natural catastrophes, and migration remains a point of contention, despite the fact that it is receiving more attention in both academic and policy circles. There must be further proof at the microeconomic level. Even in the context of a region-specific analysis, it can be challenging to establish a common pattern of migration response to climatic shocks. This is due to diverse factors such as differences in adaptation capabilities, imperfect data, differences in variable measurement, and unobserved cultural and anthropological differences. The situation becomes even more complex when there is larger heterogeneity in the space being considered. So, it is possible that this trend is situational (Mbaye, 2023).

While there are ongoing attempts to collect data over longer time periods, this information is still insufficient for understanding the long-term impacts of catastrophes. To better understand the effects of disasters on both the source and destination areas, it is helpful to have access to more data on the socio-demographics of migrants in the aftermath of shocks. It is also important to understand how people displaced by climate shocks and natural disasters differ from conventional migrants, like labor migrants. Policy actions may be better shaped with more data on the impacted people and on individuals who move within their own nations after a catastrophe.

Finally, studying return migration to disaster-stricken places is an important subject for future research. The effects of disaster-induced migration on the health of both the displaced and their communities must be better understood. Why is this so crucial? Well, for starters, it has the potential to influence migrant integration in their host nations, as well as the results of the labor markets in their home countries.

6 Study's Key Points of View

- Major causes of migration and displacement, such as natural catastrophes and climate change, will persist.
- Climate change is particularly dangerous for the poor. They are less likely to have the resources to be ready, less knowledgeable about how to react in the event of a catastrophe, and more likely to reside in high-risk locations. But they are also the ones who will have the most trouble making the move.
- Those who are compelled or choose to relocate due to climate change should be assisted by national adaptation programs. Migrants need to be warned about potential dangers and given the tools to thrive in their new environments.
- The rights of those compelled to relocate abroad must be safeguarded by international frameworks and bilateral agreements.
- The arrival of migrants may increase the strain on services and infrastructure. Addressing the needs and impacts of new migrants caused by climate change should be a priority for national initiatives.

7 Relocating as a Means of Surviving Climate Change and Natural Catastrophes

When people lose their homes and means of subsistence due to natural disasters, migration becomes an inevitable response. Because it lets impacted individuals access new opportunities and resources, migration may be seen as an adaptation strategy in the face of catastrophes. This helps lessen the unfavorable impacts of the shocks. When all other options have been exhausted, it might be used as a means of coping. Reducing the number of meals eaten each day, selling family assets, drawing on food reserves, and taking advantage of solidarity via gifts, loans, and help are all examples of possible solutions and, in some instances, survival strategies to problems like drought. The possibility of migration caused by shocks may be mitigated by having a variety of coping methods available.

The majority of the internal displacement caused by catastrophes in 2020 (91.1% worldwide) occurred in developing nations, which include low- and middle-income countries according to the World Bank's categorization. The majority of the displaced people are citizens of middle-income nations (Fig. 1). It also lends credence to the idea that calamities do not always force the poorest citizens to flee their homes, at least on a national basis. The poorest cannot afford to go, but the wealthy have alternative ways to lessen the impact of catastrophes on their lives, such as restoring their wealth or gaining access to better social services and infrastructure. Those in the center of the income spectrum may only afford to relocate if they have few other options for dealing with natural catastrophes.

The very low rates of disaster-related migration from low-income and high-income nations should not be misunderstood as a sign that no one in either group travels. The correlation between catastrophes and migration is not uniform throughout developing nations; rather, it depends on factors such as the degree to which a person is vulnerable and the magnitude and severity of the shock. Factors including political climate, social structure, asset and income diversification, poverty level, and exposure to shocks all contribute to an individual's vulnerability. For instance, compared to individuals living in calm environments, those whose political climate is unstable have less options for ensuring their own existence. Intensity is a measure of how violent the shocks caused by the tragedy are. How often, how far apart in time and space the shock occurs, and how the impacted individuals interpret it all contribute to a measure of the shock's intensity.

The number of shocks that occur within a certain time frame is called their frequency. Time between two occurrences is measured by the temporal gap. The geographical distribution describes the magnitude of the shock in a particular location; for example, it is simpler to cope with an occurrence that affects only a small portion of the country rather than the whole nation. Lastly, one's expectations and past experiences impact how they receive shocks.

From 1994 to 2010, the story of Bangladesh exemplifies the significance of vulnerability and severity. Particularly for susceptible populations like the impoverished or women, moderate flooding reduced their probability of making a long-distance migration but raised their likelihood of moving locally, in comparison to less severe flooding. In addition, families residing in regions experiencing significant crop loss were more prone to relocate, even if they were not directly impacted by the floods (Gray & Mueller, 2012). When looking at the African example, Fig. 2 effectively shows that different kinds of natural catastrophes have different impacts on migration. At 18.1 million, or 85% of the total disaster-related displacement from 2009 to 2018, floods were the leading cause of displacement during that time.

Even in industrialized nations, there is a correlation between the magnitude of a migratory reaction and the impact of a catastrophe. In the United States, data reveal that countries experience a larger net out-migration in the aftermath of major catastrophes compared to less severe ones (Boustan et al., 2020).

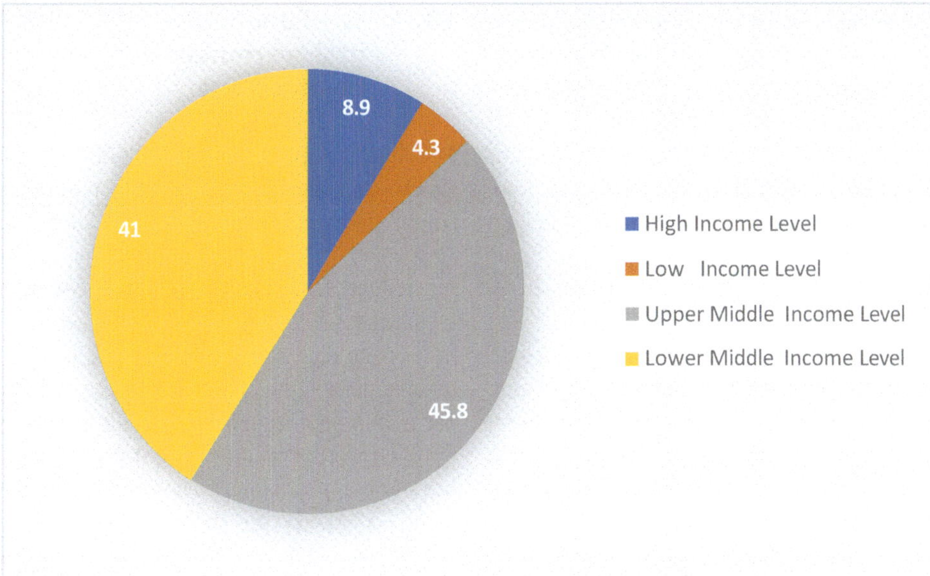

Fig. 1 By income level, the average worldwide displacement brought on by natural catastrophes in 2020. (Source: Prepared by the authors. [IDMC: Global Report on Internal Displacement (2021): Internal displacement in a changing climate. Geneva: The Internal Displacement Monitoring Centre, 2021])

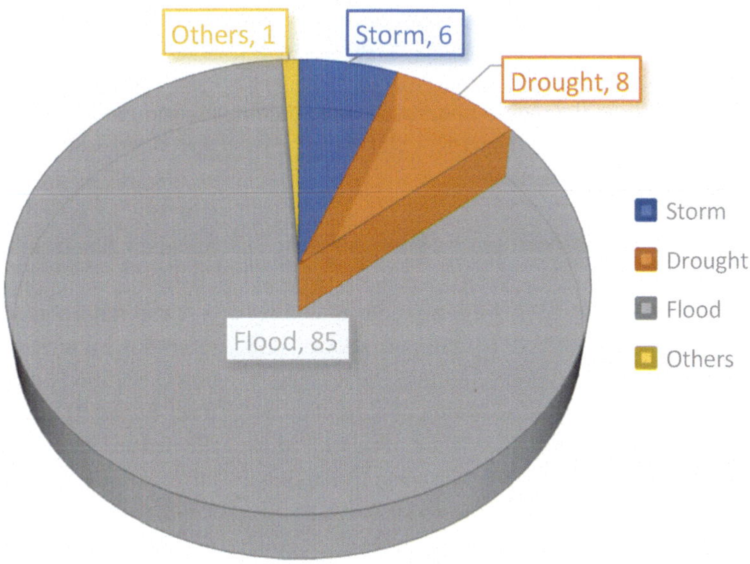

Fig. 2 Displacement in Africa categorized by hazards from 2009 to 2018. (Source: Prepared by the authors. [IDMC. Africa Report on Internal Displacement (2019). Geneva: The Internal Displacement Monitoring Centre, 2019])

8 How Natural Disasters Affect Migration in a Roundabout Way

According to some research, climate has an indirect effect on migration rather than a direct one (Beine & Parsons, 2015). Agricultural production is one potential pathway for transmission (Cattaneo & Peri, 2016; Cai RuoHong et al., 2016). Thus, climate shocks first lower agricultural production, which causes more people to leave their homes. An excellent illustration of how unfavorable climate shocks impact population mobility is shown in economic historical evidence, namely in the agriculture sector. Large dust storms brought on by extreme drought and heavy land usage in the 1930s, for example, permanently eroded soil in agriculturally reliant plains counties in the United States. As a result of people fleeing from high-erosion areas to low-erosion ones, the relative population drop helped to alleviate some of the agricultural expenditures associated with the American Dust Bowl (Hornbeck, 2012).

The impact on agriculture is one way in which natural catastrophes influence internal migration. New research, however, reveals that agricultural production is another way in which climate shocks (i.e., long-term occurrences) influence international migration. For example, from 1980 to 2010, international migration flows were found to increase for countries that rely heavily on agriculture, according to yearly migration data from 163 source countries and 42 destination countries, most of which are (Organsiation for economic co-operation and development OECD) countries (Cai RuoHong et al., 2016). As a result of climate catastrophes, company productivity drops and salaries fall, which in turn motivates people to leave the region in quest of better economic prospects (Boustan et al., 2020).

9 How Remittances Play a Part

Remittances sent back by emigrants may help communities impacted by climate shocks and natural catastrophes cope. Families are less likely to suffer catastrophic losses due to natural catastrophes and shocks when they receive remittances. Transfers made by migrants protect those left behind in the event of unexpected events. So, remittances are a great way for families to cope with the sudden loss of income that catastrophes might create. There is evidence that remittances mitigate poverty more effectively in nations that have experienced natural disasters (Mbaye & Drabo, 2017), which supports the idea that remittances serve as an insurance policy against the financial impact of such catastrophes. It is crucial to reduce the expenses of sending remittances, as this emphasizes.

10 The Function of Governmental Involvement

Reducing a population's susceptibility to climate shocks and increasing resilience require public action (Mbaye & Zimmermann, 2016). In the 1920s and 1930s, for example, some regions of the United States had net inflows of migrants from regions hit by floods, while others saw outflows from regions hit by tornadoes. Early public involvement is the main explanation for the disparity in migratory responses after tornadoes and floods. In order to make flood-prone communities more appealing as places to live, the government invested in their infrastructure and protection in advance, making them more robust to shocks. In places prone to tornadoes, such measures did not take place (Boustan et al., 2012).

Important factors in the choice to move or remain include post-migration public intervention. Again, looking at the United States as an example, there is evidence that subsidies may affect the choice of nearby families to stay in their region after a catastrophe. After Katrina, this is precisely what transpired in New Orleans as a result of the Louisiana Road Home reconstruction grant program, according to evidence based on data from 2004 to 2010 (Fu & Gregory, 2019).

Crop failure generated a tremendous surge in migration in Bangladesh between 1994 and 2010; however, floods did not trigger large migration from impacted regions. Authorities' responses to the calamities explain, once again, the differences in migratory reactions; for example, after floods, aid was supplied, but after crop losses, it was not (Gray & Mueller, 2012).

11 Climate Migration

According to Clouds' (1992) assessment, "Migration will undoubtedly occur whenever there is an ecological issue due to environmental change." Consequently, we might claim that mobility and environmental change are interchangeable. Movement can be brought on by sea level rise, volcanic emissions, bushfires, famine, beachside flooding, and the dry season (Asian Development Bank). The movie *Moana* provided the perfect illustration. Because humans contribute to climate change, there is an obvious connection between environmental change and humanity. Humans burn waste, reject the 3Rs (reduce, recycle, and reuse), and practice deforestation, among other things. In light of this apathy toward humanity, climate change and its consequences become so great that they force mankind to relocate. Therefore, a growing amount of rearranging of relocation plans is likely to occur when environmental changes cause automatic or restricted mobility, or relocation (Richardson, 1992).

As the environment changes gradually, there may be insufficient long-term planning and adjustment, and victim groups' ability to adapt will be weakened by rehabilitation efforts. Thus, it may result in the edges of (maximum) tolerance being satisfied, which over time may cause a change in migration patterns from direct to indirect changes. Similarly, with regard to environmental hazards, significant increases in the frequency, magnitude, or perception of risks may lead to indirect changes (Bronen, 2010). According to a study report, "migration is clearly and extensively used as a major adjustment to react to massive changes in the political, ecological, and financial settings across Asia and the Pacific" (Geey, 1999).

It is crucial to understand that migration affects everyone who lacks the resources to withstand it, not just the impoverished. Migration is inevitable because of the government's lack of action and insufficient resources, which render anybody unable to combat climate change. Insofar as migration and climate change are concerned, migration is mostly caused by climatic change. The main causes of migration include changes in weather patterns, bushfires, starvation, sea level rise, and coastal floods (Suphachalasai, 2014).

Both forced and voluntary migration may be impacted by climate change. In the Pacific, deliberate mobility, whether permanent or temporary, is seen as having three main advantages as a strategy for adapting to environmental change. First and foremost, deliberate migration may reduce population pressure and increase environmental sustainability in areas affected by climate change. Additionally, mobility will enhance the generation of money via remittances. In addition to boosting the local economy, remittances may assist individuals in meeting their fundamental requirements. Thirdly, moving may increase a community's access to technology and knowledge by fostering aptitude transfers between the sending and receiving countries. This is particularly the case when temporary migrants return by accident or when permanent migrants return with newly acquired skills and knowledge. Both the government and the economy will benefit from this.

People migrate forcibly when they are compelled to by external circumstances, such as rising sea levels. According to Fig. 1, the percentages of forced and voluntary immigration are 20% and 80%, respectively. They are referred to be refugees often. The United Nations Environment Programme first used the phrase "environmental refugees" in a report title in 1985. Because the consequences of climate change became more severe and intolerable, people were compelled to leave their homes and communities. This leads to other issues in the migration-affected nations, including few resources and land. Numerous people's health is also impacted by it.

There are two distinct types of migration, as shown in Fig. 3: voluntary and forced. When individuals are compelled to move as a result of climate change, a person is said to have been forced to migrate if they are forced to do so because of climate change. Many are coerced, both from inside and beyond. Rising sea levels and coastal floods have forced many to leave their homes. The worst-case scenario of coastal flooding and rising sea levels has already occurred in countries like Tuvalu and Kiribati (Farbotko & Lazrus et al., 2012). Such a movement may be short-lived or long-term. People engage in voluntary migration when they voluntarily uproot their lives and relocate from their home. As a result of climate change, they feel compelled to do so.

Fig. 3 Migration by choice and compulsion

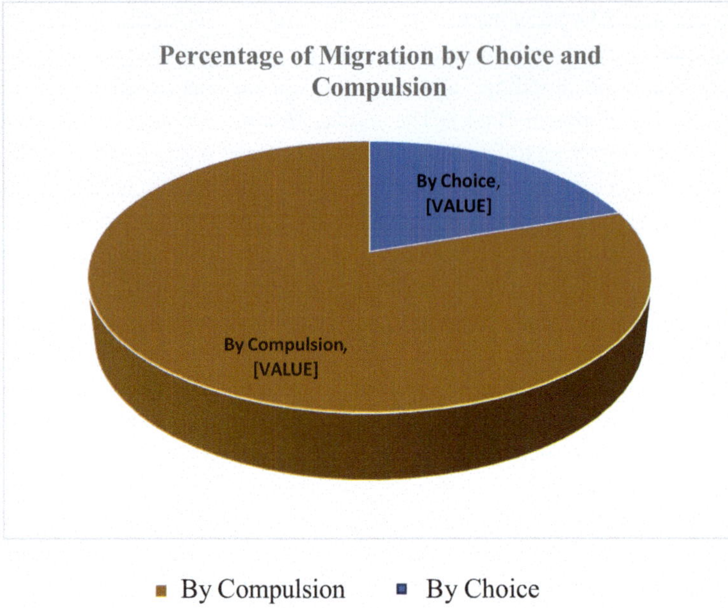

Among the causes of voluntary migration are climatic disasters that, with government assistance, cause predictable and short-term disruptions to livelihoods.

12 Climate Migration in the Pacific

Pacific climatic change and migration is abbreviated as PCCM. Through its collaboration with the (United nations development programme UNDP), the (International Labour Organisation ILO), and United Nations University (UNU), the Economic and Social Commission for Asia and the Pacific (ESCAP) and the United Nations University (UNU) are implementing this EU-funded initiative. UNU oversaw the project's research strategy, data collection, and analysis. As a result of ecological and social changes, as well as chances for education and economic advancement, Pacific Islanders have been moving. According to a United Nations (UN) assessment, nearly half a million Pacific Islanders, or nearly a quarter of the population, live abroad as a result of the heavy momentum of movement from Pacific Island states Asian development Bank (ADB).

Late (2003) states that "sea-level rise, coastal erosion, saltwater intrusion, intense cyclone, weather pattern and more frequent and worst droughts" would cause islanders to migrate more often, as seen in Fig. 2. Consequently, it is anticipated that by 2055, there would be a significant surge in migration to Kiribati and Tuvalu, in line with the usual predictions of population growth and climate change. If the trend keeps up, nations like Fiji, Australia, and New Zealand may see a dramatic influx of migrants, which is very troubling. Imagine a nation like Fiji, where resources are already few; an influx of migrants would put even more strain on the economy and the little resources that are available.

The reasons why individuals move are shown in Fig. 4. However, rising sea levels pose the greatest danger to the Pacific region and are responsible for a great deal of destruction. This has the greatest impact on the Kingdom of Tonga, Tuvalu, and Kiribati. Since saltwater intrusion is associated with rising sea levels, it also adds to migration. Extreme cyclones and droughts brought on by shifting weather patterns also have a role in forcing people to relocate.

Moving could be an option for almost 30% of families in the event that floods or rising sea levels become more severe. Due to the island

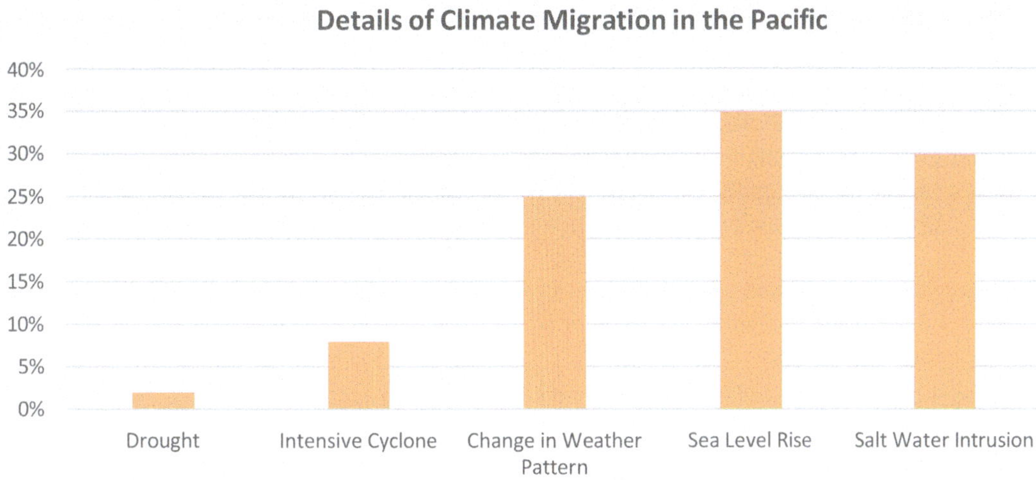

Fig. 4 Climate migration in the Pacific

nation's size and the widespread disruption caused by phosphate mining, internal migration is not a possibility in Nauru. Assuming agricultural production drops, 16% of Nauruan's think migration is unavoidable.

"There was a wide assortment of understandings of the existence, seriousness and impacts of climate change," as per a journal paper published in 2006. However, the majority of respondents were concerned about how migration and climate change will impact island traditions and independence. A small but deeply held belief among islanders is that God will save them, linked to God's promise to Noah—meaning that evacuation is seen as unnecessary. However, opinions vary on how much islanders can adjust to climate change, especially their mobility options. If the sea level continues to rise, action will be required unless the government takes measures to limit it.

Kiribati, Tuvalu, the Kingdom of Tonga, and other atoll nations often suffer from insufficient surface and groundwater as well as a scarcity of arable land, according to Kuruppu (2009). They are particularly susceptible to seawater intrusion and rely on subsistence fishing for the most part. Storm surge, king tides, natural catastrophes, and sea-level rise pose a threat to the whole population because of their concentration on coastal locations. People will inevitably relocate, and neighboring countries will need to step up to assist these countries. Recognizing their nation is especially susceptible to sea-level rise, the Kiribati administration announced in 2008 that they had set up a long-term relocation plan to encourage voluntary migration in order to prevent future forcible displacement. The reasoning behind this is that nations that are relocating will be able to retain some measure of respect and dignity if preparations for migration are made in advance. Two parts make up the vision: first, connecting Kiribati with possible destination nations; and second, providing Kiribati with the skills necessary to obtain employment and reside overseas. Still, the challenge for the current administration will be the plight of the ignorant and unskilled, who will be left behind with no available employment.

Tuvalu, Tonga, Samoa, and Vanuatu are instances where migration is functioning as a constructive development strategy and lowering susceptibility to climate change, as stated in a United Nations study (Busby, 2004). Due to the fact that migration is entirely voluntary and provides a means of generating remittances and income diversity, all four of these nations—New Zealand included—have a large immigrant population. The total yearly population growth rate of Tuvalu is approximately 0.27%, which is rather low considering the country's high fertility rate. The high per capita access to overseas migration in Tuvalu is limiting the population increase (International Social Science Council, United Nations Educational, Scientific, & Cultural Organization, 2013). However, owing to high fertility and low levels of international migration, Kiribati's population is rapidly expanding at a pace of 1.55% per year, making the island nation susceptible to the effects of climate change (International Social Science Council, United Nations Educational, Scientific, & Cultural Organization, 2013).

The nation has to take a few things into consideration before it can move. It is essential to promote relocation in a moral manner from the very beginning. In addition, islanders might benefit from preparation by increasing their content in foreign employment marketplaces. At last, more research in other developing nations with small island populations is required. Thirdly, figure out how to make sure that national capitals may grow in a reasonable fashion. Table 1 provides a list of specific initiatives that contribute to and sustain progress in response to environmental change.

13 Migration Trends in Fiji

Natural catastrophes and the consequences of climate change pose risks to Fiji in the form of tropical cyclones, flash floods, droughts, and increasing sea levels. With respective areas of about 10,429 km^2 and 5556 km^2, Viti Levu and Vanua Levu are the two biggest islands in the country. The majority of Fijians (almost 90%) call the coast home; here, you will find the country's infrastructure, agricultural output, and social hubs. The coastal resources of Fiji are anticipated to be significantly impacted by climate change. Climate change is supposedly playing out in Fiji. The most vulnerable areas of Fiji include Vanua Levu, Ba, Rakiraki, and Labasa. Many Asian nations also experience mass migration as a result of environmental crises, forcing their native populations to abandon their homes and seek refuge elsewhere (Bacolod, 2014).

Despite fierce internal opposition, Fiji and the Pacific Island Countries (PICs) are among the most at-risk nations due to climate change. This group is known as Small Island Developing States (SIDS), based on research by Barnett and Campbell (2010).

Table 1 Projects aid and sustain climate migration in the Pacific

Section	Projects	Slow onset event
Social vulnerability, adaptation, and environmental Migration (SVAEM)	Evidence for policy on migration, the environment, and climate change (MECLEP) Livelihood Resilience in Kiribati: A Resilience academy project on action research on loss and damage in vulnerable nations	Drought and climate change; coastal erosion, drought, salinization, and shifting rainfall patterns; riverbank erosion; and drought
Risk management, adaptive planning, and vulnerability assessment (RMAPVA)	Coastal areas readiness for extreme and rare disasters (PEARL) Urbanization, climate risk, and political economy in Vietnam (PEUR)	Higher water levels; lack of water
Ecosystem services and environmental vulnerability	Promoting the long-term viability of deltaic systems	Intrusion of salinity; melting glaciers
(ESEV)	Via the use of a unified model for evaluating potential threats (DELTAS)	

Source: Climate Change Response from the Pacific Islands

First and foremost, rising sea levels are a major effect of climate change. The global community is fixated on the high anticipated rates of probable sea-level rise. Particularly at risk from inundation and coastal flooding are low-lying and tiny island nations, which are concerned that their land areas would be decreased and, worse, that a large number of its citizens may be forced to relocate to neighboring countries. In light of the growing awareness of climate change and its consequences, there has been a surge in global interest in this subject, according to Mimura (2013). The second is that as sea surface temperatures rise, coral bleaching may become more common. Sediment supply, essential for maintaining coastal stability, might be reduced as a consequence of this and the delay in reef growth. Additionally, coral bleaching may harm marine ecosystems and the variety of life inside them. Third, there may be an uptick in coastal flooding and erosion if storminess patterns change, for example, tropical cyclone numbers or intensity.

Right now, the beachside zone of Viti Levu is dealing with consequences that humans have brought about. Rapid population growth has worsened urban change occurrences, watershed erosion, pollution, and the systematic exploitation of coastal holdings (both physical and biological), leaving vast stretches of shore vulnerable to erosion and flooding. Because of this, coastal frameworks are not very good at adapting to things like climate change, rising sea levels, and human activities; as a result, coastal communities and infrastructure are more likely to be hit hard by things like storm waves, tsunamis, high tides, and sea level rise. Migration from lower to higher zones, for instance, has previously been taken into account. Climate change has forced the evacuation of many cities.

On the island of Viti Levu, the Suva Peninsula is home to more people than any other metropolitan region. The Suva Peninsula and the Rewa Delta will feel the effects of rising sea levels in various ways.

1. The rise of water tables in locations with low elevations.
2. Flooding sewer syphoning frameworks and making in-ground septic systems less practical,
3. When faced with higher intense wave occurrences, the protection of the downtown Suva shoreline is exceeded under a 25-cm SLR scenario.
4. Withdrawal of mangroves from the coast in the Rewa Delta.
5. The Rewa Delta drains have become more sedimented, making them more susceptible to floods.

All of the foregoing contributed to the Suva people's desire to leave the peninsula. The environment and climate significantly impact people's quality of life. Water and food supplies, infrastructure, educational opportunities, and social security systems could all take a hit if climate change makes already-existing health problems worse. These include deaths caused by extreme weather, respiratory and cardiovascular diseases, infectious illnesses, and starvation (World Health Organization, 2015). Changing weather patterns are causing a slew of medical issues in Fiji as well. Nevertheless, individuals have relocated to many other locations. People are fleeing areas and countries hit hard by climate change-related diseases including malaria, filariasis, ciguatera, dengue fever, diarrhea, and nutrition-related ailments (Naicker, 2012; Ali, 2013). There is an increasing number of infectious tropical illnesses, such as dengue fever, typhoid fever, leptospirosis, zika, and chikungunya, that disproportionately affect Pacific islands like Fiji and add to the high mortality rate from non-communicable diseases (Moceituba & Tsang, 2015).

The potential for dengue fever to spread is highest at coastlines and decreases inland. The western and northern parts of Viti Levu are more prone to pestilences than the eastern parts. The vulnerability of Fiji to viral illness flare-ups has been exacerbated by the shifting weather patterns. Fiji had a case of the diarrheal sickness in 2011 owing to the dry season, fought off a flare-up of leptospirosis in 2012 due to the floods, and contained a dengue outbreak in 2013. Because of the increased rainfall in the summer, the danger of leptospirosis, a zoonotic disease, is greater in Fiji. In 2021, there were 160 instances of leptospirosis throughout Fiji, with over 10 fatalities. At the same time, there was one fatality from typhoid and 335 instances of dengue according to the health ministry (Xinhua News, 2021).

A lot of coastal Fijian people, like the ones in Vunidogoloa (in the province of Cakaudrove, Savu Savu), are facing similar problems right now, and moving may be their only choice. Consequently, as the dangers escalate, people keep fleeing from these regions. In 2017, Charan, Kaur, and Singh published a study. Climate change exposed the Vunidogoloa people to floods due to high tides and severe rains, which increased the river level, forcing them to evacuate. They were the first inhabitants of the country to face this threat. The life of the community was greatly affected by sea-level rise, floods, and erosion. The hamlet, which has 26 dwellings in Natewa Bay, is home to 140 people. A more modern response to the effects of climate change, designed migration is only used after all other options have been exhausted by Fiji's Ministry of Economy. Many people who are interested may find migration to be a difficult and confusing time. There are a handful of homes, insubstantial opinions on migration that can endanger identities, and a plethora of losses in terms of psychology, socioeconomic status, cognition, and culture (Dutt, 2013; Benge & Neef, 2020) (Table 2).

The following table shows the number of people displaced in different provinces of Fiji due to climate change. Other major factors that prompted people to leave their homes were coastal erosion, rising sea levels, and saltwater penetration. Moving or transferring was the only option since the arguments were too strong. The following table shows that three communities have moved completely, two have migrated partially, and two more are getting ready to leave. The majority of these people are from the provinces of Cakaudrove, Yadua Island, Kadavu, and Lau on Vanua Levu, the second-largest island in Fiji, with one exception from Ba province (Fig. 3).

Table 2 Climate change impact on migration in the villages of Fiji Islands

Successful migration of villages	Villages partially relocated	Planned relocations in initial stages
The settlement of Vunidogoloa in the province of Cakaudrove	Vunisavisavi village in Cakaudrove province	Narikoso village in Ono, Kadavu
Tukuraki village in Ba	Nagasauva hamlet in the Cakaudrove district, Udu point	Central Lakeba Waciwaci School in Lau
Denimanu village on Yadua Island		

Source: Climate Change Response from the Fiji Islands

There have been three complete town relocations in Fiji, two partial relocations, and two more in the early phases of resettlement; around 40 additional communities are proposed to be moved under the Fijian Green Growth Framework. It was said by DeLacy in 2014 that accurately assessing the social dynamics of these cultures is crucial.

Consequently, the Government of Fiji has put up a Relocation Manual that details their procedures for managing migration work in the country. The regulations are meant to serve as a helpful and brief manual for those who are looking for help with climate-related migration, as well as for the government agencies and non-governmental groups that are helping with the relocation process. Data on climate-related migration in Fiji show that rising sea levels are the primary factor causing the greatest number of departures from the country (45%). Figure 4 shows that only 2% of migrants leave their homes because of health concerns, whereas coastal floods account for 41% and violent cyclones for 12%. The gravity of the repercussions must be carefully considered.

As per the National Climate Change Policy of Fiji, the world's sea levels will rise significantly by the century's end. Since 1993, the average yearly rise in sea level in Fiji has been 6 millimeters (0.2 inch), far greater than the rest of the world. Parts of the island country have become desolate due to the rapid increase in sea levels and the saltwater interruption caused by the more intense intensity of storms that hit the coast. The island is becoming increasingly vulnerable to dangerous food- and water-borne diseases because of rising sea levels, greater temperatures, and harsher El Nino cycles. Protecting Fiji against the worst effects of climate change is a matter of national importance (Fiji 2017–2018, COP23) (Fig. 5).

Sea level rise is likely to occur, and the most severe year for it is shown in Table 3. Rising seas are a problem that this generation and the government must do something about. The government

Fig. 5 The first wave of villagers from Vunidogoloa evaded rising sea levels

Table 3 An overview of the rise in Sea levels

Year	2025	2050	2100
Scenario:			
Mid-range, best guess	11 cm	23 cm	50 cm
High-range, worst case	21 cm	43 cm	103 cm

Source: Climate Change Response from the Fiji Islands

should construct sea barriers to assist prevent coastal flooding and the effects of rising sea levels (Fig. 6). According to Table 4, the Government of Fiji has already begun constructing sea walls to protect communities at risk of coastal flooding and rising sea levels.

The Pacific Islanders' quality of life, chances of survival, and means of subsistence may all be improved via adaptation to climate change. It would still be a top priority to adapt to climate change so that vulnerable populations may lessen its impact and mitigate its impacts as much as possible. Migrating due to climate change would be less necessary if we could adapt and mitigate the effects of climate change. Consequently, traditional adaptation and mitigation strategies for climate change will not be compromised by any measures that encourage migration due to climate change.

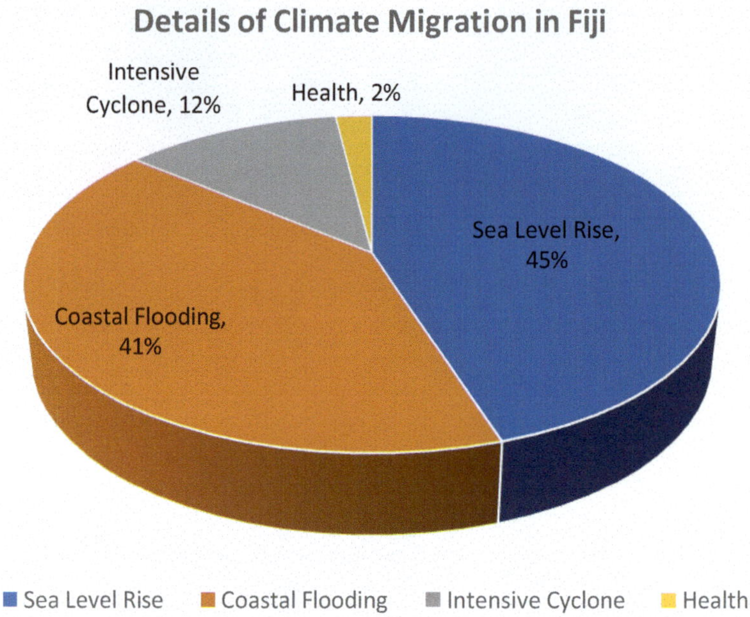

Fig. 6 Climate migration in Fiji Islands

Table 4 Villages benefited from government assistance in constructing sea barriers

Sea walls completed	Future sea wall plans
Kiuva village in Tailevu	Naivakacau in Tailevu—To curb the invasion of saltwater
Kumi village in Tailevu	Vabea village in Ono, Kadavu
Tailevu riverbank protection in the Wainibokasi basin	
Preservation of the Buretu riverbank in the Tailevu village	
The Tailevu Matainoco seawall	
Works to enhance drainage and safeguard Daku village in Tailevu from saltwater intrusion	

Source: Climate Change Response from the Fiji Islands

14 Policy Recommendations

Although natural catastrophes may happen anywhere, they disproportionately affect poor nations because of their lack of resources and ability to recover quickly. After a natural catastrophe has occurred, migration may help lessen the impact. The function is further supported by remittances, which strengthen families' ability to withstand shocks and mitigate their impact on their livelihoods. Migrants may flee their homes or even go abroad in search of safety after a natural disaster. One essential aspect of the link between catastrophes and migration is liquidity limitations, which people must be able to afford in order to relocate. The lack of migration in cases of slow-onset catastrophes might also be explained by liquidity restrictions. Not only that, but individuals may only be able to migrate briefly or engage in short-distance migrations due to liquidity restrictions, which is a common occurrence in situations with a quick start. Another way in which climate-related disasters impact migration is by lowering agricultural production, usually in the form of reduced crop yields. In conclusion, research reveals that the kind of policy interventions before, to, and during catastrophes might shed light on the varying migratory reactions to various disasters.

Private shock-coping strategies like migration and remittances should be seen as supplementary to state action rather than replacements. As a result, inequality between migrant and non-migrant families, as well as remittance recipients and non-receivers, may worsen if migration and remittances were the sole options for reducing the impact of catastrophes. This means that public and private safety nets must work together in a comprehensive strategy for catastrophe management. Governments should therefore provide assistance in remittance processing and migration channeling as needed. Additionally, government agencies may assist families in preparing for disasters by implementing measures such as stronger social protections, more efficient insurance systems, and economic diversification to lessen reliance on agriculture. After catastrophes, it is critical for governments to step in quickly and establish reliable institutions to handle the influx of help (Mbaye & Zimmermann, 2016).

While most people agree that climate change and natural disasters are major problems in the modern world, reaching an agreement on how to solve these issues remains challenging, especially when it comes to nations with different levels of development. Remittances and other private solidarity mechanisms are in operation, but international action is lacking in the magnitude that is required. This raises questions of fairness since developing nations are more susceptible to the negative impacts of climate change than wealthier nations, despite the fact that they contribute relatively less to the process compared to developed economies. Citizens' engagement in political discussions and efforts to address climate change have grown in recent years. More effective global assistance distribution to vulnerable nations and more humane migration policies toward migrants from such nations are necessary to combat climate change.

There are several reasons why people should not see migration as a danger when it follows climate shocks or natural catastrophes. One thing to keep in mind is that the majority of migrants either stay inside their own country or go to neighboring nations. Second, migrants may assist the left-behind in coping with shocks because of the possibility of remittances. Finally, in the event of a climate-related catastrophe, migration should be prioritized as a fundamental human right. In this way, everyone benefits when migrants are welcomed and assimilated into host communities.

15 Findings and Suggestions

This research found that sea levels will rise, coastal acidification would occur more often, and hot days, heavy rainfall, and overall temperature would all increase in the Pacific area. The Pacific is projected to have a sea level increase of 0.18–0.59 m by 2080–2099, which is in line with the worldwide average. Less frequent but more severe droughts are anticipated. There will likely be an increase in the average yearly rainfall. The frequency and intensity of tropical storms and cyclones are projected to decrease.

The most devastating effects of climate change for Pacific islanders will include higher sea levels, more frequent and severe droughts, coastal floods, saltwater penetration, and other similar problems.

Fiji has committed to using only renewable energy sources by 2030 and has begun a program of tree planting with the intention of sequestering carbon dioxide from the air. Fiji has been working with the UN, the Green Climate Fund, and the Global Environment Facility to identify and address climate-related risks as soon as possible.

The Government of Fiji does not have the financial resources to conduct the necessary research and evaluations on the consequences of climate change and mitigation strategies, or to implement the programs that are needed to address the current impacts of climate change. Consequently, Fiji will need short- and medium-term support from the convention's financial system, along with further donor relief. To combat climate change and reduce the need for migration, Fiji is taking all necessary measures. As a tiny nation, Fiji risks economic stagnation and disruption to its cyclical flow if its citizens begin to leave. Because of the insurmountable challenges posed by climate change, migration has become an absolute need in the other Pacific nations. It is easier to move if you are young and talented, and you may find that you fit in well in your new nation. However, the government must devise means of providing for the remaining unskilled and illiterate population. Fiji and the Pacific face a greater threat from climate change than the rest of the world due to their extreme poverty, lack of development, and, most importantly, their oceanic position. As a result, migration is obvious and inevitable, on a global and regional scale. The process of migrating is never without its pros and cons. The government can tackle climate change in several ways that will aid migration. Public education and awareness campaigns are two of the most effective tools at the government's disposal. Handling the issue of rising sea levels and climate change sea level rise (SLR) challenges requires a comprehensive national plan, which includes education programs, public awareness campaigns, and calls to action (Fiji Islands Response, Fiji Climate Change). Future efforts will concentrate on expanding climate change coverage or inclusion in the following domains. Conversely, in order to lessen the severity of climate change, Fiji plans to implement adaptation programs at the grassroots level. A project for adaptation that is supported by the Canadian government is another something it is involved in. Sustainable waste management is being promoted and endorsed by the government as a means to combat the effects of climate change, preserve Fiji's biodiversity, lessen the depletion of natural resources, educate the public, and enact the Sustainable Bill.

Another strategy to lessen the impact of coastal flooding caused by rising sea levels is to construct sea barriers. As a preventative measure against coastal flooding and rising sea levels, the Government of Fiji has begun constructing sea walls around evacuee villages.

16 Summary and Conclusion

Worldwide, natural catastrophes and climate change pose the greatest risks to present and future generations' ability to advance economically and socially. The link between climate change and human migration has likely always been there, but the exact ways in which one influences the other remain a mystery. In the wake of a tragedy, it is crucial for policymakers to understand the factors that may encourage or discourage migration. Determining the nature of the interaction between climate change, natural disasters, and migration, as well as the impact of shock intensity, is of the utmost importance. More importantly, understanding whether individuals migrate inside their own country or outside likely would not have the same impact on migrant policy management, future remittance amounts, or the economics of relocation. For these and other reasons, policymakers want scientific data to educate them on the various migratory reactions to climate change and natural catastrophes. Better policies to shield the most vulnerable people from the effects of climate change may be crafted with this information in hand.

References

Ali, M. (2013). Dengue: Factors driving the emerging epidemic. *Clinical Epidemiology, 5*, 461–463. https://doi.org/10.2147/CLEP.S54834

Australian Bureau of Meteorology (ABM), Commonwealth Scientific and Industrial Research Organisation (CSIRO). (2011). *Climate change in the Pacific: Scientific assessment and new research: Volume 1—Regional overview*.

Bacolod, E. D. (2014). Fiji 2020 agriculture sector policy agenda. *A report prepared by the Ministry of Agriculture with the Technical Assistance from the Food and Agriculture Organization of the United Nations*.

Barnett, J., & Campbell, J. (2010). *Climate change and small Island states: Power, knowledge and the South Pacific*. Earthscan.

Beine, M., & Parsons, C. (2015). Climatic factors as determinants of international migration. *The Scandinavian Journal of Economics, 117*(2), 723–767.

Benge, L., & Neef, A. (2020). Planned relocation as a contentious strategy of climate change adaptation in Fiji. In *Climate-induced disasters in the Asia-Pacific region: Response, recovery, adaptation*. Emerald Publishing Limited.

Books, M. M. (2014). *Asian Development Bank*. Asian Development Bank.

Boustan, L. P., Kahn, M. E., & Rhode, P. W. (2012). Moving to higher ground: Migration response to natural disasters in the early twentieth century. *American Economic Review, 102*(3), 238–244.

Boustan, L. P., Kahn, M. E., Rhode, P. W., & Yanguas, M. L. (2020). The effect of natural disasters on economic activity in US counties: A century of data. *Journal of Urban Economics, 118*, 103257.

Bronen, R. (2010). Forced migration of Alaskan indigenous communities due to climate change. In *In Environment, forced migration and social vulnerability* (pp. 87–98). Berlin Work Press.

Busby, N. P. (2004). *The security implications of climate change for the UN system. Environmental change and security project report.* United Nations Press.
Cai RuoHong, C. R., Feng ShuaiZhang, F. S., Oppenheimer, M., & Pytlikova, M. (2016). *Climate variability and international migration: The importance of the agricultural linkage.*
Cattaneo, C., & Peri, G. (2016). The migration response to increasing temperatures. *Journal of Development Economics, 122*, 127–146.
Charan, D., Kaur, M., & Singh, P. (2017). Customary land and climate change induced relocation—A case study of Vunidogoloa Village, Vanua Levu, Fiji. In *Climate change adaptation in Pacific countries* (pp. 19–33). Springer.
Cloud, D. (1992). *A regional geography*. Cambridge University Press.
Cloud, D. (2002). *Traversing laser locating system in migration*. York University Press.
Clouds, D. (2014). *The Leodicario show*. Dicapario Press.
COP23. (2017–18). https://cop23.com.fj/fiji-and-the-pacific/how-fiji-is-affected-by-climate- change/
DeLacy, T. (2014). *Green growth and travelism: Concept, policy and practice for sustainable tourism*. Fijian Green Growth Framework.
Dutt, L. (2013). *Relocation of the islands*. The Fiji Times.
Ekstrom, M. (2010). A framework to diagnose barriers to climate change adaptation. *Proceedings of the National Academy of Sciences, 107*(51), 125–129.
Farbotko, C., & Lazrus, H. (2012). The first climate refugees? Contesting global narratives of climate change in Tuvalu. *Global environmental change, 22*(2), 382–390.
Fu, C., & Gregory, J. (2019). Estimation of an equilibrium model with externalities: Post-disaster neighborhood rebuilding. *Econometrica, 87*(2), 387–421.
Geey, V. (1999). *Environment and planning*. Nelson Press.
Gray, C. L., & Mueller, V. (2012). Natural disasters and population mobility in Bangladesh. *Proceedings of the National Academy of Sciences, 109*(16), 6000–6005.
Hornbeck, R. (2012). The enduring impact of the American dust bowl: Short-and long-run adjustments to environmental catastrophe. *American Economic Review, 102*(4), 1477–1507.
International Social Science Council, United Nations Educational, Scientific, & Cultural Organization. (2013). World Social Science Report 2013 Changing Global Environments: Changing Global Environments. OECD Publishing.
Kuruppu, N. (2009). Adapting water resources to climate change in Kiribati: the importance of cultural values and meanings. *Environmental Science & Policy, 12*(7), 799–809.
Late, L. R. (2003). *Quaternary cold events recorded in Southern Ocean sediments in the Pacific*. Pacific Sector.
Mbaye, L. M. O. (2023). *Climate change, natural disasters, and migration*. IZA World of Labor.
Mbaye, L. M., & Drabo, A. (2017). Natural disasters and poverty reduction: Do remittances matter? *CESifo Economic Studies, 63*(4), 481–499.
Mbaye, L., & Zimmermann, K. F. (2016). *Natural disasters and human mobility*. ZEF-Center for Development Research, University of Bonn, Working Paper, 151.
Mimura, N. (2013). Sea-level rise caused by climate change and its implications for society. *Proceedings of the Japan Academy. Series B, Physical and biological sciences, 89*(7), 281–301. https://doi.org/10.2183/pjab.89.281
Moceituba, A., & Tsang, M. (2015). Averting climate change's health effects in Fiji. *World Health Organization. Bulletin of the World Health Organization, 93*(11), 746.
Naicker, M. (2012). Climate change and health in Fiji: Environmental epidemiology of infectious diseases and potential for climate-based early warning systems. *Fiji Journal of Public Health, 1*, 7–13.
Richardson. (1992). The Caribbean in the wider world. In *A regional geography* (pp. 58–69). Cambridge University Press.
Suphachalasai, A. (2014). *Assessing the costs of climate change and adaptation in South Asia*. Asian Development Bank.
United Nations. (2013). *International migration report 2013*. UN.
United Nations. (2018). *2018 revision of world urbanization prospects*.
World Health Organization. (2015). *Climate and health country profile 2015: Fiji* (No. WHO/FWC/PHE/EPE/15.35). World Health Organization.
Xinhua News. (2021). *More than 10 Fijians die from leptospirosis this year*. http://www.xinhuanet.com/english/asiapacific/2021-03/03/c_139780593.htm

Coastal Local Community Adaptation to Hazards: Empowering Resilience and Sustainability: A Case Study in the Red River Delta, Vietnam

Hue Thi Thu Ha and Hung Sy Uong

1 Introduction

Coastal areas are increasingly exposed to risks such as erosion and flooding, which escalate the costs for communities adapting to these disasters (Lincke & Hinkel, 2018). Rising sea levels, along with other hazards, impact coastal communities by affecting land, livelihoods, and the availability of natural resources such as minerals, forests, and water bodies. They also threaten intangible assets like traditions, languages, folklore, and historical sites (Noble et al., 2015). Soil erosion, rising sea levels, and flooding pose dangers to coastal communities, leading to severe damage to many coastal structures (Kumar et al., 2018).

In addition to affecting residences, roadways, and structural facilities, these hazards have the potential to destroy significant cultural treasures, including burial grounds and sacred sites (Nunn et al., 2017). The intrusion of saltwater into freshwater sources contaminates important drinking water supplies and makes land unsuitable for farming (Gingerich et al., 2017; Storlazzi et al., 2018). Depending on factors such as area size, population density, and whether the region is urban or rural, coastal adaptation may involve various strategies, such as land reclamation and on-site protection through engineering or ecosystem-based approaches (Oppenheimer et al., 2019).

In cities, adaptation measures may include complex and costly solutions, such as land reclamation, dikes, and seawalls (Aerts, 2018). Approaches to adapting to natural disasters can be either top-down or bottom-up (Bennett et al., 2016; Butler et al., 2015). Research into grassroots adaptation within small island communities has shown that local cultural and social factors significantly affect adaptive capability (Lin et al., 2017; Nunn et al., 2017; Rouse et al., 2017), as well as the availability of natural resources (Cinner et al., 2018; Warrick et al., 2017). Adaptation cost–benefit analyses typically focus only on the financial aspects of minimizing impacts on infrastructure and people, often neglecting long-term impacts on coastal environmental health, security, and the welfare of indigenous, at-risk communities (Jenkins & Jupiter, 2015).

Adapting to coastal hazards is more cost-effective and prevents greater damage in the long run than not adapting (Lincke & Hinkel, 2018). Understanding the environmental, social, and other factors influencing local adaptation contexts is crucial for developing effective and culturally appropriate adaptation strategies (Barnett & O'Neill, 2012). All countries have action plans

H. T. T. Ha (✉)
School of Interdisciplinary Sciences and Arts, Vietnam National University, Ha Noi (VNU-SIS), Vietnam

H. S. Uong
WWF, Quang Tri, Vietnam

Z. Khanam et al. (eds.), *Climate Change, Resilience and Ecosystem Services*, Sustainable Development Goals Series, https://doi.org/10.1007/978-3-031-96977-5_5

to adapt to climate change at the national level (Betzold, 2015; Robinson, 2015).

Like many other countries, Vietnam is increasingly suffering losses due to climate change and human activities. With 3260 km of coastline, Vietnam faces a high risk of rising sea levels. The impact of climate change is estimated to cause the Vietnamese economy to lose about 3.2% of its gross domestic product (GDP), with expectations for further increases in the future (World Bank Group, 2022). Climate change is causing significant salinity increases in Vietnam's coastal areas, reducing rice production and leading to other environmental consequences. Saltwater intrusion into the mainland severely affects agricultural production and domestic water sources. Understanding community awareness of climate change is important for integrating climate change response into strategies and planning to ensure future socioeconomic development

The Red River Delta (RRD), one of Vietnam's two largest deltas, is crucial to the country's socioeconomic development and is facing severe impacts from natural disasters and climate change. Previous studies on the Red River Delta have focused primarily on determining the paleoenvironment and ancient coastline during the Holocene to aid in future sea-level modeling and survival strategies (Duong et al., 2020). Other research works have addressed topics such as mangrove forest mapping (Long et al., 2021), sustainable intensive agriculture (Morton, 2020), and heavy metals and antibiotic residues in surface water (Da Le et al., 2021). However, there has been no large-scale research from central to local levels or involving the population to assess awareness of natural and man-made disasters in the coastal provinces along the Red River, and how people cope with these disasters, thereby integrating community vulnerability into policy briefs.

This study aims to address critical gaps in understanding how coastal communities can effectively adapt to environmental hazards and build resilience. Specifically, the study seeks to: (1) Identify and analyze adaptation strategies—this includes investigating both traditional and innovative approaches to hazard management and resilience building; (2) Evaluate the effectiveness of current measures—this involves evaluating the success of different strategies in enhancing community resilience and identifying areas where improvements are needed; (3) Explore socioeconomic and environmental interactions—this includes understanding how socioeconomic disparities influence adaptation capacity and how environmental changes affect community resilience; and (4) Propose sustainable adaptation frameworks—these frameworks aim to integrate community perspectives, scientific knowledge, and policy recommendations to support sustainable and resilient coastal management practices.

2 Research Methodology

2.1 Study Sites

The Red River Delta is located in the northern region of Vietnam, downstream of the Red River and Thai Binh River. It encompasses 10 provinces and is the second-largest delta in Vietnam, with a population of 21.6 million people, accounting for 22.1% of the country's population. It is also the most densely populated region in Vietnam. The Red River Delta plays a crucial role in the socioeconomic development of the country, with a coastline stretching 400 km from Hai Phong to Ninh Binh.

The Red River Delta (RRD) is significant not only for its geographical and environmental features but also for its substantial socioeconomic contributions. It supports millions of people and is vital to the national economy, particularly in agriculture, fisheries, and industry. The Red River Delta is a major agricultural zone, significantly contributing to Vietnam's rice production. Its extensive paddy fields are essential for the country's food security, and the fertile soil and abundant water resources make it ideal for growing rice, vegetables, and fruits for both local consumption and national supply.

The delta's coastal and estuary areas are crucial for fisheries and aquaculture. The region is renowned for its production of clams, shrimp, and fish, which are important for both domestic consumption and export. Aquaculture, in particular, is a key economic activity that provides livelihoods for many local families and contributes to the national economy through seafood exports. Over recent decades, the Red River Delta has experienced rapid economic growth and industrialization. Cities like Hanoi and Hai Phong have become economic powerhouses, driving industrial and commercial activities. This growth has led to job creation and increased income levels but also imposes pressures on natural resources and environmental sustainability.

The study was conducted in 6 communes and 12 villages across 3 coastal provinces of the Red River Delta: Hai Phong, Thai Binh, and Ninh Binh (see Fig. 1). In this region, farmers typically live in close-knit villages, where they are connected through land ownership, village culture, and familial relationships. They adhere to village regulations related to environmental protection and community norms.

2.2 Data Collection

Data were collected from 2020 to 2023 as part of a comprehensive study on sustainable development in the Red River Delta. The study involved conducting in-depth interviews with leaders at various levels—central, provincial, district, commune, and village—on coastal issues and local community adaptation. A total of 33 in-depth interviews were conducted, along with 36 group discussions (including groups of women, general community members, and youth) and interviews with 900 households.

All interview and group discussion formats were tested and research staff received thorough training. Each group discussion, consisting of eight to ten participants, lasted approximately two hours. Household interviews were conducted in about one hour. For household interviews, a systematic sampling method was used. The list of households in each village was obtained from village heads. The distance coefficient k: $k = N/n$ (N is total number of individuals in the population, n is the sample size (set at 75 for each village). The

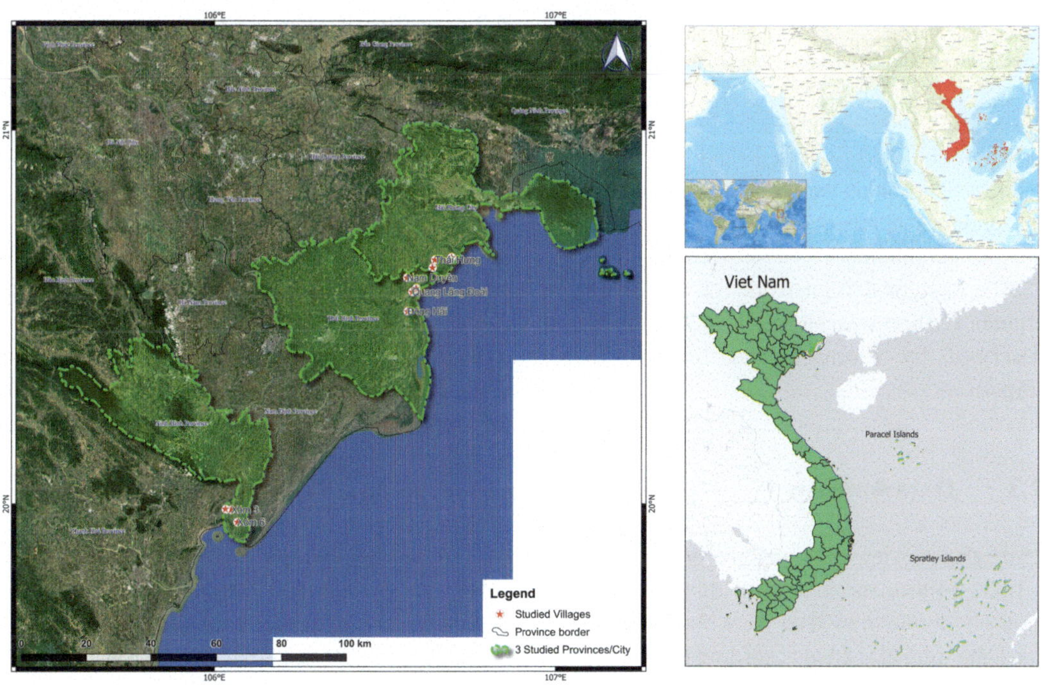

Fig. 1 Location of study site

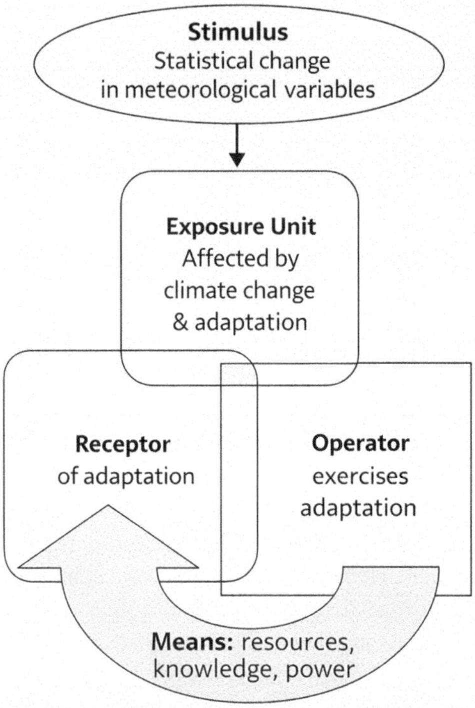

Fig. 2 The diagram illustrates key concepts of the framework. Rounded-corner boxes represent entities that can function as either actors or biophysical units, while operators exclusively function as actors. It is important to note that operators, receptors, and exposure units may not always align, as indicated by the overlapping boxes. A straight arrow denotes a causal relationship, whereas a larger arrow signifies a teleological connection

first household was selected randomly, and subsequent households were selected by adding k to the previous number. The 75th research object had the serial number $i + (75 - 1)\,k$.

The analytical framework for responding to climate change was based on the methodology described by Eisenack and Stecker (2020) (Fig. 2).

2.3 Data Analysis

In-depth interview data were recorded and transcribed to ensure accuracy and prevent distortion. Household interview data were coded for analysis. Data were classified according to a road-map: identifying hazards, affected assets, conditions required for adaptation, and community adaptation capacity. Each response was rated on a scale of 1 (Low), 2 (Average), 3 (High), and 4 (Very High). Statistical Package for the Social Sciences (SPSS) Version 20 software and Microsoft Excel were used to analyze graphs, descriptive statistics, and test parameters (Hue et al., 2018).

3 Results

3.1 Basic Information about Household Interview Participants

The interview participants were predominantly women across all three provinces (see Fig. 3). They were classified as average households according to Vietnamese standards, with most having received secondary education. Their main occupations included agriculture and aquaculture. Additionally, they engaged in alternative activities such as handicrafts (in Ninh Binh), harvesting seafood from mangrove forests, or working as hired laborers.

3.2 Identify the Hazards

Water pollution is the most frequently mentioned natural and man-made hazard across all three provinces of the Red River Delta (54.6%), followed by Coronavirus Disease-2019 (COVID-19) and diseases (including those affecting aquaculture, livestock, human health, and crops), extreme weather events (such as floods, droughts, storms, and erratic rain and sunshine), salinity, and environmental pollution (such as air and waste pollution). Figure 4 illustrates people's awareness of the top five natural and man-made hazards in the Red River Delta and the likelihood of these hazards occurring in the future.

The results of testing several hypotheses using the Independent-Samples Kruskal-Wallis Test in SPSS confirm that the hazards mentioned have a significant impact at the study site (Table 1).

Perceptions of hazards vary between provinces, as demonstrated by pairwise comparisons (Table 2).

Some of the main causes of these hazards, as reported by locals, are primarily related to human activities contributing to water pollution. According to residents, water pollution began around five to six years ago due to wastewater from factories in industrial parks being discharged directly into rivers. Additionally, people use many chemicals in agricultural production; they wash pesticide sprayers in rivers, discard pesticide and herbicide packaging into rivers, and allow wastewater from households to flow directly into rivers. There are also issues with people encroaching on canals to build houses and disposing of garbage indiscriminately into the rivers. Intensive farming households use chemicals to clean ponds, which are then discharged directly into rivers. This polluted water affects the health of aquatic ecosystems, diminishing fish and shellfish stocks, and directly impacts the fishing and aquaculture industries. Contaminated water sources also pose health risks to local communities, leading to increased medical costs and

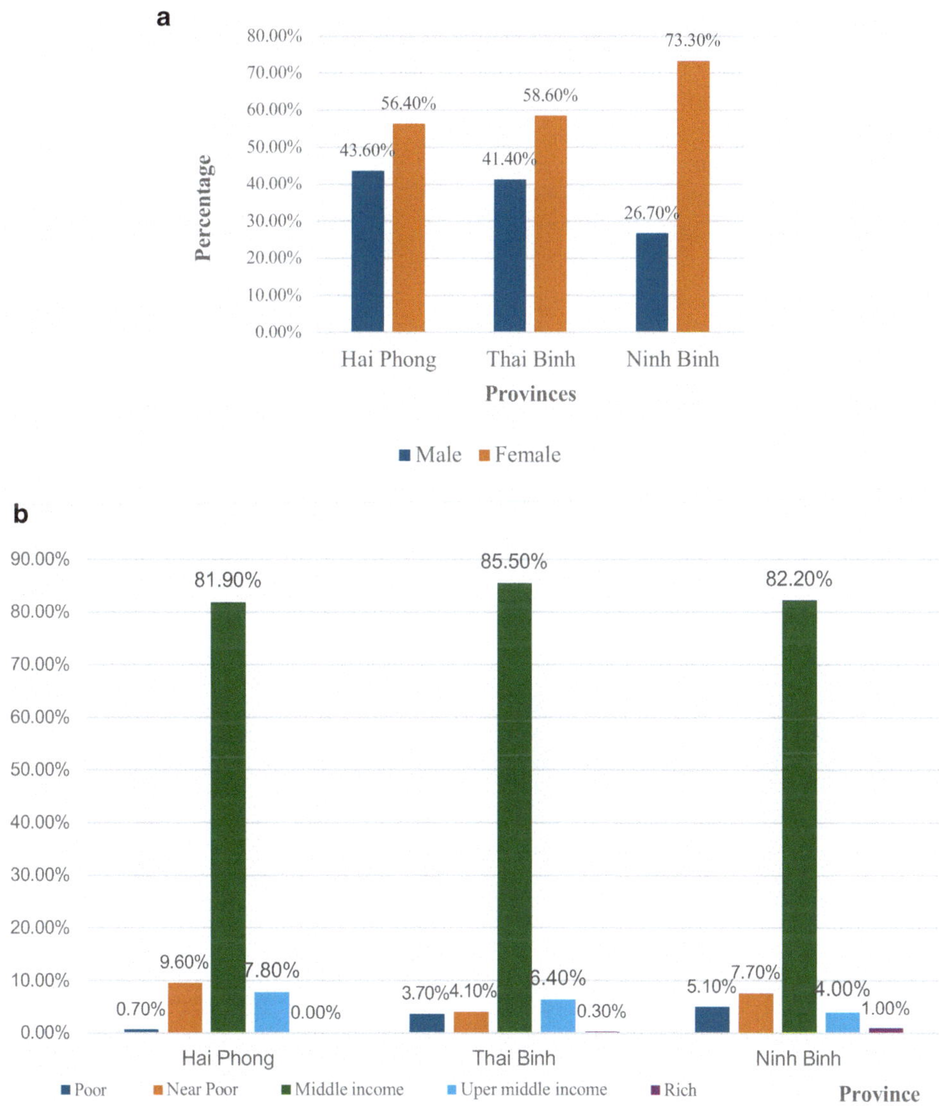

Fig. 3 General information of the respondents

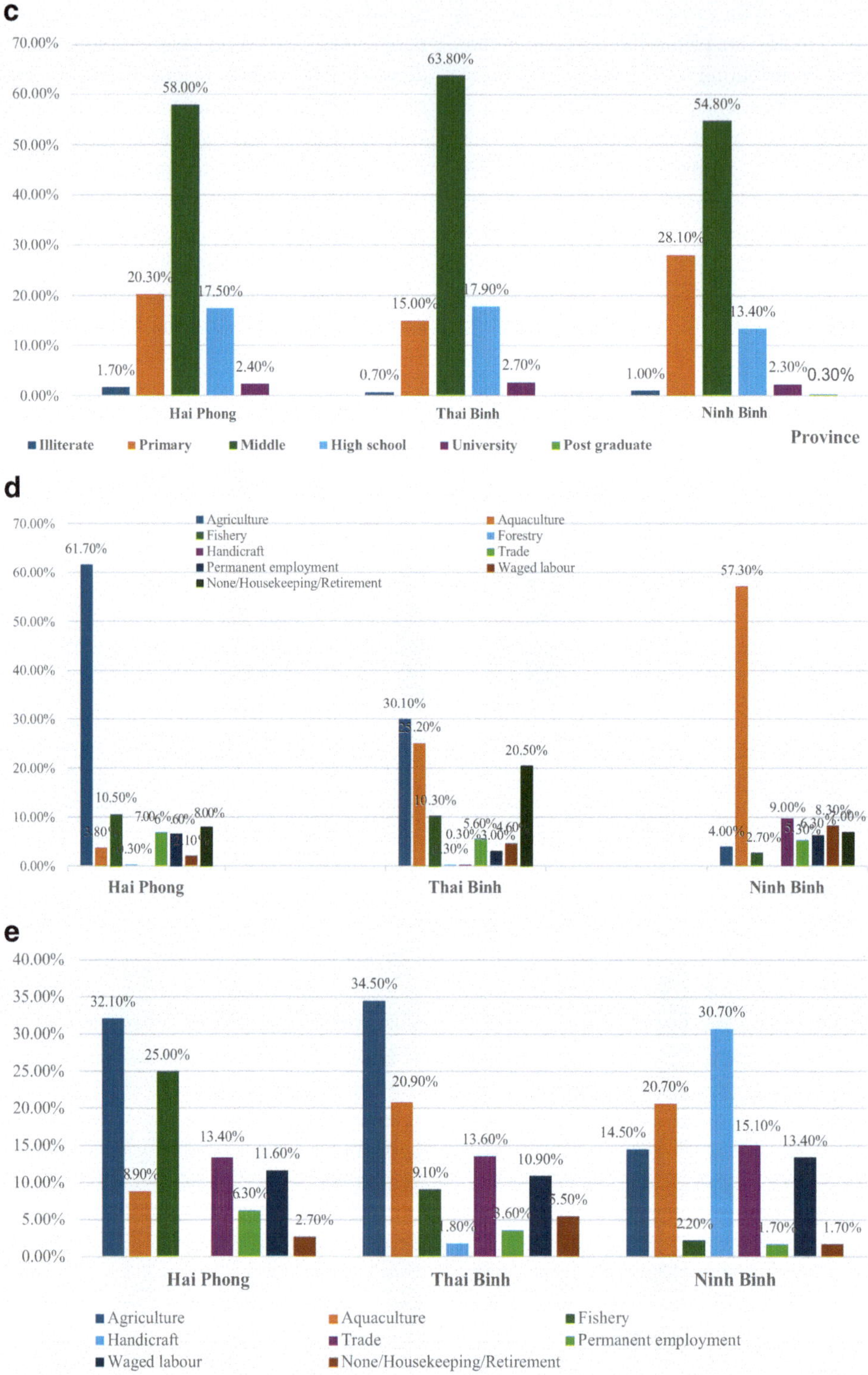

Fig. 3 (continued)

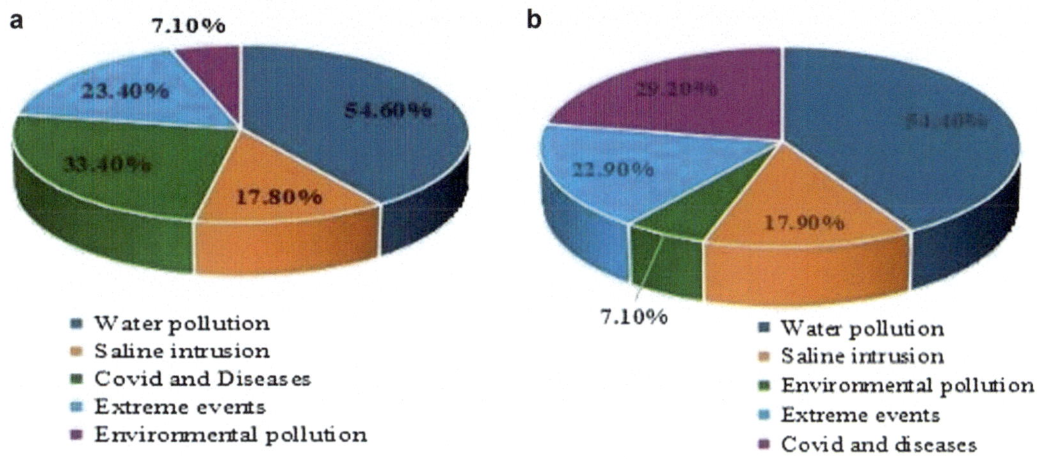

Fig. 4 Identification of hazards

Table 1 Hypothesis test summary

	Null hypothesis	Test	Significance[a]	Decision
1	The distribution of WaterPollution_Imp is the same across categories of province	Independent-Samples Kruskal-Wallis Test	0.035	Reject the null hypothesis
2	The distribution of SalineIntrusion_Imp is the same across categories of province	Independent-samples Kruskal-Wallis test	<0.001	Reject the null hypothesis
3	The distribution of ExtremeClimate_Imp is the same across categories of province	Independent-samples Kruskal-Wallis test	0.028	Reject the null hypothesis
4	The distribution of Temperature_Imp is the same across categories of province	Independent-samples Kruskal-Wallis test	0.003	Reject the null hypothesis
5	The distribution of ClimateChange_Imp is the same across categories of province	Independent-samples Kruskal-Wallis test	0.039	Reject the null hypothesis
6	The distribution of Diseases_Imp is the same across categories of province	Independent-samples Kruskal-Wallis test	0.014	Reject the null hypothesis
7	The distribution of Covid_Imp is the same across categories of province	Independent-samples Kruskal-Wallis test	0.008	Reject the null hypothesis

[a]The significance level is 0.050

loss of productivity. Extensive farming households, in particular, face mass die-offs of fish and shrimp due to polluted water. Extreme weather phenomena (such as extreme heat, heavy rain, and erratic weather patterns) are attributed to climate change (78%) (Fig. 5). However, when asked about climate change, people typically describe it as a long-term change in weather patterns.

The interactions between water pollution, saline intrusion, disease, extreme weather events, and environmental pollution create a complex web of challenges for the Red River Delta. These hazards do not act in isolation but rather amplify and compound each other's effects, leading to more severe and multifaceted impacts on local economies, ecosystems, and communities. For example, water pollution from industrial discharges and agricultural runoff can contribute to elevated salinity levels in coastal and riverine areas. When saline water intrudes into freshwater systems, it can further concentrate pollutants in the soil and water. This increased salinity exacerbates the negative impacts on soil quality and crop productivity, making it even more challenging for farmers to

Table 2 Differences in perception of hazards across the three provinces

Water pollution	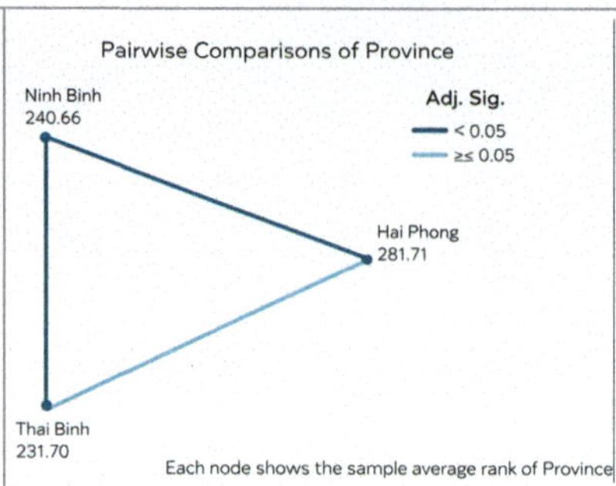	There is a difference in awareness about water pollution between people in Thai Binh and Hai Phong provinces
Saline intrusion	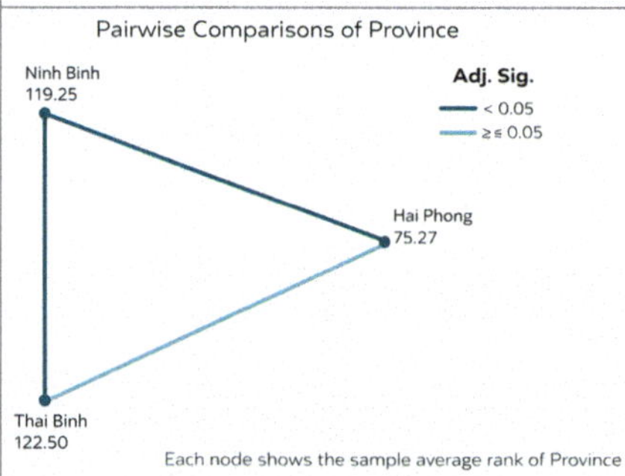	There is a difference in awareness about saline intrusion between people in Thai Binh and Hai Phong provinces
Environmental pollution	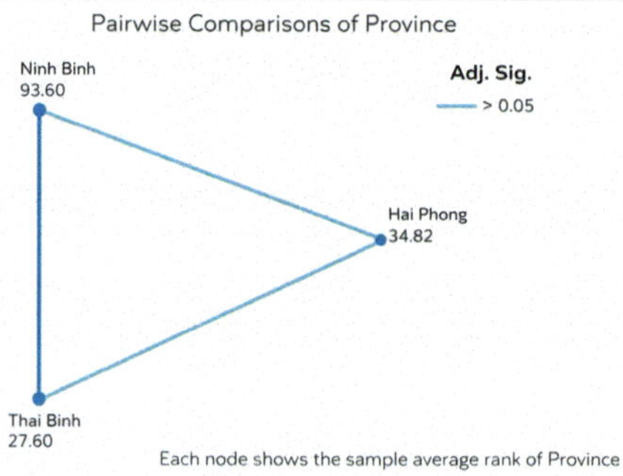	There are differences in awareness about environmental pollution among people of all three provinces
Extreme weather	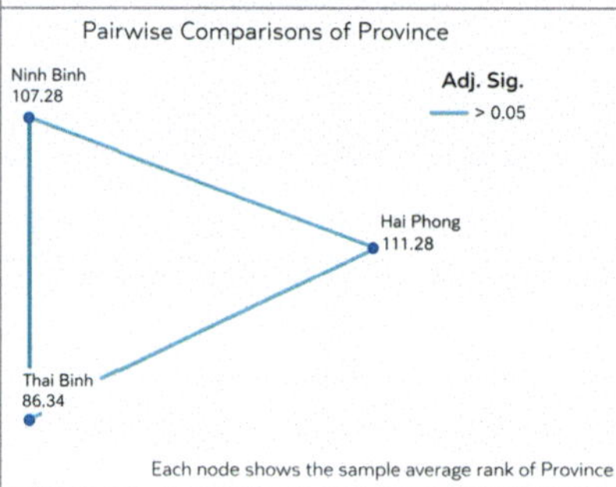	There are differences in awareness of extreme weather among people of all three provinces
Covid and disease	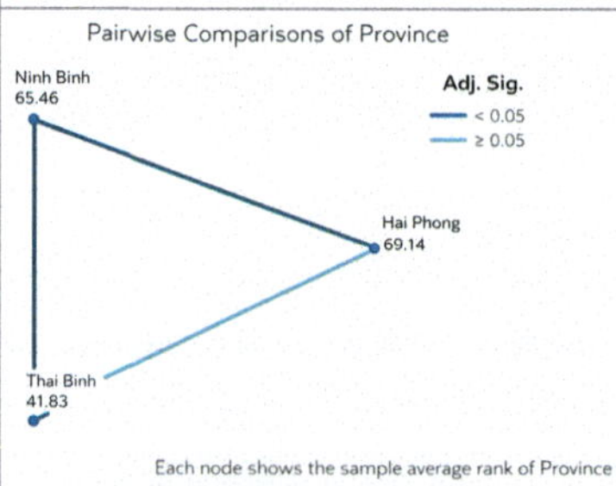	There are differences in awareness about Covid and disease between people of Thai Binh–Ninh Binh and Thai Binh–Hai Phong

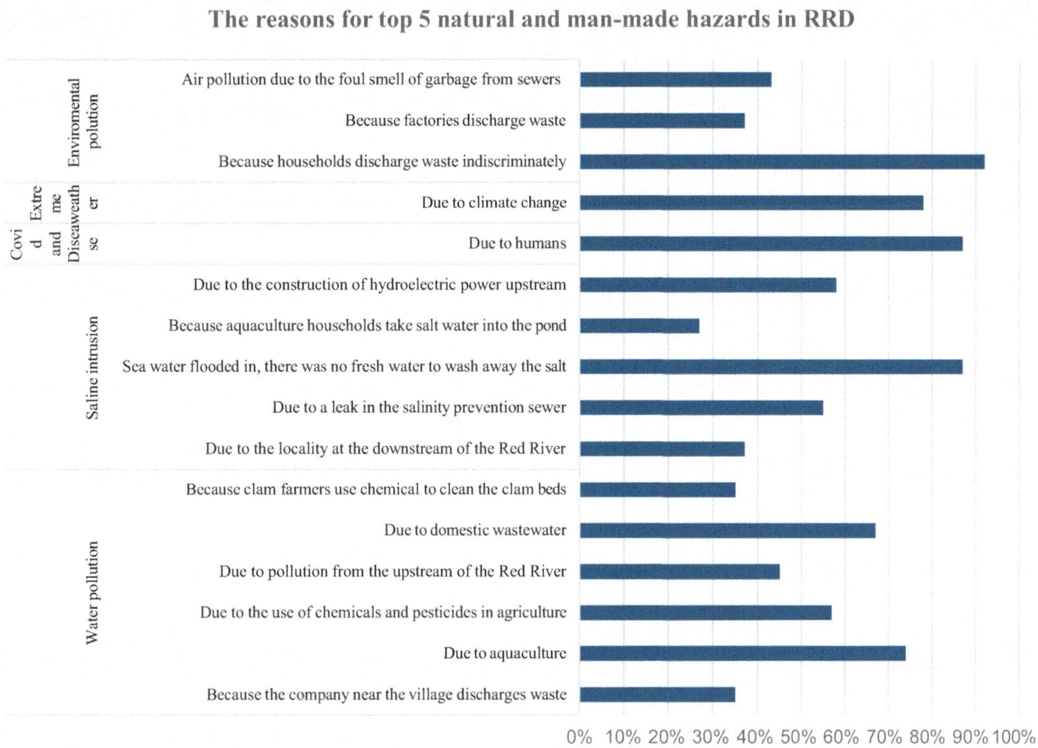

Fig. 5 Reasons for the top five natural and man-made hazards in the Red River Delta

grow rice and other essential crops. Polluted water sources that become saline are less suitable for irrigation and aquaculture. The combination of saline intrusion and water pollution can lead to more severe water quality degradation, affecting both agricultural yields and aquatic ecosystems. This compounded effect reduces the availability of clean water for both human consumption and agricultural use.

3.3 Affected Assets

Water pollution, environmental pollution, and salinity are the three hazards that cause the most damage to people in the coastal areas of the Red River Delta (Fig. 6). Salinity impacts agricultural production significantly. In the provinces of Hai Phong and Thai Binh, where agriculture is prevalent, residents grow crops such as onions, garlic, and tobacco, which are vital for their income and food supply.

When water is polluted, especially water used for aquaculture, it leads to the death of shrimp, crabs, and fish in aquaculture ponds, resulting in substantial losses. Many households are forced to abandon their ponds, rent them out, sell them, or leave them unused. In Thai Binh, locals report that due to severe water pollution, 70% of shrimp production has perished, and approximately 20% of sea bass and grouper have died compared to nearly a decade ago.

These hazards severely impact agricultural and livestock production, employment in agriculture, inland shrimp and fish farming, small-scale inshore fishing, and locally generated income. They also affect drinking water, environmental sanitation, and human health. Additionally, saline water damages household furniture and equipment. While these hazards primarily affect households, they have minimal impact on infrastructure such as transportation, health facilities, and educational institutions at the community level.

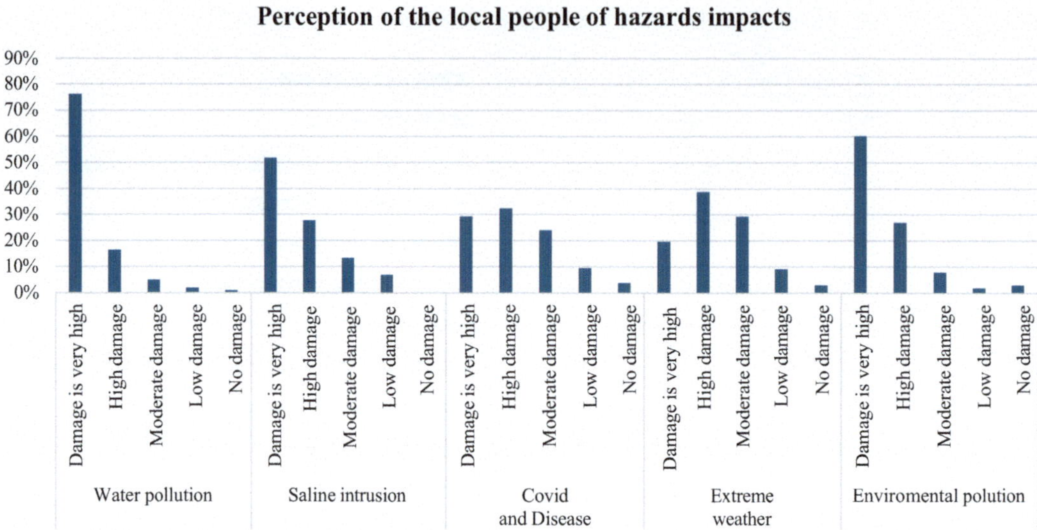

Fig. 6 Percentage of local people affected by hazards

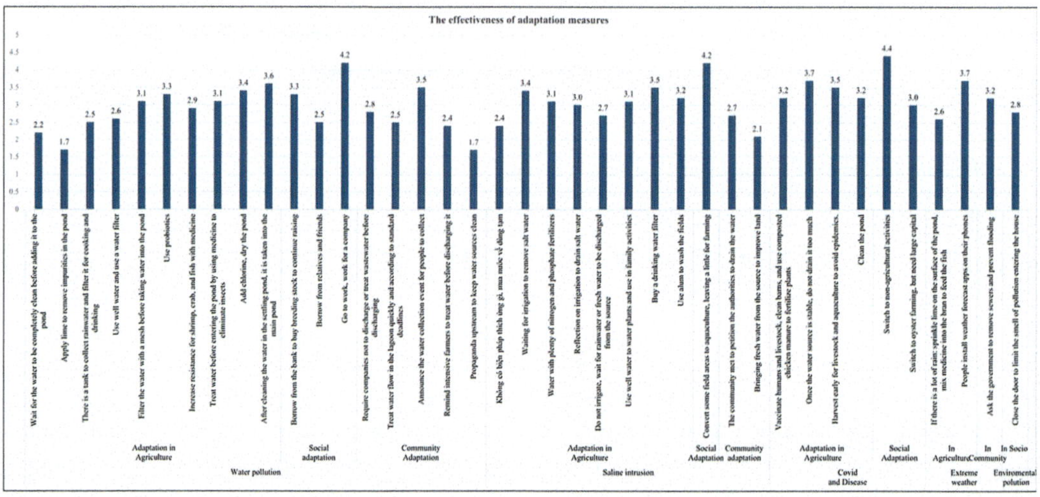

Fig. 7 Effectiveness of adaptation measures

3.4 Local People's Adaptation

When asked about adaptation measures for each hazard, the effectiveness of each type of adaptation was rated on a scale from 1 to 5 (1 = Ineffective; 2 = Slightly effective; 3 = Moderately effective; 4 = Highly effective; 5 = Very highly effective). The average scores for different types of adaptation are shown in Fig. 7. In general, people in the area lack formal measures for adapting to hazards and rely primarily on experience. When hazards impact their families, they often suffer asset loss, accumulate debt, and some households are forced to change careers. Many individuals become hired laborers, and young people in the family may move to cities or other areas to find work. Among the interview participants, 13% reported having someone working far away, only returning home every six months. They find work in cities like Hanoi, Binh Duong, and Ho Chi Minh City, while many families borrow money from banks to send their children to work abroad in countries such as Korea, Japan, and Taiwan.

When hazards affect families, 76% of respondents do not receive any support from local authorities or other organizations. In contrast, 19% received support, such as assistance during the COVID-19 pandemic in the form of rice, medicine, and food. Additionally, some households that experienced the death of all their shrimp, crabs, and fish in their ponds received support from local authorities, which included lime powder and pond disinfectant.

However, the awareness of hazards, the level of impact, and the adaptive capacity of the three coastal communities in the study locations differ slightly (Fig. 8).

Indigenous knowledge plays a crucial role in adapting to natural and environmental hazards. In the Red River Delta, residents rely not only on weather forecasts from mass media and smartphones but also on traditional knowledge and experience to deal with natural and man-made disasters. Traditional forecasting methods include observing the color of the sky, temperature, moon phases, and animal migrations. For instance, if the sky appears yellow, it is expected to be windy; if red, it is likely to rain. Observations such as wind coming from the west suggest stormy weather, while bird migrations might indicate fewer storms that year. Lightning in the east suggests heavy rain, and if dragonflies fly low, rain is expected soon. A high number of black ants crawling around may signal an impending flood, and if crabs enter the yard, a flood may be imminent.

Fishermen use calendars to determine optimal fishing days based on tides, and they follow seasonal calendars for rice planting. They adjust their practices according to weather patterns: for example, they fertilize rice with nitrogen if heavy rain is anticipated, reduce aquaculture activities, raise larger breeds, and harvest before a storm. Decisions about plants, fishing, agriculture, and aquaculture are based on weather observations.

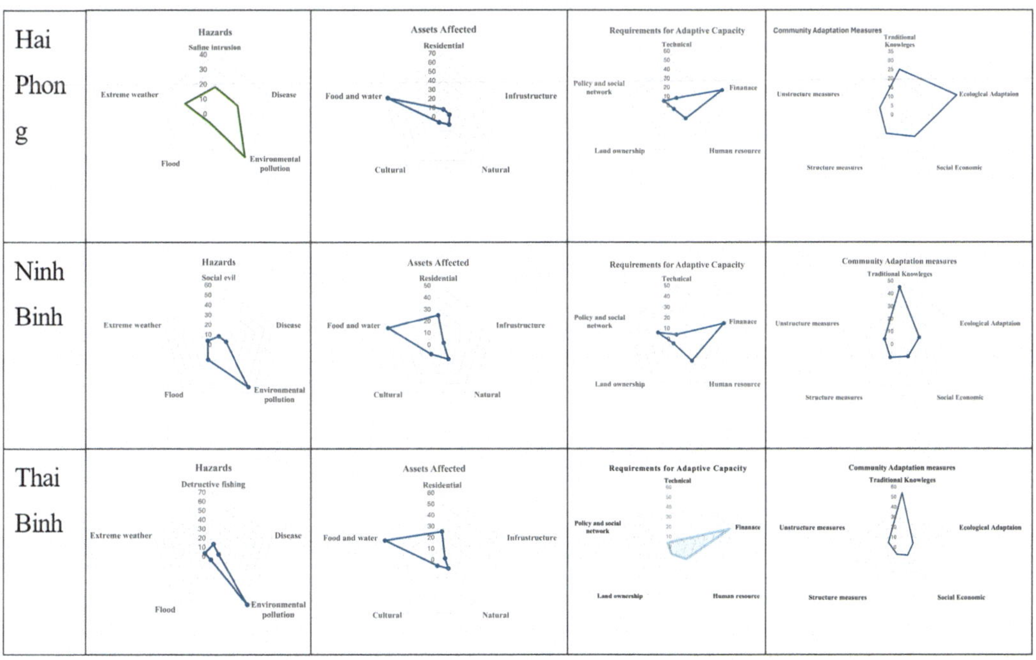

Fig. 8 Awareness of hazards, level of impact, and adaptive capacity of coastal communities in the three research provinces

However, most respondents noted that this experience-based weather forecasting was more common 15–20 years ago. Due to changing environmental conditions and phenomena, traditional methods are becoming less reliable. Young people today are less familiar with or interested in traditional agricultural, aquaculture, or fishing practices and instead rely on scientific knowledge and weather forecasts for their activities.

Adaptation measures for natural and man-made hazards in the Red River Delta (RRD) at the community level are emphasized by people as ecological solutions. Mangrove restoration is particularly important to the local population. The people in the area strictly protect mangrove forests because these forests are crucial for their livelihoods, protect their property, and provide significant income. This is especially true for residents of Thuy Hai Commune in Thai Binh, where 100% of households depend on aquaculture, fishing in mangrove forests, and salt production, and do not have agricultural land for cultivation. For coastal residents, mangrove forests are vital assets. Without them, people would feel insecure living in the area, even though construction measures are also highly valued.

Construction measures include widening and reinforcing embankments, seawalls, and breakwaters. Non-structural measures, such as improving communication facilities and early warning systems, are also highly appreciated by the community (Fig. 9).

When asked about changes in their lives over the past 5 years, 10 years, and more than 10 years, most people feel that their lives have improved significantly (Fig. 10). Previously, there were fewer mangrove forests, and the dyke system was inadequate. During storms or floods, many people's houses, property, fields, and ponds were damaged or lost. Today, living conditions have greatly improved. The electricity system, roads, schools, and public facilities are much better. The new rural development program has made the villages greener, cleaner, and more beautiful, and people are more conscious of maintaining overall hygiene.

When asked about their concerns for the future, most people responded that they do not know (Fig. 11). Their primary worry is their

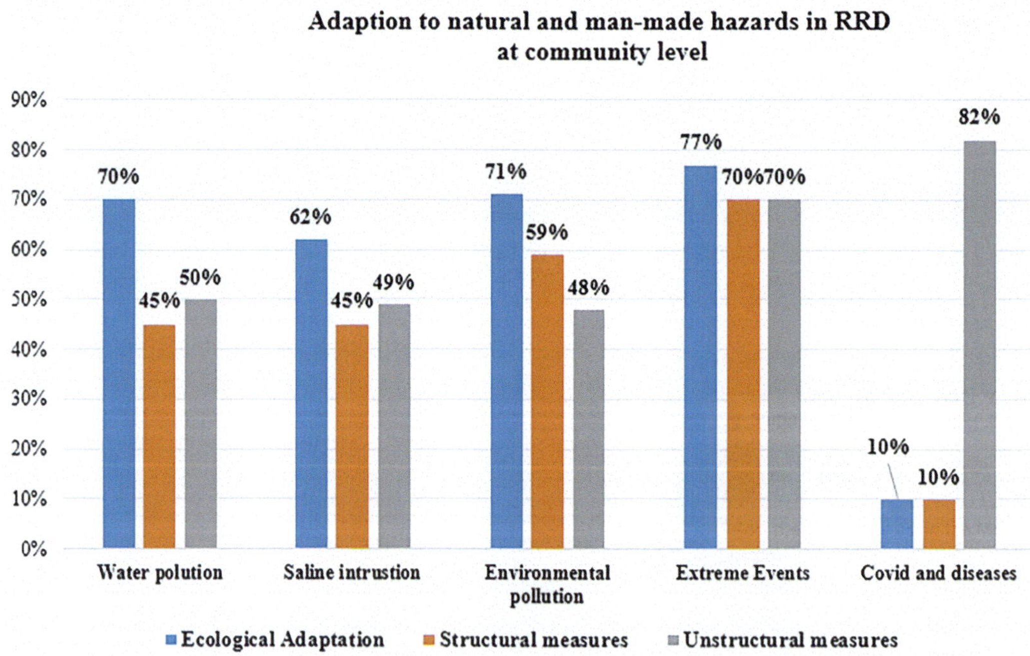

Fig. 9 Adaptation to natural and man-made hazards in the Red River Delta (RRD) at the community level

health, as the environment is increasingly polluted and no longer as fresh as it was in the past. This pollution is leading to more illnesses and an increase in cancer cases. Water pollution (17%) directly affects agricultural activities, particularly aquaculture. Over the past 10 years, severe water pollution has significantly impacted aquaculture in these three coastal provinces. Many financially stable households have shifted to industrial farming, which requires substantial capital and technology, or to oyster farming (in Ninh Binh). Young people are increasingly leaving agriculture, aquaculture, and fishing jobs to work in companies, leading to a potential labor shortage in rural areas in the near future. The local community does not have any plans for addressing future hazards.

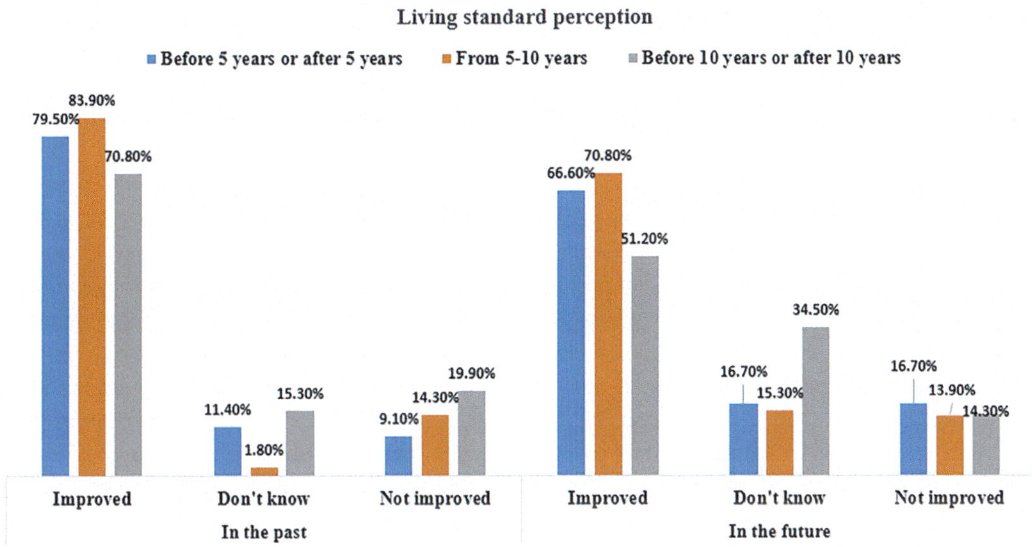

Fig. 10 Perception of living standards

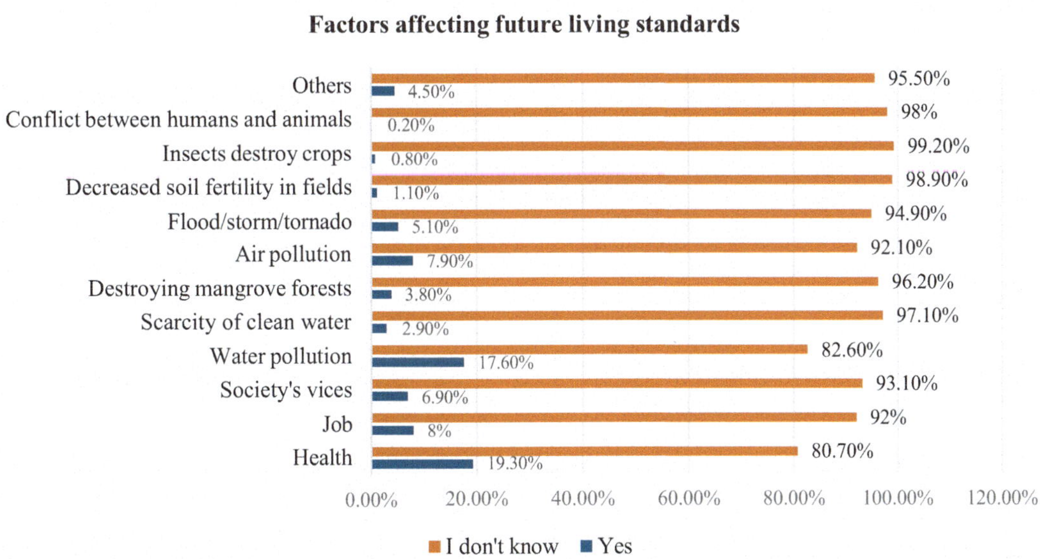

Fig. 11 Factors affecting future living standards

4 Discussions and Conclusions

This study examines local adaptive capacity in the context of natural and man-made hazards in coastal communities in the Red River Delta. By integrating data on risk identification, affected assets, and countermeasures, the study provides insight into how coastal communities in this region respond to risks. Unlike many global and regional studies, the problem of environmental pollution is identified as the most dangerous and severe for the community in the three research sites in the Red River Delta. Polluted water severely hampers aquaculture. Many households with financial means have switched from extensive to intensive industrial farming or oyster farming (in Ninh Binh), while those without the means often abandon their ponds, sell them, or rent them out. Water pollution, primarily due to inadequate treatment of wastewater from farming activities, pesticide use, and chemical fertilizers, has led to the death of shrimp and fish in lagoons. Oyster and clam farming also contribute to pollution due to chemical use in cleaning practices.

According to local residents, the last major storm occurred in 1986, a time when dikes encroached on the sea, and mangrove forests were cut down or planted. Since then, the improved mangrove forests and dyke system have greatly reduced the frequency and severity of storms and floods. Other coastal areas globally also suffer from floods caused by storms, high tides, or tsunamis (Aerts, 2018; Hinkel et al., 2014) and coastal erosion (Scyphers et al., 2019).

Similar to other global studies, residents use indigenous knowledge and local materials to adapt to hazards. They use lime powder for polluted water, dry lagoons, remove impurities, and use alum to reduce acidity. Coastal communities in some places also use local materials and knowledge, such as raising house foundations and building seawalls with wood, stone, or coral. Local expertise and resources are crucial for coastal adaptation initiatives (Narayan et al., 2020).

While some studies on coastal adaptation focus on structural measures like coastline construction (Aerts, 2018; Gittman et al., 2015), this study, along with that of Natarajan et al. (2022), finds that people still prioritize ecosystem-based measures. These measures offer more benefits by saving costs, providing significant aquatic resources, and creating livelihoods and attractive landscapes. These benefits sometimes outweigh the advantages of structural coastal protection (Lau et al., 2019; Narayan et al., 2016).

In this study, natural hazards are less significant compared to human-induced hazards. Over the past 30 years, the area has been relatively unaffected by storms or high tides, thanks to its mangrove forests and solid dyke system. Findings show a trade-off between economic development activities and the sustainable livelihood of local people. Poor wastewater management by authorities has exacerbated water pollution. Sustainable development requires equal emphasis on the economy, society, and environment (Narayan et al., 2020). In our study, land reclamation in saline areas has been challenging due to limited local capacity, expertise, funding, and resources. Converting land from agriculture to aquaculture, or vice versa, is difficult due to state land regulations.

Unlike local reactions, withdrawing from the coastline has proven unpopular and challenging in both our study and that of Narayan et al. (2020), due to deep-rooted connections with the environment, society, culture, and land ownership rights. This is similar to situations in the Solomon Islands and other locations (Albert et al., 2018). However, young people in the area often migrate to other places for work, send money back to their families, and return to their hometowns in old age. This adaptation, chosen by 13% of respondents, risks disrupting livelihoods, employment, and existing social norms (Mikami & Shibayama, 2015). Many returning young people also bring social issues and problems.

This study highlights the need to understand the critical point of risk or hazard more accurately. Further research is required to determine when the need for safety outweighs attachment to one's living place and community if change is feasible (Mcmichael et al., 2019).

Social and institutional capacity are crucial for adaptive capacity at our study site. The community values social networks, using platforms like Zalo to exchange experiences, receive early warnings, learn farming techniques, and sell products. Social networks play a vital role in adaptation (Cinner et al., 2018).

Real institutional support: For example, every year, the district organizes training courses on aquaculture and animal husbandry, which have helped the community gain more professional knowledge and resources for implementing adaptation measures (Eui et al., 2018). Therefore, even if locally appropriate adaptation measures are identified, they will not be implemented effectively without adequate institutional support.

5 Policy Recommendations Based on Findings

Based on the findings from the study, the following specific policy recommendations are proposed.

Strengthen Coastal and Water Management Policies Develop and enforce comprehensive coastal and water management policies that address saline intrusion, pollution, and flood risks. Establish and implement an Integrated Coastal Zone Management (ICZM) framework that incorporates scientific data, community input, and cross-sectoral coordination to manage coastal resources sustainably. Develop water management plans that balance freshwater allocation with protection of wetlands and estuaries to mitigate saline intrusion. Strengthen enforcement of regulations on wastewater discharge and agricultural runoff to reduce pollution in coastal waters. Facilitate community engagement in water management decisions, support local enforcement of regulations, and provide resources for implementing best practices in coastal and water management. Develop national policies and standards for integrated coastal and water management, provide funding for research and infrastructure, and oversee compliance with environmental regulations.

Enhance Disease Surveillance and Public Health Infrastructure Improve disease surveillance systems and public health infrastructure to address climate-sensitive diseases. Establish and expand disease monitoring programs that track the incidence of waterborne and vector-borne diseases, particularly in areas vulnerable to climate change. The government provide funding and technical support for disease surveillance systems and healthcare infrastructure improvements. Ensure that health policies incorporate climate change projections.

Foster Sustainable Aquaculture and Pollution Reduction Implement policies to promote sustainable aquaculture practices and reduce pollution from aquaculture and other sources. Encourage the adoption of sustainable aquaculture practices, such as organic farming and integrated pest management, through incentives and support programs. Enforce stricter regulations on the use of chemicals and waste management in aquaculture operations. Implement best practices for wastewater treatment. Develop certification programs for sustainable aquaculture products to provide market incentives for environmentally friendly practices. Local governments support local aquaculture businesses in transitioning to sustainable practices, provide training and resources, and monitor compliance with pollution control regulations. National governments develop and enforce national standards for sustainable aquaculture, provide financial incentives for adopting best practices, and support research into pollution control technologies.

Support Community-Based Adaptation and Capacity Building Strengthen community-based adaptation efforts and build local capacity for managing climate-related hazards. Involve local communities in the development and implementation of adaptation strategies, ensuring that local knowledge and needs are incorporated. Provide training and resources to enhance local capacity for disaster preparedness, environmental management, and climate adaptation. Offer grants and financial assistance to community-led projects that focus on adaptation and resilience-building. Local Governments facilitate community participation in adaptation planning, provide support for local capacity-building initiatives, and manage local adaptation projects. National Governments create and fund programs that support community-based adaptation efforts, establish frameworks for integrating local input into national policies, and promote collaboration between communities and government agencies.

6 Future Research

Develop and apply advanced climate models to project future scenarios of saline intrusion in the Red River Delta. Assess how rising sea levels and changing precipitation patterns might alter the extent and intensity of saline intrusion. Include scenarios of different greenhouse gas emission pathways to understand a range of potential impacts. Research the effectiveness of adaptive measures for agriculture in the face of increased saline intrusion. This could include studying the resilience of salt-tolerant crop varieties and the impact of soil amendments on reducing salinity effects. Investigate how integrated water management strategies can mitigate the effects of saline intrusion. This research could focus on optimizing freshwater allocation, improving irrigation techniques, and restoring wetlands to buffer against salinity changes. Explore how climate change might influence the spread and incidence of waterborne and vector-borne diseases in the Red River Delta. Develop models to predict how changes in temperature, precipitation, and humidity could affect disease vectors and pathogen survival. Analyze projections of extreme weather events, such as typhoons, heavy rainfall, and heatwaves, and their potential impacts on the Red River Delta. Research how the frequency, intensity, and duration of these events might change under different climate scenarios. Study how climate change might influence the levels and distribution of environmental pollutants in the Red River Delta. Research interactions between climate variables (e.g., temperature, rainfall) and pollution sources (e.g., industrial runoff, agricultural chemicals).

Acknowledgments We would like to thank the government and the people of Thai Binh, Hai Phong, and Ninh Binh provinces for their invaluable assistance in completing this research. We also extend our gratitude to the researchers who participated in conducting the household survey.

Declaration of Interest The authors declare that there is no conflict of interest in this paper.

References

Aerts, J. C. J. H. (2018). A review of cost estimates for flood adaptation. *Water (Switzerland), 10*(11). https://doi.org/10.3390/w10111646

Albert, S., Bronen, R., Tooler, N., Leon, J., Yee, D., Ash, J., Boseto, D., & Grinham, A. (2018). Heading for the hills: Climate-driven community relocations in the Solomon Islands and Alaska provide insight for a 1.5°C future. *Regional Environmental Change, 18*(8), 2261–2272. https://doi.org/10.1007/s10113-017-1256-8

Barnett, J., & O'Neill, S. J. (2012). Islands, resettlement and adaptation. *Nature Climate Change, 2*(1), 8–10. https://doi.org/10.1038/nclimate1334

Bennett, N. J., Kadfak, A., & Dearden, P. (2016). Community-based scenario planning: A process for vulnerability analysis and adaptation planning to social–ecological change in coastal communities. *Environment, Development and Sustainability, 18*(6), 1771–1799. https://doi.org/10.1007/s10668-015-9707-1

Betzold, C. (2015). Adapting to climate change in small Island developing states. *Climatic Change, 133*(3), 481–489. https://doi.org/10.1007/s10584-015-1408-0

Butler, J. R. A., Wise, R. M., Skewes, T. D., Bohensky, E. L., Peterson, N., Yanuartati, Y., Handayani, T., Habibi, P., Puspadi, K., Bou, N., Rochester, W., Wise, R. M., Skewes, T. D., Bohensky, E. L., & Peterson, N. (2015). Integrating top-down and bottom-up adaptation planning to build adaptive capacity: A structured learning approach integrating top-down and bottom-up adaptation planning to build adaptive capacity: A structured learning approach. *Coastal Management, 0753*. https://doi.org/10.1080/08920753.2015.1046802

Cinner, J. E., Adger, W. N., Allison, E. H., Barnes, M. L., Brown, K., Cohen, P. J., Gelcich, S., Hicks, C. C., Hughes, T. P., Lau, J., Marshall, N. A., & Morrison, T. H. (2018). Building adaptive capacity to climate change in tropical coastal communities. *Nature Climate Change, 8*(2), 117–123. https://doi.org/10.1038/s41558-017-0065-x

Da Le, N., Hoang, A. Q., Hoang, T. T. H., Nguyen, T. A. H., Duong, T. T., Pham, T. M. H., æüNguyen, T. D., Hoang, V. C., Phung, T. X. B., Le, H. T., Tran, C. S., Dang, T. H., Vu, N. T., Nguyen, T. N., & Le, T. P. Q. (2021). Antibiotic and antiparasitic residues in surface water of urban rivers in the red River Delta (Hanoi, Vietnam): Concentrations, profiles, source estimation, and risk assessment. Environmental Science and Pollution Research, 28(9), 10622–10632. https://doi.org/10.1007/s11356-020-11329-3.

Duong, N. T., Lieu, N. T. H., Cuc, N. T. T., Saito, Y., Huong, N. T. M., Phuong, N. T. M., & Thuy, A. T. (2020). Holocene paleoshoreline changes of the red River Delta, Vietnam. *Review of Palaeobotany and Palynology, 278*, 104235. https://doi.org/10.1016/j.revpalbo.2020.104235

Eisenack, K., & Stecker, R. (2020). A framework for analyzing climate change adaptations as actions. *Manuscript Appears in Mitigation and Adaptation Strategies to Global Change*, 1–22. https://doi.org/10.1007/s11027-011-9323-9

Eui, Y., Jin, J., Chang, Y. S., Jang, B., Chon, J., Choi, Y. E., Jin, J., Chang, Y. S., Jang, B., & Chon, J. (2018). Role of official development assistance in enhancing resilient coastal Community in Small Island Developing States role of official development assistance in enhancing resilient coastal Community in Small Island Developing States. *Coastal Education and Research Foundation, 85*, 1466–1470. https://doi.org/10.2112/SI85-294.1

Gingerich, S. B., Voss, C. I., & Johnson, A. G. (2017). Seawater-flooding events and impact on freshwater lenses of low-lying islands: Controlling factors, basic management and mitigation. *Journal of Hydrology*. https://doi.org/10.1016/j.jhydrol.2017.03.001

Gittman, R. K., Fodrie, F. J., Popowich, A. M., Keller, D. A., Bruno, J. F., Carolyn, A., Peterson, C. H., & Piehler, M. F. (2015). Engineering away our natural defenses: An analysis of shoreline hardening in the US. *Research Communications*. https://doi.org/10.1890/150065

Hinkel, J., Lincke, D., Vafeidis, A. T., Perrette, M., Nicholls, R. J., Tol, R. S. J., Marzeion, B., Fettweis, X., Ionescu, C., & Levermann, A. (2014). Coastal flood damage and adaptation costs under 21st century sea-level rise. *Proceedings of the National Academy of Sciences of the United States of America, 111*(9), 3292–3297. https://doi.org/10.1073/pnas.1222469111

Hue, H. T., Pradit, S., Jarunee, C., Lim, A., Nitiratsuwan, T., & Goncalo, C. (2018). Physical properties of three Songkhla lagoon fish species in the lower gulf of Thailand during and after the monsoon season. *Applied Ecology and Environmental Research, 16*(5), 6113–6127.

Jenkins, A. P., & Jupiter, S. (2015). *Natural disasters, health and wetlands: A Pacific small Island developing state perspective BT - wetlands and human health* (pp. 169–191). Springer Nature. https://doi.org/10.1007/978-94-017-9609-5_9

Kumar, L., Eliot, I., Nunn, P. D., Stul, T., Mclean, R., Kumar, L., Eliot, I., Nunn, P. D., Stul, T., & Mclean, R. (2018). *An indicative index of physical susceptibility of small islands to coastal erosion induced by climate change: An application to the Pacific islands*, 5705. https://doi.org/10.1080/19475705.2018.1455749

Lau, J. D., Hicks, C. C., Gurney, G. G., & Cinner, J. E. (2019). What matters to whom and why? Understanding the importance of coastal ecosystem services in developing coastal communities. *Ecosystem Services, 35*, 219–230. https://doi.org/10.1016/j.ecoser.2018.12.012

Lin, B. B., Capon, T., Langston, A., Taylor, B., Wise, R., Williams, R., & Lazarow, N. (2017). Adaptation pathways in coastal case studies: Lessons learned and future directions. *Coastal Management, 45*(5), 384–405. https://doi.org/10.1080/08920753.2017.1349564

Lincke, D., & Hinkel, J. (2018). Economically robust protection against 21st century sea- level rise. *Global Environmental Change, 51*(June 2017), 67–73. https://doi.org/10.1016/j.gloenvcha.2018.05.003

Long, C., Dai, Z., Zhou, X., Mei, X., & Mai Van, C. (2021). Mapping mangrove forests in the red River Delta, Vietnam. *Forest Ecology and Management, 483*, 118910. https://doi.org/10.1016/j.foreco.2020.118910

Mcmichael, C., Katonivualiku, M., & Powell, T. (2019). Planned relocation and everyday agency in low-lying coastal villages in Fiji. *The Geographical Journal, 185*(3), 325–337. https://doi.org/10.1111/geoj.12312

Mikami, T., & Shibayama, T. (2015). Chapter 5: Tsunami disasters in remote islands: 2009 Samoan and 2010 Mentawai Islands tsunamis. In M. Esteban, H. Takagi, & T. Shibayama (Eds.), *Handbook of coastal disaster mitigation for engineers and planners* (pp. 75–91). Butterworth-Heinemann. https://doi.org/10.1016/B978-0-12-801060-0.00005-8

Morton, L. W. (2020). Working toward sustainable agricultural intensification in the red River Delta of Vietnam. *Journal of Soil and Water Conservation, 75*(5), 109–116. https://doi.org/10.2489/jswc.2020.0304A

Narayan, S., Beck, M. W., Reguero, B. G., Losada, I. J., Van, B., Lange, M., & Burks-copes, K. A. (2016). *The effectiveness, costs and coastal protection benefits of natural and nature- based defences*, 1–17. https://doi.org/10.1371/journal.pone.0154735

Narayan, S., Esteban, M., Albert, S., Laurice, M., Crichton, R., Heck, N., Goby, G., & Jupiter, S. (2020). Local adaptation responses to coastal hazards in small Island communities: Insights from 4 Pacific nations. *Environmental Science and Policy, 104*(December 2018), 199–207. https://doi.org/10.1016/j.envsci.2019.11.006

Natarajan, N., Newsham, A., Rigg, J., & Suhardiman, D. (2022). A sustainable livelihoods framework for the 21st century. *World Development, 155*, 105898. https://doi.org/10.1016/j.worlddev.2022.105898

Noble, I. R., Huq, S., Anokhin, Y. A., Carmin, J. A., Goudou, D., Lansigan, F. P., Osman-Elasha, B., Villamizar, A., Patt, A., Takeuchi, K., & Chu, E. (2015). Adaptation needs and options. In *Climate change 2014 impacts, adaptation and vulnerability: Part A: Global and sectoral aspects* (pp. 833–868). Cambridge University Press. https://doi.org/10.1017/CBO9781107415379.019

Nunn, P. D., Runman, J., Falanruw, M., & Kumar, R. (2017). Culturally grounded responses to coastal change on islands in the Federated States of Micronesia, northwest Pacific Ocean. *Regional Environmental Change, 17*(4), 959–971. https://doi.org/10.1007/s10113-016-0950-2

Oppenheimer, M., Glavovic, B. C., Hinkel, J., van de Wal, R., Magnan, A. K., Abd- Elgawad, A., Cai, R., Cifuentes-Jara, M., DeConto, R. M., Ghosh, T., Hay, J., Isla, F., Marzeion, B., Meyssignac, B., & Z. S. (2019). Sea level rise and implications for low- Lying Islands, coasts and communities. In *IPCC special report on the ocean and cryosphere in a changing climate* (pp. 321–445). Cambridge University Press. https://doi.org/10.1017/9781009157964.006

Robinson, S. (2015). Climate change adaptation trends in small island developing states. *Mitig Adapt Strateg Glob Change*. https://doi.org/10.1007/s11027-015-9693-5

Rouse, H. L., Bell, R. G., Lundquist, C. J., Blackett, P. E., Hicks, D. M., & King, D.-N. (2017). Coastal adaptation to climate change in Aotearoa-New Zealand. *New Zealand Journal of Marine and Freshwater Research, 51*(2), 183–222. https://doi.org/10.1080/00288330.2016.1185736

Scyphers, S. B., Beck, M. W., Furman, K. L., Haner, J., Josephs, L. I., Lynskey, R., Keeler, A. G., Landry, C. E., Powers, S. P., Webb, B. M., & Grabowski, J. H. (2019). *A waterfront view of coastal hazards: Contextualizing relationships among geographic exposure, shoreline type, and hazard concerns among coastal residents*. Sustainability Article.

Storlazzi, C. D., Gingerich, S. B., Dongeren, A. Van, Cheriton, O. M., Swarzenski, P. W., Quataert, E., Voss, C. I., Field, D. W., Annamalai, H., Piniak, G. A., & Mccall, R. (2018). *Most atolls will be uninhabitable by the mid-21st century because of sea-level rise exacerbating wave-driven flooding*, 1–10.

Warrick, O., Aalbersberg, W., Dumaru, P., McNaught, R., & Teperman, K. (2017). The 'Pacific adaptive capacity analysis framework': Guiding the assessment of adaptive capacity in Pacific Island communities. *Regional Environmental Change, 17*(4), 1039–1051. https://doi.org/10.1007/s10113-016-1036-x

World Bank Group. (2022). *A framework for Analysing climate change adaptations as action* (CCDR series). World Bank. http://hdl.handle.net/10986/37618 License: CC BY 3.0 IGO

Heating the Planet: The Nexus of Greenhouse Gases and Global Temperature

Sana Rafi, Pragati Verma, Ashutosh Singh, Mahendra Singh, and Azizur Rahman Siddiqui

1 Introduction

The living organisms have a significant influence on the Earth system as they play a crucial role in shaping cloud formation, atmospheric dust levels and the accumulation of greenhouse gases (GHGs), mainly carbon dioxide (CO_2), methane (CH_4) and nitrous oxide (N_2O) in the air (Moiceanu & Dinca, 2021). GHG emissions behave like a blanket around the planet Earth, trapping the solar heat and elevating the temperature of Earth's surface. GHGs come usually from events such as burning oil and gases, fossil fuels, heating buildings, agriculture, forests fire and much more. The amount of GHGs released, its capacity to trap heat and its longevity determine its concentration and its relative capacity to alter radiative forcing (Mehmood et al., 2022) in the atmosphere. GHGs are also crucial for the Earth's reflectivity or albedo. Numerous climate models have consistently demonstrated a positive correlation between the increased concentration of GHGs and significant global warming. The IPCC's Sixth Assessment Report (AR6) of 2023 specified that global surface temperatures rose by 1.1°C (1.98°F) from the pre-industrial levels (the late 1800s) to the early 2000s. Further, 90% of this temperature rise is attributed to GHGs and less than 5% to natural climatic processes. The previous decade (2011–2020) was the hottest on record, and all of the past four decades have been warmer than any preceding decade since 1850. Current policies forecast a temperature rise to a 3°C by the end of this century.

Brovkin et al. (2004) utilized the CLIMBER-2 Earth system model to explore the effects of both natural phenomena (such as volcanic eruption, changes in sun's radiation, and aerosols) and anthropogenic activities (e.g. alterations in landscape cover and emissions of carbon dioxide) on global climate patterns spanning the last 150 years. Their research involved comparing the impacts of historical changes in land cover on atmospheric CO_2 levels and climate change, considering both bio-geophysical and biogeochemical aspects. Their findings suggest that the influence of land cover changes on the increase in CO_2 levels has diminished over time, decreasing from 36 to 60% between 1850 and 1960 to 4–35% from 1960 to 2000 because of fossil fuels consumption. Furthermore, if fossil fuel emissions continue to rise in the future, it is likely that the relative contribution of land cover changes to the growth of atmospheric CO_2 will

S. Rafi · A. Singh · M. Singh (✉)
Department of Geography, School of Earth Science, Hemvati Nandan Bahuguna Garhwal University, S.R.T. Campus, Tehri, Uttarakhand, India

P. Verma
Centre for Atmospheric Sciences, Indian Institute of Technology, Delhi, New Delhi, India

A. R. Siddiqui
Department of Geography, University of Allahabad, Prayagraj, Uttar Pradesh, India

Z. Khanam et al. (eds.), *Climate Change, Resilience and Ecosystem Services*, Sustainable Development Goals Series, https://doi.org/10.1007/978-3-031-96977-5_6

further decline in the twenty-first century. Human activities introduce various substances into the atmosphere, impacting climate in diverse ways. These substances vary in their longevity, their effects on a global or regional scale and whether they directly or indirectly influence climate dynamics (Tol et al., 2012).

Greenhouse gases (GHGs) significantly influence global warming and climate change (Gregory & Forster, 2008; Macedo & Madaleno, 2022). There are available theories and reasoning behind GHGs and temperature rise. One proponent theory is Anthropogenic Global Warming (AGW) Theory that explains the changes in atmospheric concentration as a possible cause of temperature rise. According to this theory, human emissions of greenhouse gases—primarily CO_2, CH_4 and N_2O—have caused a catastrophic rise in Earth's temperatures over the last 50 years. Proponents of the AGW theory attribute a range of environmental challenges to man-made carbon dioxide (CO_2). These challenges include floods, water deficiencies, crop failures, extreme weather, disease outbreaks, species extinctions, coral bleaching and other various kinds of disasters. According to their predictions, as global temperatures continue to rise, these tragedies will become increasingly frequent and severe. To mitigate these catastrophic events, urgent and substantial reductions in human emissions are essential for safeguarding our planet. In the late nineteenth century, T.C. Chamberlin proposed the Carbon Dioxide Theory. He postulated that variations in the Earth's surface and atmospheric temperatures are influenced by changes in its carbon dioxide content. This theory provides an explanation for global warming. Al Gore (2006) identified four critical factors driving global warming: (1) Population Explosion: The rapid growth of human population impacts resource consumption and emissions; (2) Scientific Revolution: Advances in scientific understanding have shaped our awareness of climate change; (3) Technological Revolution: Technological developments influence energy production and consumption patterns; (4) Fundamental Shift in Climate Crisis Perception: Our perspective on the Earth's ecosystem has evolved, emphasizing the need for urgent action. These factors have fundamentally transformed the connection between human evolution and the planet's delicate environmental balance.

Increasing levels of GHGs, particularly CO_2, in the atmosphere are causing a positive shift in climate radiative forcing, leading to an anticipated rise in surface temperatures (Bruhwiler et al., 2021). Maruyama, (2019) examined the correlation between atmospheric CO_2 and various climate indices (namely the carbon isotope ratio, El Niño-Southern Oscillation, Pacific Decadal Oscillation and North Atlantic Oscillation) using a multifractal analysis approach. The study found a significant link between CO_2 fluctuations and global temperature, with a high CO_2 growth rate observed during El Niño periods and a less pronounced multifractality in global temperature during the pause in global warming. GHGs impact climate based on their energy absorption capacity and atmospheric lifespan. The Global Warming Potential (GWP), adopted by the IPCC, measures a gas's total absorbed energy over a typical period of 100 years (GWP_{100}), relative to CO_2. A higher GWP indicates more warming caused by the particular gas (Verma & Ghosh, 2022). Among all the GHGs, sulphur hexafluoride (SF_6) has the largest warming impact (GWP_{100} = 23,900) followed by perfluorocarbons (PFCs) – GWP_{100} = 7000–9200, hydrofluorocarbons (HFCs) – GWP_{100} = 140–11700, nitrous oxide (N_2O) – GWP_{100} = 310, methane (CH_4) – GWP_{100} = 21 and CO_2 – GWP_{100} = 1 (IPCC Third Assessment Report, 2001). Each 0.5°C (0.9°F) rise will significantly increase the occurrence and severity of heat extremes, heavy rainfall and regional droughts. For instance, heatwaves that used to occur once every 10 years will likely to occur 4.1 times more often with 1.5°C (2.7°F) of heating, and 5.6 times with 2°C (3.6°F) of heating (IPCC, 2023). Increasing temperatures has also led to the increasing emanations of instable organic compounds which are the pioneers of tropospheric aerosols and ozone. Higher air temperatures are also favourable for higher concentration of ozone gas and a positive feedback loop exists between global warming and ozone depletion (Al Gore

2007). With rising temperatures and peak ozone-depleting episodes, the climate penalty is expected to increase (WMO 2018). Few studies like Davis (2017), Rivera and Khan (2012) and (Viterito, 2016) found that GHGs are not the primary driver of global temperature rise. Davis (2017) investigates the relationship between atmospheric CO_2 concentration and global temperature over the last 425 million years, using databases of stable-isotope proxies from the Phanerozoic Eon and MODTRAN modelling. Findings revealed that marginal radiative forcing exerts a minimal positive impact on temperature. Climate change is caused by an increase in the average global temperature (global warming), which causes changes in cloud cover and precipitation (particularly over land), melting ice caps and glaciers, an increase in the frequency of extreme climate events, rising ocean temperatures and acidity due to heat and carbon dioxide absorption by seawater. As a result, it incorporates any environmental changes that are likely to have negative consequences for both ecological and socioeconomic aspects of our life (WMO, climate status reports 1993–2018 and GHG bulletin 2004–2018). It is one of the primary causes of the unprecedented increase in natural disasters around the world in recent decades. Bhatti et al. (2024) explored the link between worldwide climate change and heat trapping GHG emissions over 246 states and identifies a direct positive correlation between global warming and GHG emanations. Despite G7 nations implementing various emission reduction strategies, existing economic growth strategies indicate a net rise in global GHG emissions by 2100. The effects of global climate crisis are predominantly observed through the escalating occurrence of extreme weathers like floods, heat-waves and tornadoes which are now happening more frequently all over the world (Maruyama, 2019). But, assessing the effect of a particular forcing agent on the global climate system is complex due to the interplay among different reactive gases (Thornhill et al., 2021). The alterations in climate, resulting from the rise in greenhouse gases, are predominantly irreversible and will persist for a millennium even after the cessation of emissions (Solomon et al., 2009). Assuming the Earth continues to respond as it has historically, the gradual melting of ice sheets, the presence of natural aerosols and changes in vegetation distribution will persistently push global warming beyond the 2°C goal over the ensuing millennia, irrespective of whether human-induced contributions remain constant (Ring et al., 2022). Addressing climate change requires collective action, as emissions originate from every part of the world and impact everyone (https://www.un.org/en/climatechange/what-is-climate-change). Efforts to limit global atmospheric temperature increase to not more than 1.5°C are required to eliminate the harmful climate impacts and maintain a liveable atmosphere.

Against this background, this research chapter delves into the relationship between global greenhouse gases and global warming. And it is divided into number of sections. The first section includes the introduction followed by applied materials and methods (Sect. 2). Section 3 explains the results and findings of the study, by analysing the trends of global GHGs and radiative forcing. Discussion part of the study is compiled in Sect. 4. The last section (Sect. 5) provides a further ground map for future scope and research possibilities.

2 Materials and Methods

2.1 Survey of literature and data collection

At first, high-quality research documents for investigation have been collected. For this purpose, existing literature has been extracted from the Dimensions.ai scholarly database which covers 140 million (m) publications, 260 m online citations, 1.8 m policy documents, 29 m datasets and significant amounts of patents, grants and clinical trials (https://www.dimensions.ai). The retrieved research material was extracted by applying the keywords string as ('Greenhouse gases (GHGs)' AND 'Global

Temperature' AND 'Global Warming' AND 'Climate Change' AND 'Radiative forcing') in the title and abstract fields. To select the relevant studies, publication year was limited to only two decades (2004–2024) and the research categories were limited to four fields of research, namely Earth Sciences, Atmospheric Sciences, Environmental Sciences, and Physical Geography and Environmental Geosciences. Non-English documents, PhD theses, as well as master theses were not taken into account. After implying these filtrations, 25 publications including articles, chapters and reports were finalized for the systematic review to thoroughly study the contexts, methods and findings. The flowchart of literature access is provided in Fig. 1. After detailed analysis of the selected literature, only four greenhouse gases namely CO_2, CH_4, N_2O and SF6 were selected to study their relationship with global temperature. The data for global temperature and concentration of CO_2, CH_4, N_2O and SF6 has been collected from Global Monitoring Laboratory available at https://gml.noaa.gov/.

2.2 Trend analysis

To assess the concentration of GHGs, time series analysis (trend line by least square) method is applied over the time period and further represented through line graph to show the trends of global GHGs concentration. Time series analysis stands out as a top method for extracting trends spanning a time frame. Through this approach, a linear trend is derived, represented by the equation Y = a + bX, where 'a' and 'b' are fixed values. This linear equation establishes a functional connection between the X and Y series, enabling the prediction of future values. Besides employing the least squares method for trend analysis, this technique is widely utilized for forecasting.

2.3 Regression analysis

Further, to analyse the relation in global mean atmospheric temperature and GHGs (CO_2, NH_4 and N_2O) and to predict future changes, multiple linear regression analysis was also employed through SPSS ver. 26 using Eq. 1.

$$Y_i = \beta_0 + \beta_1 X_{i1} + \beta_2 X_{i2} + \ldots\ldots\ldots + \beta_p X_{ip} + \varepsilon \quad (1)$$

where for i = n observations

Yi = dependent variable

Xi = explanatory variable

β_0 = y-intercept constant

βp= slope coefficients for each explanatory variable

ε = The model's error term also known as the residuals.

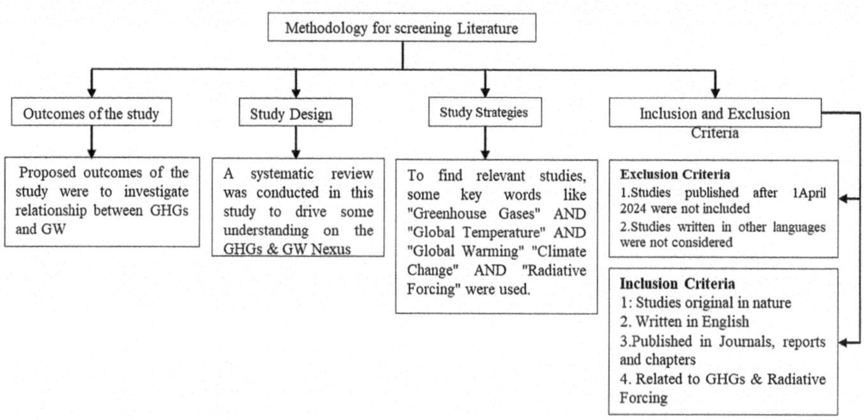

Fig. 1 Flowchart of literature access

2.4 Relative change analysis

It is also imperative to analyse how GHGs compare with past periods, for example, relative to a year ago or ten years ago or some specific period. For this purpose, absolute and relative change analysis have also been done.

Absolute change represents the difference in any indicator or two periods in time while relative change denotes the absolute change as a percentage of the value of the indicator in the earlier period, i.e.,

$$Abs.Change = \\ Indicator\ Value\ in\ Period\ 2 \\ -Indicator\ Value\ in\ Period\ 1$$

$$Rel.Change = \frac{Indicator\ Value\ in\ Period\ 2 \\ -Indicator\ Value\ in\ Period\ 1}{Indicator\ Value\ in\ Period\ 1} \times 100$$

3 Results and Discussion

3.1 Carbon Dioxide (CO_2) Concentration and Its Trend (1958–2024)

Since the 1700s, human activities particularly fossil fuel combustion (currently at about 7 gigatons of carbon per year) and deforestation (0.6-2.5 gigatons of carbon per year) – have led to a 35% increase in CO_2 concentration in the atmosphere. In the pre-industrial period, CO_2 concentration stood at 280 parts per million (ppm). It has steadily rose to 360 ppm in 1995, 379 ppm in 2005, 400 ppm in 2015, 414.89 ppm in 2021 and 422.86 ppm in 2024. Air temperature has also undergone significant changes over millennia. Approximately 18,000 years ago, it was –5°C, while 9,000 years ago, it reached 0°C. Presently, the average global temperature is around 15°C. Recent scientific studies and models confirm that this temperature increase aligns with rising CO_2 levels. T.C. Chamberlin's carbon dioxide theory supports this correlation. Unfortunately, the lower atmosphere's temperature tends to rise by 3°C with a doubling of CO_2 concentrations. Based on the CO_2 data from 1958 to 2024 collected from Global Monitoring Laboratory, it has been found that global mean monthly CO_2 has been consistently rising with two down peaks observed only in 1975 and 1984. It has reached to 422.86 ppm in 2024 from a mere concentration of 310 ppm in 1958. The prediction also showed the rising trend with concentration reaching at 475 ppm in 2038. Least square method showed a rising trend with equation $Y = 1.6303x + 302.46$ and R^2=0.935 (Fig. 2). The monthly variation in CO_2 concentration over same period is illustrated in Fig. 3a–m. Table 1 delineates the descriptive statistics of monthly global CO_2 concentration (1958–2021).

3.2 Methane (CH_4) Concentration and its Trend (1983–2024)

Methane (CH_4), in comparison to carbon dioxide, has a significantly shorter lifespan in the atmosphere, lasting approximately 12 years rather than centuries. However, during its presence in the atmosphere, methane absorbs a considerably larger amount of energy. Currently, the concentration of methane in the atmosphere stands at roughly two and a half times higher than pre-industrial levels. Estimating methane emissions involves a notable degree of uncertainty, but the latest comprehensive assessment, outlined in the Global Methane Budget, suggests that annual global methane emissions total around 580 million metric tons. These emissions comprise both natural sources, accounting for approximately 40%, and human activities, known as anthropogenic emissions, constituting the remaining 60%. Agriculture represents the largest anthropogenic source, responsible for about a quarter of methane emissions, closely followed by the energy sector, which encompasses emissions from coal, oil, natural gas and biofuels. Methane (CH_4), another potent greenhouse gas, was at approximately 700 ppb before industrialization. Due to anthropogenic emissions, it has increased by a factor of 2.5 reaching 1,784 ppb in 2003. From the data observed during 1983 to 2024, CH_4 has been consistently rising with a minimal decline from 2005 to 2015. It has reached to 1930.33 ppm in 2024 from a mere concentration of 1,636 ppm in 1983. The trend

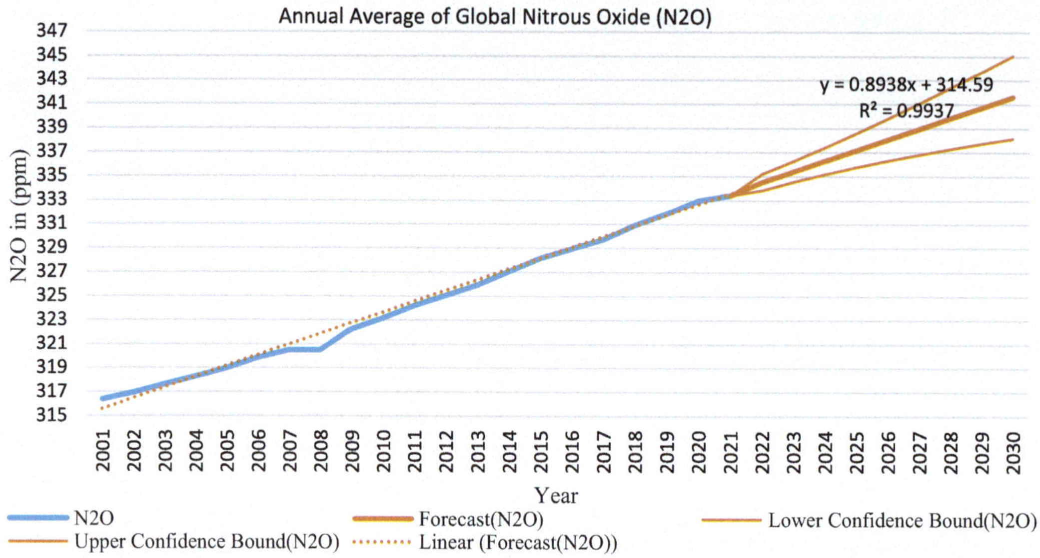

Fig. 2 Global trend of CO_2 concentration

will likely increase as per the prediction of CH_4 reaching at 1988 ppm in 2038. Linear trend gives the value (R^2 = 0.948, Y = 5.4041x + 1660.8) (Fig. 4). The monthly variation of CH_4 concentration over same period is illustrated in Fig. 5a–m. There are several studies (Kemfert & Schill, 2009; Mar et al., 2022; IEA, 2022) which indicate that due to its chemical and physical properties, methane has contributed to approximately 30% of global warming, since the industrial revolution (Methane and Climate Change—Global Methane Tracker 2022—Analysis—IEA). To curb immediate temperature increases and enhance air quality, it's crucial to decrease methane emissions consistently and swiftly. Table 2 provides the descriptive statistics of monthly global CH_4 concentration (1983–2024).

3.3 Nitrous Oxide (N_2O) Concentration and Its Trend (1997–2024)

Nitrogen oxide (N_2O) is a toxic gas with a reddish-orange hue. Its primary formation occurs at elevated combustion temperatures, such as those found in internal combustion engines and fossil fuel-fired power plants. Nitrogen oxide compounds play a significant role in indirectly influencing radiative forcing through interactions with CO_2 and methane. Additionally, N_2O serves as a precursor to ozone formation in the troposphere. When combined with aerosols in the environment, it contributes to the creation of brownish smog, reducing atmospheric visibility and adversely affecting plant growth and crop productivity. Trend line of N_2O shows a rising tendency. The concentration of N_2O was merely 270 ppb before industrialization, but it has consistently grown to 337.51 ppb in 2024 and predicted to reach 342 ppb in 2032. The least square gives the trend equation Y = 0.8938x + 314.59 and R^2 = 0.9937 (Fig. 6). The monthly variation in N_2O concentration over same period is illustrated in Fig. 7a–m. The study by Ming et al. (2016) emphasizes more on removing nitrous oxide gas than other GHGs in order to attain the pre-industrial global temperature. Table 3 gives the descriptive statistics of monthly global N_2O concentration (1997–2024).

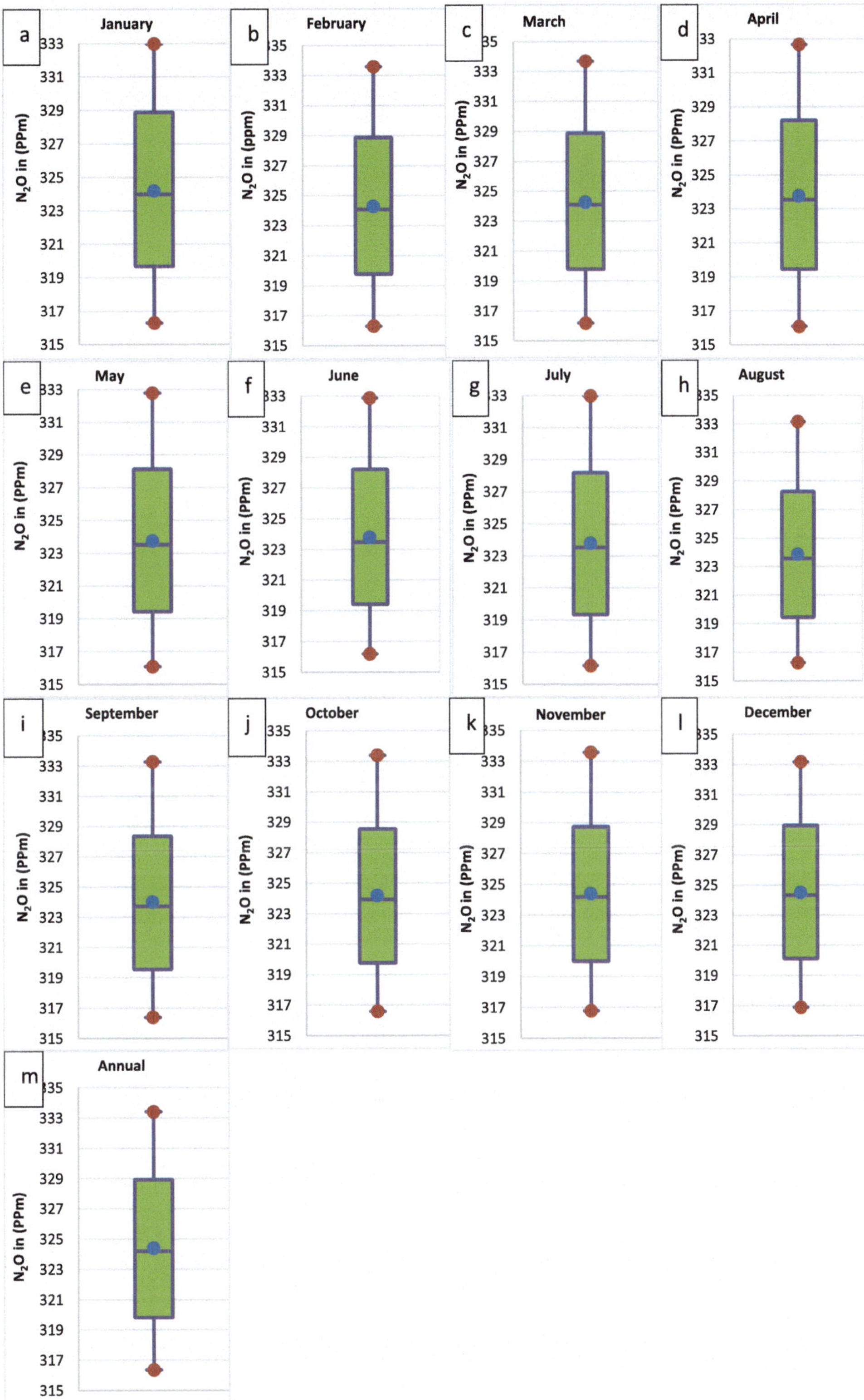

Fig. 3 (a)–(m) Global CO_2 concentration (1958–2021)

Table 1 Descriptive statistics of monthly global CO_2 concentration (1958–2021)

Statistic	Jan	Feb	March	April	May	June	July	Aug	Sept	Oct	Nov	Dec	Annual
Min	315.6	316.38	315.71	315.10	317.50	317.10	315.86	314.800	313.20	312.66	313.330	314.15	295.23
Max	415.5	416.75	417.34	416.45	417.31	416.62	414.61	412.780	411.52	411.51	413.110	414.25	416.54
Range	99.9	100.37	101.63	101.35	99.81	99.52	98.75	97.980	98.32	98.85	99.780	100.1	121.30
Mean	356.6	357.34	357.55	350.80	358.46	357.88	356.37	354.365	352.81	352.85	354.186	348.66	355.44
Variance (*n*)	868.6	871.82	884.61	4121.61	849.62	844.36	833.69	831.133	835.37	844.78	854.038	4097.53	970.06
Standard deviation (*n*)	29.5	29.53	29.74	64.20	29.15	29.06	28.87	28.829	28.90	29.07	29.224	64.01	31.15
Skewness (Pearson)	0.35	0.35	0.35	−5.39	0.36	0.36	0.35	0.360	0.37	0.36	0.361	−5.36	0.26
Kurtosis (Pearson)	−1.06	−1.06	−1.06	35.70	−1.05	−1.05	−1.06	−1.056	−1.06	−1.06	−1.055	35.42	−1.02
Standard error of the mean	3.74	3.75	3.75	8.153	3.70	3.69	3.67	3.661	3.67	3.69	3.711	8.13	3.92
Lower bound on mean (95%)	349.10	349.84	350.06	334.50	351.06	350.51	349.04	347.046	345.47	345.47	346.766	332.41	347.60
Upper bound on mean (95%)	364.06	364.83	365.04	367.10	365.86	365.26	363.70	361.684	360.14	360.22	361.605	364.91	363.29

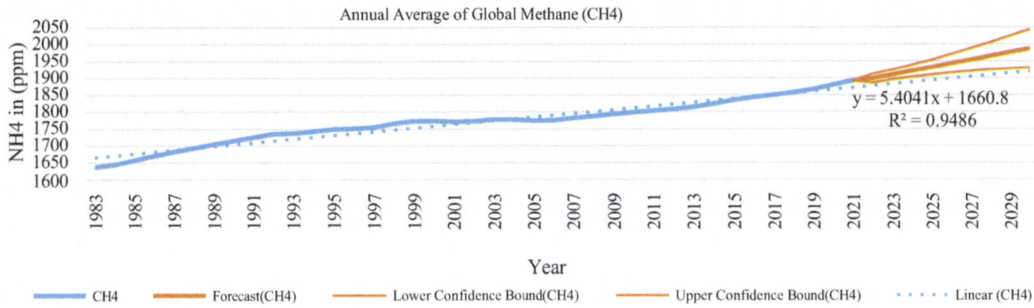

Fig. 4 Global trend of CH_4 concentration

3.4 Sulphur Hexaflouride (SF6) Concentration and Its Trend (1983–2024)

SF6 is known to be one of the most powerful greenhouse gases, with exceptional atmospheric stability and a remarkable ability to trap infrared radiation, making it far more effective at warming the Earth's atmosphere over extended periods compared to CO_2. Its impact is significantly more potent than CO_2. As the world moves towards achieving net zero carbon emissions by 2050 to address climate change, there is a growing recognition of the need to reassess the use of SF6.

Over a 100-year span, SF6 is estimated to be 23,500 times more effective at trapping infrared radiation than CO_2, implying that 1 kg of SF6 has the equivalent impact of 23,500 kg of CO_2. Once released into the atmosphere, SF6 remains for an extended period with an atmospheric lifetime of 3,200 years, leading to its accumulation without degradation for millennia. Between 2008 and 2018, annual SF6 emissions increased by 24%, largely driven by the expanding use of switchgear utilizing the gas, particularly in developing nations. In 2008, approximately 7,300 tonnes of SF6 were emitted globally, which rose to 9,040 tonnes in 2018. To put this into perspective, 9,000 tonnes of SF6 emissions are equivalent to the greenhouse gas emissions from about 44 million passenger vehicles driven for a year, or the combustion of 103 million tonnes of coal. Furthermore, it is suspected that the actual level of SF6 released into the atmosphere is significantly underestimated, particularly in developing countries, where tracking its release proves to be challenging. From 1997 to 2024 timeseries data collected from Global Monitoring Laboratory, it is revealed that the concentration of SF6 gas showed a rising trend from a mere concentration of 4 ppt in 1996 to 11.64 ppt in 2024. Least square gives the linear trend equation of $Y = 0.2739x - 543.32$ and R^2=0.9917 (Fig. 8). The monthly variation in SF_6 concentration over same period is illustrated in Fig. 9a–m. Table 4 gives the descriptive statistics of monthly global SF_6 concentration (1997–2024).

3.5 Absolute and Relative Change Analysis of Global GHGs

It is also imperative to analyse how GHGs compared with past periods for example relative to a year ago or ten years ago or some specific periods are changing. Figure 10 shows the year absolute increase in the concentration of GHGs relative to the pre-industrial period of 1750. Each data point in the graph represents the incremental concentrations in GHGs over time, providing a visual depiction of the escalating trend in atmospheric GHGs levels since the onset of industrialization. The graph portrays a clear upward trajectory, indicating a continuous and accelerating accumulation of GHGs in the Earth's atmosphere. Figure 11 shows the mean annual absolute increase in GHG concentration during the last 10 years. From the figure, the inference can be derived that absolute increase in CH_4 is very steep during 2013 (3.8%ppb) to 2016 (6.8%ppb) and after that a stability in concentration is observed. When focusing on the other GHGs,

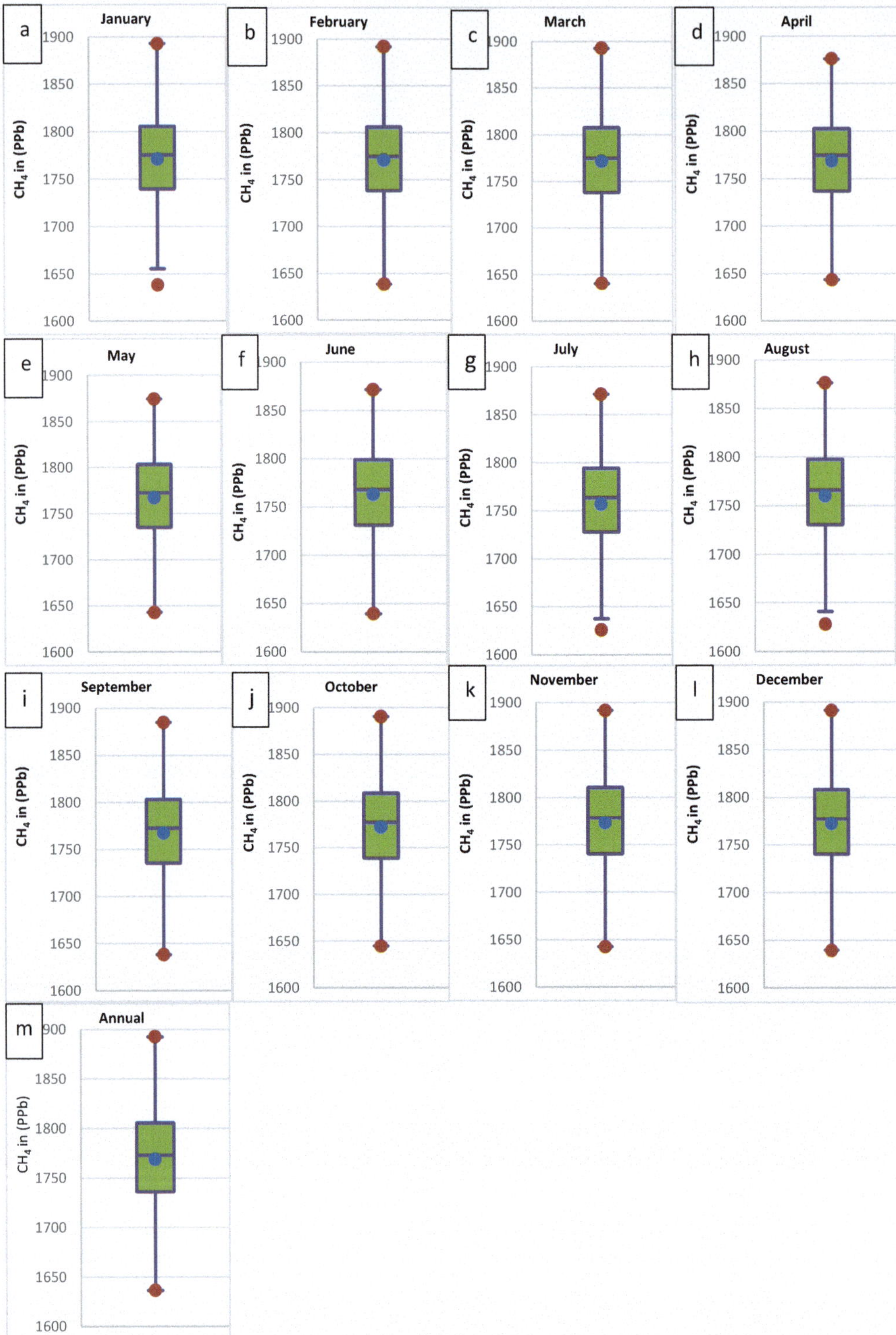

Fig. 5 (a)–(m) Global CH_4 concentration (1983–2024)

Table 2 Descriptive statistics of monthly global CH_4 concentration (1983–2024)

Statistic	Jan	Feb	Mar	April	May	June	July	Aug	Sept	Oct	Nov	Dec	Annual
Min	1638.6	1638.80	1640.70	1643.70	1642.90	1639.60	1625.90	1628.10	1638.40	1644.80	1642.60	1639.40	1636.53
Max	1892.9	1891.90	1893.00	1876.30	1874.50	1871.70	1871.50	1876.50	1884.80	1890.60	1891.90	1891.30	1892.60
Range	254.3	253.10	252.30	232.60	231.60	232.10	245.60	248.40	246.40	24,580	249.30	251.90	256.07
Mean	1771.5	1771.21	1771.97	1769.04	1767.44	1763.51	1757.12	1760.24	1767.49	1772.50	1773.66	1772.79	1768.92
Variance (*n*)	3576.7	3579.47	3571.41	3223.10	3200.68	3184.54	3588.13	3624.63	3582.83	3574.69	3619.94	3597.96	3899.75
Standard deviation (*n*)	59.81	59.83	59.76	56.77	56.58	56.43	59.90	60.21	59.86	59.79	60.16	59.98	62.45
Skewness (Pearson)	−0.20	−0.21	−0.18	−0.30	−0.26	−0.24	−0.29	−0.26	−0.27	−0.21	−0.24	−0.23	−0.19
Kurtosis (Pearson)	−0.30	−0.30	−0.32	−0.38	−0.38	−0.34	−0.35	−0.36	−0.39	−0.38	−0.34	−0.32	−0.37
Lower bound on mean (95%)	1751.60	1751.28	1752.06	1749.85	1748.32	1744.44	1737.17	1740.18	1747.55	1752.59	1753.61	1752.81	1748.41
Upper bound on mean (95%)	1791.44	1791.14	1791.88	1788.23	1786.57	1782.59	1777.07	1780.29	1787.43	1792.42	1793.70	1792.77	1789.43

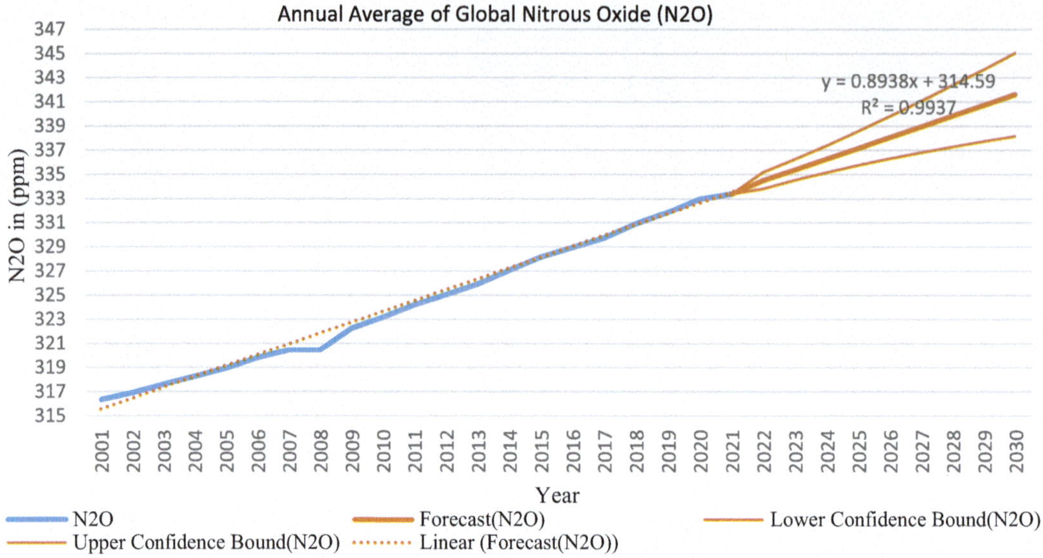

Fig. 6 Global trend of N_2O concentration

i.e., CO_2 and N_2O, the absolute increase in atmospheric concentration during past 10 years has remain almost stable. Figure 12 demonstrates the year relative increase in GHGs concentration to the year 1750. The relative concentration of GHGs has been very fluctuating compared to year 1750. The trend lines in the figure, indicating the GHGs relative concentration trajectory, are constantly upward moving which imply that concentration compared to historic eras has risen drastically and further predicted to be increased if necessary measures were not taken. These observed trends of GHGs underscore the persistent impact of human activities, particularly industrialization, fossil fuel combustion and deforestation, on altering the atmospheric composition. The steepness of the curve signifies the intensification of GHGs emissions over the years, reflecting the growing magnitude of anthropogenic contributions to climate change. The increasing rate of GHG concentration highlights the urgency of addressing emissions to mitigate the adverse effects of global warming and climate disruption. Overall, Figs. 10, 11 and 12 provide a compelling visual representation of the relentless rise in GHG concentrations since the pre-industrial era, underscoring the imperative for concerted global efforts to curb emissions and transition towards sustainable practices to safeguard the planet's future.

3.6 Radiative Forcing

The radiative forcing represents the change in forcing per unit increase in GHGs concentration. Figure 13 depicts the year-by-year absolute increase in Atmospheric Radiative Forcing (ARF), which quantifies the imbalance between incoming solar radiation absorbed by the Earth's atmosphere and surface and the outgoing long-wave radiation emitted back into space. This measure serves as a key indicator of the impact of greenhouse gases (GHGs) and other forcing agents on Earth's energy balance and climate system. The graph illustrates a notable upward trend in ARF from long lived GHGs over time, indicating a progressive enhancement in the greenhouse effect caused by rising concentrations of GHGs in the atmosphere. As GHGs such as carbon dioxide (CO_2), methane (CH_4) and nitrous oxide (N_2O) accumulate, they trap more heat within the Earth's atmosphere, leading to a positive radiative forcing and consequent warming of the planet's surface. The steepness of the curve reflects the accelerating pace at which ARF has been increasing, underscoring the exacerbating influence of human activities on temperature rise and climate change. This trend highlights the significant role of GHGs in driving radiative forcing and amplifying global warming, thereby contributing to the myriad environmental impacts asso-

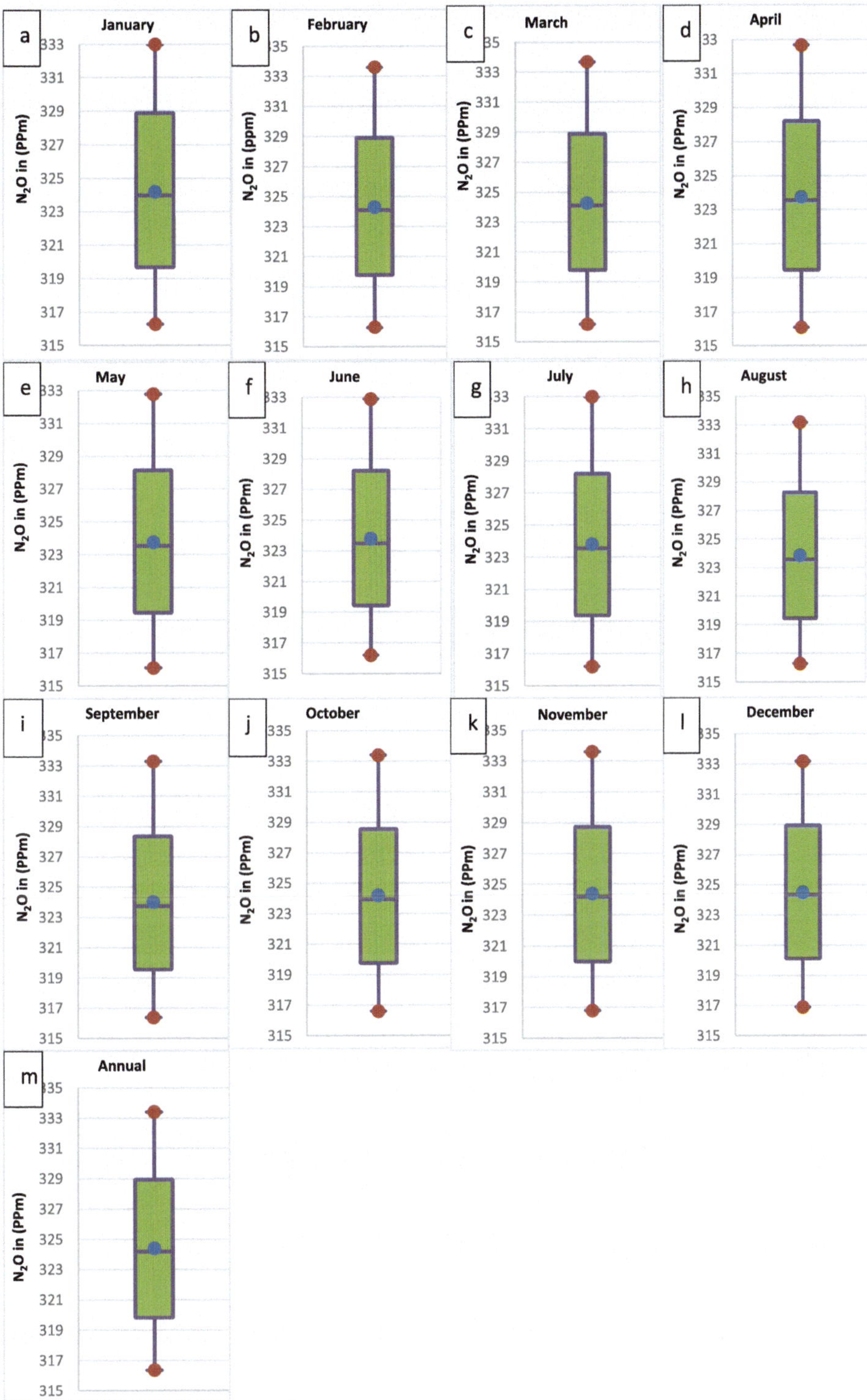

Fig. 7 (a)–(m), Global N_2O concentration (1997–2024)

Table 3 Descriptive statistics of monthly global N_2O concentration (1997–2024)

Statistic	Jan	Feb	Mar	Apr	May	Jun	Jul	Aug	Sept	Oct	Nov	Dec	Annual
Min	316.30	316.30	316.20	316.10	316.10	316.20	316.20	316.30	316.40	316.60	316.80	316.90	316.37
Max	333.00	333.60	333.70	332.70	332.80	332.90	333.00	333.20	333.30	333.40	333.60	333.20	333.43
Range	16.70	17.30	17.50	16.60	16.70	16.70	16.80	16.90	16.90	16.80	16.80	16.30	17.07
Mean	324.21	324.29	324.28	323.78	323.77	323.80	323.83	323.90	324.03	324.20	324.41	324.53	324.42
Variance (*n*)	28.46	29.11	29.25	26.40	26.54	26.83	27.14	27.39	27.18	27.18	27.12	26.70	29.48
Standard deviation (*n*)	5.34	5.40	5.408	5.138	5.151	5.180	5.210	5.233	5.214	5.213	5.208	5.167	5.429
Variation coefficient (*n*)	0.02	0.017	0.017	0.016	0.016	0.016	0.016	0.016	0.016	0.016	0.016	0.016	0.017
Skewness (Pearson)	0.14	0.16	0.17	0.17	0.18	0.18	0.19	0.19	0.19	0.18	0.17	0.14	0.15
Kurtosis (Pearson)	−1.30	−1.26	−1.26	−1.27	−1.27	−1.27	−1.27	−1.26	−1.25	−1.25	−1.24	−1.28	−1.29
Standard error of the mean	1.19	1.21	1.21	1.18	1.18	1.19	1.20	1.20	1.20	1.20	1.20	1.19	1.21
Lower bound on mean (95%)	321.7	321.77	321.76	321.31	321.30	321.31	321.32	321.39	321.52	321.69	321.90	322.04	321.89
Upper bound on mean (95%)	326.7	326.80	326.80	326.24	326.24	326.29	326.33	326.41	326.53	326.70	326.91	327.01	326.95

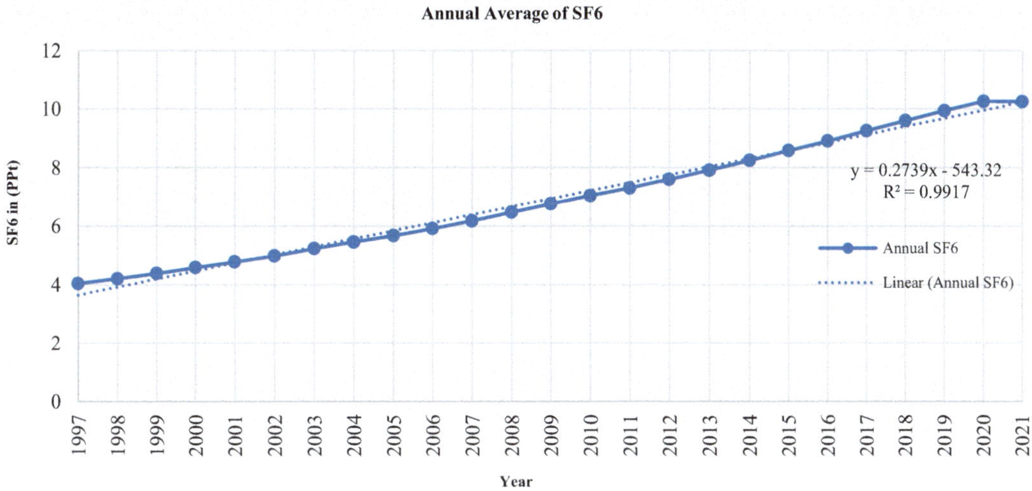

Fig. 8 Global trend of SF_6 concentration

ciated with climate change, including rising temperatures, sea level rise, altered precipitation patterns and more frequent extreme weather events. Overall, the figure underscores the critical link between GHG emissions, Atmospheric Radiative Forcing and climate change, emphasizing the urgent need for concerted global action to mitigate emissions, transition to renewable energy sources and implement adaptive measures to address the growing challenges posed by anthropogenic climate change.

3.7 Correlation and Regression

Table 5 shows the descriptive statistics of the independent variables (CO_2, CH_4 and N_2O) and dependent variable (global temperature). Table 6 shows the relationship between global CO_2 and temperature ($r = 0.79$), global CH_4 and temperature ($r = 0.50$), global N_2O and temperature ($r = 0.24$) highest being observed between CO_2 and temperature. The global mean surface temperature is found to be highly correlated with GHGs. A statistically positive association was found between CO_2 and temperature as well as CH_4 and temperature and a positive correlation between N_2O and temperature (Table 6).

A multiple regression was run to predict global temperature from global CO_2, NH_4 and N_2O. These variables statistically significantly predicted temperature, F (3, 13) = 9.444, $p < 0.005$ (i.e. the regression model is a good fit of the data), $R^2 = 0.685$ (Tables 7, 8 and 9). The coefficient of determination R^2 which is 0.685 indicates that our independent variables, i.e., CO_2, NH_4 and Nox, explain 68% of the variability in global mean temperature. CO_2 added statistically significantly to the prediction, $p < 0.05$ (Table 9). The multiple correlation coefficient R is 0.828, indicating a good level of prediction. Finally, the residuals (errors) were examined to ensure that they are roughly regularly distributed (Table 10). The general form of the equation used here to predict global mean surface temperature from independent variables namely CO_2, NH_4 and N_2O is given below:

$$\begin{aligned}\text{Predicted Global Temperature} = \\ -0.030 + (0.863 \times \text{Global}\,CO_2) \\ + (0.133 \times \text{Global}\,CH_4) \\ + (0.72 \times \text{Global}\,N_2O)\end{aligned} \tag{2}$$

It further showed that temperature would increase by the magnitude of 0.863 with increasing CO_2, ($t = 4.220$, $p < 0.05$) because CO_2, an independent variable coefficient, is statistically significantly different from 0 (zero).

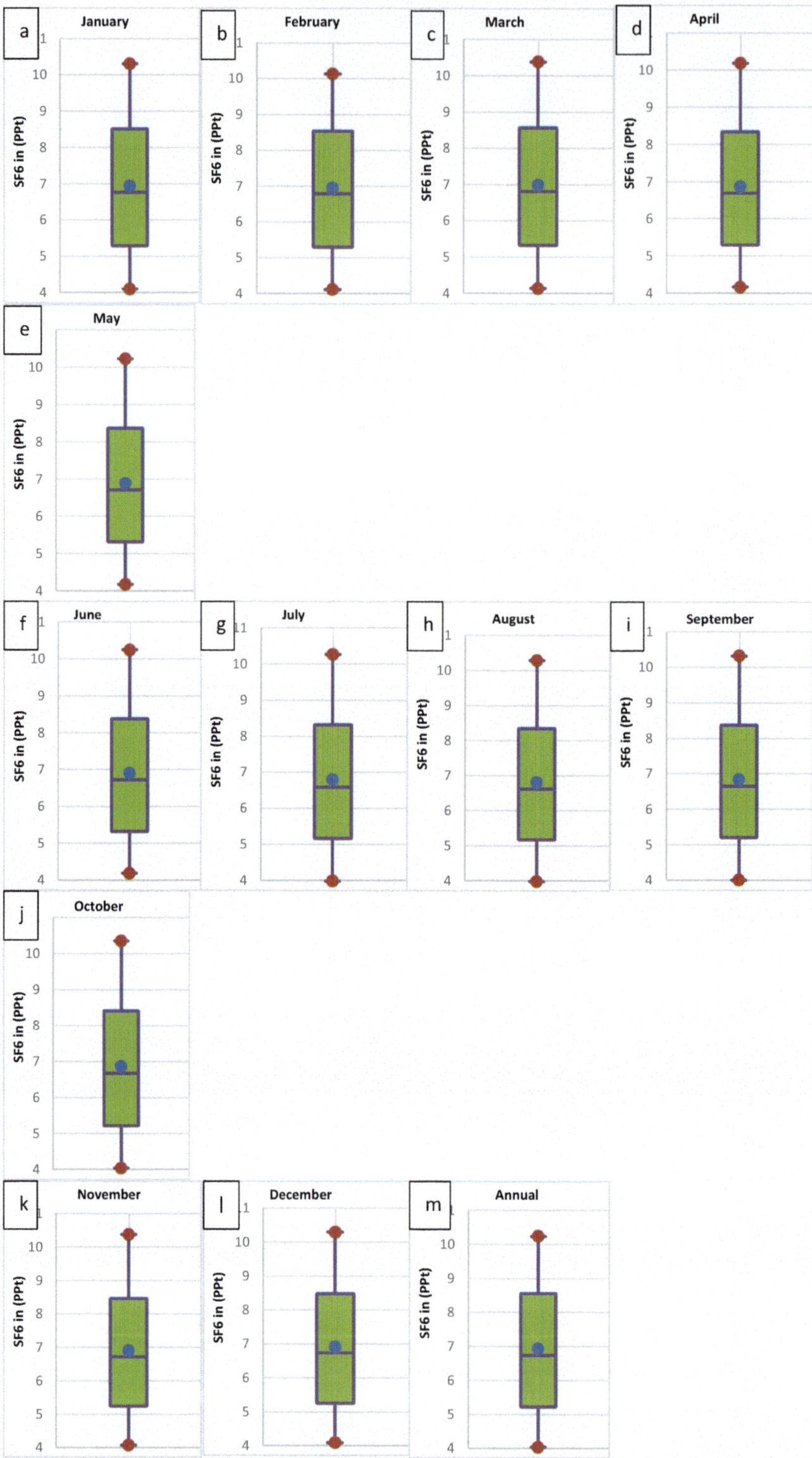

Fig. 9 (a)–(m), Global SF_6 concentration (1997–2024)

Table 4 Descriptive statistics of monthly global SF_6 concentration (1997–2024)

Statistic	Jan	Feb	Mar	April	May	Jun	Jul	Aug	Sept	Oct	Nov	Dec	Annual
Min	4.11	4.130	4.150	4.180	4.190	4.200	4.000	4.000	4.020	4.050	4.080	4.100	4.042
Max	10.32	10.150	10.400	10.200	10.230	10.260	10.280	10.300	10.330	10.360	10.390	10.310	10.259
Range	6.21	6.020	6.250	6.020	6.040	6.060	6.280	6.300	6.310	6.310	6.310	6.210	6.218
Mean	6.95	6.963	7.003	6.878	6.897	6.910	6.803	6.820	6.844	6.875	6.908	6.930	6.942
Variance (n)	3.70	3.632	3.742	3.389	3.410	3.437	3.653	3.667	3.675	3.690	3.711	3.692	3.934
Standard deviation (n)	1.93	1.906	1.934	1.841	1.847	1.854	1.911	1.915	1.917	1.921	1.926	1.922	1.983
Variation coefficient (n)	0.28	0.274	0.276	0.268	0.268	0.268	0.281	0.281	0.280	0.279	0.279	0.277	0.286
Skewness (Pearson)	0.23	0.19	0.22	0.24	0.24	0.24	0.26	0.25	0.25	0.25	0.245	0.231	0.216
Kurtosis (Pearson)	−1.21	−1.25	−1.21	−1.18	−1.18	−1.18	−1.17	−1.17	−1.17	−1.18	−1.18	−1.20	−1.24
Standard error of the mean	0.40	0.40	0.40	0.39	0.39	0.40	0.40	0.40	0.40	0.40	0.40	0.40	0.41
Lower bound on mean (95%)	6.12	6.14	6.17	6.06	6.08	6.09	5.98	5.99	6.02	6.05	6.08	6.10	6.11
Upper bound on mean (95%)	7.78	7.79	7.84	7.69	7.71	7.73	7.63	7.65	7.67	7.70	7.74	7.76	7.78

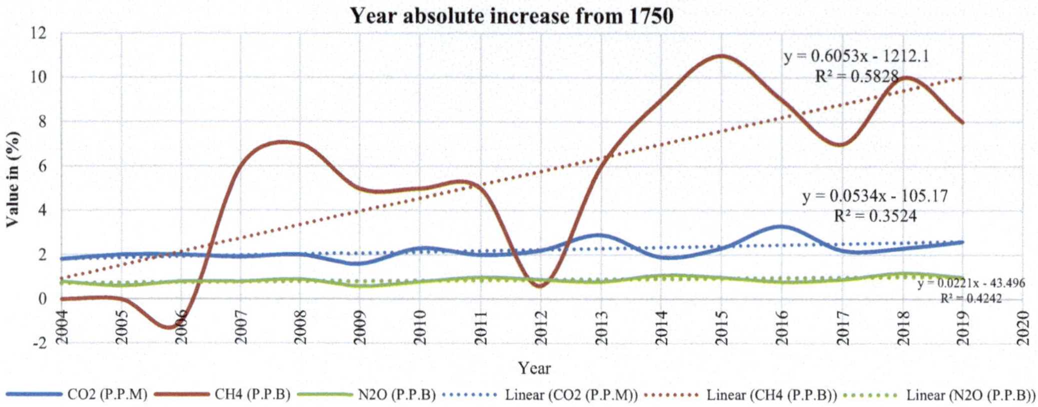

Fig. 10 Year absolute increase in GHG concentration from 1750

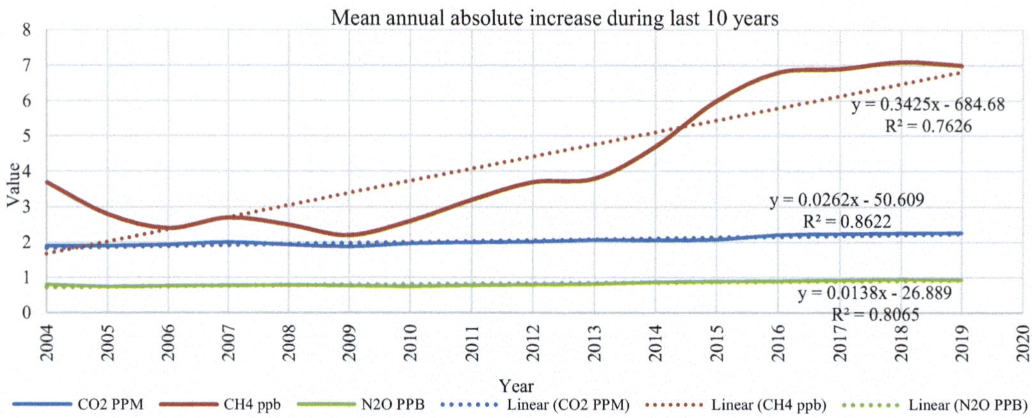

Fig. 11 Mean annual absolute increase in GHG concentration during the last 10 years

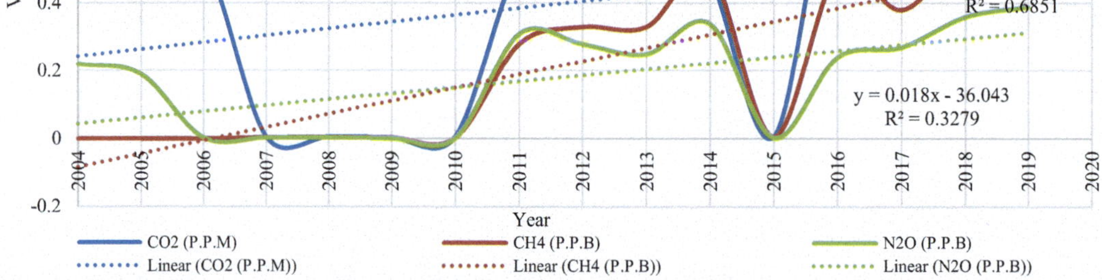

Fig. 12 Year relative increase in GHGs concentration to the year 1750

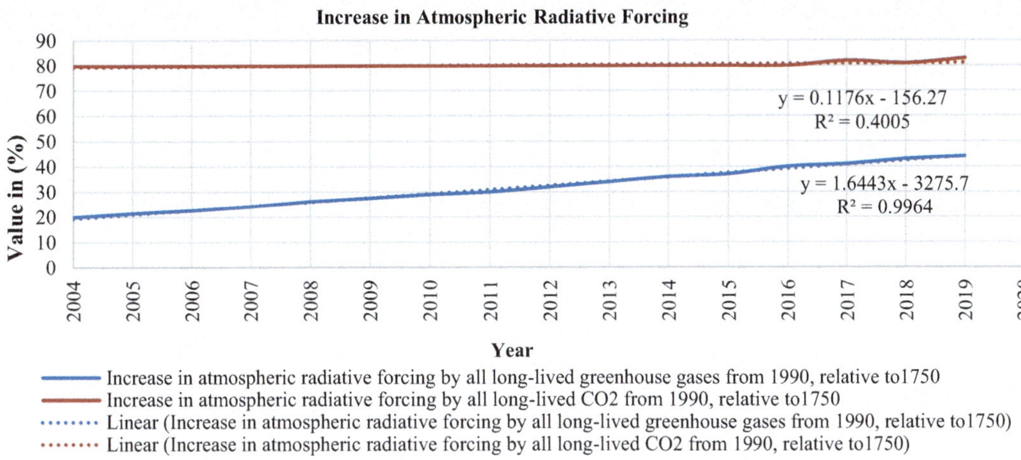

Fig. 13 Increase in atmospheric relative increase. Source: Prepared by the author

Table 5 Descriptive statistics of dependent and independent variables used in the regression analysis

	Mean	Std. deviation	*N*
Global mean surface temperature	0.5176	0.12814	17
CO_2	0.5647	0.10513	17
CH_4	0.3053	0.22336	17
N_2O	0.2806	0.06179	17

Table 6 Correlation between GHGs and temperature

Global mean surface temperature	*r*	*p*
Global carbon dioxide (CO_2)	0.793**	0.000
Global methane (CH_4)	0.501*	0.041
Global nitrous oxide (N_2O)	0.241	0.352

**Correlation is significant at the 0.01 level (2-tailed)
*Correlation is significant at the 0.05 level (2-tailed)

Table 7 Regression model statistics summary

R	R square	Adjusted R square	Std. error of the estimate
0.828[a]	0.685	0.713	0.07972

[a]Predictors: (Constant), N_2O, CO_2, CH_4
[b]Dependent variable: Global mean surface temperature

Table 8 ANOVA

ANOVA[a]						
Model		Sum of squares	Df	Mean square	F	Sig.
1	Regression	0.180	3	0.060	9.444	0.001[b]
	Residual	0.083	13	0.006		
	Total	0.263	16			

[a]Dependent variable: Global mean surface temperature
[b]Predictors: (Constant), N_2O, CO_2, CH_4

Table 9 Coefficient statistics

Coefficients[a]

Model		Unstandardized coefficients		Standardized coefficients	T	Sig.	95.0% confidence interval for B	
		B	Std. error	Beta			Lower bound	Upper bound
1	(constant)	−0.030	0.153		−0.198	0.846	−0.361	0.301
	CO_2	0.863	0.205	0.708	4.220	0.001	0.421	1.305
	CH_4	0.133	0.118	0.231	1.124	0.281	−0.122	0.388
	N_2O	0.072	0.401	0.035	0.179	0.861	−0.795	0.939

[a]Dependent variable: Global mean surface temperature

Table 10 Residual statistics

Residuals statistics[a]

	Minimum	Maximum	Mean	Std. deviation	N
Predicted value	0.3829	0.7589	0.5176	0.10609	17
Residual	−0.17001	0.10165	0.00000	0.07186	17
Std. predicted value	−1.270	2.274	0.000	1.000	17
Std. residual	−2.133	1.275	0.000	0.901	17

[a]Dependent variable: Global mean surface temperature

4 Discussion

Greenhouse gases like carbon dioxide (CO_2), methane (CH_4) and nitrous oxide (N_2O) are pivotal in regulating the temperature of Earth's atmosphere through the greenhouse effect. Since sunlight penetrates to the Earth's exterior, some of it is captured, and hence heating the planet. Subsequently, the Earth discharges this captured energy back into the atmosphere as infrared waves. GHGs in the air observe some amount of this released infrared waves, stopping it from dissolving into space. This mechanism functions akin to a blanket, maintaining the global surface at a warmer temperature than it could otherwise be. In the absence of GHGs, the average global temperature would be highly colder, rendering it uninhabitable for sustaining lives of living beings as we know it. Nonetheless, human interferences, such as the burning of fossil fuels, forest fires and forests loss and industrial emissions, have remarkably heightened the accumulation of GHGs in the atmosphere. This exaggerated greenhouse effect tends to huge retention of heat, culminating in global temperature and subsequent variation in climate. This study discusses the relationship between global GHGs and global temperature since pre-industrial era (1750). The findings of the study reveal an increasing trend in both absolute and relative concentration of GHGs. Among all GHGs, CO_2 has shown the highest contribution in raising global mean temperature. The correlation and regression analysis disclose a high association between global CO_2 concentration and global temperature ($r = 0.79$).

A number of scientific evidences, encompassing data from climate models and empirical observations, overwhelmingly corroborate the association between intensifying GHGs accumulations and the ascent of global atmospheric temperature. Findings reported from the Intergovernmental Panel on Climate Change (IPCC) consistently underscore this nexus, attributing a substantial portion of recent global warming to human disturbances, particularly the burning of fossil fuels. As concentrations of GHGs continue to rise, driven primarily by human activities, global temperatures are projected to increase further, with potentially severe harmful impact on ecosystems, human societies, seasonal patterns and sea levels. The research done by Solomon et al. (2009) reported that the alterations in climate, resulting from the rise in greenhouse gases, are predominantly irreversible and will persist for a millennium even after the cessation of emissions. Assuming the Earth continues to respond as it has historically, the gradual melting of ice sheets, the presence of natural aerosols and changes in vegetation distribution will persistently push global warming beyond the 2 °C goal over the ensuing millennia, irrespective of whether human-induced contributions remain constant (Ring et al., 2022). Addressing the root causes of GHG emissions is therefore essential to reduce the climate change and combating the associated impacts on the planet. This includes moving towards renewable energy sources, refining energy efficiency, preserving forests and practicing sustainable land use mechanism. Hence, identifying natural CO_2 reservoirs, devising strategies to boost their absorption capacity and formulating techniques to construct synthetic reservoirs for deliberate CO_2 removal are critical future research areas (Saini & Saini, 2022). Lenton (2011) suggests a restructured policy objective to limit the total intensity, rate of alteration and spatial variations of human-induced radiative forcing, and the ensuing climate change, by limiting emissions of human-made aerosols, land use patterns, and levels of both short-term and long-term greenhouse gases. Efforts are underway to develop eco-friendly substitutes for SF6 to curb the release of this harmful gas into our atmosphere. By adopting these new alternatives, we can not only save on equipment replacement costs but also lessen our carbon footprint.

5 Conclusion

The interconnection between greenhouse gases (GHGs) and temperature increase is fundamental to understand climate change. Climate change impacts extend beyond temperature rise; they include intense droughts, water scarcity, severe fires, rising sea levels, flooding, melting polar ice, catastrophic storms and declining biodiversity. People experience climate change in diverse ways, affecting health, food security, housing, safety and livelihoods. Vulnerable communities, such as those in small island nations and developing countries, are particularly at risk. This chapter delves into the intricate relationship between GHGs emissions and their impact on global temperature and climate change. Global temperature and GHGs data from global monitoring observatory were used to run through time series analysis by least square method and multiple regression analysis. Outcomes of the study showed that the CO_2 concentration has consistently been rising from a mere concentration of 310 ppm in 1958 to 433.30 ppm in 2024 and its highest accumulation is observed in April Month, while CH4 has increased from 1636 ppm in 1982 to 1930 ppm in 2024, of which the highest concentration was observed in January, and N_2O has rose from 316 to 334 ppb in 2024 which is predicted to reach 342 ppb in 2032 being highest in March, SF6 gas showed a rising trend from a mere concentration of 4 ppt in 1996 to 11.86 ppt in 2024 highest being observed in September month. A statistically positive association was found between CO_2 and temperature as well as CH_4 and temperature and a positive correlation between NOx and temperature. The prediction model (independent variables, namely CO_2, CH_4, N_2O and dependent variable temperature) gives the value of multiple correlation coefficient (R) equal to 0.828, indicating a good prediction level. The value of coefficient of determination (R^2) is 0.685, which indicates that our explanatory variables, i.e., CO_2, CH_4 and N_2O, explained 68% of the variability in temperature with $F(3, 13) = 9.444$, $p < 0.005$, $R^2 = 0.685$. CO_2 is found to be statistically significant ($p < 0.05$) to the prediction. While CO_2 has been identified as the major contributor to global temperature increase, other GHGs such as methane (CH_4) and nitrous oxide (N_2O) also play significant roles. Future research should conduct a detailed longitudinal analysis of these gases to understand their specific impacts and interactions with CO_2.

References

Bhatti, U. A., Bhatti, M. A., Tang, H., Syam, M. S., Awwad, E. M., Sharaf, M., & Ghadi, Y. Y. (2024). Global production patterns: Understanding the relationship between greenhouse gas emissions, agriculture greening and climate variability. *Environmental Research, 245*, 118049. https://doi.org/10.1016/J.ENVRES.2023.118049

Brovkin, V., Sitch, S., von Bloh, W., Claussen, M., Bauer, E., & Cramer, W. (2004). Role of land cover changes for atmospheric CO_2 increase and climate change during the last 150 years. *Global*

Change Biology, 10, 1253–1266. https://doi.org/10.1111/j.1365-2486.2004.00812.x

Bruhwiler, L., Basu, S., Butler, J. H., Chatterjee, A., Dlugokencky, E., Kenney, M. A., McComiskey, A., Montzka, S. A., & Stanitski, D. (2021). Observations of greenhouse gases as climate indicators. *Climatic Change, 165*, 1–18. https://doi.org/10.1007/s10584-021-03001-7

Davis, W. J. (2017). The relationship between atmospheric carbon dioxide concentration and global temperature for the last 425 million years. *Climate, 5*, 76. https://doi.org/10.3390/CLI5040076

Gore, A. (2006). *An inconvenient truth: The planetary emergency of global warming and what we can do about it*. Rodale.

Gregory, J. M., & Forster, P. M. (2008). Transient climate response estimated from radiative forcing and observed temperature change. *Journal of Geophysical Research—Atmospheres, 113*, 1–15. https://doi.org/10.1029/2008JD010405

IEA. (2022). *Global methane tracker 2022*. International Energy Agency. https://www.iea.org/reports/global-methane-tracker-202

Intergovernmental Panel on Climate Change [IPCC]. (2023). *Climate change 2023: Synthesis report. Contribution of working groups I, II and III to the sixth assessment report of the intergovernmental panel on climate change* (H. Lee & J. Romero, Eds.). IPCC. https://doi.org/10.59327/IPCC/AR6-9789291691647

IPCC Third Assessment Report. (2001). *Climate change 2001: Synthesis report. A contribution of working groups I, II, and III to the third assessment report of the intergovernmental panel on climate change* [Watson, R.T. and the Core Writing Team (eds.)]. Cambridge University Press, pp. 398. Available from www.ipcc.ch

Kemfert, C., & Schill, W.-P. (2009). *Mitigation of methane emissions: A rapid and cost- effective response to climate change*. Discussion paper. DIW Berlin.

Lenton, T. M. (2011). Beyond 2°C: Redefining dangerous climate change for physical systems. *Wiley Interdisciplinary Reviews: Climate Change, 2*, 451–461. https://doi.org/10.1002/wcc.107

Macedo, P., & Madaleno, M. (2022). Global temperature and carbon dioxide nexus: Evidence from a maximum entropy approach. *Energies, 16*, 277. https://doi.org/10.3390/EN16010277

Mar, K. A., Unger, C., Walderdorff, L., & Butler, T. (2022). Beyond CO_2 equivalence: The impacts of methane on climate, ecosystems, and health. *Environmental Science & Policy, 134*, 127–136. https://doi.org/10.1016/J.ENVSCI.2022.03.027

Maruyama, F. (2019). Relationship between the atmospheric CO_2 and climate indices by wavelet-based multifractal analysis. *Journal of Geoscience and Environment Protection, 07*, 38–51. https://doi.org/10.4236/gep.2019.71004

Mehmood, T., Hassan, M. A., Li, X., Ashraf, A., Rehman, S., Bilal, M., Obodo, R. M., Mustafa, B., Shaz, M., Bibi, S., & Shakoor, A. (2022). Mechanism behind sources and sinks of major anthropogenic greenhouse gases. In *Climate change alleviation for sustainable progression* (pp. 114–150). Taylor & Francis. https://doi.org/10.1201/9781003106982-8

Ming, T., de Richter, R., Shen, S., & Caillol, S. (2016). Fighting global warming by greenhouse gas removal: Destroying atmospheric nitrous oxide thanks to synergies between two breakthrough technologies. *Environmental Science and Pollution Research, 23*, 6119–6138. https://doi.org/10.1007/s11356-016-6103-9

Moiceanu, G., & Dinca, M. N. (2021). Climate change-greenhouse gas emissions analysis and forecast in Romania. *Sustainability, 13*, 12186. https://doi.org/10.3390/SU132112186

Ring, S. J., Mutz, S. G., & Ehlers, T. A. (2022). Cenozoic proxy constraints on earth system sensitivity to greenhouse gases. *Paleoceanography and Paleoclimatology, 37*, 1–21. https://doi.org/10.1029/2021PA004364

Rivera, P. C., & Khan, T. M. A. (2012). Discovery of the major mechanism of global warming and climate change. In *Greenhouse gases: Sources, sinks and mitigation* (Vol. 8). Springer. https://doi.org/10.6000/1927-5129.2012.08.01.29

Saini, M., Saini, S.K., 2022. Identification of major sinks of greenhouse gases. Greenh. Gases Sources, Sink. Mitig. 36–62. https://doi.org/10.1007/978-981-16-4482-5_3.

Solomon, S., Plattner, G. K., Knutti, R., & Friedlingstein, P. (2009). Irreversible climate change due to carbon dioxide emissions. *Proceedings of the National Academy of Sciences of the United States of America, 106*, 1704–1709. https://doi.org/10.1073/pnas.0812721106

Thornhill, G. D., Collins, W. J., Kramer, R. J., Olivie, D., Skeie, R. B., O'Connor, F. M., Luke Abraham, N., Checa-Garcia, R., Bauer, S. E., Deushi, M., Emmons, L. K., Forster, P. M., Horowitz, L. W., Johnson, B., Keeble, J., Lamarque, J. F., Michou, M., Mills, M. J., Mulcahy, J. P., Myhre, G., Nabat, P., Naik, V., Oshima, N., Schulz, M., Smith, C. J., Takemura, T., Tilmes, S., Wu, T., Zeng, G., & Zhang, J. (2021). Effective radiative forcing from emissions of reactive gases and aerosols-a multi-model comparison. *Atmospheric Chemistry and Physics, 21*, 853–874. https://doi.org/10.5194/acp-21-853-2021

Tol, R. S. J., Berntsen, T. K., Oneill, B. C., Fuglestvedt, J. S., & Shine, K. P. (2012). A unifying framework for metrics for aggregating the climate effect of different emissions. *Environmental Research Letters, 7*, 044006. https://doi.org/10.1088/1748-9326/7/4/044006

Verma, P., & Ghosh, P. K. (2022). REDD+ strategy for forest carbon sequestration in India. *The Holistic Approach to Environment, 12*, 117–130. https://doi.org/10.33765/thate.12.3.4

Viterito, A. (2016). The correlation of seismic activity and recent global warming. *Journal of Earth Science and Climatic Change, 7*. https://doi.org/10.4172/2157-7617.1000345

WMO Statement on the state of the global climate in 2018 (WMO-No. 1233).

Climate Change, Urbanization and Ecosystem Services: A Complex Systems Approach

Valuation of Ecosystem Services Amidst Land Use and Land Cover Change Dynamics in Kedarnath Wildlife Sanctuary, Uttarakhand, India

Azizur Rahman Siddiqui and Pragati Verma

1 Introduction

The connection between humanity and the ecological world is increasingly evident as we collectively strive to address global objectives such as preserving biodiversity, combating climate change, fostering economic growth and enhancing human welfare (UN, 2015). Every aspect of human existence is intertwined with, and reliant upon, ecosystems. These ecosystems offer valuable contributions, known as 'ecosystem services', to humanity. Conversely, the actions and behaviours of individuals and societies as a whole have a substantial influence on the quality and functioning of these ecosystems (Leviston et al., 2018). Hence, a comprehensive and profound investigation into the correlation between ecosystem services and human welfare has emerged as a crucial scientific matter within the domain of ecological economics (Xu et al., 2019). Enhancing our understanding of the intricate interdependencies between ecological and socioeconomic systems is paramount for the continuous and sustainable operation of both anthropogenic systems and ecosystems (Millennium Ecosystem Assessment, 2003). Millennium Ecosystem Assessment (2005) offers new insights into the connections between socioeconomic systems and ecosystem services. This assessment was undertaken from 2001 to 2005 with the aim of evaluating the impact of changes in ecosystems on human welfare (Balasubramanian & Sangha, 2023). It sought to develop a scientific foundation for measures required to improve the preservation and sustainable utilisation of ecosystem services. Ecosystem service can be described as a human-centred utilitarian concept, where the value of the benefits provided by ecosystems depends on the usefulness that individuals gain from their use, either directly or indirectly‖ (Goldenberg et al., 2017; UNEP, 2011). These services encompass provisioning services, regulating services, habitat services, cultural services and others (Li et al., 2022; Sharma et al., 2023). Ecosystem service value (ESV) is a comprehensive method used to quantify the monetary values of various ecological units that often involve invaluable ecosystem services contributing significantly to the socioecological importance of naturally occurring processes integral in producing natural capital and enhancing sustainable green economy of the area (Millennium Ecosystem Assessment, 2005; Sannigrahi et al., 2020). In essence, ESV approach quantifies the financial value of natural assets or ecosystem services from an ecological standpoint, employing both economic and environmental research methodologies (Costanza et al., 1997).

Ecosystem services, despite their substantial contribution to both natural processes and human

A. R. Siddiqui
Department of Geography, University of Allahabad, Prayagraj, Uttar Pradesh, India

P. Verma (✉)
Department of Economics, University of Allahabad, Prayagraj, Uttar Pradesh, India

Z. Khanam et al. (eds.), *Climate Change, Resilience and Ecosystem Services*, Sustainable Development Goals Series, https://doi.org/10.1007/978-3-031-96977-5_7

sustainability and survival, have experienced a significant global degradation across temporal and spatial dimensions (Gong et al., 2022). In the preceding half-century, anthropogenic influences have precipitated alterations in ecosystem services at a pace and range unparalleled in recorded human history (Scolozzi et al., 2012). Among the most significant consequences of anthropogenic disturbances on varied ecosystems is the modification in terrain utilisation. This not only brings transformation in the configuration of vegetation and topography but also influences the structure and operational dynamics of ecosystems (Hasan et al., 2020; Wang et al., 2022), consequently leading to transformations in ecosystem classifications, a progressive deterioration of Ecosystem Services and a reduction in biodiversity, subsequently instigating alterations in ecosystem service values (ESVs). Therefore, comprehending the influence and interplay of alterations in terrestrial use on the benefits of ecosystem services is of paramount importance (Yuan et al., 2019). Optimising land use patterns to align with changes in ecosystem service values (ESVs) and holistically augmenting ecosystem services holds substantial importance for the preservation of environmental systems and ecological diversity (Wang et al., 2022). While global-level assessments of land use and land cover (LULC) diversions exist, these models often lack the specificity needed to capture critical landscape modifications occurring at more granular spatial resolutions, like sporadic tropical forest loss and forest regeneration. Models of a more refined scale are necessitated for the exploration of the responses of land cover transformations on environmental benefits at local and regional levels (Hu et al., 2008). Concurrent with socioeconomic progression, the overutilisation of terrestrial resources by humans has remarkably damaged regional ecological systems (Tang et al., 2024). The degree and direction of LULC fluctuations have the potential to influence ecological phenomena such as energy transference, hydrological equilibrium and biogeochemical rotations within and interconnecting ecosystems. These dynamics can create additional transformations in ESVs, increasing certain aspects while diminishing others or potentially leading to ecological deterioration (Polasky et al., 2011). That is why, research focused on ESVs that are predicated on land utilisation can provide the spatiotemporal impact of regional changes in land use. This understanding can clarify the repercussions these changes have on ESVs, thereby providing a theoretical foundation and benchmark for ecological reformation (Zuo et al., 2023).

Ecosystem services represent the actual flow from ecological systems to socioeconomic structures, which are utilised within a defined time and space. In this context, various elements of ecological and socioeconomic structures are essential to evaluate the actual transfer of services and understand the temporal changes (Hein et al., 2016; Vallecillo et al., 2019). The significance of ecological advantages can be noticed by the quantification of ecosystem services in economic units, as the management of entities without any assigned value creates intricate challenges. There is an escalating need to numerically evaluate the role of ecological systems in enhancing human welfare and economic prosperity (Vallecillo et al., 2019). There are a number of direct and indirect use methods to estimate and allocate the economic value to terrestrial ecosystem services, such as contingent valuation method 'hedonic price method', 'opportunity cost method', 'inferred willingness-to-pay', 'price- based market valuation', revealed preference technique and 'the benefit transfer method (BTM)' (Barbier et al., 2009; Chopra et al., 2022; Marta-Pedroso et al., 2018; Verma & Ghosh, 2023). Within the diverse array of methodologies for the quantification of ecosystem service value, the benefit transfer method (BTM) has garnered extensive utilisation due to its operational practicability and inherent simplicity (Costanza et al., 2014; Dammag et al., 2024).

This study has selected Kedarnath Wildlife Sanctuary (KWLS), the largest protected area (PA) in Indian Himalayan Region (IHR), situated in Uttarakhand, as the study area due to its ecological significance and fragile mountainous nature. In the last few decades, KWLS has witnessed extensive land transition due to natural and human disturbances such as extreme flood, landslides, forest fire, exploitative tourism and excessive resource extraction (Manral, 2018). Studying the variations in monetary value of KWLS's ecosystem services in response to the transition in LULC is crucial for raising awareness and implementing strict regulations by both national and regional managing authorities in order to stabilise climatic, socioeconomic and cultural vulnerabilities. This study covering a longitudinal period of over two decades (2001–2024) utilises Geographic Information System (GIS) and remote sensing technique and the BTM approach to assess the ESV of KWLS and encompasses the following objectives:

(i) To prepare the LULC classification map of Kedarnath Wildlife Sanctuary (KWLS) for the years 2001, 2014 and 2024 using random forest (RF) technique.

(ii) To conduct change detection analysis from 2001 to 2024.

(iii) To estimate the ecosystem service value (ESV) of KWLS for the period 2001, 2014 and 2024.

1.1 Study Area Description

The Kedarnath Wildlife Sanctuary, nestled in the majestic Himalayan range of Uttarakhand, India, encompasses an area of approximately 975.20 km². Established under the Wildlife Protection Act of 1972, the sanctuary's primary purpose is to protect the endangered Himalayan musk deer. Its geographical coordinates range from approximately 30°25′–30°41′ North latitude and 78°55′–79°22′ East longitude (Fig. 1). Located in the Garhwal region of Uttarakhand,

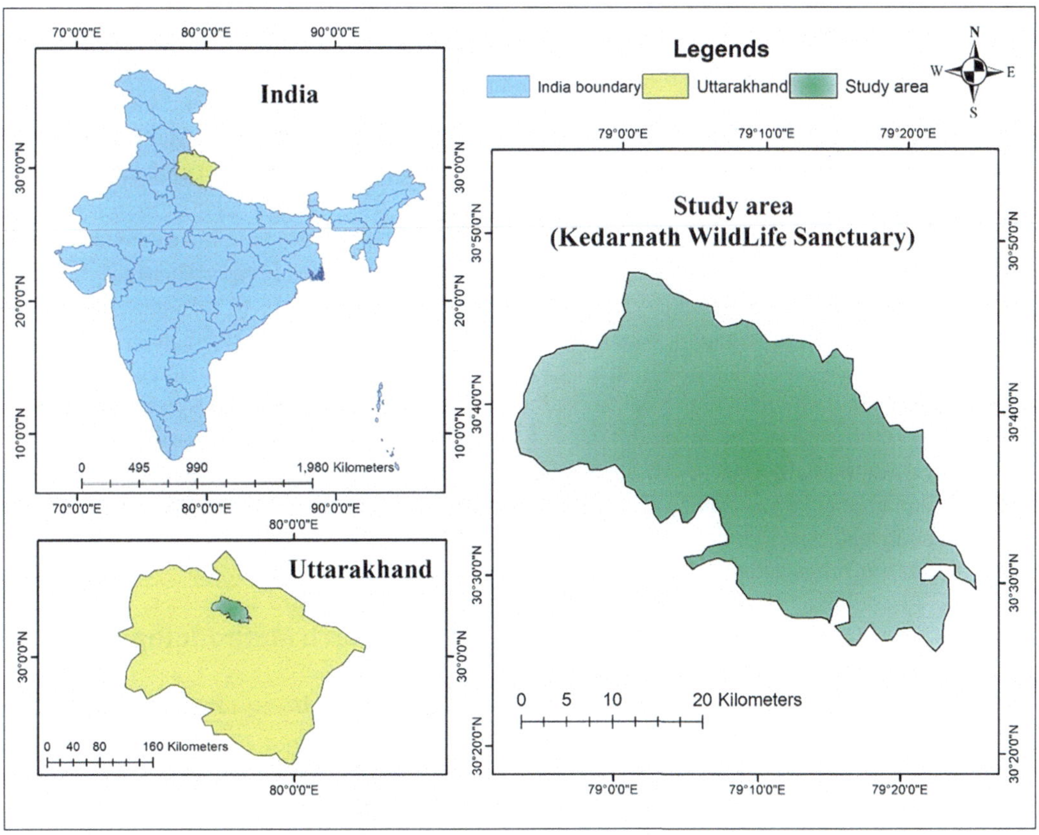

Fig. 1 Location map. Source: Generated in ArcMAP 10.7.1

Table 1 Satellite data sources information

Sl. no.	Landsat data	Satellite ID	Year	Spatial resolution	Cloud cover (%)	Data source
1.	Landsat 05 TM	LANDSAT/ LT05/C02/T1_TOA	2001	30 m	1	USGS Earth Explorer
2.	Landsat 08 OLI/TIRS	LANDSAT/LC08/C02/T1	2014	30 m	1	USGS Earth Explorer
3.	Landsat 08 OLI/TIRS	LANDSAT/LC08/C02/T1	2024	30 m	1	USGS Earth Explorer

the sanctuary stretches across the districts of Rudraprayag and Chamoli. Situated at altitudes ranging from 1160 m (near Phata) to 7553 m (Chaukhamba peak) above sea level, this sanctuary is a haven for diverse flora and fauna, offering a unique ecosystem for scientific exploration and conservation efforts.

The sanctuary experiences a sub-tropical to alpine climate, characterised by cold winters and pleasant summers. Annual precipitation is about 3000 mm, with the majority occurring during the monsoon season from June to September. Snowfall is common in higher elevations (alpine and temperate regions) in the winter season (December–February) (Thakur et al., 2011). The sanctuary's topography varies from gentle slopes to steep ridges and valleys, with the presence of numerous glaciers and perennial streams that serve as lifelines for the region's biodiversity. The main high peak mountains are Kedarnath (6940 m), Mandani (6193 m) and Chaukhamba (7553 m). Kedarnath Wildlife Sanctuary boasts a diverse array of vegetation due to its wide altitudinal range. It has vast alpine pastures in the northern region and numerous dense oak mixed forests in the southern area (Malik et al., 2014). The lower elevations are dominated by subtropical forests comprising species such as oak, pine trees and rhododendron. As elevation increases, these forests transition into alpine meadows, coniferous forests and eventually rocky bare landscapes. The reserve also serves as the origin of numerous significant rivers and their tributaries, such as the Mandakini, Alaknanda and Bhagirathi Rivers (https://wandersky.in/kedarnath-wildlife-sanctuary/). Alpine meadows adorn the higher altitudes, adding to the sanctuary's ecological richness. Owing to its unique geographical and ecological features, the Kedarnath Wildlife Sanctuary serves as an invaluable site for scientific research and conservation initiatives.

1.2 Landsat Image Retrieval in Google Earth Engine (GEE)

Data acquired through remote sensing has been known as an important source of information for cartographic representation of terrestrial cover and for tracing changes in land cover over temporal scales, with Landsat serving as the predominant data repository. Google Earth Engine (GEE), a web-based cloud interface, provides access to archived Landsat data for users, including data from Landsat 5 TM (Thematic Mapper) covering the years 1985 to 2011, Landsat 7 ETM+ (Enhanced Thematic Mapper Plus) from the period of 1999 to 2014, and Landsat 8 OLI/TIRS (Operational Land Imager/Thermal Infrared Sensor) data from 2013 to till date. The spatial resolution of these Landsat images is 30 m and the original source of image acquisition is USGS (United States Geological Survey) Earth Explorer website (https://earthexplorer.usgs.gov/). The present research work has utilised Landsat series images 05 TM and 08 OLI/TIRS to analyse terrestrial changes from 2001 to 2024 (Table 1).

2 Research Methodology

2.1 Satellite Image Processing

The selection of the satellite images was dependent on their availability, spatial resolution and overall general quality, especially those with less cloud interference or obstructions in the scene.

To minimise the impact of seasonal fluctuations on vegetation and reduce cloud cover interference, all images were chosen from the dry season (January–March) (Admasu et al., 2023). January to March is generally considered the dry season in Uttarakhand, India. During this time, the region experiences lower levels of precipitation compared to the monsoon season, which typically occurs from June to September. The dry season is characterised by clearer skies, cooler temperatures and lower humidity levels, making it an ideal time for LULC assessment. Prior to the classification process, the images were transformed into an RGB colour composite. This was done to enhance the visibility of surface characteristics and to facilitate their clear identification during the classification stage (Kazakeviciute-Januskeviciene et al., 2020). For uniformity and compatibility, every image was projected to the UTM zone 44 N, using World Geodetic System 84 (WGS84) as the reference datum. After that, sub-setting and resampling process was performed in ERDAS IMAGINE and ArcGIS 10.7.1 software, respectively, to match the extent and cell-size of all the three images (2001, 2014 and 2024).

2.2 LULC Classification Using Machine Learning Classifier

Multi-temporal Landsat images for 2001, 2014 and 2024 were extracted to prepare the LULC map for KWLS. Before implying any method of classification, i.e., supervised/unsupervised, understanding the nature and intricate structure of the study area is an important step (Shivakumar & Rajashekararadhya, 2018). Various band composites were examined to determine the spectral distinctiveness of multiple land use features, but some classes overlapped due to having similar radiometric or spectral values, known as digital numbers (DN). Classification entails careful identification of spectral patterns for the extraction of good quality and accurate training samples, also referred to as signature gathering (Verma et al., 2024). The study area's landscape was segmented into seven possible major LULC categories: forests, rangelands, cropland, waterbody, settlements, bare land and snow cover. To recognise the LULC class features in remotely accessed Landsat images, all the sample data were accumulated based on band combinations, manual visual interpretation and high-resolution Google Earth imagery (Baidoo & Obeng, 2023). Average pixel values of the spectral signatures were employed in Google Earth Engine (GEE) to generate the sample training data. The signature samples were collected employing two band combinations, i.e., FCC (false colour composite), intended for forest, cropland, rangelands and bare land recognition. In contrast, the natural colour composite (NCC) was intended for settlements and waterbody class identification (Sharma et al., 2023). For FCC, Landsat 08 used band composites: B5, B4, B3 and Landsat 05 used band composites: B4, B3, B2. Similarly, B5, B4, B1 band combinations were used to access 'nature-like' rendition in Landsat 05 and B7, B5, B3 bands combinations were used to access natural colour image with atmospheric removal. For each type of LULC classes, over 30–40 spectral signatures or values were collected. The sample size for each LULC type was decided based on its area proportion. The training data must adequately represent the whole area under study (Belay et al., 2022). Further, by utilising GEE functions, the sample dataset was divided into 70% for training purposes and 30% for testing. Additionally, these sample datasets were utilised as a training test in order to classify the LULC features using machine learning procedures such as random forest (RF).

2.2.1 Random Forest (RF) Classifier

A suitable classification model can establish the inherent relationship between the features (for instance, band data) and the targets (like the seven categories in this study) through supervised learning from training data. Once the model is trained, it can predict and assign class labels to new, unseen data. Various classification algorithms have been employed in scholarly works for supervised learning, such as random forests (RF), Support Vector Machines (SVM), nearest neighbour methods, decision trees and neural networks. RF is frequently employed as an algorithm for categorising land cover based on remote sensing data (Breiman, 2001).

RF is a classification algorithm based on ensemble learning which creates numerous decision trees during the training phase. The ultimate outcome is a composite of the classifications derived from each separate decision tree. This approach can prevent overfitting and is significantly more reliable than a single decision tree. Previous studies also indicate that the RF technique can deliver high precision, strong robustness and reduced computational burden (Belay et al., 2022; Zhang et al., 2021). That is why, this study utilised the RF algorithm in GEE to perform LULC classification using selected sample features. The LULC classification was executed by running the ee. smileRandomForest function within the GEE. In RF approach, some hyperparameters, i.e., Ntree (number of trees) and Mtry (variablesPerSplit), are required to be tuned so that a robust classification performance could be achieved. In this study, Mtry parameter was kept at default setting and only Ntree was tuned, where iterations were ranged from 10 to 200 with an interval of 10 from that the Ntree value giving the maximum overall accuracy (OA) was selected for each year. Figure 2 reflects the optimal Ntree values that are (a) 70, 80, 150, 160, 170 and 190 for the LULC 2001, (b) 10 for the LULC 2014 and (c) 30 for the LULC 2014.

To determine the accuracy of classification, various accuracy statistics were calculated, which include overall accuracy (OA), producer's accuracy (PA), user's accuracy (UA) and overall Kappa accuracy (KIA). This study has achieved OA values of 93.87%, 92.59%, and 95.65% for the years 2001, 2014 and 2024, respectively, and KIA statistics of 0.92, 0.91, and 0.91 for the years 2001, 2014 and 2024, respectively (Table 2) which fall within the acceptable range of accuracy assessment. The LULC maps are presented in Fig. 3 for all the specified years.

2.3 Assessment of Ecosystem Service Valuation

The well-being of humanity and the smooth operation of the global economy rely on the amenities provided by ecosystems. However, these services are at risk due to the complicated relations between people and the ecological system, leading to the deterioration of ecosystems and a decrease in biodiversity (Muche et al., 2023). Most of the Earth's environmental benefits have been varied because of the interference of human activities. These activities include deviations in landscape cover, deforestation and exploitation of natural resources, which have significantly increased and are now unavoidable. Among all the factors causing changes in ESVs, LULC is observed to be the most influential (Sannigrahi et al., 2020). Moreover, quantifying ecological benefits in financial value is becoming increasingly common. This approach aims to promote awareness among individuals, offer evidence for decision-makers, calculate the potential restoration expenses and ease the process of compensating for ecosystem services. The assessment of environmental benefits and their values has been in progress since the 1970s. However, accurately evaluating the values of the Earth's environmental services is challenging due to the absence of a related theoretical framework and methodology. Lately, the assessment of ecosystem service functionality and value has emerged as a focal point of research.

Ecosystem valuation model was first established by Costanza et al., 1997; Millennium Ecosystem Assessment, 2005. Costanza et al. (1997) offer significant information and methodologies for the valuation of ecosystem services by calculating 17 ecological services supplied by 16 global biomes with an average annual value exceeding US$33 trillion. This incentivise a comprehensive exploration of ecosystem service value (ESV) across various levels by global and

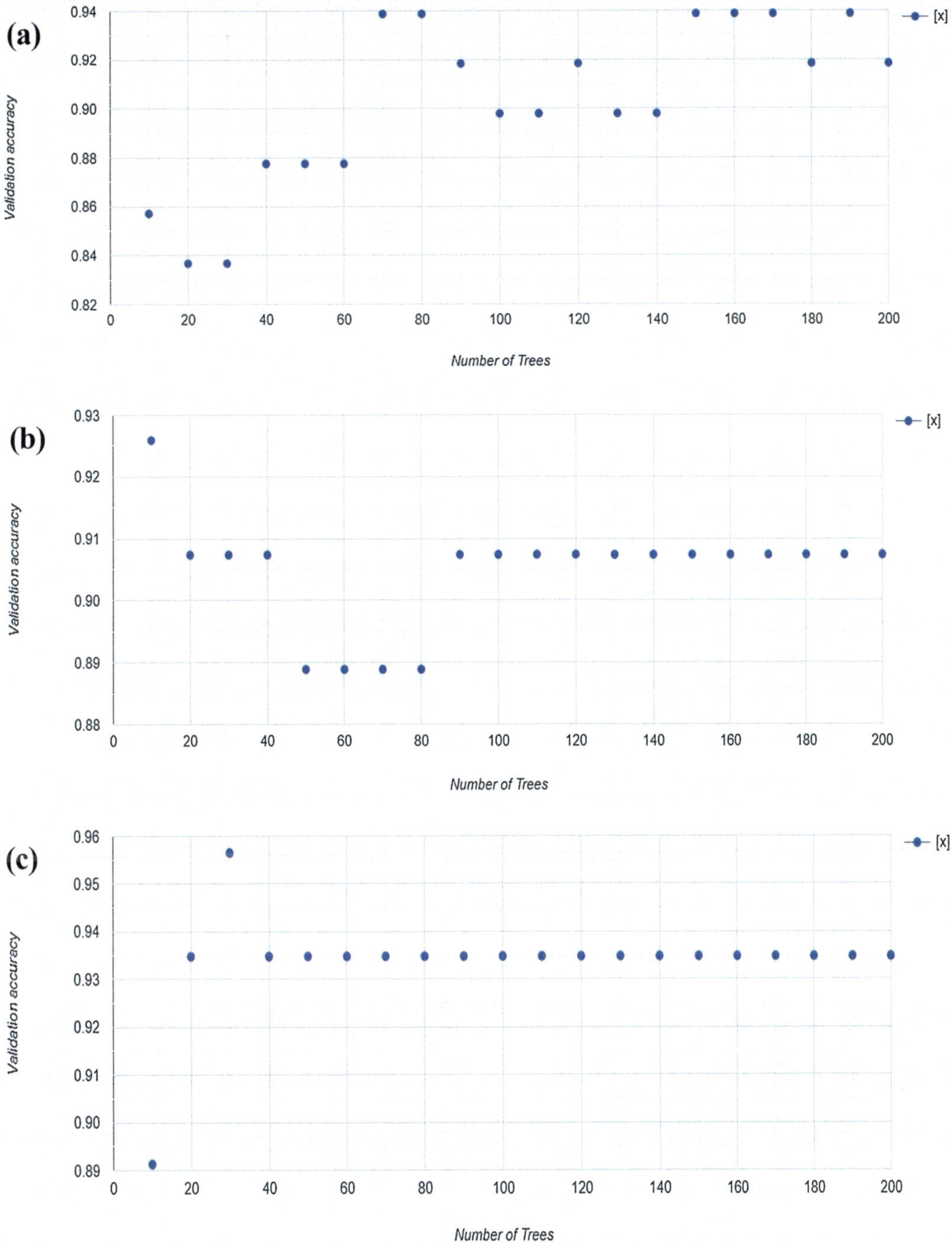

Fig. 2 Validation accuracy-based hyperparameter tuning for the number of trees: (**a**) 2001, (**b**) 2014, (**c**) 2024

Table 2 LULC accuracy assessment

LULC classes	200		2014		2024	
	PA (%)	UA (%)	PA (%)	UA (%)	PA (%)	UA (%)
Forests	100	100	100	100	100	100
Rangelands	90	90	88	100	100	100
Cropland	100	83.3	100	100	100	100
Waterbody	100	85.7	100	90	100	83.3
Settlements	77.7	100	80	67	75	75
Bare land	100	100	71	100	83.3	100
Snow cover	100	100	100	83.3	100	100
Overall Accuracy (OA)	93.87%		92.59%		95.65%	
Kappa Coefficient (KIA)	0.92		0.91		0.91	

local researchers. Nonetheless, the findings put forth by Costanza and his colleagues in 1997 have faced criticism and their valuation approach has been questioned due to its inherent uncertainties and limited applicability at diverse scales (Kindu et al., 2016; Kusi et al., 2023). In response to this Costanza et al. (2014) revised and improved the global estimates of ecosystem benefits. They took into account an extensive array of case studies (more than 300) from around the globe to tackle some major uncertainties. Upon examining multiple locations, Costanza et al. (2014) asserted that the models and statistical information utilised in their assessment have the potential to be implemented at diverse scales for

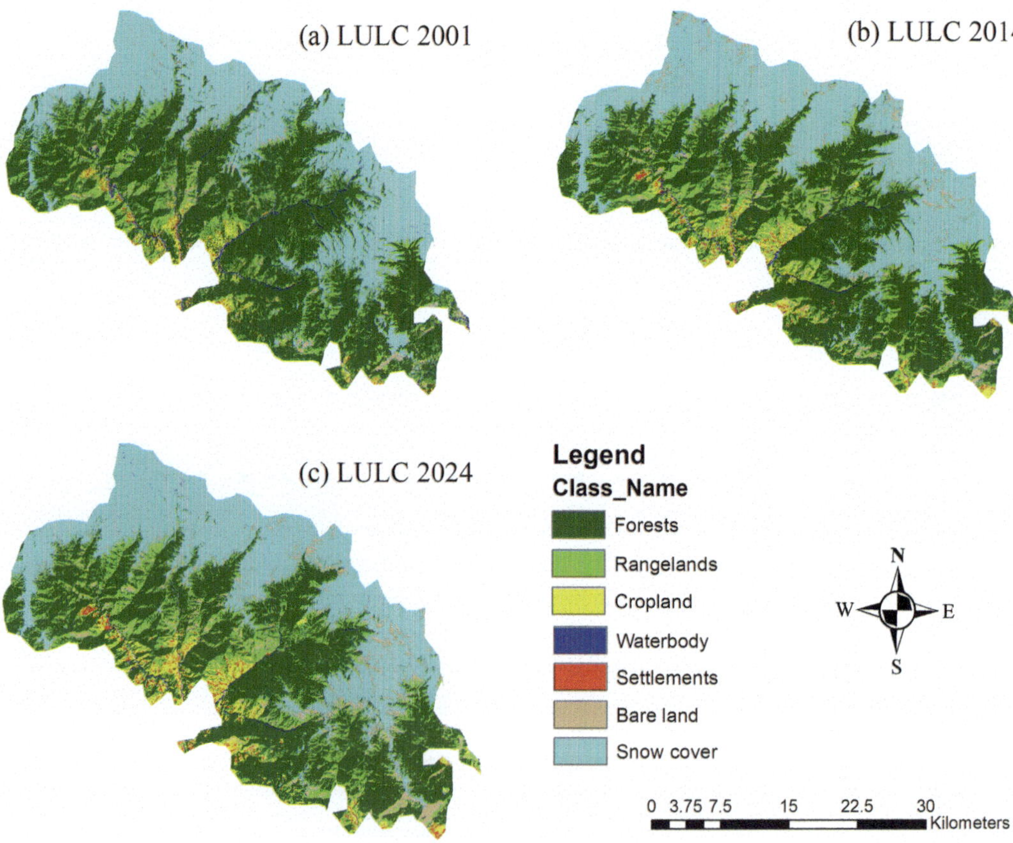

Fig. 3 LULC maps of Kedarnath Wildlife Sanctuary for the specified years: (**a**) 2001, (**b**) 2014 and (**c**) 2024

evaluating changes in ESV. There are several studies (Kindu et al., 2016; Tolessa et al., 2017; Xie et al., 2003) which have also employed modified monetary value estimated by Costanza et al. (1997, 2014). Xie et al. (2003) performed the ecosystem service valuation assessment per unit area of the various land use classes based on Costanza's global ecological service valuation technique in the Tibetan Plateau in which an updated benefit transfer method (BTM) was adopted to evaluate the variations in the value of ecosystem benefits due to changes in land uses on ESVs.

The benefit transfer method (BTM) presumes uniformity across space and allows the transfer of an equivalent value coefficient from one location to other regions that are similar in terms of ecology and socioeconomic dynamics (Kindu et al., 2016; Sannigrahi et al., 2020). The BTM approach, which relies on global value coefficients or modified value coefficients, has been a commonly employed strategy, particularly in regions lacking sufficient data (Munthali et al., 2023). This study adopts the equivalent coefficient table provided by Xie et al. (2003), which has also been applied in studies by Rai et al. (2018) and Zhilong et al. (2017) as they encompass biomes nearly similar to those in this study area.

The most indicative biome was utilised as a proxy for each LULC class including (1) forests for tropical forests, (2) rangelands for Shrubland (3) cropland for cropland (4) waterbody for lakes/ rivers (5) settlements for urban (6) bare land for desert and (7) snow cover for snow/glacier (Table 3).

The ecosystem service values (ESVs) for each LULC category were determined by multiplying the area (in hectares) of each LULC type by its respective value coefficients, as per the formula provided below in Eq. 1 and the total ESV of the landscape for each reference year was calculated by adding up the values for the LULC types in each respective year.by using Eq. 2.

Table 3 Ecosystem services value (ESV) coefficient respective to each LULC classes

LULC classes	Equivalent biome	Ecosystem services value (ESV) in USD/ ha/year
Forests	Tropical forests	2168.84
Rangelands	Shrubland	1089.19
Cropland	Cropland	699.37
Waterbody	Lakes/ Rivers	6552.97
Settlements	Urban	0
Bare land	Desert	0
Snow cover	Snow/ glacier	0

Source: Xie et al. (2003)

$$ESV_t = \left(A_k * VC_K\right) \qquad (1)$$

$$ESV_t = \Sigma\left(A_k * VC_K\right) \qquad (2)$$

where,

ESV_k = Estimated ecosystem service value for individual LULC class

ESV_t = Total ecosystem service value

A_k = The area (in ha) for LULC type k

VC_k = The value coefficient (in US$/ha/year) for LULC type k.

3 Data Analysis and Results

3.1 Change Detection Analysis (2001–2024)

Land use pattern and formation have experienced significant diversification in the past 23 years. To analyse the trajectories of LULC transitions, a change analysis of seven LULC classes has been performed based on study period 2001, 2014 and 2024. The LULC year-wise area proportion (in ha) of all the seven major classes (forests, rangelands, cropland, waterbody, bare land, settlements and snow cover) is presented in Fig. 4 and their change dynamics from 2001 to 2024 are provided in Table 4 (2001–2014) and Table 5

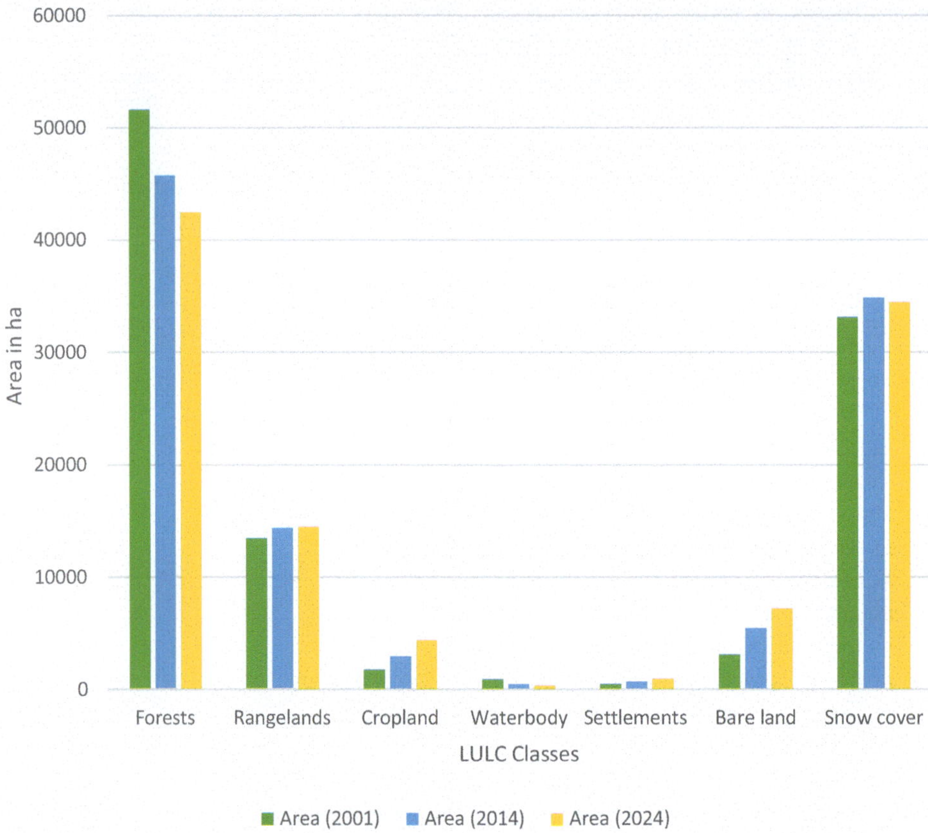

Fig. 4 LULC year-wise area proportion (in ha)

Table 4 LULC change matrix from 2001 to 2014 (in ha)

LULC classes	Area (2001)	Area (2014)	Area change dynamics (2001–2014)	
			Difference in ha	% difference
Forests	51640.11	45779.49	−5860.62	−11.35
Rangelands	13438.89	14352.3	913.41	6.80
Cropland	1805.76	2948.4	1142.64	63.28
Waterbody	944.28	488.52	−455.76	−48.27
Settlements	506.34	706.32	199.98	39.5
Bare land	3133.89	5453.28	2319.39	74.01
Snow cover	33176.61	34917.57	1740.96	5.25

Source: Author's own calculations

(2014–2024) to explore the variations occurred among all the LULC types over time. The land use classes like forests, rangelands, waterbody and snow cover are the naturally formed landscapes, while the other remaining three classes represent human-modified landscapes resulting from the transformation of forests, rangelands, waterbody and snow cover into cropland, settlements and bare land. The overall land coverage of LULC classes in all the reference periods shows significant domination of forests coverage followed by snow cover, rangelands, bare land, cropland, waterbody and settlements. Despite of being the dominant land use class, forested area has witnessed downfall over the time, i.e., 51640.11 ha in 2001 > 45779.49 ha in 2014 > 42511.14 ha in 2024 with the area difference of −5860.62 ha (2001–2014) and

Table 5 LULC change matrix from 2014 to 2024 (in ha)

LULC classes	Area (2014)	Area (2024)	Area change dynamics (2014–2024)	
			Difference in ha	% difference
Forests	45779.49	42511.14	−3268.35	−7.14
Rangelands	14352.3	14531.4	179.1	1.25
Cropland	2948.4	4407.03	1458.63	49.47
Waterbody	488.52	391.77	−96.75	−19.8
Settlements	706.32	976.41	270.09	38.24
Bare land	5453.28	7267.77	1814.49	33.27
Snow cover	34917.57	34567.56	−350.01	−1.01

Source: Author's own calculations

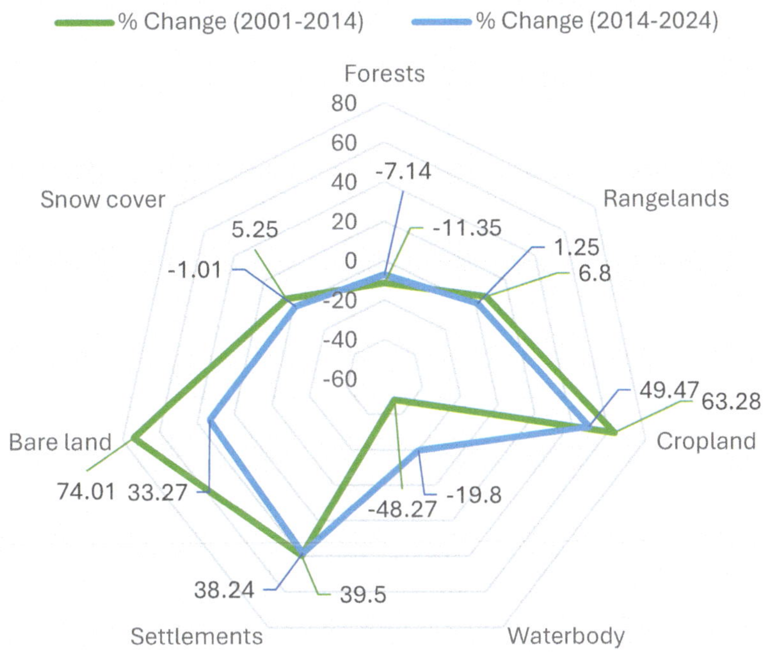

Fig. 5 Percentage change in LULC classes from 2001 to 2024

−3268.35 ha (2014–2024); however, this downfall has happened with decreasing rate as demonstrated in Fig. 4, Tables 4 and 5. The second most dominant landscape feature is snow cover which indicates up and down trajectory since it has increased from 33176.61 ha in 2001 to 34917.57 ha in 2014 with the area difference of 1740.96 ha which makes 5.25% increment in the class cover which further falls with the area difference of −350.01 ha, an decrement of −1.01%. The third most occupied land cover rangelands have shown increment in area coverage from 2001 (13438.89 ha) to 2014 (14352.3 ha) that is 6.80% increase but it has remained almost persistent during 2014–2024 with percentage increase of only 1.25. Bare land has witnessed only increment over time with the area difference of 2319.39 ha (in 2001–2014) and 1814.49 ha (in 2014–2024). Cropland has also drastically increased with the percentage difference of 63.28 (in 2001–2014) and 49.47 (in 2014–2024). Waterbody and settlements have the least area coverage where waterbody has shown decreasing trend and settlements have shown increasing trend from 2001 to 2024.

Figure 5 demonstrates percentage difference in LULC classes from 2001 to 2024. In this figure, green line indicates percentage change from 2001 to 2014 and blue line indicate percentage change from 2014 to 2024. The most fluctuated

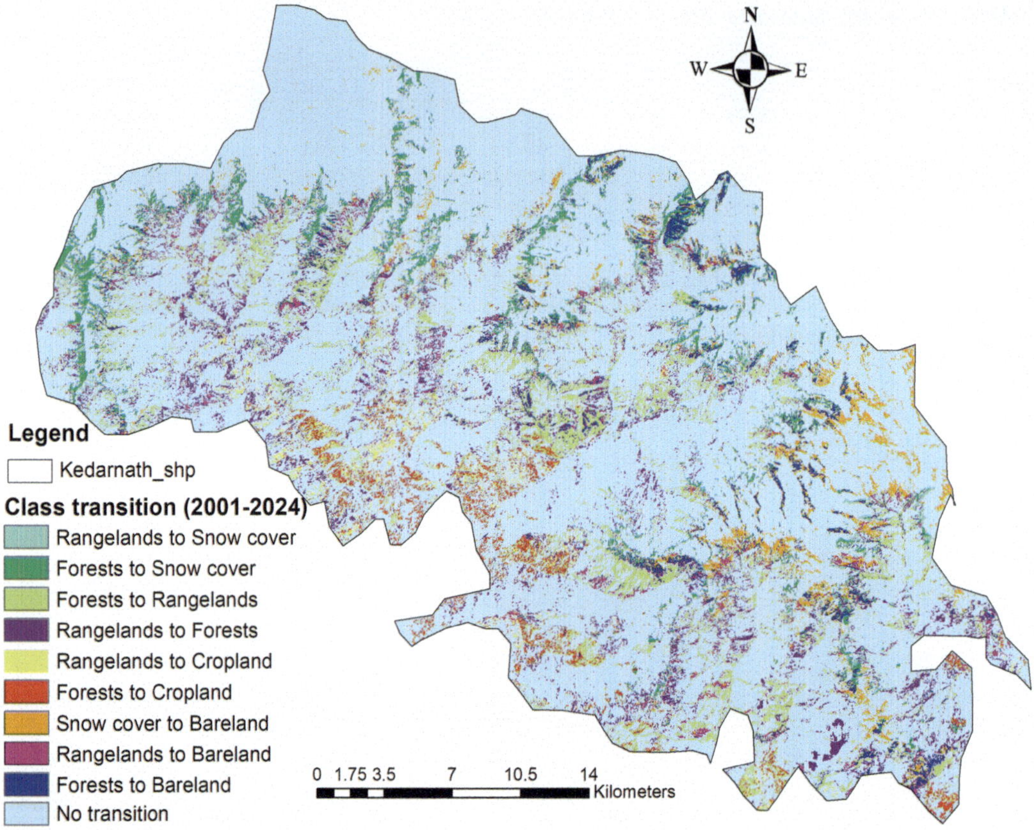

Fig. 6 Spatial transition map of major LULC classes from 2001 to 2024

LULC classes during 2001 to 2014 are bare land, cropland, waterbody and settlement. Among them bare land has the highest positive percentage difference (74.01%), followed by cropland (63.28%) and settlements (39.5), while waterbody has shown drastic negative percentage difference of −48.27. Time span (2014–2024) has shown high positive fluctuation in cropland (49.47%) followed by settlements (38.24%) and bare land (33.27%). Least change has been observed in rangelands and snow cover, while the most dominant land use class forests has witnessed moderate change. Figure 6 displays spatial transition map of KWLS in LULC classes during 2001–2024. It reflects conversion of highly occupied land cover, i.e., forests, rangelands and snow cover into other land use classes such as forests has converted into rangelands, cropland, bare land and snow cover. Rangelands have converted into forests, cropland, bare land and snow cover, while the snow cover has majorly transformed into bare land.

3.2 Estimation of Ecosystem Service Value Over Time from the Year 2001 to 2024

In the present assessment, the benefit transfer method (BTM) was implied to measure the ESVs in monetary terms as a response to the spatiotemporal changes in diverse LULC classes for the periods 2001, 2014 and 2024. Drawing on ecosystem services quantified by Xie et al. (2003) (Table 3), this study has adopted and calculated the ESV for the Kedarnath Wildlife Sanctuary (KWLS) between 2001 and 2024. The classified LULC map of KWLS was utilised to assess the variations in the value of ecological services over time.

Table 6 Calculated ecosystem services value (ESV) in USD/ha/year

LULC classes	2001	2014	2024
Forests	111999136.2	99288389.09	92199860.88
Rangelands	14637504.6	15632381.64	15827455.57
Cropland	1262894.371	2062022.508	3082144.571
Waterbody	6187838.512	3201256.904	2567257.057
Settlements	0	0	
Bare land	0	0	
Snow cover	0	0	

Source: Author's own calculations

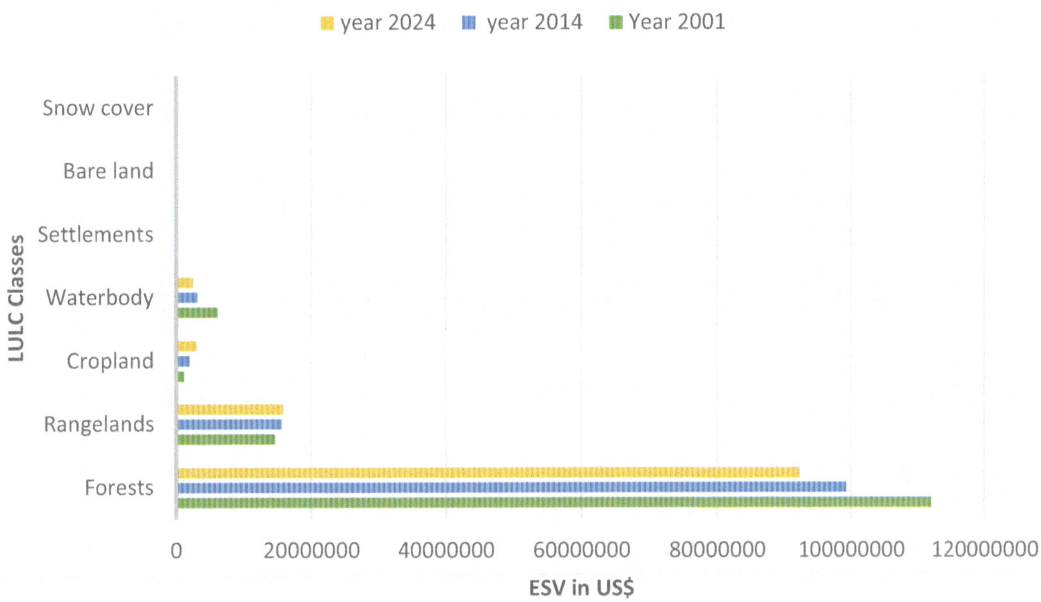

Fig. 7 Ecosystem service value of each LULC classes in the year 2001, 2014 and 2024 (Source: Author's own creation)

Table 6 and Fig. 7 depict economic quantification of ESVs all land use types, and Fig. 7 provides the graphical representation of the total ESV for each reference study period. ESV for all land cover classes was computed by multiplying the land cover (in ha) of land use types to their corresponding adopted coefficient factor. In this study, land use classes forests, rangelands, cropland and waterbody were assigned their respective coefficient value; however, no coefficients were assigned to settlements, bare land and snow cover classes since these classes lack in vegetation (Muche et al., 2023). In the year 2001, forests land cover class indicate the highest value of ecosystem service (~112.00 million dollars), followed by rangelands (~14.64 million dollars), waterbody (~6.19 million dollars) and cropland (~1.26 million dollars). Reference year 2014 again shows the similar trends of LULC class contribution in ESV as the year 2001; however, the ESV of forests (~99.29 million dollars) and waterbody (~3.20 million dollars) has decreased and ESV of rangelands (~15.63 million dollars) and cropland (~2.06 million dollars) has increased. The year 2024 has demonstrated change in ESV dynamics due to the changes in land coverage of LULC classes. In 2024, cropland (~3.08 million dollars) has surpassed the contribution of waterbody (~2.57 million dollars) in ESV, while forests class still has the highest contribution (~92.20 million dollars) followed by rangelands (~15.83 million dollars). From Fig. 6,

Fig. 8 Year-wise total estimated ESV in USD/ha/year. (Source: Author's own creation)

it is apparent that total ESV of forests and waterbody has decreasing trend, while rangelands and cropland are showing increasing trend during 2001 to 2024.

The total ESV (ESV_t) for the Kedarnath sanctuary was computed by adding the calculated ESVs from all the LULC classes. ESV_t for the year 2001, 2014 and 2024 is presented in Fig. 8. From the figure, a declining trend of ESVs has been observed for the study periods as indicated by: ESV_{2001} = ~134.09 million dollars, ESV_{2014} = ~120.08 million dollars and ESV_{2024} = 113.68 million dollars. The ESV was estimated using the areas of seven LULC classes, based on equivalent proxy coefficients; that is why, the spatial changes and distribution of each LULC class and total ESVs for the selected reference years are similar to that of the LULCs. Decline in total ESV has occurred due to the unprecedented transformation in the structure of terrain cover in the last few decades. The sanctuary has undergone substantial LULC transitions due to human activities, such as the expansion of agricultural areas and urbanisation. Forests land use class caused substantial impact to the reduction in ESVs, and its ESV has constantly been declining following the drastic reduction in area coverage over time.

Due to the conversion of most vegetated surface, i.e., forests and rangelands into cropland, snow cover and bare land which produce less ecosystem service, the green service across the region has massively deteriorated. Therefore, it is imperative to preserve the integrity and extent of natural environments within the study area since it is essential for the sustenance of diverse ecosystem services and benefits, and to ensure their long-term provision at the research location.

4 Discussion and Conclusion

Assigning a monetary value to ecosystem services is an effective method for increasing public consciousness about the limited availability of natural resources and the advantages they offer, many of which are not commercially traded. This study has conducted an assessment of the effects of changes in land use patterns on the value of environmental services within the Kedarnath wildlife sanctuary (KWLS) using Landsat images for the year 2001, 2014 and 2024. Benefit transfer method (BTM) was used to calculate the monetary value of ecosystem services using the GIS and remote sensing technique which is cost-effective, and less time taken method to study the data scarce region. Random forest (RF) classifier was employed in Google Earth Engine (GEE) to prepare the LULC map which further was utilised to estimate the ESV of each LULC class. The result reveals a massive decline in total ESV during (2001–2024) resulting from reduction in natural vegetation and rise in artificial landscape.

The wildlife sanctuary, in spite of being a protected area (PA), witnessing rapid transitions in land use resulting into the deterioration of whole landscape. A considerable reduction in forests and waterbody LULC classes has been observed across the study region from 2001 to 2024. Forests offer a wide array of ecosystem benefits, including carbon sequestration, climate regulation, provision of food, fuel, animal feed and wood. They are instrumental in sustaining the water cycle, safeguarding and preserving land-based biodiversity and aiding in the prevention of land deterioration and salinisation (Verma & Ghosh, 2022). The aquatic ecosystems such as rivers and lakes also provide many crucial ecosystem services, i.e., water provisioning and recreation, water retention, water purification and climate regulation (Grizzetti et al., 2016). These two land use types hold high ESV factor in the study region that is 2168.84 for forests and 6552.97 for waterbody but from them forests have the highest contribution in total estimated ESV which are continuously being reduced. There are a collection of studies experiencing the loss of forest cover with same pattern over time (Muche et al., 2023; Munthali et al., 2023; Sannigrahi et al., 2018). The main reason of declination might be attributed to changes in government policies, human interference and variations in climatic conditions. Cropland, settlements and bare land are swiftly encroaching on landscape cover, replacing natural forests as well as rangelands. Number of studies, i.e., (Admasu et al., 2023; Baidoo & Obeng, 2023) report the same findings that degradation of natural vegetation and increment in less-vegetation land cover such as settlements and bare land are causing reduction in total environmental services.

The conversions in landscape cover are characterised by their dynamism and non-linearity. This implies that the transition from one type of land use to another doesn't adhere to a consistent pattern. This variability can be attributed to both natural and human-induced factors such as changes in policy, fluctuations in population and a decline in land productivity. In contrast to the findings of this study, several studies revealed improvement in total ecosystem services resulting from increased open forests, cultivated land and grassland over the time (1985–2016) (Solomon et al., 2018) and increase in total ESV resulting from the growth of natural vegetation, forests, farmland and urban cover during (1992–2015) (Kusi et al., 2023). This improvement was possible due to the intervention of several environment protection, natural ecosystem management, soil and water conservation as well as participation of local communities in the protection of natural resources. Many incentives can be taken from such studies applying environment and land use policy implications to save the integrity and long-term resilience of Himalayan protected areas. Integrating LULC and ecological service quantification data can assist in recognising regions that are most susceptible to variations in ecosystem services at a landscape scale and offer an initial foundation for potential future land management strategies. Hence, this current research lays the groundwork for future studies into forecasting LULC dynamics in relation to ecosystem service valuation (ESV) using remote sensing data, land suitability modelling, as well as developing a more precise and reliable value coefficients.

References

Admasu, S., Yeshitela, K., & Argaw, M. (2023). Impact of land use land cover changes on ecosystem service values in the Dire and Legedadi watersheds, central highlands of Ethiopia: Implication for landscape management decision making. *Heliyon, 9*(4), e15352. https://doi.org/10.1016/j.heliyon.2023.e15352

Baidoo, R., & Obeng, K. (2023). Evaluating the impact of land use and land cover changes on forest ecosystem service values using Landsat dataset in the Atwima

Nwabiagya North, Ghana. *Heliyon, 9*(11), e21736. https://doi.org/10.1016/j.heliyon.2023.e21736

Balasubramanian, M., & Sangha, K. K. (2023). Valuing ecosystem services applying indigenous perspectives from a global biodiversity hotspot, the Western Ghats, India. *Frontiers in Ecology and Evolution, 11*(March), 1–12. https://doi.org/10.3389/fevo.2023.1026793

Barbier, E. B., Baumgärtner, S., Chopra, K., Costello, C., Duraiappah, A., Hassan, R., Kinzig, A. P., Lehman, M., Pascual, U., Polasky, S., & Perrings, C. (2009). The valuation of ecosystem services. In *Biodiversity, ecosystem functioning, and human wellbeing: An ecological and economic perspective*. Oxford University Press. https://doi.org/10.1093/ACPROF:OSO/9780199547951.003.0018

Belay, T., Melese, T., & Senamaw, A. (2022). Impacts of land use and land cover change on ecosystem service values in the Afroalpine area of Guna Mountain, Northwest Ethiopia. *Heliyon, 8*(12), e12246. https://doi.org/10.1016/j.heliyon.2022.e12246

Breiman, L. (2001). Random forests. *Machine Learning, 45*(1), 5–32. https://doi.org/10.1023/A:1010933404324/METRICS

Chopra, B., Khuman, Y. S. C., & Dhyani, S. (2022). Advances in ecosystem services valuation studies in India: Learnings from a systematic review. *Anthropocene Science, 1*(3), 342–357. https://doi.org/10.1007/S44177-022-00034-0

Costanza, R., D'Arge, R., de Groot, R., Farber, S., Grasso, M., Hannon, B., Limburg, K., Naeem, S., O'Neill, R. V., Paruelo, J., Raskin, R. G., Sutton, P., & van den Belt, M. (1997). The value of the world's ecosystem services and natural capital. *Nature, 387*(6630), 253–260. https://doi.org/10.1038/387253a0. https://royalroads.on.worldcat.org/oclc/4592801201; https://www-nature-com.ezproxy.royalroads.ca/articles/387253a0.pdf

Costanza, R., de Groot, R., Sutton, P., van der Ploeg, S., Anderson, S. J., Kubiszewski, I., Farber, S., & Turner, R. K. (2014). Changes in the global value of ecosystem services. *Global Environmental Change, 26*(1), 152–158. https://doi.org/10.1016/J.GLOENVCHA.2014.04.002

Dammag, A. Q., Dai, J., Cong, G., Derhem, B. Q., & Latif, H. Z. (2024). Assessing and predicting changes of ecosystem service values in response to land use/land cover dynamics in Ibb City, Yemen: A three-decade analysis and future outlook. *International Journal of Digital Earth, 17*(1). https://doi.org/10.1080/17538947.2024.2323174

EarthExplorer. (n.d.). Retrieved April 30, 2023, from https://earthexplorer.usgs.gov

Goldenberg, R., Kalantari, Z., Cvetkovic, V., Mörtberg, U., Deal, B., & Destouni, G. (2017). Distinction, quantification and mapping of potential and realized supply-demand of flow-dependent ecosystem services. *Science of The Total Environment, 593–594*, 599–609. https://doi.org/10.1016/j.scitotenv.2017.03.130

Gong, Y., Cai, M., Yao, L., Cheng, L., Hao, C., & Zhao, Z. (2022). Assessing changes in the ecosystem services value in response to land-use/land-cover dynamics in Shanghai from 2000 to 2020. *International Journal of Environmental Research and Public Health, 19*(19). https://doi.org/10.3390/ijerph191912080

Grizzetti, B., Lanzanova, D., Liquete, C., Reynaud, A., & Cardoso, A. C. (2016). Assessing water ecosystem services for water resource management. *Environmental Science & Policy, 61*, 194–203. https://doi.org/10.1016/J.ENVSCI.2016.04.008

Hasan, S. S., Zhen, L., Miah, M. G., Ahamed, T., & Samie, A. (2020). Impact of land use change on ecosystem services: A review. *Environmental Development, 34*(April), 100527. https://doi.org/10.1016/j.envdev.2020.100527

Hein, L., Bagstad, K., Edens, B., Obst, C., De Jong, R., & Lesschen, J. P. (2016). Defining ecosystem assets for natural capital accounting. *PLoS One, 11*(11), e0164460. https://doi.org/10.1371/JOURNAL.PONE.0164460

Hu, H., Liu, W., & Cao, M. (2008). Impact of land use and land cover changes on ecosystem services in Menglun, Xishuangbanna, Southwest China. *Environmental Monitoring and Assessment, 146*(1–3), 147–156. https://doi.org/10.1007/s10661-007-0067-7

Kazakeviciute-Januskeviciene, G., Janusonis, E., Bausys, R., Limba, T., & Kiskis, M. (2020). Assessment of the segmentation of RGB remote sensing images: A subjective approach. *Remote Sensing, 12*(24), 4152. https://doi.org/10.3390/RS12244152

Kedarnath Wildlife Sanctuary in Uttarakhand. (n.d.). Retrieved May 12, 2024, from https://wandersky.in/kedarnath-wildlife-sanctuary/

Kindu, M., Schneider, T., Teketay, D., & Knoke, T. (2016). Changes of ecosystem service values in response to land use/land cover dynamics in Munessa-Shashemene landscape of the Ethiopian highlands. *Science of the Total Environment, 547*, 137–147. https://doi.org/10.1016/j.scitotenv.2015.12.127

Kusi, K. K., Khattabi, A., & Mhammdi, N. (2023). Analyzing the impact of land use change on ecosystem service value in the main watersheds of Morocco. *Environment, Development and Sustainability, 25*(3), 2688–2715. https://doi.org/10.1007/s10668-022-02162-4

Leviston, Z., Walker, I., Green, M., & Price, J. (2018). Linkages between ecosystem services and human wellbeing: A Nexus Webs approach. *Ecological Indicators, 93*(May), 658–668. https://doi.org/10.1016/j.ecolind.2018.05.052

Li, Y., Li, J., & Chu, J. (2022). Research on land-use evolution and ecosystem services value response in mountainous counties based on the SD-PLUS model. *Ecology and Evolution, 12*(10), 1–18. https://doi.org/10.1002/ece3.9431

Malik, Z. A., Bhat, J. A., & Bhatt, A. B. (2014). Forest resource use pattern in Kedarnath wildlife sanctuary and its fringe areas (a case study from Western

Himalaya, India). *Energy Policy, 67*, 138–145. https://doi.org/10.1016/j.enpol.2013.12.016

Manral, U. (2018, September). Kedarnath wildlife sanctuary: Nature-culture linkages in a sacred landscape in Indian Himalayan region. *Journal of World Heritage Studies*.

Marta-Pedroso, C., Laporta, L., Gama, I., & Domingos, T. (2018). Economic valuation and mapping of ecosystem services in the context of protected area management (Natural park of Serra de São Mamede, Portugal). *One Ecosystem, 3*. https://doi.org/10.3897/oneeco.3.e26722

Millennium Ecosystem Assessment. (2003). *A report of the conceptual framework working group of the millennium ecosystem assessment*. Island Press.

Millennium Ecosystem Assessment. (2005). *Ecosystems and human well-being: Synthesis*. Island Press. https://doi.org/10.11646/zootaxa.4892.1.1

Muche, M., Yemata, G., Molla, E., Adnew, W., & Muasya, A. M. (2023). Land use and land cover changes and their impact on ecosystem service values in the north-eastern highlands of Ethiopia. *PLoS One, 18*(September), 1–24. https://doi.org/10.1371/journal.pone.0289962

Munthali, M. G., Kindu, M., Adeola, A. M., Davis, N., Botai, J. O., & Solomon, N. (2023). Variations of ecosystem service values as a response to land use and land cover dynamics in Central Malawi. *Environment, Development and Sustainability, 25*(9), 9821–9837. https://doi.org/10.1007/s10668-022-02461-w

Polasky, S., Nelson, E., Pennington, D., & Johnson, K. A. (2011). The impact of land-use change on ecosystem services, biodiversity and returns to landowners: A case study in the state of Minnesota. *Environmental and Resource Economics, 48*(2), 219–242. https://doi.org/10.1007/S10640-010-9407-0/METRICS

Rai, R., Zhang, Y., Paudel, B., Acharya, B. K., & Basnet, L. (2018). Land use and land cover dynamics and assessing the ecosystem service values in the trans-boundary Gandaki River Basin, Central Himalayas. *Sustainability (Switzerland), 10*(9), 1–22. https://doi.org/10.3390/su10093052

Sannigrahi, S., Bhatt, S., Rahmat, S., Paul, S. K., & Sen, S. (2018). Estimating global ecosystem service values and its response to land surface dynamics during 1995–2015. *Journal of Environmental Management, 223*(December 2017), 115–131. https://doi.org/10.1016/j.jenvman.2018.05.091

Sannigrahi, S., Chakraborti, S., Banerjee, A., Rahmat, S., Bhatt, S., Jha, S., Singh, L. K., Paul, S. K., & Sen, S. (2020). Ecosystem service valuation of a natural reserve region for sustainable management of natural resources. *Environmental and Sustainability Indicators, 5*(November 2019), 100014. https://doi.org/10.1016/j.indic.2019.100014

Scolozzi, R., Morri, E., & Santolini, R. (2012). Delphi-based change assessment in ecosystem service values to support strategic spatial planning in Italian landscapes. *Ecological Indicators, 21*, 134–144. https://doi.org/10.1016/J.ECOLIND.2011.07.019

Sharma, S., Hussain, S., & Singh, A. N. (2023). Impact of land use and land cover on urban ecosystem service value in Chandigarh, India: A GIS-based analysis. *Journal of Urban Ecology, 9*(1), 1–12. https://doi.org/10.1093/jue/juac030

Shivakumar, B. R., & Rajashekararadhya, S. V. (2018). Investigation on land cover mapping capability of maximum likelihood classifier: A case study on North Canara, India. *Procedia Computer Science, 143*, 579–586. https://doi.org/10.1016/j.procs.2018.10.434

Solomon, N., Pabi, O., Annang, T., Asante, I. K., & Birhane, E. (2018). The effects of land cover change on carbon stock dynamics in a dry Afromontane forest in northern Ethiopia. *Carbon Balance and Management, 13*(1). https://doi.org/10.1186/s13021-018-0103-7

Tang, H., Halike, A., Yao, K., Wei, Q., Yao, L., Tuheti, B., Luo, J., & Duan, Y. (2024). Ecosystem service valuation and multi-scenario simulation in the Ebinur Lake Basin using a coupled GMOP-PLUS model. *Scientific Reports, 14*(1), 1–17. https://doi.org/10.1038/s41598-024-55763-6

Thakur, A. K., Singh, G., Singh, S., & Rawat, G. S. (2011). Impact of pastoral practices on Forest cover and regeneration in the outer fringes of Kedarnath wildlife sanctuary, Western Himalaya. *Journal of the Indian Society of Remote Sensing, 39*(1), 127–134. https://doi.org/10.1007/s12524-011-0079-1

Tolessa, T., Senbeta, F., & Abebe, T. (2017). Land use/land cover analysis and ecosystem services valuation in the central highlands of Ethiopia. *Forests Trees and Livelihoods, 26*(2), 111–123. https://doi.org/10.1080/14728028.2016.1221780

UN (United Nations). (2015). *Transforming our world: The 2030 agenda for sustainable development A/RES/70/1*. 41 pp. https://sustainabledevelopment.un.org/content/documents/21252030%20Agenda%20for%20Sustainable%20Development%20web.pdf

UNEP. (2011). *Towards a green economy: Pathways to sustainable development and poverty eradication—a synthesis for policy makers*. www.unep.org/greeneconomy

Vallecillo, S., La Notte, A., Ferrini, S., & Maes, J. (2019). How ecosystem services are changing: An accounting application at the EU level. *Ecosystem Services, 40*(November), 101044. https://doi.org/10.1016/j.ecoser.2019.101044

Verma, P., & Ghosh, P. K. (2022). REDD+ Strategy for forest carbon sequestration in India. *The Holistic Approach to Environment, 12*(3), 117–130. https://doi.org/10.33765/thate.12.3.4

Verma, P., & Ghosh, P. K. (2023). The economics of forest carbon sequestration: A bibliometric analysis. In *Environment, development and sustainability*. Springer. https://doi.org/10.1007/s10668-023-02922-w

Verma, P., Siddiqui, A. R., Mourya, N. K., & Devi, A. R. (2024). Forest carbon sequestration mapping and economic quantification infusing MLPnn-Markov chain and InVEST carbon model in Askot Wildlife Sanctuary, Western Himalaya. *Ecological Informatics, 79*(August 2023), 102428. https://doi.org/10.1016/j.ecoinf.2023.102428

Wang, Z. J., Liu, S. J., Li, J. H., Pan, C., Wu, J. L., Ran, J., & Su, Y. (2022). Remarkable improvement of ecosystem service values promoted by land use/land cover changes on the Yungui Plateau of China during 2001–2020. *Ecological Indicators, 142*(July), 109303. https://doi.org/10.1016/j.ecolind.2022.109303

Xie, G., Lu, C., Leng, Y., Zheng, D., & Li, S. (2003). Ecological assets valuation of the Tibetan Plateau. *Journal of Natural Resources, 18*(2), 189–196. https://doi.org/10.11849/ZRZYXB.2003.02.010

Xu, Z., Wei, H., Fan, W., Wang, X., Zhang, P., Ren, J., Lu, N., Gao, Z., Dong, X., & Kong, W. (2019). Relationships between ecosystem services and human well-being changes based on carbon flow—A case study of the Manas River Basin, Xinjiang, China. *Ecosyst Serv, 37*(May), 100934. https://doi.org/10.1016/j.ecoser.2019.100934

Yuan, K., Li, F., Yang, H., & Wang, Y. (2019). The influence of land use change on ecosystem service value in Shangzhou district. *International Journal of Environmental Research and Public Health, 16*(8). https://doi.org/10.3390/ijerph16081321

Zhang, T., Su, J., Xu, Z., Luo, Y., & Li, J. (2021). Sentinel-2 satellite imagery for urban land cover classification by optimized random Forest classifier. *Applied Sciences, 11*(2), 543. https://doi.org/10.3390/APP11020543

Zhilong, Z., Xue, W., Yili, Z., & Jungang, G. (2017). Assessment of changes in the value of ecosystem services in the Koshi River Basin, Central High Himalayas based on land cover changes and the CA-Markov model. *Journal of Resources and Ecology, 8*(1), 67–76. https://doi.org/10.5814/j.issn.1674-764x.2017.01.009

Zuo, Z., Yang, Y., Wang, R., Li, J., & Zhang, P. (2023). Analysis of the gains and losses of ecosystem service value under land use change and zoning in Qiqihar. *Frontiers in Ecology and Evolution, 11*(June), 1–14. https://doi.org/10.3389/fevo.2023.1192952

Land Use and Land Cover (LULC) Dynamics and Their Impact on Land Surface Temperature (LST) Changes: A Spatio-temporal Analysis of Araria Town of Bihar, India

Fouzia Rahman

1 Introduction

The world's towns and cities are facing a dire situation due to unplanned urban expansion and land use and land cover (LULC) transformations. Localised microclimate and surface energy budget have been altered by the transformation of naturally occurring land into impervious zones. It is quite a challenge to document and comprehend how humans affect the environment globally. Research is required on population growth and its projected upward trends, which are causing the atmosphere to deteriorate, resulting in resource depletion, ecological circle destruction, hydrological process destructions and altered climate conditions (Dhar et al., 2019). It is anticipated that urbanisation processes would persist around the globe. According to UN 2014, more than half of current population, i.e., two-thirds of the world's population by 2050 is expected to reside in cities. Urban areas and new settlements expand significantly as a result of the growing demand for housing and the strain these places on natural resources (Haas, 2016). Converting terrestrial surface to constructed settings is one of the most permanent impacts caused by human activity on the global ecosystem.

The surface temperature of earth is a significant factor in land use and environmental studies. For applications in the meteorological, hydrological and climatological fields at the micro- as well as macro-levels, mechanisms of surface temperature are used (Zhang et al., 2016; Bian et al., 2017). The physical characteristics of land features may also be obtained using geospatial technology, which is flexible enough to map patterns in urban expansion and track land use and cover transformation (Pathan et al., 2021).

In recent years, the effects of shifting land use pattern on natural processes have drawn attention in recent years. Analysing these shifts can help with environmental analysis, temperature change analysis and urban/rural planning initiatives. Plenty of studies have shown that land surface temperature (LST) mapping is a crucial factor in comprehending how LULC affects the temperature of the surface (Weng et al., 2004; Peng et al., 2014; Bright et al., 2017). LST is defined by Srivastava et al. (2010), as the surface luminescence heat (spectral radiance) that corresponds to the sensor's instant range of view. LST is an important consideration in land cover appraisal, energy budget assessment and other research pertaining to earth surface features. This aids in improving comprehension of environmental

F. Rahman (✉)
P.G Department of Geography, Veer Kunwar Singh University, Ara, Bihar, India

Z. Khanam et al. (eds.), *Climate Change, Resilience and Ecosystem Services*, Sustainable Development Goals Series, https://doi.org/10.1007/978-3-031-96977-5_8

studies and categories of land use and land cover (LULC) in general (Mallick et al., 2012; Biro et al., 2013). In urban contexts and places, it may be applied to appraise and analyse the spatial link between land surface temperature and various land use and land cover (Amiri et al., 2009). LULC changes are the result of human adaptation which has a major influence on the micro, spatial and worldwide environments (Weng, 2001; Pei et al., 2007). Natural climate and ecosystem functioning are all significantly impacted by LULC changes, and these effects go across spatiotemporal dimensions (Hao & Ren, 2009).

In 2016, Babalola and Akinsanola examined land use and land cover changes and their impact on land surface temperature during a 29-year period for the Nigerian metropolis of Lagos using images from Landsat 5, 7 and 8. Pal and Ziaul in 2017 used Landsat 5 and 8 pictures during a 23-year period of analysis in northern India to confirm that LST mean fluctuates according to many LULC unit types. Guangrong Shen et al. (2015) presented a study on chronological changeability of thermal data of green cover and LULC of Chongming Island of Shanghai, China using temperature and vegetation index. In 2020, Sandipta Das & P. Angadi presented a study which leads to LULC transformation, and its correlation with LST of Barrackpore subdivision of West Bengal for 26 years, in which they demonstrated that impact of LST, is very less on wetland or extreme humid region.

In 2021, Edan et al. propounded the study, which reveals the impact of LULC on LST in Al Kut, Iraq. In 2022, Ramzan et al. analysed a temporal changes in LST and LULC in urban areas of Lahore city of Pakistan. It was a remote sensing-based prediction of changes in LULC and LST. In 2024, Jagtap et al. explored the effect of LULC and LST modifications in Pune, and also anticipated the LULC for 2030. He suggested in his paper that, planners should focus on policies based on land use type, regulations of zoning and strategies for smart urban development. Saha et al. (2024) presented a 15 years study of Kishoreganj, Bangladesh, which examines the seasonal dynamics of LULC and LST from 2008 to 2023.

In this work, the study area is situated in the catchment area of the Kosi River, known as sorrow of Bihar. The study area has a good average annual rainfall (59.29 mm) but, at the same time it shows a sound rise in surface temperature in last few decades. Therefore, it would be interesting to analyse the spatiotemporal changes in surface temperature in last two decades of the study area.

1.1 Study Area

Araria town has been chosen as the study area for this study with geographical extent of 26°11′0″ N to 26°15′0″ N latitude and 87°40′0″ E to 87°48′0″ longitude. The town is located at the southern part of Araria district of the state of Bihar (Fig. 1). Araria town is classified as class II town which consists of 11 wards. As per district census handbook 2011, the total population of the town is 79,021 with population density of 2551 persons per sq.km. In 2001, the population of the town was 60,861 and increased by 18,160 in 2011 (Fig. 3). The research area receives 59.29 cm annual rainfall with average air temperature of 27.43 °C. The idea behind choosing Araria town as the study area was that the town is situated near by the Nepal Himalayas (Tarai region) and enjoys a good amount of annual precipitation. In spite of it changes in land usage and rising temperatures have been the outcome of the problem of the town's growing population in recent years, which was not adequately planned and executed. Decadal population growth of the town was approximately 30% during 2001 to 2011. Rapidly increasing population plays a crucial role as a catalyst for increasing land use issues in terms of expansion of builtup area and fallow land, and, shrinking of vegetation and water bodies. So that, the advantage of being situated in Tarai region of Himalaya becomes pale here with the noticeable effects of land use in surface temperature in the town. Additionally, the town is lying in the catchment area of Kosi River, which is also known as 'Sorrow of Bihar' for its disastrous effects of flood in the northern part of Bihar. The development of urban areas and the gradual depletion of agricultural land, floodplains,

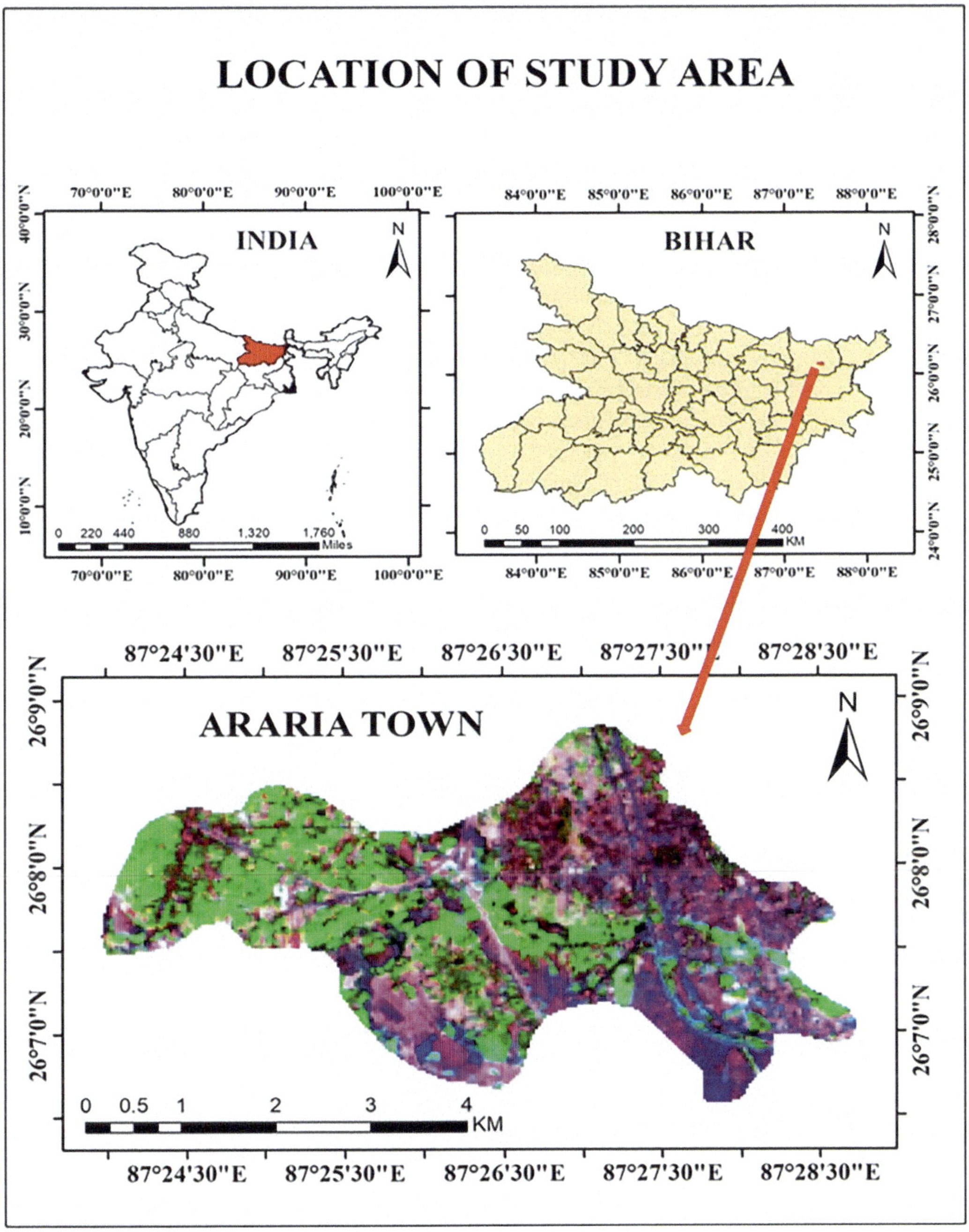

Fig. 1 Location of Araria Town

waterbodies and wetlands due to rapid population increase have profound effects on flood dynamics. Furthermore, the expansion of cities and infrastructure is linked to alterations in ecological and hydrological systems, the disappearance of drainage networks and a heightened vulnerability of certain regions to flooding. The government must increase the amount of land that is farmed in order to accommodate population expansion as well ecological balance.

There are various studies that have been done for establishing relationship between land use and land cover and land surface temperature. Mostly, these studies were focused on metropolitan cities or big cities having multistoried buildings, factories and other urban facilities. This study is trying to represent a small town, which is far away from national and state capitals and slightly urbanised. The work will present a very new kind of study in this particular small town, which is almost untouched by research work in this field by highlighting the changing pattern in land use and surface temperature. In this order for showing land use and land cover changes in 2003, 2013 and 2023, this study selects the Landsat TM and Operational Land Imager/ Thermal Infrared Sensor (OLI/TIRS) images from 2003, 2013 and 2023 and explores the land surface temperature of Araria town.

This study is basically based on secondary sources of data such as satellite imageries and other statistical data. Upcoming studies can be done by using primary sources of data like ground verification of land use changes and accuracy assessment. Furthermost, the actual change in temperature can be determined by collecting primary data in the study area. At the same time, R programming which is an open-source programming language for data analysis and visualisation can be used for more appropriate results.

1.2 Objectives

This study aims to investigate the spatiotemporal analysis of land use and land cover (LULC) changes and its impact on land surface temperature (LST) in Araria town of Bihar using multispectral satellite data. In this order, the study first identifies the different types of land use and land cover and dynamically monitors their spatial and temporal variations in 2003, 2013 and 2023 using Landsat TM imagery and other additional data and then estimates land surface temperature for the years 2003, 2013 and 2023. To this end, for retrieval of LST imageries of Landsat Thematic Mapper (TM) of three time periods, i.e., November 2003 (TM), November 2013 (TM) and November 2023(TM) have been used.

2 Materials and Methods

On the basis of spatial and non-spatial data, this work has been analysed. The spatial data is derived from remotely sensed imageries of three different years; however, non-spatial data is collected from census handbook of Araria district, 2011. Classification of land use and land cover has been determined by multispectral satellite images of Landsat 5 (LT5 TM) and 8 (LT8 OLI/ TIRS) using GIS and Remote Sensing software ArcGIS and ERDAS IMAGINE 10.4. For the year 2003 Landsat 5 imageries have been used, whereas for 2013 and 2023 images of Landsat 8 have been considered. Methodological flowchart for the study has been shown in Fig. 2.

Non-Spatial Data (Fig. 3)

Spatial Data

For this study, the spatial data has been used in the form of multispectral satellite data of LANDSAT 5 and LANDSAT presented in Table 1. There are two types of multispectral data, i.e., Landsat 5 TM and Landsat 8 OLI/ TIRS, which were used for the path 139 and row 042. Table 1 is presenting the date, sensor and thermal band, spatial resolution and band name with different wavelength of the images.

2.1 Creation of LULC Map

The world's population depends upon land resources to fulfil their most fundamental requirements. Changes in land resources during the past few decades have been mostly influenced by human activity or anthropogenic activities. The usage of land has transformed due to unplanned and uncontrolled population increase on a micro to global scale (Hegazy & Kaloop, 2015). This study categorises satellite imageries into six categories (Table 2) for better understanding of the features of the study area.

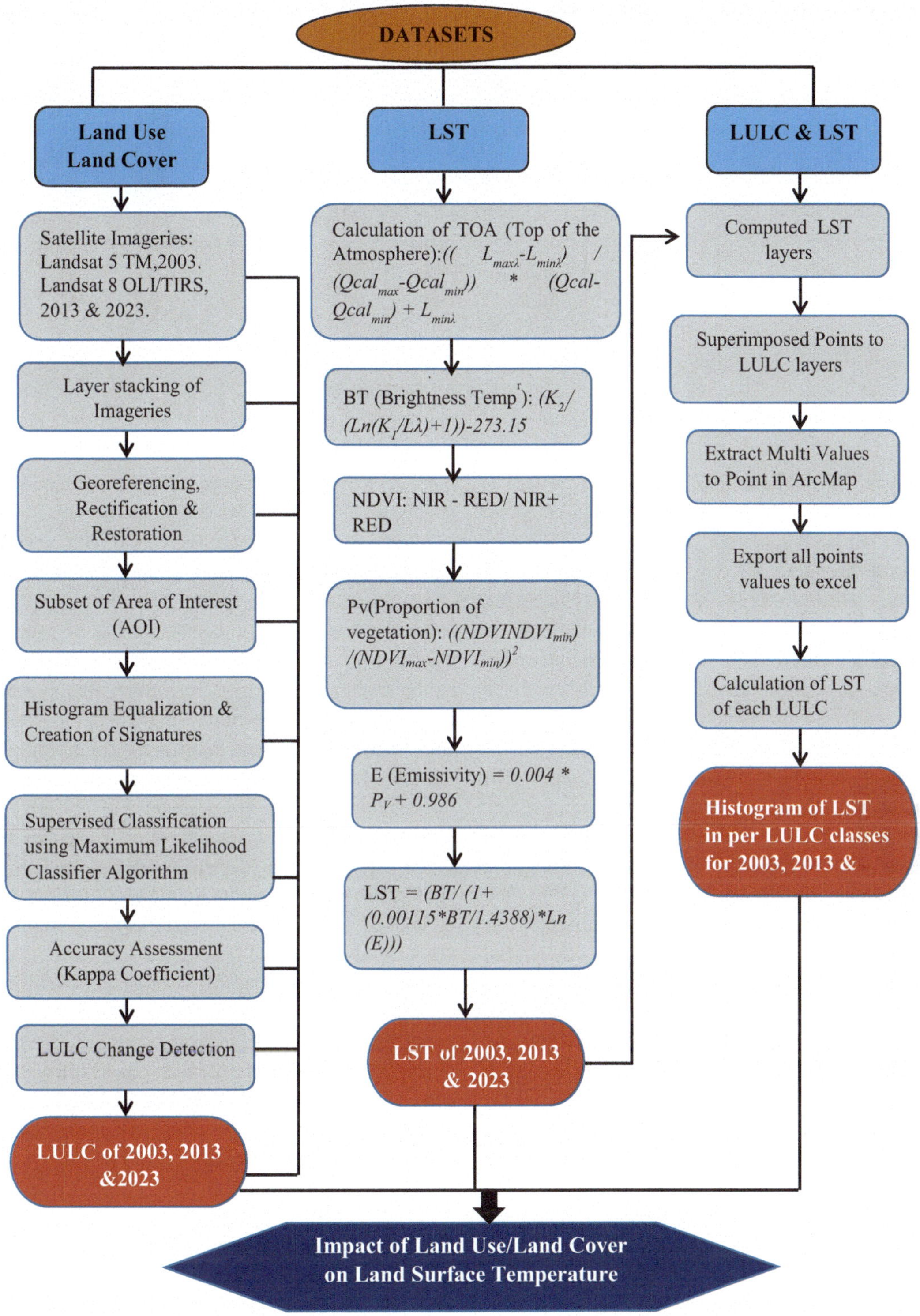

Fig. 2 Methodological flowchart

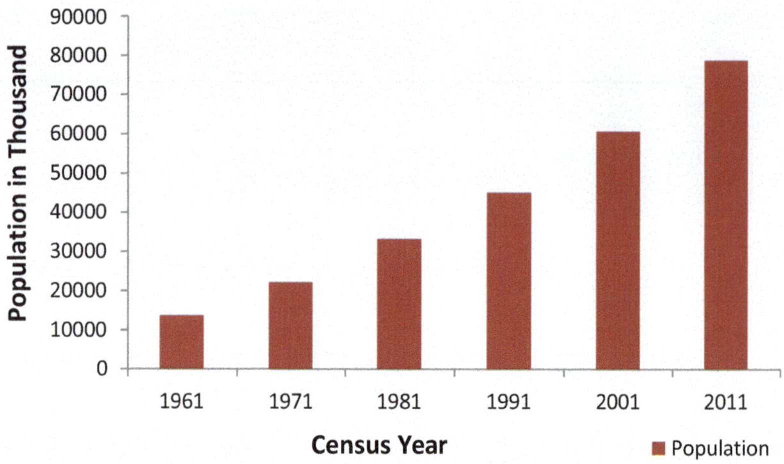

Fig. 3 Decadal growth of population in Araria town

The USGS Earth Explorer website, which is free access, is the source of Landsat imagery. Multitemporal remotely sensed satellite imageries for three referred years 2003, 2013 and 2023 were collected during the month of November, which is a cloud-free and rain-free month in northern India. During the pre-processing stage, images were georeferenced and rectified with the help of World Geodetic System (WGS) 84 datum and the UTM Zone 45 N (Universal Transverse Mercator) projection. In order to register the image with the best feasible overlay accuracy, 30 ground control points (GCP) were generated using the closest neighbour resampling technique.

Histogram equalisation and a subset of the area of interest were used to enhance the Landsat satellite image after it had been rectified and restored using the closest neighbour resampling approach. Now the land use and land cover map of the study area has been created using the supervised image classification technique.

2.2 Supervised Image Classification Method

Following the subset of area of interest (AOI), 30 sample data from each class—builtup, agricultural, vegetation, water body, fallow land and wetland, were collected through Google Earth and a topographical map in order to create a signature using the supervised classification technique.

The accuracy rate of non-parametric algorithm-based image categorisation techniques is greater (Hasan et al., 2020). Six land use types for the region have been classified using constraint-based method, i.e., supervised classification techniques and the maximum likelihood classifier (MLC) algorithm in this study. The MLC method has been considered to employ a variety of classification evaluations employing possibility and cost utilities (Duda et al., 2000). It is predicated on the resemblance of the targeted pixels' spectral signatures to a certain class (Li et al., 2014). The complexity of LULC attributes is reduced and spectral information-based analysis is suitably assessed (Li et al., 2014; Dewan & Yamaguchi, 2009). The MLC has been used in several previous studies for LULC classification as well as mapping of landscape. It is still in use today because it offers a greater possibility of extracting information from the surface landscape while lowering the requirement for data (Sudhira et al., 2003;Kumar et al., 2007).

2.3 Accuracy Assessment

Evaluating accuracy is supposed to most important parts of the procedure of classification. The purpose of accuracy assessment is to quantifiably measure the efficiency of the sampled pixels into

Table 1 Multitemporal Landsat satellite images

Date	Satellite	Sensor/thermal band	Path/Row	Spatial resolution (metre)	Band name/wavelength (μm)
25/11/2003	Landsat 5	TM/thermal band 6	139/042	30 m	Band 1: 0.45–0.52 (Blue) Band 2: 0.52–0.60 (Green) Band 3: 0.63–0.69 (Red) Band 4: 0.77–0.90 (NIR) Band 5: 1.55–1.75 (SWIR1) Band 7: 2.09–2.35 (SWIR2) Band 6: 10.40–12.50 (TIR)
20/11/2013	Landsat 8	OLI/TIRS/thermal band 10	139/042	30 m	Band 1: 0.43–0.45 (Coastal) Band 2: 0.45—0.51 (Blue) Band 3: 0.53—0.59 (Green) Band 4: 0.64—0.67 (Red) Band 5: 0.85—0.88 (NIR) Band 6: 1.57–1.65 (SWIR) Band 7: 2.11–2.29 (SWIR) Band 8: 0.50–0.68 (Panchromatic) Band 9: 1.36–1.39 (Cirrus) Band 10: 10.3–11.3 (LWIR)
16/11/2023	Landsat 8	OLI/TIRS/thermal band 10	139/042	30 m	Band 1: 0.43–0.45 (Coastal) Band 2: 0.45—0.51 (Blue) Band 3: 0.53—0.59 (Green) Band 4: 0.64—0.67 (Red) Band 5: 0.85—0.88 (NIR) Band 6: 1.57–1.65 (SWIR) Band 7: 2.11–2.29 (SWIR) Band 8: 0.50–0.68 (Panchromatic) Band 9: 1.36–1.39 (Cirrus) Band 10: 10.3–11.3 (LWIR)

Source: https://earthexplorer.usgs.gov

Table 2 Classification of different LULC classes

1.	Agriculture	Cultivated land, seasonal and temporal cropland
2.	Builtup	Residential, commercial, transportation, roads, industrial and other man-made structures etc.
3.	Fallow land	Exposed soils, landfills, bare land, excavated regions now under excavation, sand etc.
4.	Water body	River, canal, pond etc.
5.	Wetland	Permanent and seasonal wetlands, marshy land, swamps
6.	Vegetation	Grassland, mixed forest, deciduous forest, plantation, rooftop plants, urban green space

Source: Roy et al. 2016

Table 3 Types of agreement of kappa coefficient

Value of kappa coefficient (K^)	Types of agreement
0	Less than chance agreement
0.01–0.02	Slight agreement
0.21–0.40	Fair agreement
0.41–0.60	Moderate agreement
0.61–0.80	Substantial agreement
0.81–0.99	Almost perfect agreement
1	Perfect agreement

Source: Anthony and Ruther (2007)

the relevant land cover groups. For accuracy valuation, a total of 256 truth points were generated for each classified Landsat image using ERDAS IMAGINE 2014 software.

The accuracy of image classification has been examined for the current study by using two methods, i.e., Overall accuracy and Cohen's kappa coefficient (K^). Overall accuracy is determined by dividing the total number of categorised pixels by the total number of sample points. The formula for calculating the percentage of overall accuracy was -.

$$\frac{\text{Total number of accurate samples}}{\text{Total number of samples}} \times 100$$

The Kappa statistic considers both the overall accuracy of image classification and the accuracy of each individual category as a measure of the truly agreement between classification and observation. It is calculated by using this formula mentioned below.

Cohen's kappa coefficient can be defined by using this formula.

$$(K\wedge) = \frac{n\sum_{i=1}^{r} xii - \sum_{i=1}^{r}(xi + x_{+i})}{n^2 - \sum_{i=1}^{r}(xi + x_{+i})}$$

where

r = number of rows in matrix,
Xii = number of observations in row i and column i xi = marginal total row of i.
x_{+i} = marginal total of column i,
n = total number of observations (samples/pixels).

Kappa has a range of values from 0 to 1, where 0 characterises confirmation that is entirely probability based. Meanwhile, 1 demonstrates that the two data sets completely agree with one another (Cohen, 1968). According to Anthony and Ruther (2007), different values of the coefficient depict different category of agreement as shown in Table 3.

2.4 Retrieval of Land Surface Temperature (LST)

There are two methods extracting land surface temperature from satellite imagery, i.e., split window and mono-window algorithms. These are useful for utilising Visible- Near infrared (VNIR) and Thermal Infrared (TIR) sensors. An essential component in estimating LST is land surface emissivity (Qin et al., 2001; Dwivedi & Khire, 2018).

For this study, multitemporal satellite imageries of LT 5 TM and LT 8 OLI/TIRS were utilised. For drawing out of LST from band 6 of Landsat 5 TM, the mono-window algorithm technique has been used. The split-window algorithm was used for extraction LST from band 10 from Landsat 8 OLI/TIRS. Mono-window or single-channel algorithm is useful for retrieving LST from the sensors having single thermal band, i.e., TM or ETM+. However, split-window or multichannel algorithm is useful for retrieving LST from sensors having multiple thermal bands, i.e., LT 8 TIRS (Käfer et al., 2020). Therefore, there are two different kind of algorithms that have been used for two different satellite images for retrieving LST for this study.

To derive LST from Landsat 5 and Landsat 8 imageries, following equations are required:

Step I

Equation 1 may be used to estimate radiance (Lλ) for the TM and OLI sensor bands, which transforms the energy received at the top of the atmosphere from the DN (digital number).

$$\mathrm{TOA}(L\lambda) = \left(\left(L_{\max\lambda} - L_{\min\lambda}\right) / \left(\mathrm{Qcal}_{\max} - \mathrm{Qcal}_{\min}\right)\right) ^{*} \left(\mathrm{Qcal} - \mathrm{Qcal}_{\min}\right) + L_{min\lambda} \quad (1)$$

where $L\lambda$ = sensor radiance, $L_{\max\lambda}$ = maximum radiance of band 6, $L_{\min\lambda}$ = minimum radiance of band 6, Qcal = quantised calibrated pixel value in DN, $QCal_{max}$ = maximum quantised calibrated pixel value in DN, $QCal_{min}$ = minimum quantised calibrated pixel value in DN.

Step II

The radiance was converted to brightness temperature at the top of the atmosphere (ToA) using Eq. (2), which correlated to the surface temperature that was brought to the sensor by atmospheric interference.

$$\mathrm{BT}(\text{Brightness temperature}) = \left(\mathrm{K}_{2/} / \left(\mathrm{Ln}\left(\mathrm{K}_1 / L\lambda\right) + 1\right)\right) - 273.15 \quad (2)$$

where K1 = band specific thermal conversion constant from the metadata (K1_CONSTANT_BAND_*x*, where x is thermal band number), K2 = band specific thermal conversion constant from the metadata (K2_CONSTANT_BAND_x, where *x* is thermal band number) and Lλ = TOA.

Step III

Further Normalised Difference Vegetation Index (NDVI) is calculated to quantify proportion of vegetation using Eqs. 3 and 4 respectively.

$$NVDI = \frac{(NIR - RED)}{(NIR + RED)} \quad (3)$$

where NIR (near infrared) band for Landsat 5TM is band 4 and RED is band 3, whereas for Landsat 8 OLI, NIR is band 5 and RED is band 4.

Step IV

NDVI methods can be used to estimate proportion of vegetation.

$$\mathrm{P_V}\left(\text{Proportion of Vegetation}\right) = \left(\left(\mathrm{NDVI} - \mathrm{NDVI}_{\min}\right) / \left(\mathrm{NDVI}_{\max} - \mathrm{NDVI}_{\min}\right)\right)^2 \quad (4)$$

Step V

It is necessary to estimate the surface emissivity values precisely in order to retrieve the LST of multispectral satellite data of the thermal zone (Prata et al., 1995). Several methods have been employed to estimate ground object emissivity utilising satellite sensors. One of the easiest and most trustworthy methods for projecting emissivity across the surface is the NDVI-based emissivity methodology. LSE was computed using the assumption that emissivity was 0.99 for pixels with an NDVI > 0.5 and 0.98 for pixels with an NDVI < 0.2, according to Wang et al. (2016). The ability of the surface of a substance to emit heat radiation is known as its emissivity. If the emissivity of an item is known, its radiant temperature may be used to infer its real kinetic temperature.

Equation 5 is the formula for emissivity. It can be derived using following formula mentioned in Eq. 4.

$$\mathrm{E}(\text{Emissivity}) = 0.004^{*}\,\mathrm{P_V} + 0.986 \quad (5)$$

Step VI

Land surface temperature has been derived by using following formula mentioned in Eq. 6.

$$\text{LST} = \left(\text{BT} / \left(1 + \left(0.00115^{*}\text{BT} / 1.4388\right) * \text{Ln}\left(\text{E}\right)\right)\right) \quad (6)$$

3 Result and Discussion

3.1 Accuracy Assessment and Validation of Map

Accuracy assessment of classified satellite imageries has been done with the help of overall accuracy and Cohen's kappa coefficient (K^). As stated earlier that, kappa has a range of values from 0 to 1, where 0 characterises confirmation that is entirely probability based. Meanwhile, 1 demonstrates that the two data sets completely agree with one another (Cohen, 1968).

Table 4 presents overall accuracy and Kappa coefficient for the classified multitemporal satellite imageries.

3.2 Land Use and Land Cover Changes from 2003 to 2023

The ecological mechanism and sociocultural process are directly influenced by land use and land cover changes or transformations (Gessesse et al., 2019). Similarly food supply chain, climate vulnerability to hazards and natural calamity, climate change, fertility and quality of soil are also another aspects of impacts of LULC change (Geist et al., 2006). Figure 4 (a–c) shows land use and land cover map of Araria town for the year 2003 (Fig. 4a), 2013 (Fig. 4b) and 2023 (Fig. 4c). After classifying and creating the LULC of satellite imageries, area of all classes (Table 5) has been compiled with the help of ArcGIS 10.7.1. It was found that in 2003 area under agriculture was 5.73 km^2 (33.13%). It was 5.11 km^2 (29.71%) and 4.19 km^2 (24.36%) in 2013 and 2023 respectively. This figure depicts that the area under agriculture is continuously decreasing during the referred period (2003–2023) with the transformation rate of 26.5%. On the other hand, builtup area is showing increasing trend during 2003–2023. In 2003, area under builtup category was 2.93 km^2 (17.03%). It was 3.98 km (23.14%) and 4.73 km^2 (27.50%) area in 2013 and 2023 respectively, as shown in Figs. 3 and 4. Builtup area has increased during 2003–2023 with the rate of 58.83%, due to increasing population in the study area followed by fallow land. In 2003, it was accounted for 0.89 km^2 (5.17%) increased by 2.16 and acquired 3.05 km^2 area (17.73%) in 2013 and 3.03 km^2 (17.61%) in 2023. In 2003, area under vegetation was 2.93 km^2 (17.03%) which decreased in 2013 and accounted for 2.60 km^2 (15.11%) area, but in 2023 it increased and covered 3.10 km^2 (18.02%) area. The reason is that during the period of 2003 and 2023 the area under waterbody and wetland has been decreased and transformed in the form

Table 4 Accuracy assessment report of different LULC classes

Land use classes	2003		2013		2023	
	PA	UA	PA	UA	PA	UA
Agriculture	82.98%	100.00%	94.29%	84.62%	91.67%	89.19%
Builtup	84.21%	66.67%	90.00%	90.00%	87.18%	94.14%
Vegetation	68.18%	93.75%	91.67%	68.75%	91.30%	72.41%
Fallow land	90.00%	100.00%	77.78%	96.55%	83.33%	78.95%
Waterbody	91.30%	75.00%	71.43%	71.43%	100.00%	60.00%
Wetland	94.12%	76.19%	90.00%	90.00%	61.54%	80.00%
	Overall accuracy = 91.41%		Overall accuracy = 92.58%		Overall accuracy = 91.80%	
	Kappa coefficient = 0.88		Kappa coefficient = 0.90		Kappa coefficient = 0.89	

Source: ERDAS IMAGINE 2014

Note: PA = Producer's accuracy, UA = User's accuracy

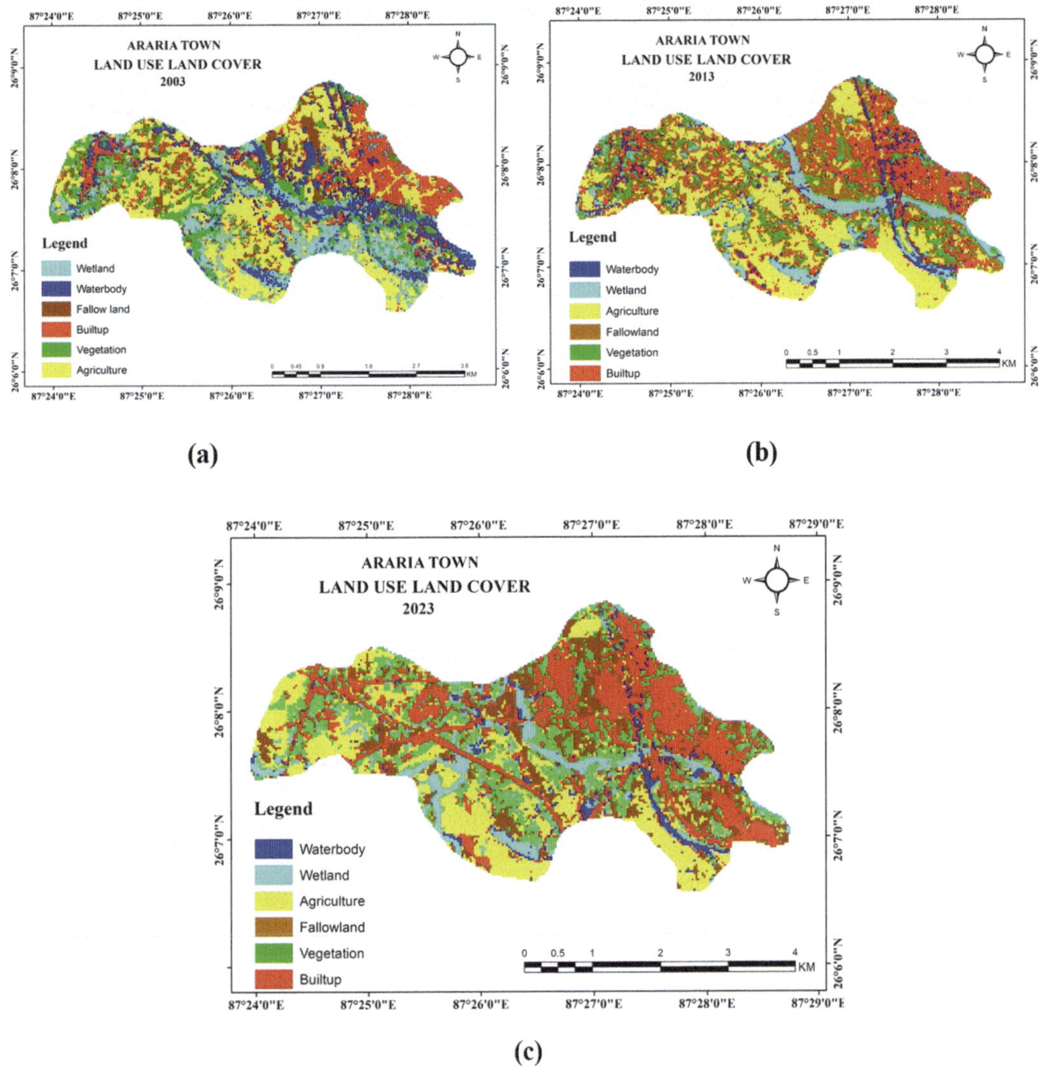

Fig. 4 LULC map of Araria town (**a**) 2003, (**b**) 2013 and (**c**) 2023

Table 5 LULC changes from 2003 to 2023 in Araria town

	2003		2013		2023	
LULC classes	Area (km^2)	Area (%)	Area (km^2)	Area (%)	Area (km^2)	Area (%)
Agriculture	5.73	33.13	5.11	29.71	4.18	24.36
Builtup	2.93	17.03	3.98	23.14	4.73	27.50
Fallow land	0.89	5.17	3.05	17.73	3.03	17.61
Vegetation	2.93	17.03	2.60	15.11	3.10	18.02
Waterbody	2.33	13.54	0.88	5.11	0.61	3.54
Wetland	2.39	13.89	1.58	9.18	1.54	8.95

Source: Compiled by researcher

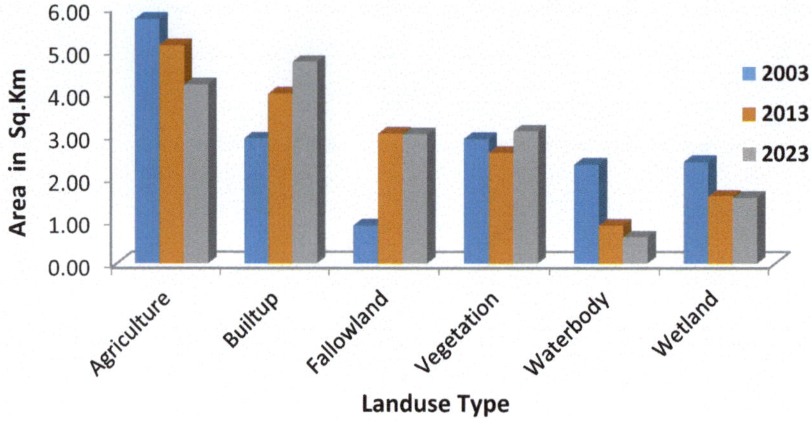

Fig. 5 Illustration of LULC changes during 2003–2023

of aquatic plants and other floating vegetation. Water body is showing decreasing trend, that is, 2.33 km^2 (13.54%) in 2003, 0.88 km^2 (5.11%) area in 2013 and 0.61 km^2 (3.54%) in 2023. Similarly, wetland is also showing decreasing trend as it was 2.39 km^2 (13.89%) in 2003, 1.58 km^2 (9.18%) area in 2013 and 1.54 area sq. km (8.95%) in 2023. The total transformation rate of wetland during 2003 and 2023 was 37.7%. Due to climate change and increasing temperature, waterbody and wetland are decreasing (Fig. 5).

3.2.1 Land Use and Land Cover Transformation During 2003 and 2023

Table 6 a–c shows land use and land cover change matrix for 2003 and 2013, 2013 and 2023, 2003 and 2023 respectively. It is clear from the Table 6a 2003 and 2013 that most of the land use features have been transformed into builtup area and fallow land. In this regard, major portion of agriculture has been transformed as builtup area (1.18 km^2) and barren land (1.13 km^2).

Change matrix represents the change magnitude, which shows the increase or contraction of LULC size (Kafi et al., 2014). In this work, matrix table between 2013 and 2023 is also shows the similar pattern of transformation. During the year 2003 and 2023 (Table 6c), out of 5.73 km^2 of agriculture 2.18 km^2 area remained unchanged, 1.45 km^2 area has been transformed into builtup area followed by fallow land (1.15 km^2) and vegetation (0.72 km^2). Major portion (0.35 km^2) of fallow land was unchanged in this period. Due to increasing population demand of food grains increased, resulting into deforestation and developing farm land for agriculture practices. Table 6c is showing the transformation of vegetation into agriculture area that is, out of 2.93 km^2 of vegetation has been transformed into 0.66 km^2 of agriculture area. Area under waterbody has decreased over the last 20 years and transformed into builtup area and wetland as well. However, wetland has been transformed into agriculture (0.67 km^2) followed by builtup (0.48 km^2) and vegetation (0.40 km^2).

3.3 Land Surface Temperature Changes from 2003 to 2023

Figure 6 (a–c) shows spatial pattern of land surface temperature of study area for the year 2003, 2013 and 2023. In 2003, maximum temperature was 26.58 and minimum temperature was 21.66 °C with coefficient of variation (CV) 3.19% as shown in Table 7. The LST of 2013 validates that the span of temperature was 31.05 °C (maximum)–23.55 °C (minimum) with CV 4.31%. In this order the year 2023 continues

Table 6 Land use and land cover transition matrix of Araria town

		Agriculture	Builtup	Fallow land	Vegetation	Water body	Wetland	Total
(a) Between 2003 and 2013								
2003								
2013	Agriculture	2.93	0.37	0.22	0.42	0.21	0.95	5.11
	Builtup	1.18	1.53	0.10	0.36	0.40	0.41	3.98
	Fallow land	1.13	0.26	0.44	0.69	0.27	0.25	3.05
	Vegetation	0.33	0.49	0.10	0.94	0.49	0.25	2.60
	Waterbody	0.11	0.21	0.01	0.15	0.22	0.19	0.88
	Wetland	0.04	0.07	0.00	0.37	0.75	0.34	1.58
	Total	5.73	2.93	0.89	2.93	2.33	2.39	17.20
(b): Between 2013 and 2023								
2013								
2023	Agriculture	2.40	0.41	0.70	0.44	0.12	0.13	4.18
	Builtup	0.78	2.25	0.68	0.55	0.33	0.15	4.73
	Fallow land	1.08	0.49	0.95	0.35	0.04	0.12	3.03
	Vegetation	0.46	0.60	0.61	0.97	0.13	0.32	3.10
	Waterbody	0.14	0.10	0.03	0.06	0.14	0.14	0.61
	Wetland	0.26	0.13	0.08	0.24	0.12	0.72	1.54
	Total	5.11	3.98	3.05	2.60	0.88	1.58	17.20
(c): Between 2003 and 2023								
2003								
2023	Agriculture	2.18	0.32	0.14	0.66	0.22	0.67	4.18
	Builtup	1.45	1.54	0.20	0.45	0.60	0.48	4.73
	Fallow land	1.15	0.33	0.35	0.52	0.29	0.39	3.03
	Vegetation	0.72	0.58	0.17	0.74	0.49	0.40	3.10
	Waterbody	0.08	0.09	0.01	0.10	0.18	0.15	0.61
	Wetland	0.16	0.08	0.02	0.45	0.55	0.30	1.54
	Total	5.73	2.93	0.89	2.93	2.33	2.39	17.20

Source: Based on LULC data

with increased temperature range, i.e., 34.69 °C (maximum)–26.74 °C (minimum) with coefficient of variation 4.15%.

It is clear from Table 7 that mean LST in 2003 was 23.50 °C which has grown to 2.9 °C and reached to 26.40 °C in 2013. In 2023, the mean LST was 29.83 °C. During the previous 20 years, the mean land surface temperature has raised by 6.33 °C in the town.

3.4 Relation Between Land Use and Land Cover and Land Surface Temperature

Table 8 depicts the land surface temperature of different land use and land cover classes for the referred years. It is clear from the Fig. 7 that builtup area and fallow land classes are showing high temperature with respect to maximum, minimum and mean LST values in all the referred years, followed by agricultural land and vegetation. Builtup area is rapidly increased over the last 20 years from 17.03% in 2003 to 27.0% in 2023. Land surface temperature reflected from builtup area ranged from 23.7 °C to 25.7 with a mean 24.7 °C in 2003. It was increased and ranged from 31.5 to 33.0 °C with mean temperature of 30 °C in 2023. Overall mean rise in temperature recorded was from 24.7 °C in 2003 to 30 °C in 2023. In comparison to other land use classes, builtup area is characterised by more temperature, i.e., 25.7 °C in 2003 and 33.0 °C in 2023, showing 7.3 °C increment in LST of this class between 2003 and 2023. After builtup area, fallow land is showing high temperature rise during the referred period of time, i.e., maximum temperature has grown

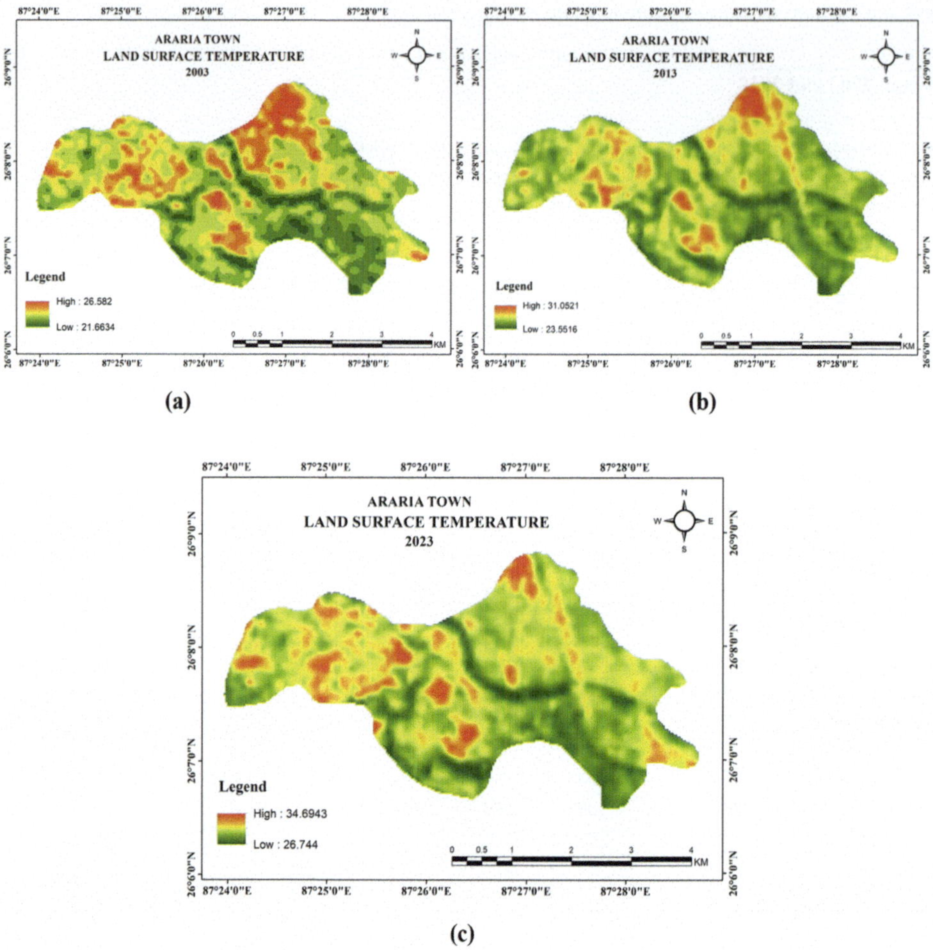

Fig. 6 Land surface temperature of Araria Town (**a**) 2003, (**b**) 2013 and (**c**) 2023

Table 7 LST coefficient of variation for different periods

LST	2003	2013	2023
Maximum temp. (°C)	26.58	31.05	34.69
Minimum temp. (°C)	21.66	23.55	26.74
Mean temp. (°C)	23.50	26.40	29.83
Standard deviation (SD)	0.74	1.14	1.24
Coefficient of variance (CV)	3.19	4.31	4.15

Source: Compiled by researcher

Table 8 LST of different land use classes of Araria town (2003, 2013 and 2023)

LULC Classes	LST(°C) 2003			LST (°C) 2013			LST(°C) 2023		
	Max	Min	Mean	Max	Min	Mean	Max	Min	Mean
Agriculture	23.9	23.2	23.5	27.1	25.1	26.1	29.5	27.7	28.6
Builtup	25.7	23.7	24.7	30.0	26.9	28.0	33.0	31.5	30.0
Fallow land	25.2	24.3	24.8	30.0	25.2	27.6	31.0	29.1	30.2
Vegetation	23.4	23.0	23.2	26.2	25.3	25.9	29.6	28.5	29.2
Waterbody	23.4	22.1	22.8	26.9	24.2	25.4	30.2	27.3	29.0
Wetland	23.4	22.1	22.6	26.2	23.8	25.2	30.0	27.7	29.1

Source: Compiled by researcher using ArcGIS 10.7.1

from 25.2 to 31 °C with mean temperature of 24.8 and 30.2 °C in 2003 and 2023 respectively. In comparison to builtup and fallow land, vegetation cover and agricultural land or land cover with moisture (water body and wetland) are less affected by surface temperature (Das & Angadi, 2020). It is revealed from Table 8 and Fig. 7 a–c) that LST over agricultural land was 23.9 °C in 2003 and 29.5 °C in 2023 with mean 23.5 °C in 2003 and 28.6 °C in 2023. However, vegeta-

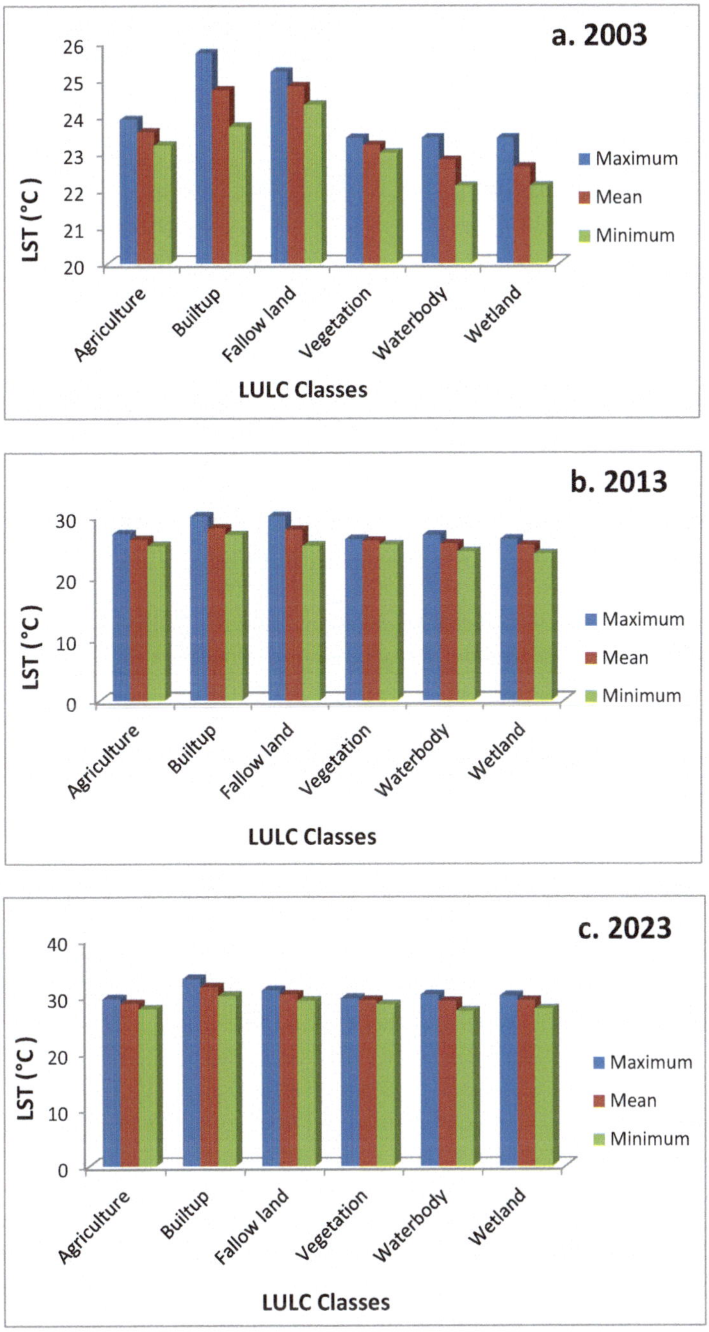

Fig. 7 Histogram of land surface temperature (°C) in per LULC types

tion cover is showing LST of 23.4 to 29.6 °C during 2003 and 2023.

In comparison to other LULC classes, water body and wetland are enjoying lesser land surface temperature during the previous 20 years in the study area. Overall mean temperature of waterbody varies from 26.9 to 30.2 °C during 2003 and 2023 respectively. LST on wetland was 23.4 and 30 °C with mean value 22.6 and 29.1 °C in 2003 and 2023 respectively.

4 Conclusion

This study revealed that over the last 20 years, there is a noticeable increasing population trend can be seen in the study area. Consequently, demand of habitats, economic development, educational and health facilities, trading and other recreational activities have grown rapidly. As a result, agricultural land and vegetation cover have been transformed into housing colonies and other infrastructural setup. It can be clearly seen in Table 5 that percentage of builtup area has increased (17.03–27.50%) during 2003 and 2023, whereas the area under agriculture, water body and wetland has decreased during this period. The study area lies in the catchment area of the Kosi River which is known for shifting its path frequently. So that after shifting waterbodies, builtup and vegetation have grown in that particular site resulting into increase in these classes during the last two decades. But at the same time due to climate change wetland and waterbodies are shrinking and transformed into builtup and fallow land. Land surface temperature is showing close relationship with these land use classes. Population growth and urban development always become the cause of increase of builtup area with concrete structure which is a good conductor of heat and temperature. Therefore, as the builtup area has grown it has increased the surface temperature. Major portion of agricultural land has been transformed into fallow land which is also a reason behind the increase in temperature. Since water body and wetland are decreasing in the study area, consequently they are reflect low LST during last two decades. But these areas are declining rapidly in the study area. Waterbody has shrinked from 13.54 in 2003 to 3.54% in 2023, whereas wetland was 13.89% in 2003 and 8.95% in 2023. This spatial distribution of land use classes affects the distribution of land surface temperature in the study area.

It is advised that rapid land transformation measures be put into place in the area of study region, particularly with relation to the conversion of fallow land which during a 20 years period collectively occupied 18% of the total area. Similarly, time conservation of waterbodies and wetland should be the inevitable part of the policymaking so that surface temperature can be maintained in the study area. Furthermore, it was concluded that ongoing observation of LULC dynamics is necessary to direct viable land use strategies that support both economic growth and environmental preservation in this area.

References

Amiri, R., Weng, Q., Alimohammadi, A., & Alavipanah, S. K. (2009). Spatial–temporal dynamics of land surface temperature in relation to fractional vegetation cover and land use/cover in the Tabriz urban area, Iran. *Remote Sensing of Environment, 113*(12), 2606–2617.

Anthony, G., & Ruther, H. (2007). Land cover mapping: Optimizing remote sensing data for SVM classification. *Journal of Remote Sensing-Beijing, 11*(5), 694.

Babalola, O. S., & Akinsanola, A. A. (2016). Change detection in land surface temperature and land use land cover over Lagos Metropolis, Nigeria. *Journal of Remote Sensing & GIS, 5*(3), 1–7.

Bian, J., Li, A., Zhang, Z., Zhao, W., Lei, G., Yin, G., & Huang, C. (2017). Monitoring fractional green vegetation cover dynamics over a seasonally inundated alpine wetland using dense time series HJ-1A/B constellation images and an adaptive endmember selection LSMM model. *Remote Sensing of Environment, 197*, 98–114.

Biro, K., Pradhan, B., Buchroithner, M., & Makeschin, F. (2013). Land use/land cover change analysis and its impact on soil properties in the northern part of Gadarif region, Sudan. *Land Degradation & Development, 24*(1), 90–102.

Bright, R. M., Davin, E., O'Halloran, T., Pongratz, J., Zhao, K., & Cescatti, A. (2017). Local temperature response to land cover and management change driven by non-radiative processes. *Nature Climate Change, 7*(4), 296–302.

Cohen, J. (1968). Weighted kappa: Nominal scale agreement provision for scaled disagreement or partial credit. *Psychological Bulletin, 70*(4), 213–220.

Das, S., & Angadi, D. P. (2020). Land use-land cover (LULC) transformation and its relation with land surface temperature changes: A case study of Barrackpore Subdivision, West Bengal, India. *Remote Sensing Applications: Society and Environment, 19*, 100322.

Dewan, A. M., & Yamaguchi, Y. (2009). Land use and land cover change in Greater Dhaka, Bangladesh: Using remote sensing to promote sustainable urbanization. *Applied Geography, 29*(3), 390–401.

Dhar, R. B., Chakraborty, S., Chattopadhyay, R., & Sikdar, P. K. (2019). Impact of land-use/land-cover change on land surface temperature using satellite data: A case study of Rajarhat Block, North 24-Parganas District, West Bengal. Journal of the Indian Society of Remote Sensing, 47(2), 331-348.

Duda, R. O., Hart, P. E., Stork, D. G., & Ionescu, A. (2000). *Pattern classification, chapter nonparametric techniques* (pp. 177–178). Wiley-Interscience Publication.

Dwivedi, A., & Khire, M. V. (2018). Application of split-window algorithm to study urban heat island effect in Mumbai through land surface temperature approach. *Sustainable Cities and Society, 41*, 865–877.

Edan, M. H., Maarouf, R. M., & Hasson, J. (2021). Predicting the impacts of land use/land cover change on land surface temperature using remote sensing approach in Al Kut, Iraq. *Physics and Chemistry of the Earth, Parts A/B/C, 123*, 103012.

Geist, H., McConnell, W., Lambin, E. F., Moran, E., Alves, D., & Rudel, T. (2006). Causes and trajectories of land-use/cover change. In *Land-use and land-cover change: Local processes and global impacts* (pp. 41–70). Springer Berlin Heidelberg.

Gessesse, B., Bewket, W., & Bräuning, A. (2019). Land use and land cover transformation and its implication on land degradation: The case of Modjo watershed, Ethiopia. In *Extreme hydrology and climate variability* (pp. 503–518). Elsevier.

Haas, J. (2016). *Remote sensing of urbanization and environmental impacts*. Doctoral dissertation, KTH Royal Institute of Technology.

Hao, H. M., & Ren, Z. Y. (2009). Land use/land cover change (LUCC) and eco-environment response to LUCC in farming-pastoral zone, China. *Agricultural Sciences in China, 8*(1), 91–97.

Hasan, M. E., Nath, B., Sarker, A. R., Wang, Z., Zhang, L., Yang, X., et al. (2020). Applying multi-temporal Landsat satellite data and markov-cellular automata to predict forest cover change and forest degradation of Sundarban reserve forest, Bangladesh. *Forests, 11*(9), 1016.

Hegazy, I. R., & Kaloop, M. R. (2015). Monitoring urban growth and land use change detection with GIS and remote sensing techniques in Daqahlia governorate Egypt. *International Journal of Sustainable Built Environment, 4*(1), 117–124.

Jagtap, A. A., Shedge, D. K., & Mane, P. B. (2024). Exploring the effects of land use/land cover (LULC) modifications and land surface temperature (LST) in Pune, Maharashtra with anticipated LULC for 2030. *International Journal of Geoinformatics, 20*(2).

Käfer, P. S., Rolim, S. B. A., Diaz, L. R., Rocha, N. S. D., Iglesias, M. L., & Rex, F. E. (2020). Comparative analysis of split-window and single-channel algorithms for land surface temperature retrieval of a pseudo-invariant target. *Boletim de Ciências Geodésicas, 26*(02).

Kafi, K. M., Shafri, H. Z. M., & Shariff, A. B. M. (2014, June). An analysis of LULC change detection using remotely sensed data; a case study of Bauchi City. In *IOP conference series: Earth and environmental science* (Vol. 20, 1st ed., p. 012056). IOP Publishing.

Kumar, A., Uniyal, S. K., & Lal, B. (2007). Stratification of forest density and its validation by NDVI analysis in a part of western Himalaya, India using remote sensing and GIS techniques. *International Journal of Remote Sensing, 28*(11), 2485–2495.

Li, M., Zang, S., Zhang, B., Li, S., & Wu, C. (2014). A review of remote sensing image classification techniques: The role of spatio-contextual information. *European Journal of Remote Sensing, 47*(1), 389–411.

Mallick, J., Singh, C. K., Shashtri, S., Rahman, A., & Mukherjee, S. (2012). Land surface emissivity retrieval based on moisture index from LANDSAT TM satellite data over heterogeneous surfaces of Delhi city. *International Journal of Applied Earth Observation and Geoinformation, 19*, 348–358.

Pal, S., & Ziaul, S. K. (2017). Detection of land use and land cover change and land surface temperature in English Bazar urban centre. *The Egyptian Journal of Remote Sensing and Space Science, 20*(1), 125–145.

Pathan, S. A., Ashwini, K., & Sil, B. S. (2021). Spatio-temporal variation in land use/land cover pattern and channel migration in Majuli River Island, India. *Environmental Monitoring and Assessment, 193*(12), 811.

Pei, H., Qin, Z., Fang, S., Xu, B., Zhang, C., Lu, L., & Gao, M. (2007). Impacts of land use/cover change on spatial variation of land surface temperature in Urumqi, China. In *Geoinformatics 2007: Remotely Sensed Data and Information* (Vol. 6752, pp. 1306–1314).

Peng, S. S., Piao, S., Zeng, Z., Ciais, P., Zhou, L., Li, L. Z., et al. (2014). Afforestation in China cools local land surface temperature. *Proceedings of the National Academy of Sciences, 111*(8), 2915–2919.

Prata, A. J., Caselles, V., Coll, C., Sobrino, J. A., & Ottle, C. (1995). Thermal remote sensing of land surface temperature from satellites: Current status and future prospects. *Remote Sensing Reviews, 12*(3–4), 175–224.

Qin, Z., Karnieli, A., & Berliner, P. (2001). A mono-window algorithm for retrieving land surface temperature from Landsat TM data and its application to the Israel-Egypt border region. *International Journal of Remote Sensing, 22*(18), 3719–3746.

Ramzan, M., Saqib, Z. A., Hussain, E., Khan, J. A., Nazir, A., Dasti, M. Y. S., & Niazi, N. K. (2022). Remote sensing-based prediction of temporal changes in land surface temperature and land use-land cover (LULC) in urban environments. *Land, 11*(9), 1610.

Roy, P. S., Meiyappan, P., Joshi, P. K., Kale, M. P., Srivastav, V. K., Srivasatava, S. K., Behera, M. D., Roy, A., Sharma, Y., Ramachandran, R. M., Bhavani, P., Jain, A. K., & Krishnamurthy, Y. V. N. (2016). Decadal land use and land cover classifications across India, 1985, 1995, 2005. *ORNL DAAC.*

Saha, J., Ria, S. S., Sultana, J., Shima, U. A., Seyam, M. M. H., & Rahman, M. M. (2024). Assessing seasonal dynamics of land surface temperature (LST) and land use land cover (LULC) in Bhairab, Kishoreganj, Bangladesh: A geospatial analysis from 2008 to 2023. *Case Studies in Chemical and Environmental Engineering, 9*, 100560.

Shen, G., Ibrahim, A. N., Wang, Z., Ma, C., & Gong, J. (2015). Spatial–temporal land- use/land-cover dynamics and their impacts on surface temperature in Chongming Island of Shanghai, China. *International Journal of Remote Sensing, 36*(15), 4037–4053.

Srivastava, A., Kumar, S. N., & Aggarwal, P. K. (2010). Assessment on vulnerability of sorghum to climate change in India. Agriculture, ecosystems & environment, 138(3-4), 160-169.

Sudhira, H. S., Ramachandra, T. V., & Jagadish, K. S. (2003). Urban sprawl pattern recognition and modeling using GIS. *Map India, 2003*, 28–31.

Weng, Q. (2001). A remote sensing? GIS evaluation of urban expansion and its impact on surface temperature in the Zhujiang Delta, China. *International Journal of Remote Sensing, 22*(10), 1999–2014.

Wang, W., Zhang, C., Allen, J. M., Li, W., Boyer, M. A., Segerson, K., & Silander Jr, J. A. (2016). Analysis and prediction of land use changes related to invasive species and major driving forces in the state of Connecticut. Land, 5(3), 25.

Weng, Q., Lu, D., & Schubring, J. (2004). Estimation of land surface temperature–vegetation abundance relationship for urban heat Island studies. *Remote Sensing of Environment, 89*(4), 467–483.

Zhang, F., Tiyip, T., Kung, H., Johnson, V. C., Maimaitiyiming, M., Zhou, M., & Wang, J. (2016). Dynamics of land surface temperature (LST) in response to land use and land cover (LULC) changes in the Weigan and Kuqa river oasis, Xinjiang, China. *Arabian Journal of Geosciences, 9*(7), 499.

Assessment of Growth and Development of Ayodhya Municipal Corporation Area Using Geospatial Approach

Sadaf

1 Introduction

Today, one of the primary consequences of human activity around the world is urbanisation. Thus, urbanisation has a major impact on both the sustainability of urban growth and the quality of urban living. Urbanisation is a process of socioeconomic transformation either due to rapid rural–urban migration or natural growth of population (Velmurugan & Sajjad, 2009; Nourqolipour et al., 2016; Jat et al., 2008). Since over half of the global population currently lives in urban areas, urban planners, urban geographers and academicians in developing countries frequently discuss the use of geospatial technologies to monitor and analyse urban expansion (Bose & Chowdhury, 2020; Kang et al., 2020; Henderson & Turner, 2020; Roy et al., 2021). The population of cities in developing nations is growing at an uncontrollable rate, and the land use of these areas is changing quickly from rural to urban. The past few decades have seen a significant increase in land use as well as land cover change areas due to quick urban expansion, which makes it more difficult to accomplish sustainable urban development. Generally, population in cities and towns is considered the important factor for an uncontrolled and unorganised urban growth rate. As the urban population increases, the need for infrastructural facilities such as housing, transportation, safe drinking water and schools also increases, and this mostly leads to a new phenomenon, i.e., urban expansion (Sumari et al., 2017; Fenta et al., 2017; Ujoh et al., 2019). So, cities are expanding without any proper planning and inadequate provision of basic services, resulting in unsystematic, unplanned and uncontrolled urban growth (Grimm et al., 2008; Cohen, 2006). So, the land beyond the municipal boundary of the cities loses its rural characteristics and transforms into non-agricultural land due to urban growth. Rapid population growth demands for residential areas beyond the city limits require urban expansion (Cavailhès & Wavresky, 2003), and this process leads to the shrinkage of agricultural lands around urban centres (Livanis et al., 2006). So, rural–urban land use transformation poses unprecedented challenges in terms of hunger, food insecurity, malnutrition and agricultural production. There could be pressure on vital services including waste management, water supply and transportation due to the built-up area's rapid expansion (Ahmad et al., 2024).

Nowadays, many kinds of models and theories for assessing urban growth have been developed due to advances in remote sensing and geographic information systems (GIS) technology, which have resulted in their widespread use for mapping changes over the years (Al-Shalabi et al., 2013; Awasthi et al., 2011; Haack & Rafter,

Sadaf (✉)
PG Department of Geography, Veer Kunwar Singh University, Ara, Bihar, India

Z. Khanam et al. (eds.), *Climate Change, Resilience and Ecosystem Services*, Sustainable Development Goals Series, https://doi.org/10.1007/978-3-031-96977-5_9

2006). Therefore, geospatial technologies are very useful for governments, scholars and urban planners to analyse the pattern of growth in urban areas in any given urban centre (Sumari et al., 2019; Huang et al., 2018). Using GIS and remote sensing techniques, numerous studies have analysed information on urban expansion patterns and their effects on environmental conditions. Remote sensing data and GIS techniques are very useful to analyse the land use pattern, urban centres' growth and direction, deforestation, agricultural production, alterations to the ecosystem and cultivation (Ligate et al., 2018; Fu et al., 2013).

With the use of geospatial technology, a few studies have examined the expansion of built-up areas and their correlation with population growth. Farooq and Ahmad (2008) determined the size, rate and pattern of expansion of the Aligarh City area using data from Landsat, IRS and Quick Bird. The urban area grew by nearly three times between 1971 and 2006, according to their analysis. Up to the 1980s, the pace of land consumption for urban uses was largely moderate; nevertheless, in the 1990s, there was a notable increase in land consumption relative to population growth. Jamal and Ali (2022) examined the spatiotemporal analysis of land use and land cover patterns in Kolkata Municipal Corporation with the help of geospatial technology. The city saw a rapid built-up expansion from 1990 to 2020 as a result of the city's growing population and economic development. Regarding Ayodhya Municipal Corporation, no such study has been conducted.

Therefore, it is critical to comprehend the spatiotemporal pattern of urban expansion utilising geospatial technologies. This research investigates the spatiotemporal urban growth development of Ayodhya Municipal Corporation, focusing on built-up area growth patterns and processes over a 20-year period from 1955 to 1975, 1975 to 1995, 1995 to 2015 and 2015 to 2024. The present research examines how the city's built-up area has grown over an extended period of time. The study also highlights the association between Ayodhya Municipal Corporation's urban population and its built-up areas'growth, providing valuable insights for decision-making and better city management.

2 Database and Research Methodology

This research paper is based on satellite imagery, topographic maps, Google Earth imagery and municipal maps. A topographic map (NG 44-7, Series U502) prepared by the Army Map Service Survey was used for identifying the extent of built-up area in 1955 (Table 1). Topographic map sheet number 63 J/I 1975 (surveyed 1971, published 1975) was used for mapping the extent of built-up area for 1975. These topographic maps were georeferenced and digitised by using ArcGIS 10.7. Landsat 5 Thematic Mapper data (path 143, rows 41 and 42) acquired on 20/03/1995 from Earth Explorer-UUSGS were used to map the extent of built-up area for 1995.

Table 1 Secondary sources of the data of the research methodology

Spatial data	Data	Sources of data
Satellite image	Landsat 5 Thematic Mapper (TM)	USGS Earth Explorer Path: 143 and rows 41 and 42
Topographic sheet of 1955	Toposheet number (NG 44-7, Series U502) (scale 1:250,000)	Prepared by the Army Map Service Survey
Topographic sheet of 1975	Toposheet number 63 J/I 1975 (Surveyed 1971, published 1975) (scale 1: 50,000)	Survey of India
Google Earth Imagery	KML file	Google Earth Pro
Administrative data	Municipal corporation map	Master plan report prepared by Ayodhya Development Authority
Non-spatial data	Calculation of projected urban population of 1955, 1975, 1995, 2015 and 2024 based on the population 1951, 1071,1991, 2011 Census of India)	

ERDAS 9.2 software was used for layer stacking, rectification, image mosaicking and georeferencing. The land use classes were interpreted and delineated using visual interpretation methods of satellite data. For fulfilment of the objectives, only built-up area has been identified rather than other land use classifications. Roads, roads, railways and built-up areas were highlighted using Google Earth imagery for the years 2015 and 2024. For the conversion of the KML file into a polygon, ArcGIS 10.7 software was used. The municipal boundary map of Ayodhya Municipal Corporation for 2019 was obtained from a master plan report prepared by the Ayodhya Development Authority. Non-spatial data such as demographic details for 1951, 1961, 1971, 1981, and 1991 were obtained from the Census of India (1951,1971,1991, and 2011), Directory of Census Operations, Uttar Pradesh, Government of India. Projected population was used for analysing the growth rate of population for 1955, 1975, 1995, 2015 and 2024. The population of 1951, 1971 and 2011 was taken as a base year for calculating the projected urban population of 1955, 1975, 2015 and 2024, respectively.

The limitation of this study is that the built-up area has been identified from different sources because the built-up growth has been taken from 1955 and the built-up area that has been identified from satellite imagery has not been assessed for accuracy, which could be a limitation of this study. Google Earth imagery has been used for both the years 2024 and 2015 because in this study only the built-up area was to be analysed.

3 Research Methodology

3.1 Built-up Area Growth

The overlapping process was used for identifying the built-up area growth of Ayodhya Municipal Corporation. Municipal area with no development and area outside the limit has been calculated with the help of cut polygon and calculated geometry with the help of ArcGIS 10.7 software.

3.2 Regression Analysis

In this study, the association between the independent variable (urban population) and the dependent variable (built-up area) was determined using the linear regression method. Regression analysis has been used to determine the relationship between the projected urban population and builtup area between 1955 and 2024. Regression analysis is expressed by the following formula:

$$Y = a + b_1 X_1 + b_2 + X_2 + b_3 \quad (1)$$

where

$b1$ represents the slope (beta coefficient) for $X1$, a (alpha) is the constant or intercept and Y represents the dependent variable (built-up area) value, what is being predicted or explained,

$X1$ represents first independent variable (urban population) that takes into account Y's variance The slope (beta coefficient) for $X2$ is denoted by $b2$.

$X2$ is the second independent variable that accounts for the variation in Y For $X3$, and $b3$ represents the slope (beta coefficient).

$X3$ is the third independent variable describing the variance in Y.

3.3 Calculation of Projected Population

The population projection of Ayodhya Municipal Corporation for 1955, 1975, 1995, 2015 and 2024 based on the population of 1951, 1971,1991 and 2011 has been calculated by the arithmetic method by using this formula:

$$A = P(1 + r/100)^n \quad (2)$$

where n = number of years, r = annual growth rate, P = present population and A = projected population (Fig. 1).

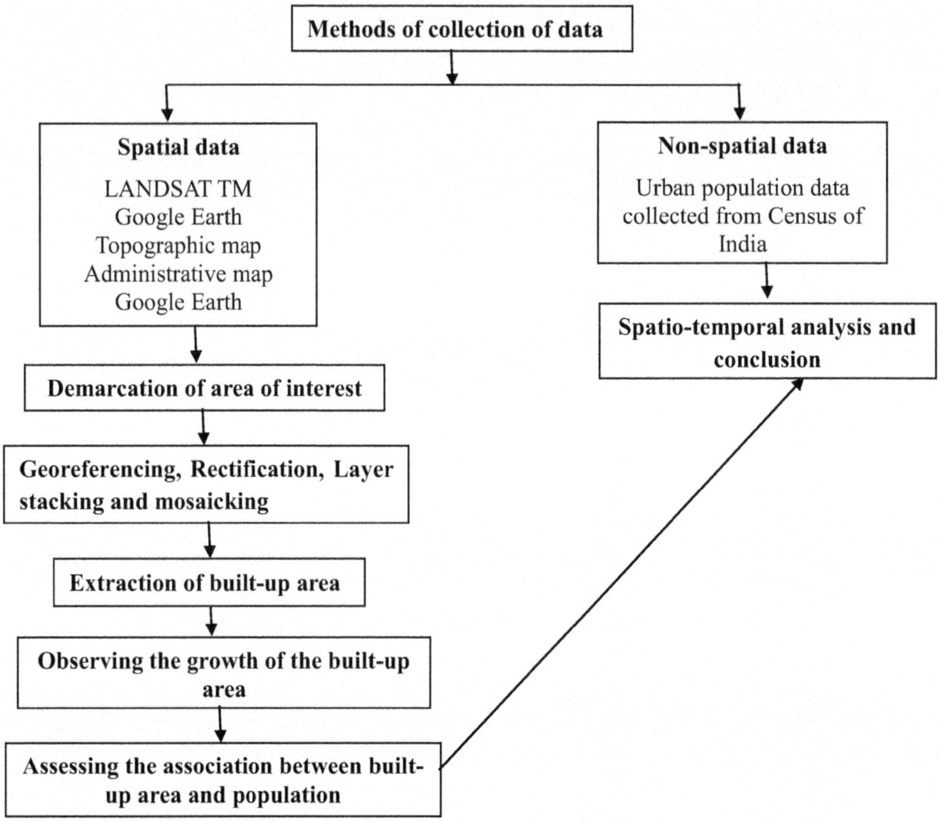

Fig. 1 Schematic diagram illustrating the research methodology

3.4 Study Area

The study area (Fig. 2) is situated at latitude 26°47′59.08″N and longitude 82°12′18.59″E in the Ayodhya District in eastern Uttar Pradesh (Ayodhya Development Authority, 2022), on the south bank of the Ghaghara River, commonly known as Saryu. The cities of Lucknow, Kanpur, Gorakhpur and Varanasi are all easily accessible from Ayodhya via road and rail. The city of Ayodhya can be accessed by National Highway 27, which links Nepal with New Delhi. By acquiring 600 acres of land, the city was able to establish an international airport that connects it to other regions both domestically and outside. The city is well-known as a significant Hindu pilgrimage site in India. Earlier, both Ayodhya City and Faizabad City became separate municipalities on 7 January 1978, and both cities started to provide their services to citizens.

The Government of Uttar Pradesh issued an order to merge again the Ayodhya Nagar Palika with Nagar Palika Faizabad, subsequently, in May 2017, the twin cities of Ayodhya and Faizabad reunited to establish the Ayodhya Municipal Corporation. Name of the Faizabad city and district was changed to Ayodhya in November 2018. For this reason, Ayodhya has been referred to as Ayodhya and Faizabad as Faizabad (Ayodhya) in this study. Prior to 17 December 2019, the Ayodhya Municipal Corporation only had nine villages inside its municipal boundaries; however, after that day, 56 villages were added to the municipal corporation limit (Ayodhya Nagar Nigam, 2024). Ayodhya Cantonment is a cantonment town near the city of Ayodhya in the Ayodhya District of Uttar Pradesh, India. It was founded in 1858 and is currently a Class I cantonment. The Government of India changed the official name from Faizabad

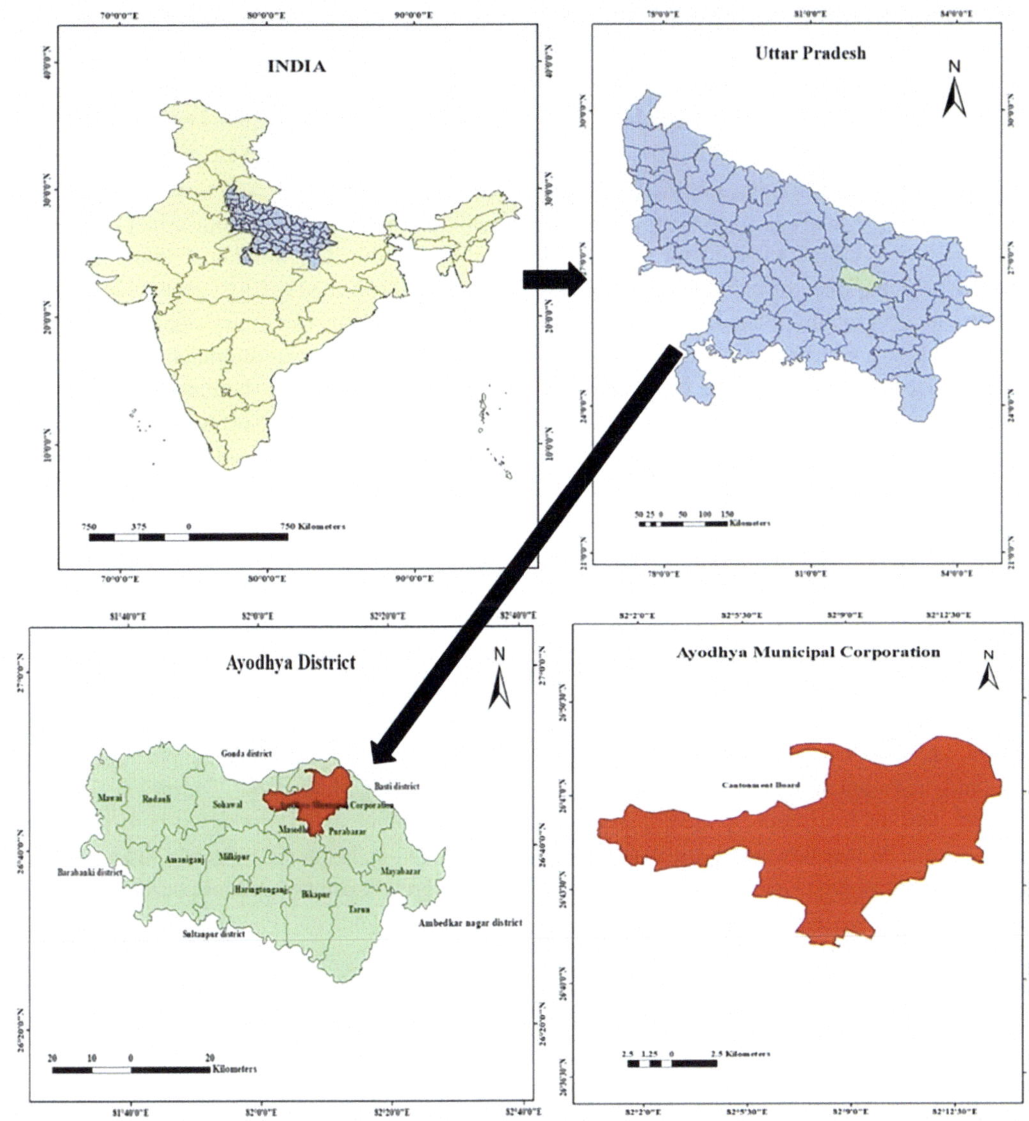

Fig. 2 Ayodhya Municipal Corporation (study area). (Source: Census of India, 2011; Ayodhya Municipal Corporation, 2024)

Cantonment to Ayodhya Cantonment in October 2022 (The Free Press Journal, 2022).

There are 2,21,118 people living in Ayodhya Municipal Corporation as of the 2011 Census. The city is well-known for being a great religious and historical centre that attracts visitors from all over the nation and the world. The city is expanding quickly compared to prior decades because of migration and population increase. The city's southern portion indicates the possible future growth pattern of the municipal corporation. Due to the Saryu River, which runs through the northern part of the city, the northern sector of the municipal corporation is unable to expand.

4 Result and Discussion

4.1 Built-up Area Growth: A Spatiotemporal Analysis

The dynamics of any city are identified through the spatiotemporal analysis of the growth of the built-up area. It is crucial to detect the spatial and temporal patterns, rates of change and trends of built-up area growth because it helps to understand how any urban area grows and develops in diverse socioeconomic and environmental conditions.

The 1955 built-up area mapping formed the basis for spatiotemporal assessment of built-up area growth in this study. Mainly because this study focuses on the long-term growth of the city's built-up area. Since data on built-up area was available from 1955 to 2024, it is possible to determine how the built-up area has evolved after 1955, and it is essential to understand about any urban centre. Figure 3 illustrates that the total built-up area of Faizabad (Ayodhya) City is 5.29 km^2., the cantonment board covered only 2.46 km^2. and the total built-up area of Ayodhya City is 4 km^2. At that time, only the core part of the cantonment board, Faizabad (Ayodhya) and Ayodhya City is visible on the map. It is very clear that the cantonment board, Ayodhya and Faizabad (Ayodhya) were separate urban centres and their boundaries are detached from each other's. These urban centres represent a very small built-up area, and even they have not touched their municipal boundary. Any kind of built-up development outside the municipal boundary is not available. Faizabad (Ayodhya) City had a larger built-up area in comparison to Ayodhya City and cantonment board. All three urban centres were separated from each other and confined within the municipal boundary in 1955.

Figure 4 shows that the built-up area of Ayodhya Municipal Corporation was 18 km^2 in 1975. Ayodhya (Faizabad) and Cantonment Board are attached to each other during the span of 20 years. It means that both the cities must have grown towards each other; hence, both appear to be one. We are not able to find out the clear-cut demarcation between the cantonment board and Ayodhya (Faizabad) City. Both twin cities, i.e., Ayodhya (Faizabad) and Ayodhya, appear connected to each other through a narrow strip of built-up area along the National Highway 27. Built-up area growth of Ayodhya City and Cantonment Board is not visible beyond the municipal boundary, and both urban centres are confined within the municipal boundary. Only Ayodhya (Faizabad) City grew beyond the municipal boundary in a very small extent during 1955–1975. Linear development can be seen beyond the municipal boundary of Faizabad (Ayodhya) City only. So, Ayodhya Municipal Corporation is characterised by very low development of built-up area outside the municipal boundary in 1975.

Figure 5 shows the growth pattern of the built-up area of Ayodhya Municipal Corporation in 1995, and the total built-up area of Ayodhya Municipal Corporation is 27.17 km^2 at this time. A linear pattern of built-up area growth started along the major roads such as Lucknow Road, Raebareli Road, Prayagraj Road and Sultanpur Road outside the municipal boundary. During 1975–1995, Ayodhya Municipal Corporation built-up growth was following the previous decade pattern of built-up area growth pattern. Because during this period also only Ayodhya (Faizabad) City expanded beyond the municipal boundary, while Cantonment Board and Ayodhya City were confined within the municipal boundary. A low level of urban growth development can be observed in 1995.

Figure 6 shows the built-up area of Ayodhya Municipal Corporation was 42 km^2 in 2015. Based on an analysis of Fig. 6, it is noticeable that Ayodhya (Faizabad) City has experienced a rapid growth of its built-up area that has taken place in a south and south-west direction along the transport routes such as Lucknow, Prayagraj, Sultanpur, Raebareli and Varanasi City. Slight growth of built-up area around the Ayodhya City can be seen in 2015. While the cantonment board remained unaffected without any built-up activity, it is confined within the boundary. The pattern of built-up area expansion is transformed from low to fast built-up area expansion around Ayodhya (Faizabad) in 2015. The northern part of Ayodhya Municipal Corporation remained

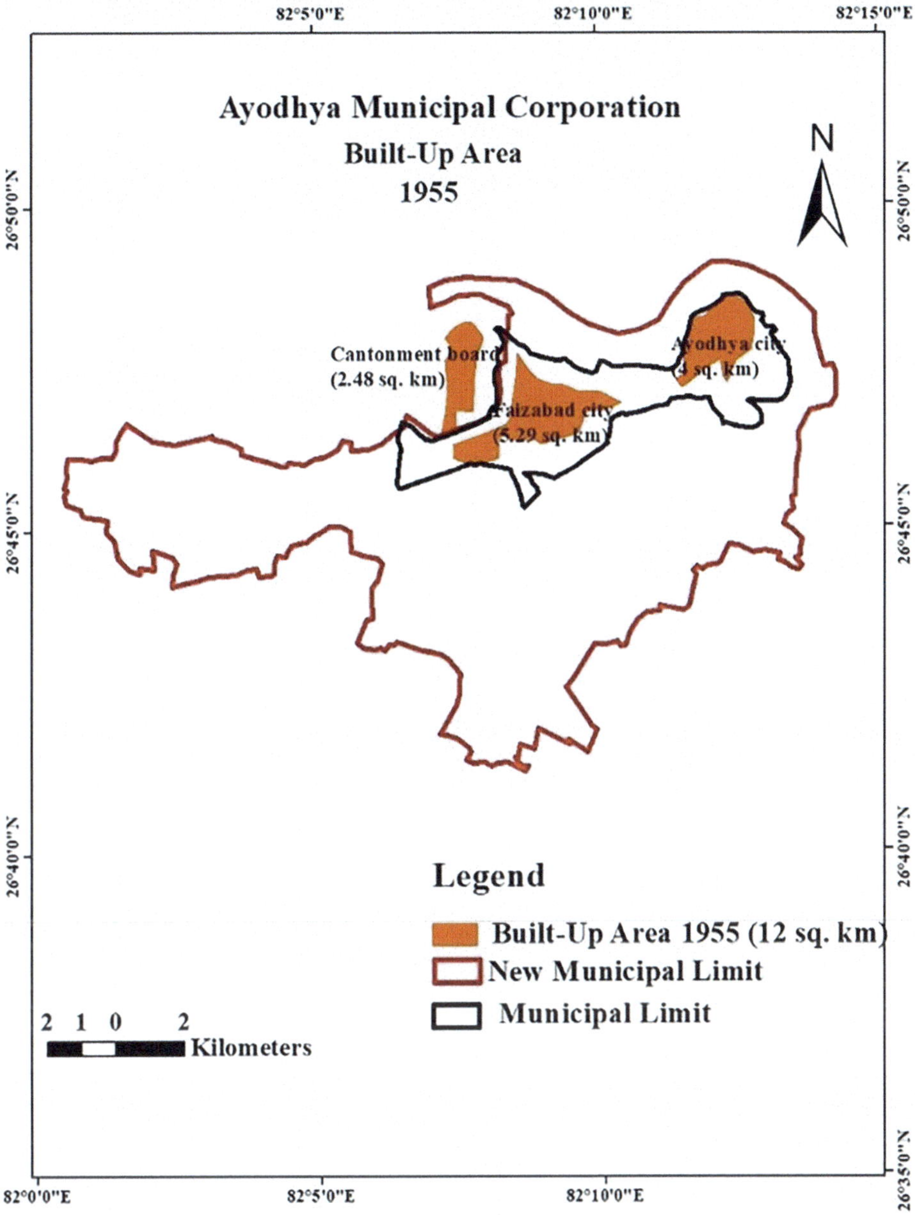

Fig. 3 Built-up area of Ayodhya Municipal Corporation, 1955. (Source: Toposheet NG 44-7, Series U502)

unchanged without built-up area growth because the Saryu River is flowing, which constrains built-up area expansion in this direction.

Figure 7 clearly gives an overview of built-up area expansion of Ayodhya Municipal Corporation in 2024. Total built-up area of Ayodhya Municipal Corporation is 76.30 km^2 in 2024. Based on analysis, it can be shown that the Ayodhya Municipal Corporation area witnessed rapid growth in built-up area between 2015 and 2024. The cantonment board area has remained unchanged, but the other two urban centres, Ayodhya (Faizabad) City and Ayodhya City, are growing very fast and have significantly expanded beyond the municipal boundary.

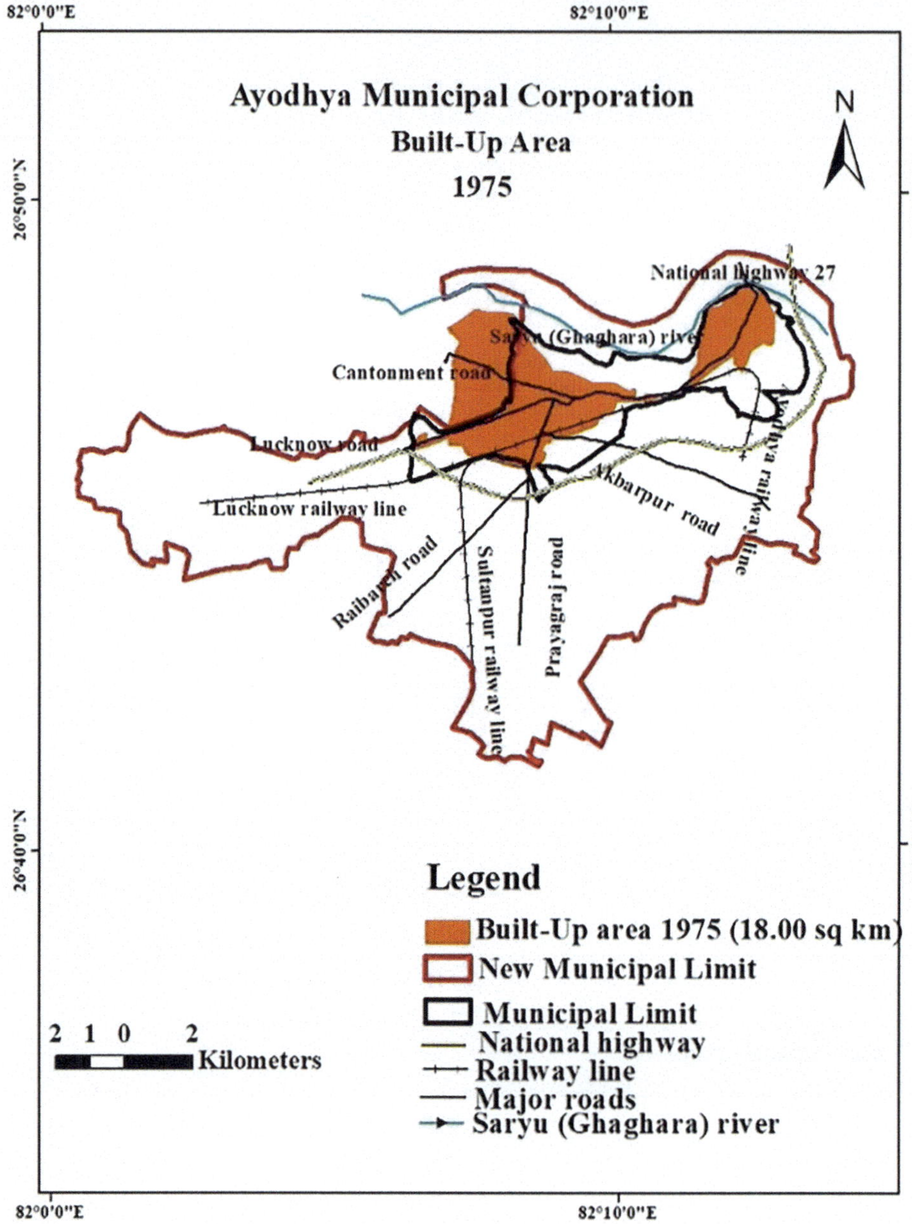

Fig. 4 Built-up area of Ayodhya Municipal Corporation, 1975. (Source: Toposheet 63 J/I)

Figure 8 illustrates that the built-up area is not growing much in the north or north-west direction. It is a known fact that Ayodhya Municipal Corporation is situated on the Saryu River's bank, which prevents further built-up area growth from expanding. Therefore, it is not feasible for the city to grow in this direction. The cantonment board acts as a sociocultural barrier in a north-west direction, which constrains the built-up area expansion.

Study shows that linear urban growth development can be seen along the transport routes. The growth of the built-up area is mainly seen along routes like Ayodhya-Lucknow, Ayodhya-

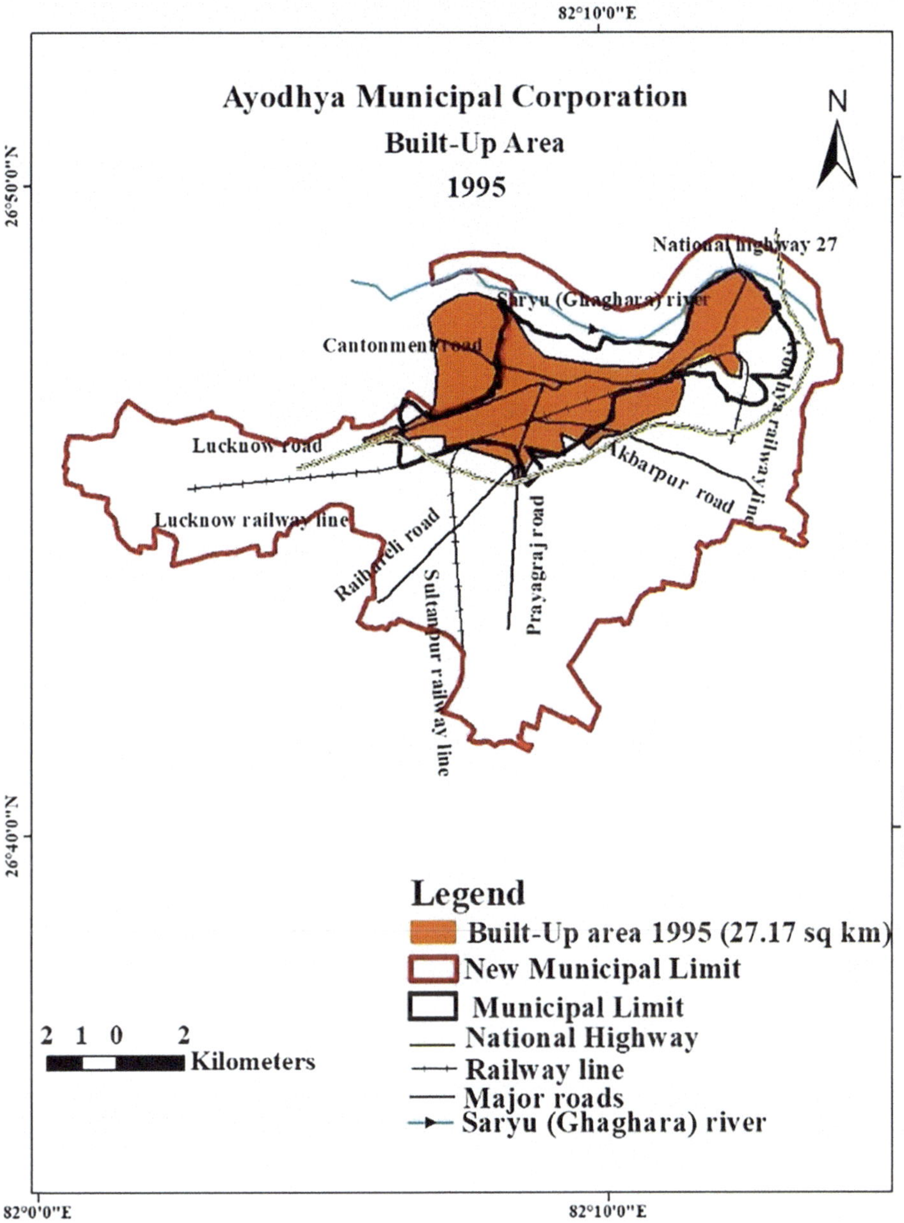

Fig. 5 Built-up area of Ayodhya Municipal Corporation, 1995. (Source: LANDSAT 5 TM)

Raebareli, Ayodhya-Prayagraj and Ayodhya-Azamgarh. Ambedkar Nagar Road, Sultanpur Road and Lucknow's commercial, residential and industrial development are the reasons for the quicker rate of development along these routes. On the basis of all these maps, it is clearly visible that Ayodhya Municipal Corporation is expanding quickly, and possible future growth of the city would be further in the southern direction of Ayodhya Municipal Corporation.

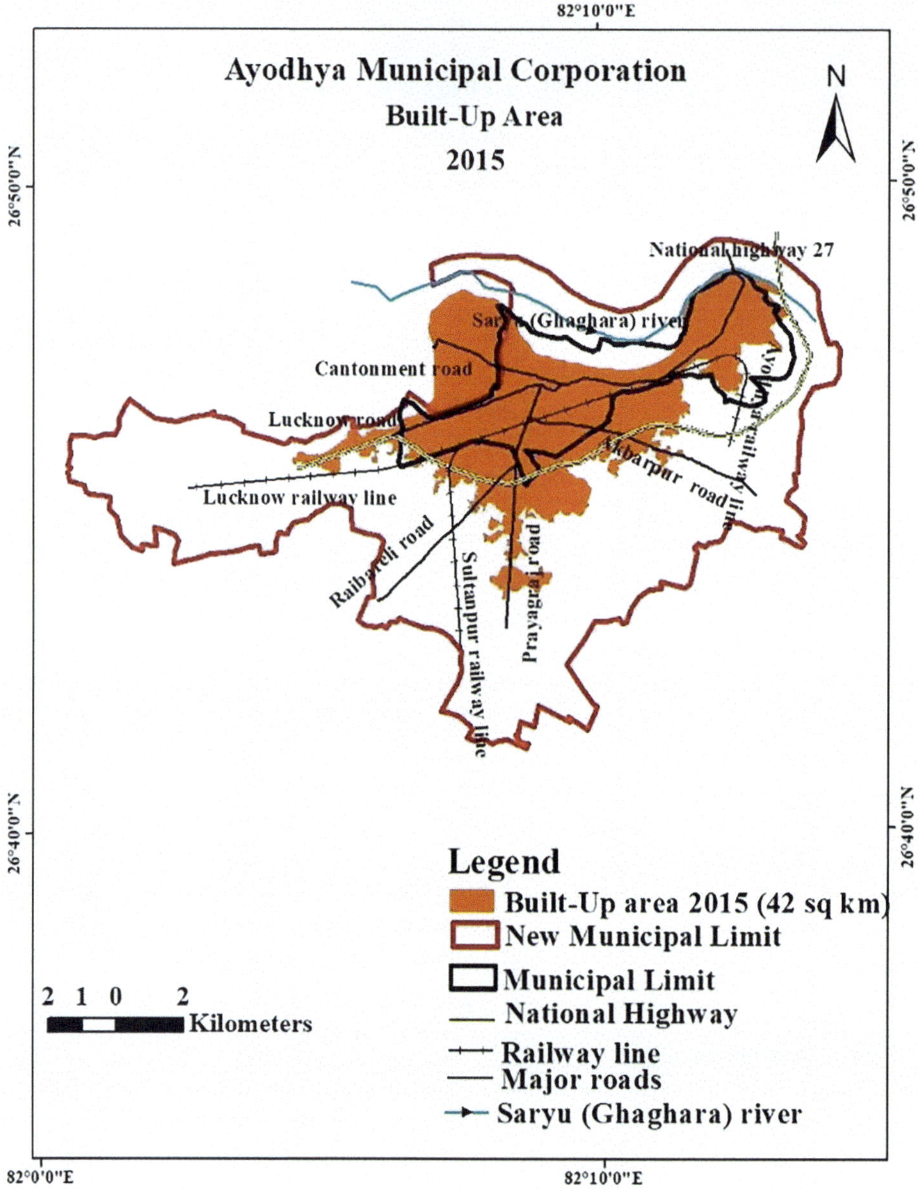

Fig. 6 Built-up area of Ayodhya Municipal Corporation, 2015. (Source: Google Earth Imagery)

5 Built-up Area Growth Metrices

Table 2 depicts the built-up area metrics and total projected population of the Ayodhya Municipal Corporation. The built-up area of Ayodhya Municipal Corporation was 12 km², with 12,4987 people living there as of 1955. During the period between 1955 and 1975, the Ayodhya Municipal Corporation's total built-up area grew from 12 square km to 18 square km, and its total population increased from 94,987 to 119,987. The total built-up area grew to 6 km² from 12 km² to 18 km² in two decades during 1955–1975. It is clearly observed that the Ayodhya Municipal Corporation recorded 50% built-up area growth during 1955–1975. The net built-up area increased by

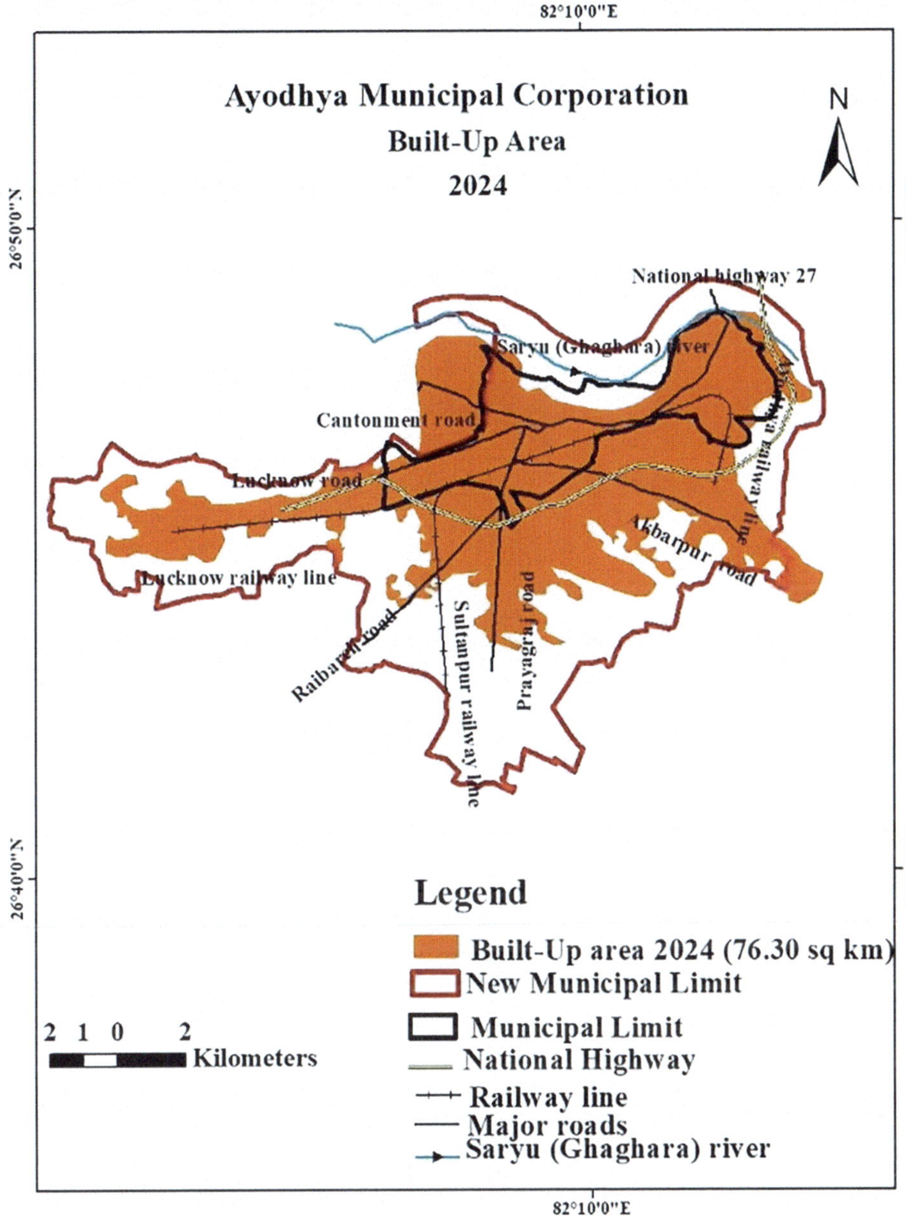

Fig. 7 Built-up area of Ayodhya Municipal Corporation, 2024. (Source: Google Earth Imagery)

8.72% from 18 km^2 (1975) to 27.17 km^2 (1995). During 1975 and 1995, the percentage of built-up area growth grew significantly from 50% to 50.94%. The total projected population of Ayodhya Municipal Corporation is rapidly increased from 1,19,987 to 1,92,386 during 1975–1995. The total projected urban population has increased by 2,78,792 persons from 1,92,386 during 1995–2015. The built-up area grew by 42% from 27.17 km^2 to 42 km^2 during 1995–2005. The highest growth rate of built-up area (81.66%) was observed during 2015–2024. The total projected urban population has increased from 2,78,792 to 3,62,000 during this period.

The total municipal area of Ayodhya Municipal Corporation is 25 km^2 till 2019 (Table 3). Now,

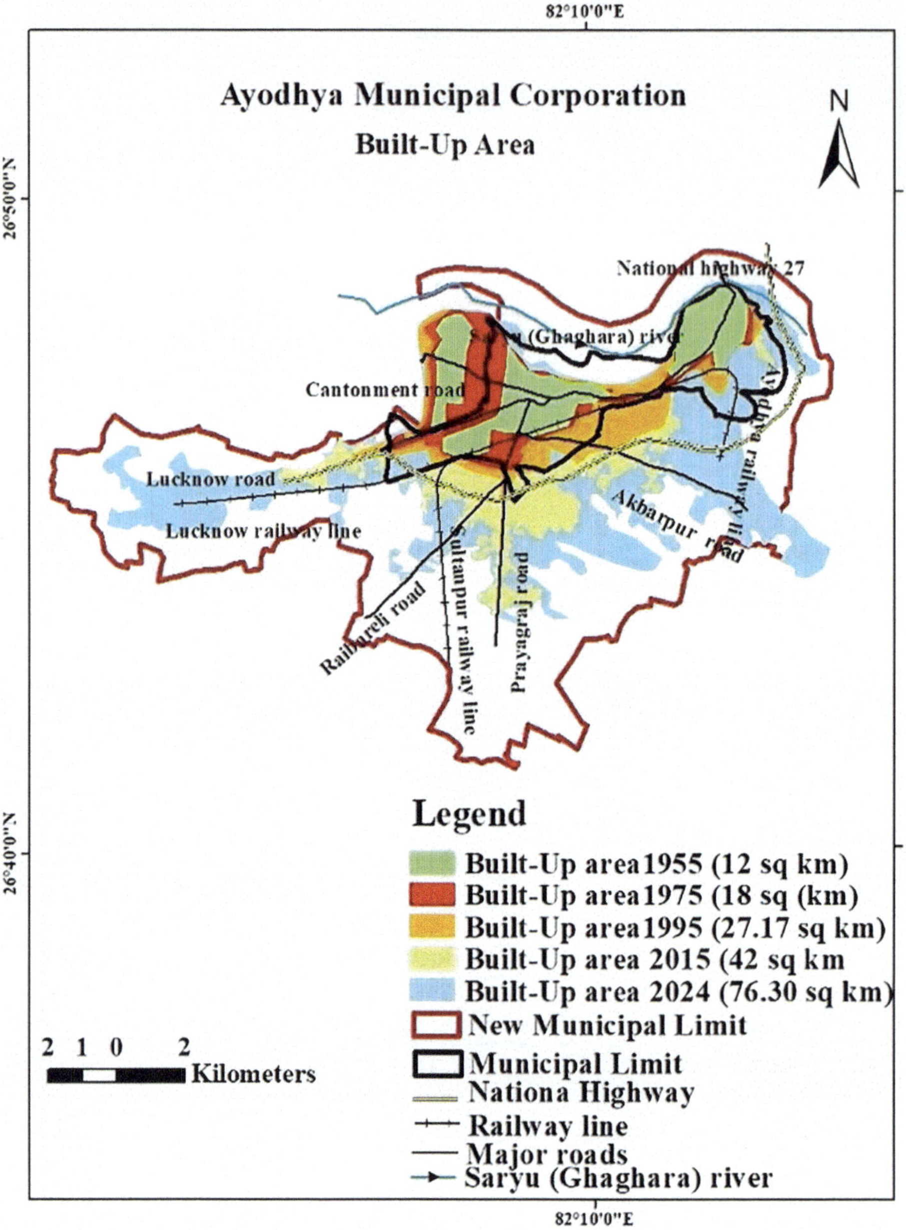

Fig. 8 Built-up area growth of Ayodhya Municipal Corporation, 1955–2024. (Source: Toposheet NG 44-7, Series U502, Toposheet 63 J/I, LANDSAT 5TM, Google Earth Imagery)

the total municipal area of Ayodhya Municipal Corporation is 142.18 km^2. Development of built-up area outside the municipal limits was totally absent in 1955. There was only 5.30 km^2 of built-up area observed beyond the municipal limit in 1975, and it has increased from 5.30 km^2 to 8.92 km^2 during 1975–1995. The built-up area beyond the municipal limit has rapidly increased from 8.92 km^2 to 21 km^2 during 1995–2015. Municipal area with no development has decreased from 13.00 km^2 to 4.00 km^2 during 1955–2015. Further, built-up area beyond the municipal limit is only 7.38 km^2 in 2024 due to a change in the municipal boundary limit.

Table 2 Built-up area metrices and population of Ayodhya Municipal Corporation

Year	Urban population (projected)	Built-up area (km^2)	Built-up area	
			Growth from previous decades (km^2)	Growth in percentage
1955	94,987	12.00	–	–
1975	1,19,987	18.00	6.00	50.00
1995	1,92,386	27.17	8.72	50.94
2015	2,78,792	42.00	14.83	54.58
2024	3,62,000	76.30	34.30	81.66

Source: Computed by researcher, based on Topographic map (NG 44-7, Series U502), Toposheet 63 J/I, LANDSAT 5 TM, Google Earth imagery and projected population for 1955, 1975, 1995, 2015 and 2024 based on Census of India 1951, 1971, 1991, 2011

Table 3 Relationship between built-up area and municipal limit of Ayodhya Municipal Corporation (km^2 area)

Year	Municipal limit area	Built-up area	Area beyond the municipal boundary	Municipal corporation area with no development
1955	25.00	12.00	0.00	13.00
1975	25.00	18.00	5.30	12.70
1995	25.00	27.17	8.92	6.75
2015	25.00	42.00	21.00	4.00
2024	142.18	76.30	7.38	68.92

Source: Computed by researcher, based on topographic map (NG 44-7, Series U502), Ayodhya Municipal Corporation, Toposheet 63 J/I, LANDSAT 5 TM, Google Earth imagery, 2015 and 2024

Table 4 Population and built-up area growth in Ayodhya Municipal Corporation

Year	Population		Growth of built-up area	
	Net growth	Percentage	Net growth (km^2)	Percentage
1955–1975	25,000	26.31	6.00	50.00
1975–1995	72,399	60.33	8.72	50.94
1995–2015	86,406	44.91	14.83	54.58
2015–2024	83,208	29.84	34.30	81.66

Source: Computed by researcher, based on Topographic map (NG 44-7, Series U502), Toposheet 63 J/I, LANDSAT 5 TM, Google Earth imagery and projected population for 1955, 1975, 1995, 2015 and 2024 based on Census of India 1951, 1971, 1991, 2011

Table 4 reveals that the net population growth of Ayodhya Municipal Corporation has been recorded as 25,000 during 1955–1975 and as 72,399 during 1975–1995. The net growth of urban population of Ayodhya Municipal Corporation is recorded at 86,406 during 1995–2015. The net urban population growth improved from 1955 to 2015, while the percentage of urban population growth declined during the same period. It is clearly observed that the net population growth declined from 86,406 to 83,208 during 2015–2024, while the net built-up area has improved enormously from 14.83 km^2 to 34.30 km^2 during 2015–2024 due to rapid changes in land-use patterns in surrounding areas. So, the main environmental effects of built-up growth include increased air and noise pollution, difficulties managing solid waste and the loss of green spaces. The results also showed that the expansion of built-up areas significantly consumes agricultural land and open-space water bodies. The Ayodhya Municipal Corporation will face challenges in fulfilling the needs of its expanding urban population in terms of housing and basic utilities like electricity, sewerage and water.

6 Relationship Between Urban Population and Built-up Area

The association between built-up area growth and population has been examined using regression analysis. Both scatterplots and linear regression have been used to analyse the association between population growth and the growth of built-up area.

Regression analysis findings indicate an upward trend (Fig. 9) with a coefficient of determination of (R2 = 0.9408) and a strong positive association (94.08%) between built-up area growth and population. This suggests that the main factor contributing to the study's rapid built-up expansion is population growth and how the population growth is causing Ayodhya Municipal Corporation to extend beyond its boundary at a faster rate. The population growth resulted in a rapid increase of built-up areas to

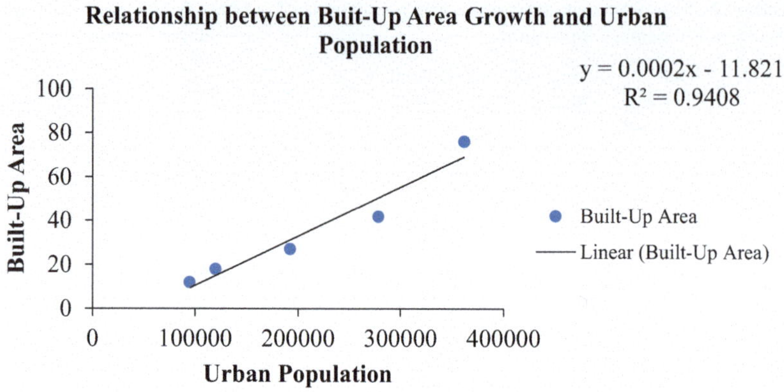

Fig. 9 Relationship between built-up area growth and urban population. (Source: Topographic map (NG 44-7, Series U502), Toposheet 63 J/I, LANDSAT 5 TM, Google Earth Imagery and projected population for 1955, 1975, 1995, 2015 and 2024 based on Census of India 1951, 1971, 1991 and 2011)

accommodate residential, administrative and commercial facilities as well as employment prospects.

7 Conclusion

The study analysed the spatiotemporal analysis of the growth of Ayodhya Municipal Corporation's built-up area, which has experienced rapid urban expansion from 1995 to 2024 due to rapid population growth and economic progress with the help of statistical techniques and geospatial technology. The analysis revealed that built-up area has been expanded rapidly from 12 km^2 in 1955 to 76.30 km^2 in 2024. Mainly built-up expansions have taken place in a south and south-west direction along the transport routes. Furthermore, the study reveals a significant and positive relationship (94.08%) with a coefficient of determination of (R2 = 0.948) between the total population and the built-up area growth during 1955 and 2024. In addition to being a centre for administration and education, the city is known for having a rich spiritual heritage, which attracts the people of surrounding areas. The study concludes that the main cause of the study area's rapid built-up area growth is population growth (resulting from migration and natural growth). Public utility services, waste disposal management and infrastructure could all be impacted by the built-up area's rapid expansion. In order to ensure Ayodhya Municipal Corporation's sustainable development, these issues must be resolved. Inclusive urban planning is necessary to ensure sustainable urban growth. Regular monitoring of built-up area growth of Ayodhya Municipal Corporation may help for future initiatives. Zoning regulations and extensive land use planning are necessary for ensuring sustainable urban growth. To prevent urban expansion and restrict the conversion of agricultural land, efforts should be undertaken to support compact and mixed land use development patterns. The study's conclusions will be helpful to the town planners, government and administrators for identifying the trend and direction of built-up area growth of the Ayodhya Municipal Corporation for making a sustainable city. Maps and diagrams of study will help town planners, administrative authorities and academicians for future planning of the city.

References

Ahmad, W. S., et al. (2024). Spatiotemporal analysis of urban expansion in Srinagar city, Kashmir. *Discover Cities, 1*(8). https://doi.org/10.1007/s44327-024-00009-3

Al-shalabi, M., Billa, L., Pradhan, B., Mansor, S., & Al-Sharif, A. A. (2013). Modelling urban growth evolution and land-use changes using GIS based cellular automata and SLEUTH models: The case of Sana'a metropolitan city, Yemen. *Environmental Earth Sciences, 70*(1), 425–437.

Awasthi, A., Chauhan, S. S., & Goyal, S. K. (2011). A multi-criteria decision-making approach for location planning for urban distribution centers under uncertainty. *Mathematical and Computer Modelling, 53*(1–2), 98–109.

Ayodhya Development Authority (2022). *Ayodhya Master Plan Draft, 2023.* https://www.ayodhyalive.com/wp-content/uploads/2022/12/AyodhyaMasterPlan2031v1-1.pdf

Ayodhya Municipal Corporation. (2024). *Map of Ayodhya Nagar Nigam*. https://www.nagarnigamayodhya.in/Map (Accessed on December, 2023).

Ayodhya Nagar Nigam. (2024). *About Nagar Nigam Ayodhya*. https://www.nagarnigamayodhya.in/about (Accessed on January 2024).

Bose, A., & Chowdhury, I. R. (2020). Monitoring and modelling of spatio-temporal urban expansion and land-use/land cover change using Markov chain model: A case study in Siliguri metropolitan area, West Bengal, India. *Modelling Earth Systems and Environment, 6*(4), 2235–2249. https://doi.org/10.1007/s40808-020-00842-6

Cavailhès, J., & Wavresky, P. (2003). Urban influences on peri urban farmland prices. *European Review of Agricultural Economics, 30*(3), 333–357.

Census of India. (1951,1971,1991, and 2011). Directory of census operations, Uttar Pradesh. Government of India.

Cohen, B. (2006). Urbanization in developing countries: Current trends, future projections, and key challenges for sustainability. *Technology in Society, 28*(1–2), 63–80.

Farooq, S., & Ahmad, S. (2008). Urban sprawl development around Aligarh City: A study aided by satellite remote sensing and GIS. *Journal of the Indian Society of Remote Sensing, 36*, 77–88.

Fenta, A. A., Yasuda, H., Haregeweyn, N., Belay, S., Hadush, Z., & Gebremedhin, M. A. (2017). The dynamics of urban expansion and land use/land cover changes using remote sensing and spatial metrics: The case of Mekelle City of Northern Ethiopia. *International Journal of Remote Sensing, 38*(14), 4107–4129. https://doi.org/10.1080/01431161.2017.1317936

Fu, Y., Lu, X., Zhao, Y., Zeng, X., & Xia, L. (2013). Assessment impacts of weather and land use/land cover (LULC) change on urban vegetation net primary productivity (NPP): A case study in Guangzhou, China. *Remote Sensing, 5*(8), 4125–4144. https://doi.org/10.3390/rs5084125

Grimm, N. B., Faeth, S. H., Golubiewski, N. E., Redman, C. L., Wu, J., Bai, X., & Briggs, J. M. (2008). Global change and the ecology of cities. *Science, 319*(5864), 756–760.

Haack, B. N., & Rafter, A. (2006). Urban growth analysis and modelling in the Kathmandu Valley Nepal. *Habitat International, 30*(4), 1056–1065.

Henderson, J. V., & Turner, M. A. (2020). Urbanization in the developing world: Too early or too slow? *Journal of Economic Perspectives, 34*(3), 150–173. https://doi.org/10.1257/jep.34.3.150

Huang, X., Wang, C., & Li, Z. (2018). A near real-time flood-mapping approach by integrating social media and post-event satellite imagery. *Annals of GIS, 24*(2), 113–123. https://doi.org/10.1080/19475683.2018.1450787

Jamal, S., & Ali, M. B. (2022). Landuse assessment of Kolkata municipal corporation by using Landsat data. *The Deccan Geographer, 60*(4), 77–88.

Jat, M. K., Garg, P. K., & Khare, D. (2008). Monitoring and modelling of urban sprawl using remote sensing and GIS techniques. *International Journal of Applied Earth Observation and Geoinformation, 10*(1), 26–43.

Kang, Y., Zhang, F., Gao, S., Lin, H., & Liu, Y. (2020). A review of urban physical environment sensing using street view imagery in public health studies. *Annals of GIS, 26*(3), 261–275. https://doi.org/10.1080/19475683.2020.1791954

Ligate, E. J., Chen, C., & Wu, C. (2018). Evaluation of tropical coastal land cover and land use changes and their impacts on ecosystem service values. *Ecosystem Health and Sustainability, 4*(8), 188–204. https://doi.org/10.1080/20964129.2018.1512839

Livanis, G., Moss, C. B., Breneman, V. E., & Nehring, R. F. (2006). Urban sprawl and farmland prices. *American Journal of Agricultural Economics, 88*(4), 915–929.

Nourqolipour, R., Shariff, A. R. B. M., Balasundram, S. K., Ahmad, N. B., Sood, A. M., & Buyong, T. (2016). Predicting the effects of urban development on land transition and spatial patterns of land use in Western Peninsular Malaysia. *Applied Spatial Analysis and Policy, 9*(1), 1–19.

Roy, S., Bose, A., Singha, N., Basak, D., & Chowdhury, I. R. (2021). Urban waterlogging risk as an undervalued environmental challenge: An integrated MCDA-GIS based modelling approach. *Environmental Challenges, 4*, 100194. https://doi.org/10.1016/j.envc.2021.100194

Sumari, N. S., Shao, Z., Huang, M., Sanga, C. A., & Van Genderen, J. L. (2017). Urban expansion: A geo-spatial approach for temporal monitoring of loss of agricultural land. *International Archives of the Photogrammetry, Remote Sensing and Spatial Information Sciences - ISPRS Archives, 42 XLII-2/W7*(2W7), 1349–1355. https://doi.org/10.5194/isprs-archives-XLII-2-W7-1349-2017

Sumari, N., Xu, G., Ujoh, F., Korah, P. I., Ebohon, O. J., & Lyimo, N. N. (2019). A geospatial approach to sustainable urban planning: Lessons for Morogoro municipal council, Tanzania. *Sustainability, 11*(22), 6508. https://doi.org/10.3390/su11226508

The Free Press Journal (2022). Faizabad Cantt to be renamed as Ayodhya Cantt, Defence Minister approves the proposal. *Asian News International*. 4 October 2022.Assessed on 19/09/2024.

Ujoh, F., Igbawua, T., & Ogidi Paul, M. (2019). Suitability mapping for Rice cultivation in Benue state, Nigeria using satellite data. *Geo-spatial Information Science, 22*(4), 332–344. https://doi.org/10.1080/10095020.2019.1637075

Velmurugan A, Sajjad H (2009).The study of land transformation and land degradation in Dehradun District,Uttrakhand. Deccan Geographer, 48. (ISSN-0011-7269)

The Interplay of Legislation and Judicial Intervention in Safeguarding India's Forests: A Critical Analysis

Deepti Khubalkar

1 Introduction

The forest policy of any nation stands as a paramount directive, outlining fundamental principles that govern the utilization and conservation of its invaluable forest resources. Historically, India's forest policies were primarily centered around timber production, emphasizing the economic significance of these lush expanses. However, the contemporary landscape reveals a dynamic shift, where diverse and often competing interests vie for a share of the forest's bounty. In the present context, forest policies are evolving to accommodate a multitude of demands, reflecting a nuanced understanding of the intricate interplay between environmental conservation, economic imperatives, and social considerations. This transformation underscores the need for adaptive and responsive forest policies that address the multifaceted challenges posed by the modern world. In this era, a comprehensive examination of a nation's forest policy becomes essential to discern how it navigates the intricate balance between sustainable resource management and the diverse array of interests seeking to benefit from this vital natural asset.

D. Khubalkar (✉)
Symbiosis Law School, Nagpur Campus, Symbiosis International (Deemed) University, Pune, India
e-mail: deeptik@slsnagpur.edu.in

2 Historical Overview of Forest Policy in India: From Imperial Roots to National Strategies

India's forest policy has a rich history, dating back to the inception of the National Forest Policy in 1894 during the era of British colonial rule. This pioneering policy marked the official entry of the government into the realm of forestry, shaping the trajectory of forest management in the country. The National Forest Policy of 1894 laid the foundation for a custodial and timber- oriented approach to forest administration. Its primary objectives were to manage forestry resources for the overall benefit of the nation. The policy recognized the critical role of maintaining an appropriate forest cover in preserving the climatic and physical conditions of the country, as well as meeting the fundamental needs of its people. Under this policy, certain conditions were set to strike a balance between developmental needs and environmental conservation. First and foremost, priority was given to permanent cultivation over forests, emphasizing the importance of sustainable land use. Additionally, the policy emphasized the satisfaction of local residents' needs, ensuring that their requirements were met at noncompetitive rates, and in some cases, even for free. This commitment to local communities took precedence over revenue-related considerations. Moreover, the recognition of maximum income from forest

Z. Khanam et al. (eds.), *Climate Change, Resilience and Ecosystem Services*, Sustainable Development Goals Series, https://doi.org/10.1007/978-3-031-96977-5_10

resources was contingent upon the fulfillment of the aforementioned conditions. In essence, the National Forest Policy of 1894 established a framework that sought to harmonize economic development with ecological sustainability and community welfare (Joshi et al., 2010).

Over time, this foundational policy underwent revisions, with a significant update occurring in 1952. The post-independence era saw an increased emphasis on expanding forests beyond their customary bounds, reflecting a growing awareness of the ecological importance of forest ecosystems. The evolution of India's forest policies continued with the Indian Forest Policy of 1952, representing a progression from its colonial predecessor. While still rooted in the need for revenue generation from forests, this policy marked a shift in consciousness towards the importance of increasing the forest cover to one-third of the total land area (Pratap, 2010).

During this period, the imperative for maximum annual revenue from forests was seen as a vital national need. The aftermath of the two World Wars, defense requirements, and large-scale developmental projects such as river valley projects and industries like pulp, paper, and plywood heavily relied on forest produce for national interest. Consequently, substantial areas of forests were cleared to generate revenue for the state. The National Forest Policy of 1952 sought to address crucial national needs, including the development of a system of complementary and balanced land use. It recognized the significance of maintaining control over issues such as deforestation in hilly areas, the invasion of sea sand along the coastline, riverbank erosion, and the overall improvement of physical and climatic circumstances (*National Forest Policy, 1952*, 2019). Furthermore, the policy aimed to secure an expanded supply of pasture, small wood, and fuel, while meeting the demand for forest products essential for defense and communication sectors. A key aspect of the policy was the requirement to fulfill all these criteria while realizing the greatest possible financial returns indefinitely (Kulkarni, 1987).

The National Forest Policy of 1988 marked a pivotal shift in India's approach to forestry, transitioning from a primarily commercial focus to a more comprehensive emphasis on the ecological role of forests and participatory management. This policy, guided by fundamental goals, recognized the imperative of balancing environmental conservation with the socio- economic needs of local communities. The key goals of the National Forest Policy of 1988 were as follows: the policy aimed to protect and restore the ecological balance that had been adversely affected by severe forest degradation. It acknowledged the critical role of forests in maintaining environmental stability. Recognizing the significance of the nation's surviving natural forests, the policy sought to safeguard their diverse flora and fauna (Bheemaiah, n.d.). These forests were acknowledged as symbols of India's tremendous biological diversity and genetic riches. The policy aimed to prevent soil erosion and denudation in the catchment regions of rivers, lakes, and reservoirs. This effort was crucial for conserving soil and water, reducing the impact of floods and droughts, and delaying reservoir silting. One of the primary objectives was to significantly increase the country's forest and tree cover. This would be achieved through widespread afforestation and social forestry programs, particularly on deforested, degraded, and unproductive lands. The policy recognized the importance of meeting the needs of rural and tribal communities for firewood, food, minor forest produce, and small timber. This reflected a commitment to the socioeconomic well-being of these communities. The policy aimed to enhance forest production to meet critical national demands. This included optimizing wood replacement and promoting sustainable forestry practices. Recognizing the importance of efficient resource utilization, the policy emphasized the promotion of effective use of forest products and optimizing wood replacement. An innovative aspect of the policy was the organization of a large, women-led people's movement. This participatory approach aimed to actively involve local communities in achieving the goals of afforestation, ecological restoration, and sustainable forest management, thereby reducing the strain on existing forests (*National Forest Policy Review-India*, n.d.). The National

Forest Policy of 1988 demonstrated a holistic understanding of the role of forests in ecological stability and socioeconomic development. By incorporating participatory management and addressing the needs of local communities, this policy set the stage for a more sustainable and inclusive approach to forestry in India.

Moving forward in time, the Draft National Forest Policy of 2018 marked a significant step in revamping the forest policy, addressing contemporary challenges such as climate change, animal-human conflict, and declining green cover. This draft policy emphasized scientific interventions and stringent regulations to safeguard forested areas. One notable goal was to bring one-third of India's population under the protection of forests. The draft policy also stood out for its global perspective on climate change and proposed public–private partnership models for afforestation and reforestation initiatives. Moreover, it advocated for the strict restraint on diversion of forest land for nonforestry purposes and emphasized strict oversight on compliance with conditions to ensure the preservation and sustainable use of forest resources.

The trajectory of India's forest policies reflects a continual adaptation to evolving national needs and global concerns, demonstrating a commitment to balancing economic development, ecological sustainability, and social well-being. The evolution of India's forest policies reflects a dynamic interplay between economic, ecological, and social considerations. From its origins in the colonial era to subsequent revisions, these policies have aimed to strike a balance between sustainable resource management, community welfare, and the broader interests of the nation. As India continues to navigate contemporary environmental challenges, the legacy of its forest policies remains integral to shaping a sustainable and resilient future.

3 Research Methodology

This study is primarily qualitative, focusing on the analysis of legislative texts, judicial decisions, and secondary sources such as academic articles, reports, and news coverage. The research has compared the legal framework before and after the 2023 amendments, as well as analyze relevant judicial pronouncements, government policy especially those related to forest conservation.

4 Balancing Development and Conservation: Environmental Impact Assessment in India

The foundation of Environmental Impact Assessment (EIA) in India lies in the internationally recognized "Precautionary Principle" of environmental jurisprudence. This principle acknowledges the irreversible nature of environmental harm, advocating for minimizing harm rather than attempting remedies. The legislative framework aims to strike a balance between sustainable development and environmental protection, acknowledging the significance of large- scale projects in the country's growth.

Introduced in 1978 initially for river valley projects, EIA expanded its scope to encompass various developmental procedures. It has become mandatory for over 30 classes of projects. The Environmental Protection Rules, 1986, empower the Central, State, or Union Territory Environmental Impact Assessment Authority (EIAA) to impose restrictions on specific projects, necessitating prior approval (Diwan & Rosencranz, 2001).

Projects are categorized into A and B based on scale and impact. Category A projects require approval from the Ministry of Environment and Forests (MoEF), advised by an Expert Appraisal Committee (EAC). Category B projects need approval from a State EIAA, guided by a State Expert Appraisal Committee (SEAC).

EACs and SEACs play pivotal roles in screening and determining project scopes. Comprising professionals from diverse fields, these committees work on the principle of "Collective Responsibility" and are constituted for a term of three years. Environmental clearance is required to follow few stages called as screening, scoping, and public hearing.

For Category B projects, SEAC methodically screens projects to determine the need for a detailed Environmental Impact Assessment (EIA) report. Projects are classified into B1 (requiring an EIA report) and B2 (not requiring an EIA report). EACs and SEACs formulate detailed terms of reference for Category A and B1 projects, addressing environmental concerns. These are displayed on the Ministry of Environment and Forests (MoEF) website. It also involves public hearings to address concerns of affected stakeholders. Exceptions exist for Category B2 projects and specific cases outlined in the notification.

EACs and SEACs scrutinize applications, EIA reports, public consultation outcomes, and public hearing proceedings. Recommendations are then made to the regulatory authority for granting or rejecting environmental clearance. Clearances are made public on government portals, with copies submitted to local bodies. Proponents must submit compliance reports every six months.

The EIA 2020 (Amendment) aimed to enhance transparency and applicant-friendliness, but it faced criticism for potential dilution of environmental regulations

It permits Ex Post Facto Clearance. This amendment has introduced a route for projects to proceed without clearance, subject to fines for violations. This has raised concerns about the legality and environmental implications. Amendment has reduced public consultation time. The time for the public to file concerns regarding projects was reduced from 30 to 20 days, potentially limiting public participation. It has strategic considerations like some category exempt from public consultations, raising questions about transparency and public awareness. It further provides monitoring relaxation. Amendment has changed the reporting frequency for compliance from half-yearly to annually, increasing the risk of unnoticed hazardous implications. While the EIA in India has been a crucial tool for environmental protection, ongoing amendments and potential dilutions raise concerns about the balance between development and conservation, emphasizing the need for comprehensive and inclusive environmental governance. Moreover, the result of stringent government forest policy and statutes has significantly shown the increase in forest cover in order to achieve our goal of National Forest Policy, 1988 (Table 1).

Table 1 Forest cover of India according to 2021 Survey

Class	Area	Percentage of geographical area
Very dense forest	99,779	3.04
Moderately dense forest	3,06,890	9.33
Open forest	3,07,120	9.34
Total forest cover	7,13,789	21.71
Scrub	46,539	1.42
Nonforest	25,27,141	76.87
Total geographical area	32,87,469	100.00

Source: India State of Forest Report (*Press Information Bureau*, 2022)

The National Forest Policy of India, 1988, sets an objective to have 33% of the nation's land covered by forests and trees. The Forest Survey of India, which operates under the Ministry of Environment, Forest and Climate Change (MoEFCC), has provided an updated report on the current state of India's forests. This report includes estimates on forest cover, tree cover, growing stock, the extent of trees outside forests, mangrove cover, bamboo resources, and forest carbon stock assessment. According to the latest data, the combined forest and tree cover amounts to 24.62% of the country's total area. Specifically, the forest cover alone is 7,13,789 square kilometers, making up 21.71% of the country's geographical area, while the tree cover accounts for 2.91%.

Compared to the India State of Forest Report (ISFR) 2019, the latest assessment reveals a national-level increase of 0.28% in the combined forest and tree cover. Specifically, the forest cover has risen by 0.22%, and the tree cover has grown by 0.76%. The new changes in the forest conservation law may affect around 28% of the forest cover which is not recorded as forest land in the government record (Sayeda, 2023). Indian government has taken significant steps to address the Sustainable Development Goals (SDGs) and combat climate change through a range of policies and programs that tackle social, economic, and environmental issues. The Swachh Bharat Mission, or Clean India Mission, is one such initiative that aimed to eliminate open defecation by 2019. Through this program, over 39.5 million household toilets were constructed, and cities and villages were ranked based on their cleanliness (From Swachh Bharat to Sundar Bharat via Swasth Bharat: An Analysis of the Swachh Bharat Mission, 2019). The National Clean Air Programme (NCAP) is another key effort, focusing on reducing air pollution by creating Clean Air Action Plans for 132 cities across the country. The government's Renewable Energy Strategy aligns with the Make in India initiative, setting an ambitious target of achieving 175 GW of renewable energy capacity by 2022, spearheaded by the Ministry of New and Renewable Energy. Together, these initiatives reflect a comprehensive approach to sustainable development and climate action (UN India Digital Library, 2024).

5 Forest Legislation and Policy in India and Changing Perspectives

The evolution of forest legislation in India traces its roots back to the colonial era, reflecting the complex interplay of economic, administrative, and environmental considerations. The initiation of formal forest laws can be attributed to the establishment of the Imperial Forest Department in 1864, a pivotal moment that aimed to consolidate British control over India's vast forest resources. The ensuing Indian Forest Act of 1865 marked the beginning of a systematic approach to forest management. Under the provisions of the 1865 Act, the British government applied the authority to declare any land covered with trees as a government forest and formulate rules for its governance. This marked a decisive step towards centralizing control over the country's forested landscapes. Subsequently, the Indian Forest Act of 1878 expanded the jurisdiction of the British Administration to incorporate wastelands, a category that included forests. This legislation enabled the demarcation of reserved and protected forests, categorizing them into distinct types with varying degrees of access for local communities.

The Act of 1878 introduced a classification system, dividing forests into reserved forests, protected forests, and village forests. This categorization aimed at regulating the collection of forest produce by forest dwellers, imposing penalties for prohibited activities, and reinforcing state control. The legislation granted the administration the power to deny local rights in protected forests, and it established the foundation for the government to exert control over forest resources (Chaudhary, 2000).

Post-independence, India witnessed the designation of vast expanses of forest land as reserved and protected forests. The Indian Forest Act of 1927 emerged as a significant milestone, focusing on regulating the movement of forest produce and defining procedures for declaring areas as Reserved Forest, Protected Forest, or Village Forest. Reserved Forests, the most restricted category, granted limited access to local populations, emphasizing stringent control by Forest Officers. Protected Forests, on the other hand, allowed the state to exercise control over nonreserved areas with revenue-generating potential. Village Forests introduced the concept of assigning rights to village communities over specific lands, a step towards recognizing local participation in forest management. However, the period also witnessed shortcomings and controversies. The forest legislation often left out significant forested areas, and the revenue- centric approach overshadowed environmental values like biodiversity conservation and soil erosion prevention. The discretionary powers granted to the forest bureaucracy resulted in the marginalization of forest dwellers, particularly nomadic and tribal communities, depriving them of their traditional rights.

The historical journey of forest legislation in India reflects a narrative of evolving priorities—from colonial-era revenue generation and control to a more contemporary understanding that seeks a balance between economic interests, environmental conservation, and social justice. The journey underscores the ongoing challenge of crafting forest policies that harmonize the diverse demands placed on this critical natural resource.

The Forest Conservation Act of 1980, a landmark legislation in India, introduced crucial measures to safeguard forest ecosystems and biodiversity. Enacted with the aim of curbing deforestation and preserving wildlife, this legislation mandated that central permission be obtained for any activities, especially sustainable agro-forestry, within forest areas. The violation or absence of such permits was considered a criminal offense, underscoring the seriousness of the legal framework. The primary objectives of the Forest Conservation Act were aligned with broader conservation goals, focusing on limiting deforestation, conserving biodiversity, and protecting wildlife. By requiring central permission, the legislation sought to regulate human activities in forested regions to ensure sustainable practices and prevent irreversible environmental damage.

Despite its noble objectives and the potential it held for advancing forest conservation, the Forest Conservation Act faced challenges in achieving its intended outcomes. The complexities of enforcing regulations, competing interests, and the continuous pressures on forest resources presented obstacles to the Act's success in curtailing deforestation effectively. The Act operated in tandem with the overarching National Forest Policy, reinforcing the collective commitment to environmental stability and ecological balance (Leelakrishnan, 2004). The ultimate goal was to preserve forests as a natural heritage, recognizing their pivotal role in maintaining the equilibrium of ecosystems, sustaining biodiversity, and safeguarding the country's rich wildlife.

While dealing with a case under Forest Conservation Act, the Allahabad High Court in the case of *Nand Lal v. State of U.P., 2002* said that the word "forest" has not been defined in the Act. The court was of the opinion that "this word has been used in the general sense," the meaning of which as defined by the Food and Agriculture Organization (FAO) was stated as follows:

> All the lands bearing vegetative associations demarcated by trees of any size exploited or not, capable of producing wood or other forest products, of exerting an influence on the climate or on the water regime or providing shelter for livestock and wildlife.||

The court further extracted forest values from encyclopedia Britannica, vil.9 at page 614 (*M/s Yashwant Stone Works v. State of U.P.*, 1988).

In the case of *T.N. Godavarman Thirumalpad* (Leelakrishnan, 2010), the Supreme Court ruled that the term "forest" encompasses any land recorded as such in government documents, regardless of who owns it. The term "forest" should be interpreted based on its dictionary definition. This definition includes all legally recognized forests, whether they are classified as reserved, protected, or otherwise, in accordance with section 29(i) of the Forest (Conservation) Act. Scope and meaning of "Forest" are also widen by the court through various orders. The word forest|| has not been defined in any statute except in some state acts relating to "Private Forests." In the Indian Forest Act of 1865, "Forest" was defined as land covered with trees, bush wood or jungle but this definition was perhaps considered unnecessary and was, therefore, given up in the subsequent enactments. There is, therefore, no statutory definition of the word "forest."

In Laxman Icharam's case (Samuel, 2003), it was held that, in the absence of any definition, the word "forest" must be taken in the ordinary dictionary sense (*Laxman Icharam v. D.F.O. Raigarh*, 1953). This was further discussed and held by the Supreme Court in the year 2022 (*Narinder Singh & Ors v. Divesh Bhutani and others*, 2022).

Thus, the definition of the word "forest" is shaped up by judiciary in various judgement in absence of any definition under the Forest Conservation Act. Where in any enactment if the expression "forest" or "forest land" is not defined or the particular land is not recorded as "forest lands" in the village accounts, one has to perceive it by the popular meaning of the word "forest" bringing it within the purview of the Act.

While the Forest Conservation Act of 1980 represented a significant step towards recognizing the importance of protecting forests, its effectiveness in achieving its objectives remained a subject of scrutiny. Continuous efforts and a comprehensive approach were deemed essential to address the evolving challenges in forest conservation and to ensure the sustainable use of these vital natural resources.

6 Latest Changes in the Forest Law

The Forest (Conservation) Amendment Bill of 2023, introduced in response to evolving environmental and developmental needs, presents a set of significant provisions and concerns that require thorough consideration. The Forest (Conservation) Amendment Bill 2023 has expanded the Scope of the Act by introducing a Preamble. Additionally, it renames the Act as Van (Sanrakshan Evam Samvardhan) Adhiniyam, 1980, reflecting the potential impact of its provisions. Initially confined to notified forest land, the Act was later extended to encompass revenue forest land and lands documented as forest in government records. The amendments aim to streamline the Act's application to recorded forest lands, private forest lands, and plantations. The Bill suggests exemptions to promote afforestation and plantation outside forests. These include provisions for connectivity, security-related infrastructure, and public utility projects. Strategic projects linked to national security within 100 km of international borders are also exempt. The amended Act raises significant concerns, particularly for the northeastern states, by removing the requirement for forest clearance within 100 km of international boundaries and lines of control for strategic linear projects of national significance. This exclusion affects some of India's most biodiverse and fragile habitats. Further, the term "strategic linear projects of national importance" is vaguely defined, which could be exploited to push through infrastructure projects that harm local ecosystems.

These provisions also threaten to strip indigenous tribes of their rights to ancestral lands, rights that are safeguarded under the Forest Rights Act, also known as the Scheduled Tribes and Other Traditional Forest Dwellers (Recognition of Forest Rights) Act of 2006. The displacement of indigenous communities could lead to further degradation of the rich biodiversity and density of these forests. Existing provisions related to leasing forest land to private entities are extended to government companies, facilitating development projects and ensuring consistency in implementation. The amendments introduce activities such as infrastructure for frontline forest staff, ecotourism, zoo, and safari as recognized forestry activities for the conservation of forests. The Bill encourages activities like establishing zoos, safaris, and ecotourism, owned by the government and set up outside protected areas. This aims to enhance awareness, create livelihood opportunities, and integrate local communities with overall development.

The Bill aims to contribute to India's efforts in combating climate change by recognizing such areas as part of its forest conservation endeavors, aligning with international commitments like Net Zero Emission by 2070. Moreover, there are some concerns expressed related to the Bill.

The Act's new Hindi name faced objections for being perceived as "non-inclusive" for non-Hindi speaking populations in South India and the North-East. Concerns were raised about potential clearances of forests in ecologically sensitive regions like the Himalayan, trans-Himalayan, and northeastern areas, potentially jeopardizing the rights of indigenous communities. The Bill confines its scope to areas recorded as forests after October 1980, possibly excluding significant sections of forest land and biodiver sity hotspots from its purview. State governments argue that forest conservation falls under the Concurrent List, and the proposed amendments might shift the balance towards the center, impacting the rights and authority of state governments.

A comprehensive assessment of the proposed amendments, considering ecological, social, and environmental factors, is imperative. Meaningful consultation and dialogue with diverse stakeholders, including experts, nongovernment organizations (NGOs), tribal communities, and state governments, will ensure transparency, inclusivity, and well-informed decision-making. It is crucial to address concerns related to the Hindi name, potential impacts on ecologically sensitive areas, limited applicability, and center–state balance for the effective implementation of the amended legislation. Amendment Bill of 2023 to the Forest (Conservation) Act was challenged before the Supreme Court regarding its constitutional validity. The Supreme Court has stayed the application of the Bill and directed the government to adopt broad definition of the term "Forest" interpreted by judiciary in *T. N. Godavarman Thirumulpad v. Union of India and Ors* till the final order (Koshy, 2024).

A balanced and well-informed approach, involving all stakeholders, is essential for the successful implementation of the Forest (Conservation) Amendment Bill of 2023.

7 Role of Indian Judiciary

The courts play a vital role in defining the powers and responsibilities of administrative agencies and in balancing environmental protection with development. Principle 10 of the 1992 Rio Declaration emphasizes the need for effective access to judicial and administrative processes, including avenues for redress and remedy. Significant judicial involvement in forest-related issues began in the 1980s, marking a period notable for both administrative and legislative initiatives, as well as judicial activism in environmental law. In India, many environmental cases are pursued under Articles 32 and 226 of the Constitution. Unlike in the past, today an individual acting in good faith and with sufficient interest can approach the courts to address public concerns and enforce public duties in public interest (Parikh, 2013).

Although the expression forest and wildlife or environment has not been expressly mentioned in the Constitution, as a fundamental right, still the High courts and Supreme Court looked to the constitutional provisions through interpretation of various provisions and action linking with fundamental rights and thus, brought the healthy environment as a fundamental right guaranteed to people of India.

It took a considerable amount of time for the judiciary to clearly state that the right to life under Article 21 of the Constitution encompasses the right to live in a healthy environment. Forest is one of the great sources to keep the Earth pollution free. Forest is a source of livelihood also. Hence, conflict often arise between individual rights and environmental protection. for example, the freedom to conduct business in forest areas (Mines, use of forest produce etc.), the right to reside within forests, or the right to unrestricted access to forest resources may clash with the broader constitutional guarantee of clean and healthy environment under Article 21. these tensions highlight the ongoing challenge of balancing developmental and livelihood interests with ecological preservation and the collective right to a sustainable environment. The first indication of the right to a wholesome environment may be traced to the Dehradoon Valley case (*Rural Litigation v. Entitlement Kendra*, 1985). This was the initial case where the Supreme Court attempted to address this issue. It mandated the cessation of mining activities in specific forest regions, while allowing a gradual phase-out in other areas. Following this judgment, various High Courts across the country eagerly affirmed that the right to a healthy environment is encompassed within the right to life under Article 21 of the Constitution. In one case, the court noted that it is the courts' legitimate duty, as enforcers of constitutional objectives, to prevent any actions by the state or citizens that disrupt the environmental balance (*Subhash Kumar v State of Bihar*, 1991). Another aspect of the right to life is right to livelihood, where the court potentially checked government actions with an environmental impact that threaten to dislocate poor people and disrupt lifestyles (*Banwasi Sewa Ashram v. State of U.P. and Ors*, 1986). Whether the right to intergenerational equity can be derived from Articles 21 and 14 is decided by the court in various cases such as in Ganesh Wood Production case (*State of H.P. v. Ganesh Wood Products*, 1996). In this case, the Supreme Court referred to the National Forest Policy and the Himachal Pradesh State

Forest Policy to overturn a decision by the state industrial project authority. This authority had approved the establishment of units to produce "Katha" from the scarce Khair tree without considering the availability of Khair trees and the negative impact on state forests. The court ruled that economic liberalization policies must align with the National Forest Policy and forest laws established by the government. It cautioned government departments against ignoring forest policies, warning that such neglect would jeopardize government decisions.|| The court emphasized that government development agencies responsible for decision-making must consider ecological factors, including the environmental policies of both central and state governments. Additionally, the court banned sawmills from operating within a 10-kilometer buffer zone around forest areas (*Sushila Saw Mill v. State of Orissa*, 1995). Bench of Andhra Pradesh High Court upheld the invalidation of a government order permitting a forest to be cleared (*V. Shankar Reddy v. State of Andhra Pradesh*, 1992). Allahabad High Court endorsed prohibition of mining in the reserve area (*Pyari Devi v. State of Uttar Pradesh*, 2004). In the case filed by K.M. Chinnappa, the question was decided whether prior approval mandatory to continue mining in an area subsequently declared as a national park (*KM Chinnappa v Union of India*, 2003). On the same tune, another landmark judgement was passed in the case of *TN Godaverman*. This case is the high water mark in forest protection (*TN Godaverman v Union of India*, 2006).

Nowadays, tourism in forest area is growing as a fascination. Therefore, the question arises whether tourism will lead to disturbance in forest and wildlife (Hannam, 2004). In various government policies, tourism is always a matter of priority. Ecotourism is one of the midway to solve this problem (Baloch et al., 2022). Court in the case of *Nagarahole* (*Nagarahole Badakattu Hakku Sthapana Samiti v. State of Karnataka*, 1997) and other cases (*Forest Friendly Camps Pvt Ltd. v. State of Rajasthan*, 2002) held that the establishment of restaurant, with boarding and lodging, in the midst of national park should not be in contravention of the Forest Conservation Act 1988.

8 Global Forest Policy

Global forest policy has emerged as a critical facet of international environmental governance, shaped through deliberations in various influential fora. The development of this policy can be tracked through key milestones at various forums, including the United Nations Conference on Environment and Development (UNCED), the Intergovernmental Panel on Forests (IPF), the Intergovernmental Forum on Forests (IFF), the United Nations Forum on Forests (UNFF), the Committee on Forestry of the UN Food and Agriculture Organization and the International Tropical Timber Organization. Each of these platforms has contributed to the formulation of strategies and frameworks aimed at addressing the complex challenges facing global forests.

The United Nations Conference on Environment and Development (UNCED), also known as the Earth Summit, held in Rio de Janeiro in 1992, was a watershed moment for global forest policy. The conference recognized the vital role of forests in sustainable development and biodiversity conservation. The resulting document, Agenda 21, laid the groundwork for subsequent discussions on forests, emphasizing the need for integrated and sustainable management practices (Grebner et al., 2013).

Building upon the momentum generated by UNCED, the Intergovernmental Panel on Forests was established in 1995 to provide a comprehensive assessment of the world's forests. The IPF played a pivotal role in identifying key issues related to forest management, biodiversity conservation, and the socio-economic aspects of forestry. Its recommendations informed the subsequent creation of the Intergovernmental Forum on Forests (IFF) in 1997.

The Intergovernmental Forum on Forests became a platform for more inclusive and participatory discussions on global forest issues. It aimed to enhance cooperation and coordination among countries, fostering a shared understanding of the challenges faced by different regions. The IFF laid the groundwork for the establishment of the United Nations Forum on Forests in

2000, which assumed the role of the central international body addressing forest-related matters.

The United Nations Forum on Forests (UNFF) has since become the primary forum for the development of global forest policy. It operates as a platform for dialogue, negotiation, and cooperation among member states, emphasizing the sustainable management of all types of forests. The UNFF has played a crucial role in shaping the Non-Legally Binding Instrument on All Types of Forests (NLBI), adopted in 2007, which provides a set of principles and guidelines for sustainable forest management.

Parallel to these developments, the Committee on Forestry (COFO) of the UN Food and Agriculture Organization (FAO) has been actively involved in addressing forestry issues. The COFO serves as a technical and advisory body, facilitating collaboration and information exchange among member countries. Its contributions to global forest policy include promoting sustainable forest management practices and addressing the socioeconomic dimensions of forestry.

The International Tropical Timber Organization (ITTO) has focused on tropical forests, recognizing their unique ecological and economic significance. Through initiatives such as the ITTO Criteria and Indicators for the Sustainable Management of Tropical Forests, the organization has contributed to the development of guidelines for responsible tropical forest management.

The global forest policy reflects a collaborative and ongoing effort to address the multifaceted challenges posed by deforestation, biodiversity loss, and unsustainable forest practices. The deliberations at UNCED, IPF, IFF, UNFF, COFO, and ITTO have collectively shaped the international agenda on forests, emphasizing the importance of sustainable management, conservation, and the equitable distribution of benefits derived from forests. As the world continues to grapple with environmental challenges, these forums remain crucial in fostering global cooperation and shaping policies that promote the well-being of forests and the communities that depend on them.

9 Conclusion

In addition to the forest laws passed by center and state legislation in the past regulating the protection and management of forests and wildlife relatively restrained the central enactments namely the Wildlife (Protection) Act 1972 the Constitution (42nd Amendment) Act 1976, the Forest Conservation Act 1980 and National Forest Policy of 1988 reflect the grave concern of the Parliament and the central government to protect and managed the forests which are a renewable resource and an invaluable gift of nature. However, the role of judiciary (Hight courts and Supreme Courts) in mitigating forest offences cannot be overlooked. Though in forest matters higher judiciary has no direct jurisdiction to solve the problem, yet by wide interpretation of Articles 21, 14, and 19 judiciary has brought environmental matters within the purview of its jurisdiction to Articles 32 and 226. One of the main defects of the laws relating to forest was the word "forest" was not defined in any statute except the law, Forest Conservation Act. However, it has been removed in the latest amendment of Forest Conservation Act. Previously in the absence of definition of forest, it was difficult to bring a particular land in the purview of Forest (Conservation) Act 1980 to protect the land from using for any nonforest purposes. However, through wide interpretation of the word "forest" judiciary has widened the scope of forest in series of cases. The term "forest land" was held by the Supreme Court to be not only as understood in the dictionary sense but also pertaining to any area accorded as forests in the government record. Thus, it has widened the scope of the Forest (Conservation) Act 1980 and it became necessary for doing any prohibited activity in such area to obtain prior approval of the central government. The court followed a dual strategy in forest-related issues. Where the state was claiming its environmental agenda, the court protected the rights of the forest dwellers. While in pollution-related issues, where the state was reluctant, it cast duty in it to be concerned about the environment. Thus, in India forest law is developing through judicial decisions. However, the amendment bill to the Forest

Conservation Act was passed by the government without opposition parties as they walked out from the house in protest to Manipur violence. The bill had some shortcomings nullifying the definition of forest shaped through various decisions of the court. The bill covered only the land notified as forest under Indian Forest Act 1927. It might also violate the significant ruling of the Supreme Court in the case of *Godavarman v. Union of India*, which granted safeguards to any land meeting the dictionary definition of forests. The forest and biodiversity bills signify a significant change in India's environmental protection philosophy in three aspects (*Godavarman Thirumulkpad v. Union of India & Ors.*, 1997). This involves a shift from a focus on protection to the opening up of lands for strategic and other purposes, a reduction in the compliance burden for the government and industries, and a departure from more expansive definitions aimed at safeguarding additional forest land and supporting local communities. The previous definitions were perceived as cumbersome and challenging to legislate on. Moreover, excluding deemed forests from the scope of the bill is an unfavorable decision. The 1996 Supreme Court judgment suggests that deemed forests, meeting the dictionary definition of a forest, should also be incorporated.

10 Recommendations

To ensure the continued protection of forests, the government should revisit the legal definitions of "forest" in the Forest (Conservation) Act. Aligning these definitions with the broad, inclusive interpretations developed by the judiciary, such as those in the *Godavarman* case, will help safeguard forest ecosystems. This step is crucial for achieving SDG-15, which emphasizes the need to conserve and restore terrestrial ecosystems. Excluding "deemed forests" from the bill's scope is a particularly concerning development, as it reverses decades of judicial progress in forest conservation. The government should consider revisiting these amendments, incorporating feedback from various stakeholders, including environmental experts, local communities, and the judiciary, to ensure that India's forests remain safeguarded for future generations (*T. N. Godavarman Thirumulpad v. Union of India and Ors*, 2022).

Any amendments to forest laws should include provisions that encourage sustainable use of forest resources. This involves ensuring that the extraction of resources, where permitted, does not compromise the health of forest ecosystems. Integrating sustainable practices into the legal framework will support SDGs-1, -11, -12, and -13 by promoting responsible consumption and production patterns. Forest policies should prioritize the rights and livelihoods of local communities and forest dwellers, who are often the most affected by changes in forest laws. Ensuring that these communities have a say in forest management and benefit from forest resources can contribute to reducing poverty and addressing inequalities. Before opening up any forest lands for development or strategic purposes, it is essential to conduct thorough environmental impact assessments (EIAs). These assessments should consider the long-term effects on biodiversity, carbon sequestration, and climate change mitigation, aligning with SDG-13's goal of taking urgent action to combat climate change and its impacts. Forest conservation should be integrated into broader national development plans, ensuring that economic growth and urbanization do not come at the expense of forest ecosystems. Policies should promote green jobs and sustainable urban planning, contributing to SDG-8 and SDG-11. By aligning forest conservation efforts with the relevant Sustainable Development Goals, India can ensure that its legal and policy frameworks support both environmental sustainability and socio-economic development.

References

Baloch, Q. B., Shah, S. N., Iqbal, N., & Sheeraz, M. (2022). Impact of tourism development upon environmental sustainability: A suggested framework for sustainable ecotourism. *Environmental Science and Pollution Research International, 30*(3), 5917–5930. https://doi.org/10.1007/s11356-022-22496-w

Banwasi Sewa Ashram vs State of U.P. and Ors, (Supreme Court November 20, 1986).

Bheemaiah, V. M. (n.d.). A critique on the relevant forest laws and policies of India. *Indian Journal of Integrated Research in Law, II*(II), 1–18. Retrieved January 3, 2024, from https://ijirl.com/wp-content/uploads/2022/05/A-CRITIQUE-ON-THE-RELEVANT-FOREST-LAWS-AND-POLICIES-OF-INDIA.pdf

Chaudhary, R. N. (2000). *Law of forests in India* (1st ed.). Orient Publishing Company.

KM Chinnappa v Union of India, (Supreme Court of India 2003).

Diwan, S., & Rosencranz, A. (2001). *Environmental law and policy in India* (2nd ed.). Oxford India Paperbacks.

Forest Friendly Camps Pvt Ltd. v. State of Rajasthan, (Rajasthan High Court 2002).

From Swachh Bharat to Sundar Bharat via Swasth Bharat: An Analysis of the Swachh Bharat Mission. (2019). *Economic Survey 2018–19*, Volume 1. https://www.indiabudget.gov.in/budget2019-20/economicsurvey/doc/vol1chapter/echap08_vol1.pdf

Godavarman Thirumulpad v. Union of India & Ors, (The Supreme Court of India 1997).

T.N. Godavarman Thirumulpad vs Union of India and Ors, (Supreme Court of India 3 June, 2022).

T.N. Godaverman v. Union of India, (Supreme Court of India 2006).

Grebner, D., Bettinger, P., & Siry, J. (2013). Forest policies and external pressures. In *Introduction to forestry and natural resources* (pp. 359–383). Academic Press. https://www.sciencedirect.com/science/article/abs/pii/B9780123869012000154

Hannam, K. (2004). Tourism and Forest Management in India: The role of the state in limiting tourism development. *Tourism Geographies, 6*(3), 331–351. https://doi.org/10.1080/1461668042000269647

Joshi, A. K., Pant, P., Kumar, P., Giriraj, A., & Joshi, P. K. (2010). National forest policy in India: Critique of targets and implementation. *Small Scale Forestry, 10*(1), 83–96. https://doi.org/10.1007/s11842-010-9133-z

Koshy, J. (2024, February 25). What is the Supreme Court's interim order on the Forest Act? *The Hindu*. https://www.thehindu.com/sci-tech/energy-and-environment/what-is-the-supreme-courts-interim-order-on-the-forest-act-explained/article67882851.ece

Kulkarni, S. (1987). Forest legislation and Tribals: Comments on forest policy resolution. *Economic and Political Weekly, 22*(50), 2143–2148. https://www.jstor.org/stable/4377847

Laxman Icharam v. D.F.O. Raigarh, (Nagpur High Court 1953).

Leelakrishnan, P. (2004). *Environmental law in India* (1st ed.). Leelakrishnan, P. (2010). *Environmental law case book* (2nd ed.). Lexis Nexis Butterworth.

M/s Yashwant Stone Works v. State of U.P, (Allahabad High Court 1988).

Nagarahole Badakattu Hakku Sthapana Samiti v. State of Karnataka, (Karnataka High Court 1997).

Narinder Singh & Ors v. Divesh Bhutani and others, (The Supreme Court of India July 21, 2022). https://www.courtkutchehry.com/Judgement/Search/AdvancedV2?s_acts=Forest%20Act,%201927&section_art=section&s_article_val=34

National Forest Policy, 1952. (2019, July 26). Bare Act Live. http://www.bareactslive.com/ACA/act3449.htm?AspxAutoDetectCookieSupport=1

National Forest Policy Review-India (p. 51). (n.d.). Indian Institute of Forest Management. Retrieved December 18, 2023, from https://www.scribd.com/document/27047642/national-forest-policy-review-india

Parikh, M. (2013). The Forest conservation in India and the role of Indian supreme court: A critical analysis. *IOSR Journal of Humanities and Social Science, 13*(4), 55–61. https://www.iosrjournals.org/iosr-jhss/papers/Vol13-issue4/J01345561.pdf?id=3407

Pratap, D. (2010). Community participation and forest policy in India: An overview. *Social Change, 40*(3), 235–256. https://doi.org/10.1177/004908571004000301

Press Information Bureau. (2022, January 20). https://static.pib.gov.in/WriteReadData/specificdocs/documents/2022/jan/doc20221207001.pdf

Pyari Devi v. State of Uttar Pradesh, (Allahabad High Court 2004).

Rural Litigation v. Entitlement Kendra, (Supreme Court of India 1985).

Samuel, C. J. (2003). *Manual of Maharashtra Laws*. Shrivastava Publishing House.

Sayeda, Z. A. (2023, August 9). *Amendment to forest conservation act leaves 28% of India's forest cover vulnerable*. Scroll.in. https://scroll.in/article/1054010/amendment-to-forest-conservation-act-leaves-28-of-indias-forest-cover-vulnerable#:~:text=The%20new%20amendment%20exempts%20land,of%20forests%2C%20as%20we%20said

V. Shankar Reddy v. State of Andhra Pradesh, (High Court of A.P. 1992)

State of H.P. v. Ganesh Wood Products, (Supreme Court of India 1996).

Subhash Kumar v. State of Bihar, (Supreme Court 1991).

Sushila Saw Mill v. State of Orissa, (Supreme Court 1995).

UN India Digital Library. (2024). Retrieved from United Nations India: https://india.un.org/en/sdgs

Spatiotemporal Analysis of Human Encroachment on Wetlands: A Micro-Level Analysis of Aligarh City

Rukhsar Anjum, Sk Ajim Ali,
Mansoor Alam Siddiqui, Rashid Aziz Faridi,
Farhana Parvin, and Asefa Khatun

1 Introduction

Year after year, most sections of the world become more urbanized as people migrate from rural areas to the city core. According to a United Nation's (United Nations, 2018) estimate, 30% of the world's inhabitants were segregated and settled in cities in 1950, this number has risen up to 55% in 2018. Prognostic models showed that by 2050, the proportion will rise up to 68%, with developing countries having the greatest rate (United Nations, 2018). In most of the developing and industrialized worlds, the ever-expanding built-up expansion towards environmental territories is a typical occurrence (Yanan et al., 2011; Milan & Creutzig, 2016). Rapid inhabitants' expansion, in combination with uncontrolled building construction and other economic activity, has the potential to accelerate urban sprawl and land area extension, resulting in a serious environmental crisis (Yue et al., 2013). India's urbanization, like that of many other emerging countries, has exploded in recent decades (Anjum et al., 2023a). The pace of urbanization in India was gradual for the period of the nineteenth century and the early decades of the twentieth century, but the situation altered dramatically following India's independence, with the startling expansion of cities and townships. Unlike in the past, when urban growth was centered on streams, water bodies, and cultivated zones, today's urban growth is centered on already existing big cities, with most of them having little similarity to past trends (Dembélé et al., 2018; Mascarenhas et al., 2019). This is largely owing to technical breakthroughs that can create the basic necessities of human living wherever on the planet, regardless of the configuration and utilities of life-sustaining systems. The benefits and drawbacks of such advances must be verified and proven over time.

Water bodies are the most delicate and vulnerable ecosystems on the planet (Anjum et al., 2023b). Despite the fact that they are the most delicate and fragile natural resources, they are also the most mismanaged and neglected. The blessing of human activities from the surrounding regions causes the majority of the challenges experienced by such sensitive resources (UN-Habitat & United Nations Human Settlements Program, 2008). The upward growth of residents in the urban areas is an unavoidable circumstance, but cities with expeditious urbanization, especially in developing countries, have been growing with improper planning and relationships among the urban setup (built-up environment). This generates an unending source of social and environmental degradation (World Health Organization, 2012). During the past 5 decades, the urban population in India has surged

R. Anjum · S. A. Ali (✉) · M. A. Siddiqui ·
R. A. Faridi · F. Parvin · A. Khatun
Global Development Institute, The University of Manchester, Manchester, UK

Z. Khanam et al. (eds.), *Climate Change, Resilience and Ecosystem Services*, Sustainable Development Goals Series, https://doi.org/10.1007/978-3-031-96977-5_11

by more than six times (El-Asmar et al., 2013). The total urban population of India is projected to be approximately 481,980,332. It is more than the overall population of various nations. India also has slightly over 10% of the entire urban population of this planet. India had 20% of its total population in urban areas in 1971, which has increased up to 34.9% in the year 2020. There are 5161 urban centers in our country, a thousand increased compared to the year 1981 (Sivaramakrishna et al., 2005).

Remote sensing and geographic information system (GIS) are commonly used to designate water bodies or to estimate variables of hydro-meteorological processes such as precipitation, temperature, soil moisture, vapor transpiration, or land surface features (Demuth & Radojevic, 2011). Remote sensing has a significant ability to recognize changes throughout time for putting together the necessary planning for efficient performance of lands, both urban and rural (Meshesha et al., 2016; Arulbalaji & Gurugnanam, 2014; Ali et al., 2023). One of the primary worries of folks who remain deeply invested in decision-making on major climatic variations and sustainable progress is the land use and land cover (LU/LC) fluctuations throughout time (Yirsaw et al., 2017). Because of the equilibrium between human interferences, local climate, and agricultural practice, LU/LC practices in rural and isolated places have remained nearly unchanged for lengthy periods of time (Mathew et al., 2016; Naserikia et al., 2019). However, in metropolitan regions, where a fast population increase and economic development have resulted in substantial changes in LU/LC, the situation is quite different. These anthropogenic activities are altering the terrestrial environment through massive land/landform changes that are typically occurring at unprecedented speeds and scales, and the procedure is usually permanent in nature (Song et al., 2014; Sun et al., 2012). All of these alterations, in one way or another, can put the ecosystem's carrying capacity to the test, affecting services such as food production, the preservation of freshwater and ecological resources, climatic and air circulation, and so on (Mukherjee et al., 2019; Choudhury et al., 2019).

In past, many scholars attempted to understand the land use dynamics of the study area. Anjum et al., 2023a, c; Ajmal & Jamal, 2021; Singh & Asgher, 2005 attempted to study the LU/LC of Aligarh City along with groundwater quality. Farooq & Ahmad, 2008 identified the urban sprawl in and around the Aligarh City. Saha et al., 1990 utilized remote sensing to plot the wasteland in the Aligarh District. Khalil & Khan, 2009 showed how the solid waste was managed in the city and explained the lack of management that turned wetlands into a dump yard. So far, no work was aimed to find out about the status of existing wetlands, let alone the relation between built areas expansion with its impact on wetland. The objective of this study is to evaluate the consequences of increasing built-up area on the prevalence of surface water bodies. Here in this study, the concerns of encroachment of existing land cover in Aligarh City and their impact on wetlands within the city were focused. The study was initiated to gather information regarding the association between urban encroachment and the existence of water bodies. The methodological principles used for the current analysis are based on Landsat 5 and 8 satellite data from 1990 to 2020, then "Normalized Difference Built-up Index" and "Normalized Difference Water Index" were applied to generate the spatial relation between built-up and water index for the study area.

2 Study Area

Aligarh, another Class-I Indian city in Uttar Pradesh, is located at 27° 88′N latitude and 78° 08′E longitude, and the research area encompasses 40.43 km^2 (Fig. 1). The Aligarh City serves as the administrative epicenter for the Aligarh District and is located in the northwest of the Indian state of Uttar Pradesh (Anjum et al., 2023c). The city is 178 meters above sea level on average. The Aligarh City is well connected by road and rail and is located around 140 km to the south-west and 305 km to the north-west of the capital cities of New Delhi and Kanpur, correspondingly. By connecting with NH-91, the GT

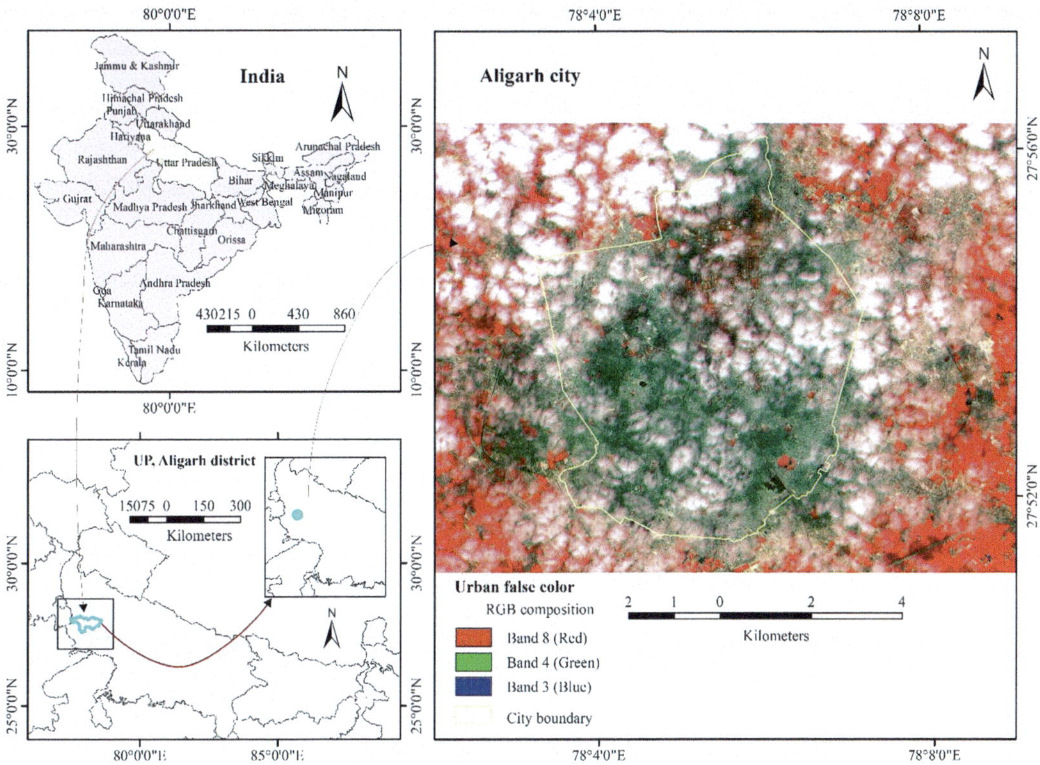

Fig. 1 Location of the study area

Table 1 Expansion of the city from 1971 to 2021

	Population		Urban area	
Period	Net area increase	Percentage increase	Net increase (km^2)	Percentage increase
1971–81	68,547	27.06	4.78	27.59
1981–91	1,59,659	49.76	7.02	31.76
1991–2001	1,88,567	39.24	2.442	8.38
2001–2011	2,03,413	30.4	4.43	14.03
2011–2021	3,17,875	36.43	29.55	82.08

Source: Farooq and Ahmad (2008)

road runs through the city. During the decades from 1990 to 2000s, another phase of evolution started when the quantity of establishments extended the encroached areas of the city (Table 1). The study focuses on the spatial architecture and norms of the city by observing the trends of its component parts and the mechanisms of their progression.

Here, we observe a subtropical humid climate affected by the monsoon. A 227 mm average rainfall and temperatures between 28 °C and 38 °C are the norm. The city is well-known for its lock and brass manufacturing industries as well as several enterprises that process and manufacture agricultural products. In 2001, the total number of population of the study area was 676,000, which shot up to 874,408, respectively, with 461,772 men and 412,636 women (2001 census). The most recent (census 2011) indicates that there are roughly 147,363 households in Aligarh with a literacy rate of 59.15% (2011 census). Aligarh City has been expanding steadily for decades due to population growth. A significant amount of formerly undeveloped or agricultural land has been developed into populated regions, increasing the number of wards. As a result, there is a noticeable modification in the city's land cover throughout time. Numerous environmental problems, reduction of vegetation, water pollution, flooding, indoor and outdoor air pollution in cities, incorrect solid waste disposal, etc., have become more prevalent due to rising population and urbanization.

3 Database and Methodology

3.1 Database

Remote sensing data are basically prime sources to obtain information for the purpose of analyzing environmental processes on both the local and global levels. In order to visualize and calculate the surface water bodies along with existing wetlands, satellite imageries from Landsat 5 for 1990, 2000, and 2010, and Landsat 8 for 2020 with 30 m resolution were selected. These satellite imageries were associated with studying the chronological as well as spatial variations in the classification of land cover in and around wetlands. All images were acquired with 0.1% or 0% cloud coverage, mostly during the winter season, so that the seasonal variation cannot affect the process of classification, which can impact the quality of image classification (Table 2).

Moreover, all images fall under the tier 1 collection (L1T), which means they are previously modified by using ground control points (GCPs) and digital elevation model (DEM) for the correction of view angle and terrain correlation. Every time series is basically rectified as well as georeferenced into Universal Transverse Mercator (UTM). Zone 36 and datum World Geodetic System (WGS) 1984.

This study used 40 years data to analyse and compare the existing situation of wetlands in the study area. Apart from surface water bodies, the wetland area includes areas that contain high moisture. They are an important aspect of ecosystem and a potential groundwater recharge point, capable enough to hinder the near-infrared (NIR) wavelength hence considered for the study area. Four cloud-free images were obtained with a time interval of 10 years.

Built-up area shows its direct impact on the encroachment of surface water bodies and agricultural land. Hence to identify the reason for changes in water bodies, Normalized Difference Built Index (NDBI) was also considered. Water bodies and wetland mapping were done with the help of the "Normalized Difference Water Index" (NDWI).

Satellite image data is not sufficient for knowing the actual reasons; hence, in order to verify the results obtained from remote sensing imageries, a total of 11 surface water bodies were chosen for the field visit. Their details are mentioned below (Table 3).

3.2 Methodology

The NDWI, proposed by McFeeters (1996), was evaluated as a drought indicator, and it is utilized to identify and monitor the humidity status of vegetation canopies over broad regions. (NDWI) is a tool for distinguishing water from arid ground or, more specifically, for water body mapping. In the visible infrared wavelength region, water bodies contain minimal radiation and high absorbability. Based on the occurrence, the NDWI utilizes near-infrared (NIR) and green bands of remote sensing images. In the vast majority of circumstances, it can effectively

Table 2 Obtained Landsat satellite imageries properties for wetland mapping in Aligarh City

Satellite ID	Sensor ID	Path/row	Acquisition date	Spatial resolution	Quality
Landsat 5	TM	146/41	30/05/1990	30 m	9
Landsat 5	TM	146/41	25/05/2000	30 m	9
Landsat 5	TM	146/41	21/05/2010	30 m	9
Landsat 8	OLI_TIRS	146/41	07/06/2020	30 m	9

Table 3 Water body locations for field visit

Sl. no.	Place	Latitude	Longitude
1	Ambedkar Nagar water body	27.872930°	78.087808°
2	Kali Deh	27.870999°	78.079724°
3	Loco pond	27.908338°	78.061976°
4	Alam pond	27.912354°	78.060862°
5	Chautal	27.913977°	78.067645°
6	Achal Tal	27.879370°	78.075918°
7	Nai Basti pond	27.897924°	78.066734°
8	Laal Diggi water body	27.902309°	78.076362°
9	Near Abdllah hall swamp	27.900825°	78.081700°
10	Mulla Para water body	27.861378°	78.060515°
11	Unnamed water body	27.869864°	78.086660°

recover water information. It's subtle in land development and frequently manifests itself in exaggerated water bodies. The NDWI was calculated by using the Eq. 1.

$$NDWI = \frac{(NIR - SWIR)}{(NIR + SWIR)} \quad (1)$$

To highlight man-made built-up areas, the "Normalized Difference Built-up Index" (NDBI) engages the NIR and SWIR bands. It is ratio based to pay off for changes in topographic luminance as well as ambient influences. The NDBI was intended to work using Landsat TM bands 5 and 4 when it was first released. Nevertheless, any multispectral sensor with a SWIR band of 1.55–1.75 m and an NIR band of 0.76–0.9 m will operate with it. Built-up features are extracted using the "Normalized Difference Built-up Index" (NDBI), where indices range from −1 to 1. This index indicates urban areas where the shortwave-infrared (SWIR) reflectance is usually greater than the near-infrared (NIR) reflectance (He et al., 2010). Watershed runoff estimates and land-use planning are two instances of applications. The NDBI was calculated by using Eq. 2.

$$NDBI = \frac{(SWIR - NIR)}{(SWIR + NIR)} \quad (2)$$

where,

NIR = Band 5 with the wavelength of 0.851–0.879.

SWIR = Band 6 with the wavelength of 1.566–1.651.

The bands used for the NDWI and NDBI vary based on the Landsat version. For Landsat 5 (TM) satellite: 2, 3, and 5 bands denote Green, Red, and NIR, while in Landsat 8 (OLI) satellite: 3, 4, and 5 bands depict Green, Red, and NIR accordingly.

The NDBI and NDWI values differ from −1 to 1 based on various factors. For the NDBI, negative values denote water bodies, where upper values denote built areas, and for the NDWI, values above 0.5 signify water bodies and minimum values signify the presence of vegetation and other features. (Donia, 2018). The NDBI and NDWI equations (Eqs. 1 and 2) were performed by using the raster calculator in ArcGIS, then an unsupervised classification was used to represent the values on the map.

4 Results

4.1 Land Cover Classification Analysis

There are various scholars who discovered unsupervised classification being better than supervised classification in terms of the classification of wetland areas (Rezaeianzadeh et al., 2018), but when the concern of recognizing the small water bodies and marshlands, the overlapping occurs in unsupervised classification. Hence, in this study, for the purpose of better reorganization of wetlands and surface water bodies, supervised classification has been used (Cawkwell, et al., 2007).

4.2 Relation with Built-up Area

Human need is an unending requirement, and they continue to fulfill it since the beginning of civilization. As time passes, people prefer areas with more opportunities and better living, hence, the inflow of population towards urban areas is a natural phenomenon. In the context of Indian cities, as per the 2011 census report, there are over a thousand Class-I cities in India, and the figure of metropolitan cities has also been amplified since 2001. This rapid urbanization trend was observed in Aligarh City as well. As per the ADA report, the total number of households were 32,787 in 1971, which increased up to 1,43,363 in 2011, and the projected number of households surpassed over a million population within the city in 2021. After increased housing, the requirement for other infrastructure has also become a necessity, like flyovers, commercial areas, schools, and hospitals, which as a result succumb the surrounding available land that is being used by other activities like agriculture, vegetation, and wetlands (Figs. 2 and 3). Here in this study, the NDBI has been used to exhibit the existing built-up regions along with the previous records from 1991 (Tables 4 and 5).

4.2.1 Built-up Area Coverage in 1990

The total built-up zone has concealed 44.43% of the whole area of 36 km^2, where the highly built-up area has only 10.70% coverage only. If we remove highly built-up areas, then only 33.73% areas have been covered by areas that were transforming into built-up areas.

4.2.2 Built-up Area Coverage in 2000

The total built-up areas have the occupancy of 63.44% and in terms of densely built-up areas, 31.75% areas have been accounted. Apart from the densely built-up areas, 31.70% areas have been recognized as the low built-up or transforming zone. Compared to the previous decade, 21.05% areas have been increased in terms of highly urbanized areas, and regarding total built-up area, 19.01% areas have been added to the accountability. Compared to the previous decade, 21.05% areas have been increased in terms of highly built-up areas, and in terms of total built-up area, 19.01% areas have been added to the accountability. In this year, mostly high built-up areas have been increased, but the transformation of low built-up areas has shown negative growth of 2.04%.

4.2.3 Built-up Area Coverage in 2010

The total built-up zone has covered 66.97% areas, where in terms of highly built-up areas, 31.17% areas have been covered. Apart from highly built-up areas, 35.80% areas have recorded conversion from other classifications to built-up areas. in comparison to the previous decade, 3.53% areas have been increased in terms of total built-up areas, and in terms of dense built-up areas, 0.58% areas have shown negative growth. The areas transformed into built-up areas have covered 4.10% areas.

4.2.4 Built-up Area Coverage in 2020

About 70.45% of the total area has been covered by total built-up areas, and in terms of highly built-up areas, 40.44% areas have been covered. The low built-up area has covered 30.00% areas out of the total area. In contrast to the prior decade, 3.48% areas have been increased in terms of total built-up areas, where dense built-up areas have shown positive growth of 9.27%. The total transformation of other land cover classification into built area has been 5.79% less compared to the previous decade.

4.3 Existence of Wetlands

The results acquired from the classification of wetland for the years 1990, 2000, 2010, and 2020 using a supervised classification method showed a reasonable result by using NDWI (Figs. 4 and 5; Tables 6 and 7). The study of four decades showed that both positive and negative growth took place over the year.

4.3.1 Wetland Area Coverage in 1990

The total water bodies were 2.13% of the entire city area, but in the wetland area, the NDWI has occupied 13.86% of the total area. If we exclude surface water bodies, the total 11.74% of the area falls under other wetland regions except water bodies. Since they are very important aspects of ecosystem, hence their reorganization has been considered.

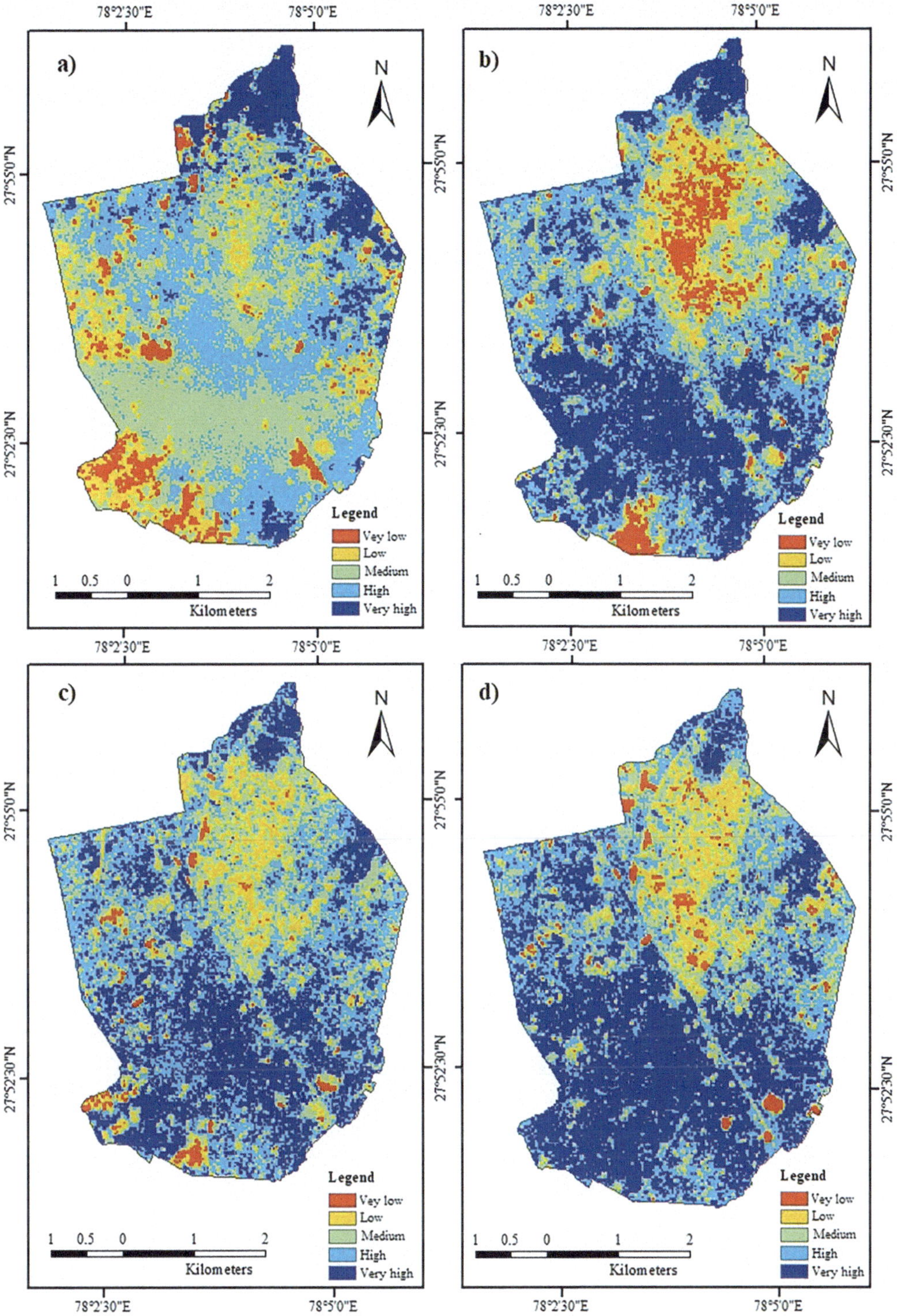

Fig. 2 Normalized difference built index where: (**a**) 1990, (**b**) 2000, (**c**) 2010, and (**d**) 2020

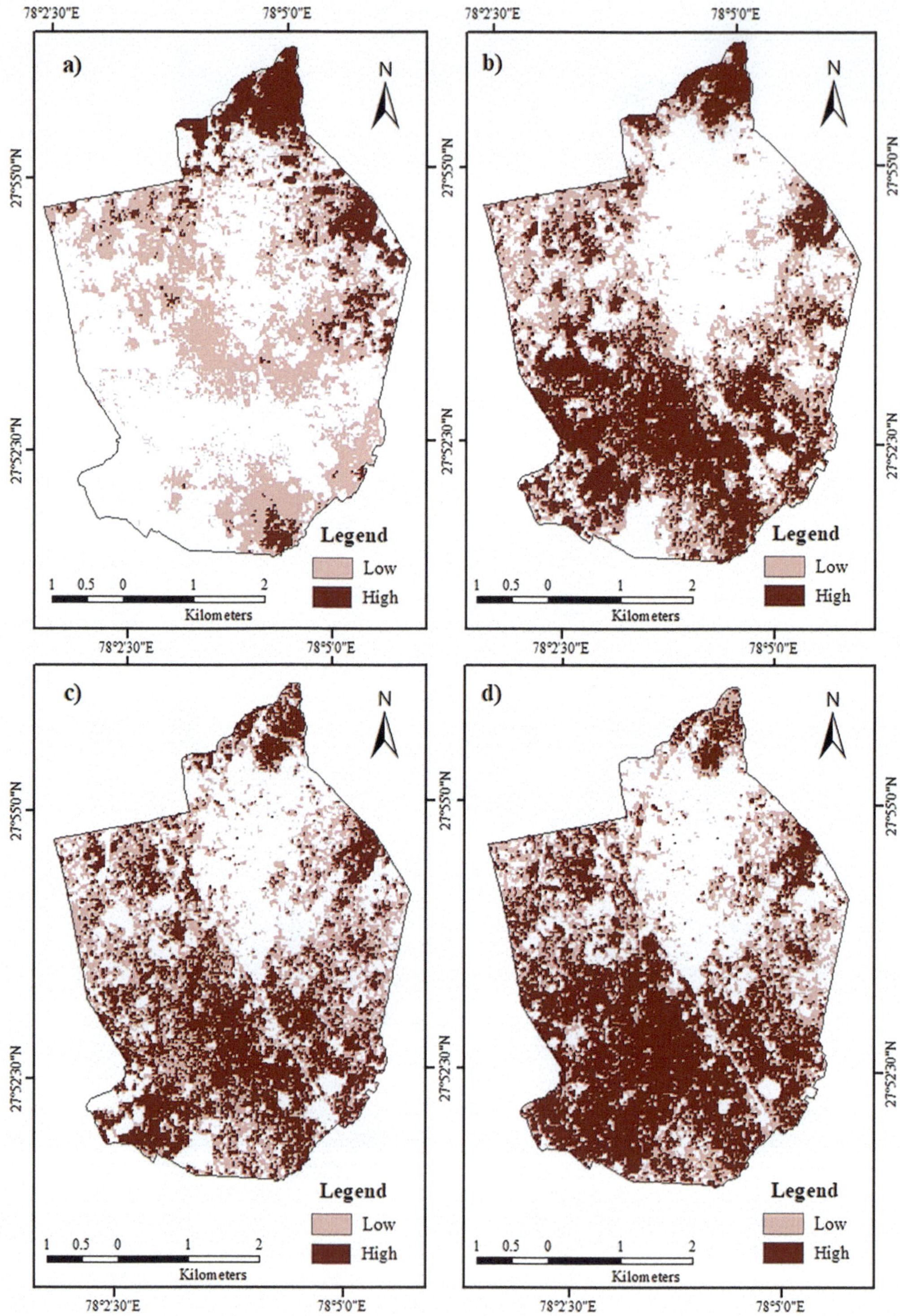

Fig. 3 Normalized difference built index density where: (**a**) 1990, (**b**) 2000, (**c**) 2010, and (**d**) 2020

Table 4 NDBI of four decades

1990		2000		2010		2020	
Pixels	Percentage (%)	Pixels	Percentage (%)	Pixels	Percentage (%)	Pixels	Percentage (%)
1908	4.86	2117	5.40	597	1.52	825	2.10
5642	14.38	4766	12.15	4327	11.03	4170	10.63
14,246	36.32	7456	19.01	8031	20.47	6594	16.81
13,230	33.73	12,431	31.70	14,041	35.80	11,769	30.00
4198	10.70	12,454	31.75	12,228	31.17	15,866	40.44

Table 5 Built-up area coverage

Class name	Year	Percentage (%)	Year	Percentage (%)	Year	Percentage (%)	Year	Percentage (%)
High built-up area	1990	10.7	2000	31.75	2010	31.17	2020	40.44
Total built-up area	1990	44.43	2000	63.44	2010	66.97	2020	70.45

4.3.2 Wetland Area Coverage in 2000

The total of 3.46% of the region was categorized under surface water bodies in the year 2000, and the total wetland of Aligarh City has covered 10.29% of the area. Out of the total wetland area, apart from surface water bodies like ponds, *Taals*, the wetland area (excluding water bodies) covers 6.83% of the total study area. Compared to the last decade, 4.9% of the wetland has decreased, but in terms of total wetland area, 3.49% area has shown a reduction in their size, but in surface water, 1.42% area has shown a positive increase over a decade.

4.3.3 Wetland Area Coverage in 2010

Total surface water bodies in Aligarh City have been 2.34% coverage out of the total area, whereas in terms of total wetland, 13.88% area has been categorized. Out of water bodies, only wetland coverage is 11.54% out of the total study area. Compared to the last decade, 4.71% area has shown a positive increase in other wetland areas (excluding water bodies) where in surface water bodies, 1.12% area showed negative growth in Aligarh City. On overall wetland area, 3.59% area has shown positive growth over a decade.

4.3.4 Wetland Area Coverage in 2020

Total surface water bodies in the city have been 2.07% of the total area, where the total wetland area covered 13.05% of the region. Excluding surface water bodies, only other wetland areas cover 10.98% of the entire city region. In comparison to the last decade, other wetland areas (excluding water bodies) covered 0.56% less area in 2020, where surface water bodies cover showed 0.27% negative growth from the last decade. The overall wetland area in 2020 covered 0.83% less area compared to 2010.

4.4 Change Detection

Digital change detection is a process that is utilized to estimate the changes that took place in the land cover of a specific region over a specific period of time interval with the help of various multispectral image data (Figs. 6 and 7). The concept has been built on the comparison between two or more images of a specific area at various times, like post-classification or pixel-based comparison.

Here in this study, post classification comparison has been selected to identify the change digitally on images dated 1990, 1999, 2010, and 2020. Post classification technique has been categorized by the capability to show the quantity and location in each and every land cover classes and the percentage of changes into other classes (Table 8).

Based on the year 2020, change detection has been performed to show the changes took place for the temporal years, i.e., 1990–2020, 2000–2020, and 2010–2020 of NDWI indices

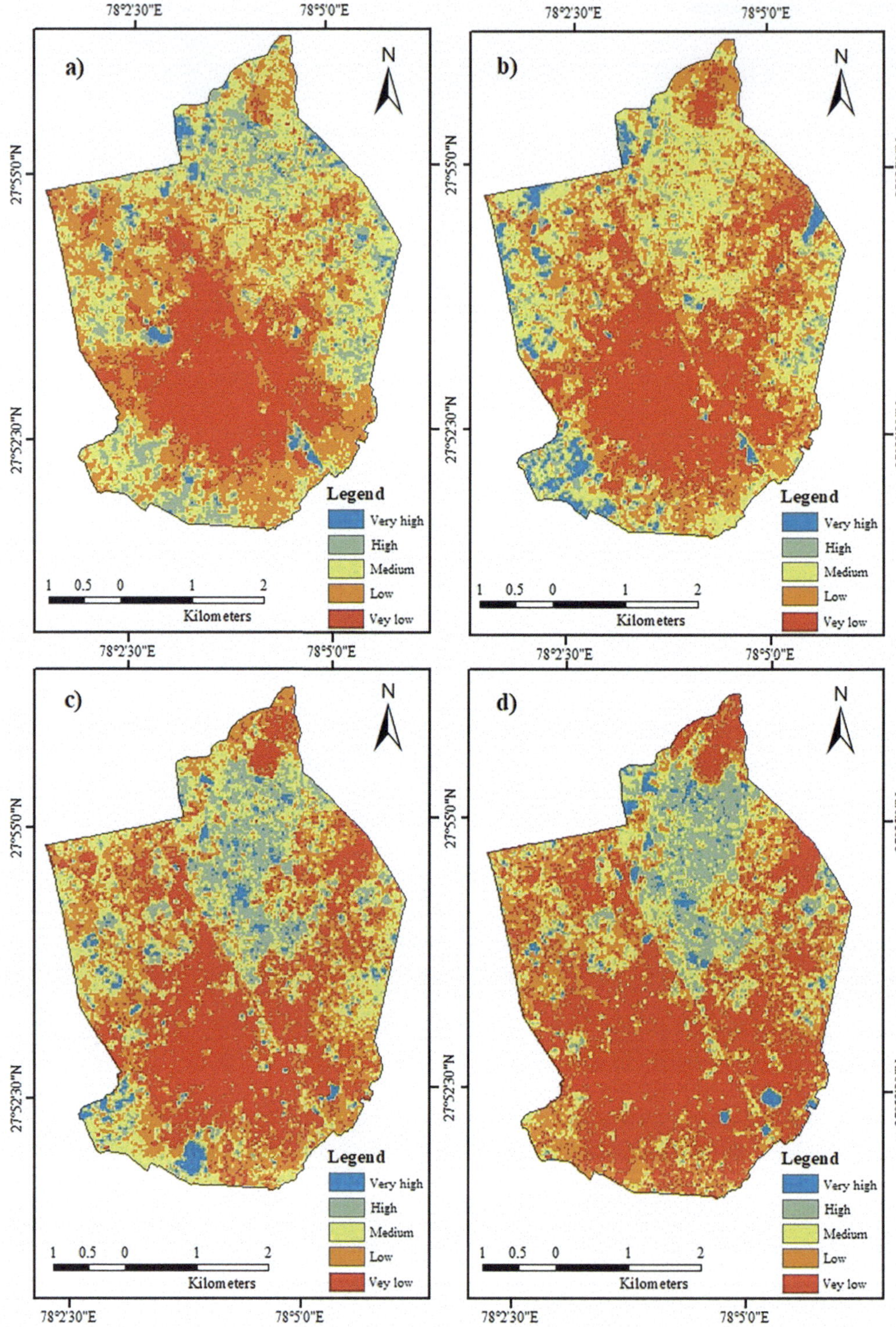

Fig. 4 "Normalized Difference Water Index" where: (**a**) 1990, (**b**) 2000, (**c**) 2010, and (**d**)2020

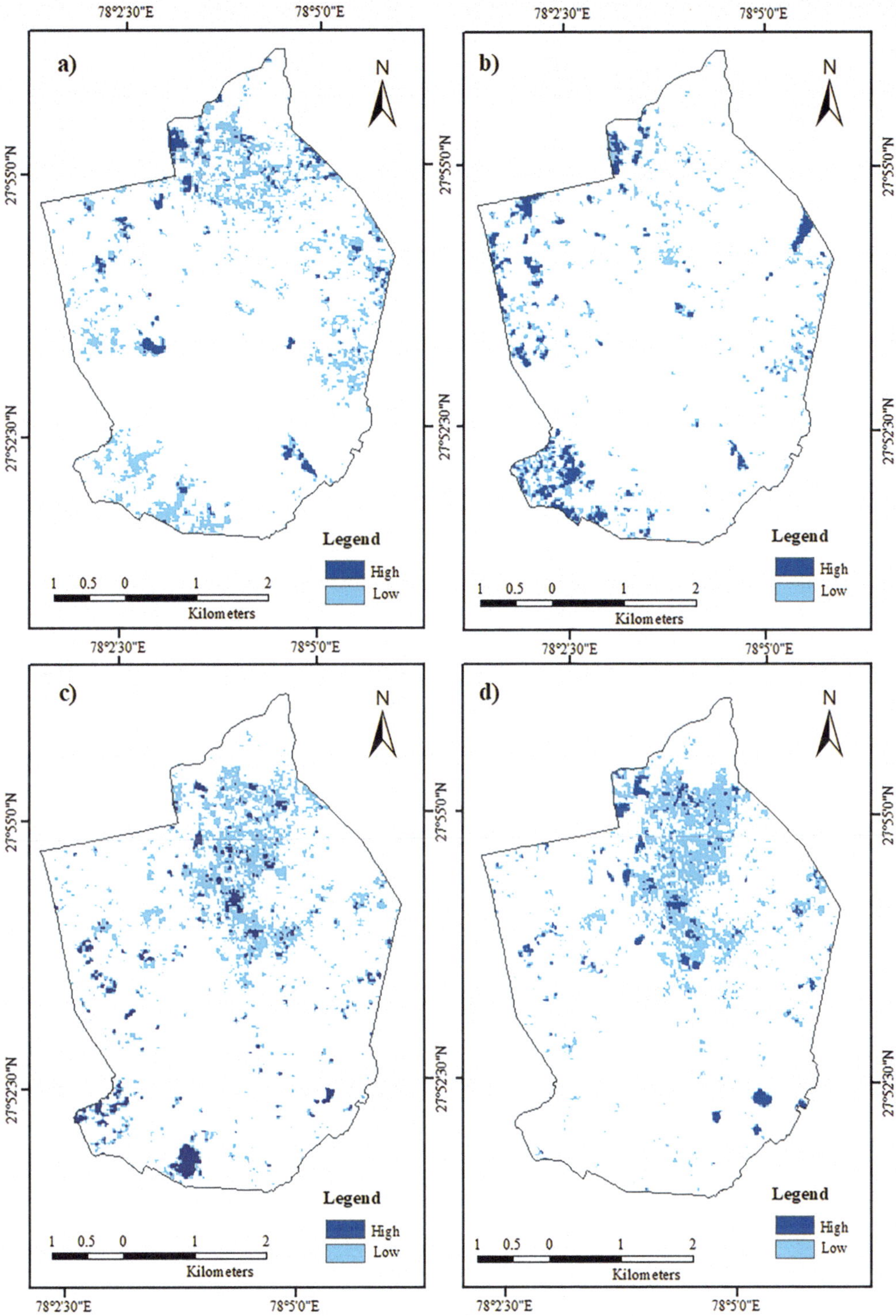

Fig. 5 Existence of water bodies where: (**a**) 1990, (**b**) 2000, (**c**) 2010, and (**d**) 2020

Table 6 NDWI of four decades

1990		2000		2010		2020	
Pixels	Percentage (%)	Pixels	Percentage (%)	Pixels	Percentage (%)	Pixels	Percentage (%)
835	2.1287	1389	3.5411	1152	2.9v369	923	2.3531
4603	11.7351	2679	6.8300	4917	12.5356	4792	12.2170
10,721	27.3327	10,900	27.7891	9355	23.8501	6802	17.3414
13,907	35.4553	13,378	34.1066	12,664	32.2863	12,058	30.7413
9158	23.3479	10,878	27.7330	11,136	28.3907	14,649	37.3470

Table 7 Surface water and total wetland area coverage

Class name	Year	Percentage (%)	Year	Percentage (%)	Year	Percentage (%)	Year	Percentage (%)
Surface water bodies	1990	2.13	2000	3.46	2010	2.34	2020	2.07
Total wetland area	1990	11.74	2000	6.83	2010	11.54	2020	10.98

shown on Fig. 5. The result of the study exposed that there has an obvious degradation of surface water bodies in Aligarh City while the built up area has an exponential increasing rate on wetland shown on Table.4.

1990–2020: The total surface water body extent was 2.13% of the total pixel values, where in 2020, the total surface water bodies were 2.07%. Based on the percentage, it is clear that changes occurred up to some extent during the four decades. However, in terms of total wetland areas, 13.05% areas were covered in 2020, but 13.86% area were considered as wetland areas in 1990. Hence, in four decades 0.76% wetland area has been reduced, and in surface water body areas, 0.06% area has been decreased.

2000–2020: The surface water body has surprisingly increased in the year 2000 compared to 1990. In 2000, 3.46% area fell under surface water bodies where 2.07% area fell under surface water bodies in 2020. In terms of overall wetland area, 10.29% area comes under this category, whereas 13.05% area comes under this category in the year 2020. In two decades, the surface water body area had a 1.39% reduction, but in terms of overall wetland areas assessment, 4.15% area has been increased in the year 2020.

2010–2020: In the last decade, the surface water body rate was 2.34% area whereas in 2020, it was 2.07% area but in terms of wetland area, 11.54% area comes under this category in 2010, and 10.98% area comes under this category in the year 2020. In terms of surface water body change, 0.27% space was consumed by built-up area, and in terms of the whole wetland area, 0.56% area has been reduced and transformed into other indices.

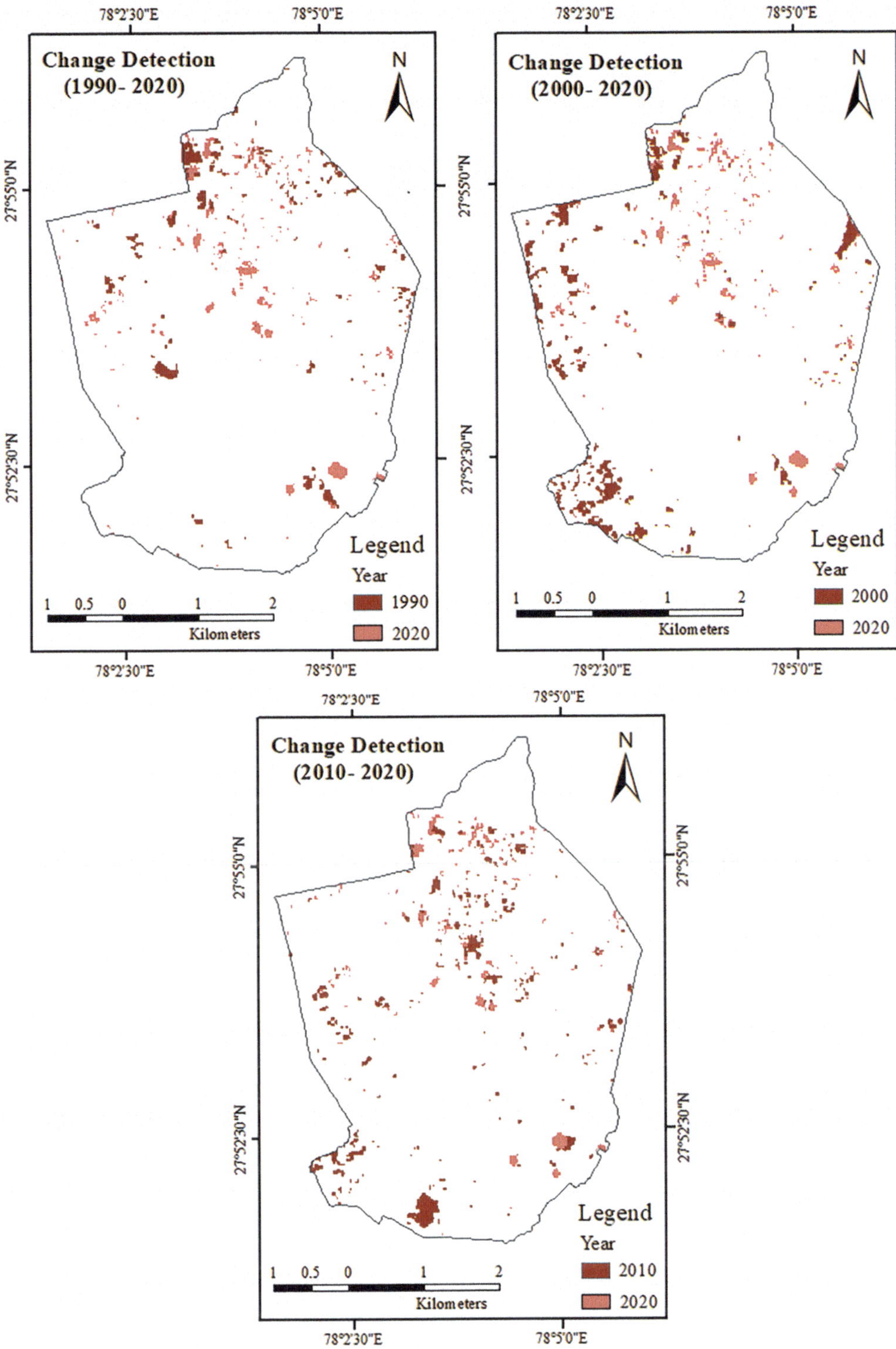

Fig. 6 Changes in water bodies in comparison to 2020

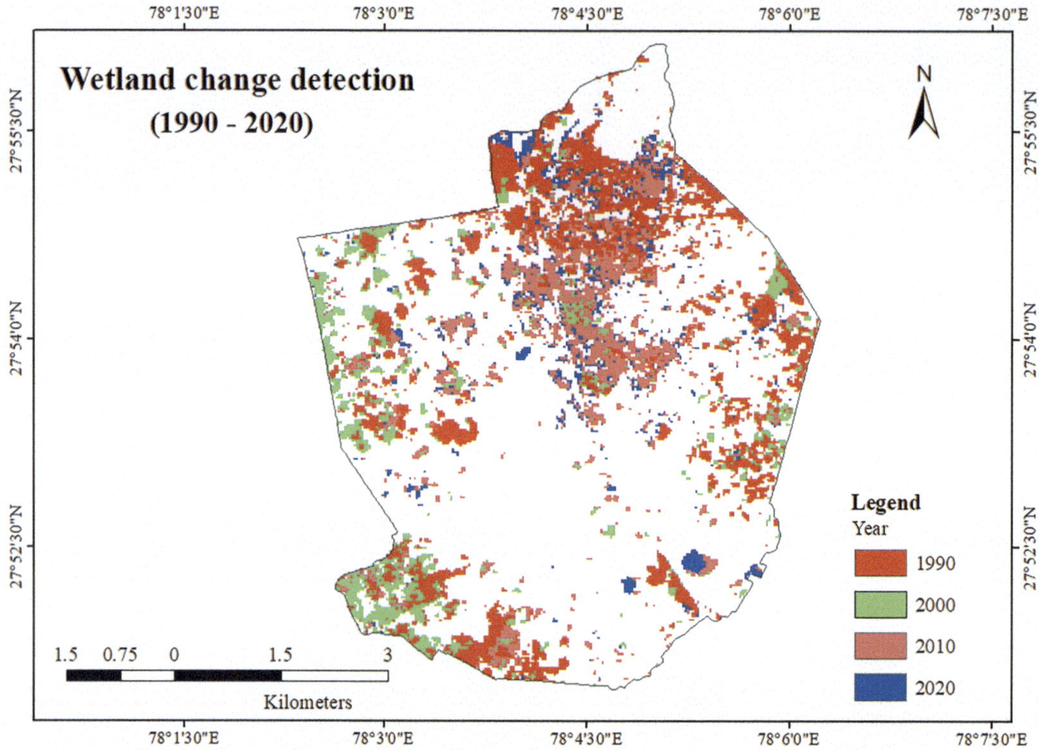

Fig. 7 Decadal changes in wetland

Table 8 Change detection of each year with 2020

Class name	Year	Percentage (%)	Year	Percentage (%)	Year	Percentage (%)
Surface water bodies	1990–2020	−0.06	2000–2020	−1.39	2010–2020	−0.27
Total wetland area	1990–2020	−0.76	2000–2020	+4.15	2010–2020	−0.56

5 Discussions

Fast urban concrete area extension was acknowledged as an essential cause for a considerable damage to natural habitats around the globe (Bassi et al., 2014; Cosgrove & Loucks, 2015; El-Hennawy, 2010; Heiden et al., 2012). Valuation of wetland area is becoming a fundamental ingredient for the purpose of management planning processes for surface water bodies or wetlands (Holland et al., 1995; Peng et al., 2020; Debanshi & Pal, 2020; Dong et al., 2022). After the construction of buildings, flyovers, and other commercial areas, inundation frequency and the magnitude of change were observed. It caused the destruction of wetland habitat and vulnerability of potential groundwater recharge (Jusuf et al., 2007; Verpoorter et al., 2012; Ouma & Tateishi, 2006). In past, numerous scholars tried to comprehend the land use dynamics of the study area.

In this study, this matter was addressed by generating land use alteration data of four decades to demonstrate the influence of built-up area extension on wetland coverage loss. In Aligarh City, hydrological alteration because of the interference of built-up area is found to become a major stimulating factor to expose the surface water body's vulnerability. To strengthen the empirical observation, we analyzed the historical land cover record of Aligarh from 1990 to 2020 to discover the straight and subsidiary influences of built-up area extension on existing wetlands (Fig. 6).

Moreover, irregular spilling felicitated humans to invade into wetland for concretization. It refers to the potential reclamation of wetland, more specifically surface water bodies, into built-up areas, and this is evident to see from the detection of change. Apart from converting wetland into cultivable lands, this mechanism is more highly responsible for making the wetland physically vulnerable. The growing fragmentation of wetland into small patches of buildings or roads is the primary indicator of degradation of wetland. In the present work, the number of residential households increased from 86,475 to 1,43,363, and the number of population also increased from 6,69,087 to 8,72,500 from 2001 to 2011 (2011 census).

This work assessed the condition of wetlands, more specifically surface water bodies, in a span of four-decade time series from 1990 to 2020 based on the NDWI and NDBI. All the decadal outcomes showed that 44.43–70.45% area was transformed in the span of four decades, which sums up as 26.02% area falls under total changes in built-up area, and 13.86–13.05% of the total wetland was decreased by the built-up area in four decades, which sums up as 0.81% area change.

The maximum value of each pixel in the map shows the surface water bodies because their size and depth absorb more near infra-red radiation, and the rest of the areas shows medium and low values, indicating the wetland areas altogether. The demarcation of surface water bodies showed a sharp decrease in its number from 2.13% in 1990 to 2.07% in 2020, and the validation of the result was proved by the NDBI values, where the maximum value 10.70% was identified in the year 1990, but in 2020 the value rose up to 40.44%, respectively. This showed a total 29.74% of growth in built areas from 1990 to 2020. Insufficient rainfall, the squeezing of active wetland areas, and encroachment of residential areas due to public demand are some prime causes behind augmented wetland change detection. Based on the field visit, it was evident that not only the urban encroachment, but also the improper solid waste dumping resulted in the disappearance of water bodies. If this trend of unplanned urban growth and transformation of surface water bodies into a dumping ground continues, the consequences will become havoc to the ecosystem and mankind. The forecasting shows that surface water bodies will reach up to 1.09% which had the 2.12% coverage in 1990, and other wetland areas, which had 11.73% area coverage, will reach up to 10.06% in the year 2050 (Table 9). If precautions are not taken on time, the repercussions will be too massive to control (Photo Plate 1).

Table 9 Estimation of water indices

Class name	Year						
	1990	2000	2010	2020	2030*	2040*	2050*
Surface water bodies	2.12	3.46	2.34	2.07	1.47	1.26	1.09
Total wetland area	11.73	6.83	11.54	10.98	10.83	10.42	10.06

*Predicted

Photo Plate 1 Physical condition of some selected water bodies

6 Conclusions

The study can be summarized as:

(i) Aligarh City had a noteworthy change in built-up area cover from 1990 to 2020 as a result of fast residential expansion, climate change, and increased demand for settlements, among other factors.
(ii) The cumulative impact of these operations has resulted in a decrease in water body capacity, deterioration of water quality, and loss of aesthetic value of Aligarh City's water bodies as evident by the NDWI record from 1990 to 2020.
(iii) The field visit shows that there is an urgent need to conserve the study area's depleting water sources and their environment. Sewage, industrial effluents, agricultural runoff, pesticides, insecticides, siltation, expanding population, improper solid waste management, and a long list of other issues are to blame for the deteriorating and diminishing conditions of these water bodies.

7 Recommendations

After the foregoing discussion and conclusion, the following recommendations can be made:

(i) The city administration must keep an eye on unauthorized settlements and urban encroachment.
(ii) Proper monitoring and maintenance of surface water bodies should be done by the administration.
(iii) In order to maintain the green infrastructure, the government can transform water bodies into public parks with a nice pond view. It will not only maintain the condition of water bodies but also maintain the aesthetic value of the park.
(iv) Strict action should be taken for improper garbage disposal at the site.

Consent for Publication Not applicable.

Availability of Data and Materials The data that support the findings of this study are available from the corresponding authors upon reasonable request.

Competing Interests The authors declare that they have no conflict of interest.

Funding No funding was received from any sources.

References

Ajmal, U., & Jamal, S. (2021). Analyzing land-use land-cover change and future urban growth with respect to the location of slaughterhouses in Aligarh city outskirts. *Environmental Challenges, 5*, 100331.

Ali, S. A., Parvin, F., & Anjum, R. (2023). GIS-based disaster risk analysis of floods using certainty factor (CF) and its ensemble with deep learning neural network (DLNN): A case study of Dima Hasao District of Assam, India. In *Emerging technologies for water supply, conservation and management* (pp. 209–234). Springer.

Anjum, R., Ali, S. A., & Siddiqui, M. A. (2023a). Assessing the impact of land cover on groundwater quality in a smart city using GIS and machine learning algorithms. *Water, Air, & Soil Pollution, 234*(3), 182.

Anjum, R., Parvin, F., & Ali, S. A. (2023b). Machine learning applications in sustainable water resource management: A systematic review. In *Emerging Technologies for Water Supply, Conservation and Management* (pp. 29–47).

Anjum, R., Ali, S. A., Siddiqui, M. A., Parvin, F., Khan, Z., Khan, N., et al. (2023c). Hydro-geochemical assessment of ground water for drinking and agricultural purposes and potential human health risk in Aligarh city, India. *Chemical Engineering Journal Advances, 16*, 100547.

Arulbalaji, P., & Gurugnanam, B. (2014). Geospatial science for 16 years of variation in land use/land cover practice assessment around Salem District, South India. *Journal of Geosciences and Geomatics, 2*(1), 17–20.

Bassi, N., Kumar, M. D., Sharma, A., & Pardha-Saradhi, P. (2014). Status of wetlands in India: A review of extent, ecosystem benefits, threats and management strategies. *Journal of Hydrology: Regional Studies, 2*, 1–19.

Cawkwell, F. G., Dwyer, N., Bartlett, D., Ameztoy, I., O'Connor, B., O'Dea, L., … & Thomas, I. (2007). *Saltmarsh habitat classification from satellite imagery*.

Choudhury, D., Das, K., & Das, A. (2019). Assessment of land use land cover changes and its impact on variations of land surface temperature in Asansol-Durgapur Development Region. *The Egyptian Journal of Remote Sensing and Space Science, 22*(2), 203–218.

Cosgrove, W. J., & Loucks, D. P. (2015). Water management: Current and future challenges and research directions. *Water Resources Research, 51*(6), 4823–4839.

Debanshi, S., & Pal, S. (2020). Wetland delineation simulation and prediction in deltaic landscape. *Ecological Indicators, 108*, 105757.

Dembélé, A., Ye, X., & Touré, A. (2018). Analysis of land surface temperature change based on MODIS data, case study: Inner Delta of Niger. *Natural Hazards and Earth System Sciences Discussions*, 1–20.

Demuth, S., & Radojevic, B. (2011). Global change and its impact on water resources: The role of UNESCO's international hydrological programme. *Water Research and Management, 1*(1), 7–15.

Dong, L., Wan, R., Li, B., Tan, Z., Yang, S., & Zhang, T. (2022). Spatiotemporal dynamics of lake wetland in the Wanjiang Plain of the Yangtze River basin, China during the recent century. *Ecological Indicators, 142*, 109295.

Donia, N. (2018). NDWI based change detection analysis of Qarun Lake Coastal Area, El-Fayoum, Egypt. In *Conference of the Arabian journal of geosciences* (pp. 121–124). Springer.

El-Asmar, H. M., Hereher, M. E., & El Kafrawy, S. B. (2013). Surface area change detection of the Burullus lagoon, north of the Nile Delta, Egypt, using water indices: A remote sensing approach. *The Egyptian Journal of Remote Sensing and Space Science, 16*(1), 119–123.

El-Hennawy, M. T. (2010). *Ecology and Management of Wetland Ecosystem: Wadi El-559 Rayan, Western Desert, Egypt*. Botany Departement, Faculty of Science, Ain Shams Univesity. Dissertation.

Farooq, S., & Ahmad, S. (2008). Urban sprawl development around Aligarh city: A study aided by satellite remote sensing and GIS. *Journal of the Indian Society of Remote Sensing, 36*, 77–88.

He, C., Shi, P., Xie, D., & Zhao, Y. (2010). Improving the normalized difference built-up index to map urban built-up areas using a semiautomatic segmentation approach. *Remote Sensing Letters, 1*(4), 213–221.

Heiden, U., Heldens, W., Roessner, S., Segl, K., Esch, T., & Mueller, A. (2012). Urban structure type characterization using hyperspectral remote sensing and height information. *Landscape and Urban Planning, 105*(4), 361–375.

Holland, C. C., Honea, J., Gwin, S. E., & Kentula, M. E. (1995). Wetland degradation and loss in the rapidly urbanizing area of Portland, Oregon. *Wetlands, 15*, 336–345.

Jusuf, S. K., Wong, N. H., Hagen, E., Anggoro, R., & Hong, Y. (2007). The influence of land use on the urban heat Island in Singapore. *Habitat International, 31*(2), 232–242.

Khalil, N., & Khan, M. (2009). A case of a municipal solid waste management system for a medium-sized Indian city, Aligarh. *Management of Environmental Quality: An International Journal, 20*(2), 121–141.

Mascarenhas, A., Haase, D., Ramos, T. B., & Santos, R. (2019). Pathways of demographic and urban development and their effects on land take and ecosystem services: The case of Lisbon Metropolitan Area, Portugal. *Land Use Policy, 82*, 181–194.

Mathew, A., Khandelwal, S., & Kaul, N. (2016). Spatial and temporal variations of urban heat Island effect and the effect of percentage impervious surface area and elevation on land surface temperature: Study of Chandigarh city, India. *Sustainable Cities and Society, 26*, 264–277.

McFeeters, S. K. (1996). The use of the Normalized Difference Water Index (NDWI) in the delineation of open water features. *International Journal of Remote Sensing, 17*(7), 1425–1432.

Meshesha, T. W., Tripathi, S. K., & Khare, D. (2016). Analyses of land use and land cover change dynamics using GIS and remote sensing during 1984 and 2015 in the Beressa Watershed Northern Central Highland of Ethiopia. *Modeling Earth Systems and Environment, 2*(4), 1–12.

Milan, B. F., & Creutzig, F. (2016). Municipal policies accelerated urban sprawl and public debts in Spain. *Land Use Policy, 54*, 103–115.

Mukherjee, N., Siddique, G., Basak, A., Roy, A., & Mandal, M. H. (2019). Climate change and livelihood vulnerability of the local population on Sagar Island, India. *Chinese Geographical Science, 29*(3), 417–436.

Naserikia, M., Asadi Shamsabadi, E., Rafieian, M., & Leal Filho, W. (2019). The urban heat Island in an urban context: A case study of Mashhad, Iran. *International Journal of Environmental Research and Public Health, 16*(3), 313.

Ouma, Y. O., & Tateishi, R. (2006). A water index for rapid mapping of shoreline changes of five East African Rift Valley lakes: An empirical analysis using Landsat TM and ETM+ data. *International Journal of Remote Sensing, 27*(15), 3153–3181.

Peng, K., Jiang, W., Deng, Y., Liu, Y., Wu, Z., & Chen, Z. (2020). Simulating wetland changes under different scenarios based on integrating the random forest and CLUE-S models: A case study of Wuhan Urban Agglomeration. *Ecological Indicators, 117*, 106671.

Rezaeianzadeh, M., Kalin, L., & Hantush, M. M. (2018). An integrated approach for modeling wetland water level: Application to a headwater wetland in coastal Alabama, USA. *Water, 10*(7), 879.

Saha, S. K., Kudrat, M., & Bhan, S. K. (1990). Digital processing of Landsat TM data for wasteland mapping in parts of Aligarh District (Uttar Pradesh), India. *International Journal of Remote Sensing, 11*(3), 485–492.

Singh, A. L., & Asgher, M. S. (2005). Impact of brick kilns on land use/landcover changes around Aligarh city, India. *Habitat International, 29*(3), 591–602.

Sivaramakrishna, C., Rao, C. V., Trimurtulu, G., Vanisree, M., & Subbaraju, G. V. (2005). Triterpenoid glycosides from Bacopa monnieri. *Phytochemistry, 66*(23), 2719–2728.

Song, C., Huang, B., Ke, L., & Richards, K. S. (2014). Remote sensing of alpine lake water environment changes on the Tibetan Plateau and surroundings: A review. *ISPRS Journal of Photogrammetry and Remote Sensing, 92*, 26–37.

Sun, Q., Wu, Z., & Tan, J. (2012). The relationship between land surface temperature and land use/land cover in Guangzhou, China. *Environmental Earth Sciences, 65*(6), 1687–1694.

UN-Habitat, & United Nations Human Settlements Programme. (2008). *State of the world's cities 2008/9: Harmonious cities*. Earthscan.

United Nations. (2018). *World Urbanization Prospects 2018*. https://population.un.org/wup/Publications/Files/WUP2018-Highlights.pdf

Verpoorter, C., Kutser, T., & Tranvik, L. (2012). Automated mapping of water bodies using Landsat multispectral data. *Limnology and Oceanography: Methods, 10*(12), 1037–1050.

World Health Organization. (2012). *UN-water global annual assessment of sanitation and drinking-water (GLAAS) 2012 report: The challenge of extending and sustaining services*.

Yanan, L., Yuliang, Q., & Yue, Z. (2011). Dynamic monitoring and driving force analysis on rivers and lakes in Zhuhai city using remote sensing technologies. *Procedia Environmental Sciences, 10*, 2677–2683.

Yirsaw, E., Wu, W., Shi, X., et al. (2017). Land use/land cover change modeling and the prediction of subsequent changes in ecosystem service values in a coastal area of China, the Su-Xi-Chang region. *Sustainability, 9*, 1–17. https://doi.org/10.3390/su9071204

Yue, W., Liu, Y., & Fan, P. (2013). Measuring urban sprawl and its drivers in large Chinese cities: The case of Hangzhou. *Land Use Policy, 31*, 358–370.

Importance of Urban Green Spaces for Sustainable Planning: A Case Study of Noida City

Rashid Aziz Faridi and Amit Sharma

1 Introduction

Urban green spaces are essential elements of urban ecosystems, providing environmental benefits and enhancing social and economic well-being within urban settings. This chapter explores these spaces' crucial role in sustainable urban ecosystems, mainly focusing on their integration into the broader urban planning framework.

Urban green spaces, which include parks, gardens, greenways, nature reserves, and other vegetated areas within urban environments, serve as critical components of sustainable urban planning. They contribute in numerous ways to urban areas' ecological, social, and economic dimensions.

Urban green spaces are vital ecological hubs within city landscapes. They support biodiversity by providing habitats for various species and serve as crucial nodes in ecological networks, facilitating species movement and ecological processes (Tzoulas et al., 2007). These spaces are instrumental in maintaining ecological balance, enhancing landscape connectivity, and preserving natural habitats in rapidly urbanizing environments (Chiesura, 2004). Green spaces play a significant role in climate change mitigation and adaptation strategies. They help reduce the urban heat island effect by cooling the air through shade and evapotranspiration (Bowler et al., 2010). Furthermore, they sequester carbon, helping to offset some of the carbon emissions typical in urban areas (Nowak et al., 2006). Green spaces can absorb significant amounts of water during heavy rains, reducing flooding and decreasing surface runoff, thereby mitigating water pollution (Gill et al., 2007). Vegetation in green spaces captures airborne pollutants, including particulate matter and nitrogen dioxide, thus improving air quality. This role is particularly critical in urban settings, where air pollution poses significant health risks to human populations (Escobedo et al., 2011).

Green spaces help recharge groundwater supplies and improve water quality through natural filtration processes by facilitating infiltration (Ferguson, 2005). This contributes to the sustainability of urban water resources, making cities more resilient to water-related issues.

Green spaces often serve as corridors that connect different parts of a city, promoting safer and more pleasant routes for pedestrians and cyclists. This connectivity enhances mobility and encourages more active lifestyles among urban residents (Forsyth & Musacchio, 2005).

R. A. Faridi
Department of Geography, Aligarh Muslim University, Aligarh, India

A. Sharma (✉)
Department of Geography, Aligarh Muslim University, Aligarh, India

New Okhla Industrial Development Authority, Noida, India

Z. Khanam et al. (eds.), *Climate Change, Resilience and Ecosystem Services*, Sustainable Development Goals Series, https://doi.org/10.1007/978-3-031-96977-5_12

Incorporating green spaces into urban planning is not merely an aesthetic choice but a strategic approach to enhancing the resilience and sustainability of urban environments. Effective planning and management of these areas are paramount in fostering ecological stability and improving the quality of life for city dwellers (Jim & Chen, 2006). As urban areas grow, integrating green spaces into the urban fabric will be crucial to sustainable urban development, making cities more livable, environmentally friendly, and socially inclusive (Haaland & van den Bosch, 2015).

1.1 The Role of Urban Green Spaces in Sustainable Urban Ecosystems

Urban green spaces are crucial for the vitality and sustainability of urban areas. These spaces, encompassing parks, gardens, greenways, and nature reserves, not only enrich the ecological diversity of cities but also significantly improve the quality of urban life (Tzoulas et al., 2007). They provide essential habitats for numerous species, acting as sanctuaries amid urban sprawl and maintaining biodiversity by serving as critical links within ecological networks (Chiesura, 2004). This role is significant as cities expand, potentially encroaching on natural habitats and disrupting local ecosystems.

The benefits of urban green spaces extend well into areas of climate regulation. By offering shade and facilitating evapotranspiration—where water is transferred from the land to the atmosphere by evaporation from soil and other surfaces and by plants—these spaces help moderate urban temperatures (Bowler et al., 2010). This is especially crucial in combating the urban heat island effect, where urban regions experience much warmer temperatures than their rural surroundings. Additionally, green spaces act as carbon sinks, absorbing carbon dioxide, a significant greenhouse gas, thus contributing to efforts to mitigate urban carbon emissions (Nowak et al., 2006).

During rainfall, urban green spaces play a pivotal role in water management. Their permeable surfaces allow rainwater to soak into the ground, reducing runoff and decreasing the likelihood of flooding, which is often a problem in heavily paved areas (Gill et al., 2007). This infiltration also helps to recharge groundwater supplies and naturally filters water, enhancing the overall quality of urban water resources (Ferguson, 2005).

Regarding air quality, the vegetation in urban green spaces can significantly reduce pollution levels. Trees and plants capture airborne pollutants—including harmful particulates and gases like nitrogen dioxide— through their leaves and bark, purifying the air and creating healthier living environments (Escobedo et al., 2011). This aspect of green spaces is critical in densely populated urban areas where air quality can often be poor due to traffic and industrial emissions (Jim & Chen, 2006).

Beyond environmental benefits, urban green spaces are vital for social connectivity and community well-being. They provide natural gathering places that foster social interactions and community engagement, which are crucial for mental health and social inclusion (Haaland & van den Bosch, 2015). These areas offer peaceful retreats from the urban hustle and opportunities for recreation. They also promote physical health by encouraging active lifestyles through walking, cycling, and other outdoor activities (Forsyth & Musacchio, 2005).

Integrating green spaces into urban planning is beneficial and necessary for sustainable urban development. As cities grow, including these spaces can make urban environments more resilient, sustainable, and livable (Jim & Chen, 2006). They address environmental and health-related challenges and enhance the aesthetic appeal of urban landscapes, contributing to a higher quality of life for all city residents (Chiesura, 2004). As such, urban planners and policymakers must prioritize preserving and expanding green spaces to ensure that cities remain vibrant, healthy, and sustainable places for future generations (Haaland & van den Bosch, 2015).

1.2 Overview of Noida's Urban Landscape and Planning Framework

Noida, officially known as the New Okhla Industrial Development Authority, is a systematically planned city under India's National Capital Region (NCR). Its planning and development is a prime example of urban modernity designed to accommodate the rapid growth associated with industrial and commercial activities. This detailed overview explores Noida's urban landscape's various facets and comprehensive planning framework.

1.2.1 Geographic and Demographic Context

Noida is strategically located in the Gautam Buddha Nagar District of Uttar Pradesh, bordering the National Capital of Delhi. The city's proximity to Delhi has significantly influenced its urban trajectory, making it a prominent hub for businesses and residential developments. Geographically, it is characterized by flat terrain, which has been advantageous in its systematic layout and construction. The Yamuna River borders the city on the west, providing critical water resources but also challenges related to flood management.

1.2.2 Urban Planning and Development

From its inception in the 1970s, Noida was conceived as a planned urban area to offload the burgeoning industrial growth from Delhi. The city's master planning has been handled by the New Okhla Industrial Development Authority, which oversees land use planning, infrastructure development, and residential and commercial space allocation. Noida's master plans have been periodically updated to adapt to changing needs and growth patterns, with the most recent revisions focusing on sustainable development, the incorporation of green spaces, and improved public transport systems.

1.2.3 Infrastructure and Connectivity

Noida's infrastructure development is notable for its emphasis on connectivity and accessibility. The city boasts a well-developed network of roads, including several significant highways connecting it to the rest of the NCR and major cities across North India. The DND Flyway over the Yamuna River is directly linked to Delhi, facilitating economic exchanges and commuting. Regarding public transport, the Noida Metro Rail Corporation (NMRC) operates the Aqua Line metro, which connects Noida to Greater Noida, enhancing intra-city and inter-city mobility.

1.2.4 Industrial and Commercial Growth

Noida is known for its vibrant industrial sector, hosting numerous national and international corporations, particularly in the information technology and biotechnology fields. The city has been developed into several sectors, each designated for specific activities such as industrial, residential, commercial, and institutional uses. This segregation helps manage urban sprawl and optimize infrastructure utilization.

1.2.5 Residential Developments

The residential sectors in Noida are well-planned, featuring a mix of high-rise apartments, gated communities, and individual housing units. These sectors are interspersed with public amenities such as schools, hospitals, and shopping centers, making them self-sufficient in many respects. The planning authority has also focused on providing affordable housing options to cater to the diverse socioeconomic backgrounds of its residents.

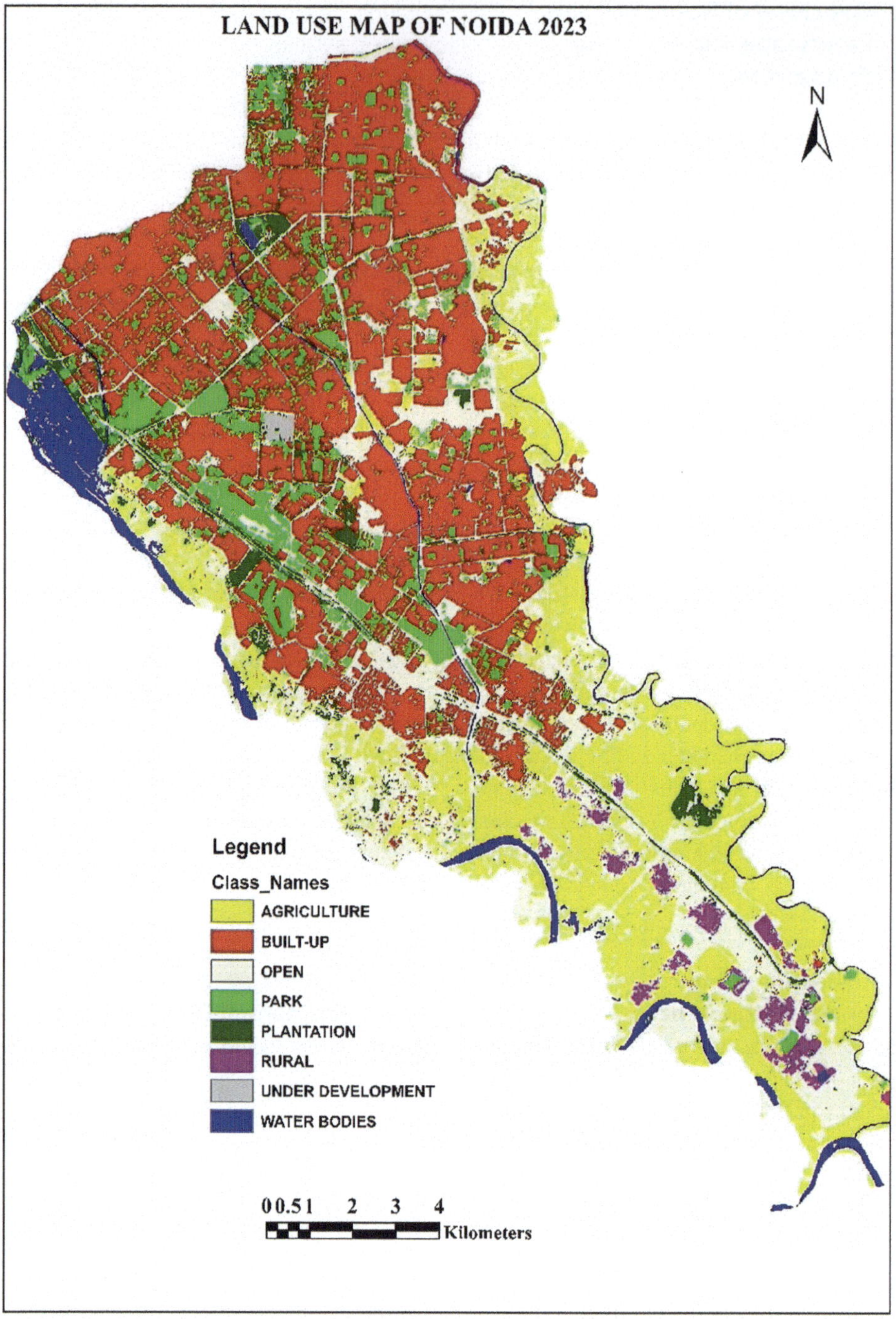
LAND USE MAP OF NOIDA 2023
N
Legend
Class_Names
AGRICULTURE
BUILT-UP
OPEN
PARK
PLANTATION
RURAL
UNDER DEVELOPMENT
WATER BODIES
0 0.5 1 2 3 4
Kilometers

1.2.6 Environmental Planning and Green Spaces

Recognizing the importance of sustainable urban environments, Noida's planning framework includes significant provisions for green spaces. Parks, green belts, and recreational areas are integral to every sector. The city planners have also incorporated eco-friendly practices in waste management, water conservation, and energy efficiency measures into the urban planning norms.

1.2.7 Challenges and Future Directions

Despite its well-planned infrastructure, Noida faces challenges typical of rapidly growing urban areas, including traffic congestion, pollution, and resource management. Plans involve expanding the metro network and other public transport options to reduce personal vehicle dependency, enhance resource use efficiency, and further integrate green technologies into the urban fabric.

Inclusively, Noida's urban landscape and planning framework highlight its evolution from an industrial township to a dynamic urban area that strives to balance growth with sustainability. The city continues to adapt and innovate to meet the demands of its growing population and expanding economy.

2 Theoretical Framework

2.1 Conceptualizing Green Spaces in Urban Settings

The theoretical framework for understanding green spaces in urban settings is deeply rooted in environmental psychology, ecology, and sustainable urban planning. The concept of green spaces encompasses a broad array of natural and semi-natural areas within urban environments, including parks, gardens, green roofs, street trees, and riverbanks, among others. These spaces are critical in shaping urban environmental, social, and economic landscapes (Tzoulas et al., 2007). Green spaces provide numerous ecological benefits, such as air purification, climate regulation, and biodiversity conservation (Gill et al., 2007). Moreover, they offer significant social advantages by promoting physical and mental well-being, facilitating social interactions, and enhancing the overall quality of life in urban areas (Hartig et al., 2014). Economically, green spaces contribute to urban resilience and sustainability by increasing property values, attracting tourism, and reducing healthcare costs (Wolch et al., 2014).

- **Environmental Psychology**: This perspective emphasizes the restorative effects of natural environments on human psychology. Studies have consistently shown that interaction with green spaces can reduce mental fatigue, enhance mood, and lower stress levels (Kaplan & Kaplan, 1989; Ulrich et al., 1991). The mere presence of greenery in urban settings has been linked to improved mental health, increased physical activity, and enhanced cognitive functioning (Berman et al., 2008; Pretty et al., 2005). The Attention Restoration Theory (ART) posits that natural environments facilitate recovery from mental fatigue by providing a setting that is both fascinating and away from daily stressors (Kaplan, 1995). The Stress Reduction Theory (SRT) also suggests that natural settings can reduce stress by eliciting positive emotional responses and promoting physiological recovery (Ulrich, 1984).
- **Urban Ecology**: From an ecological viewpoint, green spaces are vital components of urban ecosystems. They contribute to biodiversity, support wildlife, and improve ecosystem services (Dearborn & Kark, 2010). Green spaces act as natural cooling systems, mitigate urban heat islands, and improve air quality by filtering pollutants and sequestering carbon dioxide (Bowler et al., 2010; Nowak et al., 2006). Moreover, these areas play a crucial role in urban hydrology by facilitating water infiltration and reducing stormwater runoff, thus improving water quality and reducing the likelihood of flooding (Gill et al., 2007; Pugh et al., 2012).
- **Sustainable Urban Planning**: Including green spaces is increasingly recognized as a

cornerstone of sustainable urban development. Urban planners advocate for integrating green spaces into city landscapes to promote sustainable living environments resilient to environmental pressures like climate change (Beatley, 2011). The strategic placement of green spaces can also foster social interactions and create inclusive public spaces that cater to diverse populations, thus enhancing the social fabric of urban areas (Gehl, 2010; Forsyth & Musacchio, 2005). Additionally, green spaces contribute to urban resilience by providing ecosystem services, enhancing biodiversity, and mitigating the impacts of urbanization (Bolund & Hunhammar, 1999; Kabisch et al., 2016).

Several theoretical models underpin the integration of green spaces into urban planning. The "**Green Infrastructure**" concept treats green spaces as part of a broader network that includes other environmental features like water bodies and trails, emphasizing connectivity and multifunctionality (Benedict & McMahon, 2006; Ahern, 2007). Another influential concept is "**Biophilic Design**," which suggests that humans have an innate tendency to seek connections with nature and other life forms, a principle that can be harnessed to design more livable and appealing urban environments (Kellert et al., 2008; Beatley, 2016).

In summary, conceptualizing green spaces in urban settings involves understanding their multifaceted roles and their diverse benefits. This theoretical grounding provides a solid foundation for advocating for and implementing green spaces in urban areas, ensuring they contribute effectively to the sustainability and livability of urban environments.

3 Environmental Benefits of Urban Green Spaces

Urban green spaces in Noida offer a range of environmental benefits that are critical for maintaining the ecological balance and enhancing the quality of urban life. These benefits include biodiversity conservation, air and water purification, and mitigating urban heat island effects. Each of these aspects contributes significantly to Noida's sustainability and livability.

3.1 Biodiversity Conservation

In Noida, green spaces such as city parks, roadside plantings, and small forest patches serve as vital habitats for local wildlife. These areas provide refuge and resources for various species, supporting biodiversity in an urban setting. For instance, parks like the Okhla Bird Sanctuary are crucial for avian species, providing nesting sites and stopover points for migratory birds. Moreover, these green spaces contribute to ecological connectivity, allowing species to move and interact across what would otherwise be an inhospitable urban environment. By maintaining and enhancing these habitats, Noida helps preserve a wide range of plant and animal species, essential for ecological diversity and resilience.

3.2 Air and Water Purification

Urban green spaces in Noida play a significant role in improving air and water quality. Vegetation in parks and natural areas helps filter out pollutants from the air, including particulate matter and harmful gases such as nitrogen oxides and sulfur dioxide. Trees and plants absorb these pollutants through their leaves and bark, significantly reducing the overall concentration of airborne contaminants. This natural filtration process is crucial in Noida, where industrial and vehicular emissions contribute to air pollution (Fig. 1).

Similarly, green spaces enhance water quality through natural filtration systems. The soil and vegetation in these areas act as filters, capturing and breaking down pollutants from surface runoff before they can enter water bodies. This is particularly important in Noida, where rapid urbanization has increased the risk of water pollution. Additionally, preserving wetlands and other natural water bodies within urban parks helps regulate water cycles and maintains the health of aquatic ecosystems.

Fig. 1 Google Earth Image showing the Sector-57 park stretching over 77 hectares of area, a total of dense vegetation and artificial water bodies act as carbon sinks for the city

3.3 Mitigation of Urban Heat Island Effects

The urban heat island effect, where urban regions experience higher temperatures than their rural surroundings, is a significant issue in rapidly urbanizing cities like Noida. Green spaces play a crucial role in mitigating this effect. Plants and trees in urban areas help lower surface and air temperatures through shading and evapotranspiration. In Noida, large green areas like the Noida Golf Course and the Botanical Garden provide critical cooling effects, significantly reducing the need for air conditioning in nearby buildings, lowering energy consumption, and reducing greenhouse gas emissions.

Moreover, strategic placement of green spaces in urban planning can enhance these cooling effects. For example, creating green corridors and ensuring sufficient trees along streets and in public spaces can maximize shade and reduce ambient temperatures throughout the city.

In summary, the environmental benefits of urban green spaces in Noida are manifold, impacting biodiversity, air and water quality, and temperature regulation. These benefits underscore the importance of preserving and expanding green spaces as part of sustainable urban planning efforts in Noida, ensuring that the city remains livable and sustainable as it continues to grow and develop.

4 Social Dimensions of Urban Green Spaces

Urban green spaces in Noida play a pivotal role in enhancing the city's social fabric by improving physical and mental health, providing recreational opportunities, and fostering community well-being and social inclusion. These spaces serve as communal hubs, promoting social interaction and engagement among residents. Here's a detailed look at each of these social dimensions:

4.1 Enhancing Physical and Mental Health

The green spaces in Noida, such as parks and sports facilities, contribute significantly to the community's physical health by providing areas for exercise, sports, and other physical activities. Regular access to such environments encourages a physically active lifestyle, which can reduce the prevalence of diseases like obesity, cardiovascular disease, diabetes, and more. Almost all main parks situated in residential sectors have open gyms in Noida. Moreover, the mental health benefits of interacting with nature are well-documented. Natural settings have a calming effect, helping reduce stress and anxiety. For example, places like the Noida Botanical Garden, Okhla Bird Sanctuary, and City Park offer peaceful retreats from the urban hustle, where people can rejuvenate and enjoy the tranquility of nature. Villages situated in Noida City do not have an adequate number of green spaces. That is why villagers used to go to nearby residential sectors to enjoy the benefits of the green spaces (Figs. 2 and 3).

4.2 Recreational Opportunities and Community Well-being

Green spaces in Noida offer a variety of recreational opportunities that cater to all age groups and interests. These range from structured environments like playgrounds and sports complexes to more naturalistic settings like walking trails and picnic areas. Such diversity enhances physical health and contributes to emotional and psychological well-being by providing residents with places to relax, socialize, and engage in leisure activities. The recreational use of these spaces significantly enhances the overall quality of life in the city, making it more attractive for current and prospective residents (Fig. 4).

Fig. 2 Okhla Bird Sanctuary (Source: Internet)

Fig. 3 People exercising in an open gym in the park

Fig. 4 The residents of Sector 47, Noida, doing physical activities in the sector's Netaji Subhash Chandra Park

4.3 Social Inclusion and Community Engagement

Urban green spaces in Noida also play a crucial role in promoting social inclusion and community engagement. These areas are typically open to everyone, regardless of socioeconomic status, age, or ethnicity, making them essential social interaction and integration venues. Community events, cultural festivals, and sports activities in these green spaces allow diverse groups to unite, fostering a sense of community and belonging. Additionally, participatory activities in maintaining and developing these spaces encourage civic engagement and empower residents to contribute to the urban environment.

The presence of well-maintained and accessible green spaces in Noida thus not only enhances the aesthetic appeal of the city but also strengthens the social bonds among its residents. By providing safe, inclusive, and engaging environments, these spaces enhance community resilience, promote healthy lifestyles, and contribute to the social sustainability of the urban area. As Noida continues to grow, ensuring the integration and enhancement of green spaces will be crucial for maintaining and improving the quality of urban life for all its residents.

5 Economic Impacts of Green Spaces

Urban green spaces in Noida significantly impact the local economy through various channels, including influencing property values and attracting tourism and investment. These effects contribute to the city's overall economic vitality and sustainability, making it a desirable place to live, work, and visit. Here is a detailed examination of these economic impacts:

5.1 Influence on Property Values

In Noida, well-maintained green spaces have been shown to increase the value of nearby properties. Properties near parks, gardens, and other green areas typically command higher prices than those without such amenities. This price premium stems from the enhanced aesthetic appeal, recreational opportunities, and improved environmental quality of green spaces. For example, residential areas near Noida's parks and recreational zones, like those near the Botanical Garden or Noida Golf Course, are often sought after, reflected in their higher property values. This effect benefits property owners and contributes to increased property tax revenues for the city, which can be reinvested into further urban development and maintenance of public spaces. Plots that are park-facing or green bent-facing cost 5% more on the premium as per the rates of the Noida Authority.

Following Table 1 shows the circle rate of the sectors of Noida.

Table 1 The circle rate-wise category of the sectors in Noida

Categories	Sectors	Circle rate per meter
A+	14A, 15A, 44 Block A and B	175,000 Rs
A	14, 16, 16A, 16B, 17, 18, 19, 30, 35,36,38,38A, 39, 44, 47,50, 51,52, 93B, 94, 124,	92,950 Rs
B	15, 20, 21, 23, 24, 25, 26, 27, 28, 29, 31, 32, 33, 34, 37, 40, 41, 46, 48, 49,53, 55, 56, 61, 62, 82,92,93, 96, 97, 98, 99, 100, 105,108,122	64,790 Rs

Sectors falling in categories A+ and A are the sectors that have more area under green space land use comparatively.

5.2 Attraction of Tourism and Investment

Green spaces in Noida also play a crucial role in attracting tourists and investors to the city. From a tourism perspective, parks and natural attractions offer visitors a chance to experience leisure and recreational activities, which can significantly enhance the city's appeal as a travel destination. Attractions such as the Okhla Bird Sanctuary and the Noida Flower Show draw visitors from the local region, across the country, and internationally, contributing to the local hospitality and service industries. Ved Van Park in Noida is India's first Vedic theme park. It attracts lots of tourists due to its uniqueness (Fig. 5).

From an investment standpoint, companies often look for locations that offer a high quality

Fig. 5 Ved Van Park, Sector 78, Noida

of life to attract and retain skilled employees. Noida's commitment to maintaining and expanding its green spaces makes it an attractive option for businesses, particularly those in the technology, biotech, and service sectors, which prioritize environmental quality and employee well-being. This is evident from the growing number of businesses setting up operations in the city, drawn by its urban amenities and green credentials.

Moreover, the city's green spaces can spur investment in green technologies and infrastructure, such as sustainable buildings, renewable energy installations, and eco-friendly transportation solutions. This fosters a green economy, promoting sustainable practices across business operations and contributing to the broader environmental goals of the city.

In conclusion, the economic impacts of urban green spaces in Noida are profound and multifaceted. By enhancing property values and attracting tourism and investment, these spaces contribute significantly to the economic development and attractiveness of the city. As Noida continues to evolve, its green spaces'strategic development and enhancement will remain crucial for promoting economic sustainability and making the city a more competitive and appealing urban center.

6 Challenges and Opportunities in Urban Green Space Management

Urban green space management in Noida faces several challenges and opportunities critical to the city's sustainable development. These issues revolve around land use pressures, financial constraints, and stakeholder and community involvement dynamics. Each of these factors plays a significant role in shaping the effectiveness and sustainability of urban green spaces in Noida.

6.1 Land Use Pressures and Urban Development

Noida's rapid urbanization has brought significant land use pressures, with high demands for residential, commercial, and industrial developments. These demands often compete directly with the need to conserve and expand green spaces. The city's strategic location within the National Capital Region makes it a prime area for development, which can lead to the encroachment of built-up areas on existing green spaces. Managing these pressures requires careful zoning and robust urban planning, prioritizing green space amidst the city's growth. Balancing development with environmental sustainability is a continual challenge for city planners, who must ensure that urban expansion does not compromise green spaces'ecological and social benefits.

6.2 Funding Constraints and Financial Planning

Funding is another critical issue facing the management of urban green spaces in Noida. Maintaining parks, gardens, and natural areas requires substantial financial investment, often competing with other urban priorities such as infrastructure, healthcare, and education. Limited budgets may lead to inadequate maintenance of green spaces, which can diminish their quality and the benefits they provide. It is essential to develop innovative financial strategies, such as public-private partnerships, grants from environmental NGOs, or green bonds, as it can provide alternative funding streams for developing and upkeep green spaces. Additionally, effective financial planning must ensure that investments in green spaces are considered integral to urban development, not secondary considerations.

6.3 Stakeholder and Community Involvement

Managing urban green spaces in Noida also hinges significantly on stakeholder and community involvement. Engaging local communities, businesses, and environmental groups in the planning and maintaining green spaces can lead to more sustainable and widely supported urban environments. Community involvement helps ensure that green spaces meet the needs and preferences of residents, fostering a sense of ownership and responsibility towards these areas. Moreover, stakeholders such as local businesses and environmental organizations can offer resources, expertise, and advocacy for green space initiatives. However, achieving effective community engagement requires transparent communication, inclusive planning processes, and opportunities for ongoing participation in decision-making.

6.4 Opportunities for Improvement

Despite these challenges, there are substantial opportunities to enhance the management of green spaces in Noida. For example, integrating green spaces into the broader urban infrastructure—such as green roofs, walls, and permeable surfaces—can help mitigate the effects of urbanization while maximizing ecological benefits. Advances in technology also offer new ways to monitor and maintain green areas efficiently, such as using geographic information system (GIS) for spatial analysis and drone technology for regular monitoring.

In conclusion, while Noida faces significant challenges in managing its urban green spaces amid rapid urban development and financial constraints, there are also numerous opportunities to innovate and improve. Effective management strategies incorporating financial sustainability, stakeholder collaboration, and integrated urban planning will ensure that Noida's green spaces continue enhancing the city's livability and environmental resilience.

7 Policy Recommendations and Future Directions

In addressing the challenges and leveraging the opportunities associated with urban green spaces in Noida, strategic policy recommendations and future directions can guide policymakers and urban planners toward sustainable urban growth. Here is a detailed exploration of these aspects:

7.1 Strategic Recommendations for Policymakers and Urban Planners

Policymakers and urban planners in Noida need to adopt a comprehensive and strategic approach to integrate green spaces effectively into the urban landscape. This includes:

- **Developing Integrated Green Space Plans**: Policymakers should prioritize creating integrated green space plans embedded within the broader urban development strategies. This involves mapping existing green spaces, identifying potential areas for new green spaces, and ensuring that these are distributed equitably across the city. Such plans should also consider the multifunctional role of green spaces, incorporating elements like stormwater management, biodiversity corridors, and public health amenities.
- **Implementing Zoning Regulations**: Enforcing stringent zoning laws that protect existing green spaces from encroachment and ensure that all new developments include specific allocations for green space. These regulations should be rigorously applied and monitored to prevent the fragmentation and reduction of green areas.
- **Promoting Sustainable Development Practices**: Encourage adopting sustainable development practices that minimize environmental impact and enhance green space functionality. This includes promoting green building certifications, incentivizing the inclusion of green roofs and walls, and integrating sustainable transport options that connect green spaces.

7.2 Enhancing the Functionality and Accessibility of Green Spaces

To maximize the benefits that green spaces offer to urban residents, Noida must focus on enhancing their functionality and accessibility:

- **Improving Physical Accessibility**: Ensure that all green spaces are accessible to the public without physical barriers. This can be achieved by improving infrastructure such as sidewalks, crosswalks, and bike lanes that connect residential areas to parks and natural areas.
- **Designing Inclusive Spaces**: Design green spaces that cater to diverse groups, including children, elderly citizens, and people with disabilities. This involves incorporating features like playgrounds, walking trails, and seating areas that are universally accessible.
- **Increasing Functional Diversity**: Diversify the uses of green spaces to include community gardens, sports facilities, outdoor gyms, and cultural performance areas. This approach increases recreational opportunities and encourages more frequent and varied use of green spaces, promoting greater community interaction.

7.3 Sustaining Urban Growth Through Green Spaces

Green spaces are vital for sustainable urban growth in Noida, and their strategic development can propel the city towards environmental sustainability and enhanced livability:

- **Linking Green Spaces With Urban Growth**: Integrate green space planning with urban growth initiatives. For example, as new residential or commercial areas are developed, ensure that green spaces are a core component of these developments. This could be legislated as part of the planning approval process.
- **Leveraging Green Spaces for Economic Benefits**: Utilize green spaces as catalysts for economic development. This includes developing eco-tourism, hosting cultural and environmental festivals, and encouraging health and wellness industries to capitalize on the proximity to these verdant areas.
- **Fostering Public-Private Partnerships**: Encourage partnerships between the govern-

ment, private sector, and community organizations to fund and maintain green spaces. These partnerships can bring additional resources, innovative management practices, and greater community buy-in.

In conclusion, by implementing these strategic recommendations and future directions, Noida can enhance the quality, functionality, and accessibility of its urban green spaces, which are crucial for sustaining urban growth and improving the quality of life for its residents. These efforts will help ensure that the city remains a vibrant, healthy, and attractive place to live and work, now and in the future.

References

Ahern, J. (2007). Green infrastructure for cities: The spatial dimension. In V. Novotny & P. Brown (Eds.), *Cities of the future: Towards integrated sustainable water and landscape management* (pp. 267–283). IWA Publishing.

Beatley, T. (2011). *Biophilic cities: Integrating nature into urban design and planning*. Island Press.

Beatley, T. (2016). *Handbook of biophilic city planning & design*. Island Press.

Benedict, M. A., & McMahon, E. T. (2006). *Green infrastructure: Linking landscapes and communities*. Island Press.

Berman, M. G., Jonides, J., & Kaplan, S. (2008). The cognitive benefits of interacting with nature. *Psychological Science, 19*(12), 1207–1212. https://doi.org/10.1111/j.1467-9280.2008.02225.x

Bolund, P., & Hunhammar, S. (1999). Ecosystem services in urban areas. *Ecological Economics, 29*(2), 293–301. https://doi.org/10.1016/S0921-8009(99)00013-0

Bowler, D. E., Buyung-Ali, L., Knight, T. M., & Pullin, A. S. (2010). Urban greening to cool towns and cities: A systematic review of the empirical evidence. *Landscape and Urban Planning, 97*(3), 147–155. https://doi.org/10.1016/j.landurbplan.2010.05.006

Chiesura, A. (2004). The role of urban parks for the sustainable city. *Landscape and Urban Planning, 68*(1), 129–138. https://doi.org/10.1016/j.landurbplan.2003.08.003

Dearborn, D. C., & Kark, S. (2010). Motivations for conserving urban biodiversity. *Conservation Biology, 24*(2), 432–440. https://doi.org/10.1111/j.1523-1739.2009.01328.x

Escobedo, F. J., Kroeger, T., & Wagner, J. E. (2011). Urban forests and pollution mitigation: Analyzing ecosystem services and disservices. *Environmental Pollution, 159*(8–9), 2078–2087. https://doi.org/10.1016/j.envpol.2011.01.010

Ferguson, B. K. (2005). *Porous pavements*. CRC Press.

Forsyth, A., & Musacchio, L. (2005). *Designing small parks: A manual for addressing social and ecological concerns*. Wiley.

Gehl, J. (2010). *Cities for people*. Island Press.

Gill, S. E., Handley, J. F., Ennos, A. R., & Pauleit, S. (2007). Adapting cities for climate change: The role of the green infrastructure. *Built Environment, 33*(1), 115–133. https://doi.org/10.2148/benv.33.1.115

Haaland, C., & van den Bosch, C. K. (2015). Challenges and strategies for urban green-space planning in cities undergoing densification: A review. *Urban Forestry & Urban Greening, 14*(4), 760–771. https://doi.org/10.1016/j.ufug.2015.07.009

Hartig, T., Mitchell, R., de Vries, S., & Frumkin, H. (2014). Nature and health. *Annual Review of Public Health, 35*, 207–228. https://doi.org/10.1146/annurev-publhealth-032013-182443

Jim, C. Y., & Chen, W. Y. (2006). Recreation-amenity use and contingent valuation of urban green spaces in Guangzhou, China. *Landscape and Urban Planning, 75*(1–2), 81–96. https://doi.org/10.1016/j.landurbplan.2004.04.004

Kabisch, N., Qureshi, S., & Haase, D. (2016). Human-environment interactions in urban green spaces—A systematic review of contemporary issues and prospects for future research. *Environmental Impact Assessment Review, 50*, 25–34. https://doi.org/10.1016/j.eiar.2014.08.007

Kaplan, S. (1995). The restorative benefits of nature: Toward an integrative framework. *Journal of Environmental Psychology, 15*(3), 169–182. https://doi.org/10.1016/0272-4944(95)90001-2

Kaplan, R., & Kaplan, S. (1989). *The experience of nature: A psychological perspective*. Cambridge University Press.

Kellert, S. R., Heerwagen, J. H., & Mador, M. L. (Eds.). (2008). *Biophilic design: The theory, science, and practice of bringing buildings to life*. Wiley.

Nowak, D. J., Crane, D. E., & Stevens, J. C. (2006). Air pollution removal by urban trees and shrubs in the United States. *Urban Forestry & Urban Greening, 4*(3–4), 115–123. https://doi.org/10.1016/j.ufug.2006.01.007

Pretty, J., Peacock, J., Sellens, M., & Griffin, M. (2005). The mental and physical health outcomes of green exercise. *International Journal of Environmental Health Research, 15*(5), 319–337. https://doi.org/10.1080/09603120500155963

Pugh, T. A. M., MacKenzie, A. R., Whyatt, J. D., & Hewitt, C. N. (2012). Effectiveness of green infrastructure for improvement of air quality in urban street canyons. *Environmental Science & Technology, 46*(14), 7692–7699. https://doi.org/10.1021/es300826w

Tzoulas, K., Korpela, K., Venn, S., Yli-Pelkonen, V., Kaźmierczak, A., Niemela, J., & James, P. (2007). Promoting ecosystem and human health in urban areas using Green Infrastructure: A literature review.

Landscape and Urban Planning, 81(3), 167–178. https://doi.org/10.1016/j.landurbplan.2007.02.001

Ulrich, R. S. (1984). View through a window may influence recovery from surgery. *Science, 224*(4647), 420–421. https://doi.org/10.1126/science.6143402

Ulrich, R. S., Simons, R. F., Losito, B. D., Fiorito, E., Miles, M. A., & Zelson, M. (1991). Stress recovery during exposure to natural and urban environments. *Journal of Environmental Psychology, 11*(3), 201–230. https://doi.org/10.1016/S0272-4944(05)80184-7

Wolch, J. R., Byrne, J., & Newell, J. P. (2014). Urban green space, public health, and environmental justice: The challenge of making cities 'just green enough'. *Landscape and Urban Planning, 125*, 234–244. https://doi.org/10.1016/j.landurbplan.2014.01.017

Comparative Analysis of Wetland Shrinkage in Urban and Rural Settings: Understanding Driving Forces and Implications

Manal Ahmad, Mohd Atahar, Syed Naushad Ahmad, Wani Suhail Ahmad, Mohd Saqib, and Aakib Yaqoob Mir

1 Introduction

Wetlands are transitional ecosystems where water is the primary factor determining the environment, fostering diverse flora, fauna, and ecological functions (Batzer & Sharitz, 2014). They are characterized by the presence of water either seasonally or permanently, representing one of the most diverse and productive ecosystems on the planet (Mitsch & Gosselink, 2015). Wetlands are referred to as a biological "supermarket" and "kidneys of the landscape" for their various functions such as water purification, storage, carbon processing, nutrient processing, shoreline stabilization, and support for aquatic life (Jamal & Ali, 2021). Wetland ecosystems can fulfil a number of human requirements, and are frequently referred to as "wealth-lands" (Marambanyika & BeckedahL, 2016). It also plays a vital role in providing nutrients and water for both biological productivity and human survival (Schuyt, 2005). Wetlands can restrict the emission of harmful greenhouse gases like carbon dioxide, methane, and nitrous oxide (IPCC, 2007). Wetlands provide habitat for fish and wildlife, supporting a rich biodiversity, including the many threatened and endangered species (Stacy and Marvin, 2002). In addition to supplying firewood, medicinal plants, grazing, tourism, and water supply, wetlands also offer significant socioeconomic benefits (Jamal et al., 2023a) and are valued for their recreational opportunities and aesthetics (Barbier et al., 1994).

Wetlands are also acknowledged as biodiversity hotspots, supporting a myriad of plant and animal species uniquely adapted to their hydrological conditions. According to recent estimates, wetlands cover approximately 6% of the Earth's land surface and harbor around 40% of the world's known species, making them indispensable reservoirs of biological diversity (Davidson et al., 2015). It tends to play a significant role in religious, sociocultural, and economic activities (Asgher et al., 2021). Globally, people interact with wetland ecosystems for services like fuelwood, building materials, fishing, and hunting. They also benefit from herbs production, agriculture, livestock, and industrial purposes (Ahmad et al., 2024). However, despite their ecological significance, wetlands remain one of the most threatened ecosystems globally, facing extensive degradation and loss due to human activities such as urbanization, agriculture, and industrialization.

Natural and man-made stresses are among the factors threatening the wetland's survival (Schuyt, 2005). Most aquatic habitats in urban

M. Ahmad · M. Atahar (✉) · S. N. Ahmad · M. Saqib
A. Y. Mir
Department of Geography, Aligarh Muslim University, Aligarh, India

W. S. Ahmad
Department of Geography, University of Ladakh, Kargil, India

Z. Khanam et al. (eds.), *Climate Change, Resilience and Ecosystem Services*, Sustainable Development Goals Series, https://doi.org/10.1007/978-3-031-96977-5_13

and rural areas are decreasing or disappearing due to increased anthropogenic activities (Ahmad et al., 2024). The lives and livelihoods of the people living in the wetland are in danger due to these environmental consequences of urbanization. Almost 50% of worldwide wetlands were lost during the twentieth century, with experiencing the greatest amount of loss (Jamal et al., 2023a). The degradation of wetlands occurred due to the rapid population growth, urban expansion, and industrialization in recent decades (Imdad et al., 2023). More than half of the world's wetlands have vanished since 1900 (Barbier, 1993). Most wetlands have declined due to several factors such as pollution, water use, infrastructural development, and agriculture (Xu et al., 2019). The structure and function of wetland ecosystems vary geographically due to changing land use and land cover (LULC) patterns, which are often caused by human activities and can negatively affect ecosystem function (Jamal et al., 2024). The transformation of wetlands into agriculture and infrastructure has modified the land and water (Schuyt, 2005). Most of which are caused by human activity, and are causing the Earth's surface to continuously change (Chu et al., 2009). The causes of wetland alterations are several. For instance, changes in temperature and precipitation brought on by climate change will impact hydrological and biochemical cycles, which in consequence alter the distribution of wetlands (Wolfgang et al., 2013). The vast majority of aquatic ecosystems are mismanaged and discarded natural resources, despite the fact that they are scarce on Earth and very vulnerable to outside influences (Jamal et al., 2024). Understanding the complex interplay of biotic and abiotic factors within wetland environments is essential not only for ecological conservation but also for sustainable resource management and human well-being (MEA, 2005). Moreover, wetlands provide a multitude of ecosystem services vital for both natural and human systems; these services include water purification, flood regulation, carbon sequestration, and habitat provision, among others (Costanza et al., 1997). The economic value of these complex services is substantial, with estimates ranging from billions to trillions of dollars annually, underscoring the importance of wetland conservation and restoration efforts (MEA, 2005).

The estimated 15.26 million hectares of wetlands in India have experienced substantial degradation due to human interference (Ahmad et al., 2024). Since 1971, wetlands have been lost in Chennai by 90%, Vadodara by 30%, Hyderabad by 55%, Mumbai by 71%, Ahmedabad by 57%, Bengaluru by 56%, Pune by 37%, and Delhi-National Capital Region by 38%, mostly as a result of eutrophication from pollution and construction (Burgess, 2015). Bihar has a lot of wetlands, but only three of them have been classified as Ramsar sites, which show that there is a lack of awareness and action regarding the preservation of these important ecosystems (Ramsar, 2024). However, Bihar's wetlands are rapidly disappearing and becoming contaminated as a result of simultaneous pollution and reduction problems.

Significant ecological implications result from the conversion of wetlands into agricultural, urban, and other land uses, especially at the local level (Mishra, 2020; Srinivasan, 2010). Hence, the wetlands of Bihar are getting lost and encroached day by day due to human intervention. Similarly, the selected wetland, which is Moti Jheel in Motihari, Bihar, is also undergoing a continuous degradation due to human activities, being a major economic and sociocultural site. Also, this wetland is not listed in the Ramsar list.

Remote sensing and the geographic information system (GIS) are the best available technologies for analyzing wetlands (Sushma and Dey, 2012). It is one of the main sources for studying, analyzing environmental processes at the local or global level, data from satellites like Landsat, Cartosat series, Sentinel, and others are valuable for area analysis, categorization, and visualization (Kshetri, 2018). High spatial resolution satellite and aerial photography have been used recently to identify wetland species, categorize wetlands, and evaluate the accuracy of other classifications or maps (Guo et al., 2017). Even in this study, the advantage of satellite imagery of the Landsat Series for the classification of all the thematic maps of the wetlands has been taken. As the present study is based on microlevel analysis, these satellite data fulfilled our objectives for this particular research. As sustainable management of wetlands is crucial for maintaining the long-term health and well-being of all populations (Schuyt, 2005). But despite their importance, wetlands are often changed or exploited for financial and commercial gain; however, wetlands offer numerous advantages and benefits (Schuyt, 2005). Hence, the main objective of the paper is to assess the level of shrinkage in wetlands lying in urban and rural settings. The second objective is to understand the driving forces for the changes in different LULC indices in and around the wetlands in rural and urban areas during the selected time periods.

2 Study Area: Moti Jheel & Kararia Lake, East Champaran, Bihar

This study focuses on Moti Jheel, located in the East Champaran District of Bihar, India, renowned for its scenic beauty and cultural significance. Geographically, it spans from 26°38′00'‘N to 21°20′02'‘N and from 84°55′00″ E to 81°41′50″ E (see Fig. 1), making it one of the crucial oxbow lakes in Bihar. The wetland covers an area of 130 hectares within the urban

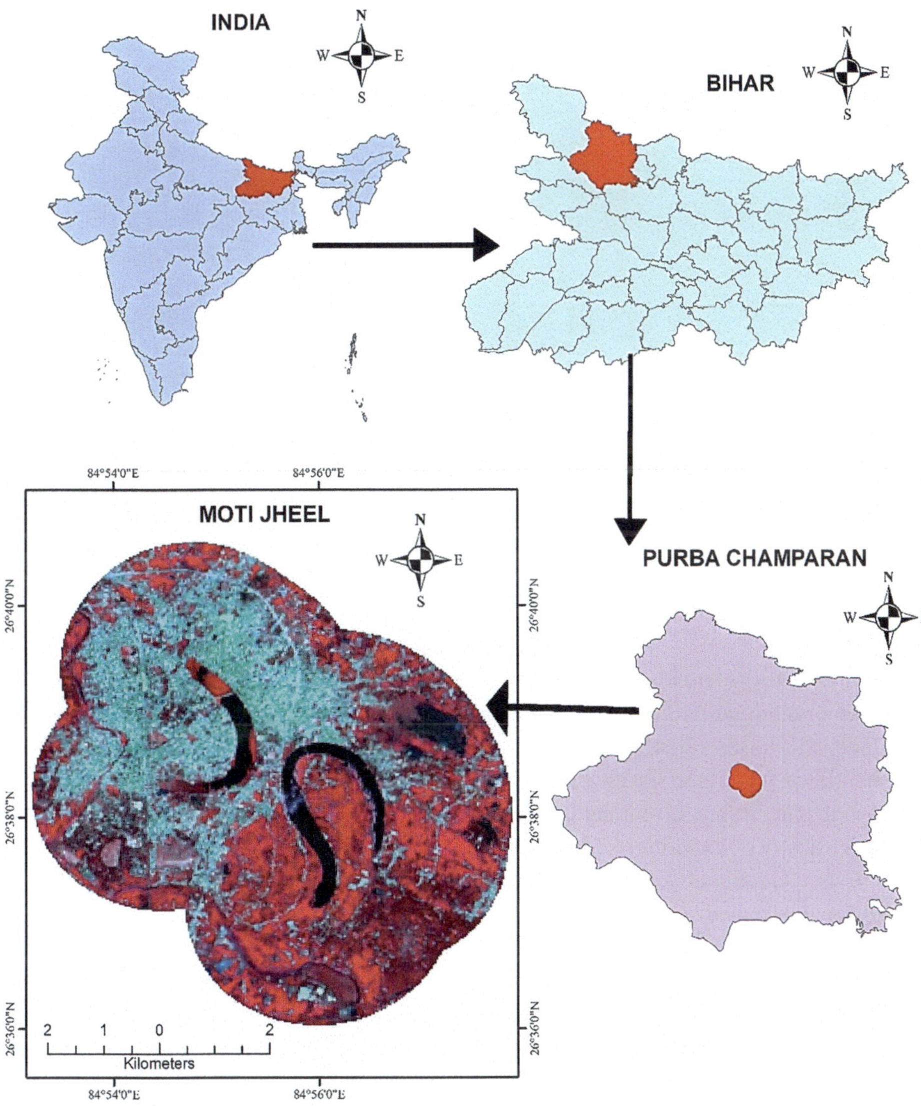

Fig. 1 Moti Jheel and Kararia Lake, East Champaran, Bihar

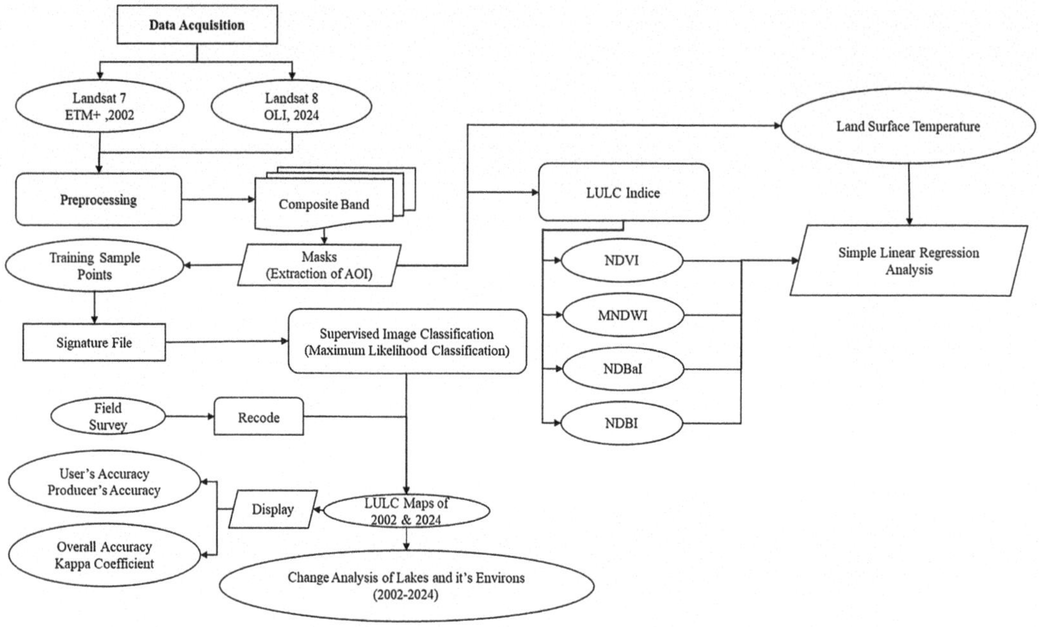

Fig. 2 Methodology

perimeter of the city. Another significant oxbow lake, Kararia, lies between the coordinates 26°39′49''N and 85°35′03″ E. This natural, perennial lake, nearly U-shaped in extension (see Fig. 1), is situated approximately 2 km east of Motihari town in a rural setting at an elevation of 66.141 m above sea level, with a water spread area of around 120 h. Kararia Lake is hydrologically connected to the Dhanauti River on one side and Moti Jheel on the other, via a canal at Bariyarpur. During the rainy season, Kararia Lake discharges excess floodwater into the Dhanauti River through an outlet. Thus, Kararia and Moti Jheel lakes are often referred to as "Twin Lakes" (isca.in). These lakes support a rich diversity of ichthyofauna, although only a small portion is used for pisciculture. Additionally, Moti Jheel offers recreational activities for local residents. The region, however, faces challenges such as pollution, siltation, and encroachment, which threaten the health of these lakes. Understanding the ecological dynamics and human impact on these Twin Lakes is crucial for developing effective conservation strategies. Sustainable management practices are necessary to preserve these water bodies for future generations (Fig. 2).

3 Methodology

The primary source of information, which is proving a boon nowadays, is satellite data. Hence, we have used the data of Landsat 7 ETM+ for 2002 and Landsat 8 for 2024 for showing the transformation taken place near the wetlands within the span of two decades. These data have been acquired from USGS site (https://earthexplorer.usgs.gov/). LULC maps have been produced using Arc GIS 10.5 software.

4 Image Classification

Land transition modelling requires the necessity of LULC maps. To prepare the maps, training samples were first collected from satellite images. These samples were then categorized into six major classes: built-up, water body, bare land, swamp, vegetation, and cropland. Multi-temporal

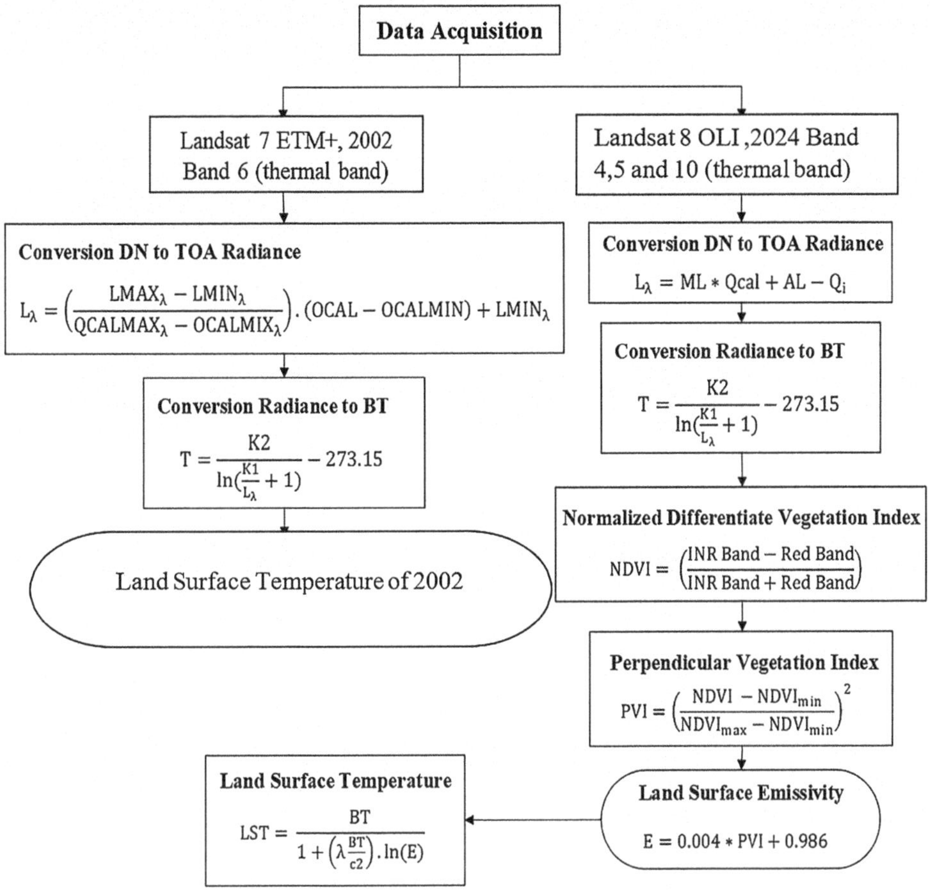

Fig. 3 Methodology for the calculation of LST

Landsat images from 2002 and 2024 were classified using the Maximum Likelihood (ML) algorithm (Fig. 3). The classified images from 2002 and 2024 were validated by comparing them with topographic sheets and previous LULC maps, respectively. Additionally, Google imagery from 2024 was used as a reference image for the LULC map of 2024.

5 Accuracy Assessment

The classification accuracy was assessed using the Kappa measure (Congalton, 1991).

Kappa Coefficient We have calculated the Kappa coefficient to assess the classification accuracy beyond chance agreement. Kappa coefficient is given by the formula:

$$k = \frac{N \cdot \Sum(\text{Diagonal}) - \Sum(\text{RowTotal} \times \text{Column Total})}{N^2 - \Sum(\text{RowTotal} \times \text{Column Total})}$$

Where N is the total number of samples. This coefficient measures the agreement between the observed and predicted classifications, regulating for the agreement that could occur by chance.

Producer's Accuracy For calculating the producer's accuracy for each class as the ratio of correctly classified samples in that class to the total number of reference samples for that class.

$$\text{Producer's Accuracy} = \frac{\text{Total Reference Samples for the Class}}{\text{Correctly Classified Samples}}$$

User's Accuracy We have calculated the user's accuracy for each class as the ratio of correctly classified samples in that class to the total number of classified samples for that class.

$$\text{User's accuracy} = \text{Correctly classified samples} \,/\, \text{Total classified samples for the class}$$

Overall Accuracy We have calculated the overall accuracy as the ratio of correctly classified samples to the total number of samples.

$$\text{Overall Accuracy} = \frac{\sum \text{Diagonal Elements}}{\text{Total Number of Samples}}$$

6 Land Transformation Dynamics

Land change dynamics over the two decades (2002–2024) were calculated using the cross-tabulation method. The land change statistics reveal significant transitions, from wetland to built-up, agriculture to built-up, and vegetation to agriculture changes being more prominent on the urban side (Moti Jheel). Conversely, transitions from wetland to agriculture and agriculture to vegetation were observed near the rural side (Kararia Maun), as shown in Fig. 6 and Table 2.

7 Calculation of Indices

For understanding the environmental implications occurring due to changes in the LULC status of the wetland, a few indices were used. These indices were calculated using specific formulas based on the reflectance values in different spectral bands (Saqib et al., 2024):

7.1 Normalized Difference Built-up Index (NDBI)

The NDBI is used to map built-up areas and is calculated using the following formula:

$$\text{NDBI} = \frac{(\text{SWIR} + \text{NIR})}{(\text{SWIR} - \text{NIR})}$$

where
SWIR: Shortwave infrared band (e.g., Landsat Band 5)
NIR: Near infrared band (e.g., Landsat Band 4).

7.2 Normalized Difference Vegetation Index (NDVI)

The NDVI is used to measure and monitor plant growth, vegetation cover, and biomass production. It is calculated using the following formula:

$$\text{NDVI} = \frac{(\text{NIR} + \text{Red})}{(\text{NIR} - \text{Red})}$$

where.
NIR: Near infrared band (e.g., Landsat Band 4)
Red: Red band (e.g., Landsat Band 3).

7.3 Normalized Difference Bareness Index (NDBI)

The NDBI is used to identify bare land areas and is calculated using the following formula:

$$\text{NDBaI} = \frac{(\text{Red} + \text{SWIR})}{(\text{Red} - \text{SWIR})}$$

where
Red: Red band (e.g., Landsat Band 3).
SWIR: Shortwave infrared band (e.g., Landsat Band 5).

7.4 Modified Normalized Difference Water Index (MNDWI)

The MNDWI is used to improve the presence of open water features while suppressing the presence of the built-up land, vegetation, and soil. It is calculated using the following formula:

$$\text{MNDWI} = \frac{(\text{Green} + \text{SWIR})}{(\text{Green} - \text{SWIR})}$$

where
Green: Green band (e.g., Landsat Band 2).
SWIR: Shortwave infrared band (e.g., Landsat Band 5).

8 Calculation of LST

The formula used to calculate the LST of 2002 and 2024 has been shown in Fig. 3 (Ali et al., 2024).

9 Correlation Between LST and LULC Indices

Calculated the Pearson correlation coefficient between LST and each index using the following formula as used by Jamal et al., (2023b):

$$r = \frac{n\sum xy - (\sum x)(\sum y)}{\sqrt{\left[n\sum x^2 - (\sum x)^2\right][n\sum y^2 - \left(\sum y^2\right)]}}$$

where.
n is the number of data points,
$\sum x$ is the sum of all x-values,
$\sum y$ is the sum of all y-values,
$\sum xy$ is the sum of the product of corresponding xxx- and y-values,
$\sum x^2$ is the sum of the squares of the xxx-values,
$\sum y^2$ is the sum of the squares of the y-values.

We have interpreted the correlation coefficients to understand the strength and direction of the relationship between LST and each index. Here, r values range from −1 to +1, where +1 indicates a perfect positive correlation, −1 indicates a perfect negative correlation, and 0 indicates no correlation.

10 Result and Discussion

11 LULC Dynamics

Figure 4a, b shows the scenario of water bodies of Motihari City in 2002 and 2024. Land use and land cover analysis of Moti Jheel and Kararia Maun has been done in order to detect the change in different categories that has taken place from 2002 to 2024. A total of six categories were identified in the district, namely, bare land, built-up, cropland, swamp, vegetation, and water body. From Fig. 4a and Table 1, it is found that the total area under the category cropland near the Jheels in 2002, which is considered to be the fundamental form of activity by man, was about 1510.55 ha, and accounts for about 33.41% of the total area. Water bodies, which are termed as the matrix of life as being the substantive part of all forms of living systems, constitute an area of about 248.51 ha in 2002, i.e., (5.50%) of the study area. The category vegetation (including forest, plantation, orchards) covers an area of about 1390.30% (30.75%) in 2002. Similarly, built-up, which is the basic requirement of man on Mother Planet, acquires an area of about 1011.65 ha (22.38%). As for as the open category is considered, which includes barren land, occupies an area of about 282.55 ha (6.25%) and is found to be the second largest category in the study area after agriculture. The swamp constitutes an area of about 77.54 ha in 2002, i.e., (1.72%) of the study area. This very class is unfortunately found to be occupying the least area in the study area (Fig. 5).

The LULC map for the year 2024 clearly reveals that all categories of class show discernible change from 2002 to 2024. From Fig. 4b and Table 1, we can see that there is a remarkable increase in class built-up as it became 1827.38 ha (40.42%) of the total geo-

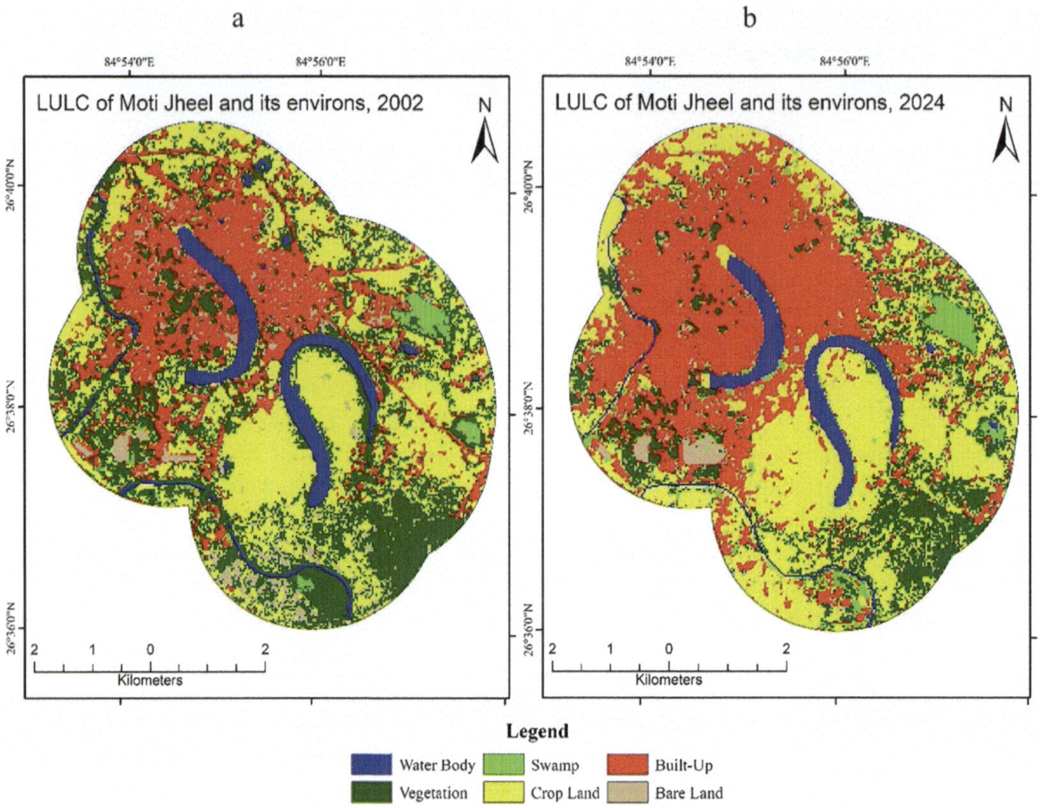

Fig. 4 LULC map of Moti Jheel: (**a**) 2002 and (**b**) 2024

Table 1 Growth and area under different LULC classes of Moti Jheel (2002–2024)

LULC class	2002 Area in ha	2002 Area (%)	2024 Area in ha	2024 Area (%)	Net change	Growth (%)
Bare land	282.55	6.25	232.46	5.14	−50.10	−17.73
Built-up	1011.65	22.38	1827.38	40.42	815.74	80.63
Crop land	1510.55	33.41	1422.28	31.46	−88.27	−5.84
Swamp	77.54	1.72	145.27	3.21	67.73	87.35
Vegetation	1390.30	30.75	660.87	14.62	−729.43	−52.47
Water body	248.51	5.50	232.83	5.15	−15.68	-6.31

graphical area. Also, the increase in the swamps category as compared to 2002 is 145.27 ha (3.21%). However, when cropland is concerned, it covers an area of about 1422.28 ha, i.e., (31.46%) of the total geographical area. This clearly indicates a significant decrease from 2002 to 2024. Also, classes like vegetation and bare land occupy an area of about 660.87 ha (14.62%), and 232.46 ha (5.14%), respectively. However, when it comes to water bodies, it shows that the area under this category in 2024 is only about 232.83 ha, i.e., (5.15%) of the

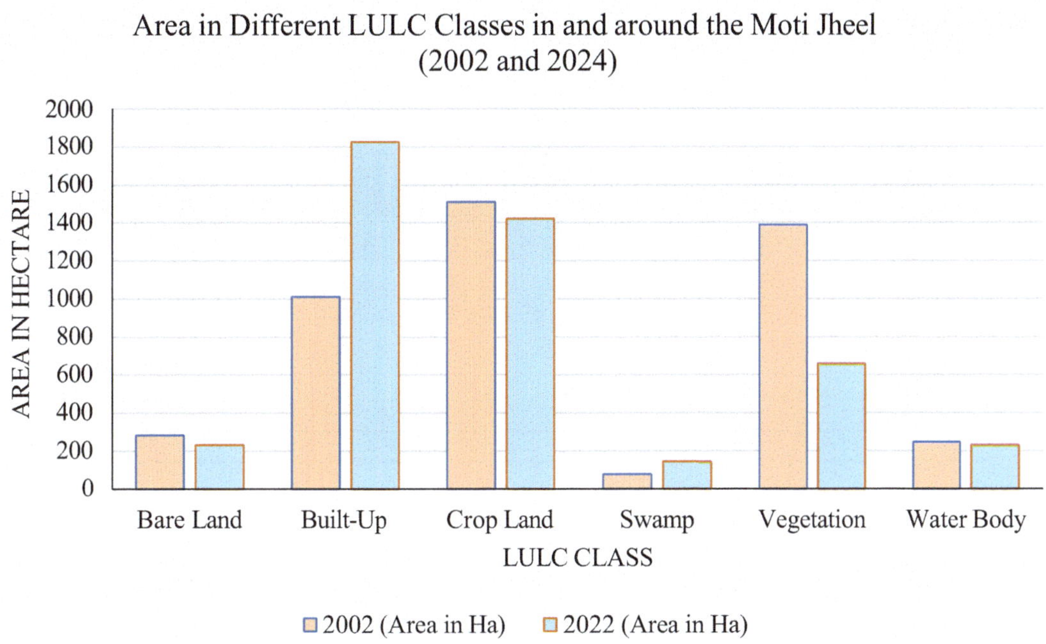

Fig. 5 Area in different LULC classes in and around the Moti Jheel (2002–2024)

total geographical area. From the present scenario of water bodies in Motihari City, clearly reveals that from 2002 to 2024, a number of water bodies/wetlands came to the verge of extinction. The water body category in the study area lost an area of about 15.68 ha from 2002 to 2024. Thus, registering a negative growth rate of about −6.31%, which is the maximum among all categories of land use and land cover in the study area. The study also revealed that Moti Jheel and Kararia Maun were having ample day-to-day needs like irrigation, etc. These sources of water also played an important role in maintaining the weather conditions in their surroundings.

12 Change Detection Analysis

Table 2 clearly reveals notable changes in various land use and land cover (LULC) categories around the wetlands of Moti Jheel and Kararia Maun over the 22-year period from 2002 to 2024. The data indicate both increasing and decreasing trends among these categories, highlighting significant land transformation dynamics. The built-up area, a major human activity indicator, expanded substantially from 1510.55 ha (22.38%) in 2002 to 1827.38 ha (40.42%) in 2024, reflecting a significant increase of 815.74 ha (80.63%). Conversely, the cropland category, which dominated the LULC in 2002, decreased from 1510.55 ha (33.41%) to 1422.28 ha (31.46%) in 2024, showing a reduction of 88.27 ha (− 5.84%). Water bodies experienced a notable decline, shrinking from 248.51 ha (5.50%) in 2002 to 232.83 ha (5.15%) in 2024, representing a loss of 15.68 ha (−6.31%), the highest decrease among all LULC categories in the study area. Vegetation cover also diminished significantly, decreasing from 1390.30 ha (30.75%) in 2002 to 660.87 ha (14.62%) in 2024, a reduction of 729.43 ha (−52.47%). Similarly, bare land

Table 2 Land transformation of Moti Jheel and its environs from 2002 to 2024

LULC class		2024 Bare land	Built-up	Crop land	Swamp	Vegetation	Water body	Row total
2002	Bare land	38.22	126.45	84.86	5.91	26.56	0.55	282.55
	Built-up	0.00	1011.65	0.00	0.00	0.00	0.00	1011.65
	Crop land	107.62	279.06	909.91	21.04	187.60	5.32	1510.55
	Swamp	1.56	16.87	8.45	41.30	7.78	1.57	77.54
	Vegetation	83.44	378.53	398.96	65.50	429.98	33.90	1390.30
	Water body	1.62	14.83	20.10	11.51	8.94	191.50	248.51
	Column total	232.46	1827.38	1422.28	145.27	660.87	232.83	4521.09

Table 3 Accuracy assessment

Year	Class	W	C	B	BR	V	S	RT	UA (%)
2002	W	9	1	0	0	1	0	11	81.82
	C	0	20	0	0	4	1	25	80.00
	B	0	0	13	5	0	0	18	72.22
	BR	0	0	2	18	0	0	20	90.00
	V	0	2	0	0	11	2	15	73.33
	S	0	1	0	0	1	9	11	81.82
	CT	9	24	15	23	17	12	100	
	PA (%)	100.00	83.33	86.67	78.26	64.71	75.00		
2024	W	9	0	0	0	1	0	10	90.00
	C	0	12	0	1	0	0	13	92.31
	B	0	1	35	1	2	0	39	89.74
	BR	0	1	2	11	0	0	14	78.57
	V	0	1	0	0	12	1	14	85.71
	S	0	1	0	0	1	8	10	80.00
	CT	9	16	37	13	16	9	100	
	PA (%)	100.00	75.00	94.59	84.62	75.00	88.89		
2002 = Number corrected 80; Classified total 100; Overall Accuracy 80%; Overall Kappa Statistics 0.7433									
2024 = Number corrected 88; Classified total 100; Overall Accuracy 88%; Overall Kappa Statistics 0.8270									

decreased from 282.55 ha (6.25%) in 2002 to 232.46 ha (5.14%) in 2024, a net decrease of 50.10 ha (−17.73%). In contrast, swamp areas expanded considerably from 77.54 ha (1.72%) in 2002 to 145.27 ha (3.21%) in 2024, indicating an increase of 67.73 ha. These changes underscore the dynamic and complex nature of land use and cover transformations in the vicinity of these wetlands over the studied period (Table 3) (Figs. 6 and 7).

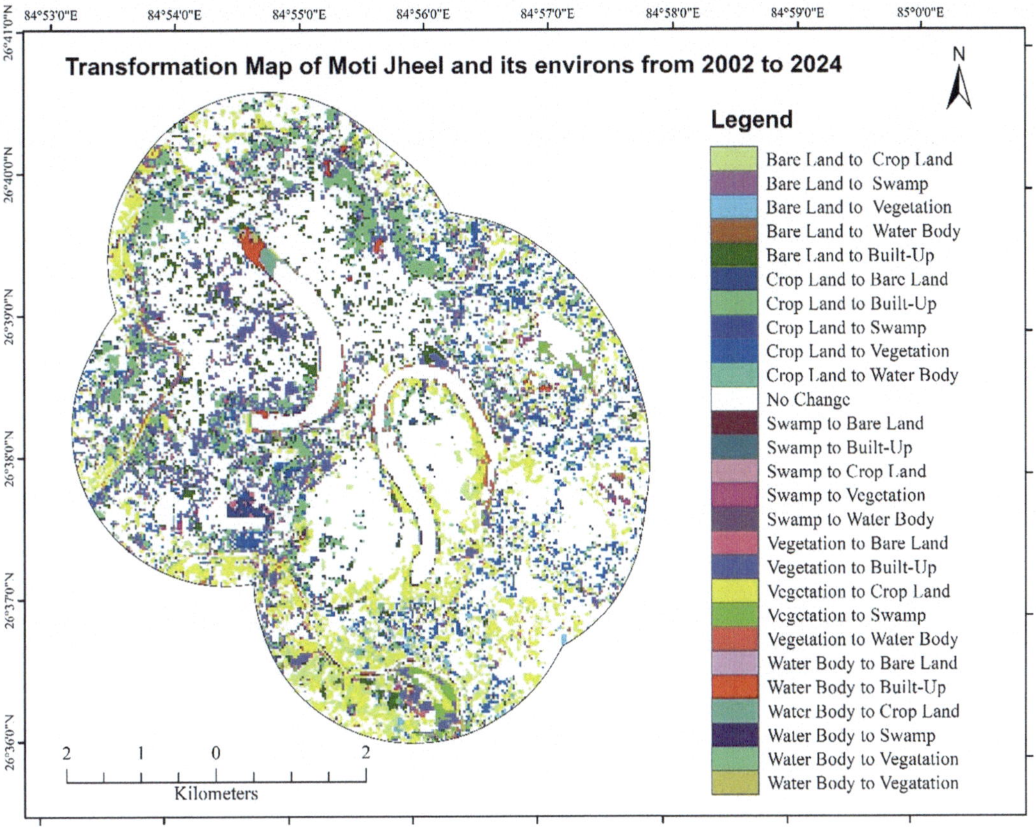

Fig. 6 Change detection analysis and transformation map of Moti Jheel and its environs from 2002 to 2024

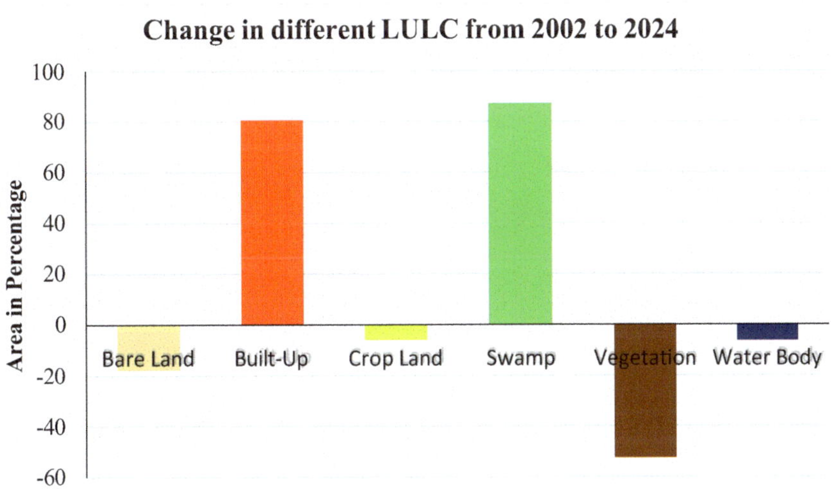

Fig. 7 Change in different LULC in the Moti Jheel (2002–2024)

13 Land Surface Temperature (LST)

Figure 8a, b shows changes in temperature distribution between two time periods: 2002 and 2024. The green color indicates lower temperatures, green to yellow color depicts moderate temperatures, while red color indicates higher temperatures. In 2002, the resulting minimum and maximum temperatures ranging from 18.69 °C to 28.34 °C. The map displays predominant of blue and purple color regions highlighting lower temperatures (water bodies, vegetation, etc.), while the green and yellow patches signify moderate temperatures such as croplands. And a few isolated red areas suggest higher temperatures, where there are built-up lands. But the conditions of 2024 are changed as the minimum and maximum temperature ranges from 19.10 °C to 26.50 °C. It is indicated by an increase in the green, yellow, and red color regions, the increase in high-temperature zones could be associated with urban expansion and increased built-up areas around the Moti Jheel. While near Kararia Maun, the temperature seems to be significantly lower because of less urbanization. There may be a few other locations as well with lower temperatures as indicated by the decreasing areas of the purple and blue color regions.

From Table 4, it is evident that there is areal increase in the very high LST zone (around Moti Jheel Wetland) from 5.10% in 2002 to 18.78% in

Table 4 LST zones of Moti Jheel from 2002 to 20,224

LST zones	2002		2024	
	Area (km^2)	Area (%)	Area (km^2)	Area (%)
Very low	424.89	9.39	912.06	20.15
Low	722.79	15.97	914.31	20.20
Medium	2288.34	50.57	918.99	20.31
High	858.6	18.97	930.24	20.56
Very high	230.76	5.10	849.78	18.78
Total	**4525.38**	**100**	**4525.38**	**100**

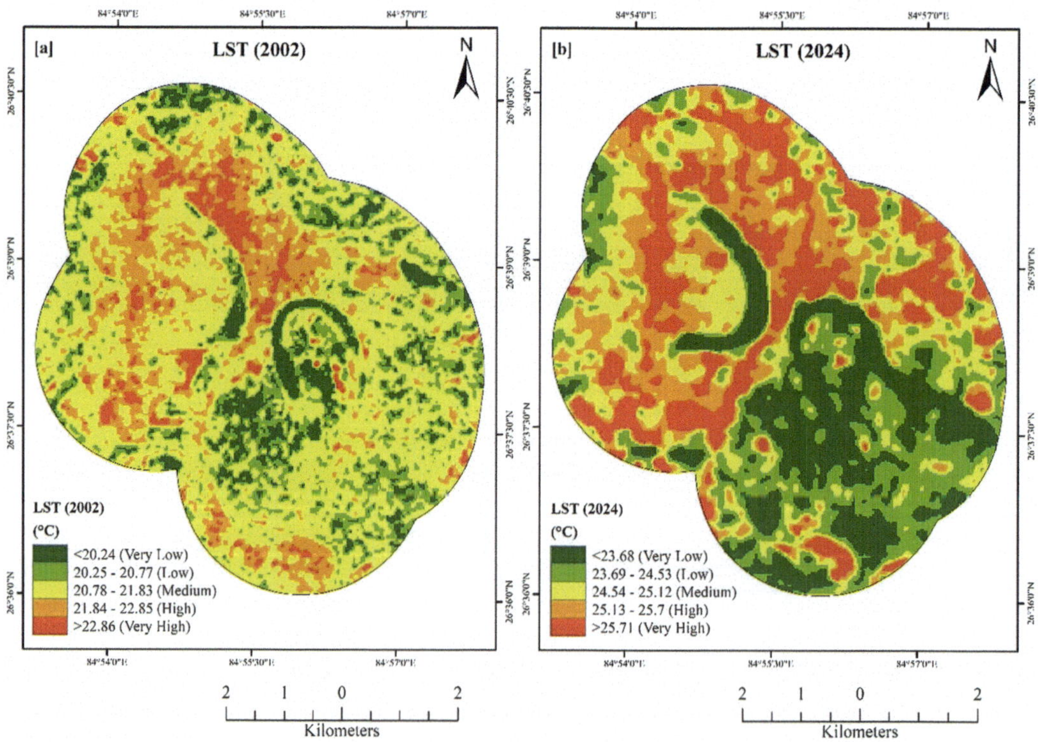

Fig. 8 Land surface temperature zones: (**a**) 2002 and (**b**) 2024

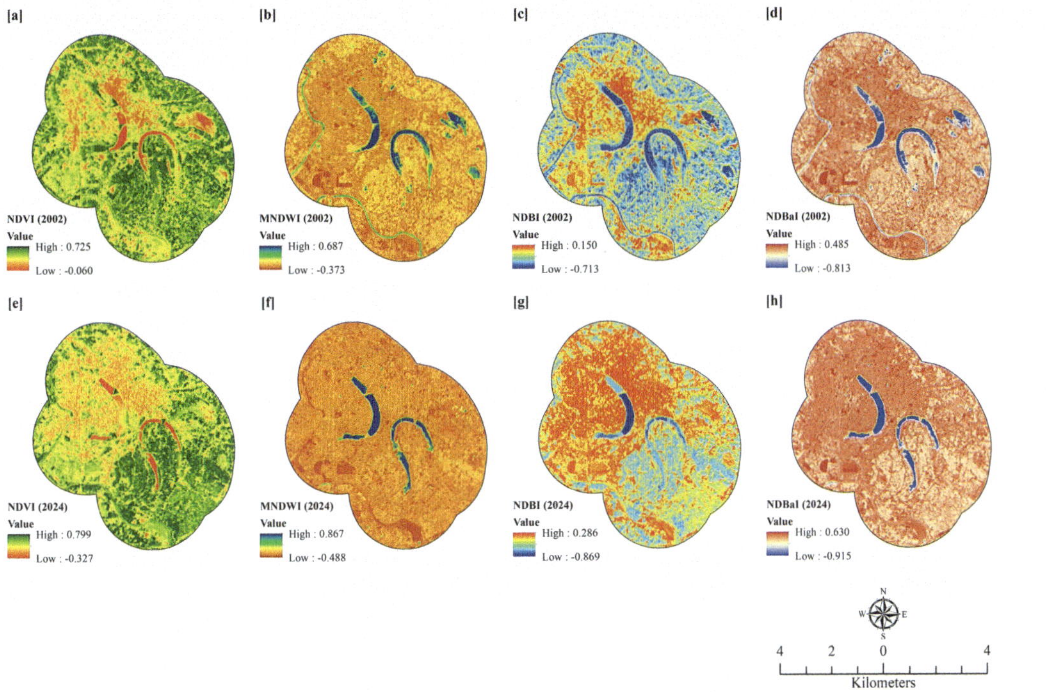

Fig. 9 LULC indices of Moti Jheel and its environs: (**a**) NDVI (2002), (**b**) MNDWI (2002), (**c**) NDBI (2002), (**d**) NDBI (2002), (**e**) NDVI (2024), (**f**) MNDWI (2024), (**g**) NDBI (2024), and (**h**) NDBI (2024)

2024 because of an increase in settlement around the Jheel. Also, the medium LST zones have undergone an areal decrease from 50.57% to 20.31% because of an increase in urbanization in the concerned Motihari City. Also, there is an areal increase in the very low LST zones from 9.39% in 2002 to 20.15% in 2024 around Kararia Maun (lying in the rural setting) because of increased cropland and vegetation in this eastern part of the city.

14 Normalized Difference Vegetation Index (NDVI) of (2002–2024)

The NDVI values ranges from −1 to +1. Higher NDVI values indicate the rich and healthy vegetative cover, while lower values indicate poor vegetation cover. As we can see the scenario of 2002, the minimum and maximum values of NDVI ranging from −0.060 to +0.725. The high value is found near Jheel, and the low value is found in water and built-up areas. But the scenario of 2024 is changed because the resulting minimum and maximum NDVI values ranging from −0.327 to +0.799 (Fig. 9a, e). The higher value areas are decreased and low value areas are increased, which means the current study's NDVI results show a prominent decrease in the croplands and vegetation areas in 2024.

15 Modified Normalized Difference Water Index (MNDWI) (2002–2024)

The MNDWI (Modified Normalized Difference Water Index) maps show changes in water bodies and wetland areas between the time periods 2002 and 2024. In 2002, the minimum and maximum MNDWI values ranges from −0.373 to +0.687. Yellow to orange color indicates lower MNDWI values, which typically correspond to areas with less water content or drier regions, mostly built-up and barren lands. The blueish green color indicates higher MNDWI values, which correspond to areas with more water content, the areas of Moti Jheel, Kararia Maun and other water bodies. There have been major changes in the value of MNDWI in 2024. The resulting minimum and maximum MNDWI values ranging from −0.488 to +0.867 (Fig. 9b, f). There appears to be a reduction in the areas covered by blueish-green patches, indicating a decrease in water content or shrinkage of water bodies. The concentration of water bodies seems to have reduced, especially when comparing the central and peripheral regions with the 2002 map.

16 Normalized Difference Built-up Index (NDBI) (2002—2024)

The NDBI (Normalized Difference Built-up Index) maps show changes in built-up areas between two time periods: 2002 and 2024. The NDBI values vary from −1 to +1. Higher NDBI values indicate a richer and healthier vegetative cover. In 2002, the resulting minimum and maximum NDBI values ranges from −0.713 to +0.150. Orange to yellow color indicates higher NDBI values, corresponding to built-up or urban areas, mostly around the Moti Jheel. There are significant patches of orange and yellow, indicating built-up areas, primarily in the central and some peripheral regions. The blue color indicates lower NDBI values, corresponding to non-built-up areas like water bodies and vegetation. The distribution of blue areas suggests the presence of water bodies, croplands, vegetation, and swampy regions. But in 2024, the resulting minimum and maximum NDBI values ranging from −0.869 to +0.286 (Fig. 9c, g). There are dynamic changes in NDBI value during the study period. The built-up and barren land area coverage have increased in the areas with higher NDBI values (orange to yellow), indicating significant urban expansion and development over the last two decades. However, the reduction in blue areas indicates a decrease in natural water bodies or non-urbanized regions.

17 The Normalized Difference Bareness Index (NDBI) (2002–2024)

Figure 9d, h shows the Normalized Difference Bareness Index (NDBI) values for two different years: 2002 and 2024. The NDBI is an index used to monitor and analyze the extent of bare land areas, with higher values indicating more bare or non-vegetated regions and lower values indicating less bare or more vegetated areas. In the map for 2002, high values +0.485 are represented by reddish-brown, while maximum values −0.813 are depicted in shades of darker-blue. In contrast, the map for 2024 shows maximum values of +0.630 and minimum values of −0.915. Comparing the maps from 2002 to 2024, there is a noticeable change in the extent and distribution of reddish-brown, suggesting a shift in the amount of bare land areas over the 22-year period. The increased intensity of brown color in certain areas points to an increase in bareness, possibly due to deforestation, land degradation, or urban expansion. The darker-blue areas might represent regions with more vegetation or less bare land, which have either increased or decreased over time. Overall, the NDBI maps illustrate significant changes in land cover from 2002 to 2024, indicating a potential increase in bare land areas. In terms of the surface temperature's spatial distribution, it is noted that the temperature of bare soil is much higher than that of agricultural areas, water bodies, and less than residentially dense places.

18 The Relationship among LST and LULC Indices

Figures 10 and 11 show the relationship between the LST and land cover indices in 2002 and 2022, respectively. The off-diagonal plots show scatter plots between pairs of variables, along with the Karl Pearson correlation coefficient (r) and the coefficient of determination (R^2). In Fig. 10, the relationship between LST and NDVI is showing that, the NDVI has moderate negative association with LST, Pearson's correlation $r = -0.21166$, $R^2 = 0.04478$ in 2002 (Fig. 10) while the value of Pearson's $r = -0.30157$, $R^2 = 0.09092$ in 2024 (Fig. 11). The LST and MNDWI also shows the moderately negative correlation, Pearson's correlation $r = -0.64679$, $R^2 = 0.41833$ in 2002, while Pearson's $r = -0.52261$, $R^2 = 0.27311$in 2024. As shown in (Figs. 10 and 11), the urbanized areas in the study area saw drastically changed throughout the study period and it could be seen that the strong positive correlations are evident between the LST and NDBI, with the Pearson's $r = 0.74457$, $R^2 = 0.55438$ in 2002 and Pearson's $r = 0.7805–9$, $R^2 = 0.60913$ in 2024. The LST and NDBI display that significant variability in LST values is reflected by the built-up areas. There is also a moderately positive relationship between the LST and NDBI, the value of Pearson's correlation coefficient $r = 0.56433$, $R^2 = 0.31845$ in 2002, and Pearson's $r = 0.67894$, $R^2 = 0.46095$ in 2024.

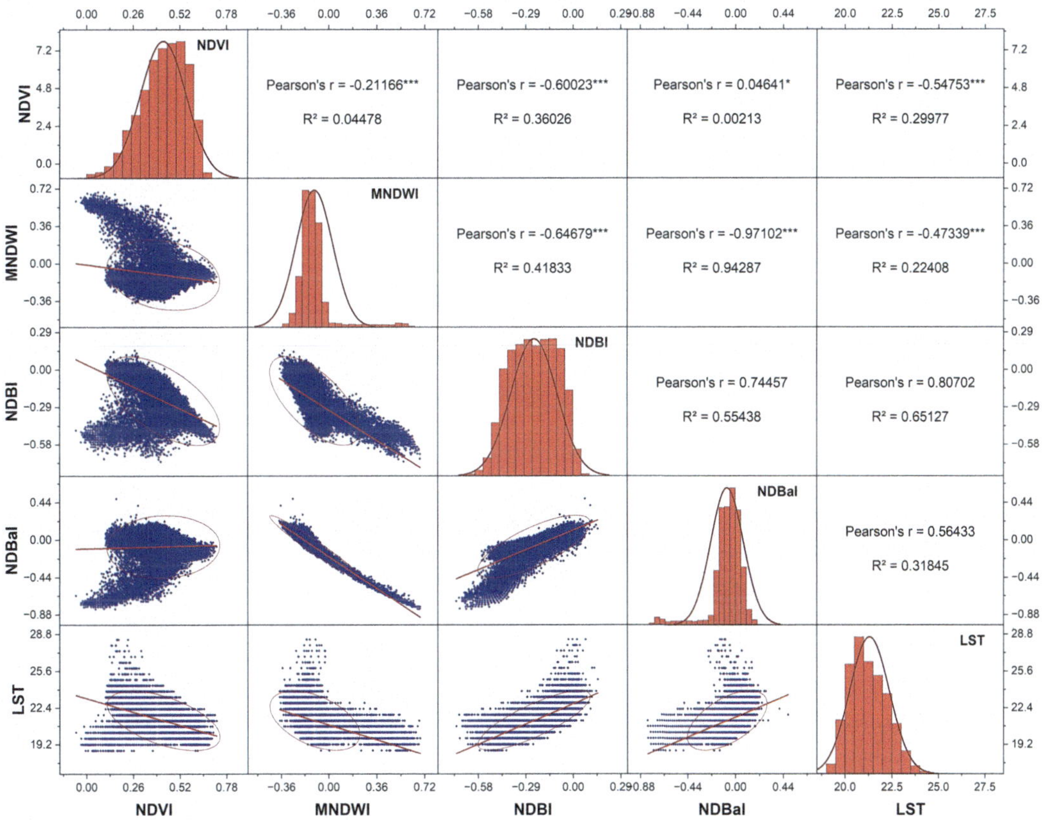

Fig. 10 Correlation between LST and LULC indices (2002)

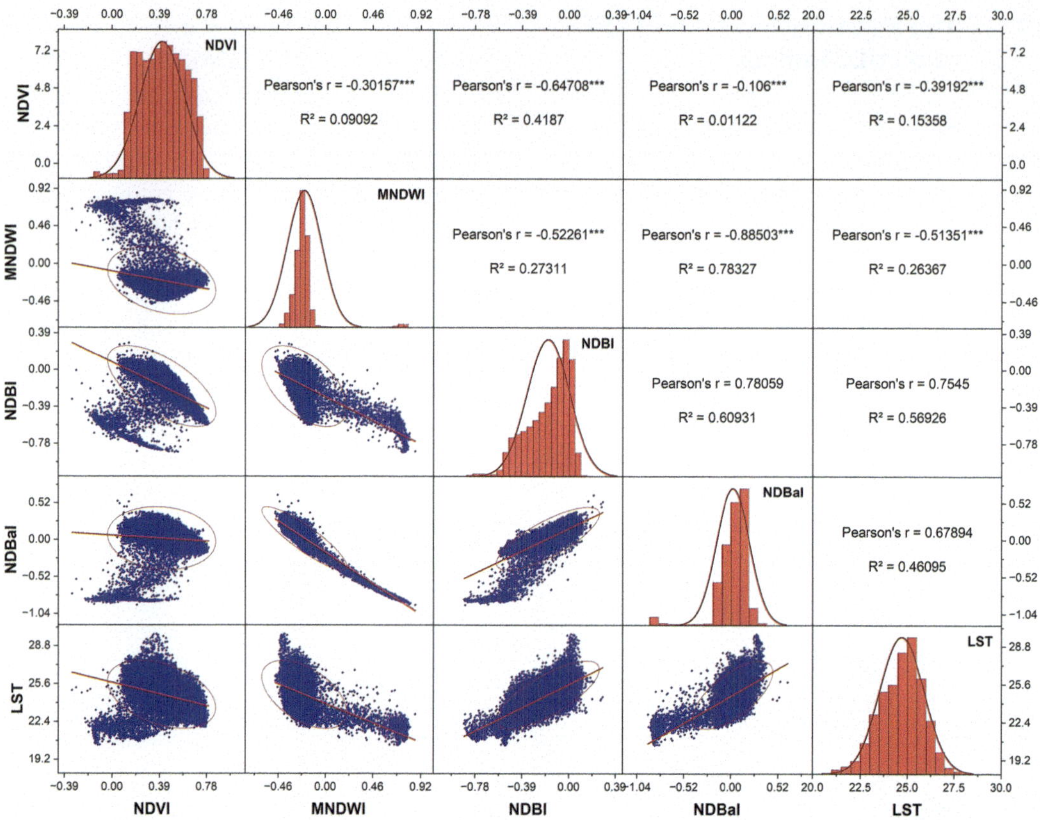

Fig. 11 Correlation between LST and LULC indices (2024)

19 Discussion

The potential reasons behind the deterioration of Moti Jheel and Kararia Maun are intricate and multifaceted, encompassing a variety of natural and anthropogenic elements. The swift pace of industrialization and the development of urban infrastructure play a significant role in the decline of Moti Jheel. Industrial facilities often release untreated effluents directly into water bodies, causing severe pollution that disrupts aquatic ecosystems. Likewise, urban development projects like road construction and building erection encroach upon water bodies, diminishing wetland areas and altering the natural hydrology. In the case of Kararia Maun, which lies in rural regions, agricultural practices are the primary source of harm, as insecticides, pesticides, erosion-induced silt, and agricultural runoff lead to considerable damage. Furthermore, nearby industrial operations can introduce chemicals and heavy metals into this water body, compounding the contamination. This study emphasizes the evolving alteration of these regions, predominantly propelled by the growth of cities, farming practices, and ecological shifts. It also demonstrates a notable rise in surface temperatures of the land, particularly noticeable in urban zones that have expanded. The standardized indicators (such as NDVI, MNDWI, NDBI, and NDBI) additionally depict the influence of these transformations on the well-being of vegetation, moisture levels, developed areas, and barren terrain.

The impact of climate change exacerbates the degradation of both water bodies, as rising temperatures and shifting precipitation patterns result in increased evaporation rates and diminished water levels. Extreme weather occurrences such as floods can trigger substantial erosion and sediment accumulation, further deteriorating these ecosystems. A lack of awareness and education regarding the significance of water bodies and wetlands results in their disregard and inappropriate use in urban and rural settings. Engaging the public through educational initiatives is vital to fostering conservation endeavors. Inadequate governance, lax environmental regulations, and ineffective enforcement of current laws all contribute to the uncontrolled exploitation of these water bodies. Sound policy formulation and robust enforcement mechanisms are crucial for safeguarding these ecosystems.

The economic pressures of urbanization often prioritize immediate profits over long-term environmental sustainability, leading to the underestimation of the ecological benefits offered by water bodies. In rural areas, the economic reliance on agriculture drives practices that harm natural water bodies. Introducing alternative livelihood opportunities and advocating for sustainable agricultural methods can help alleviate some of these pressures. The degradation of Moti Jheel and Kararia Maun has profound consequences on both plant and animal life, resulting in reduced biodiversity, disruption of habitats, and the loss of ecosystem services like water purification, flood management, and carbon storage. A comprehensive approach involving regulation of industrial activities, thoughtful infrastructure development, climate resilience strategies, public education, effective governance, and economic diversification is imperative to address the degradation and facilitate the preservation of these critical ecosystems.

20 Conclusion

So, the preservation and restoration of Moti Jheel and Kararia Maun are the need of the hour, necessitating collaborative actions from policymakers, local communities, and interested parties. Through this research, various important aspects have been put forward to support the expository and protective approaches with regard to the Moti Jheel and Kararia Maun. It points out the main causes of such abuse of these natural resources as industrial, urban, and agricultural development, and the effect of global warming. The statistical analysis investigates the degree to which boundaries defined by NDVI, MNDWI, NDBI, and NDBI reference relate to vegetative health, moisture, urbanization, and barren land existence. Further, the study reveals the fact that climate change, particularly high temperature and precipitation alternation, worsens the condition of these ecosystems by increasing evaporation and changing the hydrology of the region. It underscores the compelling need for the design of all-encompassing policies to be followed by the successful implementation of appropriate legal and institutional practices. Also, it accentuates the significance of embracing comprehensive and sustainable methodologies to tackle the obstacles introduced by urbanization, climate variations, and agricultural expansion. Through this action, we can ensure the enduring well-being and operation of these wetlands, safeguarding their ecological, climatic, and socioeconomic advantages for forthcoming generations.

References

Ahmad, M., Ahmad, W. S., Ahmad, S. N., Jamal, S., & Saqib, M. (2024). Tracing the roots of wetland degradation in India: A systematic review of anthropogenic drivers, ecological consequences and conservation strategies. *GeoJournal, 89*(1), 1–20.

Ali, M. B., Jamal, S., Ahmad, M., & Saqib, M. (2024). Unriddle the complex associations among urban green cover, built-up index, and surface temperature using geospatial approach: A micro-level study of Kolkata Municipal Corporation for sustainable city. *Theoretical and Applied Climatology, 155*(5), 4139–4160.

Asgher, M. S., Sharma, S., Singh, R., & Singh, D. (2021). Assessing human interactions and sustainability of Wetlands in Jammu, India using geospatial technique. *Modeling Earth Systems and Environment, 7*, 2793–2807.

Barbier, E. B. (1993). Sustainable use of wetlands valuing tropical wetland benefits: Economic methodologies and applications. *Geographical Journal*, 22–32.

Barbier, E. B., Burgess, J. C., & Folke, C. (1994). *Paradise lost? The ecological economics of biodiversity* (p. 267). Earthscan.

Batzer, D. P., & Sharitz, R. R. (Eds.). (2014). *Ecology of freshwater and estuarine wetlands*. Univ of California Press.

Burgess, J. (2015). WISA 2014-Water Innovations. Water SA, 41(2), 169–169.

Chu, H. J., Lin, Y. P., Huang, Y. L., & Wang, Y. C. (2009). Detecting the land-cover changes induced by large-physical disturbances using landscape metrics, spatial sampling, simulation, and spatial analysis. *Sensors, 9*(9), 6670–6700. CRA. 2011.

Congalton, R. G. (1991). A review of assessing the accuracy of classifications of remotely sensed data. *Remote Sensing of Environment, 37*(1), 35–46. https://doi.org/10.1016/0034-4257(91)90048-B

Costanza, R., d'Arge, R., de Groot, R., Farber, S., Grasso, M., Hannon, B., & van den Belt, M. (1997). The value of the world's ecosystem services and natural capital. *Nature, 387*(6630), 253–260.

Davidson, N. C., Finlayson, C. M., Milton, G. R., Prentice, R. C., & Everard, M. (2015). *Conservation of wetlands and other aquatic habitats: The role of the Ramsar convention*. Routledge.

Guo, M., Li, J., Sheng, C., Xu, J., & Wu, L. (2017). A review of wetland remote sensing. *Sensors, 17*(4), 777.

Imdad, K., Sahana, M., Ravetz, J., Areendran, G., Gautam, O., Dwivedi, S., et al. (2023). A sustainable solution to manage ecosystem health of wetlands in urban and peri-urban areas of Lucknow district, India using geospatial techniques and community based pragmatic approach. *Journal of Cleaner Production, 414*, 137646.

IPCC. (2007). *IPCC, 2007: Climate change 2007: Synthesis report. Contribution of working groups I, II and III to the fourth assessment report of the intergovernmental panel on climate change*. IPCC.

Jamal, S., & Ali, M. A. (2021). Remote sensing application for exploring land use and land cover dynamics in and around Chatra wetland, English bazar, West Bengal. *Creative Space, 9*(1), 15–27.

Jamal, S., Ali, M. A., & Ahmad, W. S. (2023a). Modelling transformation and its impact on water quality of the wetlands of lower Gangetic plain: A geospatial approach for sustainable resource management. *Geology, Ecology, and Landscapes*, 1–34.

Jamal, S., Saqib, M., Ahmad, W. S., Ahmad, M., Ali, M. A., & Ali, M. B. (2023b). Unraveling the complexities of land transformation and its impact on urban sustainability through land surface temperature analysis. *Applied Geomatics, 15*(3), 719–741.

Stacy , L. O. and Marvin , E. B. 2002 . Satellite remote sensing of wetlands . Wetlands Ecology and Management, 10 : 381–402

Sushma, N. R., & Dey, B. 2012 Assessment of land use and land cover changes of Damdama wetland, Gurugram using various change detection techniques.

Kshetri, T. (2018). Ndvi, ndbi & ndwi calculation using landsat 7, 8. *GeoWorld, 2*, 32–34.

Marambanyika, T., & Beckedahl, H. (2016). Wetland utilization patterns in semi-arid communal areas of Zimbabwe between 1985 and 2013 and the associated benefits to livelihoods of the surrounding communities. *Transactions of the Royal Society of South Africa, 71*(2), 175–186. https://doi.org/10.1080/0035919X.2016.1152520

Millennium Ecosystem Assessment (MEA). (2005). *Ecosystems and human Well-being: Wetlands and water synthesis*. World Resources Institute.

Mishra, B. K. (2020). 70 per cent of water bodies vanished in three decades in Bihar. *The Times of India*: https://timesofindia.indiatimes.com/city/patna/70-ofwater-bodies-vanished-in-three-decades/articleshow/74048278.cms

Mitsch, W. J., & Gosselink, J. G. (2015). *Wetlands*. John Wiley & sons.

Ramsar. (2024). The convention on wetlands, India names its 81st and 82nd Wetlands of International Importance on the occasion of World Environment Day, https://www.ramsar.org/news/india-names-its-81st-82nd-wetlands-international- importance-occasion-world-environment-day

Saqib, M., Jamal, S., Ahmad, M., Ali, M. A., Mir, A. Y., & Ali, M. B. (2024). Determining the impact of land use and land cover on microclimate with reference to thermal variability in Srinagar municipal corporation. In *Climate change, vulnerabilities and adaptation: Understanding and addressing threats with insights for policy and practice* (pp. 115–139). Springer Nature Switzerland.

Schuyt, K. D. (2005). Economic consequences of wetland degradation for local populations in Africa. *Ecological Economics, 53*(2), 177–190.

Srinivasan, J. T. (2010). *Understanding the Kole Lands in Kerala as a multiple use wetland ecosystem*. Centre for Economic and Social Studies, Research Unit for Livelihood and Natural Resources.

Wolfgang, J., An, S., Finlayson, C. M., & Gopal, K. (2013). Current state of knowledge regarding the world's wetlands and their; future under global climate change: A synthesis. *Aquatic Sciences, 75*, 151–167. www.isca.in

Xu, T., Weng, B., Yan, D., Wang, K., Li, X., Bi, W., ... & Liu, Y. (2019). Wetlands of international importance: Status, threats, and future protection. International journal of environmental research and public health, 16(10), 1818.

Urban Green Spaces as Catalysts for Environmental Pollution Mitigation: A Case Study of Kolkata Municipal Corporation

Md Babor Ali, Saleha Jamal, Ahmad Mujtaba Siddiqui, Arshad Amin, Mohd Saqib, and Mohd Atahar

1 Introduction

Rapid economic growth and industrial expansion in developing countries have significantly contributed to widespread ambient air pollution, impacting various societal aspects, particularly in densely populated urban areas. This increase in pollution correlates strongly with numerous diseases, prompting heightened concern over urban air quality from governments and individuals alike (Lelieveld et al., 2015; Li et al., 2015; Chen et al., 2016). Annually, more than four million premature deaths worldwide are attributed to ambient air pollution, with some regions experiencing rising pollution levels (Shaddick et al., 2020). Exposure to air pollutants elevates the risks of cardiovascular and respiratory diseases, among other health issues (Thurston et al., 2017). Notably, particulate matter (PM), varying in size and composition, is the primary health risk associated with air pollution, accounting for over half of the health burdens related to outdoor PM inhalation. PM is the fifth leading global health risk factor, underscoring the importance of reducing its concentrations for public health benefits (Cohen et al., 2017). Effective air quality management extends beyond emissions control, with research increasingly focusing on how plants and vegetation can mitigate PM effects through various mechanisms (Wolf et al., 2020).

The rapid urbanization and growing awareness of the natural environment's health impacts have driven numerous studies on outdoor land plants, often termed "green space"and "greenness" in urban planning. These nature based solutions actions inspired by, supported by, or mimicking natural processes are seen as cost-effective long-term strategies for mitigating urban air pollution (Bauduceau et al., 2015). In urban settings, blue-green infrastructure is linked to numerous benefits, from reducing air pollution to enhancing public well-being (Liu & Shen, 2014; Tzoulas et al., 2007). Trees, in particular, improve air quality by removing pollutants through direct deposition or gas absorption via stomata, and by cooling their surroundings through water vapor release (Nowak et al., 2006; McDonald et al., 2016). While the benefits of green spaces in addressing urban environmental issues are recognized, detailed information on the structural elements, composition, and management practices that enhance their air purification and climate regulation services is still needed. This requires a spatially explicit design approach with high resolution to effectively manage and plan urban green spaces.

Understanding the complex relationship between green spaces and air pollution involves various disciplines, including atmospheric sciences, plant biochemistry, and urban planning

M. B. Ali (✉) · S. Jamal · A. M. Siddiqui · A. Amin · M. Saqib · M. Atahar
Department of Geography, Faculty of Science, Aligarh Muslim University, Aligarh, Uttar Pradesh, India

Z. Khanam et al. (eds.), *Climate Change, Resilience and Ecosystem Services*, Sustainable Development Goals Series, https://doi.org/10.1007/978-3-031-96977-5_14

(Sun & Zhu, 2019). Estimating the indirect health impacts of green spaces through improved air quality is challenging due to intricate mechanisms and complex causal pathways (Janhäll, 2015; Markevych et al., 2017; Vienneau et al., 2017). Consequently, the epidemiological evidence remains limited, and the effects of green spaces on air pollution are not fully represented in current economic evaluations (Arantes & Mauad, 2019).

Furthermore, the integration of green spaces with urban structures can reduce traffic noise levels. Studies have highlighted the importance of combining green spaces with buildings, roads, and demographic features for effective noise reduction (Margaritis & Kang, 2016; Lopez & Souza, 2018). The distribution of traffic-generated noise is influenced by various components of the urban layout, including buildings and roadways (Valente-Pereira, 2014; Salomons & Pont, 2012; Silva et al., 2014). Research has examined the impact of vegetation on noise absorption and scattering, emphasizing the role of trees and different types of vegetation in mitigating traffic noise (Van Renterghem et al., 2014; Horoshenkov et al., 2013; Van Renterghem et al., 2012).

This research aims to establish correlations between urban green spaces and various parameters related to air quality and noise levels in the study area. It explores the intricate relationship between these elements to uncover patterns and connections, providing valuable insights into the potential benefits and impacts of incorporating and maintaining green spaces in urban planning and development.

2 Materials and Methods

2.1 Study Area

Kolkata, the third largest megacity in India, is situated on the bank of the Hooghly River, the largest distributary of the Ganga River. For this study, the focus is on the Kolkata Municipal Corporation (KMC) area, which is one of four municipal corporations comprising the greater Kolkata megacity. KMC, the oldest part of Kolkata, spans 206.08 km^2 and is geographically located between 22°25′0″N to 22°39′0″N latitude and 88°14′0″E to 88°28′0″E longitude (Fig. 1). The area is divided into 16 municipal boroughs and further subdivided

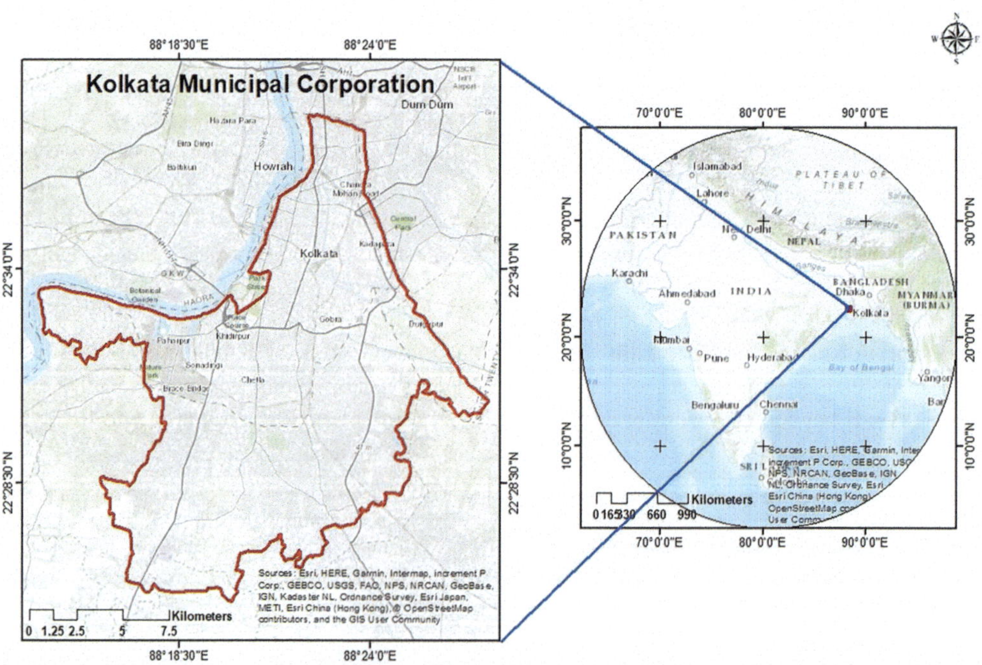

Fig. 1 Geographical location of the study area

into 144 municipal wards. According to the 2011 census, KMC has a floating population of six million and a residential population of 4.5 million, with a population density of 24,252/km^2. Kolkata has evolved into a major center for industry, trade, commerce, and other services (Dinda et al., 2021). Geographically, the study area is divided into five zones: north, south, east, west, and central. The area is part of the Bengal Basin, which evolved during the Tertiary period, and has a tropical wet and dry climate, classified as "Am" type according to Koeppen's Climatic classification. The mean annual temperature is about 25 °C, and the mean annual precipitation is around 150 cm. KMC experiences heavy rainfall during the monsoon season due to its location on the Bay of Bengal branch of the southwest monsoon, and it frequently faces cyclones. The soil is predominantly alluvial, deposited by rivers before they meet the Bay of Bengal, and is rich in tropical deciduous trees of various species. The Kolkata Municipal Corporation area hosts several urban green spaces that enhance the city's livability. Major parks like Maidan, Rabindra Sarobar, and Central Park serve as vital recreational spots and contribute to environmental well-being. These green spaces help mitigate air pollution, provide habitats for urban wildlife, and offer residents respite from the urban hustle. KMC has initiatives to expand green cover and maintain existing parks, reflecting the city's commitment to sustainability. Community gardens, tree-lined streets, and neighborhood parks collectively foster a greener, healthier urban environment in Kolkata.

3 Methods (Fig. 2)

3.1 Selection of Air Quality Monitoring Stations and Parameters

Seven monitoring stations, that is, Ballygunge, Bidhan Nagar, Thakurpukur, Jadavpur, Rabindra Bharati University, Rabindra Sarobar, and Victoria Memorial, were utilized to monitor various air quality parameters (Fig. 3). Some stations were chosen from areas with high green spaces and vegetation indices, while others were selected from areas with minimal or no green spaces. Stations in low-green areas are strategically located in zones with high vehicular traffic, industrial activities, and regions prone to natural dust or pollen emissions, as well as urban centers, to obtain accurate representations. The monitored parameters include carbon monoxide (CO), nitrogen oxide (NO), sulfur dioxide (SO_2), and particulate matter ($PM_{2.5}$ and PM_{10}), essential for evaluating pollution's environmental impact and protecting public health. Various techniques and instruments are employed for measurement and analysis. The data collected from these stations are vital for regulatory compliance, scientific research, and public health interventions, offering insights into pollution trends and identifying sources and their association with green spaces.

3.2 Selection of Noise Level Monitoring Stations

Nine noise level monitoring stations have been set up at various locations, including SSKM Hospital, Gol Park, Kolkata HQ (WBPCB), Patuli, New Market, RG Kar Hospital, Tollygunge, Bag Bazar, and Taratala (Fig. 4). Some stations were selected from areas with abundant greenery and a high vegetation index, while others were chosen from areas with minimal greenery and a low vegetation index. Stations in low-greenery areas were strategically placed in zones with heavy vehicular traffic, industrial establishments, and urban hubs to ensure accurate data representation. Both daytime and nighttime noise levels were recorded at these stations for comparative analysis.

3.3 Inverse Distance Weighting for Interpolation

Inverse Distance Weighting (IDW) is a popular spatial interpolation method used in Geographic Information Systems (GIS) to estimate values at unsampled locations based on the values of nearby sampled points. The technique is founded

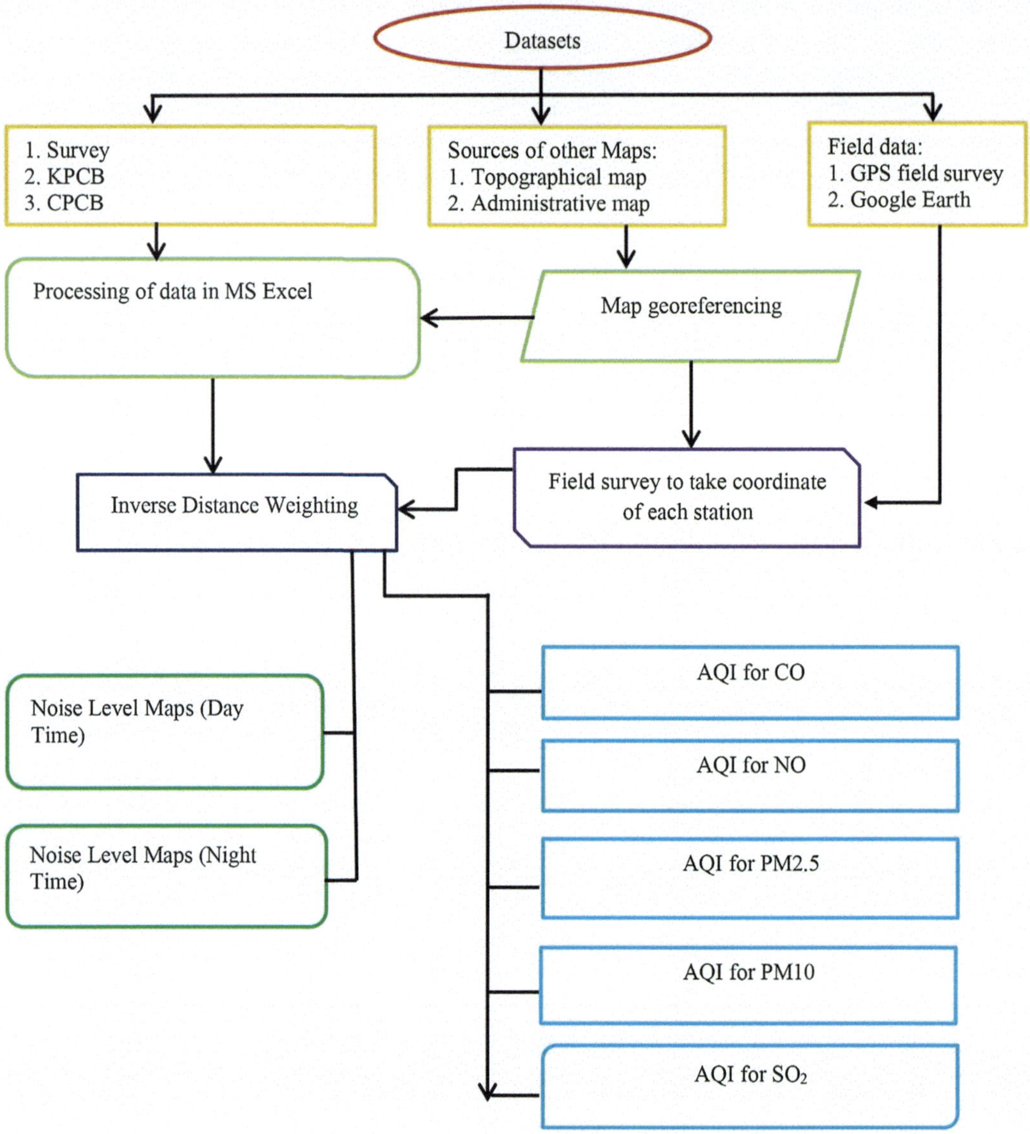

Fig. 2 Methodological flowchart of the study

on the principle that points closer to the target location exert more influence on the estimated value than those farther away.

IDW operates under the assumption that spatial datasets exhibit spatial autocorrelation, meaning that nearby locations are more likely to have similar values. This makes IDW well-suited for phenomena with a smooth and continuous spatial distribution, that is, pollution levels (Jamal et al., 2024).

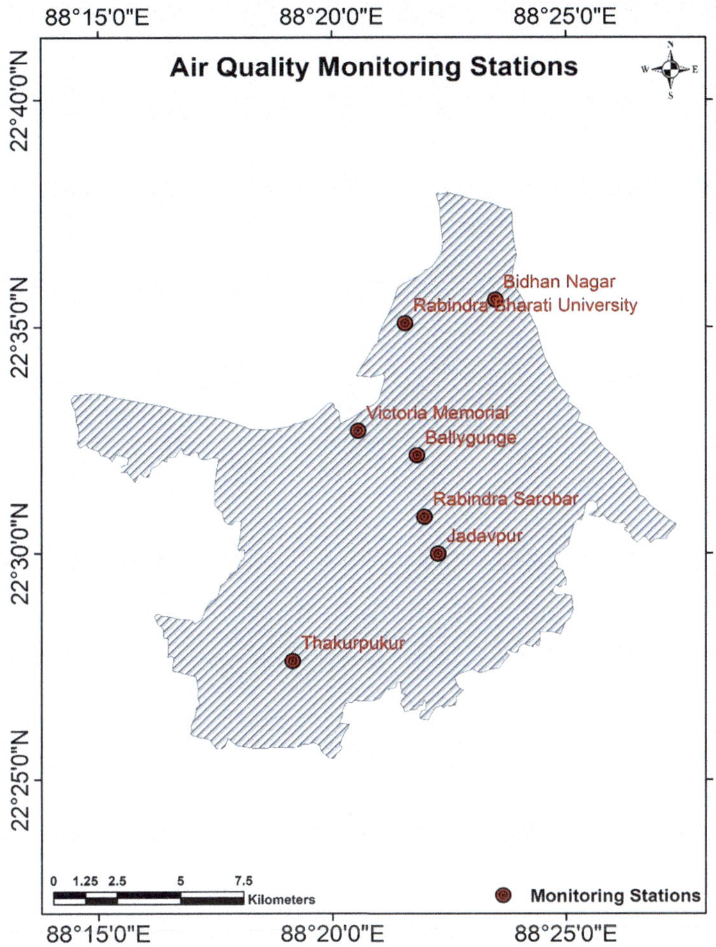

Fig. 3 Air quality monitoring stations

In IDW, each known point is assigned a weight based on its distance from the target location. These weights are calculated using a specified mathematical function, often a power-law function, where closer points receive higher weights. The power parameter controls the rate at which the influence of points decreases with distance; higher power values result in a steeper decline in influence, while lower values lead to a more gradual decrease.

Overall, Inverse Distance Weighting remains a valuable tool in GIS for its ease of implementation and its effectiveness in capturing spatial patterns and trends in various environmental and geographical phenomena.

Inverse Distance Weighting

$$vp = \frac{\sum_{i=1}^{n} \frac{1}{d} vi}{\sum_{i=1}^{n} \frac{1}{d}}$$

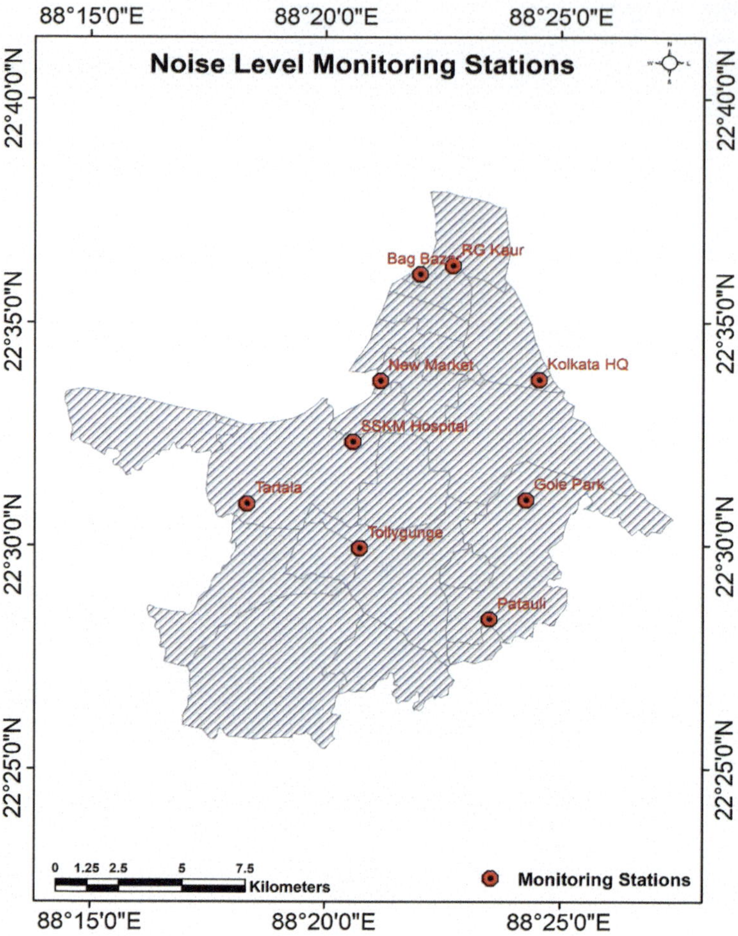

Fig. 4 Noise level monitoring stations

3.4 Normalized Difference Vegetation Index

The Normalized Difference Vegetation Index (NDVI) is employed to analyze vegetation cover and green spaces (Fig. 5). This widely used index, derived from satellite imagery, provides valuable information on vegetation distribution and health (Townshend & Justice, 1995). The NDVI is particularly effective for monitoring and mapping the distribution of vegetation over large areas (Newnham et al., 2011; Jamal et al., 2022; Ali et al., 2024). It utilizes the near infrared (NIR) and red bands of remote sensing images. The NDVI values range from −1 to +1, with values close to 1 indicating lush green cover, and values close to −1 representing barren land (Bardhan et al., 2016).

$$\text{Calculation of NDVI} = \frac{\text{Near Infrared}\left(\text{NIR}\right) - \text{Red}}{\text{Near Infrared}\left(\text{NIR}\right) + \text{Red}}$$

$$\text{NDVI for Landsat} - 8\text{OLI}, \text{NDVI} = \frac{\text{Band5} - \text{Band4}}{\text{Band5} + \text{Band4}}$$

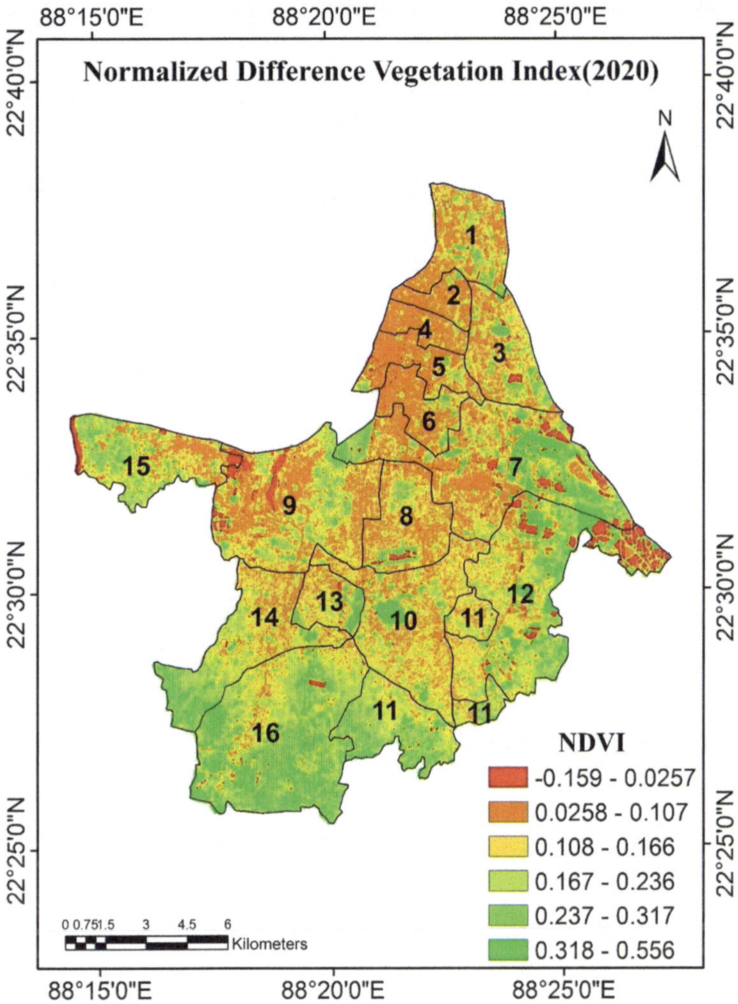

Fig. 5 Normalized Difference Vegetation Index (2020)

4 Results and Discussion

4.1 Concentration Levels of Different Parameters of Air Quality and Their Associations with Green Spaces

Table 1 and Fig. 6 provide detailed air quality data for several stations, including measurements of various pollutants such as PM2.5 (particulate matter with a diameter of 2.5 micrometers or less), PM10 (particulate matter with a diameter of 10 micrometers or less), NO (nitrogen dioxide), CO (carbon monoxide), and SO2 (sulfur dioxide), all recorded over a 24-hour period in micrograms per cubic meter (μg/m^3). Table 1 also includes the standard limits set by regulatory authorities for each of these pollutants.

Station Ballygunge shows significantly high levels of PM2.5 at 225.42 μg/m^3, far exceeding the recommended standard of 60 μg/m^3. PM10 levels are also elevated at 290.85 μg/m^3, surpassing the limit of 100 μg/m^3. Nitrogen dioxide (NO) levels are 165.1 μg/m^3, above the standard

Table 1 Different parameters-based air quality statistics

Stations	Latitude	Longitude	PM2.5	PM10	NO	CO	SO2
			Standard (μg/m³)in 24 h				
			60	100	80	4	80
Ballygunge	22.53597222	88.36352568	225.42	290.85	165.1	2.16	9.91
Bidhan Nagar	22.59316762	88.39140858	132.68	237.57	69.23	2.14	27.85
Thakurpukur	22.46026112	88.31915205	87.81	159.16	11.62	0.76	4.75
Jadavpur	22.49974494	88.37103054	26.46	119.01	28	0.89	11.67
Rabindra Bharati University	22.58454804	88.35939401	222.94	524.65	68.27	2.6	11.6
Rabindra Sarobar	22.5132309	88.36619677	124.24	167.32	53.97	1.55	29.13
Victoria Memorial	22.54496186	88.34256622	96.36	178.39	79.96	1.14	7.41

Date of data acquisition— January 13, 14, 15, 19, 20, 2020
Source: CPCB (2022)

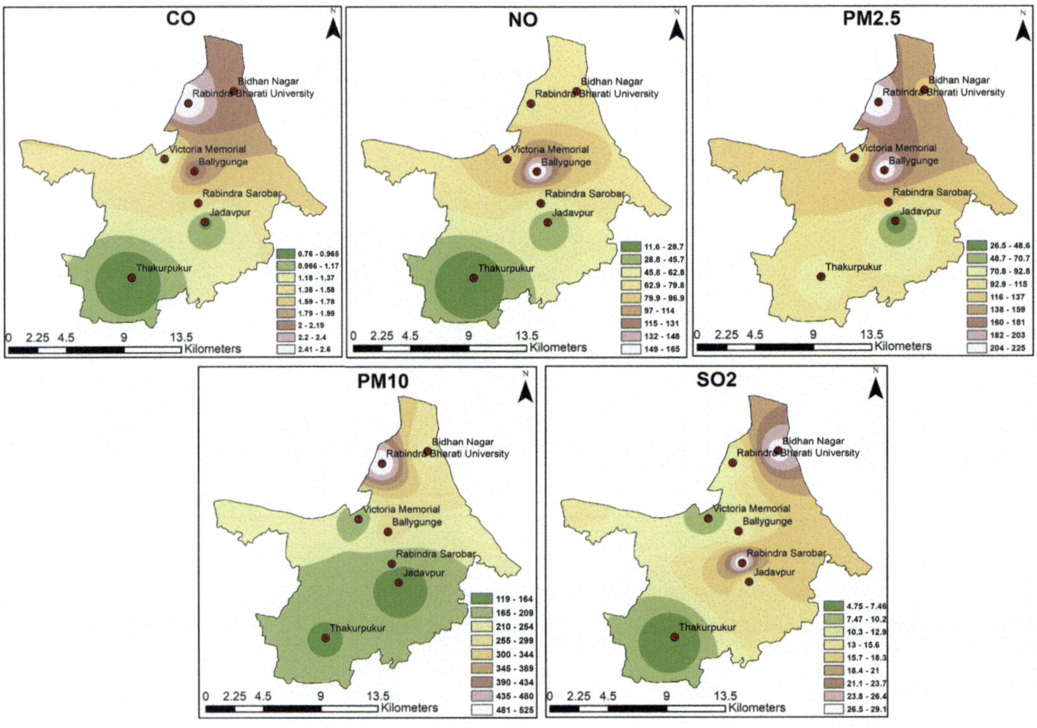

Fig. 6 Concentration levels of different parameters of air quality

of 80 μg/m³. Carbon monoxide (CO) is measured at 2.16 μg/m³, below the limit of 4 μg/m³, indicating a relatively lower presence. Sulfur dioxide (SO2) levels are 9.91 μg/m³, within acceptable ranges. The area around this station has very low vegetation cover and a low vegetation index.

At Bidhan Nagar, PM2.5 levels are recorded at 132.68 μg/m³, exceeding the recommended standard. PM10 levels are 237.57 μg/m³, also above the permissible limit. NO levels are 69.23 μg/m³, below the standard of 80 μg/m³. CO levels are 2.14 μg/m³, within the acceptable range. However, SO2 levels are high at 27.85 μg/m³, indicating a potential concern. This area is fully concretized, being located in the downtown area of the city, resulting in negligible green spaces.

Thakurpukur shows lower PM2.5 levels at 87.81 μg/m³ compared to the previous stations, still above the recommended standard. PM10 levels are 159.16 μg/m³, also exceeding the limit. NO levels are relatively low at 11.62 μg/m³, well within the acceptable range. CO levels are 0.76 μg/m^3, significantly lower than the 4 μg/m^3 standard. SO2 levels are 4.75 μg/m^3, within acceptable ranges. This area has high concentrations of green spaces and a high vegetation index.

Jadavpur exhibits the lowest PM2.5 levels at 26.46 μg/m^3 among all the stations, below the recommended standard. However, PM10 levels are 119.01 μg/m^3, above the permissible limit. NO levels are 28 μg/m^3, indicating a relatively higher presence. CO levels are 0.89 μg/m^3, indicating a lower presence. SO2 levels are 11.67 μg/m^3, within acceptable ranges. The area's high concentration of green spaces, such as Kolkata Golf Club and Kolkata Golf City, contributes to these results.

Rabindra Bharati University shows extremely high PM2.5 levels at 222.94 μg/m^3, significantly above the recommended standard. PM10 levels are very high at 524.65 μg/m^3, indicating a severe presence of particulate matter. NO levels are 68.27 μg/m^3, above the standard. CO levels are 2.6 μg/m^3, within the acceptable range. SO2 levels are 11.6 μg/m^3, within acceptable limits. This area is fully concretized, located in the downtown area of the city, and has negligible green spaces.

Rabindra Sarobar records PM2.5 levels at 124.24 μg/m^3, which is above the recommended standard. PM10 levels are at 167.32 μg/m^3, exceeding the permissible limit. NO levels are elevated at 53.97 μg/m^3, indicating a higher presence of this pollutant. CO levels at 1.55 μg/m^3 are within the acceptable range. However, SO_2 levels at 29.13 μg/m^3 are high, indicating a potential concern. This area also faces a scarcity of green spaces.

Victoria Memorial exhibits PM2.5 levels at 96.36 μg/m^3, above the recommended standard. PM10 levels are at 178.39 μg/m^3, surpassing the permissible limit. NO levels are measured at 79.96 μg/m^3, indicating a higher presence of this pollutant. CO levels at 1.14 μg/m^3 are within the acceptable range. SO2 levels at 7.41 μg/m^3 are also within acceptable limits. Compared to Rabindra Bharati University and Bidhan Nagar, this area has a higher concentration of green spaces, resulting in relatively lower pollutant levels.

Green spaces are crucial in combating air pollution. Trees and vegetation act as natural filters, capturing harmful pollutants and releasing oxygen, resulting in cleaner air. They also play a vital role in reducing the urban heat island effect and mitigating the formation of ground-level ozone (Ali et al., 2024). Additionally, green spaces support biodiversity, fostering a healthier ecosystem that aids in pollution control. In urban planning, prioritizing green spaces is essential for sustainable, thriving cities. Their impact on air quality is indispensable, making them vital in the fight against pollution and for the overall health of urban communities(Ali et al., 2024).

4.2 Concentration of Noise Levels and its Associations with Green Spaces

Noise pollution is a significant issue in urban areas, contributing to various health problems. The economic costs of noise are estimated to be between 0.2–2% of the European Union's gross domestic product. Urban green spaces in densely populated cities can significantly reduce noise levels, depending on their size, quality, and proximity to noise sources. This research highlights the importance of urban green spaces in maintaining ecological balance and ensuring environmental sustainability in cities. Tables 2 and 3 and Figs. 7 and 8 present a comprehensive dataset of noise levels, measured in Leq dB(A) during day and night, for various stations across the Kolkata Municipal Corporation over a year, broken down by month.

Noise levels at SSKM Hospital are relatively stable throughout the year, ranging from 43 dB(A) in September to 52 dB(A) in January. This station's low noise levels might be attributed to its location near the Maidan, a large green space. GolPark experiences significant monthly variations, with noise increasing from 49 dB(A) in February to 70 dB(A) in May, then gradually decreasing. This area has a low vegetation index and minimal green cover. Kolkata HQ (WBPCB) shows stable noise readings throughout the year, with minor fluctuations, indicating a consistent acoustic environment. This downtown area has very little green space. Patuli shows significant variability in noise levels, with peaks in April (62 dB(A)) and June (62 dB(A)). Lower noise levels are recorded in

Table 2 Noise level data (day time) in Leq dB(A)

Station location	Latitude	Longitude	January	February	March	April	May	June	July	August	September	October	November	December
SSKM Hospital	22.53877	88.34314	52	49	51	50	49	51	48	44	43	44	52	48
Gole Park	22.51700	88.40439	59	62	64	67	70	65	67	64	67	67	60	60
Kolkata HQ (WBPCB)	22.56185	88.40902	65	64	61	64	63	66	65	64	64	65	63	63
Patauli	22.47252	88.39159	50	44	44	62	47	62	62	62	59	50	61	67
New Market	22.56150	88.35289	66	65	61	60	61	61	62	61	60	61	63	69
RG Kar Hospital	22.60449	88.37867	56	52	45	44	45	44	45	44	43	44	44	44
Tollygunge	22.49902	88.34550	66	66	66	66	64	65	66	65	66	66	66	66
Bag Bazar	22.60128	88.36695	78	79	78	77	65	69	67	63	54	65	74	70
Tartala	22.51556	88.30533	65	65	64	65	66	66	65	66	65	64	64	65

Source: CPCB (2022)

Table 3 Noise level data (nighttime) in Leq dB(A)

Station location	Latitude	Longitude	January	February	March	April	May	June	July	August	September	October	November	December
SSKM Hospital	22.53877	88.34314	51	49	53	52	50	53	50	46	45	46	47	44
Gole Park	22.51700	88.40439	55	55	60	65	70	65	68	64	68	67	61	60
Kolkata HQ (WBPCB)	22.56185	88.40902	62	64	59	58	60	64	60	57	58	60	56	57
Patauli	22.47252	88.39159	51	44	48	63	49	58	61	64	59	50	61	69
New Market	22.56150	88.35289	67	65	61	60	62	61	62	62	60	61	60	62
RG Kar Hospital	22.60449	88.37867	57	52	45	44	45	44	45	43	43	44	44	43
Tollygunge	22.49902	88.34550	61	62	62	62	60	60	61	59	60	62	61	62
Bag Bazar	22.60128	88.36695	78	78	77	75	65	68	66	63	53	61	72	69
Tartala	22.51556	88.30533	61	61	61	62	61	62	62	61	62	60	60	60

Source: CPCB (2022)

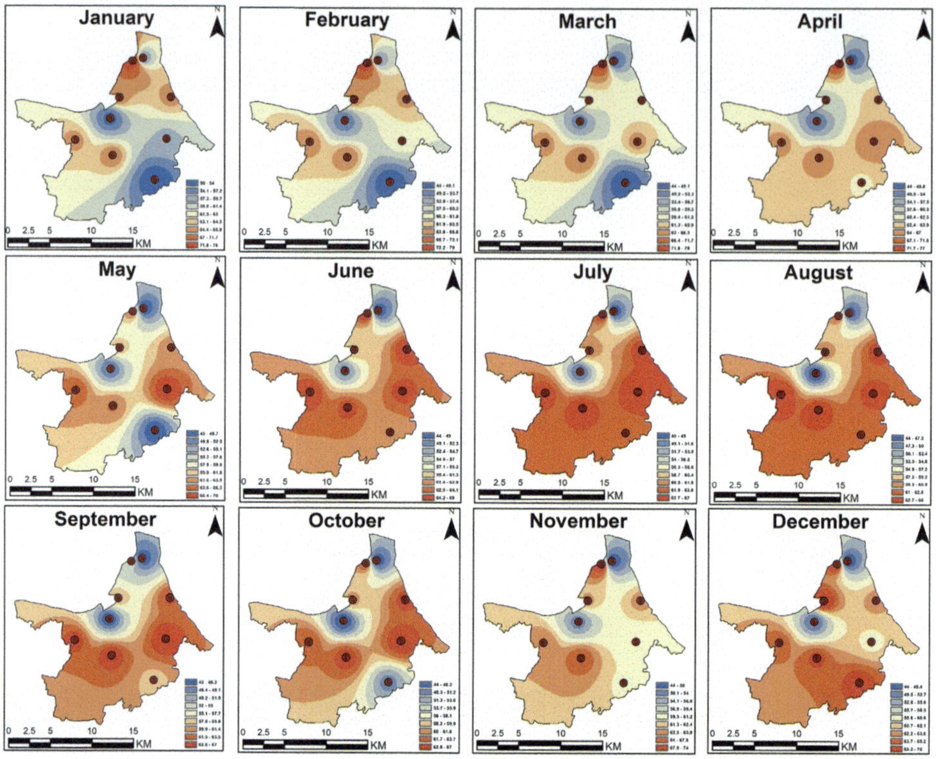

Fig. 7 Day time noise level

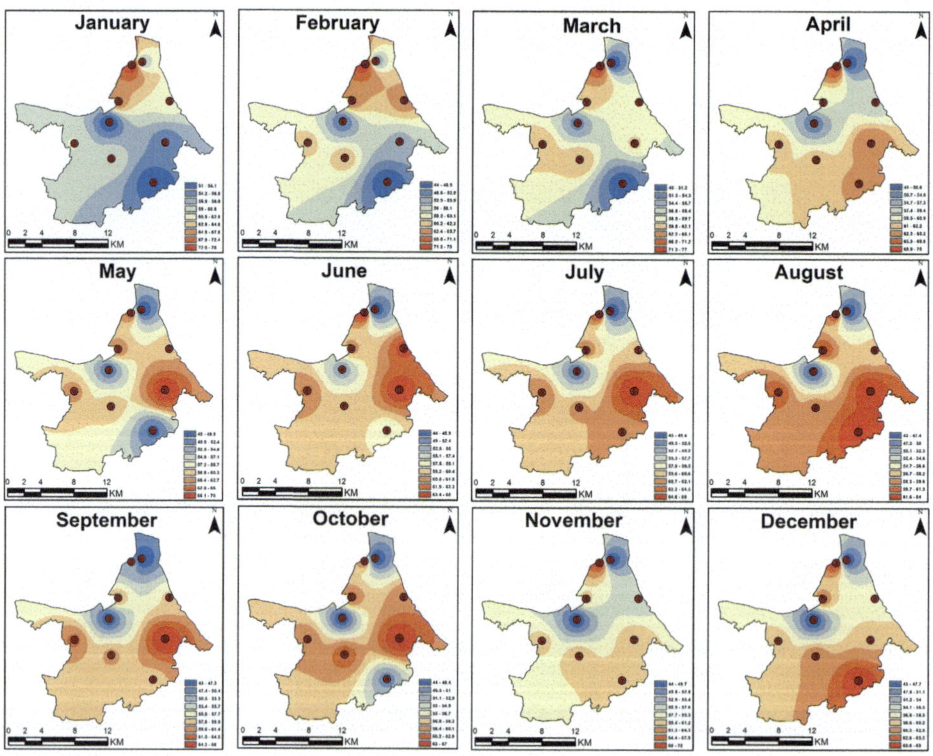

Fig. 8 Nighttime noise level

September (50 dB(A)). This station is located in an area with a moderate amount of green spaces, leading to lower noise levels from January to May and in October. New Market has a stable noise profile, with readings from 60 dB(A) in October to 69 dB(A) in December. This commercial area is heavily concretized with negligible green spaces. RG Kar Hospital records stable noise levels, ranging from 43 dB(A) in September to 56 dB(A) in January. This stability might be due to the designation of a silence zone around the hospital. Tollygunge demonstrates remarkably stable noise levels, consistently around 66 dB(A) from January to November, dropping slightly to 65 dB(A) in December. This area also lacks green spaces and experiences high noise levels throughout the year. Bag Bazar and Taratala exhibit significant noise variations, with a peak in February (79 dB(A)) and a decrease towards the end of the year. These downtown areas have negligible green spaces.

The noise level data for various stations in Kolkata show distinct patterns and variations. Understanding these trends is crucial for urban planning, environmental management, and ensuring residents' well-being. Stations located near green spaces tend to have lower noise levels, emphasizing the importance of urban greenery.

5 Conclusion

This research demonstrates a significant correlation between green spaces and both air quality and noise levels in different areas of Kolkata Municipal Corporation. The analysis indicates a clear relationship between vegetation cover and pollutant concentrations. In regions like Ballygunge and Bidhan Nagar, where green spaces are scarce, pollutant levels, particularly PM2.5 and PM10, exceed recommended standards, highlighting the potential of green spaces in reducing particulate pollution. Conversely, in Thakurpukur and Jadavpur, which have higher concentrations of greenery, pollutant levels are comparatively lower, showcasing the positive impact of vegetation on air quality. Rabindra Bharati University and Bag Bazar, located in highly urbanized, concrete-dominated areas with limited greenery, exhibit alarmingly high pollutant concentrations, reinforcing the need for integrating green spaces into urban planning to combat the adverse effects of urbanization on air quality. Victoria Memorial, situated in an area with relatively more green cover, shows comparatively lower pollutant levels, further emphasizing the importance of green spaces in mitigating air pollution. Regarding noise pollution, SSKM Hospital, surrounded by significant greenery like the Maidan, consistently records lower noise levels. In contrast, areas such as Gole Park and Bag Bazar, characterized by concrete landscapes and limited green spaces, experience higher noise levels. The data highlights that areas with more green spaces tend to have lower noise levels, underscoring the importance of preserving and creating green areas within urban environments.

To improve air quality and reduce noise pollution in Kolkata Municipal Corporation, urban planning should prioritize expanding green spaces, particularly in high-pollution areas like Ballygunge and Bidhan Nagar. Incorporating green belts, vegetative barriers, and green infrastructure in redevelopment projects will help mitigate pollution. Implementing noise reduction strategies in sensitive areas and establishing more monitoring stations are crucial. Public awareness campaigns and policy incentives for integrating green designs in buildings will further enhance environmental sustainability and health. These measures will create a healthier urban environment, benefiting Kolkata's residents.

References

Ali, M. B., Jamal, S., Ahmad, M., & Saqib, M. (2024). Unriddle the complex associations among urban green cover, built-up index, and surface temperature using geospatial approach: A micro-level study of Kolkata Municipal Corporation for sustainable city. *Theoretical and Applied Climatology, 155*(5), 4139–4160.

Arantes, B. L. D., & Mauad, T. (2019). Urban forests, air quality and health: A systematic review. *International Forestry Review, 21*(2), 167–181.

Bardhan, R., Debnath, R., & Bandopadhyay, S. (2016). A conceptual model for identifying the risk susceptibility of urban green spaces using geo-spatial techniques. *Modeling Earth Systems and Environment, 2*(3), 1–12.

Bauduceau, N., Berry, P., Cecchi, C., Elmqvist, T., Fernandez, M., Hartig, T., ... & Tack, J. (2015).

Towards an EU research and innovation policy agenda for nature-based solutions & re-naturing cities: Final report of the horizon 2020 expert group on'nature-based solutions and re-naturing cities

Chen, X., Zhang, L. W., Huang, J. J., Song, F. J., Zhang, L. P., Qian, Z. M., et al. (2016). Long-term exposure to urban air pollution and lung cancer mortality: A 12-year cohort study in Northern China. *Science of the Total Environment, 571*, 855–861.

Cohen, A. J., Brauer, M., Burnett, R., Anderson, H. R., Frostad, J., Estep, K., et al. (2017). Estimates and 25-year trends of the global burden of disease attributable to ambient air pollution: An analysis of data from the Global Burden of Diseases Study 2015. *The Lancet, 389*(10082), 1907–1918.

Central Pollution Control Board. (2022). Annual report 2020–21. New Delhi, India: CPCB.

Dinda, S., Chatterjee, N. D., & Ghosh, S. (2021). An integrated simulation approach to the assessment of urban growth pattern and loss in urban green space in Kolkata, India: A GIS-based analysis. *Ecological Indicators, 121*, 107178.

Horoshenkov, K. V., Khan, A., & Benkreira, H. (2013). Acoustic properties of low growing plants. *The Journal of the Acoustical Society of America, 133*(5), 2554–2565.

Jamal, S., Ali, M. B., Ali, M. A., & Ajmal, U. (2022). Evaluation and distribution of urban green spaces in Kolkata Municipal Corporation: An approach to urban sustainability. In *Towards sustainable natural resources: Monitoring and managing ecosystem biodiversity* (pp. 151–172). Springer International Publishing.

Jamal, S., Ajmal, U., Ali, M. B., Saqib, M., & Arfeen, S. (2024). Assessment of ambient air quality in relation to the burning of firecrackers during the festival of Diwali: A case study of Jodhpur City (India). *Environmental Monitoring and Assessment, 196*(2), 222.

Janhäll, S. (2015). Review on urban vegetation and particle air pollution–deposition and dispersion. *Atmospheric Environment, 105*, 130–137.

Lelieveld, J., Evans, J. S., Fnais, M., Giannadaki, D., & Pozzer, A. (2015). The contribution of outdoor air pollution sources to premature mortality on a global scale. *Nature, 525*(7569), 367–371.

Li, P., Yan, R., Yu, S., Wang, S., Liu, W., & Bao, H. (2015). Reinstate regional transport of PM2. 5 as a major cause of severe haze in Beijing. *Proceedings of the National Academy of Sciences, 112*(21), E2739–E2740.

Liu, H. L., & Shen, Y. S. (2014). The impact of green space changes on air pollution and microclimates: A case study of the Taipei metropolitan area. *Sustainability, 6*(12), 8827–8855.

Lopez, G. A. P., & Souza, L. C. L. D. (2018). Urban green spaces and the influence on vehicular traffic noise control. *Ambiente Construído, 18*, 161–175.

Margaritis, E., & Kang, J. (2016). Relationship between urban green spaces and other features of urban morphology with traffic noise distribution. *Urban Forestry & Urban Greening, 15*, 174–185.

Markevych, I., Schoierer, J., Hartig, T., Chudnovsky, A., Hystad, P., Dzhambov, A. M., et al. (2017). Exploring pathways linking greenspace to health: Theoretical and methodological guidance. *Environmental Research, 158*, 301–317.

McDonald, R., Kroeger, T., Boucher, T., Wang LongZhu, W. L., & Salem, R. (2016). Planting healthy air: a global analysis of the role of urban trees in addressing particulate matter pollution and extreme heat.

Newnham, G. J., Verbesselt, J., Grant, I. F., & Anderson, S. A. (2011). Relative Greenness Index for assessing curing of grassland fuel. *Remote Sensing of Environment, 115*(6), 1456–1463.

Nowak, D. J., Crane, D. E., & Stevens, J. C. (2006). Air pollution removal by urban trees and shrubs in the United States. *Urban Forestry & Urban Greening, 4*(3–4), 115–123.

Salomons, E. M., & Pont, M. B. (2012). Urban traffic noise and the relation to urban density, form, and traffic elasticity. *Landscape and Urban Planning, 108*(1), 2–16.

Shaddick, G., Thomas, M. L., Mudu, P., Ruggeri, G., & Gumy, S. (2020). Half the world's population are exposed to increasing air pollution. *npj Climate and Atmospheric Science, 3*(1), 23.

Silva, L. T., Oliveira, M., & Silva, J. F. (2014). Urban form indicators as proxy on the noise exposure of buildings. *Applied Acoustics, 76*, 366–376.

Sun, Z., & Zhu, D. (2019). Exposure to outdoor air pollution and its human-related health outcomes: An evidence gap map. *BMJ Open, 9*(12), e031312.

Thurston, G. D., Kipen, H., Annesi-Maesano, I., Balmes, J., Brook, R. D., Cromar, K., et al. (2017). A joint ERS/ATS policy statement: What constitutes an adverse health effect of air pollution? An analytical framework. *European Respiratory Journal, 49*, 1.

Townshend, J. R. G., & Justice, C. O. (1995). Spatial variability of images and the monitoring of changes in the normalized difference vegetation index. *International Journal of Remote Sensing, 16*(12), 2187–2195.

Tzoulas, K., Korpela, K., Venn, S., Yli-Pelkonen, V., Kaźmierczak, A., Niemela, J., & James, P. (2007). Promoting ecosystem and human health in urban areas using Green Infrastructure: A literature review. *Landscape and Urban Planning, 81*(3), 167–178.

Valente-Pereira, L. (2014). *Urban form definition in urban planning*. Simplíssimo.

Van Renterghem, T., Botteldooren, D., & Verheyen, K. (2012). Road traffic noise shielding by vegetation belts of limited depth. *Journal of Sound and Vibration, 331*(10), 2404–2425.

Van Renterghem, T., Attenborough, K., Maennel, M., Defrance, J., Horoshenkov, K., Kang, J., et al. (2014). Measured light vehicle noise reduction by hedges. *Applied Acoustics, 78*, 19–27.

Vienneau, D., de Hoogh, K., Faeh, D., Kaufmann, M., Wunderli, J. M., Röösli, M., & SNC Study Group. (2017). More than clean air and tranquility: Residential green is independently associated with decreasing mortality. *Environment International, 108*, 176–184.

Wolf, K. L., Lam, S. T., McKeen, J. K., Richardson, G. R., Van Den Bosch, M., & Bardekjian, A. C. (2020). Urban trees and human health: A scoping review. *International Journal of Environmental Research and Public Health, 17*(12), 4371.

Land Use and Land Cover Dynamics in Mahananda River Basin, West Bengal, India Using Geospatial Technology

Madhurima Purkait and Sujit Mandal

1 Introduction

Planning and environmental managements are the successfully dependent on land use and land cover. It provides useful data on land usage, which is crucial for assessing natural resource and biodiversity. The sustainability of ecosystems is threatened by the changes in land cover, particularly when natural vegetation is replaced by anthropogenic intervention. These changes reduce energy, water, and nutrient availability; assist exotic species invasion; and expedite the processes of natural alteration (Woldeamlak, 2002; Ojima et al., 1994; Walker & Homma, 1996). External effects beyond the development regions are also caused by the expansion of agriculture and urbanization into natural habitats. Land demand is the main driving force behind this shift (Walker & Homma, 1996). In the agricultural setting of the northeastern hilly region, land is the key component, with its requirement influenced by product and factor prices. This demand leads to productive land use, involving specific agricultural practices like tea plantations often requiring deforestation (Afify, 2011; Mitra et al., 2023; Mandal et al., 2020; Bandyopadhyay & De, 2018). Identifying land use and land cover (LULC) changes is fundamental due to their long-term impacts on the cosmopolitan settings and their contribution to climate. The fast infrastructural development sparked by the economic policies put in place in 1991 changed the basin morphology of the Mahananda River and the Terai peninsula of northeastern India. This led to noticeable vegetation loss, channel fragmentation, and widespread alteration in land cover across the floodplains. This study analyzes the effects of human activity on the land cover of the Mahananda River Basin during the last 30 years using an integrated scientific approach that incorporates remote sensing and field observations. The findings revealed that untouched vegetation-covered areas are initially transformed into cropland or grassland, and tea plantations and later predominantly into built-up areas. The study indicates that the natural vegetation class is showing negative signs of areal expansion, which could lead to a decline in the groundwater table, land loss, landslide, reduced flood intervals, and sabotage of the implanted infrastructure within the river basin (Mitra et al., 2020). Siliguri is one of the cities not only in the Darjeeling District but also the second most populated city in West Bengal, which has experienced significant growth due to rapid urbanization (Mandal & Maity, 2014). To restore this river basin's socio-hydrological and eco-hydrological balance, endorsements have been made to promote the sustainable use of available resources by localities and policymakers. Changes can be detected through

M. Purkait (✉) · S. Mandal
Department of Geography, Diamond Harbour Women's University, Diamond Harbour, India

Z. Khanam et al. (eds.), *Climate Change, Resilience and Ecosystem Services*, Sustainable Development Goals Series, https://doi.org/10.1007/978-3-031-96977-5_15

remote sensing, but precise change detection processes are necessary to retrieve details from satellite data. Remote sensing is becoming more and more important due to its unique perspective on the temporal and geographical dynamics of land use and land cover changes. Consequently, there has been substantial use of remote-sensing satellite based approaches, with good findings, to identify and track changes in land cover at various scales. Societal and environmental limitations may have a deeper imprint on changes in land use when land values rise and competition amongst alternative consumers increases. Thus, to help with planning and implementation, planners must regularly monitor these changes to assess the effects of "from-to" changes and recommend alternate land uses for further development. Additionally, by monitoring the urban augmentation, it helps to speculate the extent and futuristic distribution of inhabited sets, which is highly appreciated. This study will investigate the LULC of the Upper Mahananda River Basin in the Darjeeling area using sensor-based data analysis and Geographic Information System (GIS) approaches. It will compare images from 5 years, with a decadal interval from 1993 to the present day. To validate the data, actual ground-based surveying techniques will be used to detect changes in the area.

2 Materials and Methodology

2.1 Delineation of the Study Domain

The Upper Mahananda River Basin, located in the Darjeeling District in the foothills of the Himalayas, was the primary focus of the study. Over the past 30 years, districts of West Bengal especially the Darjeeling District have faced some of the worst catastrophic events (Sarkar & Baral, 2021). It has experienced severe landslides, earthquake/building collapse, resource shortages, monsoon-specific hazards like floods, draught, cloudbursts, and significant environmental damages. Despite so many calamities, the people living in the Darjeeling District face those hazards and disasters that frequently happen every year and are trying to survive and persevere (Vliet et al., 2011). This district is subdivided into nine blocks; Darjeeling Pulbazar, Rangli Rangliot, Mirik, Kurseong, Matigara, Jorebunglow Sukiapokhri, Naxalbari, Kharibari, and Phansidewa. Two blocks, Kurseong and Matigara, and eight watersheds make up the chosen Mahananda River Basin. The watersheds are Shiva Khola, Babu Khola and Mana Khola, Jogi Khola, Ghoramara Khola and Choplong Khola, Gulma Nala, the Mahanadi River (left bank), Panchanai Khola, and Choklong Nala. The Mahananda River rises in the eastern Kurseong block of the Darjeeling District, close to Chimli, in the Mahaldhiram Mountain (26,055′40’‘N and 88,019′12″ E). After leaving the hills, the river flows southward until Siliguri, where it shifts slightly westward, forming the border between the districts of Darjeeling and Jalpaiguri up to the Phasidewa block in the extreme southeast (O'Malley, 1907, Darjeeling District Gazetteer). It then runs through Uttar Dinajpur, Kishanganj, and Malda before entering Bangladesh close to Tentulia in Panchagarh. Subsequently, it makes its way back to India and encounters the powerful Ganga in the Malda District of West Bengal. The Mahananda River flows through the Indian states of West Bengal and Bihar for about 470 km (Bharti, 2000). The Mahananda River Basin‘s catchment area is 14,500 km^2, according to Rana (2018). A 249 km^2 region with the coordinates 26°40’ N to 26°56’ N and 88°19′ E to 88°29′ E was chosen for the study from this group. Significant topographic variation can be seen in this catchment area, where elevations range from 25 to 2234 m above the sea level. The main river that flows a distance of 47.79 km within this selected Mahananda River Basin (MRB) is the Mahananda River. The following rivers—Khola representing tiny rivers—flow into the Mahananda River within the research area: Panchanai, Rongtong, Babu, Mana, Jogi, Choklong, Gulma, Champasari Jhora, and Mahishmari. The selected region features significantly steeper slopes and landslide-prone areas in the upper reaches, extensive irrigated flood-prone regions in the lower reaches, and increasing anthropogenic influence within the entire basin. The upper portion of the study area is under the soil group of Humic Acrisols (Ah) and lower portion of the basin is under the Eutric Fluvisols (Je) soil category. Slope of the area is between 0^0 and 65^0 (Fig. 2a) and the highest elevation of the selected site is 2400 m. The cities Siliguri and Kurseong with a huge urban agglomeration are situated amid the research location. A detailed description of this Mahananda River Basin is extracted from the Toposheet No. 78 B/1 and 78 B/5 and the Digital Elevation Model (DEM) from USGS Earth Explorer which is provided in Fig. 1. Within this region, the interaction of various biophysical climate and vegetation factors with human activities, such as agriculture and livestock production, significantly influences the ecosystem transformation. The reasons behind choosing the area are as follows: (a) The Mahananda River is the lengthiest river in North Bengal; and (b) located in the foothills of the Great Himalaya, the fastest-growing city in West Bengal, Siliguri, is situated on the banks of the Mahananda (Fig. 2).

2.2 Data Acquisition and Data Preparation

The Upper Mahananda River Basin‘s LULC fluctuations were shown using GIS and remote sensing techniques. Data collection, pre-processing, supervised classification, accuracy assessment, and change detection analysis (Arulbalaji, 2019; Thakkar et al., 2016) were

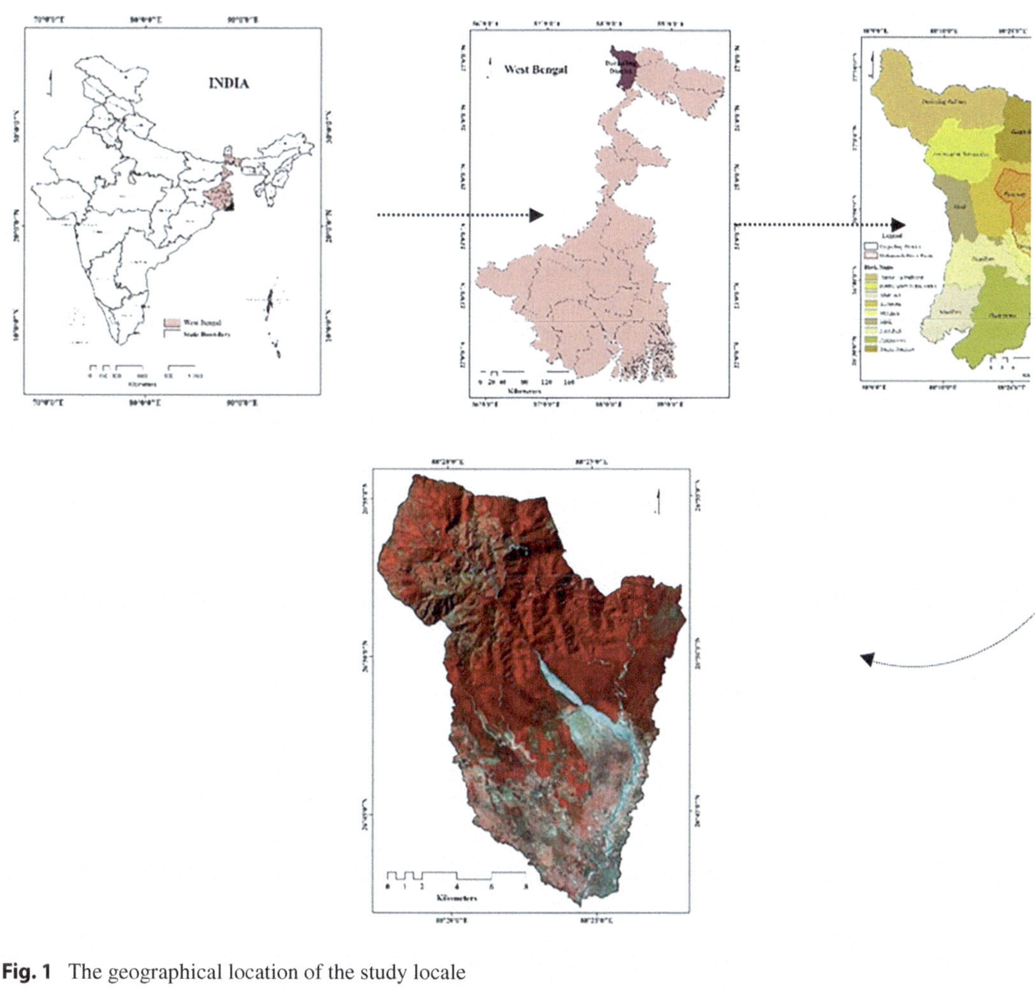

Fig. 1 The geographical location of the study locale

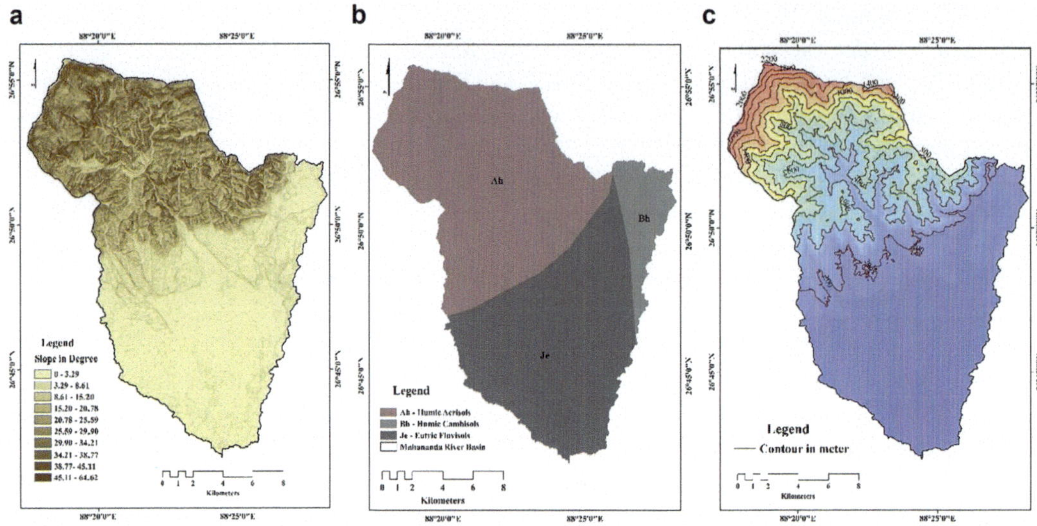

Fig. 2 (**a**) Slope map. (**b**) Soil map. (**c**) Contour map

Table 1 Geospatial datasets accustomed in the research

Data	Sensor	Date of pass	Path/row	Spatial resolution (m)	Day/night	Land cloud	Scene cloud
Landsat 4	TM	1993	139/041	30	Day	2	2
Landsat 5	TM	2003	139/041	30	Day	1	1
Landsat 8	OLI-TIRS	2013	139/041	30	Day	0	0
Landsat 9	OLI-TIRS	2023	139/041	30	Day	0	0

the phases that came after the land use and land cover change detection analysis. Data collected by remotely sensed satellites and other sources were used in this investigation. The US Geological Survey's Earth Explorer provides access to the remote sensing data. The primary tools used for data management, image interpretation, and satellite-based data processing were ArcGIS 10.8 and ERDAS Imagine 9.2. The USGS's Earth Explorer (https://earthexplorer.usgs.gov/) provided four consecutive multispectral LANDSAT files with a spatial resolution of 30 m. The Operational Land Imager (OLI)/Thermal Infrared Sensor (TIRS) and the Thematic Mapper (TM) sensor were used to collect these datasets. Exploration of geospatial data was done using these datasets (Table 1). To ensure cloud-free images, remote sensed data selection was done during the post-monsoon season (Belay et al., 2024). Satellite datasets from 1993, 2003, 2013, and 2023 were used to assess data spanning 30 years. A comprehensive overview of the demographic makeup of Siliguri Kurseong City and the neighboring regions was also provided by including census data from 1991, 2001, and 2011 for the Darjeeling and Jalpaiguri Districts.

Table 1. Geospatial datasets accustomed in the research.

2.3 Classification of Images

The "maximum likelihood classifier" option of LULC classification on ArcGIS 10.8 platform was selected to obtain precise and comprehensive geospatial information on the changing land use and land cover (LULC) using training sets derived from aerial imagery, GIS data, and field-based monitoring data in selected MRB. This algorithm is widely recognized as one of the best by modern researchers. The platform ERDAS Imagine 9.2 was used to convert the resultant output raster to vector output format, which was then imported into ArcGIS 10.8 for quantitative analysis and map production. LULC maps for a number of time series in the study area are therefore included in the output (Kafi et al., 2014; Debnath et al., 2017). All four of the chosen raster images were then categorized into six distinct LULC classifications: river, natural vegetation, tea plantation, farmland and grassland, fallow land, and built-up region. This classification was made following the National Remote Sensing Center's suggested methodology and the scientists' previous understanding of the research locale (Ganasria & Dwarakish, 2015; Arveti et al., 2016). The class tea plantation and natural vegetation was a challenging portion of the correct classification but after the detailed field investigation and reidentifying the geometric shape of the land, both the classes were identified from satellite images. The ground-truth locations were selected by stratified random method by LULC classes. For each class, minimum of 50 points were selected covering the whole river basin (Table 2).

Table 2 Explanation of land use and land cover (LULC) classes identified in Mahananda River Basin

LULC classes	Narratives
River	Geographic areas like large rivers and lakes that are submerged in water
Natural vegetation	This category comprises areas with dense tree cover, including evergreen forests, mixed forests, and deciduous forests
Tea plantation	Plantations of indigenous species of tea plants are considered here
Cropland and grassland	Crop-cultivated areas, pastures, plantations around villages, fallow sites, and scattered rural settlements were all lumped together
Fallow land	This class denotes the agricultural lands, temporarily not harvested for either environmental conditions or crop rotation systems
Built-up area	This category encompasses residential, commercial, and service areas, recreational sites, public installations, and infrastructures. Additionally, roads constructed from pavement are part of this category

2.4 Post-Classification of Corrections for Accuracy Improvement

The classified output revealed significant misclassification of non-agricultural LULC classes such as cropland and grasslands. For instance, tea plantations were erroneously categorized as cropland, and grasslands suffered a similar misclassification. Reassigning misclassified pixels and adding supplementary data (field investigation) from visual image interpretations to classified maps, as shown by Shalaby and Tateishi (2007), can improve the overall accuracy by 10%. In order to decrease classification errors between agricultural and non-agricultural LULC classes, tea plantation masks were used as auxiliary data in this study. Following the class reassignment process, the polygons representing the cropland or grassland classes of LULC and the tea plantation class were dissolved into their corresponding categories. Every processing step was carried out in the software environment of ArcGIS 10.8.

2.5 Final Output of Accuracy Assessment

The accuracy of the LULC categorization for the 1993, 2003, 2013, and 2023 images was evaluated using ground-truth observations (30%) that were chosen at random from different land cover and use classifications (Chakraborty & Saha, 2021). Each class was given at least 50 points based on the ground-truth data and visual interpretation techniques. After comparing the classification outputs with the reference points using the error matrix technique, the kappa coefficient was used to get the classification accuracy percentage. Each LULC map underwent independent validation to ensure result consistency

(Thakkar et at.; 2016, 2017). The accuracy assessed data was derived from the Landsat imagery data for 1993, 2003, 2013, and 2023, with ground-truth verification conducted for 2023. Ground-truth verification locations from 2023 were projected onto the 1993, 2003, 2013, and 2023 remote sensing images. Landscape features were visually interpreted and converted into verification points based on these locations.

2.6 Change Detection Analysis

Analyzing change detection involves measuring and summarizing variations between several images taken at the same location at different times. The long-term shifts in land use and land cover, including more urbanization and less farmland and natural vegetation, can only be fully comprehended with the use of this analytical method. Then, change detection statistics were calculated for the following intervals in the context of this study: 1993–2003, 2003–2013, 2013–2023, and 1993–2023. The purpose of this research is to shed light on the dynamics and trends of land use and land cover changes across the designated time periods by identifying key transition locations and probable environmental change sources (Tefera, 2011; Twisa & Buchroithner, 2019).

3 Outcomes and Discussion

3.1 Land Use and Land Cover Analysis

Figure 3a, 3b, 3c, 3d which is based on multi-temporal remote sensing data, displays the land use and land cover (LULC) conditions in the Upper Mahananda River Basin throughout four research periods (1993, 2003, 2013, and 2023). The study region's land cover and land use patterns may be roughly divided into six categories: built-up, cropland/grassland, natural, riverine, tea plantation, and fallow land. These comprehensive LULC classifications provide insights into the predominant land use and land cover features of the location. Figure 3a and Table 3 provide detail of the spatial extent and distribution in 1993, highlighting that approximately 3.043% (7.577 km^2) of the area was covered by the class river, 53.641% (133.567 km^2) by natural vegetation, 20.616% (51.334 km^2) by tea plantation, 2.749% (6.844 km^2) by cropland and grassland, 16.138% (40.182 km^2) by fallow land, 3.813% (9.495 km^2) by built-up zone. Figure 3b and Table 3. present the zonal distribution and scenario for the year 2003. Around 3.777% (9.404 km^2) of the area was covered by river, 51.593% (128.467 km^2) by natural vegetation, 16.24% (40.438 km^2) by tea plantation, 4.761% (11.857 km^2) by cropland and grassland, 16.343% (40.693 km^2) by fallow land, 7.286% (18.1411 km^2) by built-up area. Figure 3c and Table 3 illustrate the areal expansion and land use and land cover scenario in the year 2013. In that year, approximately 4.07% (10.135 km^2) of the area was covered by river, 55.646% (138.558 km^2) by natural vegetation, 10.839% (26.990 km^2) by tea plantation class, 3.065% (7.631 km^2) by cropland and grassland, 16.425% (40.899 km^2) by fallow lands, and 9.875% (24.587 km^2) by built-up areas. Figure 3d and Table 3 present the areal distribution and scenario for the year 2023. During this year, approximately 3.674% (9.149 km^2) of the area was covered by rivers, 57.026% (141.994 km^2) by natural vegetation, 6.974% (17.365 km^2) by tea plantations, 2.259% (5.625 km^2) by cropland and grassland, 16.655% (41. 471 km^2) by fallow land, and 13.412% (33.395 km^2) by built-up areas (Fig. 4).

In 1993, 2003, 2013, and 2023, broad woodland mostly covered the research region. These forest areas facilitate water penetration and lessen soil erosion and surface runoff, which are both vital components of the hydrodynamic process.

Throughout the years, natural vegetation has consistently covered almost 50% of the area annually. However, in 2003, this trend showed a notable decrease. After this decline, the area of natural vegetation began to recover, demonstrating a gradual increase at an annual rate of 0.281 km^2. This recovery highlights a positive trend in the restoration of natural vegetation over time.

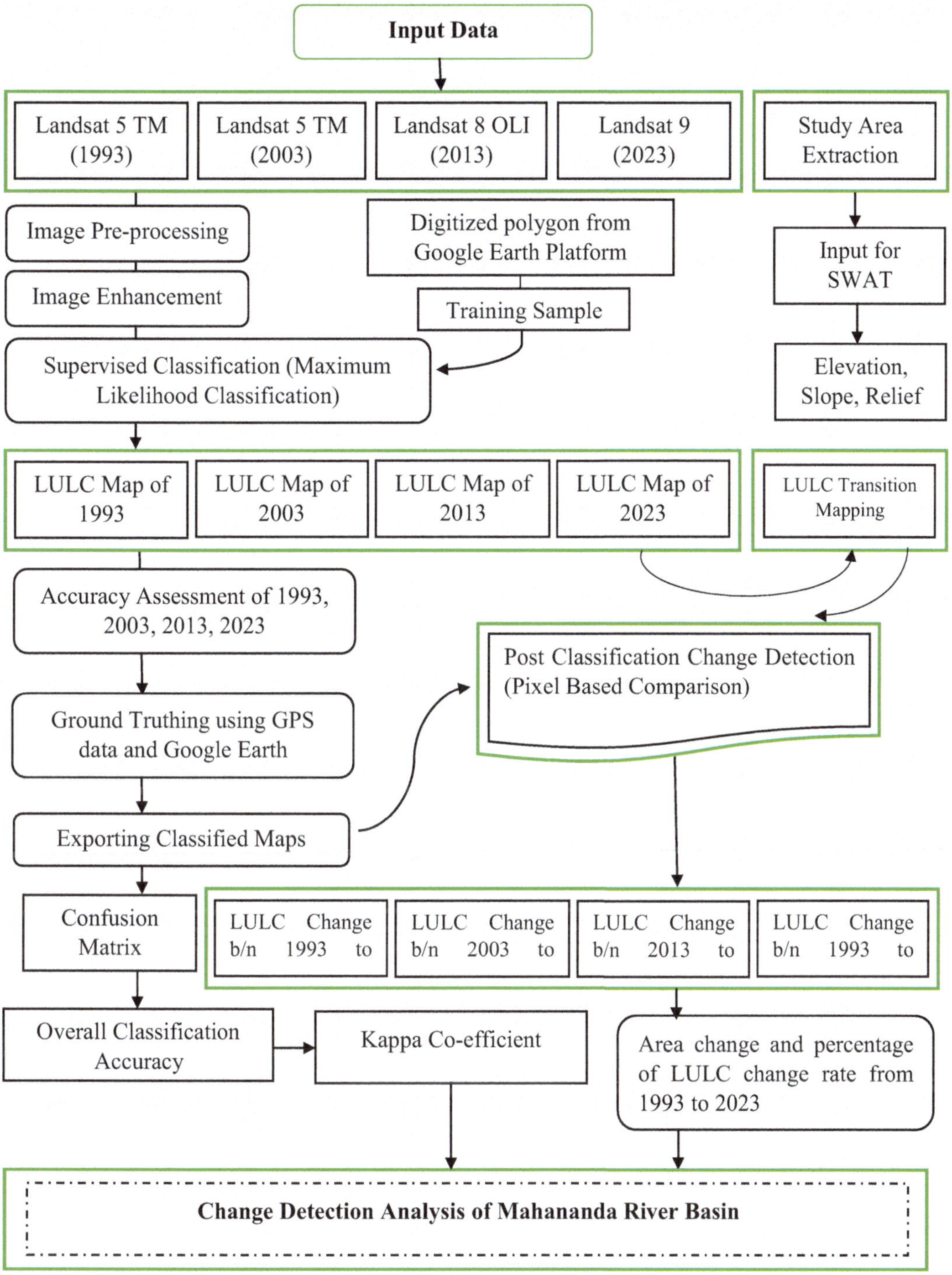

Fig. 3 Flowchart of the methodology

In contrast, starting from 1993, tea plantations experienced a significant decline, particularly evident in the years 2003, 2013, and 2023. This reduction in tea plantation area can be primarily attributed to the transformation of agricultural land for nonagricultural purposes. Factors such as increased population pressure and the subsequent demand for housing and infrastructure have driven this conversion, leading to a notable decrease in tea plantations.

Table 3 Spatial distributions of the LULC classes

Area in km^2					Area in percentage (%)			
LULC class	Year 1993	Year 2003	Year 2013	Year 2023	Year 1993	Year 2003		Year 2023
River	7.577373	9.403508	10.135454	9.14911	3.043	3.777	4.07	3.674
Natural vegetation	133.56697	128.4671	138.55831	141.99378	53.641	51.593		57.026
Tea plantation	51.333939	40.438	26.990426	17.365497	20.616	16.24		6.974
Cropland or Grassland	6.844168	11.85682	7.631417	5.625427	2.749	4.761	3.065	2.259
Fallow land	40.182327	40.69344	40.898527	41.471194	16.138	16.343	16.425	16.655
Built-up area	9.495234	18.14112	24.587784	33.395027	3.813	7.286	9.875	13.412
Total area = 249 km^2					**Percentage of area = 100%**			

The categories of river, built-up areas, and barren or fallow land have all generally increased between 1993 and 2023. This indicates a shift in land use patterns, with more land being left fallow, rivers potentially expanding or changing course, and built-up areas increasing due to urbanization and development activities.

The loss of farmland and tea plantations, as well as the growth of built-up regions, are positively correlated with population growth in the chosen MRB region. An increase in population necessitates the construction of more residences, companies, and industries. As a result of this increasing demand, once-natural and agricultural landscapes are being transformed into urban and semi-urban settings. The subsequent change has far-reaching consequences for the region's ecology, biodiversity, agricultural output, and terrain.

In sum, these shifts in land use and patches of vegetation highlight the need of sustainable land management methods in balancing development demands with environmental protection, as they show the dynamic interaction between natural processes and human influences.

3.2 Result of Accuracy Assessment

In order to identify LULC changes, accuracy analysis is vital. In order to evaluate the classification accuracy, this study used the error matrix approach to compare the categorized results with the reference data for each LULC class. Using an average of 50 reference points per class allowed the study to obtain strong validation. We calculated parameters such as producer accuracy, user accuracy, total accuracy, and kappa statistics to get a clear picture. The results of accuracy tests for 1993, 2003, 2013, and 2023 are shown in Tables 4, 5, 6 and 7, respectively, which demonstrate the consistent improvement in classification accuracy over time. Overall accuracy figures for 1993, 2003, 2013, and 2023 were 91%, 90%, 92%, and 95%, respectively (Fig. 5). The kappa coefficient of the LULC classification for 1993, 2003, 2013, and 2023 was found to be 0.9087, 0.8825, 0.9032, and 0.9415, respectively, indicating almost perfect agreement (Cohen's kappa statistics). Figure 6 shows LULC ground-truth verification, confirming the accu-

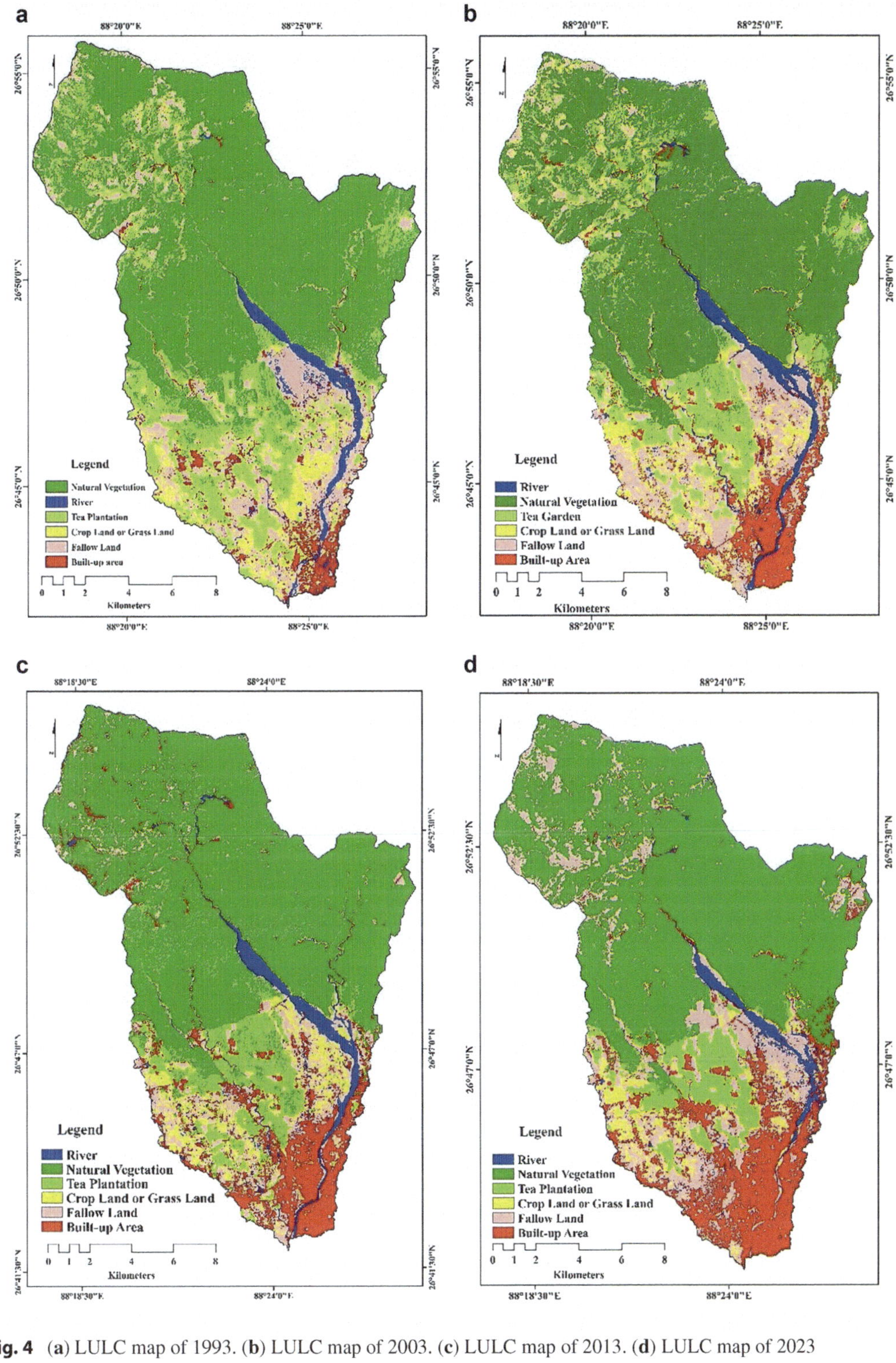

Fig. 4 (**a**) LULC map of 1993. (**b**) LULC map of 2003. (**c**) LULC map of 2013. (**d**) LULC map of 2023

Table 4 Accuracy assessment status of 1993 land use land cover map

Object Id/ Raster	Name	River	Natural vegetation	Tea Plantation	Crop land or grass	Fallow land	Built-up area	Classified total	User's accuracy %	Commission error	Producer's accuracy %	Omission error
1	River	70	0	0	0	0	3	73	95.8904	4.1096	93.3333	6.6667
2	Natural vegetation	0	101	3	1	1	0	106	95.283	4.717	98.0583	1.9417
3	Tea plantation	0	1	76	3	4	0	84	90.4762	9.5238	92.6829	7.3171
4	Crop land or grass land	0	0	0	55	2	0	57	96.4912	3.5088	91.6667	8.3333
5	Fallow land	2	1	3	1	69	5	81	85.1852	14.8148	85.1852	14.8148
6	Built-up area	3	0	0	0	5	58	66	87.8788	12.1212	87.8787	12.1212
	Reference total	75	103	82	60	81	66	467				

Overall accuracy= 91.8630%
Kappa coefficient = 0.9087

Total corrected class pixel = (70+101+76+55+69+58) = **429**

Table 5 Accuracy assessment status of 2003 land use land cover map

Object Id	Name	River	Natural vegetation	Tea plantation	Crop land or grass	Fallow land	Built-up area	Classified total	User's accuracy %	Commission error	Producer's accuracy %	Omission Error
1	River	53	0	0	0	1	2	56	94.6429	5.3508	92.9825	7.0175
2	Natural vegetation	0	84	7	0	1	0	92	91.3044	8.6956	96.5517	3.4483
3	Tea plantation	0	3	56	0	0	1	60	93.3333	6.6667	81.1594	18.8406
4	Crop land or grass land	0	0	0	40	2	0	42	95.2361	4.7639	81.6327	18.3673
5	Fallow land	3	0	5	9	85	2	104	81.7307	18.2693	92.3913	7.6087
6	Built-up area	1	0	1	0	3	65	70	92.8571	7.1429	92.8571	7.1429
	Reference total	57	87	69	49	92	70	424				

Overall accuracy= 90.3302%
Kappa coefficient = 0.8825

Total corrected class pixel = (53+84+56+40+85+65) = **383**

Table 6 Accuracy assessment status of 2013 land use land cover map

Object Id	Name	River	Natural vegetation	Tea plantation	Crop land or grass	Fallow land	Built-up area	Classified total	User's accuracy %	Commission error	Producer's accuracy %	Omission error
1	River	87	0	0	0	2	2	91	95.6044	4.3956	98.8636	1.1364
2	Natural vegetation	0	121	7	0	0	0	128	94.5313	5.4685	99.1803	0.8197
3	Tea plantation	0	1	67	1	2	1	72	93.0556	6.9444	89.3333	10.667
4	Crop land or grass land	0	0	0	32	5	0	37	86.4865	13.5135	76.1905	23.8095
5	Fallow land	1	0	1	9	71	3	85	83.5294	16.4706	83.5294	16.4706
6	Built-up area	0	0	0	0	5	87	92	94.5652	5.4348	93.5484	6.6516
	Reference total	88	122	75	42	85	93	505	Total corrected class pixel = (87+121+67+32+71+87) = **465**			

Overall accuracy= 92.0792%
Kappa coefficient = 0.9032

Table 7 Accuracy assessment status of 2023 land use and land cover map

Object Id	Name	River	Natural vegetation	Tea plantation	Crop land or grass land	Fallow land	Built-up area	Classified total	User's accuracy %	Commission error	Producer's accuracy %	Omission error
1	River	180	0	0	0	1	0	181	99.4475	0.5525	95.2381	4.7619
2	Natural vegetation	1	101	0	0	0	1	103	98.0583	1.9417	98.0583	1.9417
3	Tea plantation	0	0	101	1	1	1	104	97.1154	2.8846	100	0
4	Crop land or grass land	0	0	0	111	3	0	114	97.3684	2.6316	88.0952	11.9048
5	Fallow land	2	2	0	14	114	2	134	85.0746	14.9254	93.4426	6.5574
6	Built-up area	6	0	0	0	3	142	151	94.0397	5.9603	97.2603	2.7397
	Referenced total	189	103	101	126	122	146	787	Total corrected class pixel = (180+101+101+111+114+142) = **749**			

Overall accuracy= 95.1715 %
Kappa coefficient = 0.9415

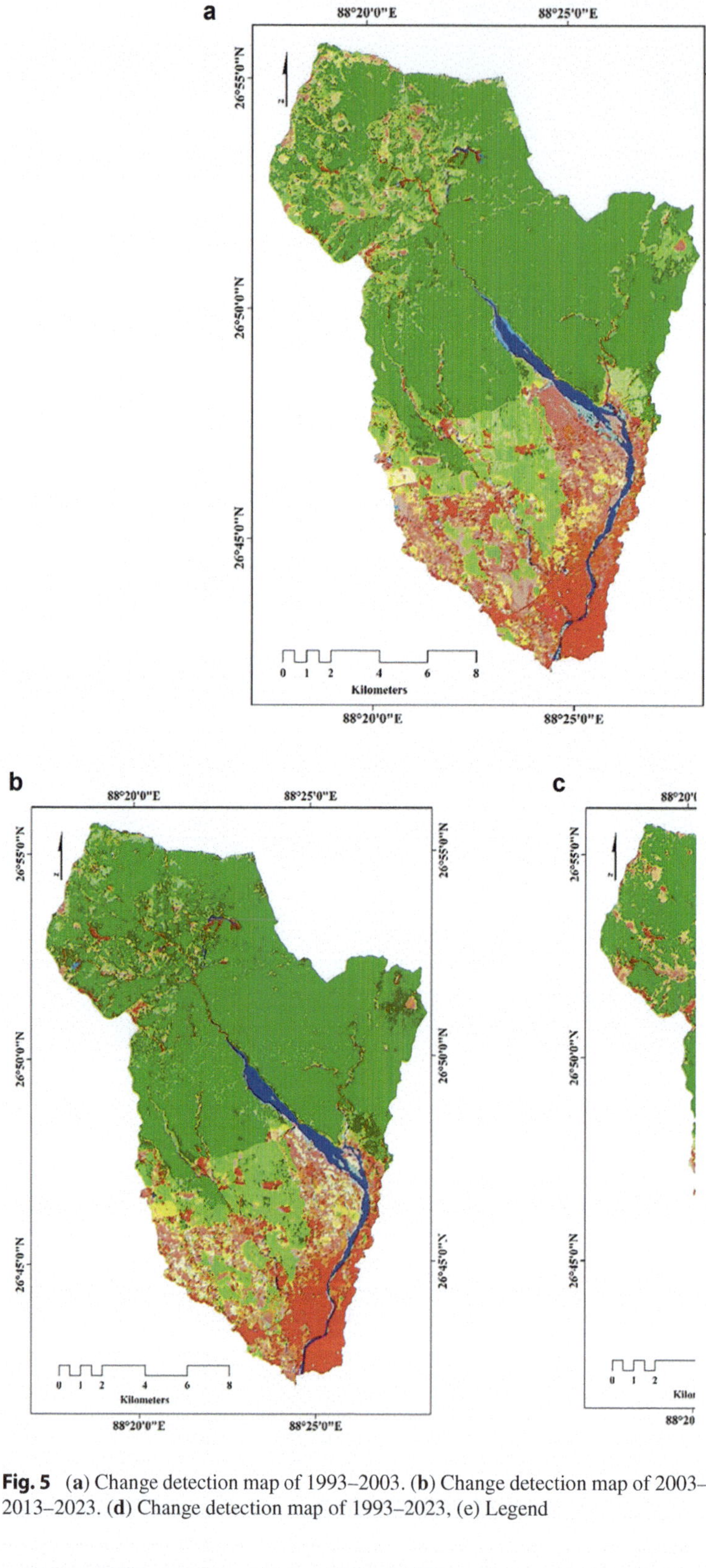

Fig. 5 (**a**) Change detection map of 1993–2003. (**b**) Change detection map of 2003–2013. (**c**) Change detection map of 2013–2023. (**d**) Change detection map of 1993–2023, (e) Legend

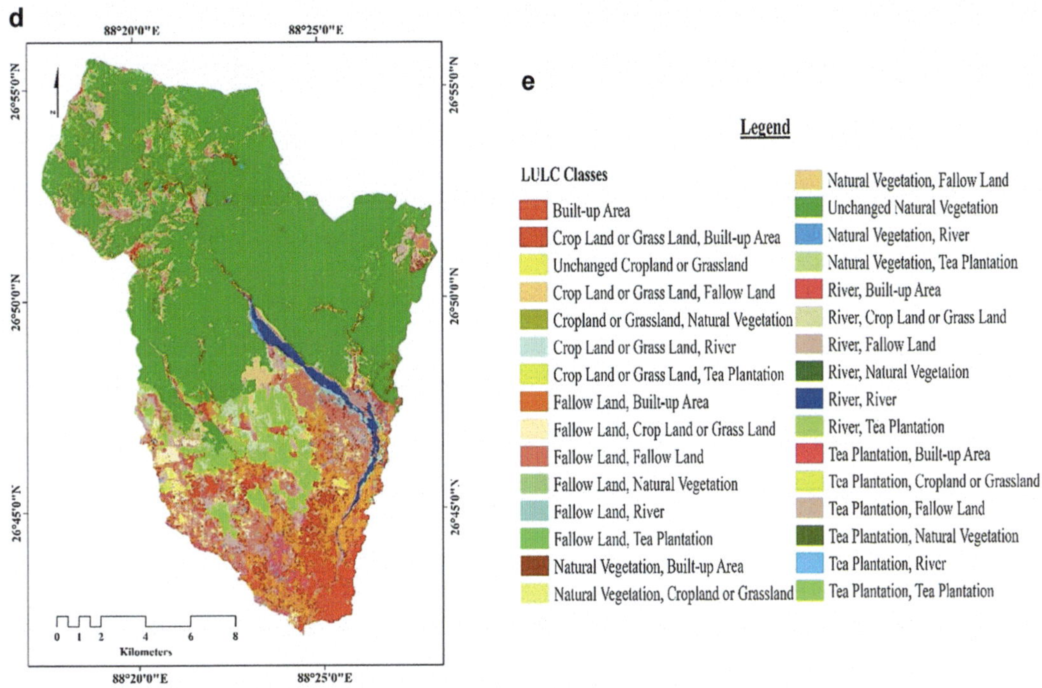

Fig. 5 (continued)

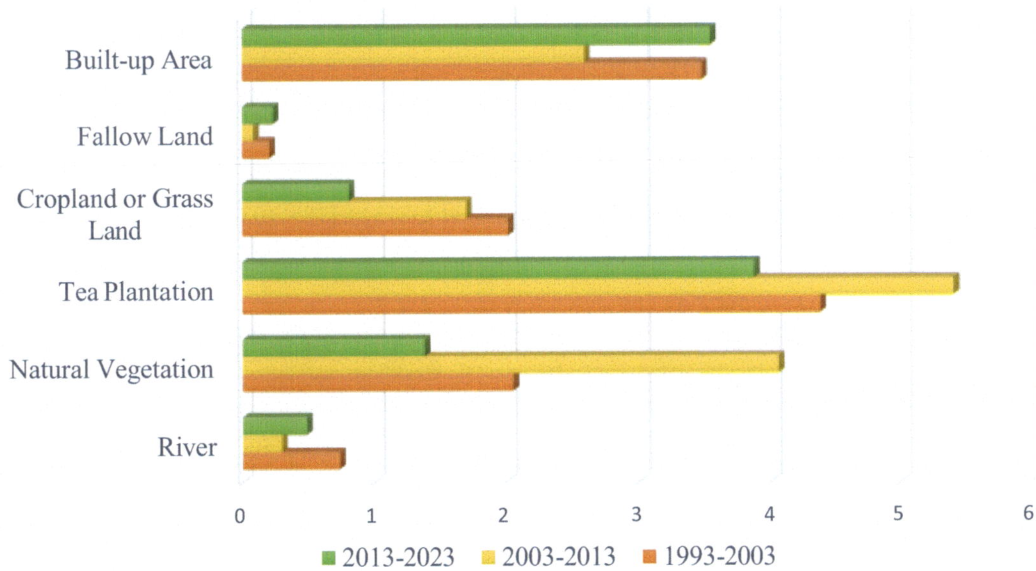

Fig. 6 Percentage of area changes of each LULC classes

racy assessment. The LULC classification accuracy exceeded expectations, highlighting the robustness and reliability of the error matrix approach and its effectiveness in ensuring high-quality classification results.

$$\text{User's accuracy} = \begin{pmatrix}\text{Corrected individual class Pixel}\\ /\text{classified individual class Pixel}\end{pmatrix}^{*} 100$$

$$\text{Producer's accuracy} = \begin{pmatrix}\text{Corrected individual class pixel}\\ /\text{reference individual class total}\end{pmatrix}^{*} 100$$

$$\text{Commission error} = \begin{pmatrix}\text{Non} - \text{classified total}\\ \text{single class pixel}\\ /\text{Total classified pixel}\end{pmatrix} * 100$$

$$\text{Omission error} = \begin{pmatrix}\text{Non} - \text{referenced total}\\ \text{single class pixel}\\ /\text{total referenced pixel}\end{pmatrix} * 100$$

The kappa coefficient (K_{hat}) takes into account all aspects of the error matrix to quantify the agreement between two maps. This is its error-matrix-based definition:

$$\text{Kappa coefficient}\left(K_{hat}\right) = \begin{pmatrix}\text{observed accuracy}\\ -\text{chance agreement}\end{pmatrix} / \left(1 - \text{chance agreement}\right)$$

Where,

Observed accuracy = Observed correct, it represents accuracy reported in the error matrix (overall accuracy).

Chance agreement = Expected correct, it represents the correct classification Chance/expected accuracy = Sum of each class's (classified total * reference total) Calculation example for the year 1993:

$$\text{Accuracy of river} = 70/73^{*}\,100.\text{So the accuracy is} = 95.8904109\%$$

$$\text{Accuracy of natural vegetation} = 101/106^{*}\,100,\text{So the Accuracy is} = 95.2830188\%$$

$$\text{accuracy of tea plantation} = 76/84^{*}\,100.\text{So the accuracy is} = 90.4761904\%$$

$$\text{Accuracy of cropland or grassland} = 55/57^{*}\,100.\text{So the accuracy is} = 96.49122807\%$$

$$\text{Accuracy of fallow land} = 69/81^{*}\,100.\text{So, the accuracy is} = 85.185185\%.$$

$$\text{Accuracy of built} - \text{uparea} = 58/66^{*}100,\text{So the accuracy is} = 87.8787878\%$$

$$\text{Total percentage of accuracy} = \begin{pmatrix}70+101+76\\ +55+69+58\end{pmatrix}^{*} 100/467.$$

$$\text{So the overall accuracy is} = 91.862955\%$$

$$\text{Chance / expected accuracy} = \left(73^{*}\,75\right) + \left(106^{*}\,103\right) + \left(84^{*}\,82\right) + \left(57^{*}\,60\right) + \left(81^{*}\,81\right) + \left(66^{*}\,66\right) = 37618$$

$$K_{hat} = (\begin{pmatrix}467(429)\\ -37618\end{pmatrix} / \begin{pmatrix}\left(467^{*}\,467\right)\\ -37618\end{pmatrix} = 0.90872284581$$

3.3 Change Detection Analysis

The results of the change detection for 1993–2003, 2003–2013, 2013–2023, and 1993–2023 are shown in Table 8. The data indicates that there may be both lower and greater LULC changes in the upper Mahananda River Basin.

Between 1993 and 2003, a significant decline was observed in tea plantations and

Table 8 The magnitude and percentage of LULC changes

LULC class	LULC area change in km²				Percentage of area change				
	1993–2003	2003–2013	2013–2023	1993–2023	1993–2003	2003–2013	2013–2023	1993–2023	
River	1.826	0.731	1.205	1.571	0.733	0.294	0.484	0.631	
Natural vegetation	−5.099	10.091	3.435	8.426	2.048	4.052	1.379	3.384	
Tea plantation	−10.895	−13.447	−9.624	−33.968	−4.376	−5.401	−3.865	−13.641	2.265
Cropland or grassland	5.012	−4.225	−2.005	−1.218	−2.013	−1.697	−0.806	−0.489	
Fallow land	0.511	0.205	0.572	1.288	0.205	0.0823	0.23	0.518	
Built-up area	8.645	6.446	8.807	23.899	3.472	2.589	3.537	9.598	
Total area change in km²	31.991	35.147	25.651	70.374	12.848	14.116	10.302	28.263	

natural vegetation by 10.895939 km^2 (−4.376%), and natural vegetation by 5.09987 km^2 (−2.048%). Conversely, there was a notable increase in built-up area by 8.645886 km^2 (+3.472%), cropland or grassland by 5.012652 km^2 (+2.013%), river by 1.826135 km^2 (+0.733%), and fallow lands by 0.511113 km^2 (+0.205%).

From 2003 to 2013, tea plantations decreased by 13.447574 km^2 (−5.401%), and croplands by 4.225403 km^2 (−1.697%). However, natural vegetation expanded by 10.09121 km^2 (+4.052%), built-up area by 6.446664 km^2 (+2.589%), water bodies by 0.731946 km^2 (+0.294%), and fallow lands by 0.205087 km^2 (+0.0823%).

Between 2013 and 2023, there was a decline in tea plantations by 9.624929 km^2 (−3.865%), and croplands by 2.00599 km^2 (−0.806%). On the other hand, natural vegetation grew by 3.43547 km^2 (+1.379%), built-up area by 8.807243 km^2 (+3.537%), water bodies by 1.20543 km^2 (+0.484%), and fallow lands by 0.572667 km^2 (+0.23%).

Over the 30 years from 1993 to 2023, the LULC change showed a decrease in tea plantations by 33.96844 km^2 (−13.641%), and croplands or grasslands by 1.218741 km^2 (−0.489%). Meanwhile, natural vegetation increased by 8.42681 km^2 (+3.384%), built-up lands increased by 23.899793 km^2 (+9.598%), and fallow land increased by 1.288867 km^2 (+0.518%).

The annual rate of changes for the periods 1993–2023 was also obtained and is presented in Table 8. The maximum rate of areal change is seen in the class of tea plantations and built-up areas.

This diagram illustrates the changing pattern of LULC classes from 1993 to 2023. The percentage of area change of river is higher between 1993 and 2003 and lowest 2003 and 2013. The built-up area and fallow land classes are also following the trend. The percentage of natural vegetation and tea plantation change was highest in 2003–2013.

3.4 Impact of Population Growth on LULC Classes

Land use and land cover (LULC) changes often accompany population growth. Essential resources including food, water, shelter, electricity, healthcare, and transportation are under high demand due to a growing population. Ecological degradation may result from this increased consumption's substantial strain on natural resources like land and water. The expansion of urban areas, deforestation for agriculture, and the conversion of natural landscapes into industrial zones are common outcomes of this growth. Consequently, these LULC changes can have profound impacts on biodiversity, climate, and the overall health of ecosystems. However, the recent decrease in the population growth rate is a positive sign (Fig. 8), suggesting potential relief from the intense pressure on natural resources. This reduction can lead to more sustainable land use practices, preservation of natural habitats, and a healthier environment overall (Figs. 7 and 8) (Table 9).

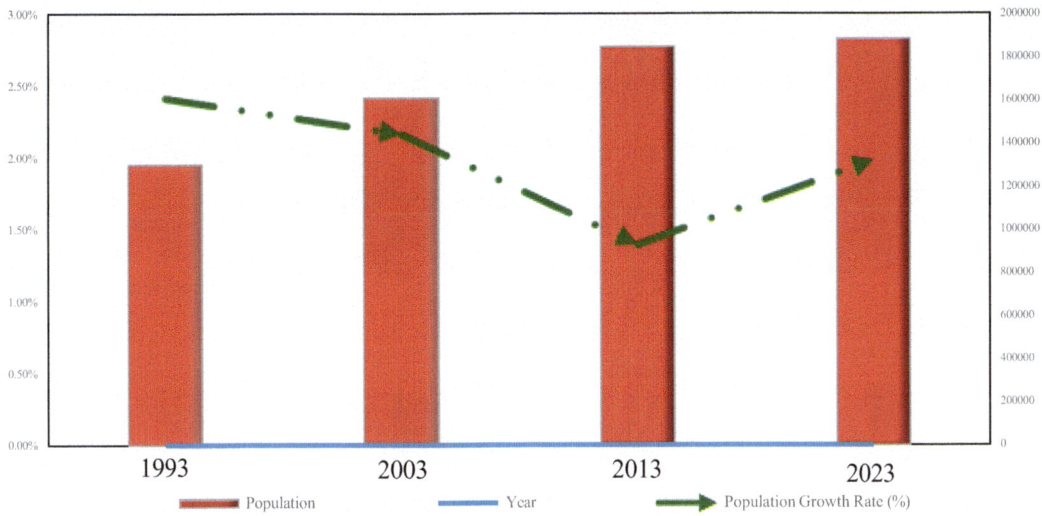

Fig. 7 Field Photographs (**a**) River. (**b**) Natural vegetation. (**c**) Tea plantations. (**d**) Cropland or grassland. (**e**) Fallow land. (**f**) Built-up area. (**a**) Mahananda River, (**b**) Downhill, (**c**) Mohorgon Tea Garden, (**d**) Bank of the Mahananda River, (**e**) Fallow land near Gulma Station, (**f**) Kurseong City

3.5 Conversion Pattern of Land Use and Land Cover Classes

The class agricultural land is mostly converted into the natural vegetation class. The percentage of coverage is 52%, while 43% of agricultural land is converted into fallow land. Only 0.1% of the area is converted into settlement (Fig. 9).

The fallow land class is substantially converted into natural vegetation (67%) and very less converted into settlement or built-up area class (1%).

The class natural vegetation is reclassed into fallow Land, which is 78% of the total changed natural vegetation class. Then it is transformed into tea plantation and settlement class.

Most of the settlement area is converted into fallow land. The reason behind this conversion is primarily by the hazards and disasters like floods and landslides. However, the amount of areal change is very low.

Overall tea plantation class is reclassified into natural vegetation (71.8%), fallow land (26%), and agricultural land (2%).

The change in fallow land class from the class waterbodies is showing the highest value. The predominant reason of this type of reclassification is expansion of river width, illegal sand mining, exposer of river bed, etc.

The most unchanged class is natural vegetation (87%) and fallow land (10.8%). The percentage of unchanged classes of agricultural land and built-up area is only 1% each and the class waterbodies and tea plantation is the highest unchanged class (0.1%) of the particular area (Fig. 9).

Fig. 8 Population and decadal growth rate in Darjeeling District (Source: District Census Handbook)

Table 9 Population growth and decadal growth of Darjeeling

Year	Population	Population growth rate (%)
1991	12,99,919	+2.41%
2001	16,09172	+2.16%
2011	18,46,823	+1.39%
2021	18,83,390	+1.98%

Fig. 9 Conversion of land use and land cover pattern of the selected area

4 Conclusion

This study is the first of its kind to examine how human activity affects changes in land cover and usage in the Upper Mahananda River Basin. By utilizing a comprehensive methodology that integrates geospatial data analysis, field investigations, and community responses, the research provides a holistic view of the environmental changes occurring in this area. By integrating geospatial data analysis, field investigations, and community responses to these three methodological approaches, the study can deliver a comprehensive picture of how anthropogenic activities are influencing morphology in the lower Darjeeling Himalaya. The combined findings provide light on the recent physical changes to the environment as well as the hidden human elements and historical background that play a role in these changes. This multifaceted approach sets a precedent for future research in similar environments, demonstrating the value of combining scientific data with local knowledge to address complex environmental issues. The changing rate of population growth of the Darjeeling District and the selected study area shows quite similar values. The recent trend of the annual rate of human interference is showing an increasing rate which is highly correlated with the decreasing level of environment. The results of this research demonstrate that cropland and grassland LULC classes, as well as tea plantations, are undergoing fast land use modification. Severe eco-hydrological and socio-hydrological threats are being caused by the intense human activities within the basin, which are affecting the natural stability of the basin and its topography. In order to preserve the standard environmental systems of the chosen field, it is vital that local authorities regulate future urbanization activities. To help groundwater recharge, newly developed regions and urban sprawls should include enough of open, unpaved space. Furthermore, the riparian zone should not be developed with concrete and should be kept in its natural form. In order to tackle and lessen the impact of possible problems, resilience should be an essential part of future plans for managing urbanization and restoration.

References

Afify, H. A. (2011). Evaluation of change detection techniques for monitoring land-cover changes: A case study in new Burg El-Arab area. *Alexandria Engineering Journal, 50*, 187–195.

Arulbalaji, P. (2019). Analysis of land use/land cover changes using geospatial techniques in Salem district, Tamil Nadu, South India. *Springer Nature SwitzerlandAG, 2019*(1), 462. https://doi.org/10.1007/s42452-019-0485-5

Arveti, N., Etikala, B., & Dash, P. (2016). Land use/land cover analysis based on various comprehensive geospatial data sets: A case study from Tirupati area, South India. *Advanced Remote Sensing, 5*, 73–82. https://doi.org/10.4236/ars.2016.52006

Bandyopadhyay, S., & De, S. K. (2018). Anthropogenic impacts on the morphology of Haora River, Tripura, India. *Geomorphologie: relief, processus, environment, 24*, 151–166. https://journals.openedition.org/geomorphologie/12019

Belay, H., Melesse, A. M., & Tegegne, G. (2024). *Scenario-based land use and land cover change detection and prediction using the cellular automata–Markov model in the Gumara watershed*. Upper Blue Nile Basin.

Bharti, R. (2000). *Rivers of India* (p. 31). NBT Publication.

Chakraborty, K., & Saha, S. (2021). Assessment of bank erosion and its impact on land use and land cover dynamics of Mahananda River basin (upper) in the sub-Himalayan North Bengal, India. *SN Applied Sciences (2022), 4*, 20. https://doi.org/10.1007/s42452-021-04904-x

Debnath, M., Ray, S., Islam, N., & Sar, N. (2017). Migration patterns and urban growth in north–East India: A study in Siliguri city. *Quest The Journal of UGC-HRDC, 11*, 118–123. https://doi.org/10.5958/2249-0035.2017.00016.X

Ganasria, B. P., & Dwarakish, G. S. (2015). Study of land use/land cover dynamics through classification algorithms for Harangi catchment area, Karnataka state, India. *Aquatic Procedia, 4*(2015), 1413–1420. https://doi.org/10.1016/j.aqpro.2015.02.183

Kafi, K. M., Shafri, H. Z. M., & Shariff, A. B. M. (2014). An analysis of LULC change detection using remotely sensed data; a case study of Bauchi City. *Earth and Environmental Science, 20*, 012056. https://doi.org/10.1088/1755-1315/20/1/012056

Mandal, S., & Maity, R. (2014). Role of lithological composition and lineaments in landsliding: A case study of Shivkhola watershed, Darjeeling Himalaya. *International Journal of Geology, Earth and Environmental Sciences, 4*, 1. January-April, pp. 126-132/Mandal and Maiti ISSN: 2277-2081.

Mandal, V. K., Raheman, S., Ahmed, R., Masroor, M., Kumar, P., & Sajjad, H. (2020). Land suitability assessment for optimal cropping sequences in Katihar district of Bihar, India using GIS and AHP. Korean Spatial Information Society 2020. *Spatial Information Research, 28*, 589–599. https://doi.org/10.1007/s41324-020-00315-z

Mitra, S., Roy, A. K., & Tamang, L. (2020). *Assessing the status of changing channel regimes of Balason and Mahananda River in the Sub-Himalayan West Bengal, India*. Earth Systems and Environment. https://doi.org/10.1007/s41748-020-00160-y

Mitra, S., Purkait, M., Chatterjee, A., Dey, J., De, S. K., & Tamang, L. (2023). Analysing the role of anthropogenic intervention in reconfiguring the hydromorphology of the sub-Himalayan river systems in India: A comparative assessment between the Mahananda–Balason and Chel–Neora systems. *Aquatic Sciences, 85*, 91. https://doi.org/10.1007/s00027-023-00988-2

O'Malley, L. S. S. (1907). *Bengal District Gazetteer: Darjeeling* (pp. 2–15). Government of West Bengal, Department of Higher Education.

Ojima, D. S., Galvin, K. S., & TurnerII, B. L. (1994). The global impact of land-use change. *Bioscience, 44*(5) Global Impact of Land-Cover Change (May, 1994), 300–304. https://doi.org/10.2307/1312379

Rana, N. K. (2018). *Analysis of Mahananda River basin using geospatial data* (pp. 239–250. ISSN 2364-6454. ISBN 978-981-10-2983-7.). Springer Hydrology. https://doi.org/10.1007/978-981-10-2984-4

Sarkar, R., Baral, M. (2021). Monitoring land use & land cover change in an urban landscape: A case study of Siliguri, India. Geospatial Science for Digital Earth Observation (GSDEO 2021). IOP Conf. Series: Earth and Environmental Science 1164 (2023) 012011 DOI: https://doi.org/10.1088/1755-1315/1164/1/012011.

Shalaby, A., & Tateishi, R. (2007). Remote sensing and GIS for mapping and monitoring land cover and land-use changes in the north–western coastal zone of Egypt. *Applied Geography, 27*, 28–41. https://doi.org/10.1016/j.apgeog.2006.09.004

Tefera, M. M. (2011). Land-use/land-cover dynamics in Nonno District, Central Ethiopia. *Journal of Sustainable Development in Africa, 13*, 1. ISSN: 1520-5509.

Thakkar, A. K., Desai, V. R., Patel, A., & Potdar, M. B. (2016). Impact assessment of watershed management programmes on land use/land cover dynamics using remote sensing and GIS. *Remote Sensing Applications: Society and Environment*. PII: S2352-9385(16)30028-3. https://doi.org/10.1016/j.rsase.2016.12.001

Thakkar, A. M., Venkappayya, R. D., Ajay, P., & Madhukar, B. P. (2017). Post-classification corrections in improving the classification of land use/land cover of the arid region using RS and GIS: The case of Arjuna watershed, Gujarat, India. *Egyptian Journal of Remote Sensing and Space Science, 20*, 79–89.

Twisa, S., & Buchroithner, M. F. (2019). Land-use and land-cover (LULC) change detection in Wami River Basin, Tanzania. *Land, 2019*(8), 136. https://doi.org/10.3390/land8090136

Vliet, J. V., Bregt, A. K., & Zanker, A. H. (2011). Revisiting kappa to account for change in the accuracy assessment of land-use change models. *Ecological Modelling, 222*(2011), 1367–1375. https://doi.org/10.1016/j.ecolmodel.2011.01.017

Walker, R., & Homma, A. K. O. (1996). Land use and land cover dynamics in the Brazilian Amazon: An overview. *Ecological Economics, 18*, 67–80. PII S0921-8009(96)00033-X.

Woldeamlak, B. (2002). Land cover dynamics since 1950s in Chemoga watershed, Blue Nile Basin, Ethiopia. *Mountain Research and Development, 22*(3), 263–269.

Disaster Risk Reduction and Management: From Preparedness to Recovery

GIS-Based Micro-Solar Power Plant Site Selection in Sundarban Delta of India: A Siting Model for Sustainable Utility of Renewable Energy

Farhana Parvin, Sk Ajim Ali, Zainab Khan, Zeba Khanam, and Rukhsar Anjum

1 Introduction

Finding possible places for a solar photovoltaic (PV) project is what a site suitability analysis for solar plants entails (Colak et al., 2020; Nebey et al., 2020). Identifying suitable sites for micro-solar plants allows the working efficiency of a solar PV power system. As a result, one of the main tactics for improving the efficiency of solar plants is to select a good location. Solar PV productivity decreases and may malfunction as a result of site inadequacy. This is why scientists have developed guidelines for the best site for solar PV systems. Natural processes provide renewable energy on a continuous basis, which is good for environment and sustainable development (Shukla et al., 2018; Khan & Arsalan, 2016).

F. Parvin · Z. Khan · R. Anjum
Department of Geography, Faculty of Science, Aligarh Muslim University, Aligarh, India

S. A. Ali (✉)
Department of Geography, Faculty of Science, Aligarh Muslim University, Aligarh, India

RA (Remote)—Leverhulme & ISPF funded Project, The University of Manchester, Manchester, UK
e-mail: skajimali@myamu.ac.in

Z. Khanam
CSRD (ICSSR), Jawaharlal Nehru University, New Delhi, India

Energy is a commonly used metric to evaluate sophistication and advancement (Colak et al., 2020). The extraction and use of energy from fossil fuels have long been deemed improper and unscientific. However, there are a number of problems that need to be handled right now, including climate change, environmental degradation, and energy shortages (Xiao et al., 2013). These problems have an impact on the future of the planet. As a result of the various governmental agencies realizing that climate change and the rapidly increasing emissions from wealthy industrialized nations posed serious environmental concerns, a great deal of emphasis was placed on adaptation and capacity building, especially for marginalized groups that depend on resources that are sensitive to climate change (Haldar, 2015). The world's main source of sustainable energy is thought to be renewable energy (Alhammad et al., 2022). With the world's population rising at an exponential rate and the range of energy resources being exploited at an even faster rate, it's time to put a stop to the unchecked use of traditional non-renewable energy sources (non-RES; Bradford, 2008; Haldar, 2015). These worries have prompted the investigation of alternative energy sources, and the use of renewable energy sources has enormous potential due to their naturalness, freedom from cost, rapid resupply, and accessibility (Colak et al., 2020). Among

Z. Khanam et al. (eds.), *Climate Change, Resilience and Ecosystem Services*, Sustainable Development Goals Series, https://doi.org/10.1007/978-3-031-96977-5_16

the choices are geothermal, biomass, wind, and solar energy. To promote a more varied energy market, some governments have set up renewable energy source (RES) portfolios. A low-carbon, safe, and sustainable future is what these portfolios seek to establish (Al Garni & Awasthi, 2017; Alhammad et al., 2022). Interest in using solar energy as a source of energy has increased since it is a new and renewable resource (Masson et al., 2013). The photovoltaic sector is currently one of the high-tech industries with the fastest rate of development and promise, because of the support of several governments (Xiao et al., 2013). In the next 30 years, solar energy is expected to provide 11% of global energy needs, according to the International Energy Agency (Colak et al., 2020).

Being the world's largest mangrove forest, the delta of Sundarban is an attraction for a diverse range of aquatic and terrestrial plants and animals. Furthermore, nearly 4.5 million people live in this environmentally fragile and climate-vulnerable region. Sunderban Delta is far from meeting the Sustainable Development Goal-7 (SDG-7), which ensures that everyone has ingress to reasonable, dependable, supportable, and contemporary energy, with special focus on production of renewable energy by 2030, due to its location in the low coastal zone. Due to its isolation, inappropriate accessibility, low cost of living, incapacity to connect to overhead wires, and inability to connect to overhead power systems, the Sundarban Region lacks adequate energy. The majority of the reclaimed villages are still disconnected from modern of electricity. Currently, decentralized distributed generation (DDG) energy systems that operate off the grid that recommend natural sources of energy (Cust et al., 2007). An ecological zone that is vulnerable to climate change catastrophes should be created and protected through practical and effective solutions.

The Sundarban Delta, renowned for its rich biodiversity and unique ecosystem, faces significant energy challenges, primarily due to its remote location and the dispersed nature of its population. Despite its abundant natural resources, including ample sunlight, the region struggles with unreliable electricity supply and frequent power outages. In this context, integration of geographic information systems (GIS) for site selection of micro-solar power plants emerges as a promising solution to address these energy inadequacies sustainably. The GIS technology enables the systematic analysis of spatial data to identify optimal sites for solar power installations, considering factors such as solar radiation, land availability, proximity to transmission lines, and environmental constraints (Colak et al., 2020). This study aims to leverage GIS-based approaches to pinpoint optimal sites for micro-solar power plants within the deltaic region, enhancing energy access while minimizing ecological impact. By prioritizing areas with the highest solar potential and the least environmental disruption, the proposed approach seeks to foster a renewable energy infrastructure that aligns with the region's conservation goals and socioeconomic needs. The successful implementation of micro-solar power plants can transform the energy landscape of the Sundarban Delta, providing reliable, clean energy to its residents and reducing dependence on conventional, polluting energy sources. This study offers a comprehensive analysis of the GIS-based site selection process, its benefits, and its potential to drive sustainable development in one of India's most ecologically sensitive regions.

2 Literature Review (Related Works)

There have been several studies on site selection in which multi-criteria decision-making approach has been applied from various angles. This approach was applied for suitable agricultural land selection and according to the majority of scholars, this approach compatible to solve complex decision-making (Azmi et al., 2017; Ennaji et al., 2018; Marinakis et al., 2017; Rezaei, 2015). Multiple competing objectives may be encircled using Multi-criteria Decision Making (MCDM) approaches in a holistic and visible way to aid energy planning as well as to identifying potential locations for renewable energy. The relevant site suitability criteria for this study are incorporated by considering recent studies. For large-scale PV power plants, sun irradiation is a significant criterion which plays a critical role in generating additional electric. Table 1 shows important choice factors in the solar plant site suitability studies.

Proximity analysis ensures adequate accessibility and helps to minimize excessive infrastructure building costs as well as negative environmental repercussions. Studies showed that certain places are suitable to avoid adverse environmental effects on urban growth and Not in My Backyard (NIMBY) resistance (Janke, 2010; Sánchez-Lozano et al., 2014; Uyan, 2013). According to other research, areas close to cities have bigger economic advantages (Aydin et al., 2013; Effat, 2013). The research area might be filtered to avoid infeasible sites that would obstruct the installation of a utility-scale PV plant, resulting in more accurate decision findings. The GIS will be used to eliminate unsuitable areas that prevent the deployment of such facilities. The most prevalent limits for solar site suitability are shown in Table 2.

Table 1 Relevant criteria considered in previous studies

Main criteria	Sub-criteria	Authors
Environmental	Land use and land cover	Janke (2010); Carrión et al. (2008); Wang et al. (2014); Massimo et al. (2014); Sabo et al. (2017)
	Distance from residential areas	Janke (2010); Sánchez-Lozano et al. (2016); Mentis et al. (2015); Uyan (2013); Sánchez-Lozano et al. (2013)
	Population density	(Janke, 2010; Massimo et al., 2014; Sabo et al., 2017; Kengpol et al. (2012))
	Distance from roads	Janke (2010); Sánchez-Lozano et al. (2016); Effat (2013); Watson & Hudson (2015); Merrouni et al. (2016)
	Distance from rivers	Ayodele et al. (2018); Nebey et al. (2020)
	Distance from protected Areas	Watson & Hudson (2015)
Economic	Land cost	Kengpol et al. (2012)
	Construction	Kengpol et al. (2012)
Climatic	Solar irradiation	Janke (2010); Sánchez-Lozano et al. (2014, 2016); Mentis et al. (2015); Carrión et al. (2008); Wang et al. (2014); Lee et al. (2015); Charabi et al. (2016)
	Global horizontal irradiation (GHI)	Al Garni & Awasthi (2017); Kasim et al. (2019)
	Land surface temperature (LST)	Sánchez-Lozano et al. (2013, 2015); Charabi & Gastli (2011); Kengpol et al. (2012); Lee et al. (2015)
Topographic	Land elevation	Sánchez-Lozano et al. (2015); Effat (2013); Charabi et al. (2016)
	Slope	Sánchez-Lozano et al. (2013, 2015)

Table 2 Restrictions applied by different researchers for solar plants site suitability

Sl. no.	Restricted areas	Studies
1	Protected areas	Sánchez-Lozano et al. (2016); Charabi & Gastli (2011); Borgogno Mondino et al. (2015)
2	Agricultural lands	Effat (2013); Massimo et al. (2014); Merrouni et al. (2016)
3	High elevated areas	Janke (2010)
4	Water bodies	Massimo et al. (2014); Sánchez-Lozano et al. (2015)
5	Cultural heritage sites	Janke (2010); Sánchez-Lozano et al. (2015); Charabi & Gastli (2011)
6	Road and railways areas	Uyan (2013); Sánchez-Lozano et al. (2015)
7	Disaster prone areas (flood, cyclone, erosion, etc.)	Kengpol et al. (2012)
8	Areas with high slope (>5°)	Watson & Hudson (2015)
9	Mountain areas	Sánchez-Lozano et al. (2015)
10	Coastal zones	Kengpol et al. (2012)

3 Materials and Methods

3.1 Description of Study Area

The Sundarban Region is a vast delta formed by the confluence of the Ganges, Brahmaputra, and Meghna Rivers. It encompasses an intricate network of tidal waterways, mudflats, and small islands, interspersed with dense mangrove forests. Covering approximately 10,000 square kilometers, the Sundarbans is renowned for its unique and diverse ecosystem, hosting the largest contiguous mangrove forest in the world. The Sundarban Region stretches between India and Bangladesh, comprising approximately 25,000 km^2, with 62% of it in Bangladesh and 38% in India (Ali et al., 2019). It is located between 21°4000″N and 22°3540″N and 88°0130″E and 89°0500″E (Fig. 1). The region is characterized by a subtropical climate, with hot and humid summers, a monsoon season with heavy rainfall and mild winters. The dynamic landscape is constantly shaped by the ebb and flow of tides, leading to frequent changes in the geography of the islands and waterways. Despite its ecological richness, the Sundarban Region faces significant challenges such as rising sea levels, coastal erosion, and frequent cyclones, exacerbated by climate change. The peoples here suffer for basic amenities.

Environmentalists have criticized the West Bengal government's decision to expand the national grid to the Sundarban Delta in order to electrify communities. The per capita consumption of electricity in the Sundarban is less than 50 kwh/year and about 96% of which is supplied through grid (Moharil & Kulkarni, 2010). Another issue with grid-based electricity is its inconsistency and unpredictability, according to the survey, with 85% of towns receiving power experiencing daily power outages lasting 4–9 h. Because the ground is soft, erecting massive, heavy transmission poles in rivers and streams may cause further erosion and alter tide patterns. These transmission

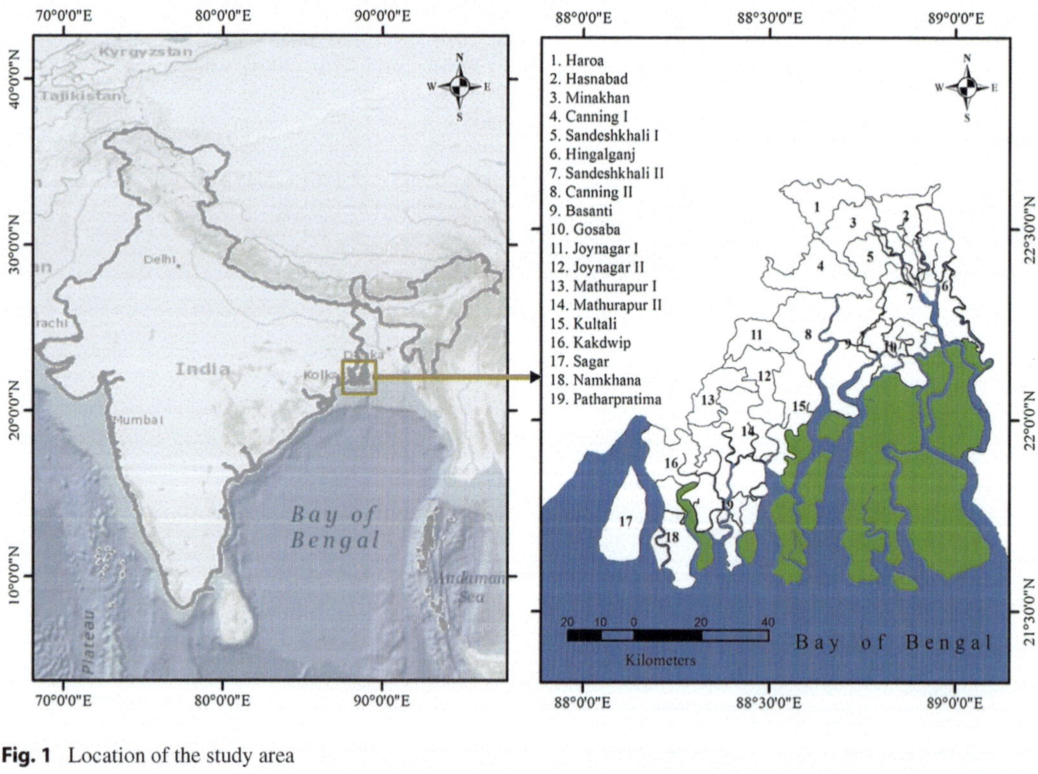

Fig. 1 Location of the study area

lines will also be vulnerable to the cyclones that impact the region on a regular basis; therefore, exploring the possibility of accessing renewable energy resources to address these issues and ensure the long-term usage of power should be considered.

3.2 GIS and MCDA-Based Framework

The geographic information system has been employed in suitability analysis for the function and visualization of geographical data in recent years (Church, 2002; Khashoggi & Murad, 2020; Lü et al., 2019). The GIS has the unique ability to determine a location's appropriateness for a certain purpose. The basic goal of appropriateness analysis is to assess how suitable or inappropriate something is in a specific context. For the GIS-based suitability study, multi-factor examination of a broad set of challenges is required. As a result, academics are becoming more interested in merging many criteria with the GIS. It is becoming simpler for decision-makers to grasp real-world difficulties as interest in the GIS and MCDM analysis grows. In circumstances involving optimal site selection and spatial decision help, suitability mapping techniques and geographic information systems (GIS) have lately become popular (Georgiou & Skarlatos, 2016).

In terms of the qualities of data that used, the GIS differs from other information systems. Such data has the capacity to address the position of things or occurrences (phenomena) on the planet (Lü et al., 2019). As a result, the positions of items or events, as well as their interactions with one another, may be depicted using a map. Real-world objects (houses, roads, mountains, and so on) are summed using the required criteria; a topographic landscape model is built and saved in the GIS as a point, line, area, or volume. The GIS technology, one of the far-flung infrastructure platforms, has progressed in recent years. The GIS stands for geographic information system, and it is a full system that allows you to view, query, and analyze geographic data and the data tables that go with it (Rahman et al., 2020; Koc et al., 2019). Because it frequently requires geographical data in relation to multiple siting laws, regulations, considerations, and limits, suitability analysis for selecting solar plant locations is a complicated procedure (Shao et al., 2020). However, due to the favorable conclusion of the site selection technique, its use has increased in popularity and piqued the curiosity of a wide range of people. MCDM when combined with the GIS may offers desire results. For the GIS-based site selection and effective map layer, various multi-criteria decision approaches such as analytical hierarchy process (AHP) and ordered weighted average, Fuzzy MCDM, Fuzzy AHP, fuzzy TODIM, and fuzzy TOPSIS have recently been used (Adedeji et al., 2020; Ali & Ahmad, 2020; Liu et al., 2021; Zhou et al., 2020). MCDMs have frequently been chosen to calculate the parameter's weights in research relating to solar energy and solar plants site selection GIS (Aly et al., 2017; Colak et al., 2020; Doorga et al., 2019; Uyan, 2017).

3.3 Selected Criteria (GIS Layers)

Technical feasibility criteria are essential in selecting the optimal site for solar power plants, significantly impacting the plant's performance. Two critical factors include the quantity of solar irradiation and the average air temperature. Additionally, the project's economic viability is influenced by the location of the solar farm. Key economic considerations encompass the site's proximity to metropolitan areas, roads, and electricity lines, as well as the land's slope and aspect (Fig. 2a–n). Below are the details of these criteria:

(C1) Land surface temperature: Air temperature influences the efficiency of solar panels. Optimal performance is achieved at moderate temperatures, as extremely high temperatures can reduce efficiency (Zhang et al., 2013; Chen et al., 2006; Silva et al., 2018). Various approaches have been proposed in the litera-

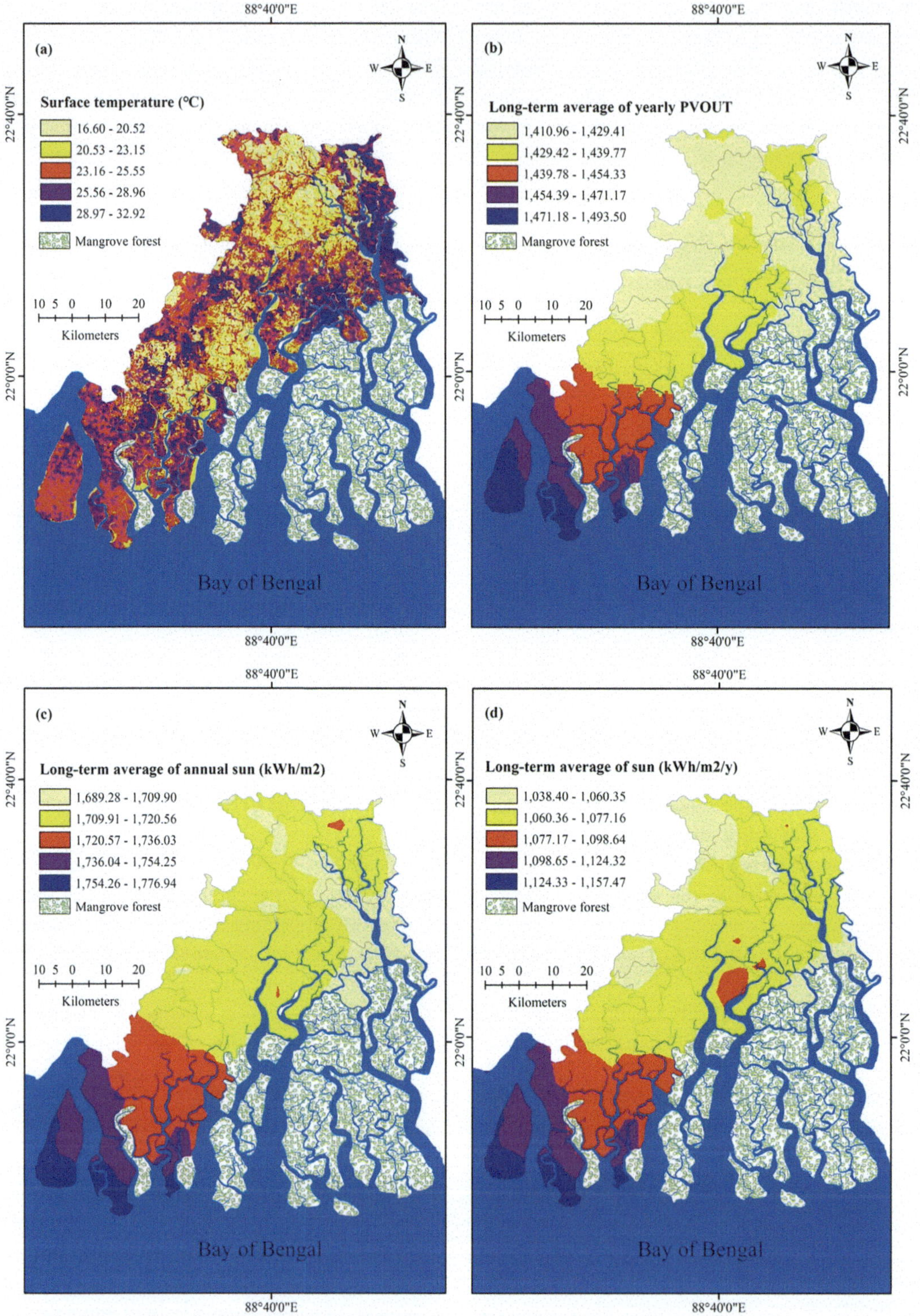

Fig. 2 Selected criteria for solar power plants site suitability: (**a**) Land surface temperature, (**b**) PVOUT, (**c**) Global Horizontal Irradiation, (**d**) Direct Normal Irradiation, (**e**) Elevation, (**f**) Slope, (**g**) NDVI, (**h**) NDBI, (**i**) Distance from Roads, (**j**) Distance from rivers, (**k**) Land cover, (**l**) Rainfall deviation, (**m**) Flood prone areas, and (**n**) Cyclone prone areas

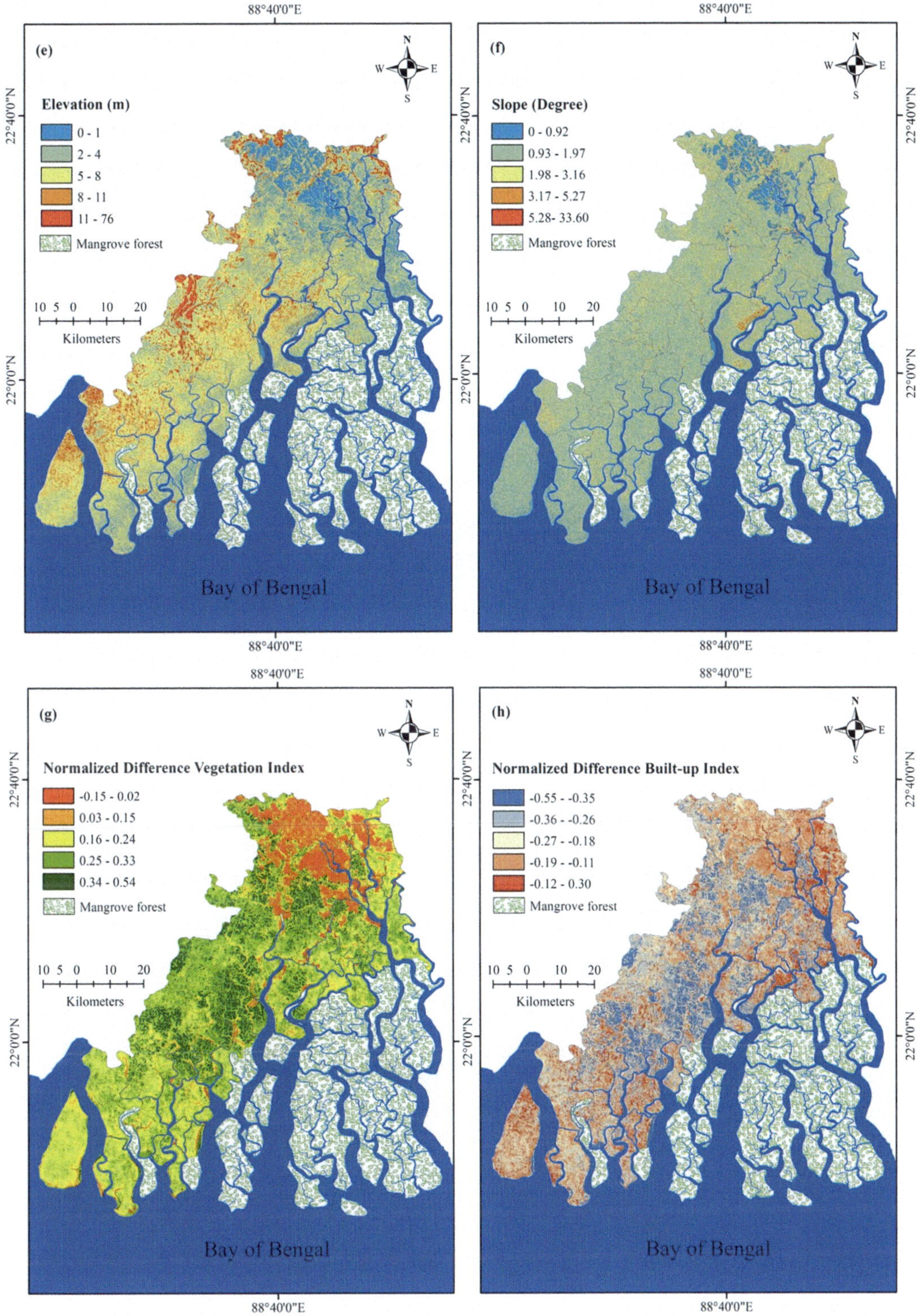

Fig. 2 (continued)

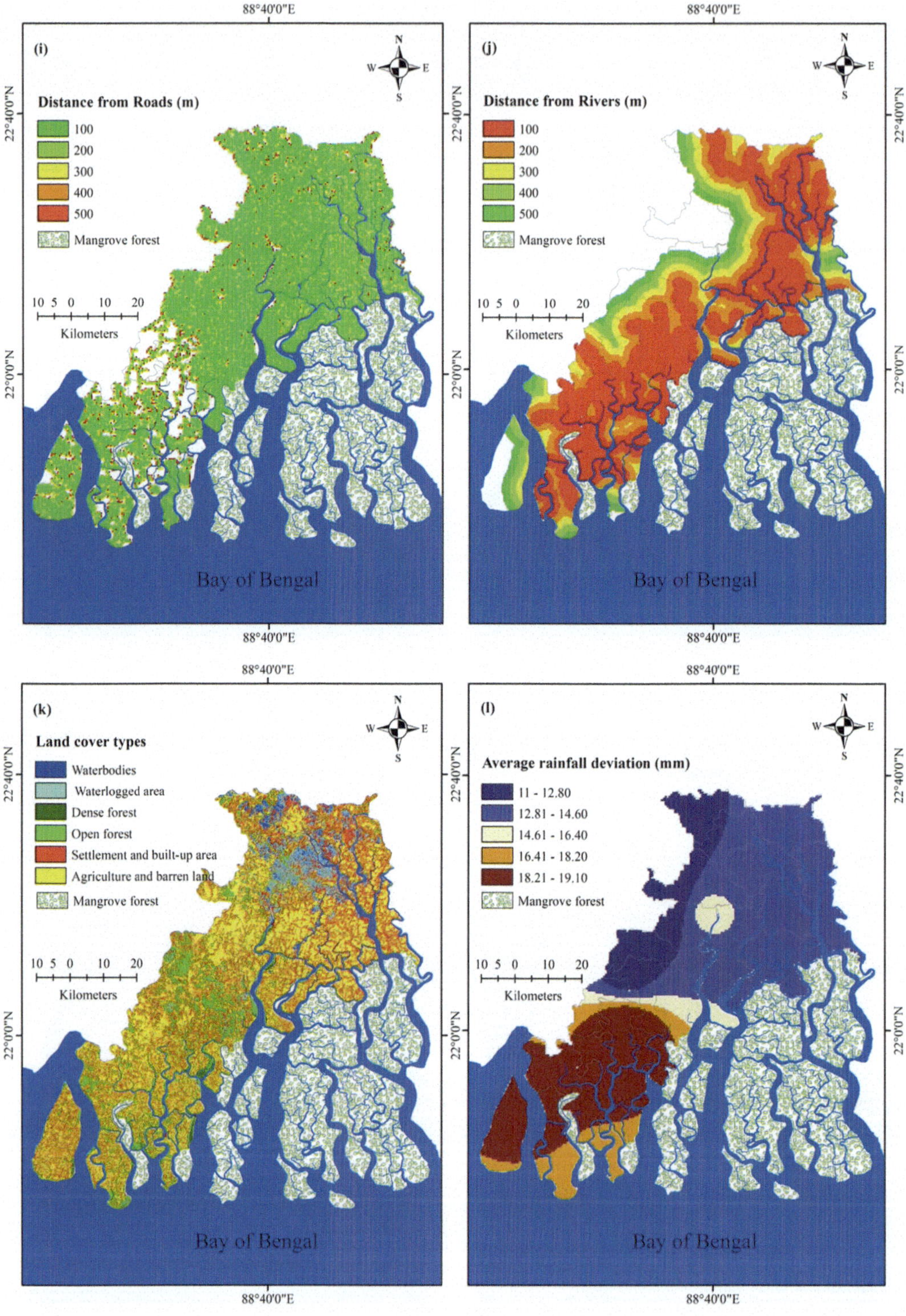

Fig. 2 (continued)

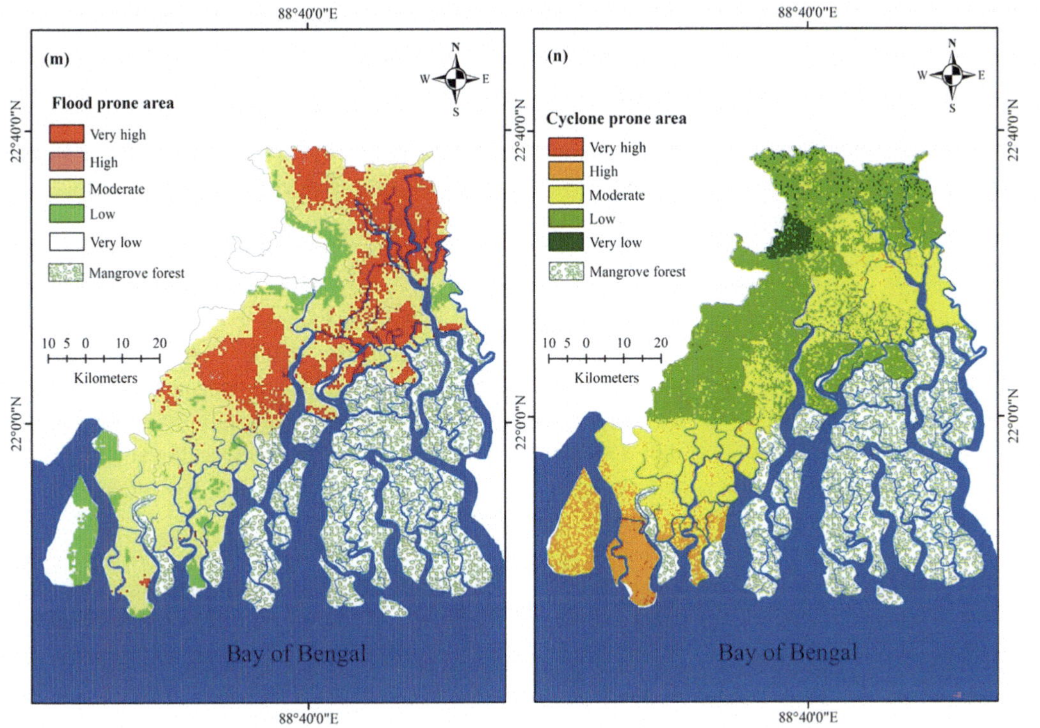

Fig. 2 (continued)

ture, including the split-window method, the radiative transfer equation-based method, the single-window (mono-window) methodology, and the single channel (single channel) method. We employed the radiative transfer equation-based technique in this work to choose solar power plants using Landsat-8's thermal band (Band 10) and calculated the results as in prior studies (Ali & Ahmad, 2019a, b). Figure 2a depicts a surface temperature map of the Sundarban Delta.

(C2) Photovoltaic Power Potential (PVOUT): The photovoltaic power potential (PVOUT) map shows the potential for solar photovoltaic (PV) power generation. The PVOUT is useful for determining the appropriateness of a solar power plant site since it offers an areal distribution of PV, with regions with a larger proportion of PV being more suited for plant placement. Figure 2b shows the PVOUT map of the Sundarban Delta.

(C3) Global Horizontal Irradiation (GHI): The amount of sunlight a location receives directly affects the energy output of solar panels. Higher solar irradiation ensures greater electricity generation. It is the long-term average of the worldwide horizontal irradiation's yearly/daily total. One of the most essential features of solar energy is sun irradiation (Kasim et al., 2019). It is linked to the amount of time spent sunbathing. The greater the temperature, the longer the sunbathing time is. Despite the fact that the polar night and day lasts for six months, the temperature remains low due to the sun's rays being slanted to the earth (Al Garni & Awasthi, 2017). In the present study, the GHI data was collected from solar GIS project. This offers country level solar data in Tiff file format. The particular data of the study area was extracted as shown in Fig. 2c.

(C4) Direct Normal Irradiation (DNI): The most essential metric for calculating energy yields and evaluating the performance of concentrating solar power (CSP) and concentrator solar photovoltaic (CPV) systems is the NDI. DNI is also crucial for calculating the global irradiation received by photovoltaic modules that are inclined or sun-tracking. The projected solar energy available for power generation and other energy purposes is summarized in this solar resource map. The research area's NDI map was created using solar GIS data and is shown in Fig. 2d.

(C5) Elevation: Solar panels located at a greater elevation will receive more solar radiation than those located at ground level, resulting in increased power output (Assouline et al., 2017; Albraheem & Alabdulkarim, 2021). Because these complete solar radiations are available, a more efficient PV system may be built than ground-mounted PV systems. The elevation layer was created using Shuttle Radar Topography Mission (SRTM) data from the Digital Elevation Model (DEM). Figure 2e depicts the elevation map of the Sundarban Delta.

(C6) Slope: The steepness or inclination of a surface in relation to the horizontal plane is measured by slope. The slope has an effect on persistent snow in the high mountains, temperature variations at different latitudes along the same longitude, and rainfall as the temperature climbs every 200 m (Al Garni & Awasthi, 2017). The slope map was created using data from the SRTM Digital Elevation Model (DEM) using the slope tool in the spatial Analyst extension. Figure 2f depicts the slope map of the Sundarban Delta.

(C7) Normalized Difference Vegetation Index (NDVI): Vegetation indices are used to improve vegetation information in remote sensing imagery by restricting a near-infrared (NIR) band to a red band. This takes use of the strong plant reflectance in the NIR spectral band as well as the high red light pigment absorption (Abd-El Monsef & Smith, 2017). Due to the influence of vegetation on the smooth operation of PV, vegetation criteria should be addressed while mapping the potentials for solar power facilities. Places with a high and low NDVI score have the lowest and greatest priority for solar power plant construction, respectively (Fig. 2g).

(C8) Normalized Difference Built-Up Index (NDBI): The NDBI was used to create the build-up land picture, which takes use of the build-up lands' distinctive spectral response, which has greater reflectance in the Mid-Infrared (MIR) wavelength range than in the NIR wavelength range (Asad et al., 2017; Xu, 2007). Figure 2h depicts the slope map of NDBI of the study area.

(C9) Distance from Roads (both national and local): In reality, the near proximity will save money on new transportation infrastructure, energy transfer, staff commutes, unit support, and environment degradation. For every power plant installation, transportation costs were the most important consideration. As a result, places remote from roadways are both economically and environmentally inappropriate. As a result, spots within 100 m of each other were chosen as very ideal solar PV sites. Locations with a distance of 200–300 m are appropriate, those with a distance of 300–400 m are fairly suitable, and those with a distance of more than 500 m are undesirable (Fig. 2i).

(C10) Distance from rivers: Rivers and water bodies have an impact on the location of solar power plants (Nebey et al., 2020). The most acceptable places were those that were the furthest away (Ayodele et al., 2018). Depending on the distance from the locations, the streams were divided into five primary types. The areas that were the furthest away were deemed more ideal, while those that were the closest were deemed unfit (Fig. 2j).

(C11) Land cover: The structure of land is what land use is all about. The quality of the land is determined by its stony or rocky surface. When the stony sections are excessive, for example,

the buildings must be utilized in conjunction with concrete feet, which raises the cost by 30% (Türk et al., 2021). The heat is dissipated more quickly on the rapidly heated surface, allowing the solar panels to warm from behind. The rear-heated solar panel loses a significant amount of heat and energy (Farrell et al., 2015; Türk et al., 2021). Figure 2k is a land-use map of the Sundarban Delta.

(C12) Rainfall deviation: This parameter is regarded as the most important one, as it plays a direct role in the hydrological process. High humidity and suspended particle concentrations, which are the most critical variables in shortwave radiation absorption or reflection, are found in areas with a high annual precipitation rate (Zoghi et al., 2017; Noorollahi et al., 2016). The map of the Sundarban Delta in terms of rainfall deviation is shown in Fig. 2l.

(C13 and C14) Flood and cyclone prone areas: The catastrophe risk affects the sites of solar power plants (Hao et al., 2021). The places farthest from disaster-prone zones (such as flood, cyclone, and erosion) were the most suited (Silva et al., 2021). Because the research location is prone to flooding and cyclones, these two aspects were taken into account. Flood and cyclone-prone locations have been divided into five groups based on their level of danger, and maps have been created based on prior studies (Ali & Ahmad, 2019a, b). The areas farthest away were deemed more ideal for solar power plants, whereas those closest to them were deemed unsuitable (Fig. 2m, n).

3.4 MCDM Techniques

In the presence of several criteria, MCDM approaches assist decision-makers in selecting the best option among several possibilities. These methods have been widely used in planning, particularly for site selection based on environmental, technological, and economic considerations. The selection of solar plant sites based just on one factor is insufficient (Løken, 2007). According to Huang et al. (1995) and Løken (2007), MCDM is the optimum method for site selection, especially for energy planning complexity. Integration of the GIS with MCDM has recently grown in popularity for a variety of siting applications, including landfill site selection (Ali & Ahmad, 2020; Ali et al., 2021; Sk et al., 2020; Mohsin et al., 2022; Şener et al., 2010), urban planning (Chandio et al., 2013; Mohit & Ali, 2006), and renewable energy resource site planning (Delivand et al., 2015; Izadikhah & Saen, 2016; Pohekar & Ramachandran, 2004).

3.4.1 Application of the Analytic Hierarchy Process (AHP)

Prof. Thomas L. Saaty invented the AHP, one of the MCDM approaches, in 1977. The AHP approach has been used to solve a variety of challenging decision-making issues. Since its conception, AHP has been used in a variety of industries to define and analyze user preferences in a variety of applications. In this process, factors are arranged in a hierarchal manner aiming at quantifying rank in a given set of alternatives. In the AHP technique, the most important task is to select relevant criteria to run the decision-making process, based on the judgments of the decision-maker.

To execute AHP in the present study, first of all, pairwise comparisons matrix (PCM) of the criteria and alternatives was established to define their relative importance as shown in Table 3. This was usually done using a scale of 1 to 9, where 1 indicates equal importance and 9 indicates extreme importance of one element over another. Then, the pairwise comparison matrix was normalized by dividing each element by the sum of its column, which is commonly known as normalized column sum (NCS). The priority vector or eigenvector was calculated by averaging the rows of the normalized matrix. Finally, the consistency of the decision matrix was checked by consistency index (CI) and consistency ratio (CR). All mathematical calculation was adopted from Ali et al. (2019).

4 Results

4.1 Determining Criteria Priorities, Preparing PCM, and Calculating Factor Weights by AHP

Because a variety of factors influence the location of solar PV power plants, it's crucial to figure out which strategy decision-makers will employ. The relevance of components in relation to one another was chosen and assessed while deciding which parameters to prioritize during application. AHP was considered to find the most significant criteria and their priority order. Because the criteria employed in this context were not examined independently, the weights and categorization of the elements among themselves were determined using the expert's view. Following that, a pairwise comparison matrix (PCM) was established (Table 3). The CR was computed for the determination of the weights to assure the correctness of the operations, and it was determined that the CR should be less than 0.1. Following this computation, the factor weights were checked to see if they were compatible with the CR to be implemented. The binary comparison matrix yielded factor weights, which are displayed in Table 4.

4.2 Optimal site Selection for Micro-Solar Power Plant

Each element was assessed independently to identify best sites for micro-solar power plants in Sundarban deltaic region of India. Saaty's rank scale, which is presented in Table 3, was used to find appropriate ranges for each parameter, and raster layers (criteria maps) were created by ranking according to the scale. The information was categorized and arranged for ranking. The weights were rated on a scale of 9 to 1, with 9 being the most acceptable and 1 being the least appropriate. All criterion maps were then divided into two categories: constraint areas (areas that are not appropriate for solar plant siting) and suitable areas (Fig. 3a–n). As a result, for further study, all appropriate and constraint areas were segregated. These raster layers were integrated in ArcGIS 10.8 with weights calculated using the AHP approach, and a map of the most ideal sites was created (Fig. 4).

Using the GIS and the AHP, a raster-based cumulative surface map with 400 dpi resolution was created in this study (Fig. 4). The pixels with the greatest numerical values were selected as the optimal solar power plant locations. The map created enabled for the optimal places for the installation of micro-solar power plants to be chosen. The most appropriate pixels are those with the greatest value, while the least appropriate pixels are those with the lowest value. The resulting map showed that there was not only an enough but also an inadequate amount of sun irradiation parameters suited for building of solar power plants. Consequently, it was determined that other factors needed to be considered.

Although different researches have been carried out on solar plants, it has been established that research has to be done in specific places, taking into account a variety of aspects in shaping the best locations for micro-solar power plants. In an ecological zone prone to climate change catastrophes, preservation and conservation should be promoted—instead, Sundarban has become a renewable energy source where every possible renewable energy experiment is carried out at the risk of destroying even more of what is left. In this light, the goal of this research (i.e., micro-power pants

Table 3 Pain-wise comparison matrix

	C1	C2	C3	C4	C5	C6	C7	C8	C9	C10	C11	C12	C13	C14
C1	1	0.50	0.33	0.50	2	3	3.03	2	0.50	3	0.50	4	0.25	0.33
C2	2	1	0.33	0.50	2	2	1	4	3	2	0.50	3	0.25	0.33
C3	3	3	1	2.00	3	2	3.03	4	3	2	0.50	3	0.25	0.20
C4	2	2	0.50	1.00	3	2	2.00	0.50	0.50	0.50	0.33	0.50	0.33	0.25
C5	0.50	0.50	0.33	0.33	1	1	0.50	0.33	0.33	0.50	1	0.33	0.50	0.33
C6	0.33	0.50	0.50	0.50	1	1	0.33	0.50	0.33	0.50	0.33	0.50	0.50	0.50
C7	0.33	1	0.33	0.50	2	3	1	0.50	0.33	0.50	0.33	0.50	0.33	0.50
C8	0.50	0.25	0.25	2	3	2	2	1	0.33	0.50	0.50	0.33	0.50	0.33
C9	2.	0.33	0.33	2	3	3	3	3	1	0.50	0.25	0.33	0.33	0.50
C10	0.33	0.50	0.50	2	2	2	2		2	1	0.33	0.33	0.50	0.33
C11	2.	2	2	3	1	3	3	2	4	3	1	0.33	0.25	0.33
C12	0.25	0.33	0.33	2	3	2	2	2	3	3	3	1	0.33	0.33
C13	4.	4	4	3	2	2	3	2	3	2	4	3	1	0.50
C14	3.	3	5	4	3	2	2	3	2	3	3	3	2	1
Column Sum	21.24	18.91	15.74	23.33	31.06	30.03	27.89	27.83	23.39	22.03	15.58	20.23	7.33	5.78

Note: C1—Land surface temperature, C2—PVOUT, C3—Global Horizontal Irradiation, C4—Direct Normal Irradiation, C5—Elevation, C6—Slope, C7—NDVI, C8—NDBI, C9—Distance from Roads, C10—Distance from rivers, C11—Land cover, C12—Rainfall deviation, C13—Flood prone areas, C14—Cyclone prone areas

Table 4 Criteria and factor weights

Criteria	Factors	Weight	Normalized weights	Rank
C1	Land surface temperature	0.0670	7.0	6
C2	PVOUT	0.0709	7.0	5
C3	Global Horizontal Irradiation	0.0953	10.0	3
C4	NDI	0.0504	5.0	10
C5	Elevation	0.0304	3.0	13
C6	Slope	0.0270	2.0	14
C7	NDVI	0.0387	4.0	12
C8	NDBI	0.0443	4.0	11
C9	Distance roads	0.0612	6.0	9
C10	Distance rivers	0.0520	5.0	9
C11	LULC	0.0888	9.0	7
C12	Rainfall	0.0776	8.0	4
C13	Flood prone areas	0.1409	14.0	2
C14	Cyclone prone areas	0.1556	16.0	1

site appropriateness) is to boost solar energy-based electricity output to greater levels. To solve this question, an applied technique was employed in this study. Moreover, solar energy production facilities are prohibited from being built on complete farmland, special product lands, sown farmland, or irrigated farmland, as well as in areas where the integrity of agricultural use would be compromised by the Regulation on the Protection, Use, and Land Consolidation of Agricultural Land. Thus, in order to carry out the best possible site selection study for the construction of small solar power plants, these characteristics were employed. The Sundarban Delta area of India was selected for the study due to its considerable solar energy production potential. In this study, pertinent geographic analyses were carried out, leading to the identification of suitable locations. The best locations were identified through the use of GIS technology and spatial research, and these weight values were then used to illustrate the locations on maps. The research led to the creation of a raster layer. The found alternative locations, which were selected using suitable pixel values, are displayed in Fig. 5.

Because these places are extremely close to mangrove forest and very susceptibility to floods and storms, the solar power plant map of the Sundarban Region reveals that solar energy is at a maximum in the northwest to southwest and at a minimum in the northeast to southeast. However, when additional characteristics were taken into account, a total of ten viable places were discovered in this region. The current land characteristics were considered to limit the number of identified locations. The regions around solar PV power facilities that are presently being developed or that are expected to be built are focused for this purpose. The majority of these sites are found in different villages of Sagar, Joynagar-I, Kakdwip, Gosaba, and Canning-II blocks (Fig. 6).

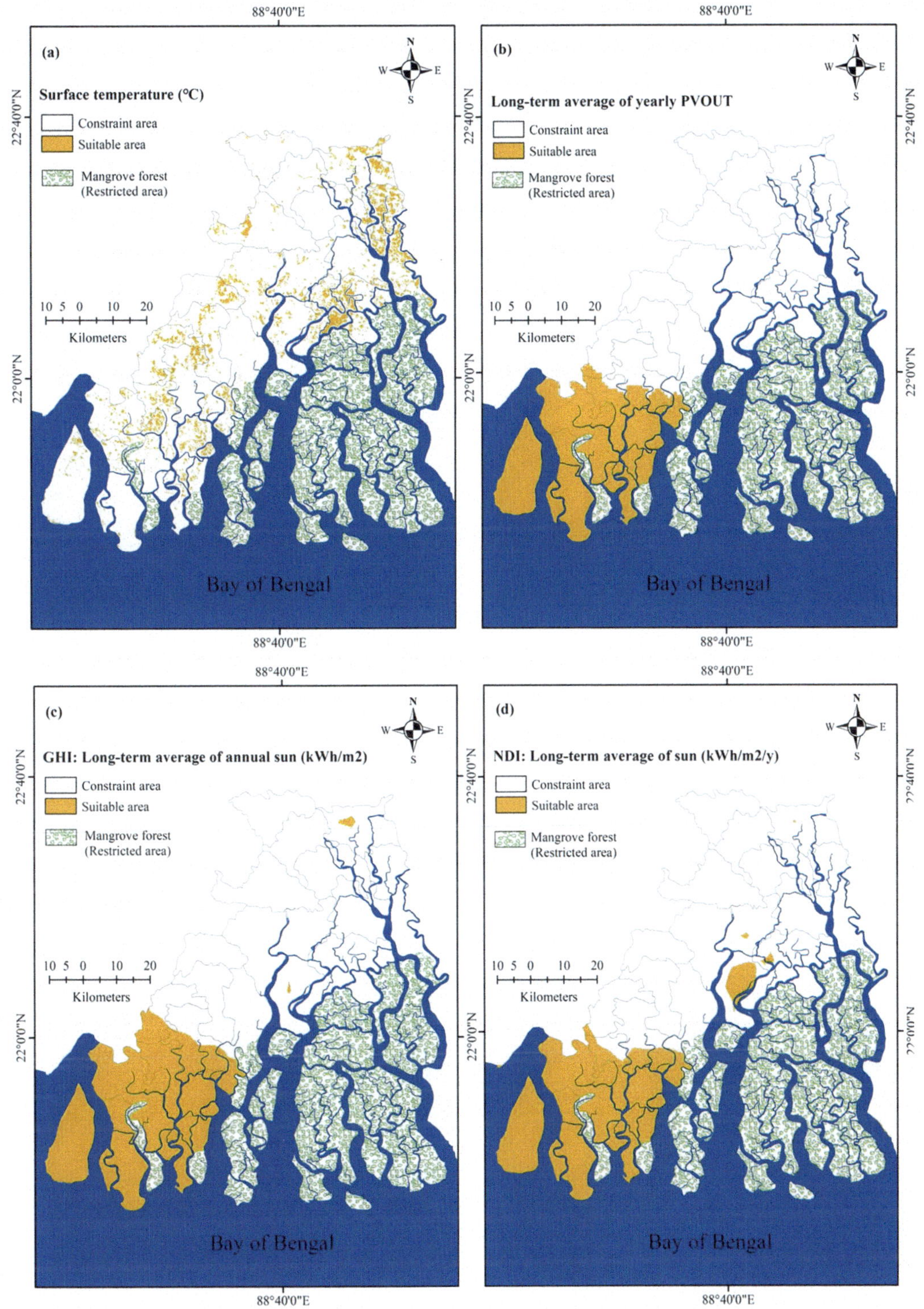

Fig. 3 Extraction of constraint and suitable areas: (**a**) Land surface temperature, (**b**) PVOUT, (**c**) Global Horizontal Irradiation, (**d**) Direct Normal Irradiation, (**e**) Elevation, (**f**) Slope, (**g**) NDVI, (**h**) NDBI, (**i**) Distance from Roads, (**j**) Distance from rivers, (**k**) Land cover, (**l**) Rainfall deviation, (**m**) Flood prone areas, and (**n**) Cyclone prone areas

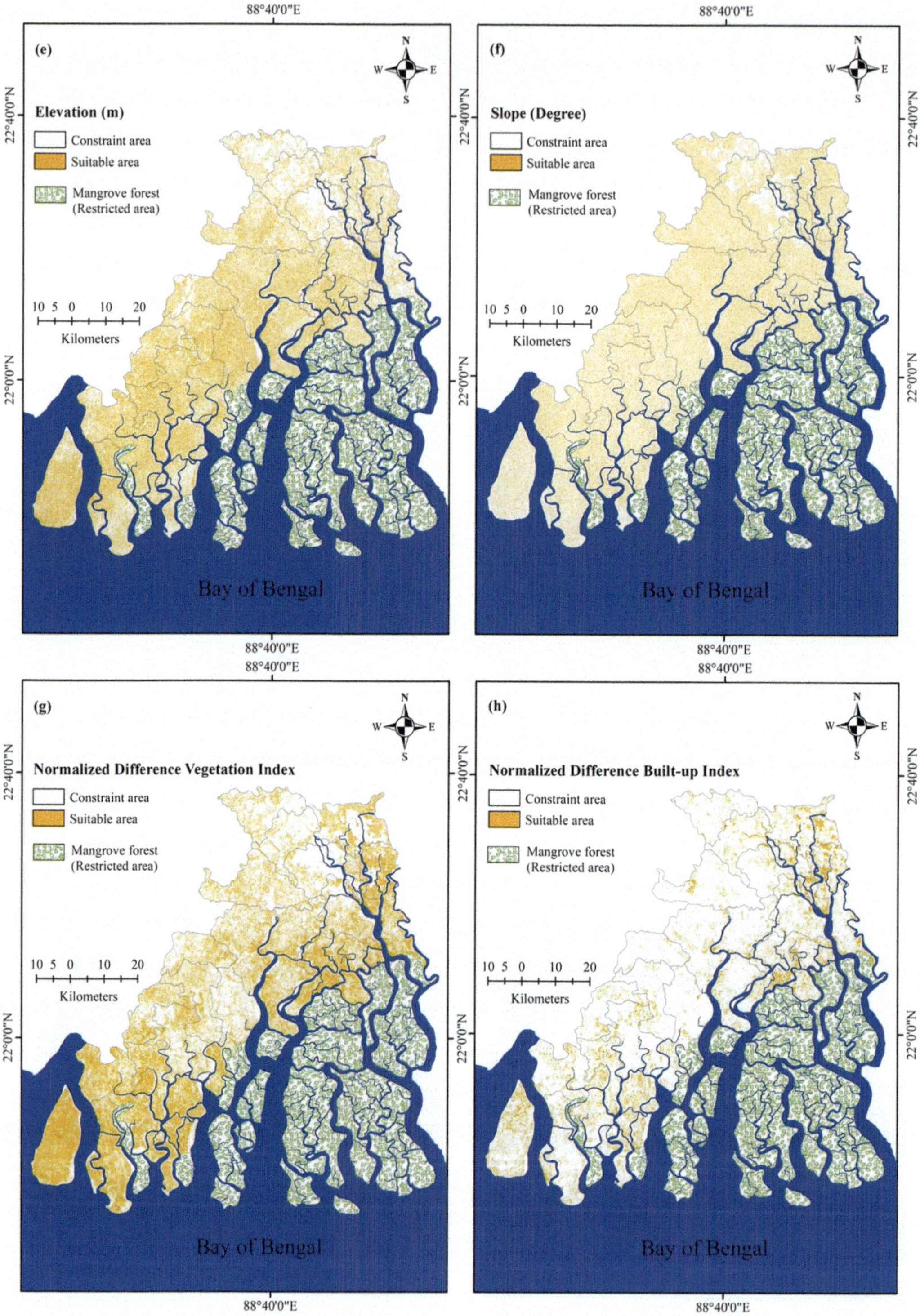

Fig. 3 (continued)

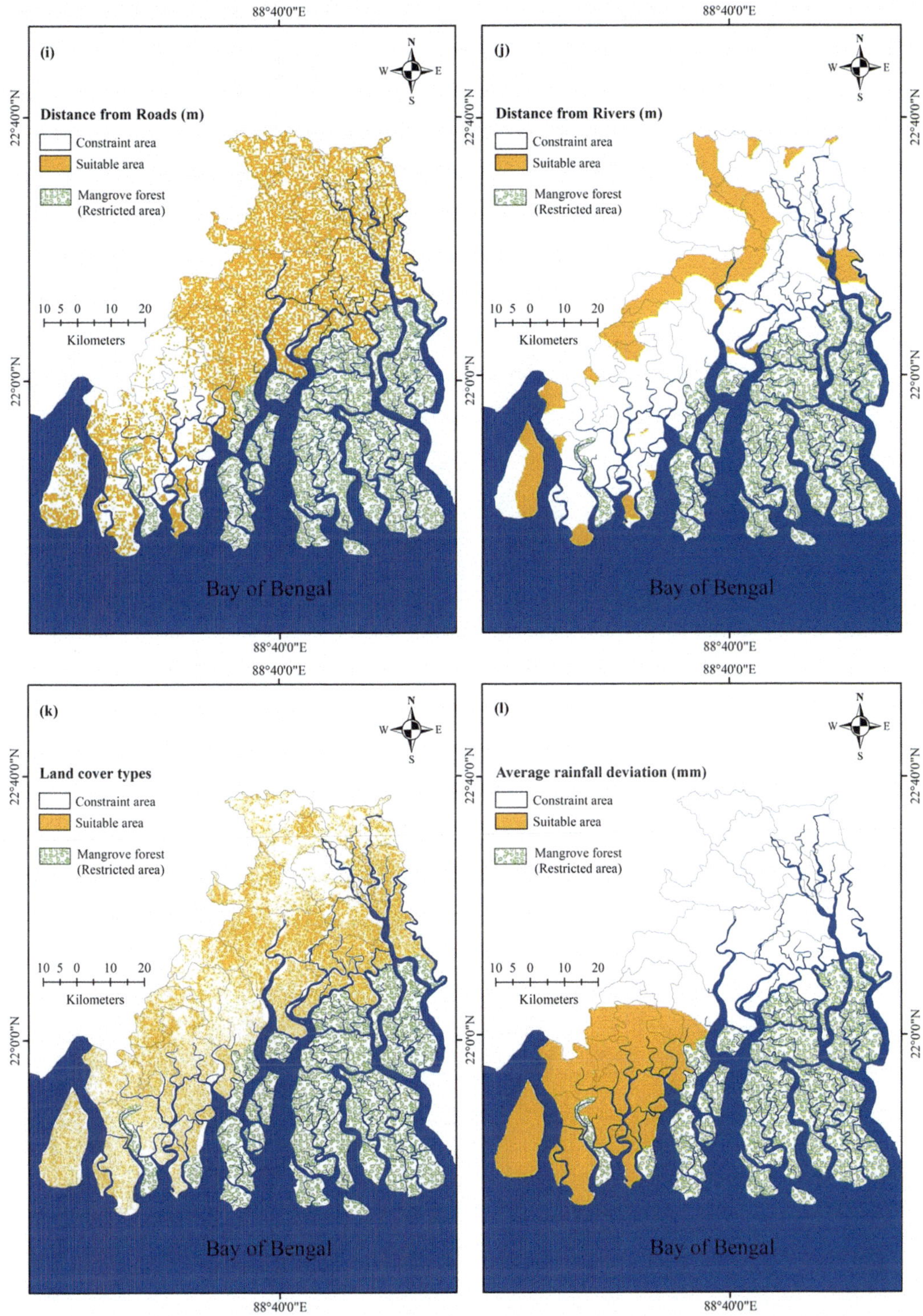

Fig. 3 (continued)

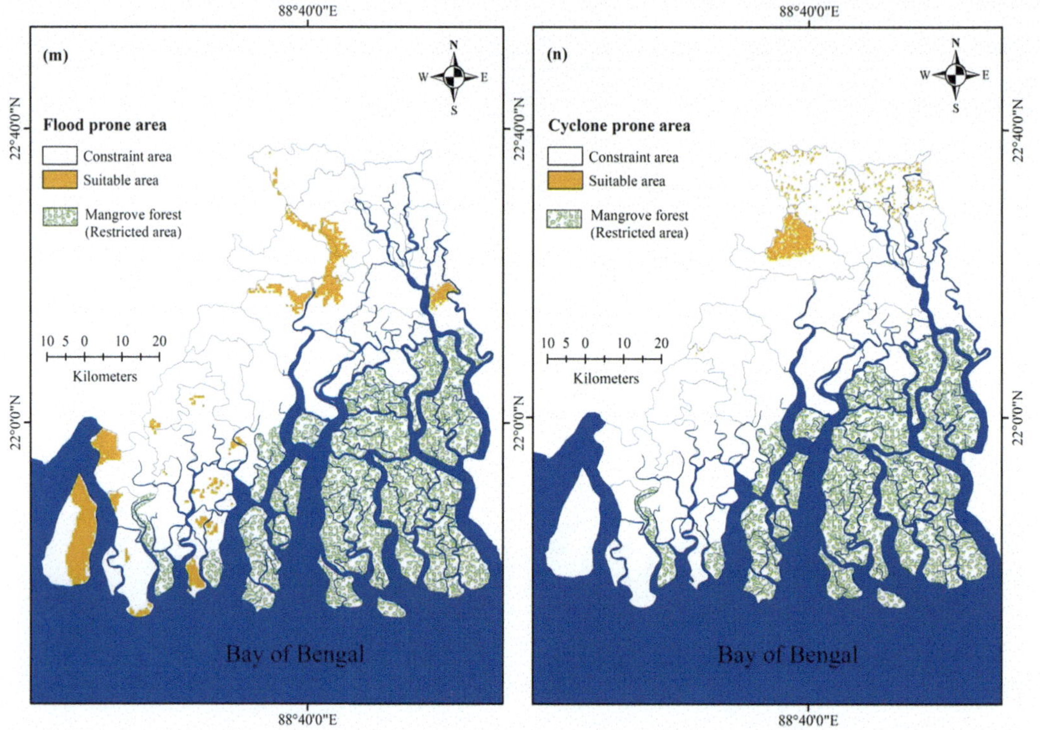

Fig. 3 (continued)

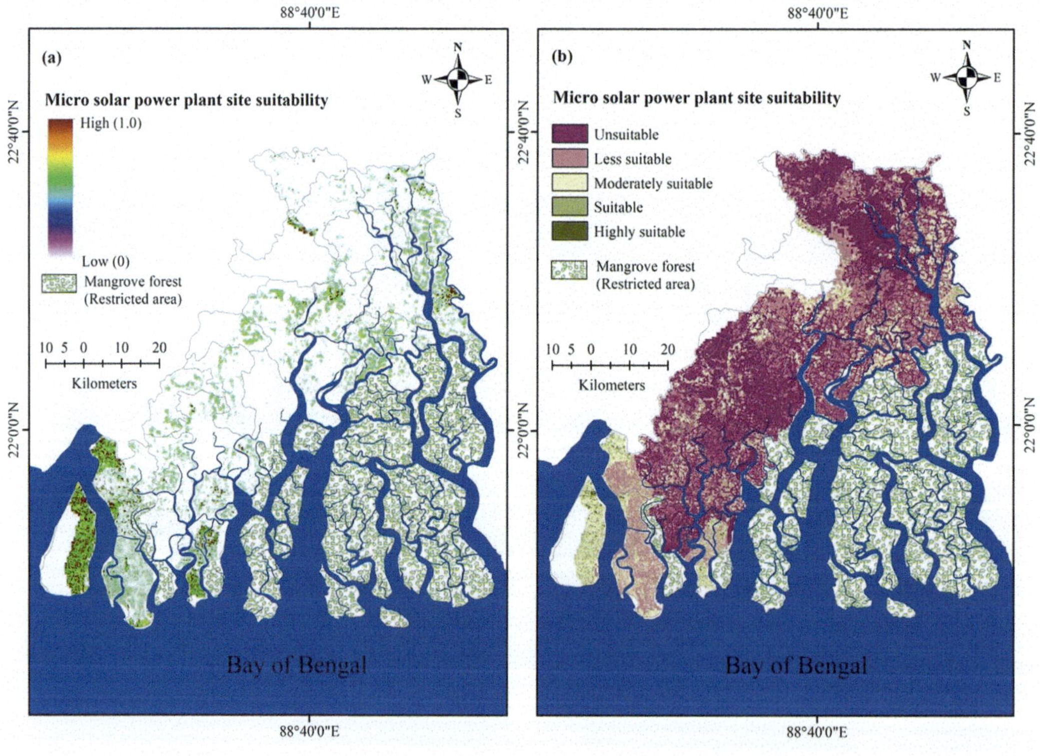

Fig. 4 Suitable sites for micro-solar power plants: (**a**) suitability score and (**b**) classified classes

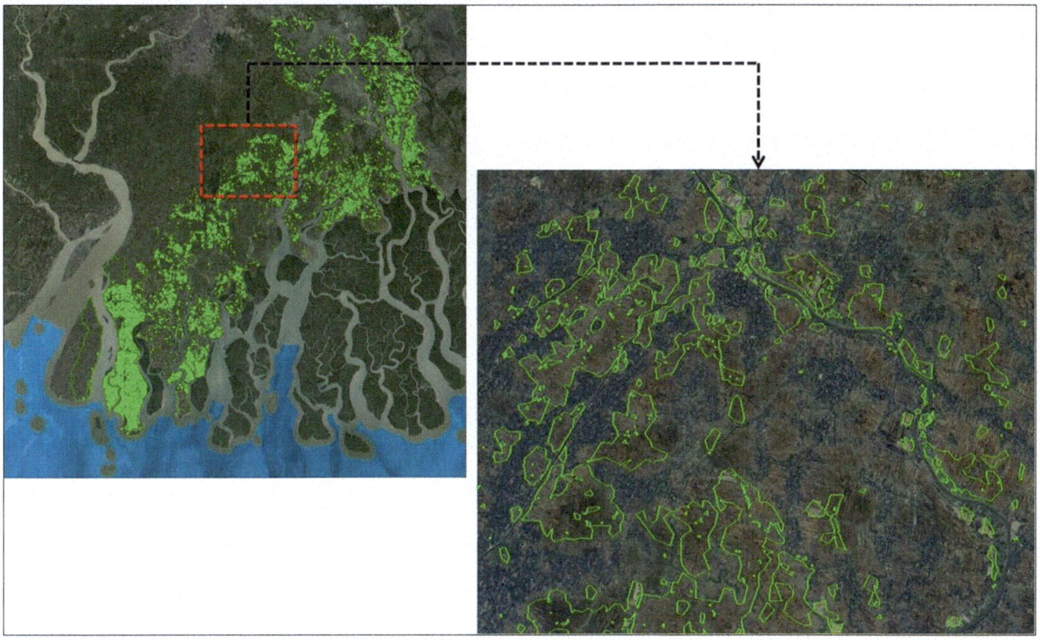

Fig. 5 Suitable sites overlaid on Google Earth to show actual ground level visualization

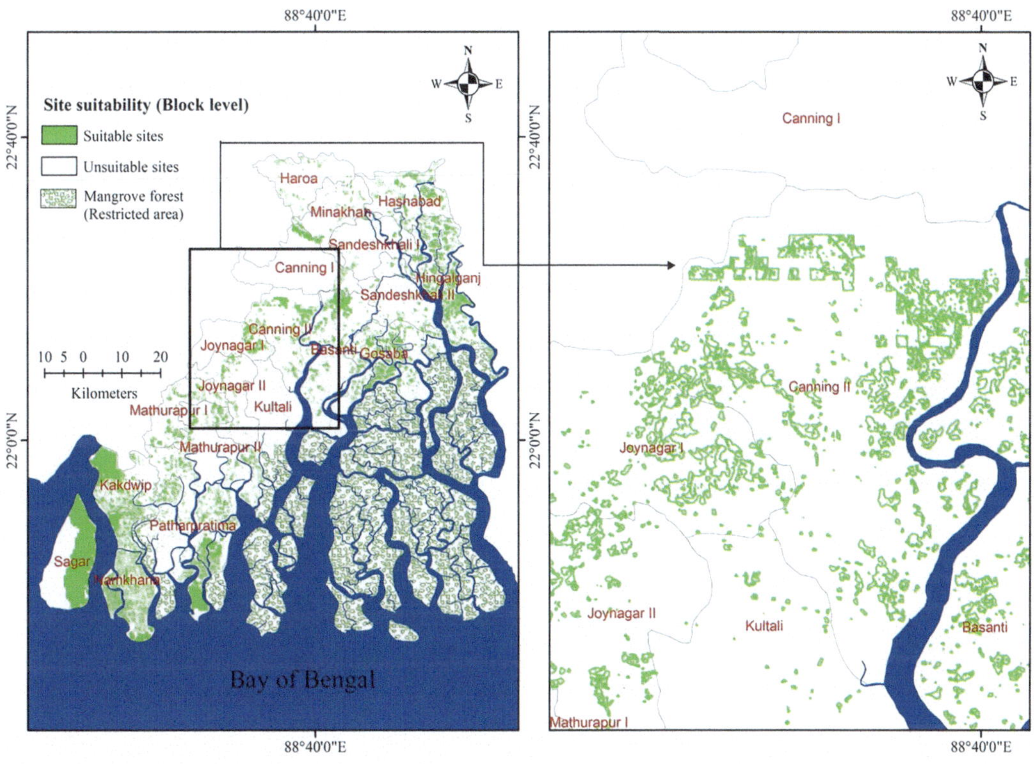

Fig. 6 Block level distribution of suitable sites for micro-solar power plants in Sundarban Delta, India

5 Discussion

The application of GIS-based techniques for selecting micro-solar power plant sites in the Sundarban Delta of India underscores the significant potential of renewable energy to transform energy accessibility in remote and ecologically sensitive regions. Our study demonstrates that GIS tools can effectively identify optimal sites by integrating multiple criteria, including solar radiation, land use, proximity to grid infrastructure, and environmental constraints. The findings reveal several high-potential areas for micro-solar power plants, particularly in regions that experience high solar insolation and have minimal environmental constraints (Khan & Rathi, 2014; Nebey et al., 2020). These areas offer the dual benefits of harnessing abundant solar energy while minimizing ecological disruption. This approach not only ensures the sustainability of the energy projects but also aligns with the broader goals of conserving the unique biodiversity of the Sundarban Delta.

Moreover, the integration of socioeconomic factors into the siting model is crucial. The Sundarban Delta, characterized by its scattered and isolated communities, faces significant challenges in terms of energy access. By prioritizing sites close to human settlements and existing infrastructure, the proposed model ensures that the benefits of renewable energy reach the most vulnerable populations. This has the potential to improve the quality of life, support local economies, and reduce the dependence on non-renewable and polluting energy sources.

Despite India's rapid and remarkable economic growth, energy remains in short supply. The country's fast-paced economic expansion is driving high energy consumption, necessitating development of new energy sources to meet this demand. Concurrently, India faces long-term developmental challenges due to population growth and environmental degradation, leading a widen gap between electricity demand and supply in near future (Arto et al., 2016). Since the mid-1990s, the Sundarban Region has had one of the world's largest solar power systems, with over 200,000 off-grid connections. However, the focus on solar power began to wane after the introduction of the mainstream electrical grid in 2012. The situation worsened when the main grid was extended to Sagar Island, and neighboring islands. The Sundarban Region is a climate change hotspot, with more extreme weather events projected in the future. In such cases, the main grid is likely to be disrupted. Therefore, it is crucial to maintain a focus on solar micro-grids as a backup and supplemental energy source. The current site suitability research aims to identify locations where small solar power plants can be established in the future, ensuring the sustainable use of solar energy as an electricity source in these remote rural communities.

The study also highlights the importance of stakeholder engagement and participatory planning in the site selection process. Involving local communities, policymakers, and conservationists in decision-making can enhance the acceptance and success of solar power projects. Furthermore, the dynamic nature of GIS allows for the continuous updating and refining of the siting model as new data becomes available or as conditions change, ensuring the long-term relevance and adaptability of the approach. The GIS-based siting model presented in this study provides a robust framework for identifying suitable locations for micro-solar power plants in the Sundarban Delta. This model not only supports the sustainable utilization of renewable energy but also promotes socioeconomic development and environmental conservation, offering a viable pathway for the region's energy future.

6 Conclusion

Today's civilizations enhance their competitiveness, expand their economies, and improve their quality of life in direct relation to their technological development levels. Technological advancements are driven by the availability of sufficient, consistent, and clean energy. The solar energy systems are the most environmentally beneficial, being free, continuous, and renewable. Consequently, countries and international organizations provide substantial incentives for solar energy. However, all energy generation, regardless of source, has various environmental repercussions. Solar plants, for instance, have a high impact on land use but very low air pollution. Thermal solar systems significantly affect water resources, while photovoltaic systems have a moderate impact. Both types can influence local ecology, biodiversity, and visual aesthetics.

Most regions in the study met the criteria for suitability analysis. The key criteria for determining solar power site suitability included solar irradiance, slope, land use, land cover, and proximity to highways, streams, and disaster-prone areas. Solar power generation potential was higher in north and south western districts of Sundarban Region. This potential can help close the energy gap in the Sundarban Delta region between demand and supply. Exploiting this high solar potential for electricity generation can bridge the energy gap in rural communities. Although solar resource studies are currently underway, our findings use MCDM techniques to explain solar site suitability in the Sundarban Delta region, an environmentally sensitive area, for the first time. The main advantage of this study is its use of natural properties to supply required power to remote populations in an environmentally friendly manner. Hybrid systems combining multiple renewable energy resources like solar PV could lead to cost-effective and technically feasible renewable energy projects in the future.

Availability of Data and Materials The data that support the findings of this study are available from the corresponding authors [Farhana Parvin and Sk Ajim Ali, skajimali.saa@gmail.com/skajimali@myamu.ac.in] upon reasonable request.

Competing Interests The authors declare that they have no conflict of interest.

Funding No fund was received from any sources.

References

Abd-El Monsef, H., & Smith, S. E. (2017). A new approach for estimating mangrove canopy cover using Landsat 8 imagery. *Computers and Electronics in Agriculture, 135*, 183–194.

Adedeji, P. A., Akinlabi, S. A., Madushele, N., & Olatunji, O. O. (2020). Neuro-fuzzy resource forecast in site suitability assessment for wind and solar energy: A mini review. *Journal of Cleaner Production, 269*, 122104.

Al Garni, H. Z., & Awasthi, A. (2017). Solar PV power plant site selection using a GIS-AHP based approach with application in Saudi Arabia. *Applied Energy, 206*, 1225–1240.

Albraheem, L., & Alabdulkarim, L. (2021). Geospatial analysis of solar energy in Riyadh using a GIS-AHP-based technique. *ISPRS International Journal of Geo-Information, 10*(5), 291.

Alhammad, A., Sun, Q. C., & Tao, Y. (2022). Optimal solar plant site identification using GIS and remote sensing: Framework and case study. *Energies, 15*(1), 312.

Ali, S. A., & Ahmad, A. (2019a). Mapping of mosquito-borne diseases in Kolkata Municipal Corporation using GIS and AHP based decision making approach. *Spatial Information Research, 27*(3), 351–372.

Ali, S. A., & Ahmad, A. (2019b). Spatial susceptibility analysis of vector-borne diseases in KMC using geospatial technique and MCDM approach. *Modeling Earth Systems and Environment, 5*(3), 1135–1159.

Ali, S. A., Khatun, R., Ahmad, A., & Ahmad, S. N. (2019). Application of GIS-based analytic hierarchy process and frequency ratio model to flood vulnerable mapping and risk area estimation at Sundarban region, India. *Modeling Earth Systems and Environment, 5*(3), 1083–1102.

Ali, S. A., Parvin, F., Al-Ansari, N., Pham, Q. B., Ahmad, A., Raj, M. S., et al. (2021). Sanitary landfill site selection by integrating AHP and FTOPSIS with GIS: A case study of Memari municipality, India. *Environmental Science and Pollution Research, 28*(6), 7528–7550.

Aly, A., Jensen, S. S., & Pedersen, A. B. (2017). Solar power potential of Tanzania: Identifying CSP and PV hot spots through a GIS multicriteria decision making analysis. *Renewable Energy, 113*, 159–175.

Arto, I., Capellán-Pérez, I., Lago, R., Bueno, G., & Bermejo, R. (2016). The energy requirements of a developed world. *Energy for Sustainable Development, 33*, 1–13.

Asad, M., Ahmad, S. R., Ali, F., Mehmood, R., Butt, M. A., & Rathore, S. (2017). Use of remote sensing for urban impervious surfaces: A case study of Lahore. *International Journal of Engineering and Applied Sciences, 4*(8), 257387.

Assouline, D., Mohajeri, N., & Scartezzini, J. L. (2017). Quantifying rooftop photovoltaic solar energy potential: A machine learning approach. *Solar Energy, 141*, 278–296.

Aydin, N. Y., Kentel, E., & Duzgun, H. S. (2013). GIS-based site selection methodology for hybrid renewable energy systems: A case study from western Turkey. *Energy Conversion and Management, 70*, 90–106. https://doi.org/10.1016/j.enconman.2013.02.004

Ayodele, T. R., Ogunjuyigbe, A. S. O., Odigie, O., & Munda, J. L. (2018). A multi-criteria GIS based model for wind farm site selection using interval type-2 fuzzy analytic hierarchy process: The case study of Nigeria. *Applied Energy, 228*, 1853–1869.

Azmi, R., Amar, H., & Kacimi, I. (2017). Photovoltaic site suitability analysis using analytical hierarchy process and sensitivity analysis methods with GIS and remote sensing in southern Morocco: Case of Draa-Tafilatet region. In *2017 international renewable and sustainable energy conference (IRSEC)* (pp. 1–5). IEEE.

Borgogno Mondino, E., Fabrizio, E., & Chiabrando, R. (2015). Site selection of large ground-mounted photovoltaic plants: A GIS decision support system and an application to Italy. *International Journal of Green Energy, 12*(5), 515–525. https://doi.org/10.1080/15435075.2013.858047

Bradford, T. (2008). *Solar revolution: The economic transformation of the global energy industry*. MIT Press.

Carrión, J. A., Estrella, A. E., Dols, F. A., & Ridao, A. R. (2008). The electricity production capacity of photovoltaic power plants and the selection of solar energy sites in Andalusia (Spain). *Renewable Energy, 33*(4), 545–552. https://doi.org/10.1016/j.renene.2007.05.041

Chandio, I. A., Matori, A. N. B., WanYusof, K. B., Talpur, M. A. H., Balogun, A. L., & Lawal, D. U. (2013). GIS-based analytic hierarchy process as a multicriteria decision analysis instrument: A review. *Arabian Journal of Geosciences, 6*(8), 3059–3066. https://doi.org/10.1007/s12517-012-0568-8

Charabi, Y., & Gastli, A. (2011). PV site suitability analysis using GIS-based spatial fuzzy multi-criteria evaluation. *Renewable Energy, 36*(9), 2554–2561. https://doi.org/10.1016/j.renene.2010.10.037

Charabi, Y., Rhouma, M. B. H., & Gastli, A. (2016). Siting of PV power plants on inclined terrains. *International Journal of Sustainable Energy, 35*(9), 834–843. https://doi.org/10.1080/14786451.2014.952298

Chen, X. L., Zhao, H. M., Li, P. X., & Yin, Z. Y. (2006). Remote sensing image-based analysis of the relationship between urban heat Island and land use/cover changes. *Remote Sensing of Environment, 104*(2), 133–146.

Church, R. L. (2002). Geographical information systems and location science. *Computers & Operations Research, 29*(6), 541–562.

Colak, H. E., Memisoglu, T., & Gercek, Y. (2020). Optimal site selection for solar photovoltaic (PV) power plants using GIS and AHP: A case study of Malatya Province, Turkey. *Renewable Energy, 149*, 565–576.

Cust, J., Singh, A., & Neuhoff, K. (2007). *Rural electrification in India: Economic and institutional aspects of renewables*. Available at SSRN 2760810.

Delivand, M. K., Cammerino, A. R. B., Garofalo, P., & Monteleone, M. (2015). Optimal locations of bioenergy facilities, biomass spatial availability, logistics costs and GHG (greenhouse gas) emissions: A case study on electricity productions in South Italy. *Journal of Cleaner Production, 99*, 129–139. https://doi.org/10.1016/j.jclepro.2015.03.018

Doorga, J. R., Rughooputh, S. D., & Boojhawon, R. (2019). Multi-criteria GIS-based modelling technique for identifying potential solar farm sites: A case study in Mauritius. *Renewable Energy, 133*, 1201–1219.

Effat, H. A. (2013). Selection of potential sites for solar energy farms in Ismailia Governorate, Egypt using SRTM and multicriteria analysis. *International Journal of Advanced Remote Sensing and GIS, 2*(1), 205–220.

Ennaji, W., Barakat, A., El Baghdadi, M., Oumenskou, H., Aadraoui, M., Karroum, L. A., & Hilali, A. (2018). GIS-based multi-criteria land suitability analysis for sustainable agriculture in the northeast area of Tadla plain (Morocco). *Journal of Earth System Science, 127*(6), 1–14.

Farrell, D. J., Sodabanlu, H., Wang, Y., Sugiyama, M., & Okada, Y. (2015). A hot-electron thermophotonic solar cell demonstrated by thermal up-conversion of sub-bandgap photons. *Nature Communications, 6*(1), 1–9.

Georgiou, A., & Skarlatos, D. (2016). Optimal site selection for sitting a solar park using multi-criteria decision analysis and geographical information systems. *Geoscientific Instrumentation, Methods and Data Systems, 5*(2), 321–332.

Haldar, A. (2015). Energy resource development in the Sundarban region of West Bengal: Perspective of climate hazards and vulnerability. *TTPP, 71*.

Hao, K., Ialnazov, D., & Yamashiki, Y. (2021). GIS analysis of solar PV locations and disaster risk areas in Japan. *Frontiers in Sustainability, 2*, 815986. https://doi.org/10.3389/frsus

Huang, J. P., Poh, K. L., & Ang, B. W. (1995). Decision analysis in energy and environmental modeling. *Energy, 20*(9), 843–855. https://doi.org/10.1016/0360-5442(95)00036-G

Izadikhah, M., & Saen, R. F. (2016). A new preference voting method for sustainable location planning using geographic information system and data envelopment analysis. *Journal of Cleaner Production, 137*, 1347–1367. https://doi.org/10.1016/j.jclepro.2016.08.021

Janke, J. R. (2010). Multicriteria GIS modeling of wind and solar farms in Colorado. *Renewable Energy, 35*(10), 2228–2234. https://doi.org/10.1016/j.renene.2010.03.014

Kasim, N. K., Hussain, H. H., & Abed, A. N. (2019). Performance analysis of grid-connected CIGS PV solar system and comparison with PVsyst simulation program. *International Journal of Smart Grid, 3*, 172–179.

Kengpol, A., Rontlaong, P., & Tuominen, M. (2012). Design of a decision support system for site selection using fuzzy AHP: A case study of solar power plant in north eastern parts of Thailand. In *2012 proceedings of PICMET'12: Technology management for emerging technologies* (pp. 734–743). IEEE.

Khan, J., & Arsalan, M. H. (2016). Solar power technologies for sustainable electricity generation–A review. *Renewable and Sustainable Energy Reviews, 55*, 414–425.

Khan, G., & Rathi, S. (2014). Optimal site selection for solar PV power plant in an Indian state using geographical information system (GIS). *International Journal of Emerging Engineering Research and Technology, 2*(7), 260–266.

Khashoggi, B. F., & Murad, A. (2020). Issues of healthcare planning and GIS: A review. *ISPRS International Journal of Geo-Information, 9*(6), 352.

Koc, A., Turk, S., & Şahin, G. (2019). Multi-criteria of wind-solar site selection problem using a GIS-AHP-based approach with an application in Igdir Province/Turkey. *Environmental Science and Pollution Research, 26*(31), 32298–32310. https://doi.org/10.1007/s11356-019-06260-1

Lee, A. H., Kang, H. Y., Lin, C. Y., & Shen, K. C. (2015). An integrated decision-making model for the location of a PV solar plant. *Sustainability, 7*(10), 13522–13541. https://doi.org/10.3390/su71013522

Liu, J., Li, Y., Xiao, B., & Jiao, J. (2021). Coupling fuzzy multi-criteria decision-making and clustering algorithm for MSW landfill site selection (case study: Lanzhou, China). *ISPRS International Journal of Geo-Information, 10*(6), 403.

Løken, E. (2007). Use of multicriteria decision analysis methods for energy planning problems. *Renewable and Sustainable Energy Reviews, 11*(7), 1584–1595. https://doi.org/10.1016/j.rser.2005.11.005

Lü, G., Batty, M., Strobl, J., Lin, H., Zhu, A. X., & Chen, M. (2019). Reflections and speculations on the progress in Geographic Information Systems (GIS): A geographic perspective. *International Journal of Geographical Information Science, 33*(2), 346–367.

Marinakis, V., Doukas, H., Xidonas, P., & Zopounidis, C. (2017). Multicriteria decision support in local energy planning: An evaluation of alternative scenarios for the Sustainable Energy Action Plan. *Omega, 69*, 1–16.

Massimo, A., Dell'Isola, M., Frattolillo, A., & Ficco, G. (2014). Development of a geographical information system (GIS) for the integration of solar energy in the energy planning of a wide area. *Sustainability, 6*(9), 5730–5744. https://doi.org/10.3390/su6095730

Masson, G., Latour, M., Rekinger, M., Theologitis, I. T., & Papoutsi, M. (2013). *Global market outlook for photovoltaics 2013–2017* (pp. 12–32). European Photovoltaic Industry Association.

Mentis, D., Welsch, M., Nerini, F. F., Broad, O., Howells, M., Bazilian, M., & Rogner, H. (2015). A GIS-based approach for electrification planning—A case study on Nigeria. *Energy for Sustainable Development, 29*, 142–150. https://doi.org/10.1016/j.esd.2015.09.007

Merrouni, A. A., Mezrhab, A., & Mezrhab, A. (2016). PV sites suitability analysis in the eastern region of Morocco. *Sustainable Energy Technologies and Assessments, 18*, 6–15. https://doi.org/10.1016/j.seta.2016.09.006

Moharil, R. M., & Kulkarni, P. S. (2010). Reliability analysis of solar photovoltaic system using hourly mean solar radiation data. *Solar Energy, 84*(4), 691–702.

Mohit, M. A., & Ali, M. M. (2006). Integrating GIS and AHP for land suitability analysis for urban development in a secondary city of Bangladesh. *Jurnal alam Bina, 8*(1), 1–20.

Mohsin, M., Ali, S. A., Shamim, S. K., & Ahmad, A. (2022). A GIS-based novel approach for suitable sanitary landfill site selection using integrated fuzzy analytic hierarchy process and machine learning algorithms. *Environmental Science and Pollution Research, 29*(21), 31511–31540.

Nebey, A. H., Taye, B. Z., & Workineh, T. G. (2020). Site suitability analysis of solar PV power generation in South Gondar, Amhara Region. *Journal of Energy, 2020*.

Noorollahi, E., Fadai, D., Akbarpour Shirazi, M., & Ghodsipour, S. H. (2016). Land suitability analysis for solar farms exploitation using GIS and fuzzy analytic hierarchy process (FAHP)—A case study of Iran. *Energies, 9*(8), 643.

Pohekar, S. D., & Ramachandran, M. (2004). Application of multi-criteria decision making to sustainable energy planning—A review. *Renewable and Sustainable Energy Reviews, 8*(4), 365–381. https://doi.org/10.1016/j.rser.2003.12.007

Rahman, M. A., Maulud, K. A., Bahri, M. S., Hussain, M. S., Oon, A. R., Suhatdi, S., et al. (2020). Development of gis database for infrastructure management: Power distribution network system. In *IOP conference series: Earth and environmental science (Vol. 540, no. 1, p. 012067)*. IOP Publishing.

Rezaei, J. (2015). Best-worst multi-criteria decision-making method. *Omega, 53*, 49–57.

Sabo, M. L., Mariun, N., Hizam, H., Radzi, M. A. M., & Zakaria, A. (2017). Spatial matching of large-scale grid-connected photovoltaic power generation with utility demand in Peninsular Malaysia. *Applied Energy, 191*, 663–688. https://doi.org/10.1016/j.apenergy.2017.01.087

Sánchez-Lozano, J. M., Teruel-Solano, J., Soto-Elvira, P. L., & García-Cascales, M. S. (2013). Geographical information systems (GIS) and multi-criteria decision making (MCDM) methods for the evaluation of solar farms locations: Case study in South-Eastern Spain. *Renewable and Sustainable Energy Reviews, 24*, 544–556.

Sánchez-Lozano, J. M., Antunes, C. H., García-Cascales, M. S., & Dias, L. C. (2014). GIS-based photovoltaic solar farms site selection using ELECTRE-TRI: Evaluating the case for Torre Pacheco, Murcia, Southeast of Spain. *Renewable Energy, 66*, 478–494. https://doi.org/10.1016/j.renene.2013.12.038

Sánchez-Lozano, J. M., García-Cascales, M. S., & Lamata, M. T. (2015). Evaluation of suitable locations for the installation of solar thermoelectric power plants. *Computers & Industrial Engineering, 87*, 343–355. https://doi.org/10.1016/j.cie.2015.05.028

Sánchez-Lozano, J. M., García-Cascales, M. S., & Lamata, M. T. (2016). Comparative TOPSIS-ELECTRE TRI methods for optimal sites for photovoltaic solar farms. Case study in Spain. *Journal of Cleaner Production, 127*, 387–398. https://doi.org/10.1016/j.jclepro.2016.04.005

Şener, Ş., Şener, E., Nas, B., & Karagüzel, R. (2010). Combining AHP with GIS for landfill site selection: A case study in the Lake Beyşehir catchment area (Konya, Turkey). *Waste Management, 30*(11), 2037–2046. https://doi.org/10.1016/j.wasman.2010.05.024

Shao, M., Han, Z., Sun, J., Xiao, C., Zhang, S., & Zhao, Y. (2020). A review of multi-criteria decision making applications for renewable energy site selection. *Renewable Energy, 157*, 377–403.

Shukla, A. K., Sudhakar, K., Baredar, P., & Mamat, R. (2018). Solar PV and BIPV system: Barrier, challenges and policy recommendation in India. *Renewable and Sustainable Energy Reviews, 82*, 3314–3322.

Silva, J. S., da Silva, R. M., & Santos, C. A. G. (2018). Spatiotemporal impact of land use/land cover changes on urban heat islands: A case study of Paço do Lumiar, Brazil. *Building and Environment, 136*, 279–292.

Silva, K., Janta, P., & Chollacoop, N. (2021). Points of consideration on climate adaptation of solar power plants in Thailand: How climate change affects site selection, construction and operation. *Energies, 15*(1), 171.

Sk, M. M., Ali, S. A., & Ahmad, A. (2020). Optimal sanitary landfill site selection for solid waste disposal in Durgapur city using geographic information system and multi-criteria evaluation technique. *KN-Journal of Cartography and Geographic Information, 70*(4), 163–180.

Türk, S., Koç, A., & Şahin, G. (2021). Multi-criteria of PV solar site selection problem using GIS-intuitionistic fuzzy based approach in Erzurum province/Turkey. *Scientific Reports, 11*(1), 1–23.

Uyan, M. (2013). GIS-based solar farms site selection using analytic hierarchy process (AHP) in Karapinar region, Konya/Turkey. *Renewable and Sustainable Energy Reviews, 28*, 11–17. https://doi.org/10.1016/j.rser.2013.07.042

Uyan, M. (2017). Optimal site selection for solar power plants using multi-criteria evaluation: A case study from the Ayranci region in Karaman, Turkey. *Clean Technologies and Environmental Policy, 19*(9), 2231–2244.

Wang, Q., M'Ikiugu, M. M., & Kinoshita, I. (2014). A GIS-based approach in support of spatial planning for renewable energy: A case study of Fukushima, Japan. *Sustainability, 6*(4), 2087–2117. https://doi.org/10.3390/su6042087

Watson, J. J., & Hudson, M. D. (2015). Regional Scale wind farm and solar farm suitability assessment using GIS-assisted multi-criteria evaluation. *Landscape and Urban Planning, 138*, 20–31. https://doi.org/10.1016/j.landurbplan.2015.02.001

Xiao, J., Yao, Z., Qu, J., & Sun, J. (2013). Research on an optimal site selection model for desert photovoltaic power plants based on analytic hierarchy process and geographic information system. *Journal of Renewable and Sustainable Energy, 5*(2), 023132.

Xu, H. (2007). Extraction of urban built-up land features from Landsat imagery using a thematic-oriented index combination technique. *Photogrammetric Engineering & Remote Sensing, 73*(12), 1381–1391.

Zhang, H., Qi, Z. F., Ye, X. Y., Cai, Y. B., Ma, W. C., & Chen, M. N. (2013). Analysis of land use/land cover change, population shift, and their effects on spatiotemporal patterns of urban heat islands in metropolitan Shanghai, China. *Applied Geography, 44*, 121–133.

Zhou, J., Wu, Y., Wu, C., He, F., Zhang, B., & Liu, F. (2020). A geographical information system based multi-criteria decision-making approach for location analysis and evaluation of urban photovoltaic charging station: A case study in Beijing. *Energy Conversion and Management, 205*, 112340.

Zoghi, M., Ehsani, A. H., Sadat, M., Javad Amiri, M., & Karimi, S. (2017). Optimization solar site selection by fuzzy logic model and weighted linear combination method in arid and semi-arid region: A case study Isfahan-Iran. *Renewable and Sustainable Energy Reviews, 68*, 986–996.

Aggregating Machine Learning and GIS for Flood Susceptibility Modeling on Agricultural Land in Banda District, Uttar Pradesh, India

Mujahid Husain, Zainab Khan, Tahreem Fatima, Syed Kausar Shamim, Ateeque Ahmad, and Sania Jawaid

1 Introduction

Globally, flooding is responsible for approximately one-third of all losses resulting from natural disasters. Research indicates a significant increase of over 40% in flood events worldwide over the past two decades (Hirabayashi et al., 2013). Between 1995 and 2015, floods affected over 100 million people annually, resulting in estimated damages of around $75 billion USD each year (Alfieri et al., 2016). The triggers of floods are multifaceted, comprising both natural and human-induced factors, including climate change, land use/land cover (LU/LC) alterations, and shifts in precipitation patterns (Kumar et al., 2022). Climate change has notably impacted flood dynamics, influencing their frequency, severity, and scope (Runhaar et al. 2017). The growing incidence of floods is often attributed to changing rainfall patterns, which are frequently linked to increased human activities (Zhang et al., 2018).

Recurring floods have an impact on millions of people's lives, infrastructures, economies, and local ecosystems (Khosravi et al., 2019; Paarcha et al., 2023). Floods are the second deadly disasters impacting agriculture sector, responsible for huge loss of crops and production that cost around USD 21 billion, between 2008 and 2018 in least developing countries and lower middle-income countries—this amounts to 19% of total loss (FAO, 2021). Agriculture is the most prevalent economic activity of India and plays an important role in the ultimate socioeconomic development of the nation (Bassi et al. 2014). India was ranked second among the worst flood affected country in the world after Afghanistan (Agarwal and Narayan, 1991). The Ganga-Yamuna Basin, i.e., world's largest area composed of modern alluvial sedimentation (Bonsor et al., 2017), is often quite susceptible to flooding, from 1986 to 2016; there were 18, 11, and 5 sites that recorded high, very high, and extreme floods, respectively (Nandargi and Aman. 2018). Banda District lying in southern part of Yamuna Basin encompasses the watersheds of Yamuna, Ken and Baghain streams draining in the area (Office of the Registrar General and Census Commissioner, India, 2011). Floods are the most

M. Husain
ARX Smart Minds | ARX Engineering Services L.L.C, Dubai, United Arab Emirates

Z. Khan (✉)
Department of Geomatics Engineering, Indian Institute of Technology, Roorkee, India

T. Fatima · S. K. Shamim · A. Ahmad · S. Jawaid
Department of Geography, Faculty of Sciences, Aligarh Muslim University, Aligarh, India

Z. Khanam et al. (eds.), *Climate Change, Resilience and Ecosystem Services*, Sustainable Development Goals Series, https://doi.org/10.1007/978-3-031-96977-5_17

frequent natural hazard in Banda disrupting its agricultural base and affecting the economy and well-being of people (Rahmati et al., 2019).

Over the past decade, Banda District has experienced frequent and severe flooding events (El Morjani et al., 2016). The Yamuna and Baghain Rivers are particularly vulnerable to flood-related issues, with research focusing on soil erosion and dam siltation within the Yamuna watershed, a flood-prone area (Sharma et al., 2017). Studies have identified the southern part of Banda District as the most susceptible to flooding. While existing flood risk assessments in this region have primarily utilized geographic information system (GIS), remote sensing, and multi-criteria analysis methods, spatial analysis techniques have proven effective in mapping flood susceptibility. Researchers have extensively applied these methods to identify, assess, and monitor flood hazards incuding (Vangelis et al., 2022). The combination of hydrological models like Soil and Water Assessment Tool (SWAT) and HYDROTEL (Jutras et al., 2009) with remote sensing and GIS has been employed to collect data and analyze spatial parameters related to flooding, offering a comprehensive approach to understanding flood risk.

Flood susceptibility mapping is crucial for flood management programs (Moradi et al., 2020a). Recently, artificial intelligence (AI)-assisted machine learning methods have demonstrated superior performance over traditional mapping methods in disaster management and mitigation (Tehrany et al., 2015). The mapping of flood susceptible areas of Tajan watershed in Mazandaran Province, Sari, was effectively done using two Machine Learning (ML) methods, i.e., Random Forest (RF) model and the Generalized Linear Model (GLM) Bayesian model (Moradi et al. 2020b). Likewise, research has demonstrated the efficacy of the gradient Boosted Regression Tree (BRT) method in modeling and examining the complex behaviors of tropical cyclones in the Western North Pacific Ocean, including re-curvature, landfall, and intensity changes (Zhang et al., 2013). Mandal et al. (2022) in his study used BRT model and found the highest area the under curve (AUC) value while predicting the landslide susceptibility in eastern Sikkim Himalayan region. Eslaminezhad et al. (2022) proposed and assessed the ability of BRT model for the evaluation of groundwater potential and found that BRT performed better than Linear Model Theory (LMT) and RF algorithms. The Boosted Regression Tree (BRT) approach combines the strengths of decision tree algorithms and boosting techniques, similar to Random Forest models. Through an iterative process, BRTs train multiple decision trees in succession, refining the model's accuracy with each iteration.

Although machine learning algorithms have proven effective in predicting natural disasters, their use in forecasting and mapping flood susceptibility in Banda is still minimal or absent. This study evaluates the effectiveness of machine learning algorithms and explores their potential to improve flood susceptibility mapping, strategic planning, and flood management in the district. Here, the agricultural land that is more vulnerable to flood is identified using "Boosted Regression Tree" (BRT) models and real-time flood has been used to validate the applicability of this model.

2 Study Area

Located in the Indo-Gangetic Plain, Banda District in Uttar Pradesh spans an area of approximately 4460 square kilometers. Geographically, it is situated between 25°00′00″ and 25°59′00″ north latitude and 80°06′00″ and 81°00′00″ east longitude (Fig. 1). The district's terrain exhibits a varied elevation, ranging from a low of 69 m to a high of 360 m above sea level (Fig. 2). The population is estimated to be 1,799,410 persons much as of census 2011. Banda District is located within the Bundelkhand plateau, characterized by older alluvial deposits and erosional terraces formed by the floodplains of the Yamuna, Ken, and Baghain Rivers which are parts of Yamuna River system (Kumar et al. 2019). Length of the Yamuna River lying in Banda is 130 km. The Yamuna River divides Banda District from Fatehpur in the north and flows west to east across the entire district. The Ken

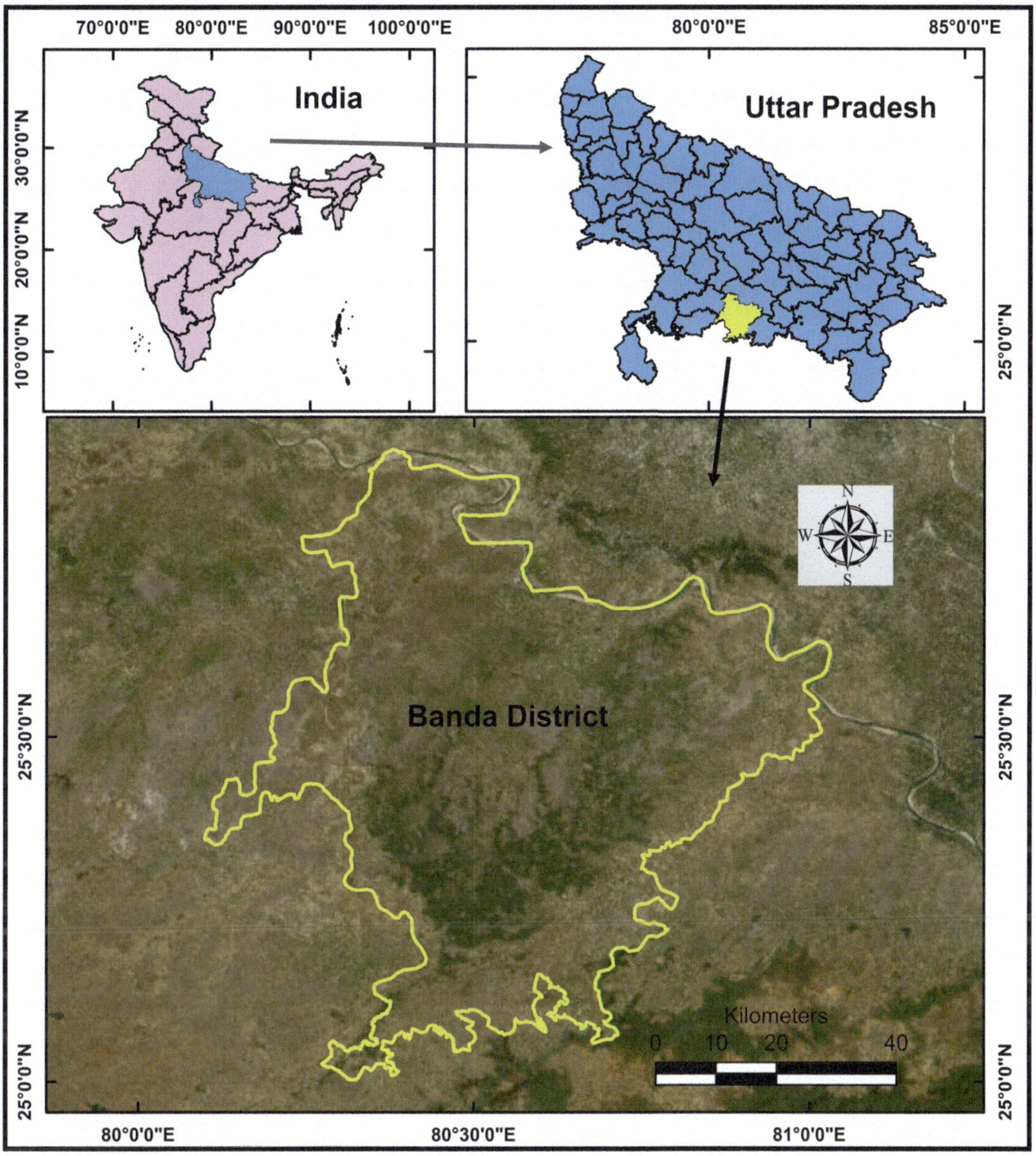

Fig. 1 Study area

River joins the Yamuna at Chilla, while the Baghain River separates Banda from Chitrakoot in the southeast. There are several other streams including Banganga, Paisuni, Chandrawal, Ranj, Karauli, Garhara, and Turi (District Statistical Handbook). Drainage density is high in the southern part, while it is moderate in the central and northern parts of the district. Out of the total, more than 98% land of the district is rural and hence agriculture is the main source of economy employing around 75% of the total population. The area employs both surface and groundwater for the purpose of irrigation. The net irrigated area is 153,804 hectares, and the net sown area is 336,000 hectares, indicating that 45.77% of the area is irrigated using ground and surface water, while the remainder relies on rainfall. The region receives approximately 80% of its annual rainfall from the southwest monsoon (2013). However, the likelihood of flooding has escalated over time

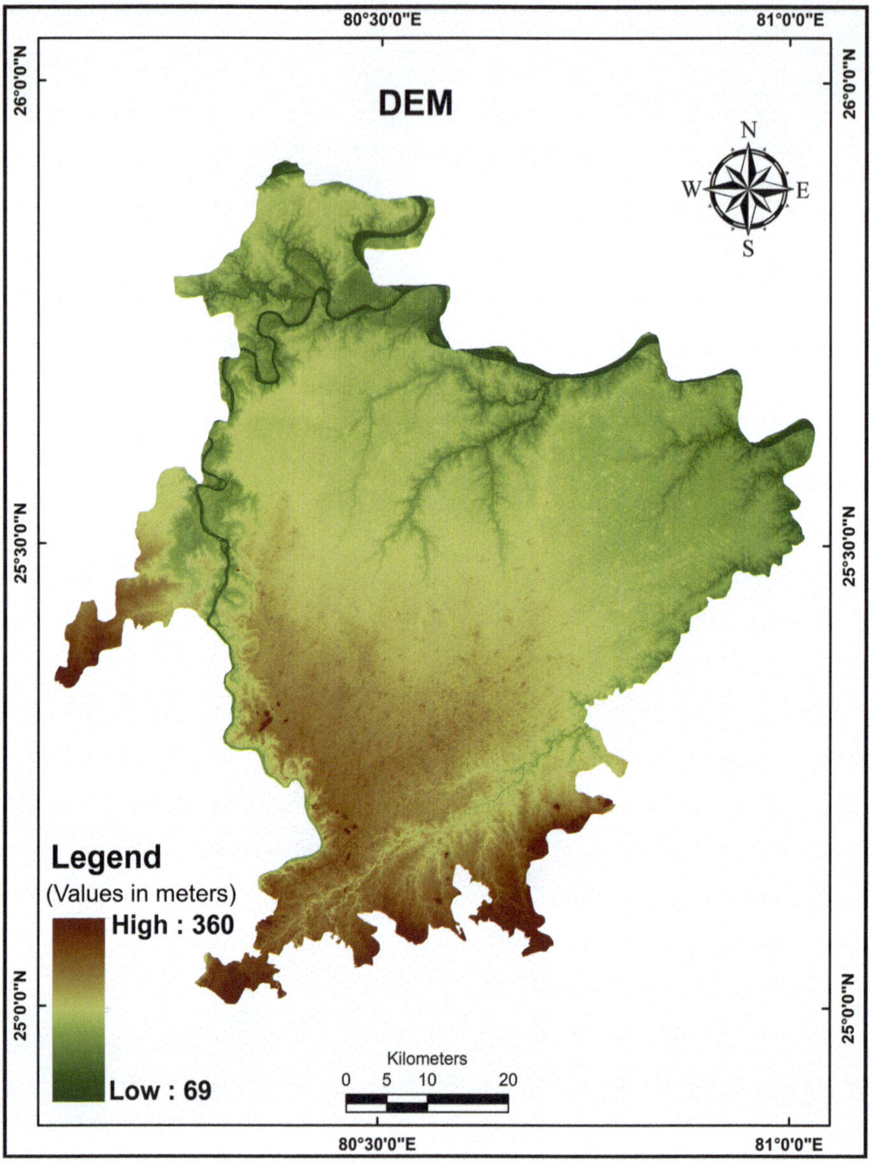

Fig. 2 DEM

due to several contributing factors, including unpredictable and irregular monsoonal rainfall patterns, excessive siltation in rivers, and the area's undulating terrain. The district's landscape is characterized by a mix of rocky outcrops and low-lying areas, which are prone to inundation during the rainy season.

3 Formulation of Database

Variable selection plays a critical role in assessing susceptibility and modeling flood events. There is ample information on factors that can influence flood conditions in a given area and increase its vulnerability to flooding. Scholars

Table 1 Datasets and ancillary data utilized for this study

Data	Data type	Year	Source
Flood	Raster	1985–2003	SEDAC, NASA
Inventory Data	Vector	NA	Dart Mouth Flood Observatory
ASTER DEM	Raster	2018	USGS
Geology	Vector	NA	Bhukosh
Toposheet	Raster	2012	Survey of India
LULC	Raster	2017–2021	NRSC

have underscored the importance of socioeconomic and demographic factors in assessing flood impacts (Vousdoukas et al., 2018; Dawson et al., 2011). Meanwhile, other researchers have explored geological and physical factors (O'Connor et al., 2002; Kourgialas & Karatzas, 2011). This research selects eight key variables for flood susceptibility mapping in Banda District: Digital Elevation Model (DEM), river proximity, curvature, gradient slope, geology, Topographic Wetness Index (TWI), Stream Power Index (SPI), and Slope Length Factor (SLF), to provide a comprehensive understanding of flood risk. These conditioning variables are modeled against the flood inventory spatial points (Table 1).

4 Processing of Database for Flood Susceptibility Mapping

It is well recognized that numerous hydrological elements affect how susceptible a region is to flood. Factors such as slope, stream development, land use, and geological structure are well-known, but their specific relationships with flood susceptibility have not been thoroughly explored. Consequently, eight input criteria were employed, featuring distance from the river as a crucial factor, widely regarded as a significant predictor of flood susceptibility. These variables were subsequently integrated with the boosted-tree and support vector machine models (Lee et al., 2017; Khan et al., 2024a), to enhance the accuracy of flood vulnerability assessments.

To accurately assess flood susceptibility, various flood-related variables are employed, each contributing to a comprehensive understanding of how different factors influence flood risks. Digital Elevation Model (DEM) data (Fig. 2), commonly used in flood modeling, provides valuable information on the topography, aiding in identifying areas prone to water accumulation (Rahmati et al., 2016). Distance from rivers (Fig. 3) is another key variable, as regions closer to rivers are more susceptible to flooding, especially during periods of high precipitation or when riverbanks overflow (Marconi et al., 2016). Similarly, the curvature of the land (Fig. 4) whether concave or convex helps in determining water accumulation areas and drainage patterns (Youssef et al., 2016).

Slope (Fig. 5) is a critical variable as it directly affects the velocity of water runoff; areas with steep slopes tend to experience faster water flow, increasing the likelihood of floods (Tehrany et al., 2013). Geological characteristics (Fig. 6) are essential to consider, as different rock and soil types affect water infiltration rates and overall landscape stability. Areas with porous rocks or well-drained soils may be less prone to flooding than regions with impermeable geology (Lawal et al., 2014). The Topographic Wetness Index (TWI) (Fig. 7) is widely used in hydrological models, as it measures the spatial distribution of wetness, indicating areas more likely to retain water and flood. The Stream Power Index (SPI) (Fig. 8) highlights areas with high water flow energy, which could exacerbate flood risks (Khosravi et al., 2016). Finally, the Slope Length Factor (SLF) (Fig. 9) is another vital variable, as it accounts for the length and steepness of a slope, influencing soil erosion and water runoff patterns. This factor is particularly important in flood-prone areas where runoff can lead to soil erosion, further complicating drainage (Borrelli et al., 2014).

Conclusively, these variables, (Figs. 2, 3, 4, 5, 6, 7, 8 and 9), provide a multi-faceted approach to flood risk assessment by considering both topographical and geological characteristics, ensuring that models incorporate the dynamic interactions between land, water, and the agrarian land.

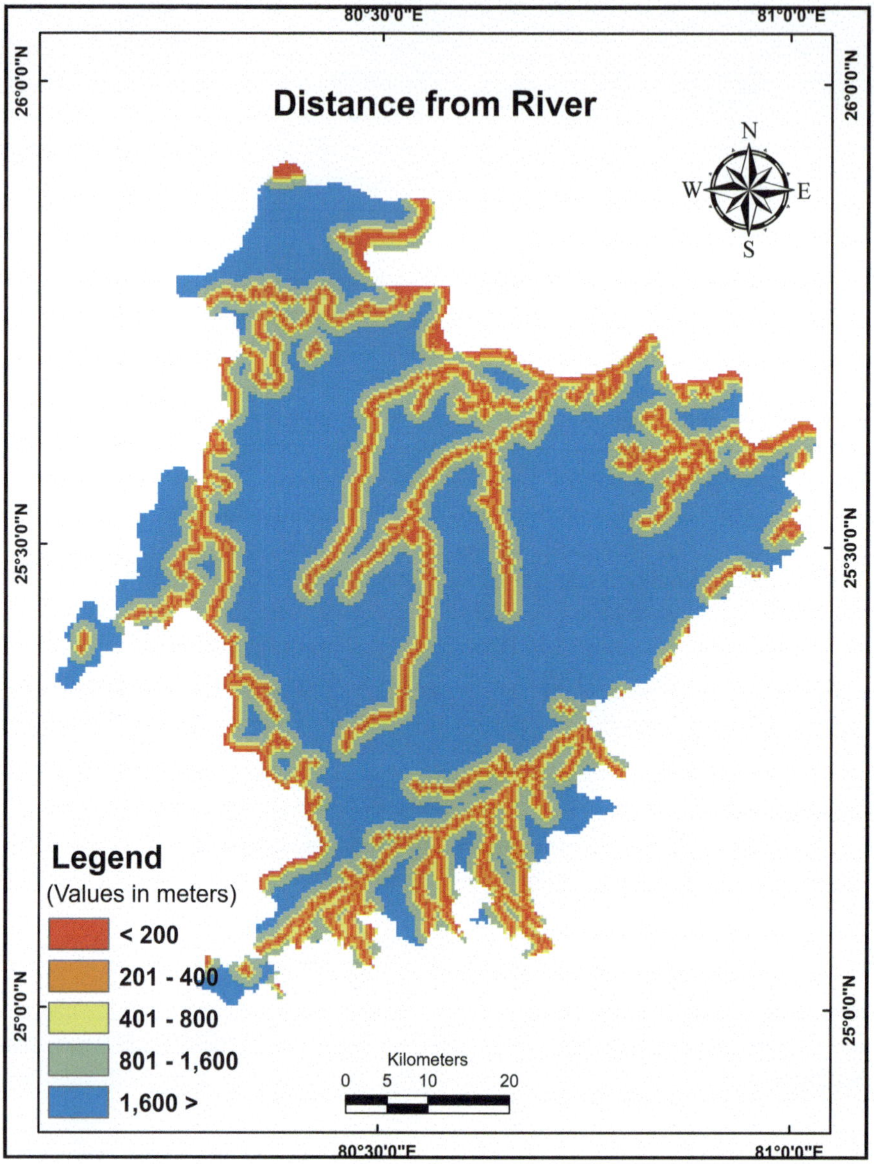

Fig. 3 Distance from river

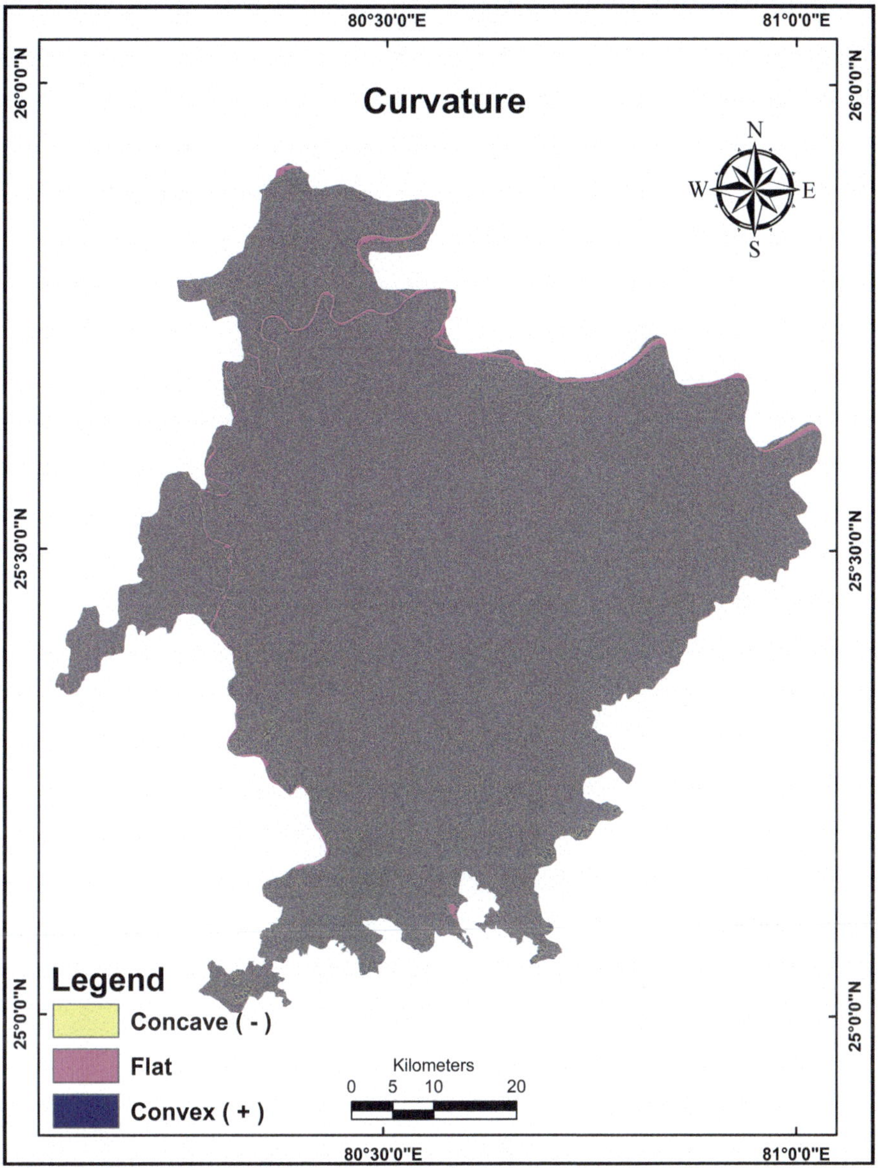

Fig. 4 Curvature

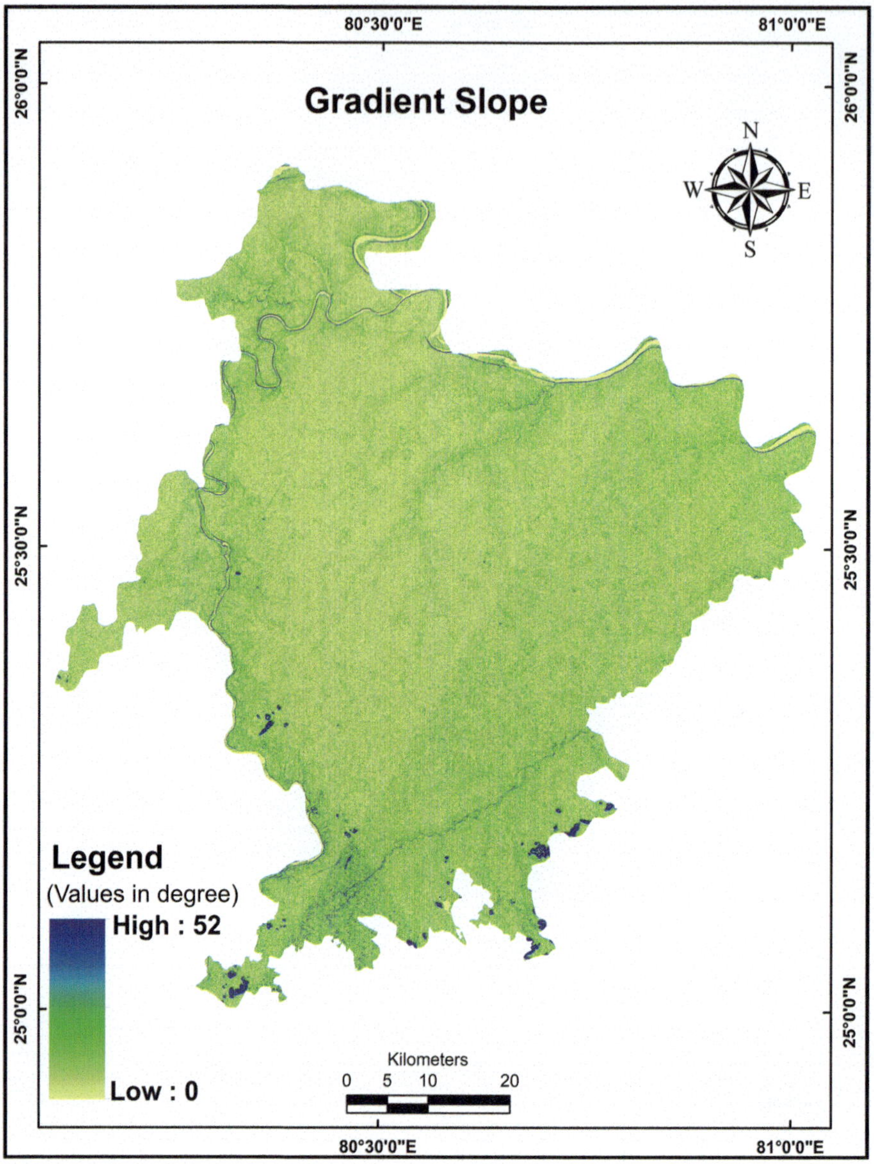
Gradient Slope
80°30'0"E
81°0'0"E
26°0'0"N
25°30'0"N
25°0'0"N
N
W
E
S
Legend
(Values in degree)
High : 52
Low : 0
Kilometers
0 5 10 20

Fig. 5 Slope

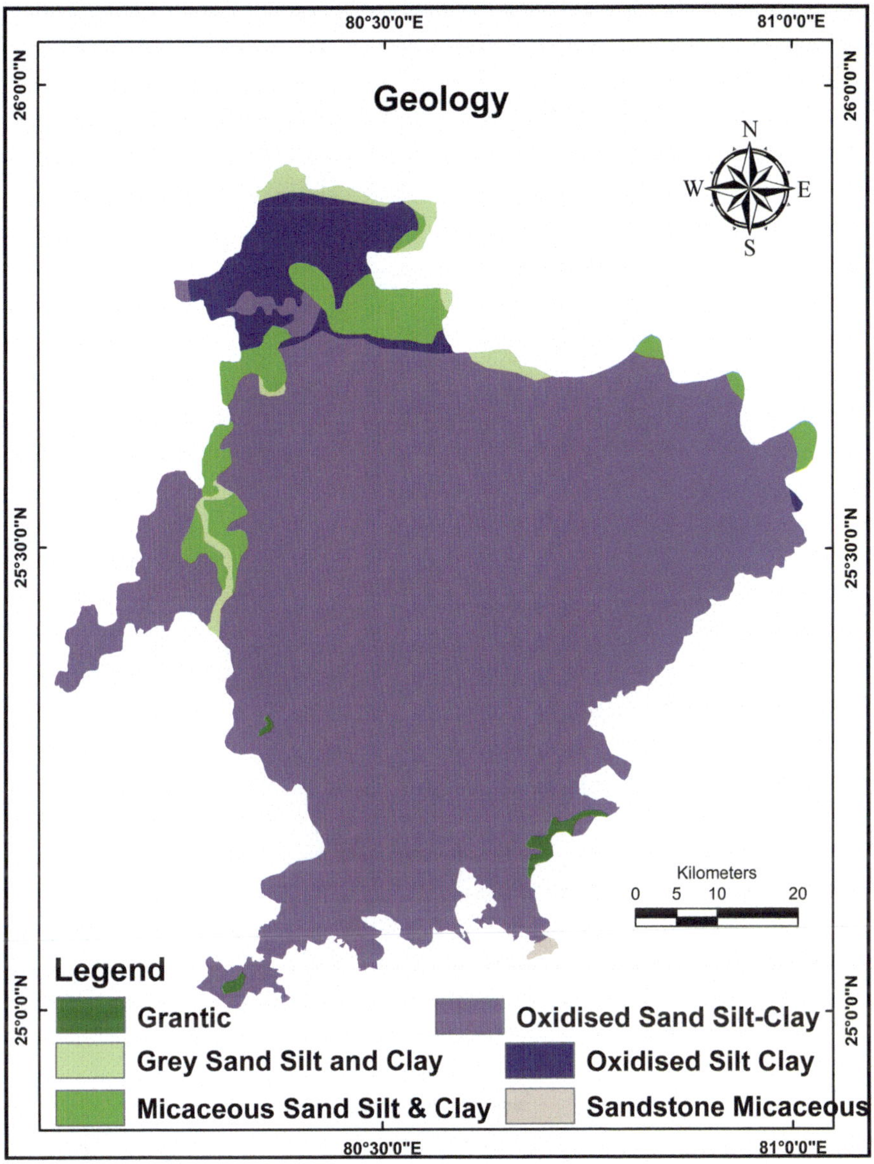

Fig. 6 Geology

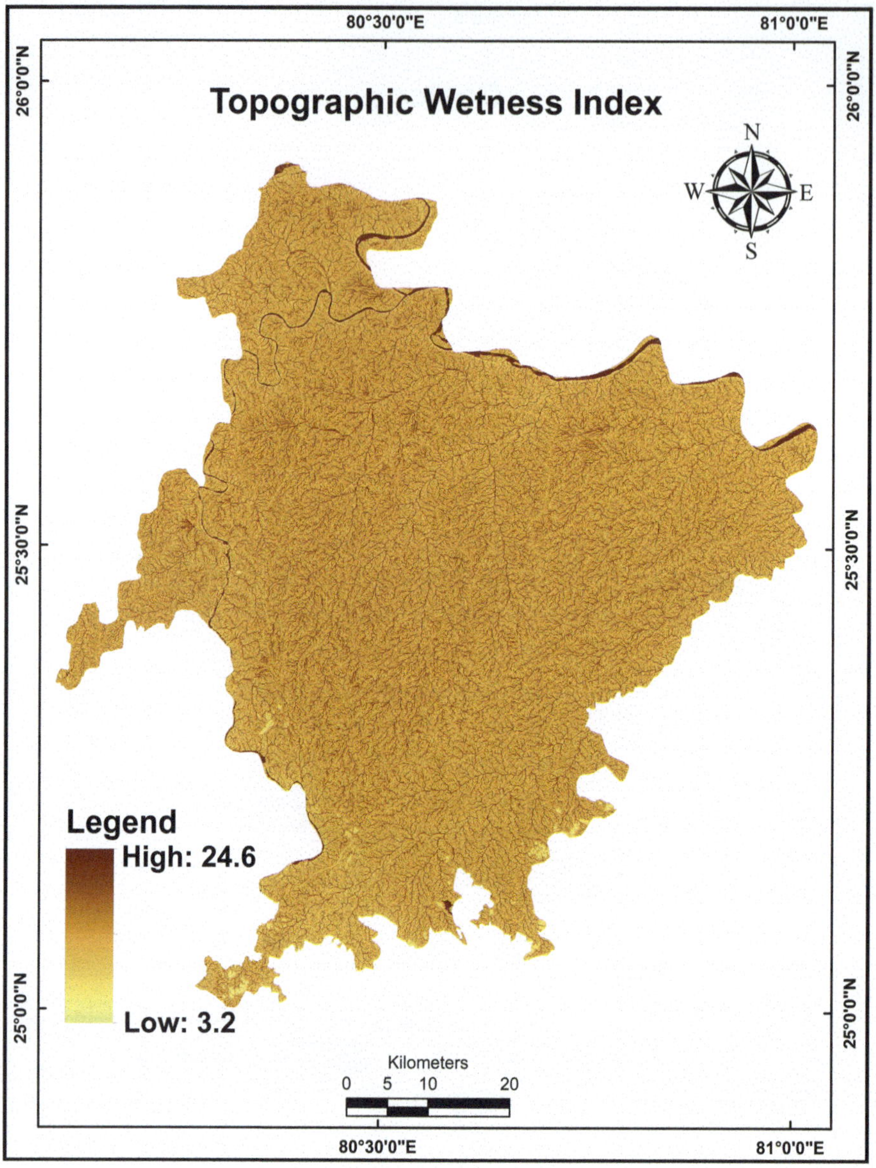
Topographic Wetness Index
80°30'0"E
81°0'0"E
26°0'0"N
25°30'0"N
25°0'0"N
N
W
E
S
Legend
High: 24.6
Low: 3.2
Kilometers
0 5 10 20

Fig. 7 TWI

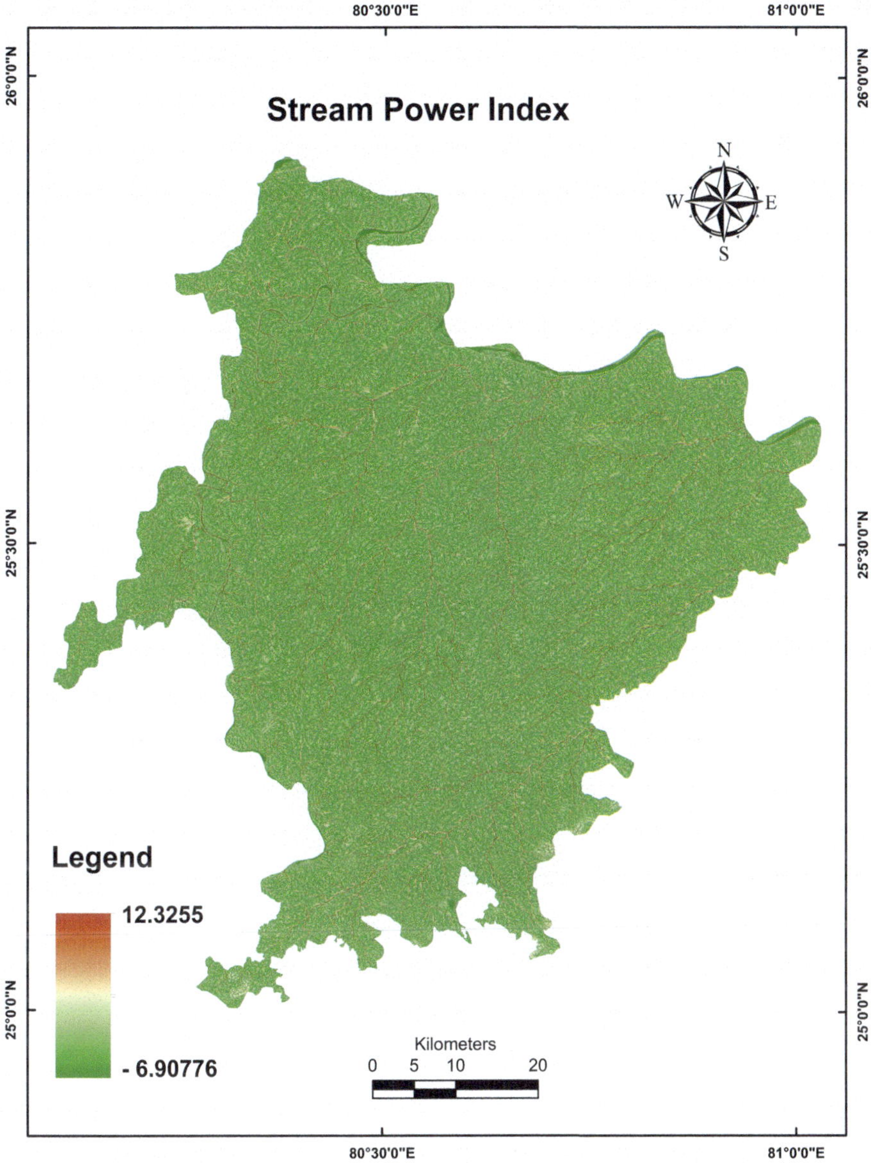
80°30'0"E
81°0'0"E
26°0'0"N
25°30'0"N
25°0'0"N
Stream Power Index
N
W
E
S
Legend
12.3255
- 6.90776
Kilometers
0 5 10 20

Fig. 8 SPI

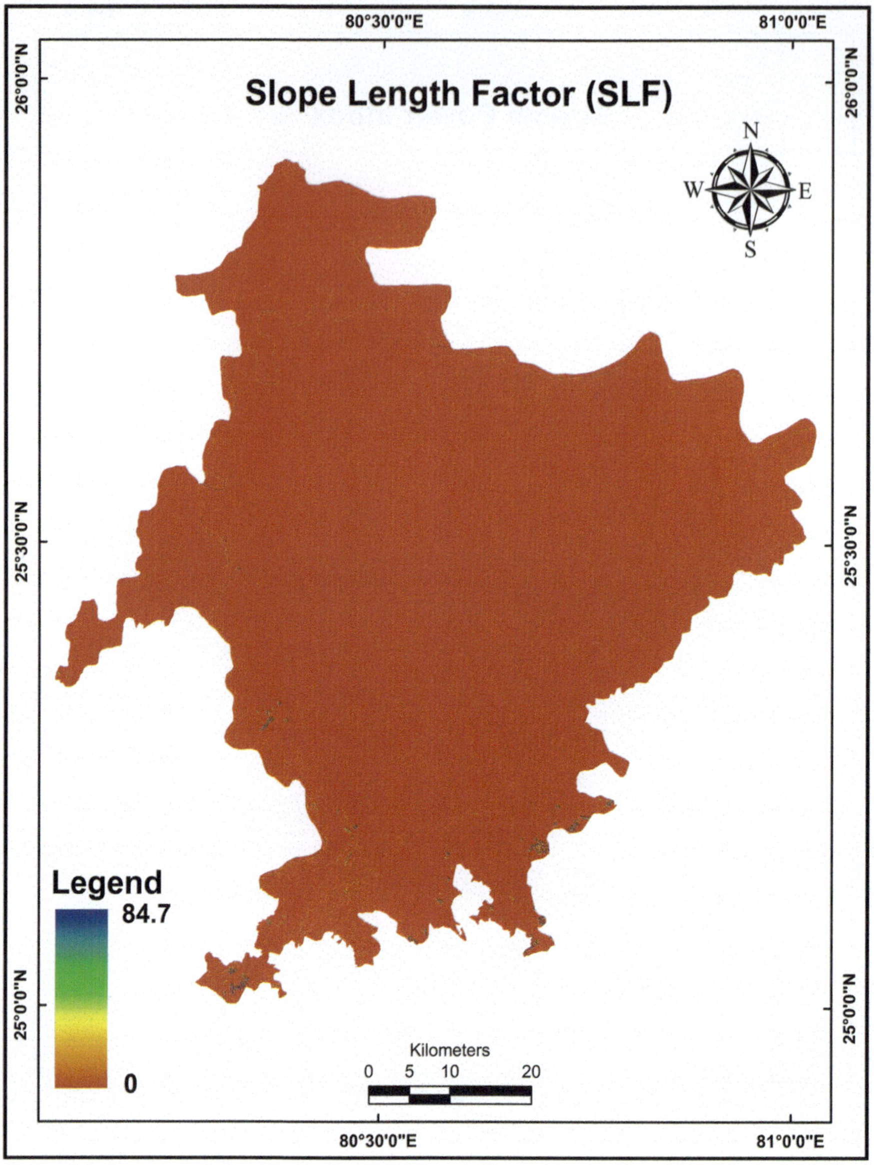

Fig. 9 SLF

5 Methodology

In the present analysis, open-access environment such as RStudio and QGIS was used for the susceptibility mapping of the flood events. The modeling process of the susceptibility raster utilized R packages such as dismo and tidyr. Flood occurrence data includes the floods caused by monsoonal rain as well as heavy rain other than monsoonal rain. Some pseudo no flood points are also plotted on the DEM raster. These points then were merged with real flood events in order to form a single feature class flood and no flood attributes. Before applying the BRT model, the conditioning variables are normalized so as to eliminate the influence of the over influence of one raster. To detect multicollinearity among the conditioning rasters, the vifstep and vifcor functions in R were utilized, with a specified threshold of 0.5. This step ensured that the variables used in the analysis were sufficiently independent and did not exhibit excessive correlation. No conditioning raster was found to be having collinearity beyond the given threshold.

6 Selection of Model

BRT algorithm is applied for the susceptibility mapping of flood—a model that is a relatively recent addition to the predictive modeling field has gained prominence over the last decade. It is adept at predicting spatial features such as snow cover, forest fires, landslides, and flood events. Modern regression trees, particularly Boosted Regression Trees (BRT), have enhanced predictive capabilities. Pérez Navarro et al. (2019) found BRT outperformed GAM, MaxEnt, and Mahalanobis in predicting extreme climatic events. BRT has become a reliable tool in flood prediction research, successfully applied in Indo-Gangetic flood modeling (Islam et al., 2022). The BRT model's probability of occurrence is represented by Eq. 1 (Elith et al., 2008):

$$P(y = 1z) = f(z), \quad (1)$$

where y is the logit and z are the covariate.

7 Accuracy Assessment

ROC The receiver operating characteristic (ROC) curve is a visualization tool used to evaluate the effectiveness of classification models. It plots the True Positive Rate (TPR), representing the proportion of correctly identified positive instances, against the False Positive Rate (FPR). The TPR is calculated using the following equation:

The True Positive Rate (TPR) is calculated as:

$$\text{TPR} = \text{TP} / \text{TP} + \text{FN}$$

where TP stands for true positives and FN denotes false negatives. The False Positive Rate (FPR) is calculated as:

$$\text{FPR} = \text{FP} / \text{FP} + \text{TN}$$

where FP represents false positives and TN stands for true negatives.

AUC The area under the ROC curve (AUC) calculates the cumulative area beneath the ROC curve, covering the entire range from the lower-left corner (0,0) to the upper-right corner (1,1), providing a comprehensive measure of the model's performance. A model with an AUC close to 1 indicates excellent separability, while an AUC close to 0 suggests poor separability (Khan et al., 2024b).

Correlation Correlation refers to the degree to which two variables are related. It measures the strength and direction of their association. Correlational studies can yield three outcomes: a positive correlation, a negative correlation, or no correlation. The formula for the correlation is

$$r = \left(\Sigma(xy)\Sigma(\Sigma x)(\Sigma y)\right) / \sqrt{\left(\left(\frac{n\Sigma x^2}{\Sigma(\Sigma x)^2}\right)\left(\frac{n\Sigma y^2}{\Sigma(\Sigma y)^2}\right)\right)}$$

where,
r = Pearson correlation coefficient,
n = number of stock pairs,
$\sum xy$ = sum of the products of paired stock scores
$\sum x$ and $\sum y$ = sums of the x and y scores, respectively
$\sum x^2$ and $\sum y^2$ = sums of the squared x and y scores, respectively (Fig. 10).

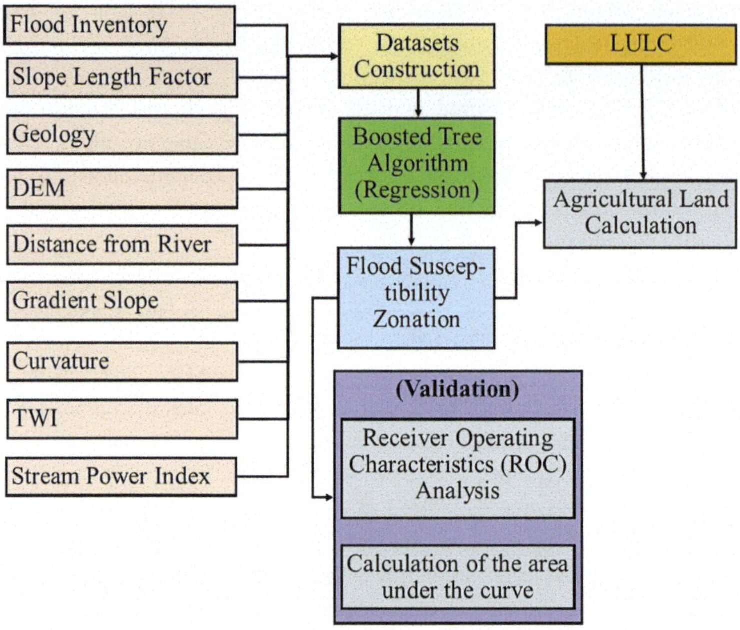

Fig. 10 Flowchart of study

8 Results

8.1 Flood Susceptibility Mapping

Table 2 Data pertaining accuracy parameters

Model	AUC	COR	Deviance
BRT	0.875	0.65	1.06

Machine learning has long been recognized as a powerful tool for evaluating the effects of flood events on the social, cultural, and economic dimensions of affected areas. In this analysis, the Boosted Regression Tree (BRT) algorithm has been employed to estimate the likelihood of flooding within the spatial confines of Banda District. This approach incorporates various factors that influence flood events, along with point data indicating both flood occurrences and areas with no flooding, to create a flood susceptibility model for Banda District.

Although machine learning models offer several benefits, their accuracy can be a subject of debate. To evaluate the performance of the BRT model, a range of metrics are employed, including the receiver operating characteristic (ROC) curve, the area under the curve (AUC), the Coefficient of Correlation (COR), the True Skill Statistic (TSS), and deviance. The ROC curve exhibits a strong performance, with an AUC value of 0.875, indicating a high degree of accuracy. The correlation coefficient (COR) of 0.65 suggests a moderate level of agreement between predicted and actual values. Furthermore, the model's deviance of 1.06 provides additional information about its goodness of fit and predictive capabilities (see Table 2).

The modeled area susceptible to the flood using BRT algorithm is mostly found in the northeastern part of the district. High probability zone of flood susceptibility is in northwestern part of the district is limited to contiguity of the river channel. The southern past remains moderately susceptible as of the Chap. 7, Fig. 1. The density curves appear near normal in terms of distribution of the flood and no flood area data (Fig. 11). The ROC curve represents the healthy accuracy with AUC being 0.875 (Fig. 12).

Figure 11 shows the density distribution of predicted probabilities for flood and no-flood areas. The blue curve represents the areas where flooding is predicted, while the red curve represents the areas where no flooding is predicted. The near-normal distribution of these curves suggests that the model effectively distinguishes between flood-prone and non-flood-prone areas, with a reason-

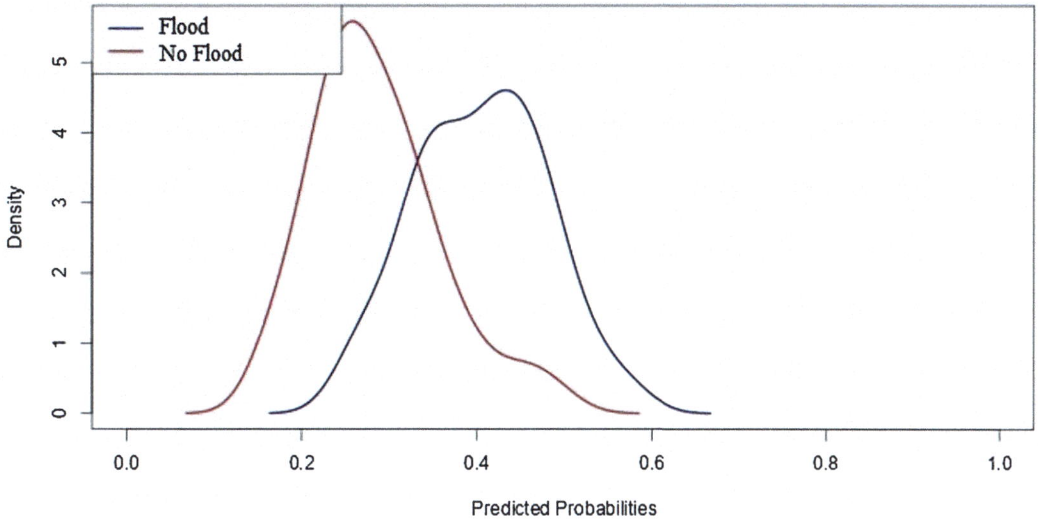

Fig. 11 Flood susceptibility map

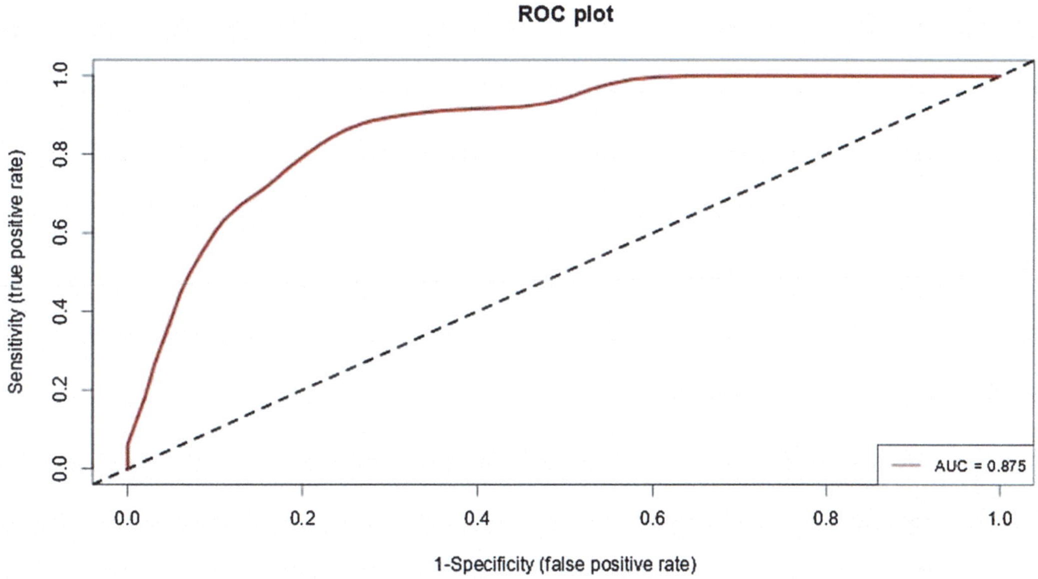

Fig. 12 ROC of the susceptibility model

able overlap that indicates a natural variability in flood susceptibility across the Banda District. This overlap is crucial for understanding the regions that fall into a transition zone, where flood risk might be moderate but not immediately apparent.

Figure 12 demonstrates the ROC curve for the BRT model, with an AUC of 0.875. The ROC curve is a graphical representation of the s ability of the model to discriminate between flood and no-flood events across different thresholds. The closer the curve follows the left-hand border and then the top border of the ROC space, the more accurate the model is. In this case, the proximity of the curve to the upper left corner indicates a high true positive rate or sensitivity and a low false positive rate or 1-specificity, confirming the model's robustness in predicting flood susceptibility. The diagonal line showing a random guess, with ROC curve for BRT

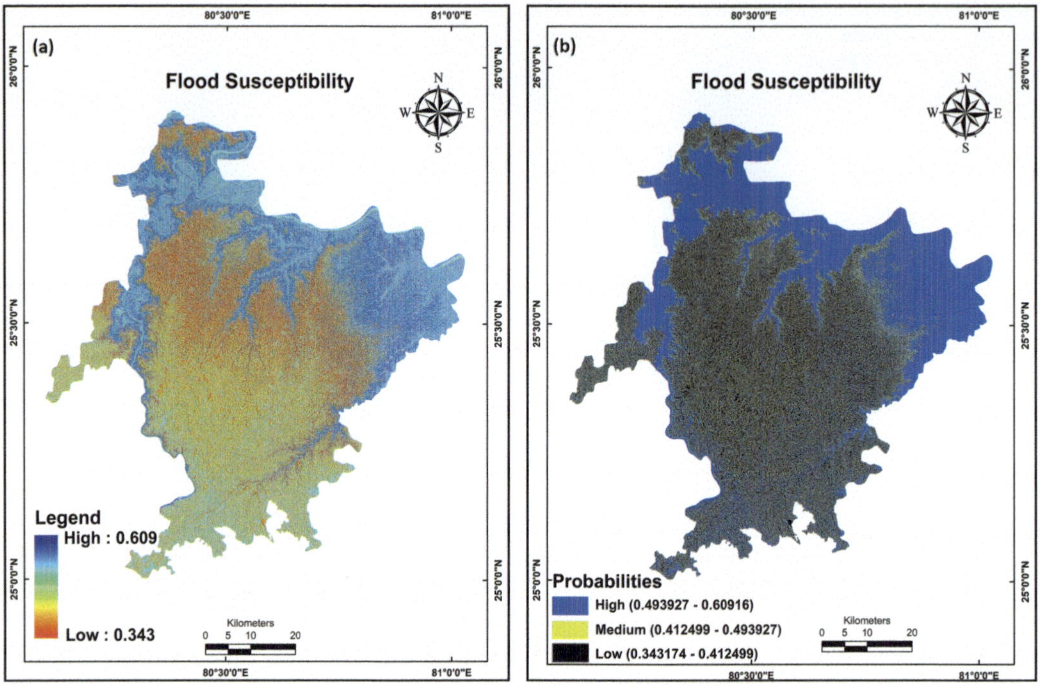

Fig. 13 Flood susceptibility of Banda District (**a**) and zonal assessments (**b**)

well above the aforesaid line, further reinforces predictive power of the model. The integration of various flood-related variables within the BRT algorithm provides a detailed and reliable flood susceptibility model for Banda District.

Figure 13a presents the spatial distribution of flood susceptibility across Banda District, depicted using a continuous gradient raster. The color scale ranges from blue, representing areas with low flood susceptibility, to red, indicating high flood susceptibility. This map visually highlights the area that are most vulnerable to flooding, particularly in the central and northeastern parts of the district, which are closely associated with the main river channels and lower-lying areas. In contrast, Fig. 13b offers a classified version of the flood susceptibility data, dividing the district into discrete zones of low, medium, and high flood risk. This classification simplifies the interpretation of the data, making it more accessible for stakeholders, decision-makers, and laymen. The high-risk areas are concentrated in the northern and northeastern regions, while medium-risk and low-risk areas are more dispersed across the central and southern parts of the district. The southeastern region, indicated in the lightest color, shows the lowest susceptibility to flooding. Together, these figures provide both detailed and practical insights into flood risk distribution.

9 Agricultural Vulnerability Assessment

The floods have been devastating for the human settlements, infrastructure, and agricultural areas. The floods not only destroy the crops but also flush away nutrients of the earth. Banda District being agricultural dominated area has been severely hit by flood in 2021.

As of Table 3, flood susceptibility in Banda District is a major problem as significant area each year gets inundated. The tabulated results highlight the total flooded area, the flooded agricultural area, and the proportion of agricultural land affected by flooding within each susceptibility category—low, medium, and high. In areas with low flood susceptibility, 957.99 square kilometers land is modeled to be prone to flooding, out of which 875.61 square kilometers is agricultural land, accounting for 91.40% of the total flood-prone area. In medium susceptibility zones,

Table 3 Flooded area of Banda in the year 2021

Susceptibility	Flooded area (sq.km)	Flooded agricultural area (sq.km)	Flooded agricultural area in total flooded area (%)
Low	957.99	875.61	91.40%
Medium	843.49	753.92	89.38%
High	2570.81	2035.33	79.17%
Total area	4372.28	3664.87	83.82%

Source: Calculated by authors

843.49 square kilometer land is predicted to be susceptible to flooding, with 753.92 square kilometers being agricultural land, representing 89.38% of the total risked area. In areas with high flood susceptibility, a significant 2570.81 square kilometers land is vulnerable to flooding, and 2035.33 square kilometers land of this area is agricultural land, which accounts 79.17% of the total flood-prone area in the aforesaid category. Overall, the total modeled flood-prone area in Banda District is 4372.28 square kilometers, with 3664.87 square kilometers being agricultural land, meaning that out of the total flood-prone area, 83.82% is agricultural land. This data clearly states the fact that flood has an utmost devastating impact on agriculture in Banda District. Therefore, the most significant consequences of flooding are borne by the agricultural sector. This is particularly concerning for Banda District, which is predominantly agrarian district. The economy and livelihoods of Banda are heavily reliant on agriculture, meaning that any disruption due to flooding has far-reaching implications, not only for food security but also for the economic stability of the region (Sarkar et al., 2019). The high percentage of prone agricultural land across all susceptibility categories reflects the vulnerability in agricultural infrastructure of Banda and highlights the need for targeted flood mitigation strategies to protect this critical sector (Pingali & Rosegrant, 1995).

10 Discussion and Conclusion

The Banda District in Uttar Pradesh has experienced an increase in flood occurrences in recent years, primarily attributed to urbanization and climate change. There is a pressing need for localized research on flood susceptibility and prediction, as it fills a significant knowledge gap. Although global and national frameworks offer valuable perspectives, there is a notable lack of localized studies, emphasizing the importance of targeted research in specific regions. Therefore, assessing flood susceptibility using specialized methods is crucial for understanding and mapping predictive flood risks in the district. The effectiveness of these specialized methods directly influences predictive performance and outcomes, making it essential to carefully choose the techniques used in such assessments. For example, Convolutional neural networks (CNNs) have shown considerable promise in mapping flood vulnerability and risk, as demonstrated by research in Shangyou County, China (Wang et al., 2020). Furthermore, the Random Naive Bayes approach has been found to be effective in evaluating flood susceptibility, underscoring its utility in assessing flood hazards. (Tang et al. 2020). Arora et al., employed Shannon entropy method for modeling flood susceptibility in the middle Ganda plain (Arora et al., 2021). Various modeling techniques have been applied across different regions globally, each contributing to a deeper understanding of flood risks.

However, there is a pressing need of localized studies in refining flood risk models and improving predictive accuracy for the Banda District. In this study, eight conditioning or causative variables sourced from Advanced Spaceborne Thermal Emission and Reflection Radiometer (ASTER), National Remote Sensing Centre (NRSC), and Bhukosh data are selected carefully and utilized as inputs for a machine learning algorithm. The model demonstrated high efficiency and accuracy in flood prediction, with an AUC of 0.875 and a COR of 0.65. The robustness of the model is evident in its ability to effectively integrate multiple data sources and accurately identify flood-prone areas, despite the complexities involved in such analyses. The high AUC score indicates the model's strong capability in distinguishing between flood-prone and non-flood-prone areas, making it a valuable tool for managing local flood risks.

Gram panchayats like Khaptiha, Piparahari, Gugauli, Jauharpur Amli in the northern part, Rampur, Jaspura, Sabada, Birkhera, Narayanpur, Sardhua, Rajapur eastern part, In the west Banda city itself lies in the high probability flood susceptibility region its nearby administrative units like Gureh, Hardauli, Kanwara, Chahitaran, Pachulla and Mawai are also lies in high probability of flood susceptibility. In the south Kanaya, Karate, Bahara these are a few regions where flood probability shows high.

Although flood susceptibility maps cannot precisely predict the exact occurrence of floods in terms of space and time, they can indicate the likelihood of flood events across different areas. Still, this study demonstrated exceptional effectiveness and accuracy in identifying flood-prone areas within the Banda District, where agriculture is intensely practiced. The dominance of agriculture activity and higher proportion of flood-prone agricultural land to is the reason for considering only agricultural land in this study. By focusing on this area, the study provides crucial insights into the risks faced by agricultural practices, which are the backbone of the local economy. The identification of these vulnerable zones is essential for formulating targeted mitigation strategies and ensuring the sustainability of agricultural productivity in the region. Moreover, understanding the specific impact of floods on agricultural land helps in prioritizing interventions that can minimize damage and safeguard livelihoods in flood-prone regions of Banda as well as areas with similar sociocultural and physical settings.

It is also noteworthy that despite the effectiveness of applied BRT model, the study is not limitation-free. The limited number of flood event data points may have affected the study's overall robustness. Additionally, the "no flood" points were based on assumptions, and the flood inventory data was amalgamated from two different sources, with the most recent flood inventory data not being available. These factors could potentially impact the accuracy of the predictions, and future studies should aim to address these limitations by incorporating more comprehensive and up-to-date data. However, compared to similar studies conducted in other regions, such as the flood susceptibility mapping in the Ganga Basin by Arora et al. (2021), which reported an AUC of 0.80, and by Talukdar et al. (2020) with an AUC of 0.85, the results from this study exhibit slightly higher predictive capability. This enhanced accuracy is likely due to the careful selection of causative factors and the specific customizing of the model to the local context. Such high accuracy ensures that the model can reliably forecast flood-prone areas, thereby significantly aiding in the development of preemptive measures to protect agricultural lands such as construction of levees.

This study is revealing several areas with a high probability of flood events. Gram Panchayats in the northern part, such as Khaptiha, Piparahari, Gugauli, and Jauharpur Amli, are at high risk due to their topography and their location. In the eastern region, Rampur, Jaspura, Sabada, Bir Khera, Narayanpur, Sardhua, and Rajapur are vulnerable because of their low-lying terrain. The western part of the region, including Banda City and its neighboring areas like Gureh, Hardauli, Kanwara, Chahitaran, Pachulla, and Mawai, faces increased flood risks distance from the Ken River. In the southern areas, Kanaya, Karatal, and Bilharka are highly flood susceptible areas. In conclusion, this study emphasizes the importance of localized research in enhancing the accuracy and reliability of flood prediction models. By focusing on agricultural lands in Banda, the research not only fills a crucial knowledge gap but also provides actionable insights that can inform effective mitigation strategies. Future research should continue to refine these models by addressing the identified limitations and incorporating additional data sources, thereby improving the predictive accuracy and overall utility of flood risk assessments in the region.

References

Agarwal, A., & Narain, S. (1991). *Floods, flood plains and environmental myths*. Centre for Science and Environment. https://csestore.cse.org.in/pub/media/catalog/product/file/content_6.pdf

Arora, A., Pandey, M., Siddiqui, M. A., Hong, H., & Mishra, V. N. (2021). Spatial flood susceptibility prediction in Middle Ganga Plain: Comparison of frequency ratio and Shannon's entropy models. *Geocarto International, 36*(18), 2085–2116.

Alfieri, L., Bisselink, B., Dottori, F., Naumann, G., De Roo, A., Salamon, P., Wyser, K., & Feyen, L. (2016). Global projections of river flood risk in a warmer world. *Earth's Future, 5*(2), 171–182. https://doi.org/10.1002/2016ef000485

Bassi, N., Kumar, M. D., Sharma, A., & Pardha-Saradhi, P. (2014). Status of wetlands in India: A review of extent, ecosystem benefits, threats and management strategies. *Journal of Hydrology Regional Studies, 2*, 1–19. https://doi.org/10.1016/j.ejrh.2014.07.001

Bonsor, H. C., MacDonald, A. M., Ahmed, K. M., et al. (2017). Hydrogeological typologies of the Indo-Gangetic basin alluvial aquifer, South Asia. *Hydrogeology Journal, 25*, 1377–1406. https://doi.org/10.1007/s10040-017-1550-z

Borrelli, P., Märker, M., Panagos, P., & Schütt, B. (2014). Modeling soil erosion and river sediment yield for an intermountain drainage basin of the Central Apennines, Italy. *Catena, 114*, 45–58. https://doi.org/10.1016/j.catena.2013.10.007

Dawson, R. J., Ball, T., Werritty, J., Werritty, A., Hall, J. W., & Roche, N. (2011). Assessing the effectiveness of non-structural flood management measures in the Thames Estuary under conditions of socio-economic and environmental change. *Global Environmental Change, 21*(2), 628–646.

Elith, J., Leathwick, J. R., & Hastie, T. (2008). A working guide to boosted regression trees. *Journal of Animal Ecology, 77*(4), 802–813.

El Morjani, Z. E. A., Seif Ennasr, M., Elmouden, A., Idbraim, S., Bouaakaz, B., & Saad, A. (2016). Flood hazard mapping and modeling using GIS applied to the Souss River watershed. In R. Choukr-Allah, R. Ragab, L. Bouchaou, & D. Barceló (Eds.), *The Souss-Massa River Basin, Morocco. The handbook of environmental chemistry* (Vol. 53). Springer. https://doi.org/10.1007/698_2016_69

Eslaminezhad, S. A., Eftekhari, M., Akbari, M., Bayat, H., & Barghi, W. (2022). Using boosted regression tree, logistic model tree, and random forest algorithms to evaluate the groundwater potential. *Watershed Researches*. https://doi.org/10.22092/wmrj.2021.356517.1440

FAO. (2021). Article Retrieved from https://www.fao.org/resources/digital-reports/disasters-in-agriculture/en/ https://banda.nic.in/statistical-report/

Hirabayashi, Y., Mahendran, R., Koirala, S., et al. (2013). Global flood risk under climate change. *Nature Clim Change, 3*, 816–821. https://doi.org/10.1038/nclimate1911

Islam, M. M., Ujiie, K., Noguchi, R., & Ahamed, T. (2022). Flash flood-induced vulnerability and need assessment of wetlands using remote sensing, GIS, and econometric models. *Remote Sensing Applications Society and Environment, 25*, 100692. https://doi.org/10.1016/j.rsase.2021.100692

Jutras, S., Rousseau, A. N., & Clerc, C. (2009). Implementation of a Peatland-Specific water budget algorithm in HYDROTEL. *Canadian Water Resources Journal/Revue Canadienne Des Ressources Hydriques, 34*(4), 349–364. https://doi.org/10.4296/cwrj3404349

Kumar, M., Sharif, M., & Ahmed, S. (2019). Impact of urbanization on the river Yamuna basin. *International Journal of River Basin Management, 18*(4), 461–475. https://doi.org/10.1080/15715124.2019.1613412

Khan, Z., Mohsin, M., Ajmal, U., & Ahmad, A. (2024a). Modelling potential zones of Gangotri glacier using GIS and ML in the wake of physico-climatic factors. In *Climate change, vulnerabilities and adaptation: Understanding and addressing threats with insights for policy and practice* (pp. 245–271). Springer Nature Switzerland.

Khan, Z., Mohsin, M., Ali, S. A., Vashishtha, D., Husain, M., Parveen, A., et al. (2024b). Comparing the performance of machine learning algorithms for groundwater mapping in Delhi. *Journal of the Indian Society of Remote Sensing, 52*(1), 17–39.

Khosravi, K., Pourghasemi, H. R., Chapi, K., & Bahri, M. (2016). Flash flood susceptibility analysis and its mapping using different bivariate models in Iran: A comparison between Shannon's entropy, statistical index, and weighting factor models. *Environmental Monitoring and Assessment, 188*, 656. https://doi.org/10.1007/s10661-016-5665-9

Khosravi, K., Shahabi, H., Pham, B. T., Adamowski, J., Shirzadi, A., Pradhan, B., et al. (2019). A comparative assessment of flood susceptibility modeling using multi-criteria decision-making analysis and machine learning methods. *Journal of Hydrology, 573*, 311–323.

Kourgialas, N. N., & Karatzas, G. P. (2011). Flood management and a GIS modelling method to assess flood-hazard areas—A case study. *Hydrological Sciences Journal–Journal des Sciences Hydrologiques, 56*(2), 212–225.

Kumar, H., Karwariya, S. K., & Kumar, R. (2022). Google earth engine-based identification of flood extent and flood-affected Paddy Rice fields using Sentinel-2 MSI and Sentinel-1 SAR data in Bihar state, India. *Journal of the Indian Society of Remote Sensing, 50*, 791–803. https://doi.org/10.1007/s12524-021-01487-3

Lawal, D. U., Matori, A. N., Yusuf, K. W., Hashim, A. M., & Balogun, A. L. (2014). Analysis of the flood extent extraction model and the natural flood influencing factors: A GIS-based and remote sensing analysis. *IOP Conference Series: Earth and Environmental Science, 18*, 012059. https://doi.org/10.1088/1755-1315/18/1/012059

Lee, S., Kim, J.-C., Jung, H.-S., Lee, M. J., & Lee, S. (2017). Spatial prediction of flood susceptibility using random-forest and boosted-tree models in Seoul metropolitan city, Korea. *Geomatics, Natural Hazards and Risk, 8*, 1185–1203. https://doi.org/10.1080/19475705.2017.1308971

Mandal, K., Saha, S., & Mandal, S. (2022). Predicting the landslide susceptibility in Eastern Sikkim Himalayan Region, India using boosted regression tree and REPTree machine learning techniques. In S. Mandal, R. Maiti, M. Nones, & H. R. Beckedahl (Eds.), *Applied geomorphology and contemporary issues. Geography of the physical environment*. Springer. https://doi.org/10.1007/978-3-031-04532-5_36

Moradi, M., Kazeminezhad, M. H., & Kabiri, K. (2020a). Integration of Geographic Information System and system dynamics for assessment of the impacts of storm damage on coastal communities—Case study:

Chabahar, Iran. *International Journal of Disaster Risk Reduction, 49*, 101665. https://doi.org/10.1016/j.ijdrr.2020.101665

Moradi, M., Delavar, M. R., & Samadzadegan, F. (2020b). Flood susceptibility mapping using ensemble machine learning models. *Journal of Environmental Water Engineering, 6*(4), 499–512. https://doi.org/10.22034/jewe.2020.220593.1351

Nandargi, S., & Aman, K. (2018). Precipitation concentration changes over India during 1951 to 2015. *Scientific Research and Essays, 13*(3), 14–26. https://doi.org/10.5897/sre2017.6540

Office of the Registrar General & Census Commissioner, India. (2011). *District census handbook: Banda, Uttar Pradesh (Part A & B)*. Government of India, Ministry of Home Affairs. https://censusindia.gov.in/nada/index.php/catalog/1180/download/3734/DH_2011_0939_PART_A_DCHB_BANDA.pdfMarconi, M., Gatto, B., Magni, M., & Marincioni, F. (2016). A rapid method for flood susceptibility mapping in two districts of Phatthalung Province (Thailand): Present and projected conditions for 2050. *Natural Hazards, 81*, 329–346. https://doi.org/10.1007/s11069-015-2082-2

O'Connor, J. E., Grant, G. E., & Costa, J. E. (2002). The geology and geography of floods. Ancient floods, modern hazards and applications of paleoflood hydrology. Water science and application. *American Geophysical Union, 5*, 359–395.

Paarcha, A., Maithani, S., Husain, M., Suhanee, N., & Azad, R. K. (2023). Monitoring urban sprawl using geo-spatial technology: A case study of Kanpur City, India. In *Advancements in urban environmental studies: Application of geospatial technology and artificial intelligence in urban studies* (pp. 87–99). Springer International Publishing.

Pérez Navarro, M. Á., Sapes, G., Batllori, E., Serra-Diaz, J. M., Esteve, M. A., & Lloret, F. (2019). Climatic suitability derived from species distribution models captures community responses to an extreme drought episode. *Ecosystems, 22*(1), 77–90.

Pingali, P. L., & Rosegrant, M. W. (1995). Agricultural commercialization and diversification: Processes and policies. *Food Policy, 20*(3), 171–185.

Rahmati, O., Zeinivand, H., & Besharat, M. (2016). Flood hazard zoning in Yasooj region, Iran, using GIS and multi-criteria decision analysis. *Geomatics, Natural Hazards and Risk, 7*, 1000–1017. https://doi.org/10.1080/19475705.2015.1045043

Rahmati, O., Samadi, M., Shahabi, H., Mohammadi, F., Ahmad, B. B., & Keesstra, S. (2019). Evaluating the efficacy of statistical and machine learning models for flood modeling in data-scarce regions. *Environmental Modeling & Software, 115*, 40–58.

Runhaar, H., Wilk, B., Persson, Å., Uittenbroek, C., & Wamsler, C. (2017). Mainstreaming climate adaptation: Taking stock about "what works" from empirical research worldwide. *Regional Environmental Change, 18*(4), 1201–1210. https://doi.org/10.1007/s10113-017-1259-5

Sarkar, A., Dasgupta, A., & Sensarma, S. R. (2019). Climate change and food security in India: adaptation strategies and major challenges. *Sustainable Solutions for Food Security: Combating Climate Change by Adaptation*, 497–520.

Sharma, V. K., Rao, G. S., Bhatt, C. M., Shukla, A. K., Mishra, A. K., & Bhanumurthy, V. (2017). Automatic procedures analyzing remote sensing data to minimize flood response time: A step towards national flood mapping service. *Spatial Information Research, 25*, 657–663. https://doi.org/10.1007/s41324-017-0132-4

Talukdar, S., Ghose, B., Shahfahad, S., Mahato, R. S., Pham, Q. B., et al. (2020). Flood susceptibility modeling in Teesta River basin, Bangladesh using novel ensembles of bagging algorithms. *Stochastic Environmental Research and Risk Assessment, 34*, 2277–2300.

Tehrany, M. S., Pradhan, B., & Jebur, M. N. (2013). Spatial prediction of flood susceptible areas using rule based decision tree (DT) and a novel ensemble bivariate and multivariate statistical models in GIS. *Journal of Hydrology, 504*, 69–79. https://doi.org/10.1016/j.jhydrol.2013.09.034

Tehrany, M. S., Pradhan, B., & Jebur, M. N. (2015). Flood susceptibility analysis and its verification using a novel ensemble support vector machine and frequency ratio method. *Stochastic Environmental Research and Risk Assessment, 29*, 1149–1165.

Tang, X., Li, J., Liu, M., Liu, W., & Hong, H. (2020). Flood susceptibility assessment based on a novel random Naïve Bayes method: A comparison between different factor discretization methods. *CATENA, 190*, 104536. https://doi.org/10.1016/j.catena.2020.104536

Vangelis, H., Zotou, I., Kourtis, I. M., Bellos, V., & Tsihrintzis, V. A. (2022). Relationship of rainfall and flood return periods through hydrologic and hydraulic modeling. *Water, 14*(22), 3618.

Vousdoukas, M. I., Mentaschi, L., Voukouvalas, E., Bianchi, A., Dottori, F., & Feyen, L. (2018). Climatic and socioeconomic controls of future coastal flood risk in Europe. *Nature Climate Change, 8*(9), 776–780.

Wang, Y., Fang, Z., Hong, H., & Peng, L. (2020). Flood susceptibility mapping using convolutional neural network frameworks. *Journal of Hydrology, 582*, 124482. https://doi.org/10.1016/j.jhydrol.2019.124482

Youssef, A. M., Pradhan, B., & Sefry, S. A. (2016). Flash flood susceptibility assessment in Jeddah city (Kingdom of Saudi Arabia) using bivariate and multivariate statistical models. *Environment and Earth Science, 75*, 12. https://doi.org/10.1007/s12665-015-4830-8

Zhang, W., Leung, Y., & Chan, J. C. L. (2013). The analysis of tropical cyclone tracks in the Western North Pacific through data mining. Part I: Tropical cyclone recurvature. *Journal of Applied Meteorology and Climatology, 52*(6), 1394–1416. https://doi.org/10.1175/jamc-d-12-045.1

Zhang, Q., Gu, X., Singh, V. P., Shi, P., & Sun, P. (2018). More frequent flooding? Changes in flood frequency in the Pearl River basin, China, since 1951 and over the past 1000 years. *Hydrology and Earth System Sciences, 22*(5), 2637–2653. https://doi.org/10.5194/hess-22-2637-2018

Analytical Hierarchy Process (AHP)-Based Cyclone Coping Capacity Index of Bhadrak District, Odisha, India Using Geospatial Techniques

Toushif Jaman, K. Dharanirajan, and Kajal Kumar Mandal

1 Introduction

Severe convective tropical cyclones are recognized as exceptionally powerful and destructive meteorological hazard which result in both losses of life and environmental damage. This is one of the most recent activities, frequent and intense. It has been said that changes are due to climatic changes, high sea surface temperature, changes in circulation in the atmosphere, and others (Knutson et al., 2010; Walsh et al., 2016; IPCC, 2014). These intensified cyclonic activities disproportionately affect coastal and adjacent inland communities, especially in the Bay of Bengal region, which has witnessed some of the deadliest cyclones on record (Ali, 1999; Mohapatra et al., 2012; Sahoo & Bhaskaran, 2016). The Indian State Odisha, particularly Bhadrak District, is highly vulnerable due to its geographic location along this cyclone-prone region (Dube et al., 2009; Mohapatra et al., 2015). Coping capacity is normally called as a coping strategies and coping skill, although strategies are commonly omitted. The word coping is generally known as a beneficial strategy which helps to minimize impact of vulnerability. It is the basis for planning and policy formulation about it. It is an important ancillary sector of disaster management science in research (Keogh et al., 2011; Gao et al., 2012; Shi et al., 2018). Indications so far included in index systems used for assessing the Chinese cop-ting ability concerning natural calamity mitigation have been casualties, economic damages and restitution measures, disaster prevention, and disaster preparedness (cutter 2003–Estember 2018). According to scholars, adaptability to natural disasters encompasses three dimensions: disaster prevention, disaster relief, and disaster recovery (Chen et al., 2020; Hu et al., 2010). Some researchers approximated the coping capacity of disaster by using the number of casualties, economic loss, and recovery time (Xie, 2006). In short, these two index systems greatly vary in their contents and indicators, yet no standardized systematic indicator system had been deployed during the assessment of natural disaster coping capacity (Jiang et al., 2016; Huang et al., 2020). The primary applied methods of analysis when measuring the capacity of disasters coping are as follows: the coping index, the fuzzy com-prehensive evaluation, as well as back propagation, otherwise referred to as back

T. Jaman (✉)
Department of Space, North Eastern Space Applications Centre (NESAC), Govt. of India, Shillong, India

K. Dharanirajan · K. K. Mandal
Department of Coastal Disaster Management, Pondicherry University, Port Blair Campus, Andamans, India

Z. Khanam et al. (eds.), *Climate Change, Resilience and Ecosystem Services*, Sustainable Development Goals Series, https://doi.org/10.1007/978-3-031-96977-5_18

propagation (BP) neural network (Ferreira et al., 2016; Jha & Gundimeda, 2019). Generally, these estimation results based on above methods are basically presented in scores or relative rank values.

The analytical hierarchy process (AHP), introduced by Saaty (1980), has emerged as a valuable tool for multi-criteria decision-making, particularly in contexts where decision-making involves evaluating complex, interdependent criteria (Vargas, 2010; Yager, 2008). AHP has been extensively applied in disaster risk management to prioritize various dimensions of vulnerability and resilience, such as in flood risk assessments (Rahman et al., 2019), earthquake vulnerability analysis (Malik & Mahmood, 2015), and wildfire risk management (Boroushaki & Malczewski, 2010). However, despite its widespread application, there is a notable lack of studies that integrate AHP with geospatial technologies like geographic information systems (GIS) and remote sensing to develop spatially explicit models of coping capacity for regions exposed to tropical cyclones (Greiving et al., 2006; Jha et al., 2016a, b; Montz et al., 2017). By structuring these criteria into a hierarchical model, AHP facilitates the quantitative comparison of different factors, thereby providing a clear framework for assessing the coping capacity of a region (Saaty, 1980). In recent years, AHP has been widely utilized in various domains, including disaster risk reduction, to prioritize and evaluate different dimensions of vulnerability and resilience (Vargas, 2010). Risk is defined in this context as an initial point of assessment, emphasizing vulnerability in the absence of considerations related to coping capacity.

Recent research highlights the importance of integrating geospatial data with multi-criteria decision-making models to enhance the accuracy and relevance of disaster risk assessments (Schneiderbauer & Ehrlich, 2006; Montz et al., 2017). However, such integrated approaches remain underexplored for cyclone-prone regions like Bhadrak District, where localized, detailed assessments are crucial for enhancing community preparedness and resilience.

In conjunction with AHP, geospatial techniques such as geographic information systems (GIS) and remote sensing have revolutionized disaster management practices by enabling precise spatial analysis and visualization of risk factors. These technologies provide critical insights into the spatial distribution of vulnerabilities and resources, which are essential for effective disaster preparedness and response (Jha et al., 2016a, b). Integrating AHP with geospatial techniques thus offers a powerful methodological framework for developing a Cyclone Coping Capacity Index (CCI) that is both comprehensive and spatially explicit. Coping capacity is defined as people's, organization's, or system's ability, by using existing skills and resources, to cope with impending, ongoing, or potential undesirable events, threats, hazards, or disasters. Capacity covers infrastructure to institutions as well as human and knowledge bases while incorporating, in collective attributes of social connections, leadership and management qualities (UNISDR, 2017). In many ways, until the Natural Disaster Prevention, Preparedness, and Mitigation 1994 guidelines were put forward, disaster coping capacity was hardly considered as a matter of great interest among international organizations or nations (IDNDR, 1994). The impact the guidelines of natural disaster coping capacity caused was tremendous long-run based on assessment, according to Maly and Suppasri (2020). The United States introduced its Capability Assessment for Readiness and State and Local Government or Federal Emergency Management Agency (FEMA) as early as 1997 (FEMA, 2009). Moreover, all states of the United States have also provided standards for the assessment of disaster coping capacity. The index system of assessment includes communication, disaster warning, training, and financial management (Chen et al., 2020). Japan assessed the natural disaster coping capacity along lines such as emergency response, reconstruction after the disasters, and training (Chen et al., 2020). Since the year 2000, studies about the evaluation of the coping capacity of China have been conducted (Zeng, 1996; Xie, 2006).

For regions like Bhadrak District in Odisha, which are highly susceptible to cyclones due to their proximity to the Bay of Bengal, there is an urgent need for a tailored approach that factors in local conditions such as distance from key infrastructure, housing types, access to emergency services, and community literacy levels. This study aims to fill this gap by creating a Cyclone Coping Capacity Index (CCI) that leverages both AHP and geospatial techniques to provide a more context-specific understanding of regional coping capacities. This approach not only helps identify the area's most at risk but also informs targeted interventions to enhance disaster preparedness and resilience at the community level.

This research aims to develop an AHP-based Cyclone Coping Capacity Index for Bhadrak District, leveraging geospatial techniques to map and analyze the region's capacity to withstand and recover from cyclone impacts. The study will identify key indicators of coping capacity, such as infrastructure resilience, socioeconomic factors, and emergency response mechanisms, and will systematically evaluate these indicators using the AHP framework. The resulting index will provide valuable insights for policymakers and disaster management authorities to enhance the region's preparedness and resilience against future cyclones.

2 Materials and Methods

2.1 Description of Study Area

The eastern region of the state of Odisha is home to the Bhadrak District. It is located between latitudes 20°43′ to 21°13′ north and longitudes 86°16′ to 87° east. The Bhadrak District is 2505 square kilometers in size. On the west lies the Kendujhar District and Bay of Bengal, on the east the Kendrapara District, on the north, the Baleshwar District, and on the south lies the Jajpur District. And encircling River Baitarani in the west is Bhadrak District. There are seven tehsils divided in the district of Bhadrak, namely Bant, Bhadrak, Dhamanagar, Chandabali, Tihidi, Basudevpur, and Bhandaripokhari for easier management of tax collections (Fig. 1). During Summer, the average daily maximum temperature is 36.40 °C, and average daily minimum temperature is 24.60 °C. The average maximum and minimum temperature for a day during winter periods lie within a range of 14–27 °C. Period for southwest monsoon starts in the middle of June to end by September. Period after monsoons are the months of October and November. The southwest monsoon season lasts from June to September. The months of October and November are the post-monsoon season. The district receives 1427.9 mm of precipitation (rainfall) on average each year. Approximately 71% of the yearly precipitation falls between June and September, which is the southwest monsoon season. The district lies directly in the trajectory of most cyclonic storms in the region. During the monsoon season, cyclones that originate in the Bay of Bengal and make landfall along the Odisha coast frequently impact this district.

2.2 Data Source

Cyclone Coping Capacity Index of Bhadrak District has been calculated using primary and secondary data. The primary data (satellite-based remote sensing) is Alos Palsar DEM, collected from ASF (Alaska Satellite Facility) data hub. The resolution of this data is 12.5 m. Road network data of the study area has collected from OSM (open-streets map) database, which is used to develop the roadsides buffer zone. The secondary data is collected from district census handbook, 2011 and HDI (Human Development Index) Bhadrak, 2006. The secondary data-based essential parameters are health facilities, types of housing, educational level, medical facilities, and shelter facilities.

2.3 Data Processing

Both primary and secondary data have processed on geospatial software (ArcGIS-Vs10.4). Relative relief map of the study area has been generated by

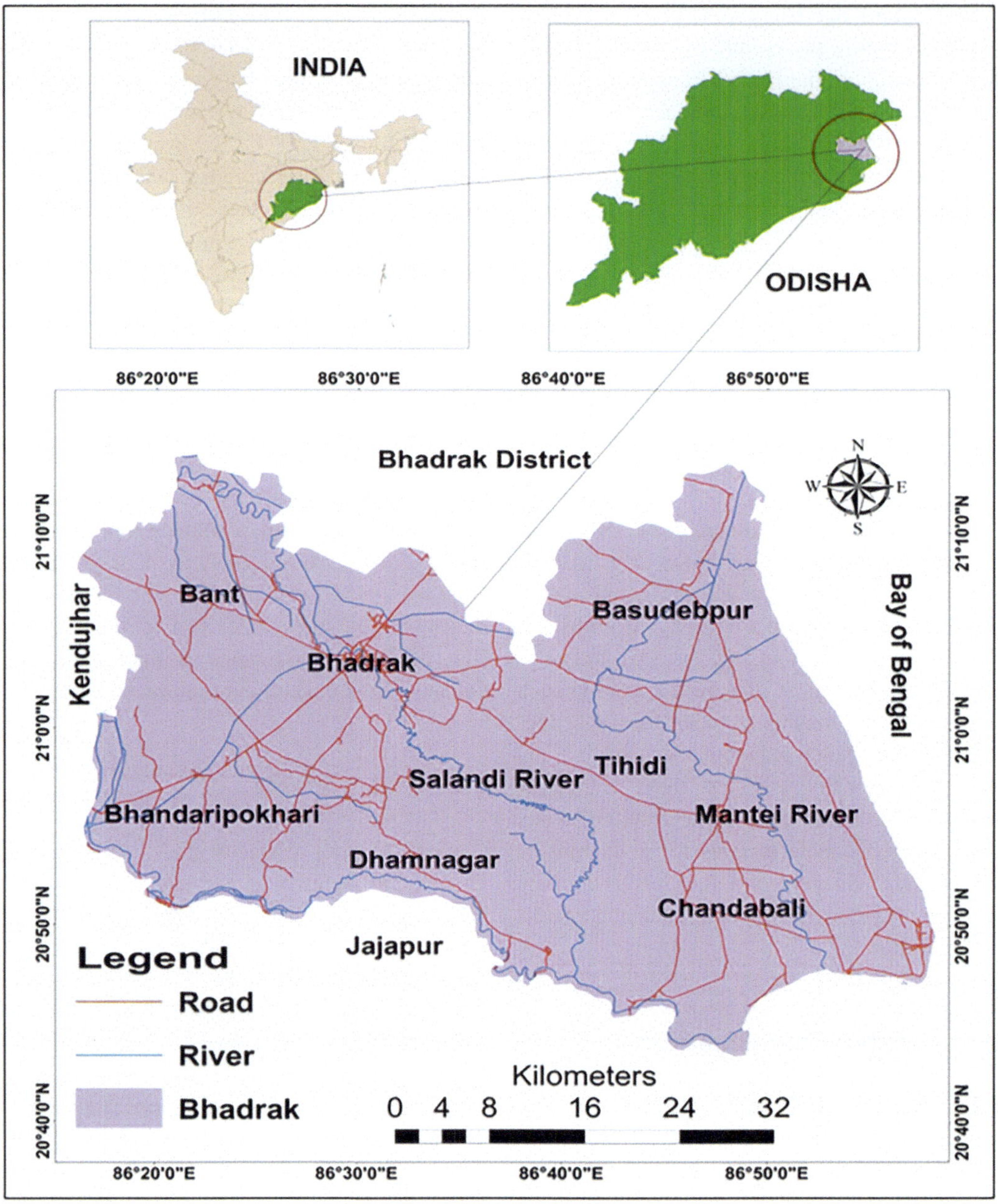

Fig. 1 Study area

surface modeling function. The river data have been extracted by applying the hydrological model on the ArcGIS platform using Alos Palsar DEM. The road network buffer zonation map was prepared on ArcGIS applying the multi-ring buffer function. The secondary data's are spatially merged with the block level database of Bhadrak District, and all data were categorized based on the density. After that, the AHP (analytical hierarchy process) qualitative techniques have used to determine each parameter's influencing value. Further using this influencing value, the weightage overlay analysis method has delineated the vulnerable Coping Capacity Index (Fig. 2).

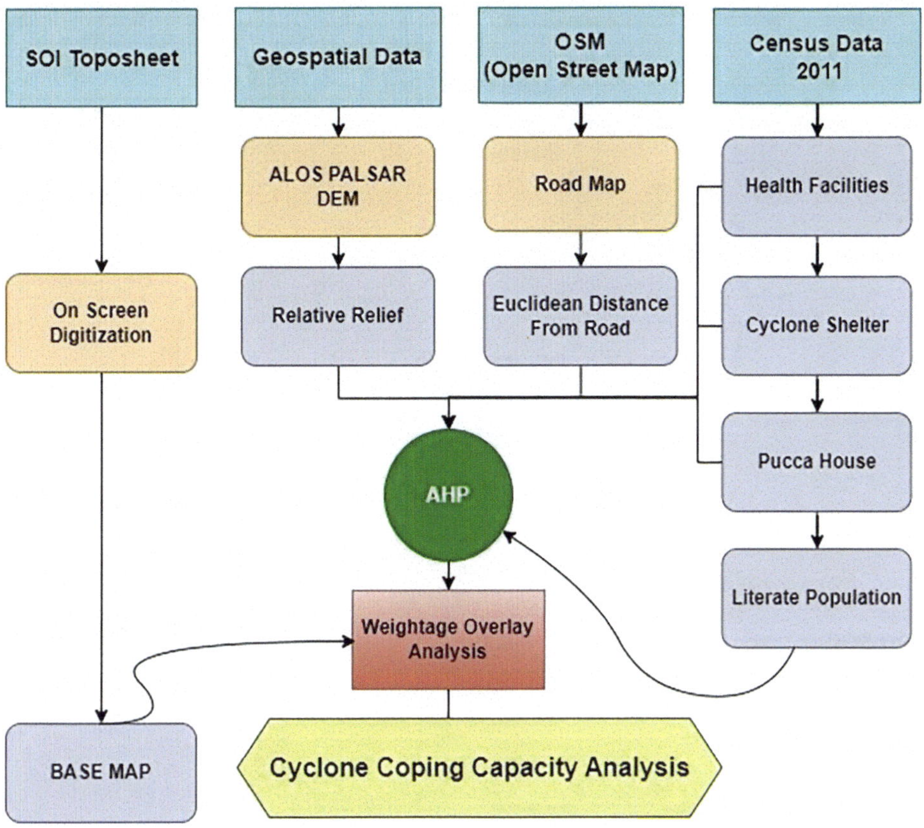

Fig. 2 Flowchart of Cyclone Coping Capacity Index

3 Cyclone Coping Capacity Variables

3.1 Relative Relief

Relative relief, or local relief, is the difference between the maximum elevation and the minimum elevation over a unit area (Mustak et al., 2012). Of the relief features of an area which might be considered quantitatively in terms other than sea level are the degree of subdivision of the features and the general assessment of the morphological profile of the landscape (Singh, 1992). It indicates the research site demarcated into five classes as shown in Fig. 3. The very low elevation class, representing areas below 150 m above mean sea level, has a minimum spatial extent of 0.55 km^2 and is predominantly located within the central and eastern hill ridges. The low elevation class, which extends up to 300 m, occupies a significantly larger area of 70.51 km^2. These elevation classes provide a clear representation of the topographic variation across the district, highlighting the transition from low-lying regions to the hillier terrains. The highest area coverage of the medium class (300–450 m) is 249.61 km^2, almost one-third (60.73%) of the research area. Besides being seen in the central, northern, and south central regions, the high class (450–600 m) is visible as an isolated patch in the western hill ranges. It covers the study area of 83.26 km^2. The class extremely high class (>600 m) occupies only an area of 6.07 km^2 and it is confined mainly to study areas northeast and southeast.

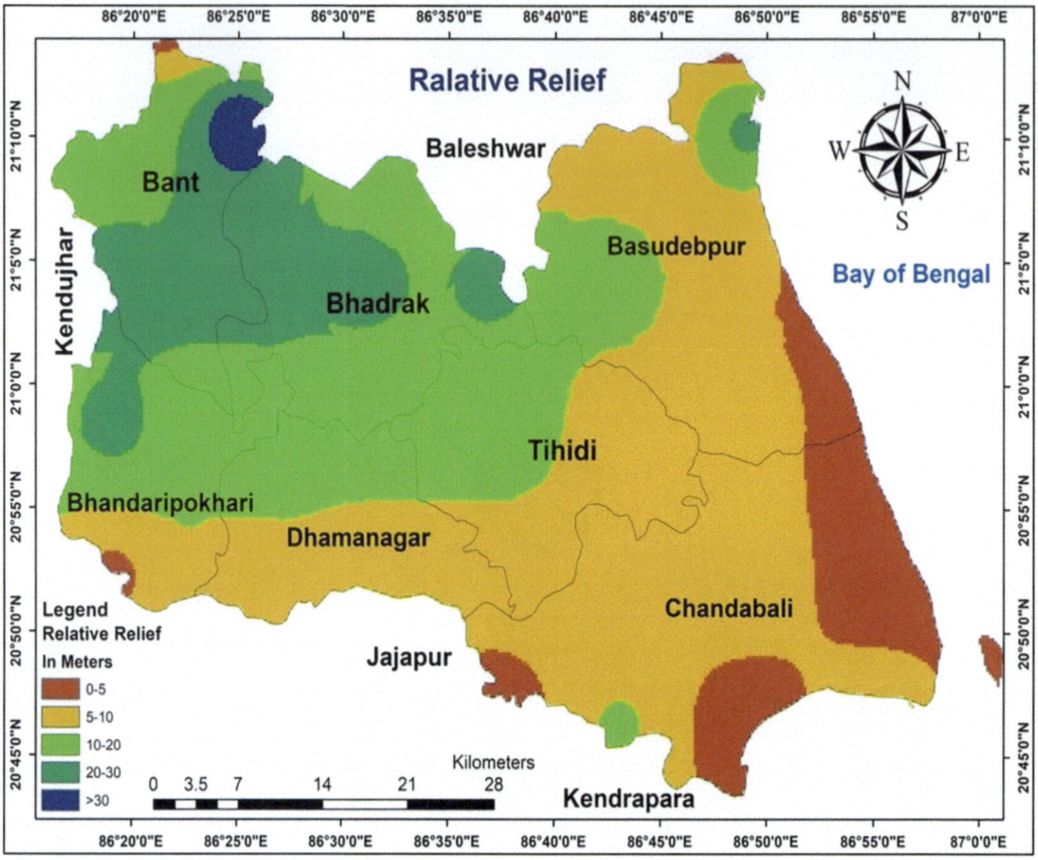

Fig. 3 Relative relief map

3.2 Distance from River

Areas closer to rivers are at a higher risk of flooding during cyclones, as heavy rainfall can lead to rivers overflowing their banks. This can exacerbate the impact of the cyclone, making it harder for communities to cope. In the event of a cyclone, evacuation routes and plans need to be more robust for areas near rivers. These regions may need more shelters and resources dedicated to moving people out of flood-prone zones. Infrastructure such as bridges, roads, and utilities near rivers might be more vulnerable to damage during cyclones. This requires better maintenance, stronger construction, and more rapid repair capabilities. The existing river networks put people and property at serious risk up to 1.5 km. Nonetheless, the category with the lowest weight was that which was farther than 1.5 km from the rivers, while the category with the most weight was that which was less than 0.5 km (Fig. 4).

3.3 Distance from Road

The road network plays a crucial role in enhancing the coping capacity of communities and regions affected by cyclones. Well-maintained and well-planned road networks facilitate quick and efficient evacuation of people from high-risk areas to safer locations. Roads provide critical access for emergency services and relief teams to reach affected areas promptly. Roads are essential for transporting food, water, medical supplies, and other necessities to cyclone-hit areas.

Efficient road networks play a crucial role in minimizing response delays and ensuring that relief and emergency assistance reaches affected populations promptly, thereby reducing significant time losses during disaster situations. Good road connectivity supports economic activities post-cyclone, helping communities to recover and rebuild faster. Properly designed road net-

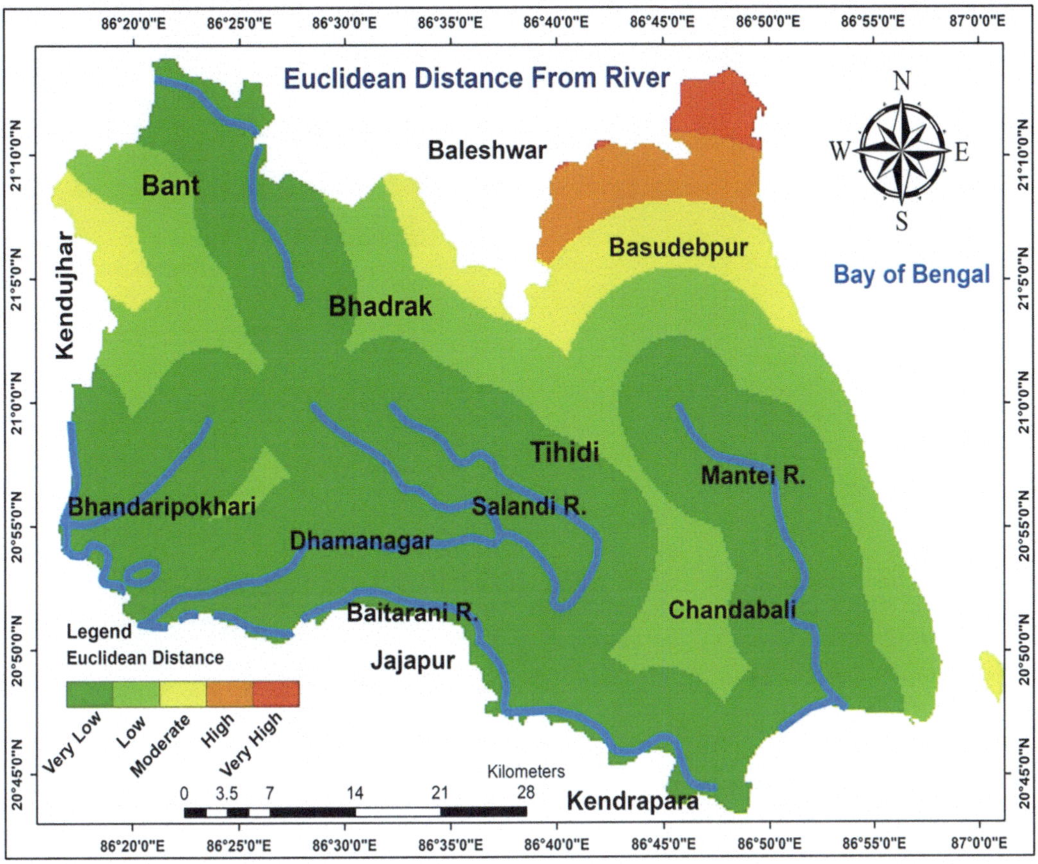

Fig. 4 Euclidean distance from river

works can help in the mitigation of risks associated with cyclones by ensuring they are not easily damaged. Roads connect rural areas with urban centers, allowing for the continuation of economic activities, which is crucial for the livelihood of the affected populations. The reconstruction and maintenance of road networks create jobs, contributing to the economic stability of affected regions. In summary, a robust road network significantly enhances the capacity of communities to cope with cyclones by ensuring efficient evacuation, timely delivery of aid, effective recovery and reconstruction, continuous communication, and long-term resilience against future cyclones (Fig. 5).

3.4 Types of Housing

The majority of rural homes are of the katcha house and those situated on plains or next to bodies of water are typically more vulnerable. Three different house types—katcha, semi-pucca, and pucca—are found in the research region. The vulnerability evaluation took these kinds of dwellings into account. Katcha dwellings were given a higher weight in the pairwise comparison matrix because of their low quality and vulnerability to severe damage from cyclones. The block-wise distribution of pucca dwellings in the Bhadrak District is shown in Fig. 6 (Table 1).

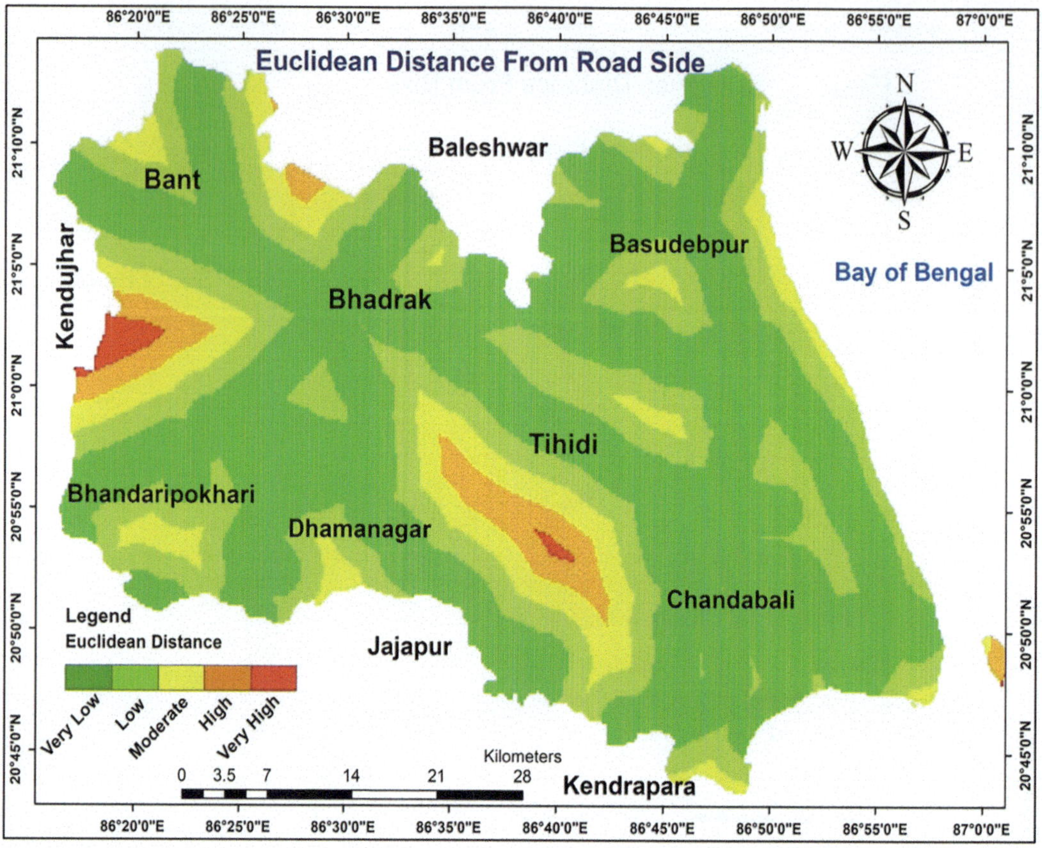

Fig. 5 Euclidean distance from road

3.5 Educational Level

The idea behind the use of the illiterate population's number (both male and female) is that those who lack literacy rely on others to help them get ready for impending disasters and leave. As a result, when weight was allocated in the decision hierarchy, the district with the highest rates of illiteracy received the greatest weight. "In general, education is a type of learning in which, via instruction, training, or research, the habits, knowledge, and abilities of a group of people are passed down from one generation to the next." According to D'Oyley & Blunt, 1994, education is an essential human resource through which one is enabled to make the appropriate decisions and make all the required efforts for survival or revival following a natural disaster. Indeed, it affects the lifetime income of an individual as well (Cutter et al., 2003). The available literature states that better educated members' households perform better than those households having lesser educated members (Paul & Routray, 2010). To test this hypothesis, overall literacy rate was used and weighted for each community to help understand their coping capacity. For example, the highest weight would be assigned to communities having the highest literacy rate both for males and females, while lowest weight would be given to those with lower literacy rate (see Table 2) (Fig. 7).

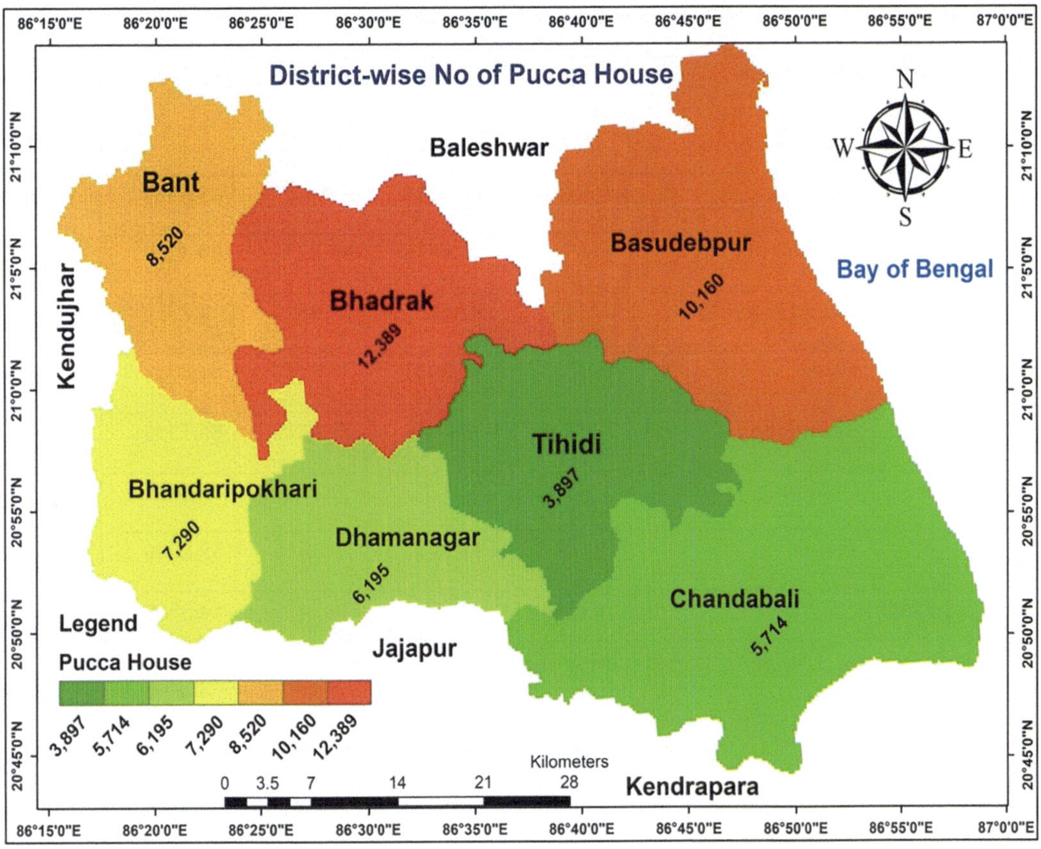

Fig. 6 District-wise distribution of pucca house

Table 1 Block-wise distribution of pucca house

Sl. no.	Block name	Number of pucca house
1	Bhadrak	12,389
2	Basudebpur	10,160
3	Bhandaripokhari	7290
4	Bant	8520
5	Chandabali	5714
6	Dhamnagar	6195
7	Tihidi	3897
Total		54,165

Source: District Disaster Management Handbook 2011

Table 2 Block-wise distribution of literacy of Bhadrak District

Sl. no.	Block name	Literate	Male literacy	Female literacy	Literacy %
1	Bhadrak	165269	91.87%	80%	85.87%
2	Basudebpur	114545	92.65%	85.82%	89.32%
3	Bhandaripokhari	84646	84.01%	64.96%	74.89%
4	Bant	120227	91.86%	77.81%	84.87%
5	Chandabali	137896	92.28%	79.85%	86.15%
6	Dhamnagar	87411	92.12%	79.88%	85.53%
7	Tihidi	51745	91.27%	79.36%	85.27%
Total		761797	90.86	78.24%	84.55%

Source: Socioeconomic Data and Applications Center (SEDAC)

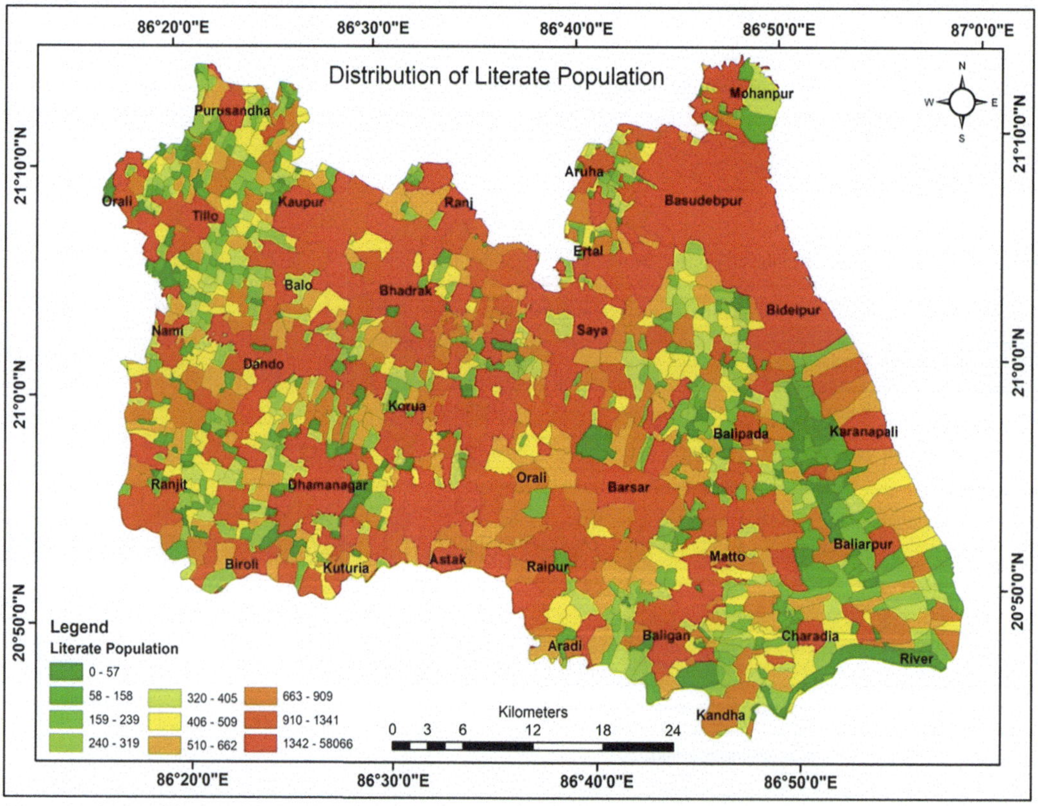

Fig. 7 Distribution of literate population

3.6 Medical Facilities

The timely supply of emergency relief and basic necessities to affected populations in cyclone-prone areas can save lives and significantly reduce the severity of damages and losses. Basically, essential materials supply like dry cyclone, drinking water facilities, and medicine are very important during the cyclone disaster which can minimize the probability of cyclone risk. The relief system facilities of the study area were serene by the different opinions and involved survey. Cyclone relief distribution becomes extremely challenging in cyclone-prone areas during and immediately after a cyclone due to the large-scale complexity of emergency operations. As a result, many affected victims are often unable to receive adequate relief supplies, despite concentrated efforts by government agencies and humanitarian organizations such as NGOs and the Red Cross. Some of the important relief materials like cyclones, drinking water, medicine, and cloth were found in the study area those are served to the victims during the cyclone situation. Among all drinking water is a vital issue for the people of Bhadrak District. Various government and non-government organizations provided relief materials to the affected population and also supported essential public health measures, including the purification of water sources, distribution of oral rehydration salts, and supply of safe portable drinking water. The collected information was utilized for model development, and a relief facility map was subsequently prepared and produced (Fig. 8) (Table 3).

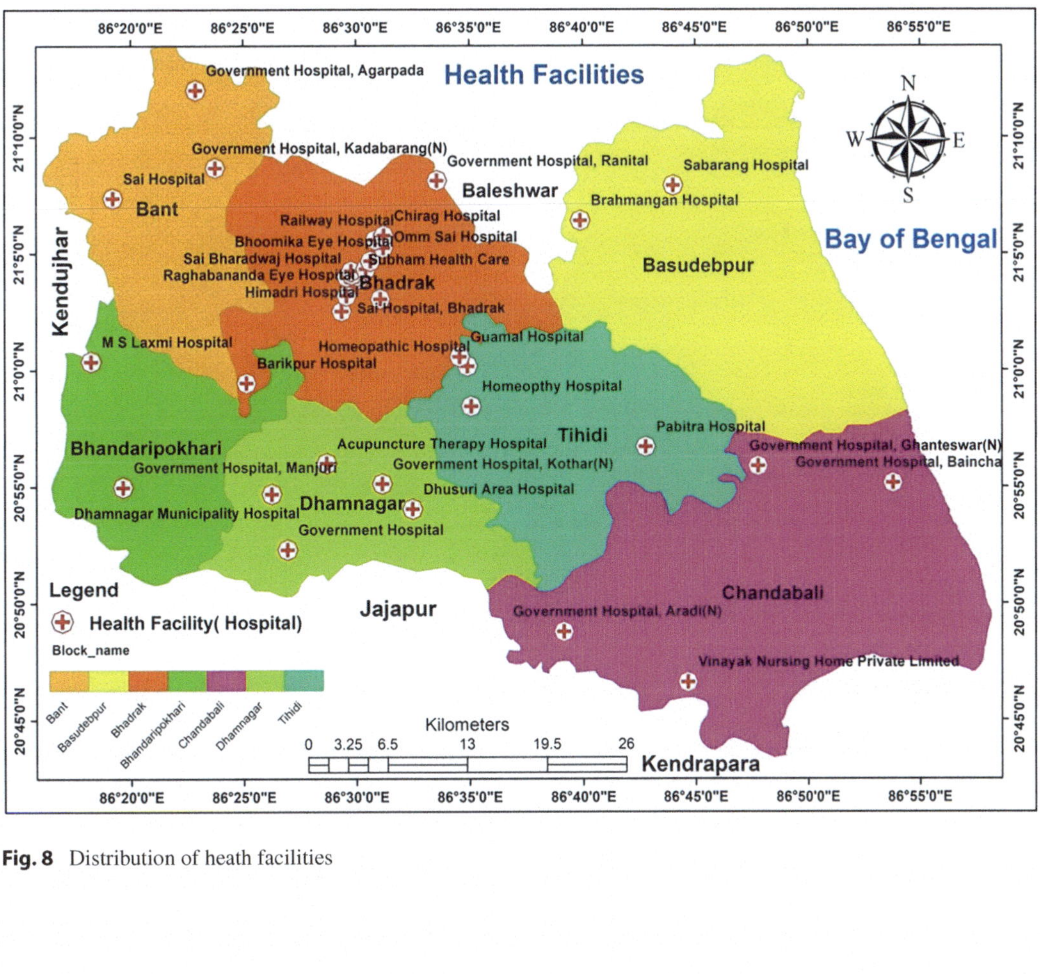

Fig. 8 Distribution of heath facilities

Table 3 Number of health facilities

Sl. no.	Block name	Number of hospital
1	Bhadrak	58
2	Basudebpur	47
3	Bhandaripokhari	28
4	Bant	31
5	Chandabali	46
6	Dhamnagar	47
7	Tihidi	41
Total		**298**

Source: District Disaster Management Handbook 2011

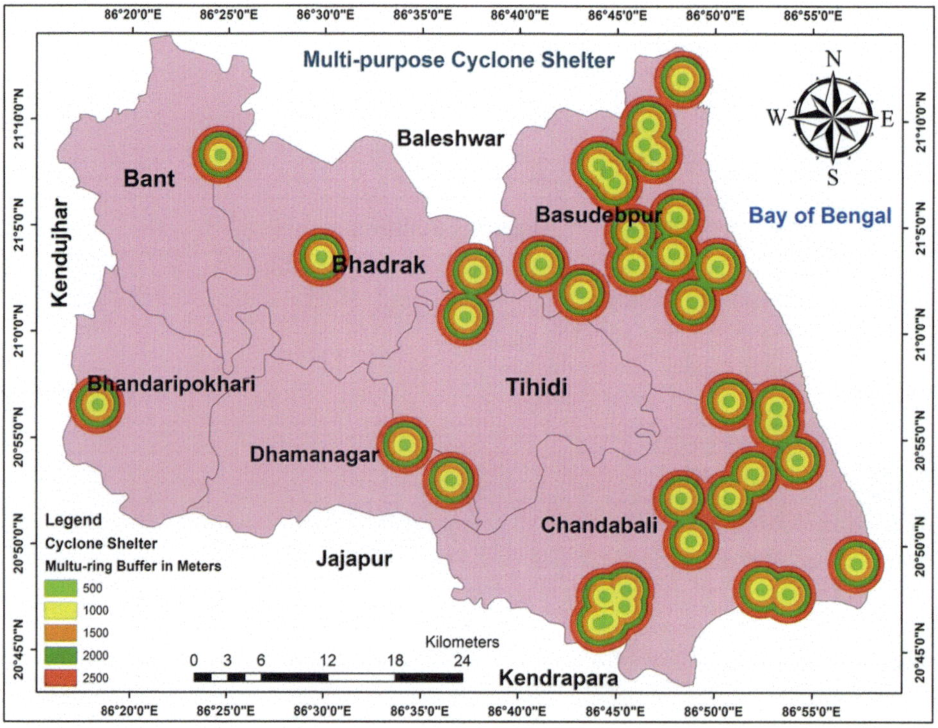

Fig. 9 Multipurpose cyclone shelter

3.7 Shelters Facility

The cyclone coping capacities depend upon the availability of cyclone shelter during and after cyclone. The government buildings, community hall, school, panchayet, and private sector buildings of flat terrain area are considered best shelter for victims. During the cyclone, people can shift into this safer place. Some places of the study area have sufficient shelters. The coastal blocks of Bhadrak District like Basudebpur and Chandabali are having a greater number of cyclone shelters. The multi-purpose cyclone shelter has been shown in the Fig. 9.

4 Application of AHP on CCI

Analytical hierarchical process (AHP) is a qualitative decision-making judgmental tool that helps to take a probabilistic, more minor error decision. This is an experience and expert knowledge-based approach, where a group of expert people can solve a single problem by contributing their decision. This technique has developed by the Thomas L. Saaty in 1977. In this study, 10 expert researcher's decision has collected to assess the cyclone vulnerable coping capacity of the Bhadrak District. After that, all expertise judgmental weights have consolidated (based on geometrical mean) in a single matrix (consolidated matrix) to develop the pairwise comparisons (DTRS 2004) matrix. The pairwise comparison matrix weights are determined based on each factor's priority and relative importance (Tables 4, 5 and 6) with coping capability and cyclone hazard. After that, the normalized Eigenvector Matrix has been generated by dividing each column's corresponding sums (Estember and Abiog 2018) value. Next, the Lambda (λ) max has been calculated by integrating the pairwise comparisons matrix's column sum value and normalized Eigen vector Matrix's normalized weights value. This Lambda (λ) max value has been used to generate the consistency ratio (CR). The CR value provides the likelihood of the matrix's judgments. The standard tolerable CR value is less than 10% (<0.1); above (>10), this not acceptable. The CR of this matrix's has calculated (Eq. 1) using the divide function between CI (Consistency Index) and RI (Random Index), where CI is calculated (Eq. 2) by using Lambda max and RI. RI is the constant value of Saaty's scale (Tables 7 and 8) (Fig. 10).

$$CR = \frac{CI}{RI} \tag{1}$$

$$CI = \frac{(\lambda_{\max} - n)}{(n-1)} \tag{2}$$

Table 4 Preference of decision-making value

Scale value	Statement
1	Same importance
2	Same, but slightly more
3	Medium importance
4	High importance
5	Very high importance
6	Extreme slightly low
7	Extreme
8	Very extreme but slightly low
9	Very extreme

Table 5 Consolidated matrix = Weighted geometric mean of participants

	Relative relief	Euclidean distance from road	Euclidean distance from river	Pucca house	Health facility	Literate population	Cyclone shelter
Relative relief	1.00	2.67	0.90	1.72	1.46	2.03	1.12
Euclidean distance from road	0.37	1.00	0.47	0.85	0.90	0.86	0.80
Euclidean distance from river	1.12	2.14	1.00	1.39	1.39	1.25	1.12
Pucca house	0.58	1.17	0.72	1.00	1.00	1.20	0.90
Health facility	0.68	1.12	0.72	1.00	1.00	1.12	1.00
Literate population	0.49	1.16	0.80	0.84	0.90	1.00	1.00
Cyclone shelter	0.90	1.25	0.90	1.12	1.00	1.00	1.00

Table 6 Normalization matrix of Cyclone Coping Capacity

	Relative relief	Euclidean distance from Road	Euclidean distance from River	Pucca house	Health facility	Literate Population	Cyclone Shelter	Normalization
Relative Relief	0.19	0.25	0.16	0.22	0.19	0.24	0.16	0.203494404
Euclidean distance from road	0.07	0.10	0.08	0.11	0.12	0.10	0.12	0.099058689
Euclidean distance from river	0.22	0.20	0.18	0.18	0.18	0.15	0.16	0.181602031
Pucca house	0.11	0.11	0.13	0.13	0.13	0.14	0.13	0.126063482
Health facility	0.13	0.11	0.13	0.13	0.13	0.13	0.14	0.128902183
Literate population	0.10	0.11	0.15	0.11	0.12	0.12	0.14	0.119374484

Table 7 Random Consistency Indexes Scale, Thomas L. Saaty

Number	1	2	3	4	5	6	7	8	9
RI	0.0	0.0	0.52	0.89	1.11	1.25	1.35	1.45	1.49

Table 8 Normalized principal Eigenvector Matrix of Coping Capacity

Matrix	Relative relief	Euclidean distance from road	Euclidean distance from	Pucca house	Health facility	Literate populatio		normalize d principal Eigen vector
Relative relief	1	2 2/3	8/9	1 5/7	1 1/2	2		20.35%
Euclidean distance from road	3/8	1	½	6/7	8/9	6/7	4/5	9.91%
Euclidean distance from river	1 1/9	2 1/7	1	1 2/5	1 2/5	1 1/4	1 1/9	18.16%
Pucca house	4/7	1 1/6	5/7	1	1	1 1/5	8/9	12.61%
Health facility	2/3	1 1/9	5/7	1	1	1 1/9	1	12.89%
Literate population	1/2	1 1/6	4/5	5/6	8/9	1	1	11.94%
Cyclone shelter	8/9	1 1/4	8/9	1 1/9	1	1		14.15%

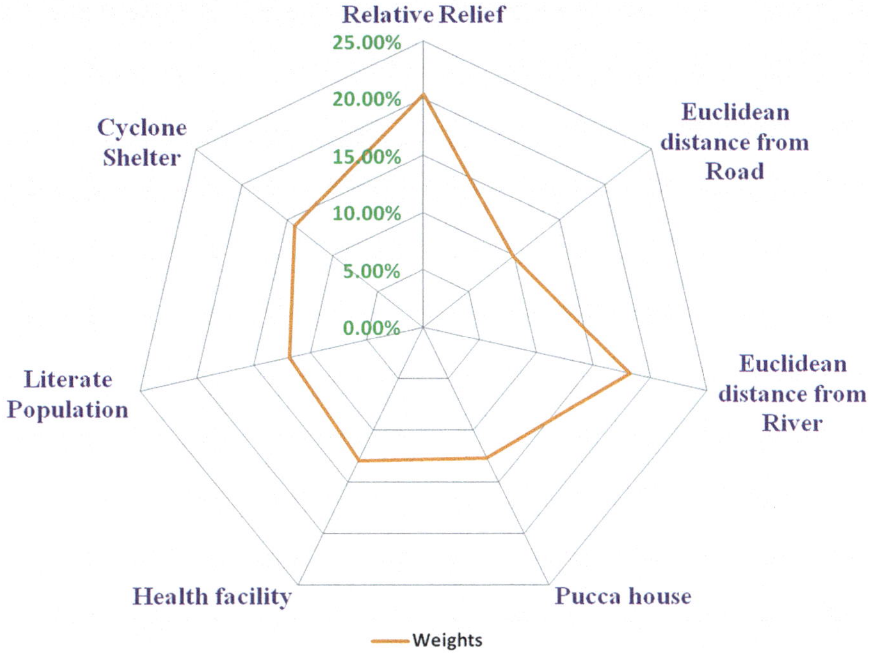

Fig. 10 Radial plot showing relative weights of the variables

5 Results

5.1 Coping Capacity Index

The Cyclone Coping Capacity Index is an essential part of cyclone risk assessment because it provides the cyclone resistance area for the entire areas. Cyclone Coping Capacity Index of Bhadrak District has been generated using geophysical and social involvement factors, which are relative relief, distance from rivers, distance from roadside, household types, emergency shelter, educational level and health facilities. The weightage of each parameter sub-class was assigned based on the individual capacity index of all considered parameters. The AHP technique has applied to generate all parameter's influencing factor, according to the priorities-based pairwise comparison matrix. After that, the weighted overlay method has been used to develop Cyclone Coping Capacity Index (CCI) of the Bhadrak District based on the weights and influencing factors. The CCI has categorized into very high CCI, high CCI zone, medium CCI zone, and low CCI zone. The results indicate that 27,645.65 hectares of land fall within the very high Cyclone Coping Capacity Index (CCI) zone, representing 11.60% of the total study area. Similarly, the high CCI zone covers 58,079.26 hectares (24.37%), while the moderate CCI zone extends over 73,115.70 hectares (30.68%). The low CCI zone accounts for the largest share, encompassing 79,409.62 hectares, which corresponds to 33.35% of the study area. These findings suggest notable spatial variation in cyclone coping capacity across the region. It has been observed (Fig. 11) that the northeastern and some part of the northwestern side of this district have fallen under the very high and high cyclone coping capacity zone, which means this is the area where less cyclone effect has been observed (Table 9). Approximately 11.60% of the district falls within the very high Cyclone Coping Capacity Zone, primarily covering the Basudebpur, Bant, and Bhadrak tehsils. Accordingly, 30.68% and 33.35% of the district fall within the moderate and low coping capacity zones, respectively, with these areas predominantly located in the Tihidi, Chandabali, and Dhamnagar tehsils. Table 10 shows block-wise distribution of cyclone coping capacity area of the study area (Fig. 12).

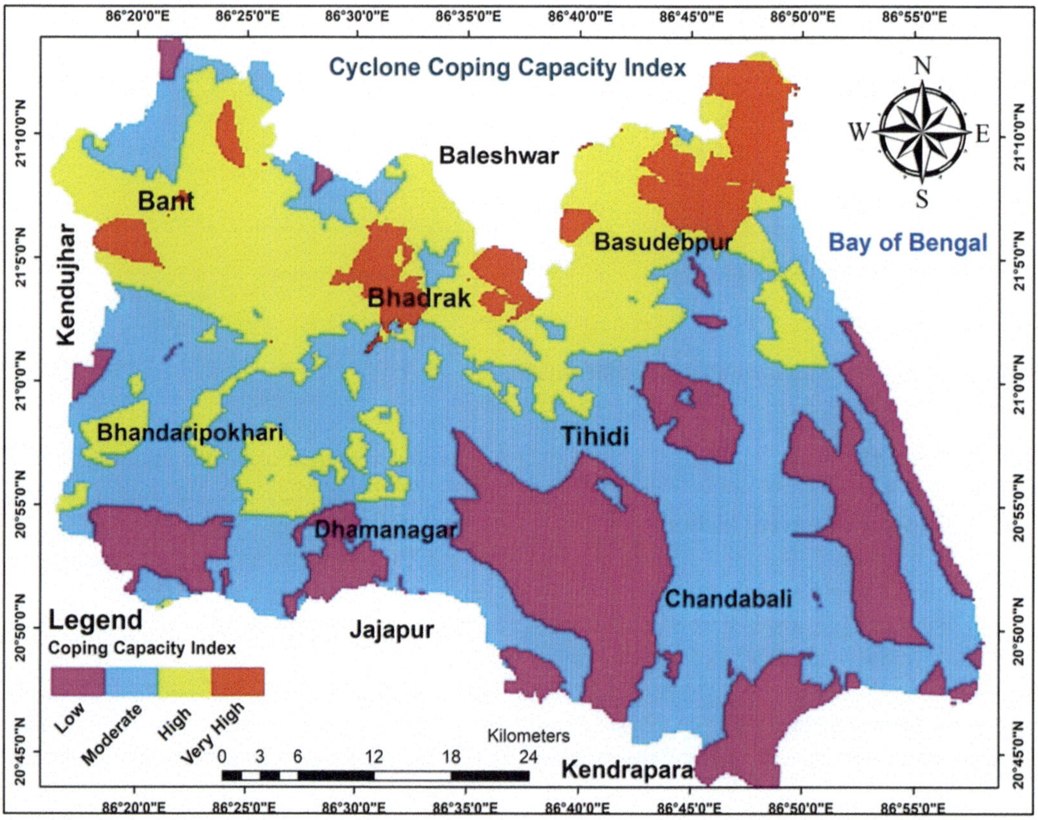

Fig. 11 Cyclone Coping Capacity Index

Table 9 Cyclone coping Capacity Zone (Area/ha)

Sl. no	Coping Capacity Index	Area/ha	(%)
1	Low	79409.62	33.35
2	Moderate	73115.70	30.68
3	High	58079.26	24.37
4	Very high	27645.65	11.60
	Total	**245105.23**	**100%**

Table 10 Block-wise Coping Capacity Index areas in hectares

Coping Capacity Index			Coping Capacity Index areas in hectares			
			Low CCI	Moderate CCI	High CCI	Very high CCI
Block name	**1**	Bant	121.36	8446.56	16787.27	1113.98
	2	Basudebpur	301.79	16114.05	16759.17	12788.75
	3	Bhadrak	1.36	704.38	20711.09	12704.46
	4	Bhandaripokhari	1979.00	17050.31	3490.75	–
	5	Chandabali	47329.77	10040.62	–	–
	6	Dhamnagar	10802.33	11564.34	2192.54	–
	7	Tihidi	18758.49	12151.90	94.66	0.73
Total 245105.23			79409.62	73115.70	58079.26	27645.65
Area in percentage 100%			33.35%	30.68%	24.37%	11.60%

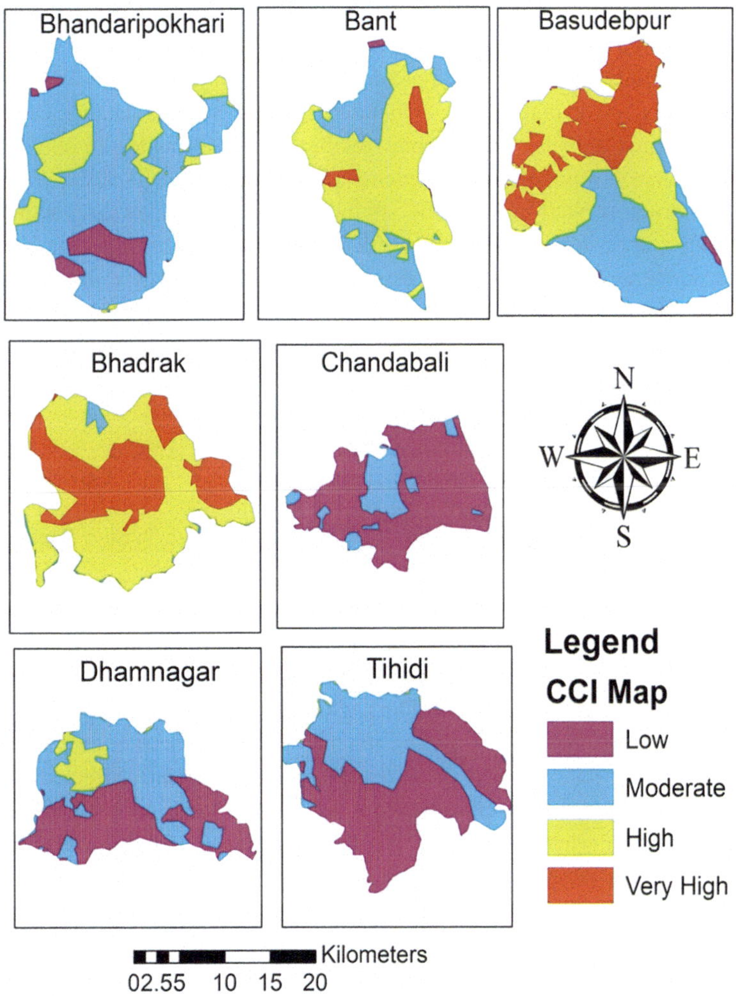

Fig. 12 Block-wise Coping Capacity Index

6 Discussion

The increasing frequency and intensity of cyclones due to climate change have highlighted the urgent need for communities to enhance their coping capacities. Human activities and unplanned development have exacerbated climate change and biodiversity loss, making it imperative to develop robust strategies for disaster management. This study focuses on the application of the analytical hierarchy process (AHP) to create a Cyclone Coping Capacity Index (CCI) for Bhadrak District, Odisha, India, using geospatial techniques. The findings indicate that 33.35% of the district falls into low coping capacity zones, primarily covering Tihidi, Chandabali, and Dhamnagar tehsils. To enhance resilience in these areas, immediate investments in critical infrastructure such as cyclone shelters, emergency response facilities, and health centers are recommended. Additionally, establishing all-weather roads to improve access to emergency services during cyclones would be beneficial.

For areas identified as having moderate coping capacities (30.68% of the district), such as parts of Basudebpur, Bant, and Bhadrak Tehsils, targeted programs focusing on community preparedness and capacity building could significantly improve resilience. Training programs on disaster preparedness, emergency response, and first aid, along with public awareness campaigns on evacuation protocols, can empower local communities to respond more effectively to cyclones. The regions classified under very high coping capacity zones (11.60% of the district) show lesser cyclone impacts due to robust infrastructure and better access to emergency services. Replicating the successful strategies from these zones—such as reinforced housing, proximity to cyclone shelters, and well-developed road networks—in lower-capacity areas could mitigate cyclone risk district-wide.

The use of geospatial technologies like GIS and remote sensing in conjunction with AHP has proven effective in identifying vulnerable areas. For continuous improvement, integrating real-time geospatial data with disaster management systems is recommended to dynamically monitor changes in cyclone patterns, land use, and population distribution, thereby allowing for more responsive and adaptive disaster planning. Developing a district-level disaster management plan that aligns with the CCI outcomes would ensure that high-risk areas receive adequate support in terms of preparedness, response, and recovery efforts.

In the long term, structural mitigation measures such as building embankments, afforestation in coastal areas, and enhancing water drainage systems could help reduce the impacts of cyclones. Incorporating these strategies into urban and rural planning will help build sustainable resilience against future cyclonic events. While this study provides a comprehensive assessment of cyclone coping capacity, future research could focus on incorporating additional socioeconomic variables, such as income levels and access to insurance, to further refine the CCI. Validation of the CCI through field surveys and historical cyclone impact data would enhance its reliability and utility for disaster management authorities.

7 Conclusion

This study Coping Capacity Index is proportionally or reciprocally related to the cyclone hazard. It explores the survival capability during the cyclone hazards situation start. The Coping Capacity Index increases the adaptation capacity of cyclone hazard. This study area is falling under Baitarani and Salandi River Basin, which is frequently affected by cyclone and flooding. For this reason, a suitable cyclone survival area (Coping Capacity Index) has produced with the help of AHP method. This study suggests that the Cyclone Coping Capacity Index of the Bhadrak District is mainly controlled by the relative heights and distance from river systems. Human activities and unplanned development are major contributors to climate change and the destruction of biodiversity globally. Cyclones, in particular, represent a significant natural disaster affecting both coastal and inland communities worldwide. Climate change has exacerbated these effects, as rising sea surface temperatures and other changing physical properties intensify the frequency and damage of tropical cyclones.

Communities that are unprepared are less capable of coping with the effects of such disasters, leading to greater suffering and damage. This study aimed to enhance the understanding of cyclone coping capacity using the analytical hierarchy process (AHP) to produce a Cyclone Coping Capacity Index (CCI) for Bhadrak District, Odisha, India, leveraging geospatial techniques. The multi-objective decision-making approach of AHP provided a quantitative evaluation of various metrics based on expert judgment. This involved designing a hierarchical framework, creating judgment matrices, determining index weights, and checking for consistency.

To combat the severe impact of cyclones, it is essential for human communities to increase their coping capacity. In this study, AHP was utilized to create the CCI, taking into account criteria such as relative relief, distance from roads and rivers, the presence of pucca houses, health facilities, and other social and geophysical factors. The parameters were weighted according to their individual capacity indices, and a weighted overlay method was used to develop the Coping Capacity Index (CCI) for Bhadrak District. Overall, the study demonstrates the utility of AHP in assessing cyclone coping capacity and highlights the importance of geophysical and social factors in mitigating the impacts of cyclones. The findings can inform disaster management strategies, ensuring better preparedness and resilience in the face of future cyclonic events.

References

Ali, A. (1999). Climate change impacts and adaptation assessment in Bangladesh. *Climate Research, 12*(2–3), 109–116.

Boroushaki, S., & Malczewski, J. (2010). Using the fuzzy majority approach for GIS-based multicriteria group decision-making. *Computers, Environment and Urban Systems, 34*(4), 322–332.

Chen, F., Jia, H., & Zhang, C. (2020). A comprehensive method for evaluating marine disaster risk reduction capacity in China. *Sustainability, 12*, 825.

Cutter, S. L., Boruff, B. J., & Shirley, W. L. (2003). Social vulnerability to environmental hazards. *Social Science Quarterly, 84*, 242–261. https://doi.org/10.1111/1540-6237.8402002

D'Oyley, V., & Blunt, A. (1994). *Education and development: Lessons from the third world.* Detseling Enterprises.

DTRS (Department of Transport & Regional Services) (2004). Natural disaster in Australia: Reforming mitigation, relief & recovery arrangements—A report to the Council of Australian Governments by a High Level O_cial'sGroup (2002–2008); Department of Transport & Regional Services: Canberra, Australia.

Dube, S. K., Jain, I., Murty, T. S., & Agnihotri, N. (2009). Storm surge modeling for the Bay of Bengal and Arabian Sea. *Natural Hazards, 51*(1), 3–27.

Estember, R. D., & Abiog, M. C. M. (2018). Vulnerability assessment of Pangasinan Province to typhoons, floodsand landslides. In *Proceedings of the 4th electronic and green materials international conference, Bandung, Indonesia, 27–28 July 2018.* AIP Conference Proceedings.

FEMA (Federal Emergency Management Agency). (2009). State capability assessment for readiness. Retrieved March 12, 2020, from https://www.docin.com/p-238379278.html

Ferreira, M. A., de Sa, F. M., & Oliveira, C. S. (2016). The disruption index (DI) as a tool to measure disaster mitigation strategies. *Bulletin of Earthquake Engineering, 14*, 1957–1977.

Gao, Z. C., Yu, M., & Ding, Z. D. (2012). Typhoon storm surge risk assessment based on GIS—A case study of Taizhou.Marine. *Environmental Sciences, 31*(439–442), 447. (In Chinese).

Greiving, S., Fleischhauer, M., & Wanczura, S. (2006). Management of natural hazards in Europe: The role of spatial planning in selected EU member states. *Journal of Environmental Planning and Management, 49*(5), 739–757.

Hu, J. F., Yang, P. G., Yang, Y. Q., & Wu, J. J. (2010). Study on evaluation index system and method for flood control and disaster reduction capacity. *Journal of Natural Disasters, 19*, 82–87. (In Chinese).

Huang, X., Song, J., Li, X., & Bai, H. (2020). Evaluation model of synergy degree for disaster prevention and reduction in coastal cities. *Natural Hazards, 100*, 933–953.

IDNDR (International Decade for Natural Disaster Reduction) (1994). *Yokohama strategy and plan of action fora safer world: Guidelines for natural disaster prevention, preparedness and mitigation.* Retrieved March 4, 2020, from https://www.unisdr.org/we/inform/publications/8241

IPCC (Intergovernmental Panel on Climate Change). (2014). Climate change 2014: Impacts, adaptation, and vulnerability. *Contribution of Working Group II to the fifth assessment report of the intergovernmental panel on climate change*. Cambridge University Press.

Jha, R. K., & Gundimeda, H. (2019). An integrated assessment of vulnerability to floods using composite index- a district level analysis for Bihar, India. *International Journal of Disaster Risk Reduction, 35*, 101074.

Jha, A., Miner, T. W., & Stanton-Geddes, Z. (2016a). *Building urban resilience: Principles, tools, and practice*. World Bank Publications.

Jha, M., Bloch, R., & Lamond, J. (2016b). *Cities and flooding: A guide to integrated urban flood risk management for the 21st century*. World Bank Publications.

Jiang, W. G., Deng, Y., Tang, Z. H., Cao, R., Chen, Z., & Jia, K. (2016). Adaptive capacity of mountainous rural communities under restructuring to geological disasters: The case of Yunnan Province. *Journal of Rural Studies, 47*, 622–629.

Keogh, D. U., Apan, A., Mushtaq, S., King, D., & Thomas, M. (2011). Resilience, vulnerability and adaptive capacity ofan inland rural town prone to flooding: A climate change adaptation case study of Charleville, Queensland, Australia. *Natural Hazards, 59*, 699–723.

Knutson, T. R., McBride, J. L., Chan, J., Emanuel, K., Holland, G., Landsea, C., Held, I., Kossin, J. P., Srivastava, A. K., & Sugi, M. (2010). Tropical cyclones and climate change. *Nature Geoscience, 3*(3), 157–163.

Malik, S., & Mahmood, S. (2015). Earthquake vulnerability assessment using analytic hierarchy process (AHP): A case study of Srinagar city, Jammu and Kashmir, India. *Journal of the Geological Society of India, 86*(1), 105–112.

Maly, E., & Suppasri, A. (2020). The Sendai framework for disaster risk reduction at five: Lessons from the 2011 greatest Japan earthquake and Tsunami. *International Journal of Disaster Risk Science, 11*, 167–178.

Mohapatra, M., Bandyopadhyay, B. K., & Tyagi, A. (2012). Best track parameters of tropical cyclones over the North Indian Ocean: A review. *Natural Hazards, 63*(3), 1285–1317.

Mohapatra, M., Bandyopadhyay, B. K., & Rathore, L. S. (2015). *Tropical cyclones over the North Indian Ocean*. Springer Science & Business Media.

Montz, B. E., Tobin, G. A., & Hagelman, R. R. (2017). *Natural hazards: Explanation and integration*. Guilford Press.

Mustak, S. K., Baghmar, N. K., & andRatre, C.R. (2012). Measurement of dissection index of Pairi River basin using remote sensing and GIS. *National GEOGRAPHICAL JOURNAL of India, 58*(2), 97–106.

Paul, S. K., & Routray, J. K. (2010). Flood proneness and coping strategies: The experiences of two villages in Bangladesh. *Disasters, 34*(2), 489–508.

Rahman, M. S., Haque, A., & Islam, M. M. (2019). Assessment of flood risk and vulnerability of communities in the northeastern region of Bangladesh using a multi-criteria decision-making approach. *International Journal of Disaster Risk Reduction, 41*, 101288.

Saaty, T. L. (1980). *The analytic hierarchy process: Planning, priority setting, resource allocation*. McGraw-Hill.

Sahoo, B., & Bhaskaran, P. K. (2016). Multi-hazard risk assessment of coastal vulnerability from tropical cyclones—A GIS based approach for the Odisha coast. *Journal of Coastal Conservation, 20*(1), 1–17.

Schneiderbauer, S., & Ehrlich, D. (2006). Risk and vulnerability mapping for the implementation of disaster risk reduction strategies. *Disasters, 30*(1), 40–58.

Shi, Y., Zhai, G., Zhou, S., Lu, Y., Chen, W., & Liu, H. (2018). How can cities adapt to a multi-disaster environment? Empirical research in Guangzhou (China). *International Journal of Environmental Research and Public Health, 15*, 2453.

Singh, S. (1992). Quantitative geomorphology of the Drainage Basin, reading on remote sensing application. In T. S. Chouhan & K. N. Joshi (Eds.). Scientific Publishers.

UNISDR. (2017). *Terminology on disaster risk reduction*. UNISDR. Retrieved March 19, 2020, from https://www.unisdr.org/files/52828_nationaldisasterriskassessmentpart52821.pdf

Vargas, L. G. (2010). Using the analytic hierarchy process (AHP) to select and prioritize projects in a portfolio. *International Journal of Project Management, 28*(5), 512–523.

Walsh, K. J. E., McBride, J. L., Klotzbach, P. J., Balachandran, S., Camargo, S. J., Holland, G., Knutson, T. R., Kossin, J. P., Lee, T. C., Sobel, A., & Sugi, M. (2016). Tropical cyclones and climate change. *Wiley Interdisciplinary Reviews: Climate Change, 7*(1), 65–89.

Xie, L. L. (2006). A method for evaluating cities' ability of reducing earthquake disasters. *Earthquake Engineering & Engineering Dynamics, 26*, 1–10. (In Chinese).

Yager, R. R. (2008). Prioritization in multi-criteria decision making. *Fuzzy Optimization and Decision Making, 7*(3), 181–199.

Zeng, L. F. (1996). Risk evaluation of typhoon disasters in the coastal areas in Guangxi. *Journal of Catastrophology, 11*, 43–47.

Navigating Urban Floods Risk Assessment: Insights from HEC-RAS and Mitigation Strategies

Azim Jamal, Shruti Kanga, Suraj Kumar Singh,
Tauseef Ahmad, and Deepak

1 Introduction

Flooding is one of the disasters that can be a natural phenomenon as well as due to the advancement of human civilization that has enabled *Homo sapiens* of creating a flood-like situation with the mismanagement of its infrastructure or an adverse situation that is caused due to failure of hydrological infrastructure. Flood in a region is mainly caused by two events, one of which is failure of human infrastructure and the other that is a natural phenomenon, i.e., precipitation with a high intensity in a very short interval of time, which results in inundation of land area around a water body (Pathan & Agnihotri, 2021). Flood is a disaster that has a global presence and has caused loss of billions of dollars every year in terms of economic loss and many human lives along with the fauna have been lost due to it. As per a report by *Statista*, the world has lost more than 2.9 trillion of dollars worth of assets including infrastructure and non-infrastructural assets such as natural resources. The World Meteorological Organization (WMO) report on disasters states that 2681 disasters have occurred only in the continent of Asia during the period from 1970 to 2012, causing a loss of more than 789 billion dollars worth of economic assets be it infrastructure or loss of resources. India is classed into the category of highly prone to disasters and as per the report said to have lost somewhere between 52 billion dollars and 100 billion dollars worth of economic assets (World Meteorological Organization, 2014). One of the worst flood-related disasters occurred in India during the month of July 1993. India incurred the wrath of floods in eight of its states. The amounts of losses were innumerable when it comes to human losses and was estimated to be 11.1 billion dollars in economic terms (World Meteorological Organization, 2014). As per the report of *Annual Disaster Statistical Review*, the world witnessed a massive total of 69 hydrological disasters of which India alone accounts for approximately 15% of the total number of hydrological disasters and ranks third on the list of worst hit countries after Republic of China and Indonesia, in the year 2016 (Guha-Sapir et al., 2016). More than 4 billion of humans currently alive on the planet were prone to natural disasters in the year 2017,

A. Jamal · S. Kanga (✉) · Deepak
Department of Geography, School of Environment and Earth Sciences, Central University of Punjab, VPO-Ghudda, Bathinda, India
e-mail: shruti.kanga@cup.edu.in

S. K. Singh
Centre for Climate Change and Water Research, Suresh Gyan Vihar University, Jaipur, India
e-mail: suraj.kumar@mygyanvihar.com

T. Ahmad
School of Computer Science Engineering, Sandip University, Madhubani, India
e-mail: tauseef.ahmad@sandipuniversity.edu.in

Z. Khanam et al. (eds.), *Climate Change, Resilience and Ecosystem Services*, Sustainable Development Goals Series, https://doi.org/10.1007/978-3-031-96977-5_19

and the number becomes 11.2 billion when counting the population affected or exposed to several disasters in the same region (Kishore et al., 2018). In India alone, more than 15 million of the population are potentially exposed to floods (Kishore et al., 2018). A more alarming account of the disasters is presented in the report by World Meteorological Department that records the total number of disasters to be 432 that India was hit by, between the years 1972 and 2012 (World Meteorological Organization, 2014).

Disaster risk assessment is one of the keys to successful mitigation of disasters, and flood is no exception to that. Flood risk assessment mandates a study of floodplain, river channel, river flow, precipitation, and meteorological as well as hydrological aspects of the area or region for flood mitigation (Chandrashekhar et al., 2003). Floods are frequent phenomenon in several parts of the country (Rafiq et al., 2016). Urban floods are modern-day phenomenon that have gripped our cities in particular, and urban areas in general. Most of the large cities in India namely Delhi, Chennai, Mumbai, Patna are troubled by flood of varying intensities (Rafiq et al., 2016).

2 Data and Methods

2.1 Study Area

The National Capital Territory (NCT) of Delhi is a union territory in India. It is divided into 11 districts, with the most recent delimitation occurring in the year 2011. The magnificent city of Indraprastha was blessed with the water of the divine Yamuna River that was brought from the divine space to earth via the seven sages and their meditation (Husain et al., 2018). At present the river trickles down through the union territory of Delhi, enveloping the capital of the country, New Delhi (Delhi Jal Board, 2016). The NCT of Delhi is enveloped by the state of Haryana from three sides and in the east of it is Uttar Pradesh (Tomar et al., 2021). The economic profile of the region is very diverse and most of the land in the region has been utilized for settlement and industries, though the outskirts of the region or the borders of the region are employed in agriculture and farming. The population profile of Delhi is very diverse; it would not be an exaggeration to call it a mini-India. As a matter of fact, the region houses population from all the corners of India as well as the world. As per the 2011 census of India, the NCT of Delhi is home to more than 16 million (1.6 crore). The sex ratio stands at 868 females per thousand males (Census, 2011). Natural disasters are ubiquitous phenomena, happening all over the globe. According to the report by Khan, Amir A. 2023, these calamities have a significant impact on the development and economy of a nation or country or region. The problem of flood is not new to the vicinity and floods have hampered the growth and development of the city from time to time. As per the report published by the Department of Flood and Irrigation of NCT of Delhi, the territory has experienced floods in the years 1967, 1971, 1975, 1976, 1978, 1988, 1995, 1998, 2011, and 2018 in the Yamuna River. The most recent of these flooding events was seen in the year 2023, during the month of July, when the water surface elevation rose to danger level and breached the mark of highest recorded level at the Old Railway Bridge that stood at 207.49 meters since 1978 at several places in the territory (IANS. 2023, July 19).

Nine major floods that occurred in 1924, 1947, 1976, 1978, 1988, 1995, 1998, 2010, and 2013, when the Yamuna River exceeded its danger level of 204.83 meters, have affected Delhi since 1900. The Yamuna River in Delhi reached 207.49 meters in 1978, resulting in 2.53 lakh cusec of water discharged at the Old Railway Bridge and 7.0 lakh cusec discharged from Tajewala. The flood caused flooding in 25 urban colonies and 130 villages in Delhi. It was the worst flood to ever occur in the city (Tomar et al., 2021) (Fig. 1).

Land use land cover (LULC) refers to the classification of land based on its current use (e.g., agriculture, urban) and the physical characteristics of its surface (e.g., forest, water bodies). It is used for environmental monitoring, urban planning, and natural resource management.

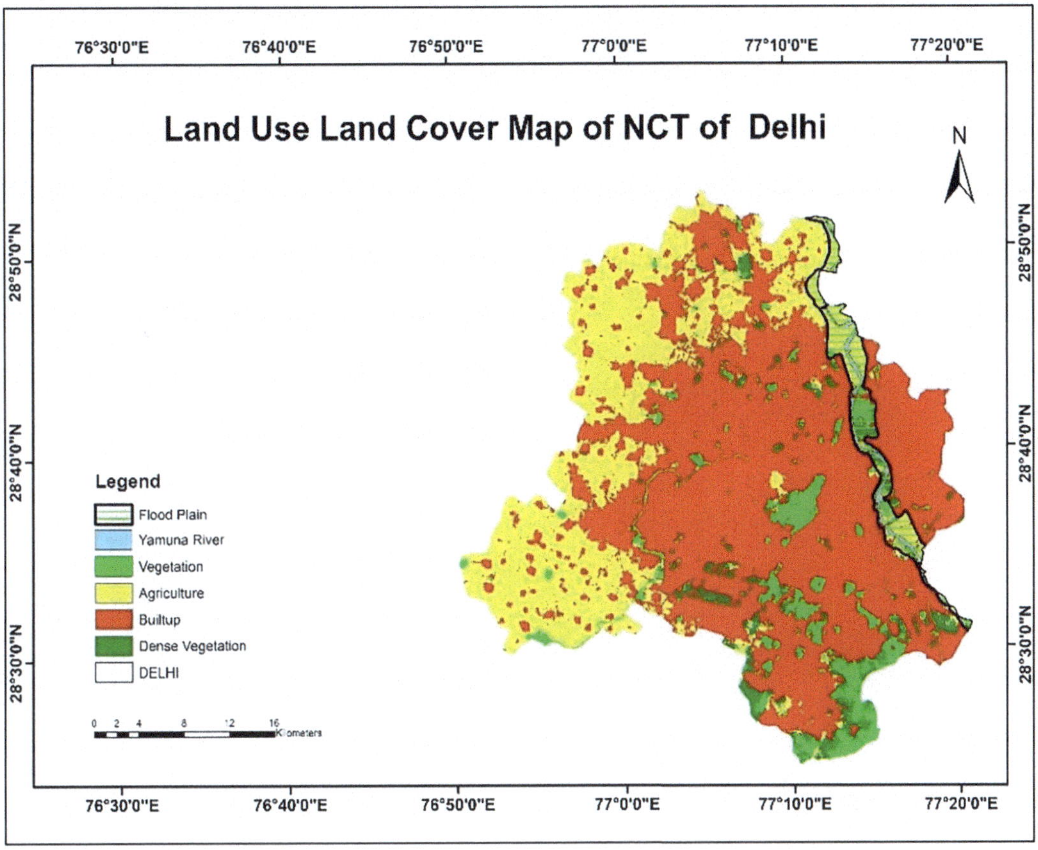

Fig. 1 Land use land cover of Delhi NCT, 2023 (Source: ESRI LULC)

2.2 Datasets

The software Hydrologic Engineering Center's River Analysis System (HEC-RAS) is developed and designed by the U.S. Army Corps of Engineer for hydrological analysis of rivers. The version employed for the study is 6.4.1. The software is used for both one-dimensional (1D) and two-dimensional (2D) modeling and for simulation of situations that inundate or spate an area with water as the lateral flow of river along its channel.

HEC-RAS software is capable of accurately simulating flood scenarios based on the input data. The datasets required for simulation can be broadly categorized into two categories, namely Geometric Data and Hydrologic Data. Geometric data includes digital elevation model (DEM) data, LULC (land use land cover data), and satellite imagery, whereas hydrologic data contains data related to discharge amount and flow hydrographs. Another important data that is required or warranted by the software are the Manning's n values. The Manning's roughness coefficient (n) is a crucial parameter in the Manning's equation, which is employed to calculate the velocity of water flow in open channels. The value of Manning's n is contingent upon the physical characteristics of the channel surface and the presence of vegetation (Table 1).

The simulation is a multi-step process that involves creation of the digital terrain model (DTM) using digital elevation model (Shuttle Radar Topographic Mission [SRTM] 30 m). Creation of 2D flow area, computation of mesh, and establishing an inflow and outflow points in the mesh are very crucial steps in the calibration for 2D simulation modeling

Table 1 Mannings *n* values (Hydrologic Engineering Center, 2024)

Land use category	Manning's *n* value
Built-up	0.013–0.017
Agriculture	0.020–0.050
Water bodies	0.025–0.060
Forest	0.050–0.110
Pasture	0.030–0.050

(Hydrologic Engineering Center, 2024). Furthermore, the simulation process requires input of hourly discharge data into the unsteady flow data section. Flow hydrograph has been used for the upstream boundary line and normal depth has been employed for the downstream boundary line (Hydrologic Engineering Center, 2024).

2.3 Methodology (Fig. 2)

The use of HEC-RAS involves several steps, from DTM creation to final unsteady flow simulation (Fig. 3).

HEC-RAS software is capable of one- and two-dimensional modeling as well as a combined 1D/2D modeling (Hydrologic Engineering Center, 2024) Following are the equations used by the software for the purpose of simulation and flood modeling (Farooq et al., 2019; Kumar et al., 2020; Mehta et al. 2014).

Continuity equation:

$$\frac{\partial h}{\partial t}+\frac{\partial p}{\partial x}+\frac{\partial q}{\partial y}=0 \tag{1}$$

Momentum equation *x* direction:

$$\frac{\partial p}{\partial t}+\frac{\partial}{\partial x}\left(\frac{p^2}{h}\right)+\frac{\partial}{\partial y}\left(\frac{pq}{h}\right)=n^2pg\frac{p^2+q^2}{h^2}-gh\frac{\partial h}{\partial x}+pf+\frac{\partial}{\partial x}\left(h_{xx}\right)+\frac{\partial}{\partial y}\left(h_{xy}\right) \tag{2}$$

Momentum equation *y* direction:

$$\frac{\partial p}{\partial t}+\frac{\partial}{\partial y}\left(\frac{q^2}{h}\right)+\frac{\partial}{\partial x}\left(\frac{pq}{h}\right)=-n^2qg\frac{p^2+q^2}{h^2}-gh\frac{\partial h}{\partial y}+qf+\frac{\partial}{\partial y}\left(h_{yy}\right)+\frac{\partial}{\partial x}\left(h_{xy}\right) \tag{3}$$

h: Height of the fluid (possibly the water depth if these are shallow water equations),
p: Momentum in the xx-direction,
q: Momentum in the yy-direction,
g: Acceleration due to gravity,
n: Manning's roughness coefficient, related to the frictional effects,
f: Coriolis parameter, accounting for the Coriolis effect if the fluid dynamics are on a rotating planet,
h_{xx}, h_{xy}, h_{yy}: Higher-order diffusion terms or stresses, possibly representing viscosity or turbulence effects in the fluid (Quirogaa et al., 2016).

3 Results

The "Depth Map of Study Area" provides a detailed visual representation of flood inundation levels along the Yamuna River in Delhi, covering districts such as New Delhi, Central, North East, East, South East, and Shahdara. Key observation points, marked with red dots, include Wazirabad Barrage Downstream, Old Railway Bridge, and Okhla Barrage Upstream; these are in order of North to South. These points are essential for monitoring water levels at various sections of the river. The map is a work of 2D simulation that is computed using the HEC-RAS software version 6.4.1 and then processing the exported raster in the Arcmap software version 10.8. The map uses a gradient of blue shades to depict water depth, with dark blue indicating high inundation levels up to 6.285 meters and light blue showing lower depths nearing 0 meters, which explicitly means that the lateral flow of the river is restricted up to that point, but it must be noted that the region also suffers from the problem of water logging. This color-coding effectively highlights the severity of flooding in different areas, making it easy to identify high-risk zones near the river and

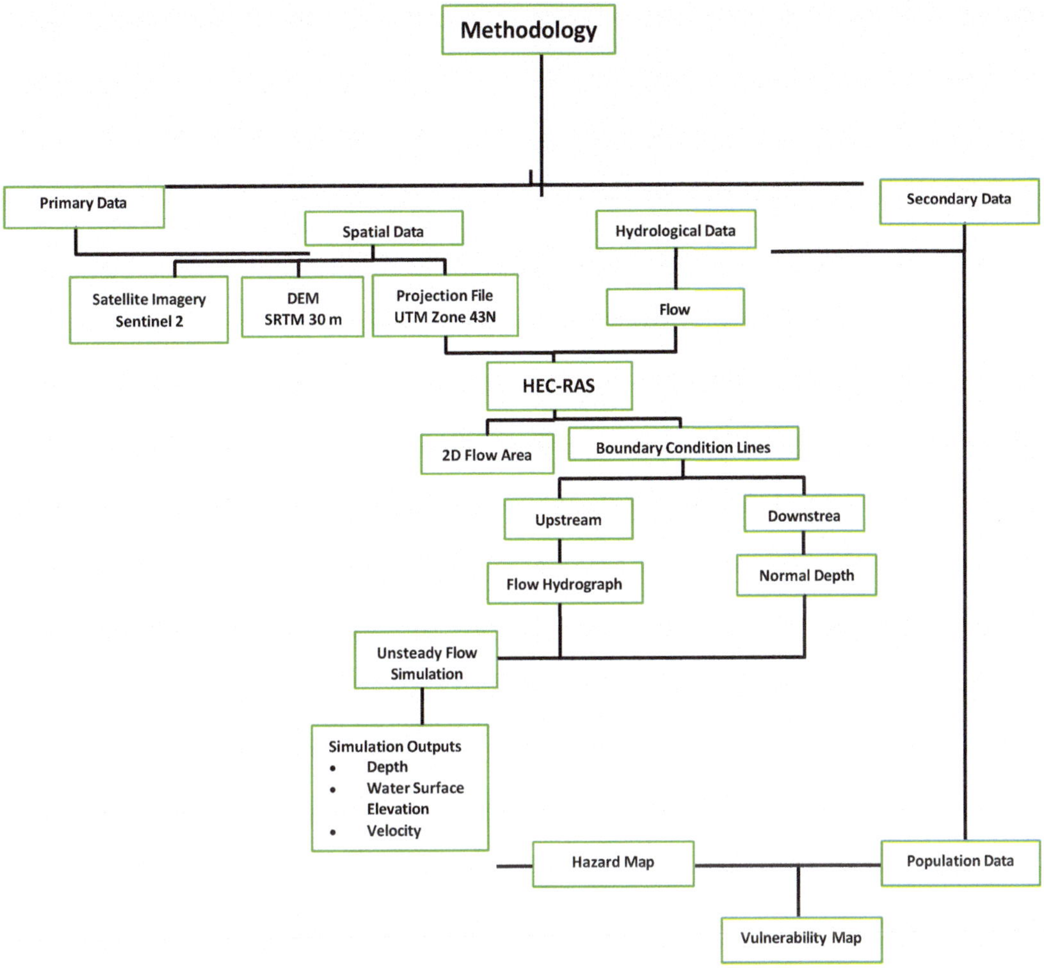

Fig. 2 Methodology flow chart

lower-risk zones further away. Flood-prone areas are clearly marked in blue, showing varying levels of risk. High-risk zones, indicated by darker shades of blue, are closer to the river, while lighter shades represent lower-risk zones situated further away. This delineation helps prioritize areas needing immediate attention and mitigation. District boundaries are outlined, providing a clear view of the impacted regions. The black line representing the Yamuna River's path is the focal point for assessing flood risk. This map is crucial for understanding flood risks, aiding urban planning, flood management, and emergency response strategies in Delhi. It helps urban planners identify vulnerable areas for infrastructural improvements, supports emergency response teams in planning evacuations and relief operations, and assists policymakers in formulating effective flood management policies. Overall, this map enhances the safety and resilience of Delhi's communities against flooding threats (Fig. 4).

The water surface elevation map displays the water surface elevation levels along the Yamuna River in Delhi, encompassing districts such as New Delhi, Central, North East, East, South East, and Shahdara. This map is a crucial tool for understanding the elevation changes and potential flood impact areas along the river, facilitating better urban planning, emergency response, and flood management strategies. Observation points are marked with red dots,

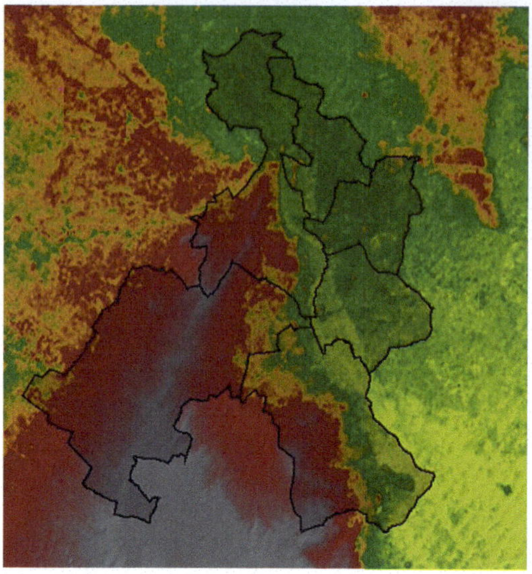

Fig. 3 Digital terrain model

highlighting significant locations including Wazirabad Barrage Upstream in the North East, Old Railway Bridge in Central Delhi, and Okhla Barrage Downstream in the South East, in order of North to South. These points are vital for monitoring the water surface elevation and understanding the hydrodynamics of the river at different segments. The map uses a gradient of colors to indicate the water surface elevation, with green representing lower elevations starting at 194.381 meters and red representing higher elevations up to 208.799 meters. This color gradient allows for easy visualization of elevation changes along the river. The areas with higher elevations are marked in red and orange, indicating regions that might be more prone to elevated water levels during peak flow conditions. Lower elevations, marked in yellow

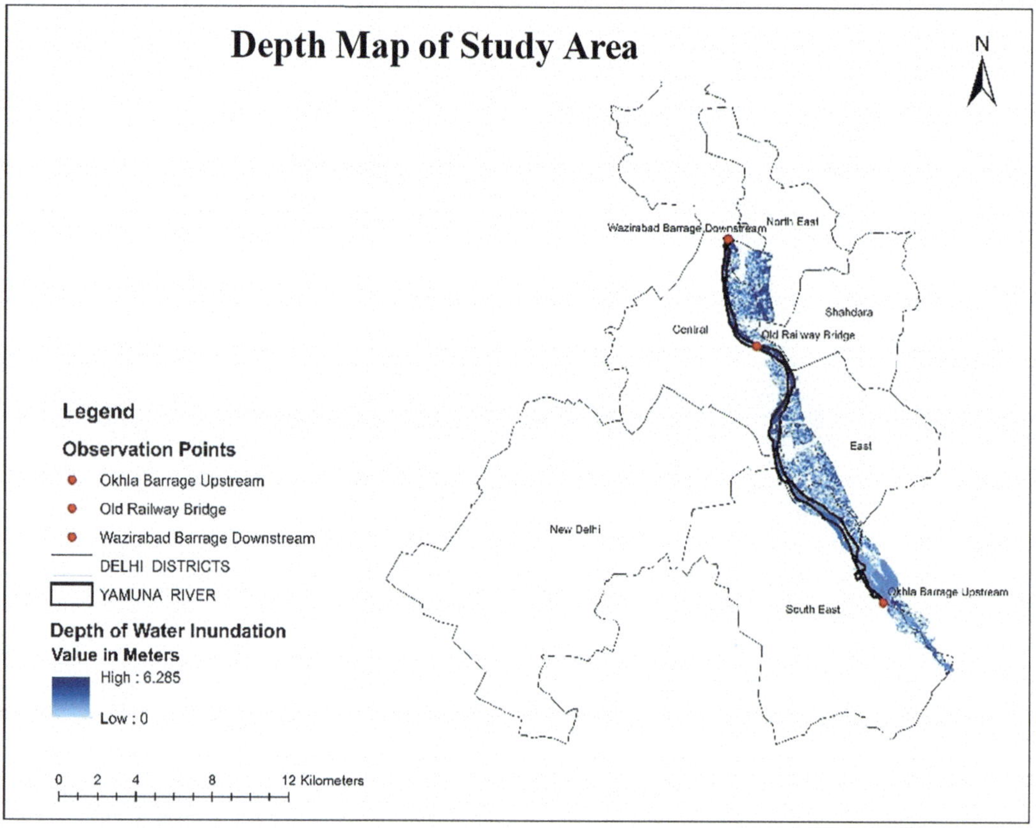

Fig. 4 Water inundation/depth map

and green, are generally downstream and may experience less dramatic elevation changes but are still significant in flood risk assessment. Flood-prone areas are clearly demarcated, with color variations indicating the degree of elevation. The higher elevation zones closer to the river, marked in red and orange, signify areas where water levels can rise significantly, posing a higher risk of flooding. Conversely, lower elevation zones marked in yellow and green represent areas that might experience less severe water level changes but are still at risk of inundation. District boundaries are outlined to provide a clear view of the regions impacted by water surface elevation changes. The black line representing the Yamuna River's path serves as a reference for assessing the elevation profile along its course. This delineation helps in understanding the geographical distribution of flood risks and the potential impact on adjacent areas. In conclusion, this "Water Surface Elevation Map of Study Area" is an essential resource for understanding the water surface elevation dynamics along the Yamuna River in Delhi. It aids urban planners in identifying vulnerable areas requiring infrastructural improvements, supports emergency response teams in planning effective evacuations and relief operations, and assists policymakers in formulating robust flood management policies. By providing detailed and visual insights into water surface elevation changes, this map enhances the preparedness and resilience of communities in Delhi against potential flood threats (Fig. 5).

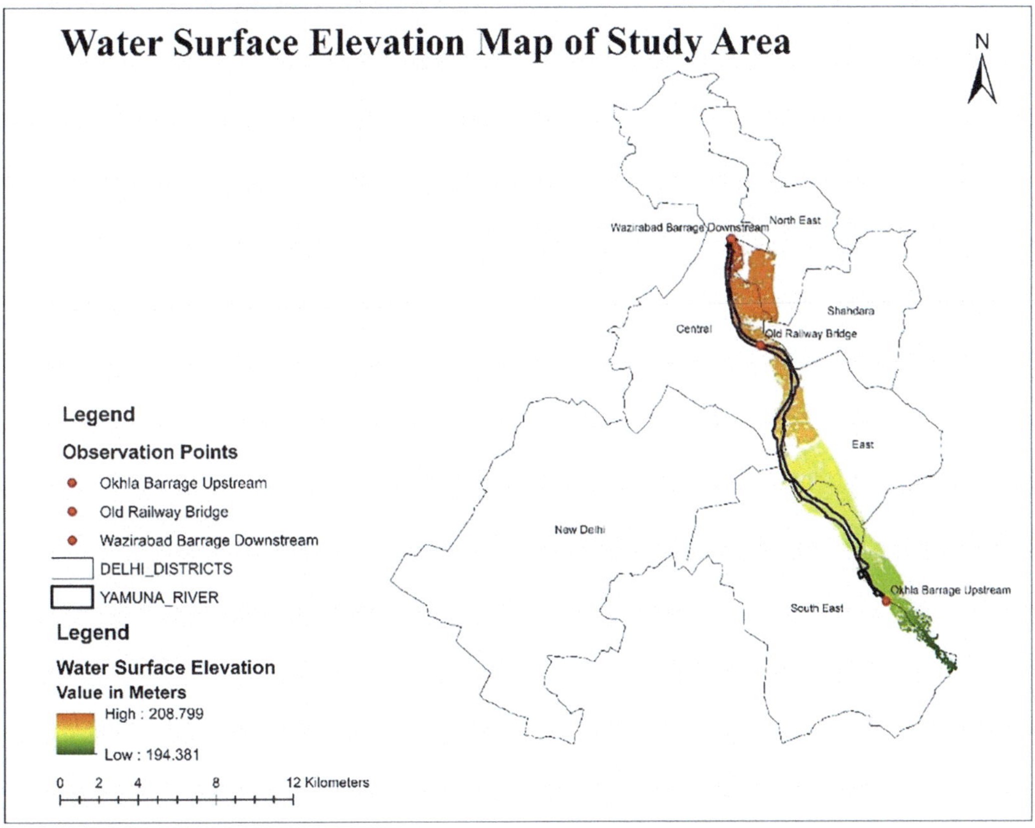

Fig. 5 Water surface elevation map

4 Discussions

The study conducted on flood dynamics along the Yamuna River in Delhi, using HEC-RAS software, highlights several critical findings with significant implications for flood risk management. The results demonstrate the strong correlation between urban expansion and increased flood risk, particularly in regions with high population density and impervious surfaces. Central and North East Delhi emerged as the most vulnerable areas, where rapid urbanization has encroached upon natural floodplains, thereby exacerbating surface runoff and reducing water infiltration capacity (Gupta et al., 2017). This aligns with existing literature that links unplanned urban development to heightened flood susceptibility. The simulation outcomes provided a detailed understanding of flood inundation patterns under different discharge scenarios (Khan et al., 2011; Rahmati et al., 2015). Areas near critical points, such as the Wazirabad Barrage and Old Railway Bridge, experienced water levels surpassing danger thresholds, leading to widespread inundation of adjacent settlements. These findings corroborate historical flood events, particularly the catastrophic 1978 flood, which remains a pivotal reference for flood management strategies in the region. The comparison of modeled inundation depths with historical data strengthens the reliability of the HEC-RAS simulation as a predictive tool for future flood events (Teng et al., 2017).

The methodological approach, which integrated high-resolution digital terrain models, discharge hydrographs, and Manning's roughness coefficients, proved effective in capturing the spatial extent and severity of flooding (Ansari et al. 2023). The use of 2D hydrodynamic modeling enabled a nuanced representation of water flow and inundation in urban settings, addressing the limitations of traditional 1D models (Ongdas et al., 2020). Moreover, the coupling of hydrometeorological data with the simulation highlighted the critical influence of precipitation intensity and river discharge on flood dynamics, reinforcing the importance of real-time data integration for accurate forecasting (Wang & Eltahir 1999). Despite the strengths of the modeling approach, challenges remain in translating these insights into actionable flood mitigation strategies. Encroachments on floodplains and limited drainage capacity continue to impede effective flood management(Klapka et al., 2010). Additionally, the increasing variability in precipitation patterns due to climate change adds complexity to predicting and managing flood risks (Chandrashekhar et al., 2003). The results underscore the need for a multifaceted approach that combines advanced modeling tools with policy interventions aimed at preserving natural floodplains, enhancing drainage infrastructure, and promoting sustainable urban planning practices. This study demonstrates the potential of HEC-RAS in identifying high-risk zones and prioritizing areas for intervention. However, its application must be complemented by robust institutional frameworks and community engagement to ensure effective flood risk reduction. The findings provide a valuable basis for integrating flood risk assessments into urban planning and disaster management frameworks, contributing to a more resilient urban environment in flood-prone regions.

5 Conclusion

The study detailed in this manuscript underscores the paramount significance of hydrodynamic modeling in comprehending and addressing urban flood vulnerabilities, particularly within highly susceptible locales such as the National Capital Territory (NCT) of Delhi. Through the utilization of HEC-RAS software version 6.4.1 to construct a meticulous 2D hydrodynamic model, our investigation meticulously simulated flood scenarios along the expansive stretch of the Yamuna River, spanning from the Wazirabad Barrage to the point of exit from the territory. Our analytical approach capitalized on a rich array of datasets, encompassing the SRTM 30 m digital elevation model (DEM), discharge records sourced from the Flood and Irrigation Department of Delhi, and pertinent census data, amalgamating to furnish a comprehensive depiction of flood dynamics. The resultant findings notably unveiled significant inundation extents across various urban areas, underscoring the imperative for the development and implementation of robust flood management strategies. Furthermore, our research elucidated the compounding influences of both natural phenomena and anthropogenic activities in amplifying flood susceptibilities. With observed water surface elevations surpassing historical benchmarks, the imperative for the integration of sophisticated hydrodynamic modeling methodologies within urban infrastructure planning and management frameworks becomes unequivocally apparent. Moreover, our study advocates for the continual monitoring and refinement of hydrological datasets to iteratively enhance the accuracy and reliability of flood models. The incorporation of diverse land use and land cover data, in conjunction with precise Manning's n values, emerged as pivotal factor in ensuring the fidelity of flood behavior simulations. In summation, the insights gleaned from our investigation possess profound implications for policymakers, urban planners, and disaster management authorities, furnishing invaluable guidance in the formulation and implementation of efficacious flood mitigation strategies. Through the adoption of a data-centric approach and the judicious utilization of advanced modeling tools like HEC-RAS, the mitigation of urban flood perils stands within reach, thereby engendering the safeguarding of lives, assets, and the overall socioeconomic resilience of flood-prone environs.

References

Ansari, M., Jabbari, I., & Sargordi, F. (2023). The effect of water resources on spatial and temporal change of soil salinity in Izdkhast playa, Fars Province, Iran. *Environmental Monitoring and Assessment, 195*(1), 63. https://doi.org/10.1007/s10661-022-10678-5

Census. (2011). Retrieved May 27, 2024, from https://des.delhi.gov.in/sites/default/files/DES/generic_multiple_files/population_of_d elhi_as_per_census_2011_1.pdf

Chandrashekhar, M. B., Singh, S., & Roy, P. S. (2003). Geospatial modelling techniques for rapid assessment of phytodiversity at landscape level in western Himalayas, Himachal Pradesh. *Current Science, 84*, 5.

Farooq, M., Shafique, M., & Khattak, M. S. (2019). Flood hazard assessment and mapping of river swat using HEC-RAS 2D model and high-resolution 12-m TanDEM-X DEM (WorldDEM). *Natural Hazards, 97*(2), 477–492. https://doi.org/10.1007/s11069-019-03638-9

IANS. (2023, July 19). Yamuna river crosses danger mark after briefly receding. The News Minute. https://www.thenewsminute.com/news/yamuna-river-crosses-danger-mark-after-briefly-receding-179923

Guha-Sapir, D., Hoyois, P., Wallemacq, P., & Below, R. (2016). Annual disaster statistical review 2016: The numbers and trends. Centre for Research on the Epidemiology of Disasters (CRED). http://www.cred.be/sites/default/files/ADSR_2016.pdf

Gupta, S. (2017). CLIMATRANS report on impact of floods in Delhi. School of Planning and Architecture, New Delhi.

Hydrologic Engineering Center. (2024). HEC-RAS river analysis system: HEC-RAS hydraulic reference manual (Version 6.6). U.S. Army Corps of Engineers.

Husain, A., Sharif, M., & Ahmad, M. L. (2018). Simulation of floods in Delhi segment of River Yamuna using HEC-RAS. American Journal of Water Resources, 6(4), 162–168.

Jongman, B., Winsemius, H. C., Aerts, J. C., Coughlan de Perez, E., van Aalst, M. K., Kron, W., & Ward, P. J. (2015). Declining vulnerability to river floods and the global benefits of adaptation. Proceedings of the National Academy of Sciences, 112(18), E2271–E2280. https://doi.org/10.1073/pnas.1414439112

Khan, S. I., Hong, Y., Wang, J., Yilmaz, K. K., Gourley, J. J., Adler, R. F., Brakenridge, G. R., Policelli, F., Habib, S., & Irwin, D. (2011). Satellite remote sensing and hydrologic modeling for flood inundation mapping in Lake Victoria Basin: Implications for hydrologic prediction in ungauged basins. *IEEE Transactions on Geoscience and Remote Sensing, 49*(1), 85–95. https://doi.org/10.1109/TGRS.2010.2057513

Kishore, N., Marqués, D., Mahmud, A., Kiang, M. V., Rodriguez, I., Fuller, A., Ebner, P., Sorensen, C., Racy, F., Lemery, J., Maas, L., Leaning, J., Irizarry, R. A., Balsari, S., & Buckee, C. O. (2018). Mortality in Puerto Rico after hurricane Maria. *New England Journal of Medicine, 379*(2), 162–170. https://doi.org/10.1056/NEJMsa1803972

Khan, Amir A. (2023): Kashmir Floods 2014: Recovery to Resilience, National Institute of Disaster Management (NIDM)

Klapka, P., Frantál, B., Halás, M., & Kunc, J. (2010). Spatial organisation: Development, structure and approximation of geographical systems. Moravian Geographical Reports, 18(3), 53–65.

Delhi Jal Board. (2016). Water Policy for Delhi 2016. Government of National Capital Territory of Delhi.

Kumar, N., Kumar, M., Sherring, A., Suryavanshi, S., Ahmad, A., & Lal, D. (2020). Applicability of HEC-RAS 2D and GFMS for flood extent mapping: A case study of Sangam area, Prayagraj, India. *Modeling Earth Systems and Environment, 6*(1), 397–405. https://doi.org/10.1007/s40808-019-00687-8

Mehta, D. J., Ramani, M., & Joshi, M. (2014). Application of 1-D HEC-RAS model in design of channels. International Journal of Innovative Research in Advanced Engineering, 1(7), 213–216.

Ongdas, N., Akiyanova, F., Karakulov, Y., Muratbayeva, A., & Zinabdin, N. (2020). Application of HEC-RAS (2D) for flood Hazard maps generation for Yesil (Ishim) River in Kazakhstan. *Water, 12*(10), 2672. https://doi.org/10.3390/w12102672

Pathan, A. I., & Agnihotri, P. G. (2021). Application of new HEC-RAS version 5 for 1D hydrodynamic flood modeling with special reference through geospatial techniques: A case of river Purna at Navsari, Gujarat, India. *Modeling Earth Systems and Environment, 7*(2), 1133–1144. https://doi.org/10.1007/s40808-020-00961-0

Quirogaa, V. M., Kurea, S., Udoa, K., & Manoa, A. (2016). Application of 2D numerical simulation for the analysis of the February 2014 Bolivian Amazonia flood: Application of the new HEC-RAS version 5. Ribagua, 3(1), 25–33. https://doi.org/10.1016/j.riba.2015.12.001

Rafiq, F., Ahmed, S., Ahmad, S., & Khan, A. A. (2016). Urban floods in India. International Journal of Scientific & Engineering Research, 7(1), 721–734.

Rahmati, O., Nazari Samani, A., Mahdavi, M., Pourghasemi, H. R., & Zeinivand, H. (2015). Groundwater potential mapping at Kurdistan region of Iran using analytic hierarchy process and GIS. *Arabian Journal of Geosciences, 8*(9), 7059–7071. https://doi.org/10.1007/s12517-014-1668-4

Statista Research Department. (2018). Economic damage caused by floods worldwide from 1900 to 2016. Statista.

Teng, J., Jakeman, A. J., Vaze, J., Croke, B. F. W., Dutta, D., & Kim, S. (2017). Flood inundation modelling: A review of methods, recent advances and uncertainty analysis. *Environmental Modelling & Software, 90*, 201–216. https://doi.org/10.1016/j.envsoft.2017.01.006

Tomar, P., Singh, S. K., Kanga, S., Meraj, G., Kranjčić, N., Đurin, B., & Pattanaik, A. (2021). GIS-based urban flood risk assessment and management—A case study of Delhi National Capital Territory (NCT), India. *Sustainability, 13*(22), 12850. https://doi.org/10.3390/su132212850

Wang, G., & Eltahir, E. A. B. (1999). Use of ENSO information in medium- and long-range forecasting of the Nile floods. *Journal of Climate, 12*(6), 1726–1737. https://doi.org/10.1175/1520-0442(1999)012<1726:UOEIIM>2.0.CO;2

Influence of Natural and Anthropogenic Factors on Mangrove Ecosystem of Malappuram–Kozhikode Districts, Kerala

P. Anusha, Shruti Kanga, Suraj Kumar Singh, Narisetty Nadha Gowrish, and Sunil Kumar

1 Introduction

Mangroves, an essential component of tropical shorelines, derive their name from the Portuguese word "Mangue" and the English term "grove." These unique ecosystems cover approximately 25% of the world's tropical coasts, ranging from the 20th parallel north to the 20th parallel south latitude (Walsh, 1974). The earliest mangrove species are believed to have originated in the Indo-Malayan region, which might explain the higher species diversity in this area compared to others.

Mangroves are a specialized group of trees and shrubs adapted to thrive in intertidal zones. These plants exhibit remarkable adaptations such as aerial roots, salt-excreting glands, and vivipary (live birth of seeds) to survive in saline, waterlogged environments. Mangrove ecosystems, found in sheltered coastlines, shallow-water lagoons, estuaries, rivers, and deltas within tropical and subtropical regions, play crucial socioeconomic and environmental roles. They support a rich biodiversity of plants, animals, and microorganisms and provide essential services such as protecting coastlines, supporting fisheries, and supplying resources like wood.

Mangrove forests, often referred to as coastal rainforests, are highly productive ecosystems that serve as a vital interface between land and sea. These "tidal forests" or "coastal woodlands" are the only tall tree forests where land, freshwater, and sea meet. Mangroves have evolved to withstand high salinity, extreme tides, strong winds, high temperatures, low oxygen, and muddy soils. Their unique features include arching and breathing roots, salt-excreting leaves, and diverse wildlife.

The biomass of mangrove ecosystems surpasses that of other aquatic systems, underscoring their ecological importance. They protect coastlines from wind, waves, and currents, prevent soil erosion and siltation, safeguard coral reefs and seagrass beds, supply wood and other forest products, provide habitats and nutrients for various organisms, and support coastal fisheries and livelihoods. Despite their significance, mangroves face severe threats due to human activities and climate change. Current global annual depletion rates of 1–2% indicate that mangrove forests

P. Anusha · S. Kanga · N. N. Gowrish · S. Kumar
Department of Geography, School of Environment and Earth Sciences, Central University of Punjab, Bathinda, India
e-mail: shruti.kanga@cup.edu.in; sunil22phd@cup.ac.in

S. K. Singh (✉)
Centre for Climate Change and Water Research, Suresh Gyan Vihar University, Jaipur, Rajasthan, India
e-mail: suraj.kumar@mygyanvihar.com

Z. Khanam et al. (eds.), *Climate Change, Resilience and Ecosystem Services*, Sustainable Development Goals Series, https://doi.org/10.1007/978-3-031-96977-5_20

may disappear by the end of the twenty-first century if present trends continue (Walton et al., 2007).

According to a report by the Food and Agriculture Organization of the United Nations (FAO), approximately 35% of global mangrove forests have already been lost, with significant losses in South Asia, including India, the Philippines, and Vietnam. Major threats to mangroves include deforestation, resource extraction, river flow alterations, coral reef destruction, pollution, and climate change (Food and Agriculture Organization of the United Nations, 2004). The increasing human population and rapid economic development have led to excessive exploitation of mangrove ecosystems, resulting in their degradation (Benton & Twitchett, 2003; Dahdouh-Guebas et al., 2005; Zhang et al., 2007).

The global status of mangroves is further exacerbated by the overexploitation and depletion of natural resources due to collective ownership and unrestricted resource access. Despite efforts through institutional, social, and legal means to promote conservation, many natural resources, including mangroves, remain at risk.

In India, 38 mangrove areas are actively managed with financial assistance from the Ministry of Environment and Forests. India's mangrove coverage spans 4639 km^2, which constitutes only 0.14% of the country's total geographical area but accounts for 3% of global and 8% of Asian mangrove reserves (Forest Survey of India, 2023). Regulatory measures such as the Forest Act of 1927, the Forest Conservation Act of 1980, the Coastal Regulation Zone (CRZ) notification of 1991, and Environmental Impact Assessment studies are crucial for protecting mangroves (FAO, 2007).

Despite these regulations, India lost about 38% of its wetlands in the 1990s (Khaleel & Jaleel, 2009). The importance of mangroves has often been underestimated, yet they provide critical ecological services, including protecting coastal areas from erosion, maintaining water quality, and serving as breeding grounds for fish and crustaceans. The loss of mangroves leads to diminished coastal water quality, reduced biodiversity, and increased vulnerability to coastal hazards such as erosion, storm surges, and tsunamis (Badola & Hussain, 2005; Ewel et al., 1998; Nagelkerken et al., 2008). Additionally, the destruction of mangrove forests results in significant carbon release, contributing to global warming and climate change.

Kerala, with over 900 km of interconnected waterways known as the backwaters, hosts a unique mangrove ecosystem. The state's mangrove forests are rapidly degrading due to demographic pressures, urbanization, conversion to coconut plantations, agriculture, and ecotourism (Hema & Devi, 2015). Kerala's mangrove area, once extensive, has significantly dwindled, with estimates suggesting it covered around 70,000 hectares in 2004 (Hema & Devi, 2015). However, up-to-date scientific data on the current extent of mangroves in Kerala is lacking.

In Kerala, 14 species of mangroves have been documented (Khaleel & Jaleel, 2009). Notable species include *Avicennia officinalis*, *Rhizophora mucronata*, and *Excoecaria agallocha*. The preservation and management of mangroves in Kerala involve community engagement, as seen in the Kadalundi–Vallikkunnu Reserve established in 2007 to conserve biodiversity with community participation. Despite these efforts, urban development projects and agricultural expansion continue to threaten mangrove habitats.

Mangrove ecosystems are invaluable for their ecological, socioeconomic, and environmental contributions. However, they face significant threats from human activities and climate change. Effective conservation and management strategies, coupled with community involvement, are essential to safeguard these vital ecosystems. Immediate action is required to halt the degradation of mangroves and ensure their survival for future generations.

2 Study Area

Kadalundi is considered one of Kerala's delicate mangrove ecosystems, located along the estuaries of Kadalundi and Vallikkunnu Panchayats in Kozhikode and Malappuram Districts, respectively. The Kadalundi River, originating from the eastern region of Karuvarakkundu Village in Kozhikode District, Kerala State, merges with two primary tributaries, Olipuzha and Veliyar, to form its main course. The river stretches over a total distance of 130 km and has a drainage basin of 1099 km^2. It flows toward Chaliyar and eventually converges with the Arabian Sea about 5 km south of the Chaliyar River. An additional small watercourse, the Pooraparamba River, spans 8 km and a catchment area of 23 km^2, with the estuary located near Kadalundi and Vallikunnu Panchayats. The overall catchment area of the basin is 1122 km^2.

3 River Source and Drainage

The Kadalundi River emanates from the Western Ghats at the western boundary of the Silent Valley, with Olippuzha originating from Cherakkombanmala and Veliyar from Erattakkombanmala. The river predominantly flows through the Malappuram District and covers a drainage basin of 1122 km^2. The river's flow and its dendritic drainage pattern are crucial for irrigating the surrounding regions.

4 Transportation and Accessibility

The region under study is an island, accessible primarily by ferry services apart from pedestrian walkways linking various small islands (Thruths). Kadalundi is located 18 km south of Kozhikode City.

5 Environmental and Ecological Significance

The combined area of waterbodies and mangrove habitats in Kadalundi and Vallikkunnu Panchayats is approximately 250 hectares. In 2007, 150 hectares of ecologically sensitive mangrove ecosystems in these panchayats were declared communal reserve forests, marking a pioneering conservation initiative in Southern India.

6 Flora and Fauna

The Kadalundi–Vallikkunnu estuarine region is biologically rich, hosting a variety of species including migratory and indigenous birds, aquatic life such as fishes and crabs, reptiles including sea turtles, mammals, and various plant species. The coconut groves, unique tidal patterns, and the estuary of the Kadalundi River add to the region's ecological value (Fig. 1).

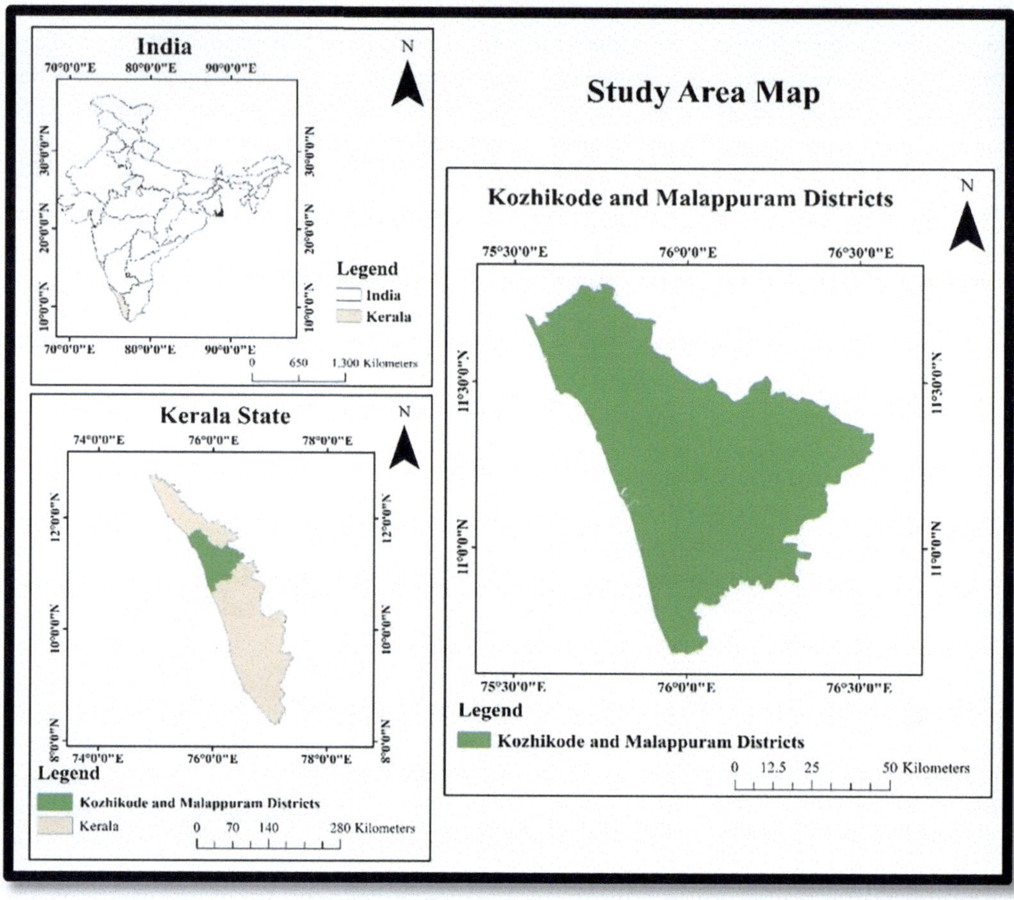

Fig. 1 Study area map

7 Community Reserve Initiative

The Kadalundi–Vallikkunnu Reserve was established by the Government of Kerala in 2007 to conserve biodiversity with active community involvement. This initiative aims to safeguard the ecosystem while maintaining the property and livelihood rights of the local communities. The reserve area includes the Kadalundi Bird Sanctuary and the mangroves, focusing on sustainable management with the participation of indigenous populations.

7.1 Geographical Features

The area comprises coastal plains with sandbars, dunes, sandy beaches, and inter-ridge depressions or basins. Notable geographical features include Kadalundi's lens-shaped channel bars, sand islands, barrier ridges, and extensive mudflats that provide habitats for various bird species.

8 Climate and Hydrology

The Kadalundi River Basin receives an average annual precipitation of 3610 mm, with 60% occurring during the southwest monsoon (June–September), 30% during the northeast monsoon, and 10% in summer. The estimated annual surface water potential of the basin is 1829 mm, creating favorable conditions for the diverse flora and fauna thriving in the Kadalundi–Vallikkunnu community reserve.

9 Conservation Challenges

Despite its ecological significance, the Kadalundi–Vallikkunnu region faces several biological disruptions such as excessive fishing, harvesting of oysters and mussels, sand, and lime extraction, and traditional coconut retting practices. The community reserve initiative aims to address these challenges by promoting conservation and sustainable use of resources.

The Kadalundi–Vallikkunnu mangrove ecosystem, with its rich biodiversity and significant ecological role, is a critical area requiring careful management and conservation efforts. The establishment of the community reserve represents a vital step towards ensuring the long-term sustainability of this unique environment while balancing the needs of local communities.

10 Methodology

The study is entirely based on both primary and secondary data. Satellite images are being used to obtain current land use information of the Kadalundi River Basin. Satellite imagery is employed to derivate current land use data within the Kadalundi River Basin. Data was obtained via personal interviews, telephone interviews, and primarily through direct observation. The supplementary secondary data was collected from various government agencies in Kerala, including the Department of Forest & Wildlife, Fisheries, as well as from published articles and reports. A method utilizing multi-temporal and multi-scale approach, utilizing remote sensing data, was employed to identify changes in the expanse of mangrove-covered area across a duration exceeding two decades. Conventional satellite data including Landsat facilitated the observation of changes in land use and land cover from the year 1993 to 2023 (Ghosh et al., 2015). The data for 1990, 2001, 2011, and 2021 were utilized in conjunction with Landsat imagery obtained from the archive of the US Geological Survey (USGS) to construct the land use and land cover (LULC) map of the specified research region. The complete collection of Landsat data, from 1990 to 2021, is currently accessible without charge on the website http://glovis.usgs; all satellite images were retrieved from the website http://earthexplorer.usgs.gov.

The research was commenced by facilitating informal dialogues with community members, authorities from the forest/agriculture/fisheries sector, representatives of local governing bodies, and senior citizens residing in the area, complemented by firsthand observations. Waterbodies were identified through a supervised classification (Giri et al., 2011). The land use map has been developed utilizing data extracted from these satellite images. The data underwent both visual and digital interpretation through the utilization of image interpretation components. In order to achieve the goals of the study, ArcGIS software made it easier to process, analyze, and integrate geographical data (Bindu & Suresh, 2015). The land represents a crucial natural asset alongside water, essential for human survival on the planet. The concept of land use pertains to how individuals are exploiting the land, whereas land cover specifically denotes the vegetation layer that extends across the land. The documentation of land cover data quantifies the extent to which an area is enveloped by natural occurrences. In contrast, LULC data delineates how individuals interact with the terrain. These datasets on land use and land cover hold immense significance for effectively managing land resources, urban planners, and policymakers. Four categories of stakeholders who relied on the ecosystem were distinguished through this methodology. These groups encompassed individuals residing in proximity to the mangrove forests and the population reliant on the services provided by the mangroves(Hema & Devi, 2015) (Fig. 2).

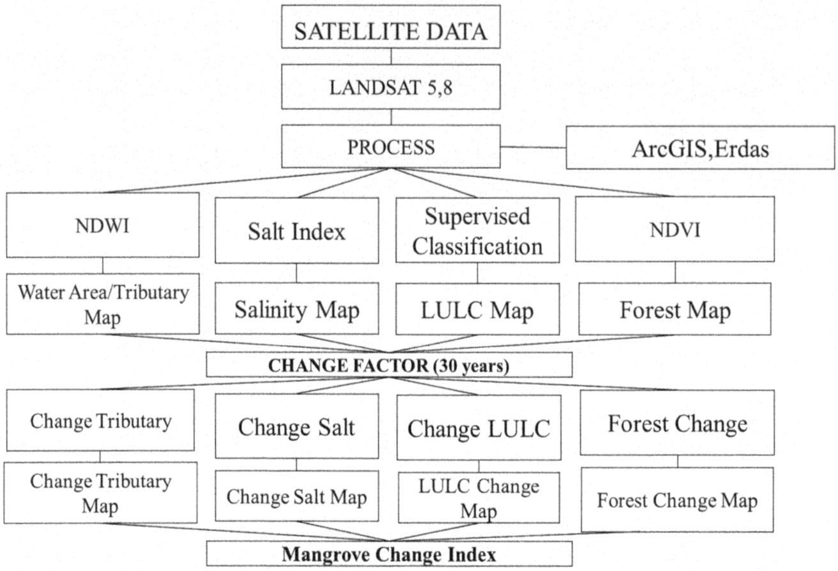

Fig. 2 Methodology

Table 1 Data used in the analysis

Datasets	Year	Spatial resolution	No. of bands	Spectral resolution
Landsat 5 TM (Thematic Mapper)	1993	30	1,2,3, 4,5,7	IR, blue, red, green
	2003	30		
Landsat 8 OLI	2013	30	1,2,3, 4,5,6,7	
	2023	30		

11 Data Used (Table 1)

12 Software

12.1 ArcGIS and ERDAS Satellite Data-Landsat (8–9 for Current Year)

The Landsat program comprises a sequence of Earth-observation satellite missions collaboratively overseen by NASA and the U.S. Geological Survey. Landsat data are used for tributary mapping, land use and land cover change detection, saltwater index, and spatiotemporal forest change detection.

The ERDAS and ArcGIS software applications shall be employed to delineate and analyze the features of land cover and use in the Kadalundi River Basin. Remote sensing and geographic information system (GIS) approaches offer more consistent and precise baseline data than any conventional survey for understanding the current pattern of land use/cover of any area. Using satellite photos, the built-up land, agricultural land, waterbodies, and forest will be prepared for thematic mapping of land use/land information. It intends to process, analyze, and integrate spatial data using ArcGIS software in order to meet the study's goals. It also plans to use image interpretation elements to digitally and visually interpret Landsat data.

13 Land Use and Land Cover (LULC)

For land use and land cover analysis, we will be using a supervised classification scheme of training samples based on the spectral properties of features. The land use map is going to be prepared based on the information obtained from GIS and satellite images. The physical and economic state of each region is reflected in its land use and land cover, which also establishes the quality of life for local residents and the availability of natural resources. Using satellite photos, the built-up land, agricultural land, waterbodies, and forest will be prepared for thematic mapping of land use/land information. We are not taken the Landsat 7 images, due to the lack of data availability and absence of Landsat images devoid of cloud cover. Land use analyses encompassed the generation of False Colour Composite (FCC) using remote sensing data (bands–green, red, and NIR), aiding in identifying heterogeneous patches within the landscape. This process also involved the selection of training polygons, which corresponded to the heterogeneous patches in the FCC and were distributed uniformly across the study area. These training polygons were then loaded, and the corresponding attribute data (land use types) for each polygon was collected from the field using GPS technology. Field investigations were conducted to compile attribute data supplemented by information from Google Earth. The training data was utilized for classification purposes, while the remaining data was allocated for validation and accuracy assessment.

14 Normalized Difference Water Index (NDWI)

Stands for the Normalized Difference Water Index will be used for water availability and spatiotemporal water availability assessment. The NDWI can assist in assessing mangrove health by indicating changes in water conditions crucial for mangrove growth.

For NDWI calculation, we used the following formula

$$\mathrm{NDWI} = \mathrm{G} - \mathrm{NIR} / \mathrm{G} + \mathrm{NIR} \quad (1)$$

15 Normalized Difference Vegetation Index (NDVI)

The Normalized Difference Vegetation Index is a measure commonly used in remote sensing to access and monitor changes in vegetation over time. The NDVI is valuable for forestry (Ragavan et al., 2019).

For the calculation of NDVI, here we have used the formula is that:

$$\mathrm{NDVI} = (\mathrm{NIR} - R) / (\mathrm{NIR} + R) \quad (2)$$

16 Normalized Difference Saltwater Index (NDSI)

It refers to the salinity level in water. It measures the concentration of dissolved salts, mainly sodium chloride. This index is crucial for assessing water quality, especially in ecosystems like mangrove forests where specific salinity levels are vital for the survival of diverse flora and fauna. The saltwater index helps ensure that the salinity remains within the suitable range for the mangrove ecosystem.

$$\mathrm{NDSI} = R - \mathrm{NIR} / R + \mathrm{NIR} \quad (3)$$

17 Results and Discussion

This chapter presents an examination of the findings and discourse. The research comprises four objectives. The first objective is the tributary mapping; second is to identify the LULC change; third is to assess the spatiotemporal saltwater index; and fourth one is to examine the spatiotemporal forest change detection in the study area. In this chapter, we discuss about the LULC change and its interpretation and analysis and the spatiotemporal changes in the NDWI and NDVI as well as the NDSI. We also analyze the trend of NDVI and the mangrove change graph.

17.1 The Spatiotemporal Dynamics and Comparison of LULC Map (Table 2)

Figures 3 and 4 show LULC map of 2013 and 2033which includes five major classifications; the vegetation or forests display a dark green hue, while the adjacent agricultural land or crop land exhibits a yellowish tone. Human settlements or built-up areas are characterized by a red appearance. The Dampier-Hodges line serves as a demarcation for the tidal impact and

Table 2 Spatiotemporal dynamics and comparison of LULC in 2013 and 2023

LULC 2013		LULC 2023	
Area (km^2)		Area (km^2)	
Classes	Area	Classes	Area
Waterbody	94.71	Waterbody	91.88
Vegetation/Forest	4716.00	Vegetation/Forest	4523.19
Cropland	175.05	Cropland	86.04
Built-up	769.99	Built-up	1054.74
Barren land	13.44	Barren land	13.33
Total area	5769.18 km^2	Total area	5769.18 km^2

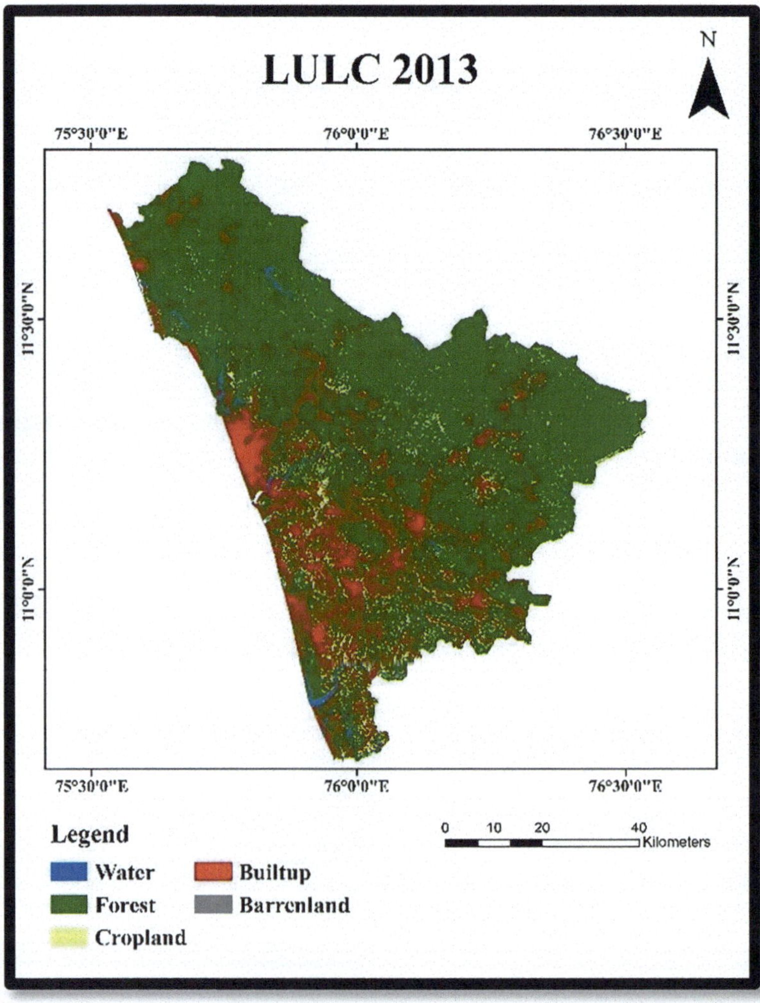

Fig. 3 LULC 2013

approximately outlines the historical reach of the mangrove forests. The waterbody appears blue, and as well as the barren lands appear ash in color. Comparing the 2013 and 2023 LULC maps of the study area, it can be seen that by 2023, the cropland will decrease relatively. Similarly, by 2023, settlements or built-ups are seen to have spread over a larger area compared to 2013. The rate of mangrove transformation exhibited temporal variability; the most substantial decline occurred between 2013 and 2023, resulting in forest reductions of 192.81 km^2, while 284.75 km^2 land was obtained through urban development. The temporal variability in mangrove change did not exhibit a consistent pattern; the most substantial decline occurred between 2013 and 2023, resulting in a loss of 192.81 km^2 of forest cover, while a gain of 284.75 km^2 was attributed to settlement expansion (Vinod et al., 2019). For the LULC

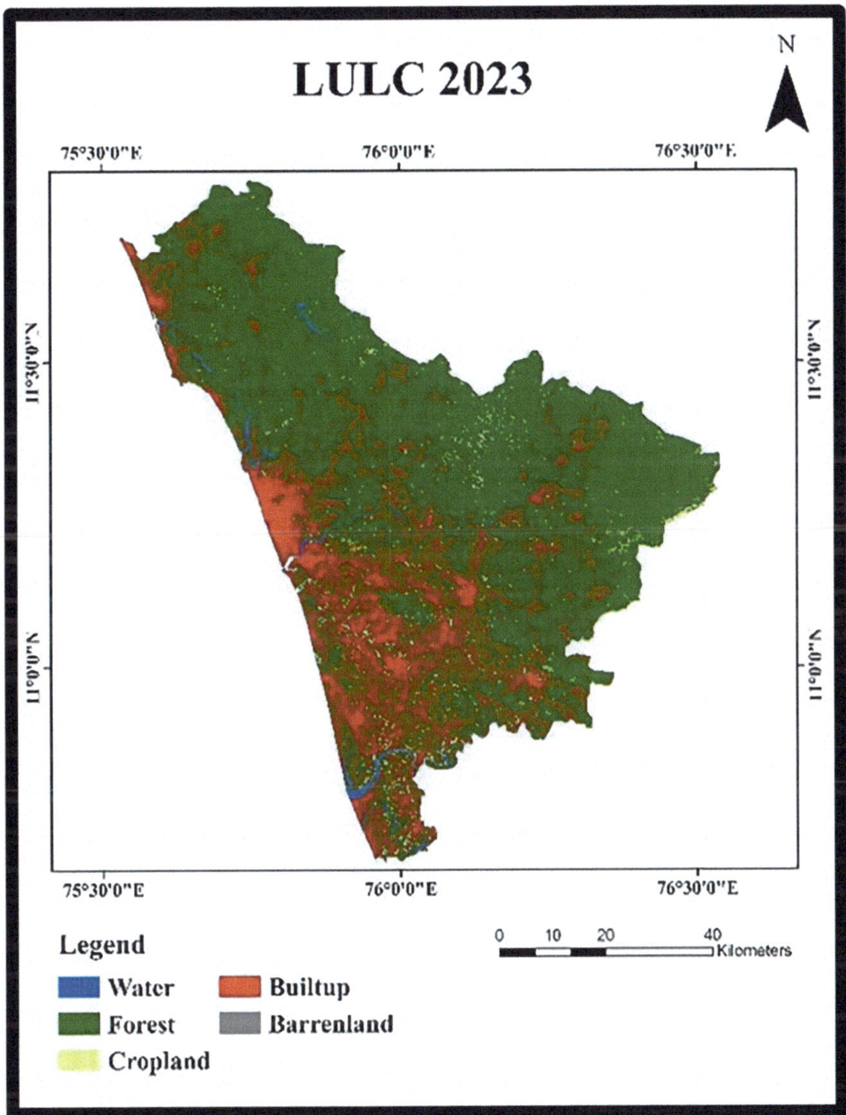

Fig. 4 LULC 2023

mapping, we mainly used Landsat 5 and Landsat 8 data and satellite images. We did not use the Landsat 7 images because of the data is not available. The change detection analysis indicates areas of land loss and land gain between 2013 and 2023 based on the Landsat data (Ghosh et al., 2015; Samanta et al., 2021).

18 Normalized Difference Water Index (NDWI)

The NDWI stands for the Normalized Difference Water Index which will be used for water availability and spatiotemporal water availability assessment. The NDWI can assist in assessing mangrove health by indicating changes in water conditions crucial for mangrove growth (Table 3).

Table 4 explains the area calculation of NDWI in four different years. Tables 3 and 4 show the huge difference between waterbody and non-waterbody. In 1993, the area of water was 225.027. After the coming years, we can see it shrinking. Comparing 1993 with 2003, we can identify a slight difference in the area of the waterbody. Year by year, the water concentration is reduced to a smaller area. Table 3 shows in 1993, 5696.9946 km^2 area was covered by non-waterbody, whereas in 2003 214.60 km^2 area was fully covered by water and 5707.41 km^2 was covered by non-waterbody. The highest concentration of water was in 1993 (Table 3), whereas the lowest concentration of water was identified in 2023. In 2013, 214.60 km^2 (Table 4) was reduced to 205.6752 km^2 (Table 4). Table 4 indicates that by 2023, approximately 138.6864 km^2 of waterbody area had been converted. Meanwhile, 5,783.3352 km^2 of nonwaterbody area has been occupied (Figs. 5, 6, 7 and 8).

Table 3 Area calculation of **NDWI** 1993 and 2003

NDWI 1993		NDWI 2003	
Total area 5922.0216 km^2		Total area 5922.0216 km^2	
waterbody	225.027	waterbody	214.6077
Non--waterbody	5696.9946	Non--waterbody	5707.4139

Table 4 Area calculation of NDWI 2013 and 2023

NDWI 2013		NDWI 2023	
Total area 5922.0216 km^2		Total area 5922.0216 km^2	
waterbody	205.6752	waterbody	138.6864
Non--waterbody	5716.3464	Non--waterbody	5783.3352

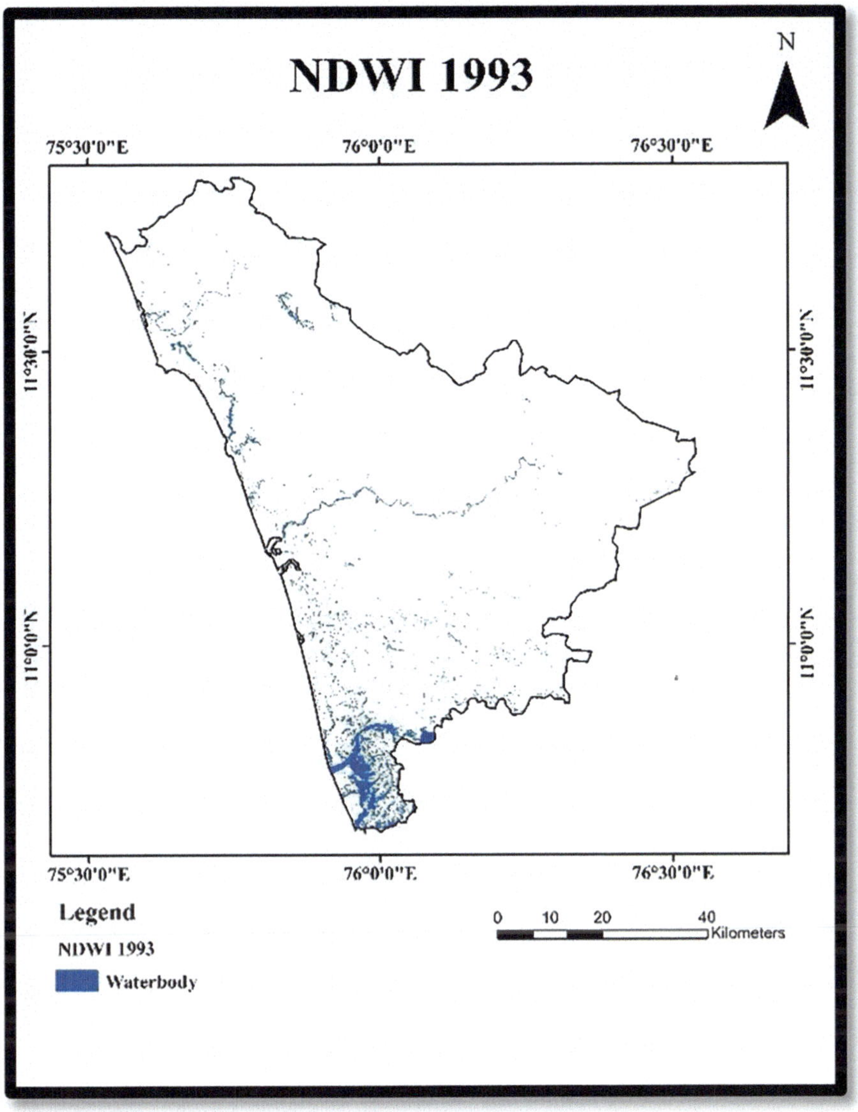

Fig. 5 NDWI 1993

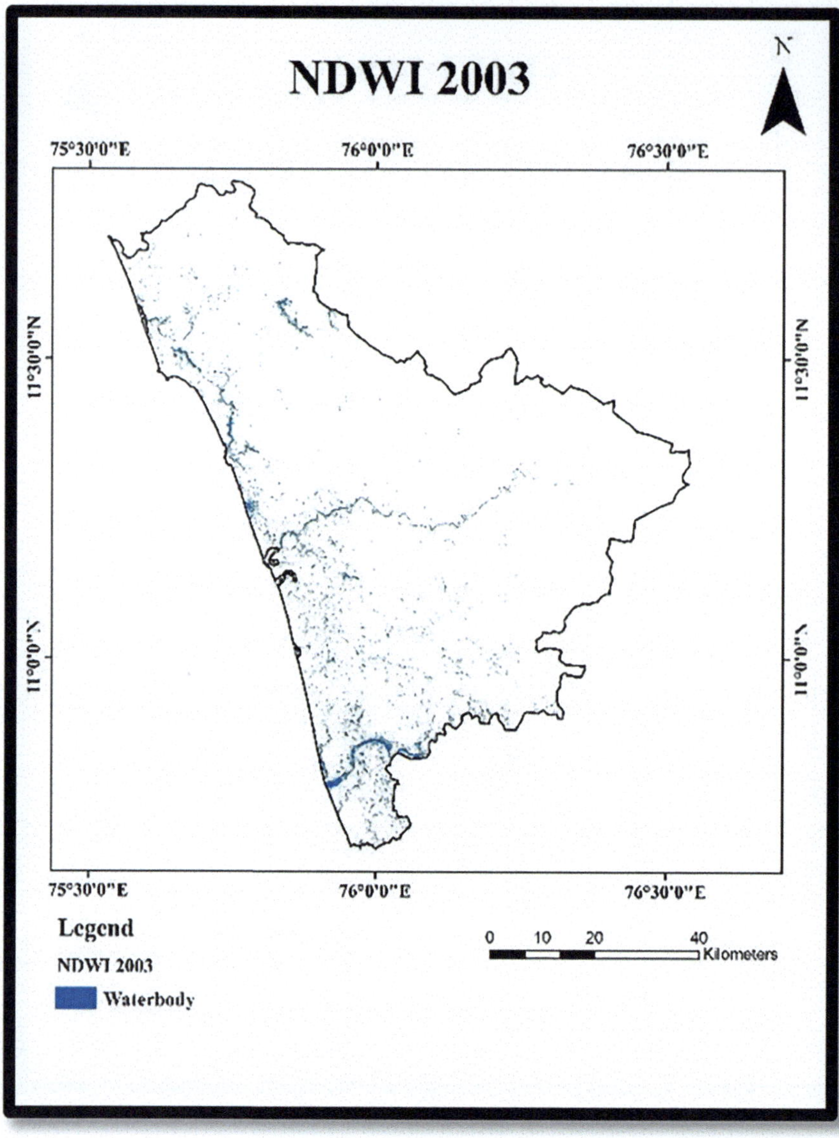

Fig. 6 NDWI 2003

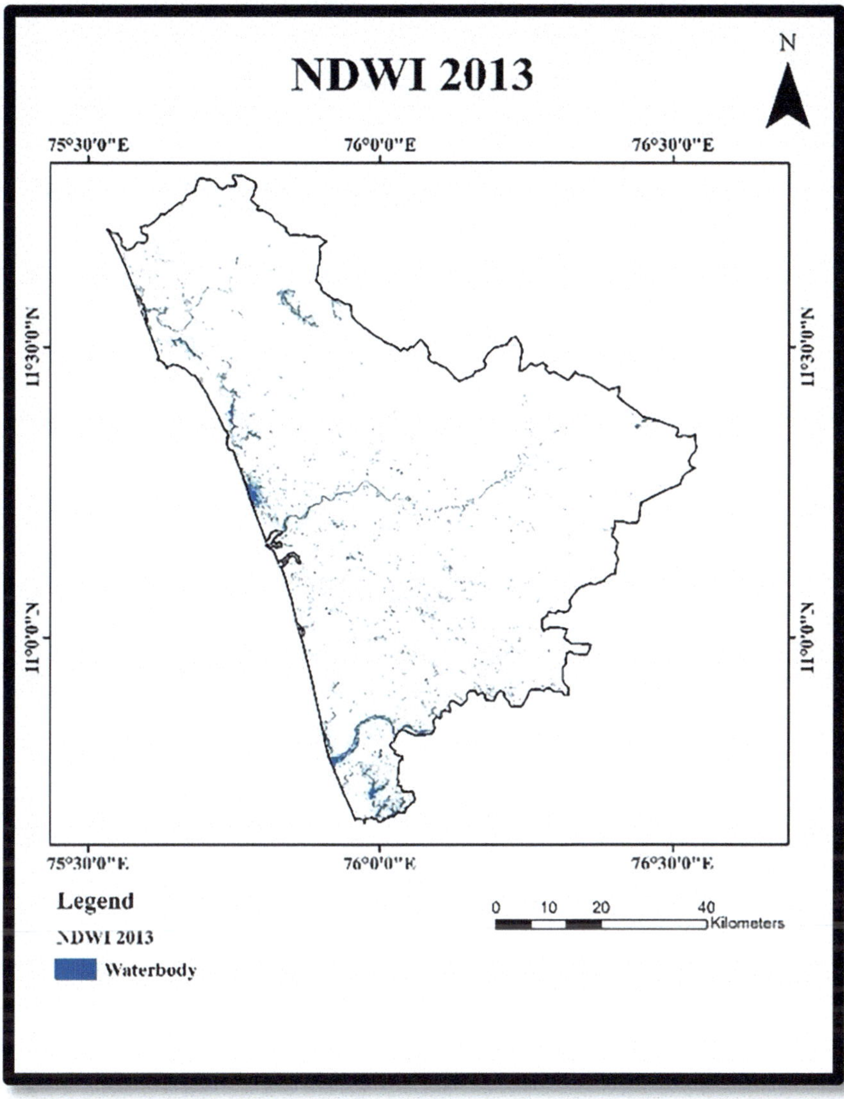

Fig. 7 NDWI 2013

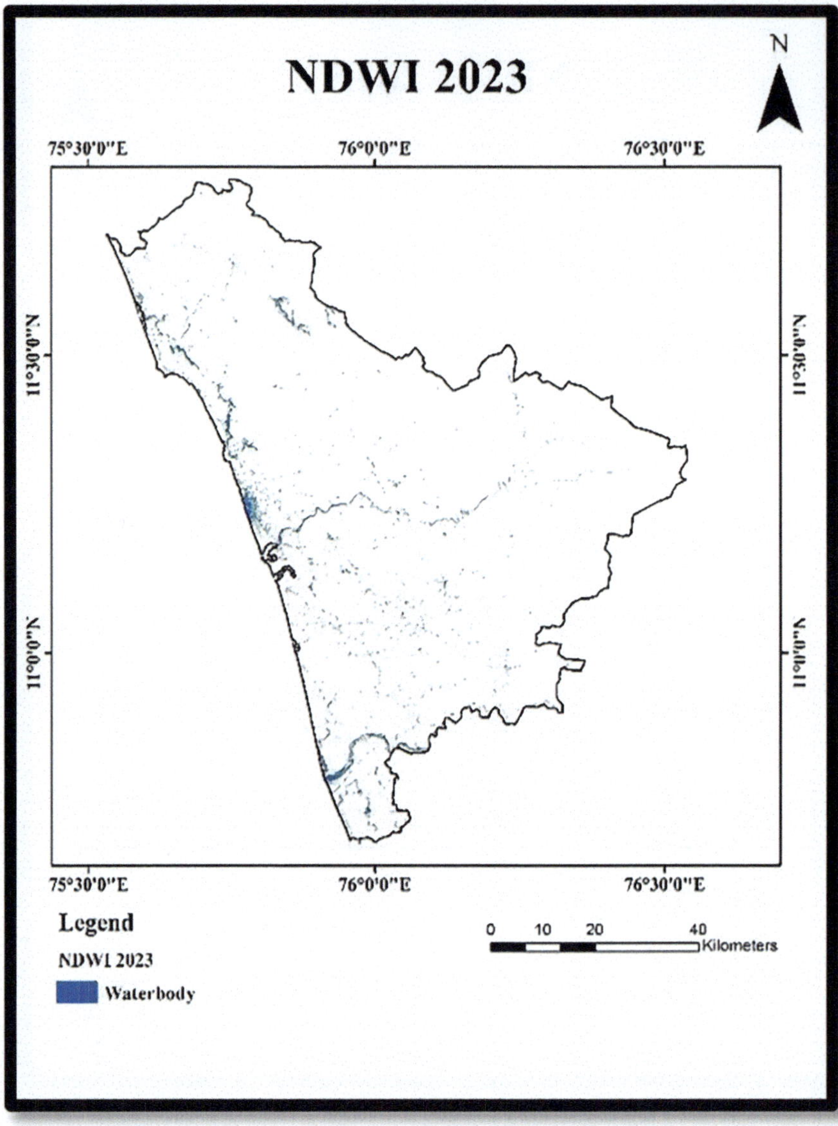

Fig. 8 NDWI 2023

19 Analyzing of NDVI Maps

Tables 5 and 6 explain the total forest cover and non-forest areas in different period of time (NDVI—Normalized Difference Vegetation Index). Table 5 shows that in 1993, the non-forest area was more than the forest cover. After 10 years, it showed a slight change in forest cover with the decreasing trend. While the non-forest area suddenly increased. Table 5 shows that the total 5623.50 km^2 area is covered by forest cover, and 298.50 km^2 is covered by non-forest areas. The post-monsoon season data and Landsat images were exclusively utilized in our analysis, while the trend of NDVI over a 20-year period was calculated based on Landsat data. A consistent reduction in and NDVI values for Landsat data during the post-monsoon season reveals a continuous decline in the overall health of the mangrove forest. This decline signifies a negative trend. Table 6 describes that in 2013, the forest areas started to decrease again, and the non-forest areas underwent a drastic change. In 2013, the forest cover occupied an area of 5,387.32 km^2, while non-forest areas covered 534.70 km^2. We can understand a huge difference in the forest cover and non-forest areas after examining the map of 2013 and 2023. After examining the area calculation, we can see that there is an inverse relation between the forest and non-forest. The forest cover is always decreasing and the non-forest area is increasing at all times. The change detection analysis highlights the areas experiencing both land loss and land gain from 1993 to 2023, using Landsat data (Figs. 9, 10, 11, 12 and 13).

Table 7 shows the change in the trend of NDVI in the study area during the last 20 years. From this table, we can observe that the class high gives the idea about the rapid change in the forest cover in the study area. We can see that 1169.5842 km^2 area is rapidly changing, while 2809.0053 km^2 area shows the moderate vegetation change in the study area. We can ascertain that there is only a slight or no change in the study area, whereas 1943.43 km^2 area comes under the low change classification.

Table 5 Area calculation of NDVI 1993 and 2003

NDVI 1993		NDVI 2003	
Total area 5922.02 km^2		Total area 5922.02 km^2	
Non-forest	274.42	Non-forest	298.50
Forest	5647.60	Forest	5623.50

Table 6 Area calculation of NDVI 2013 and 2023

NDVI 2013		NDVI 2023	
Total area 5922.02 km^2		Total area 5922.02 km^2	
Non-forest	534.70	Non-forest	645.37
Forest	5387.32	Forest	5276.65

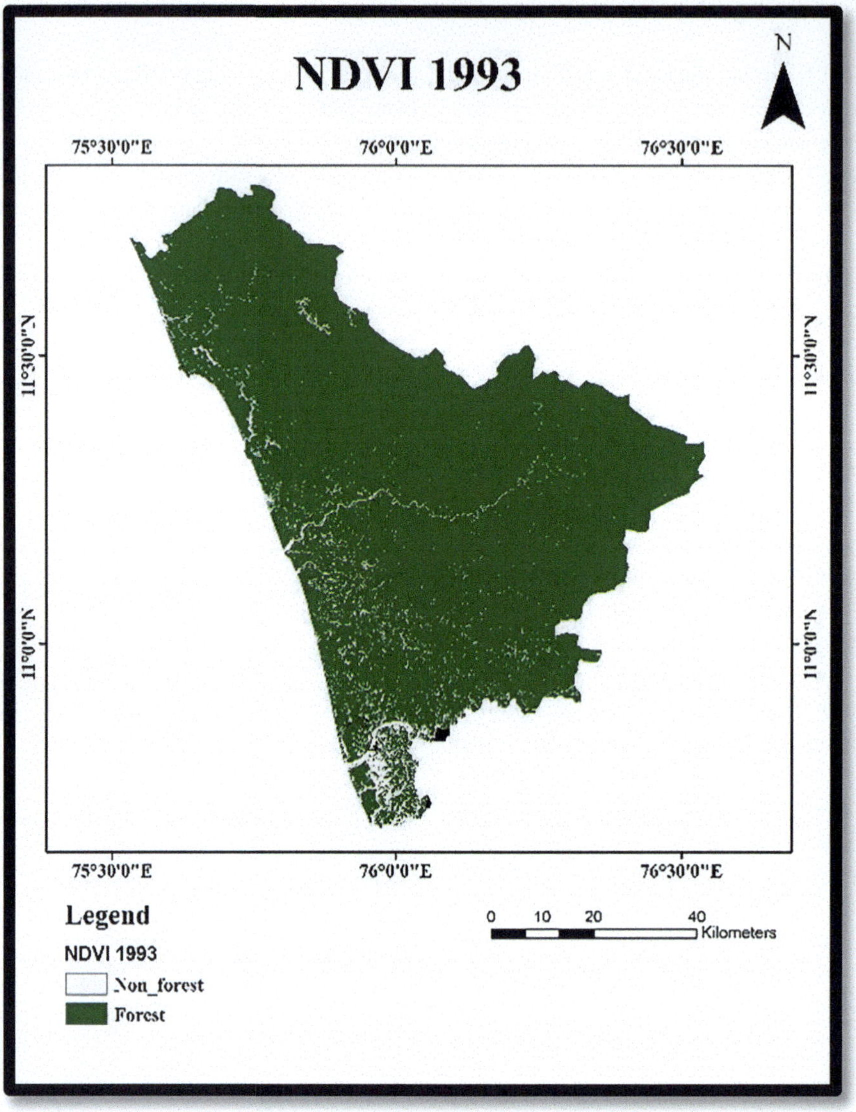

Fig. 9 NDWI 1993

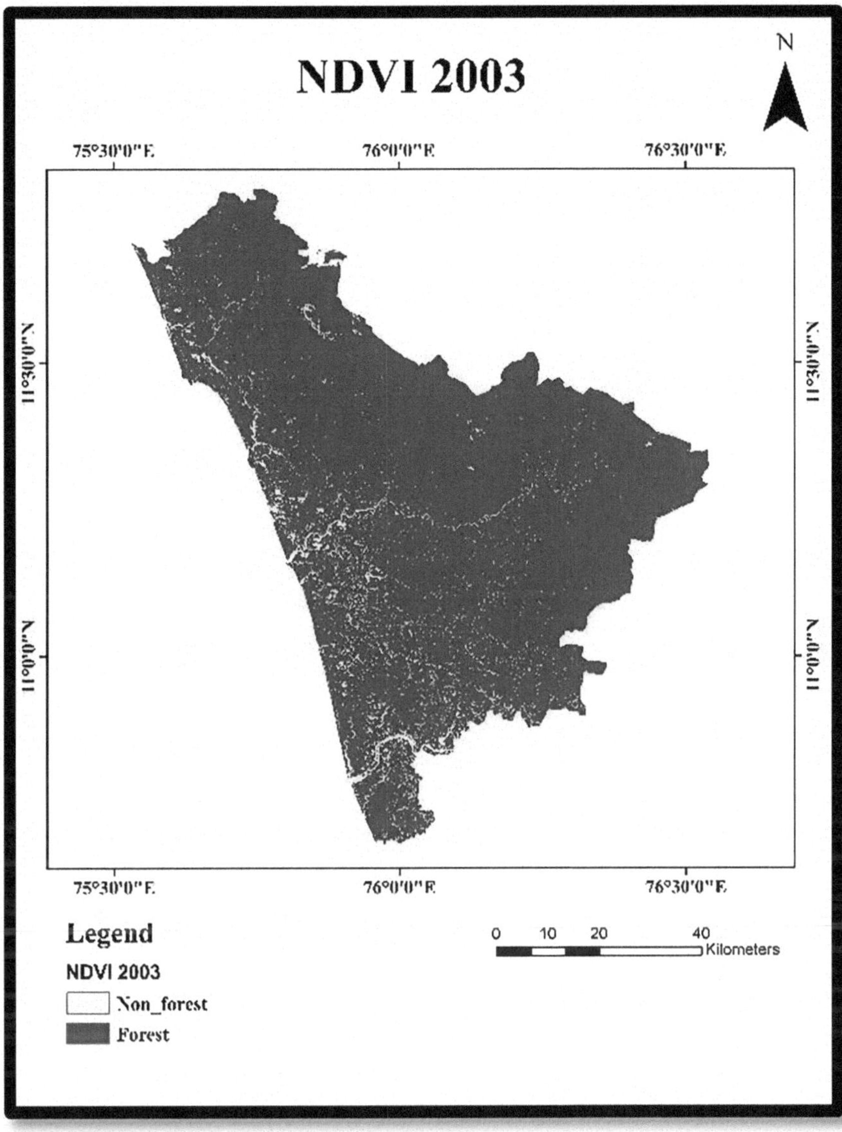

Fig. 10 NDWI 2003

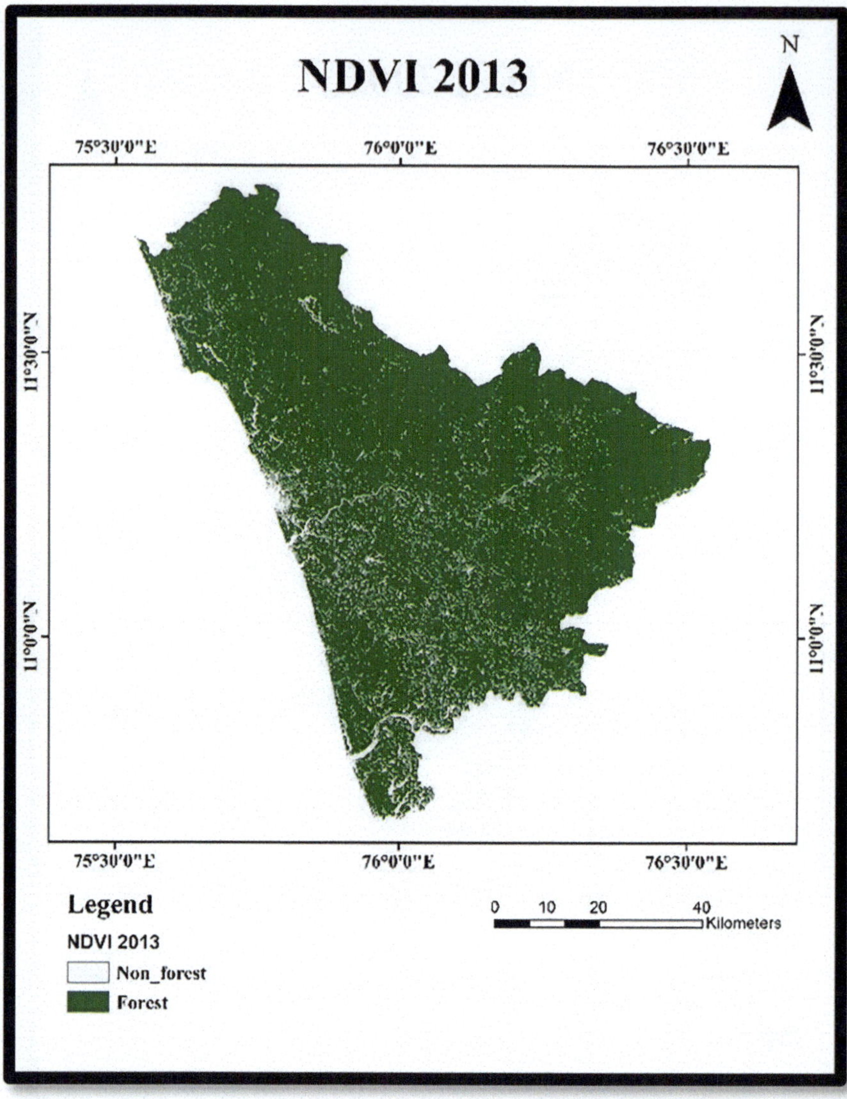

Fig. 11 NDWI 2013

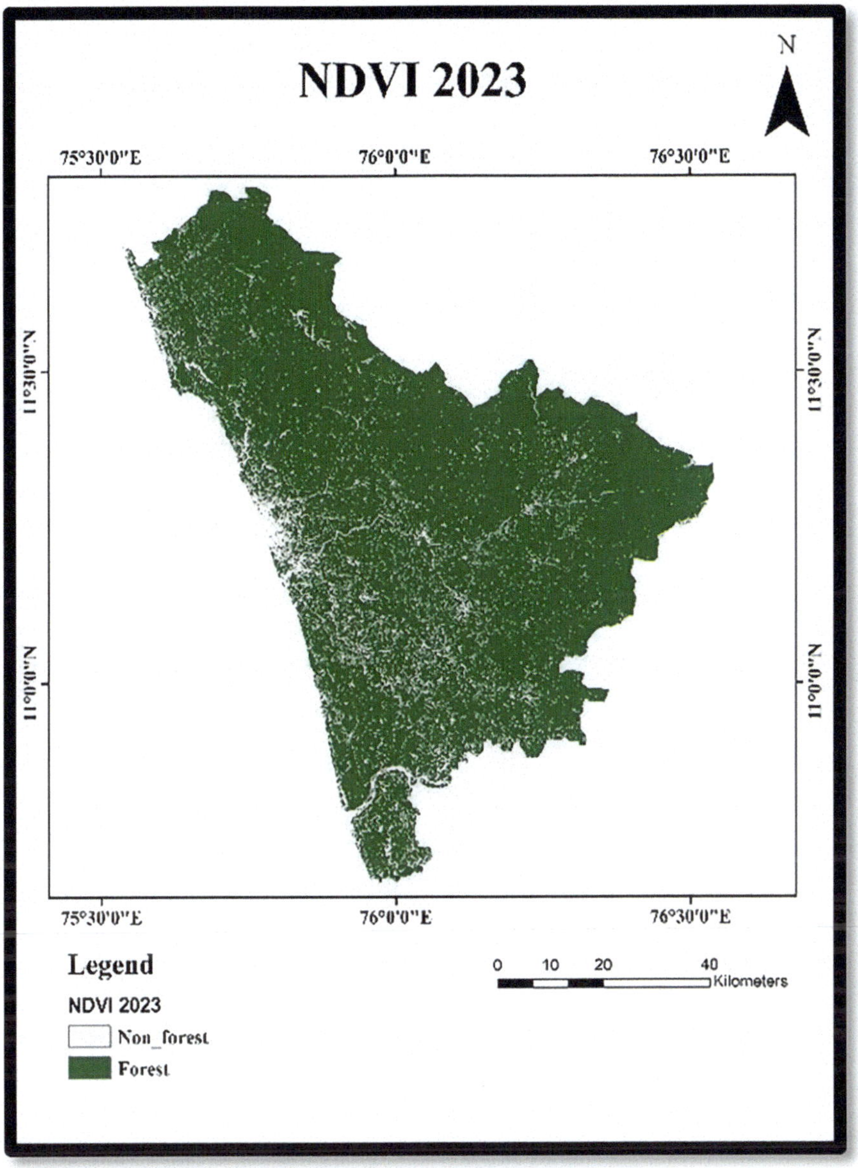

Fig. 12 NDWI 2023

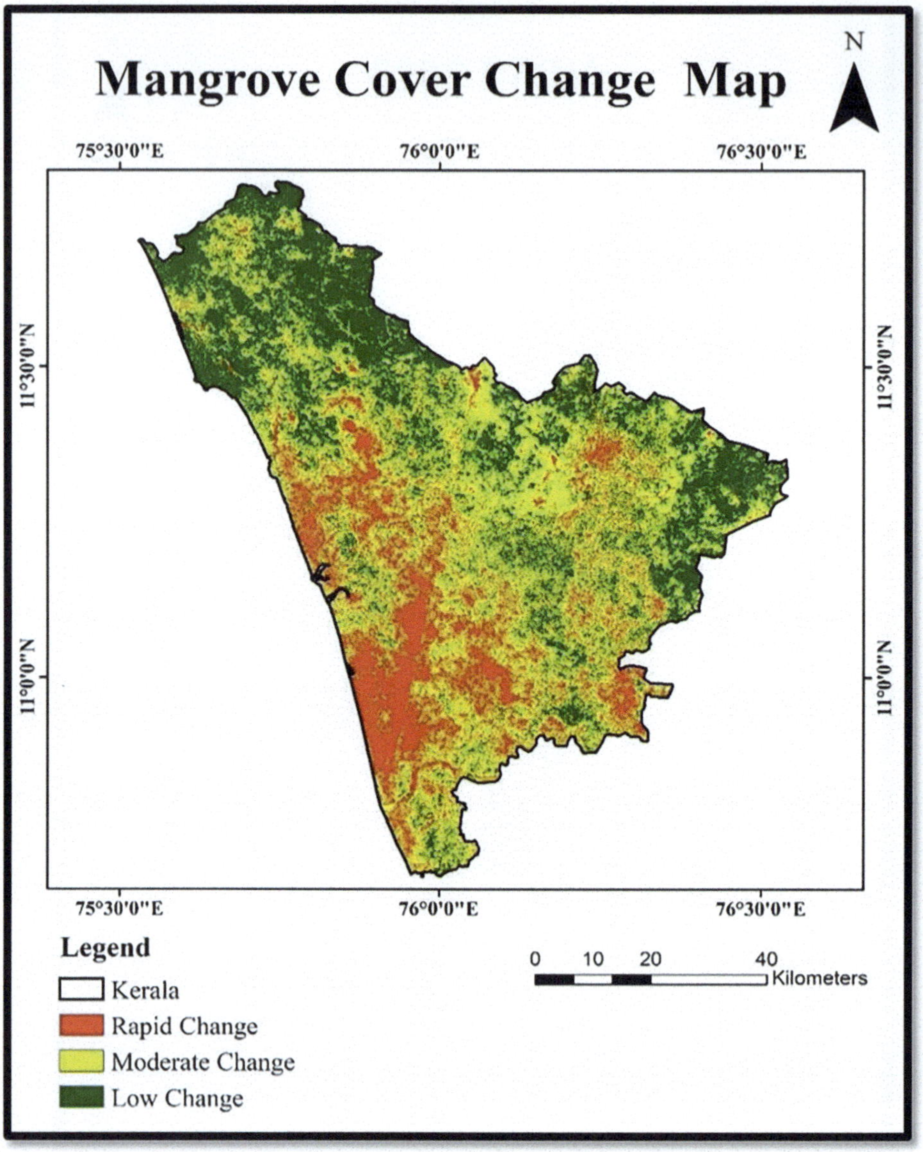

Fig. 13 Mangrove cover change map

Table 7 Mangrove NDVI trend

Total area 5922.01 km^2	
Classes	Area (km^2)
High	1169.5842
Moderate	2809.0053
Low	1943.4321

20 Mangrove Change Trend

Fig. 14 explains the change trend of the mangrove ecosystem in the study area. In Fig. 14, we can observe that in 1990 we can't see much difference in the mangrove ecosystem in that area.

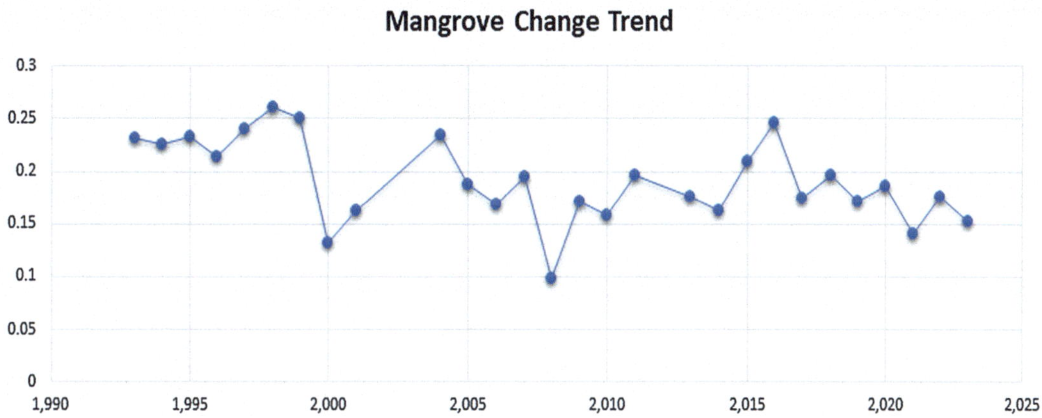

Fig. 14 Mangrove change trend graph

After five years gap, we can observe the changes. Moreover, it shows a negative trend. Then it slowly starts rising and shows a positive trend. Ater that, the graph drops sharply till 2000, then the graph rises again. In this graph, we can clearly understand that the change in the mangroves is not constant. The graph indicates that mangrove cover remains stable initially, followed by fluctuations, with periods of increase and decrease over time.

21 Analyzing and Interpreting the Normalized Difference Salinity Index (NDSI) (Table 8)

The analysis of the Normal Difference Salinity Index (NDSI) data over the years 1993, 2003, 2013, and 2023 reveals significant fluctuations in salinity levels across various classes within the study area of 5922.02 km^2. The area with very high salinity exhibited notable changes over the years. In 1993, it covered 1427.562 km^2 (24.11% of the total area) area, increasing to 1830.138 km^2 (30.89%) in 2003. However, this category saw a sharp decline in subsequent years, dropping to 957.1059 km^2 (16.16%) in 2013 and further to 501.3639 km^2 (8.47%) in 2023, representing a significant overall decrease over the 30-year period. The high salinity class followed a different pattern, starting at 2821.5234 km^2 (47.62%) in 1993, decreasing to 2269.4004 km^2 (38.32%) in 2003, then increasing to 2770.9101 km^2 (46.80%) in 2013, and reaching 2912.1588 km^2 (49.17%) in 2023, showing a slight overall increase. The medium salinity area remained relatively stable initially, with 1144.0071 km^2 (19.31%) in 1993 and 1144.5669 km^2 (19.33%) in 2003. It then increased to 1458.2988 km^2 (24.63%) in 2013 and further to 1602.7497 km^2 (27.07%) in 2023, indicating a consistent rise over the period (Figs. 15, 16, 17 and 18).

The low salinity class saw substantial fluctuations, starting at 322.7661 km^2 (5.45%) in 1993, increasing significantly to 529.1433 km^2 (8.93%) in 2003, decreasing to 451.4931 km^2 (7.62%) in 2013, and then rising again to 665.9496 km^2 (11.25%) in 2023, showing an overall increase. The very low salinity area fluctuated considerably, decreasing from 206.163 km^2 (3.48%) in 1993 to 148.7727 km^2 (2.51%) in 2003, then increasing to 284.2137 km^2 (4.80%) in 2013, and slightly decreasing to 239.7996 km^2 (4.05%) in 2023, showing a modest overall increase. These trends suggest dynamic changes in salinity

Table 8 Area calculation of **NDSI** 1993–2023

Year	Very high (km^2)	High (km^2)	Medium (km^2)	Low (km^2)	Very low (km^2)
1993	1427.562	2821.5234	1144.0071	322.7661	206.163
2003	1830.138	2269.4004	1144.5669	529.1433	148.7727
2013	957.1059	2770.9101	1458.2988	451.4931	284.2137
2023	501.3639	2912.1588	1602.7497	665.9496	239.7996

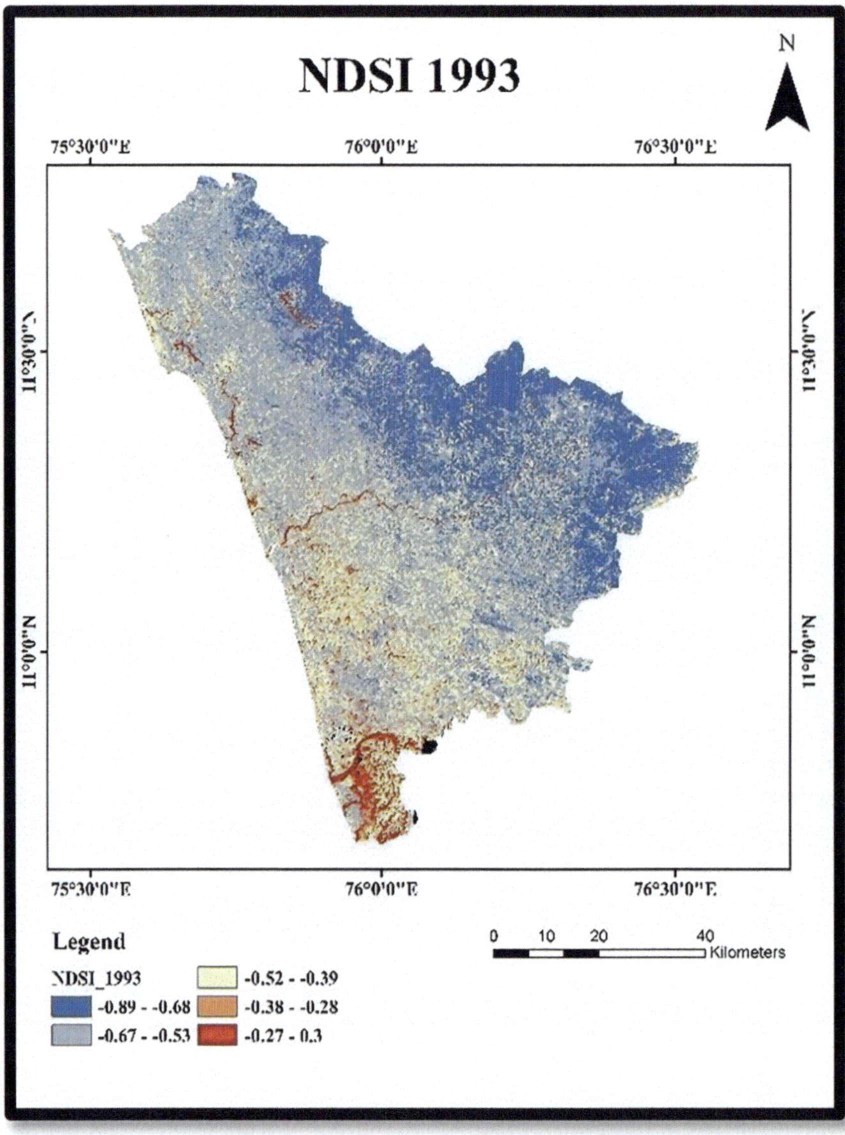

Fig. 15 NDSI 1993

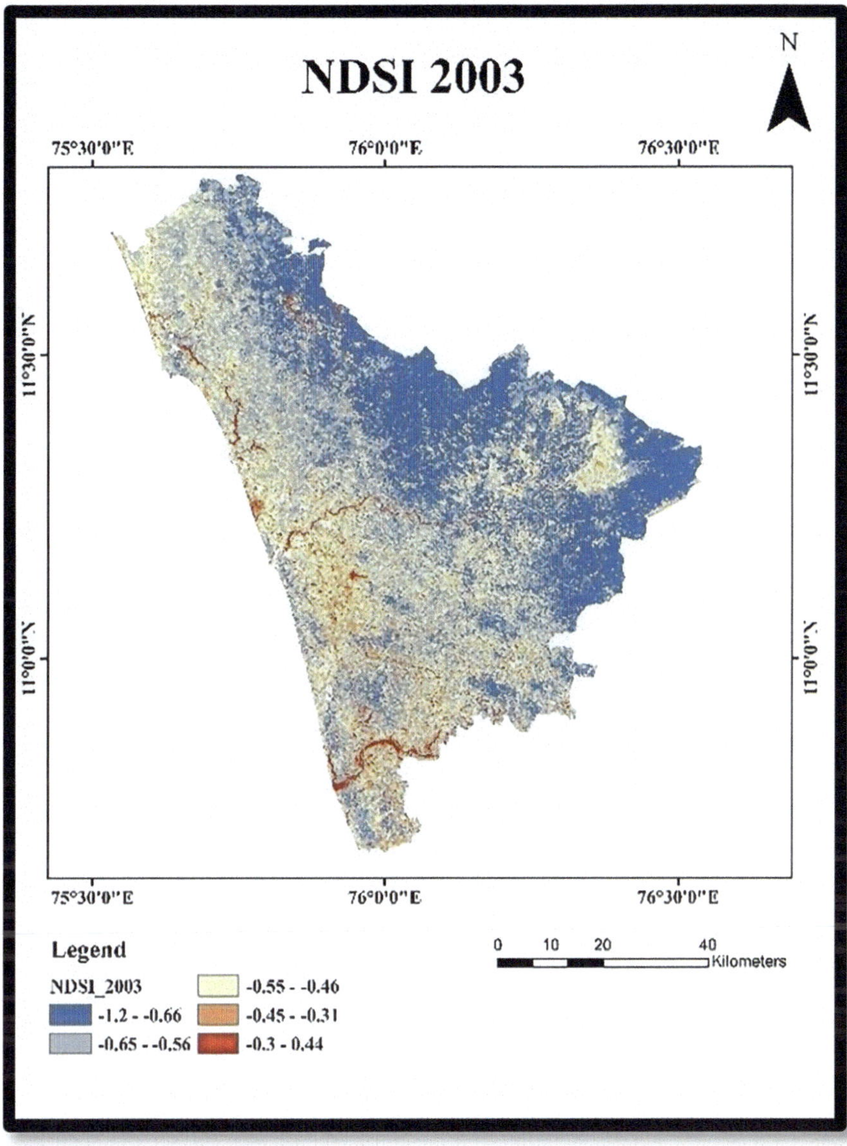

Fig. 16 NDSI 2003

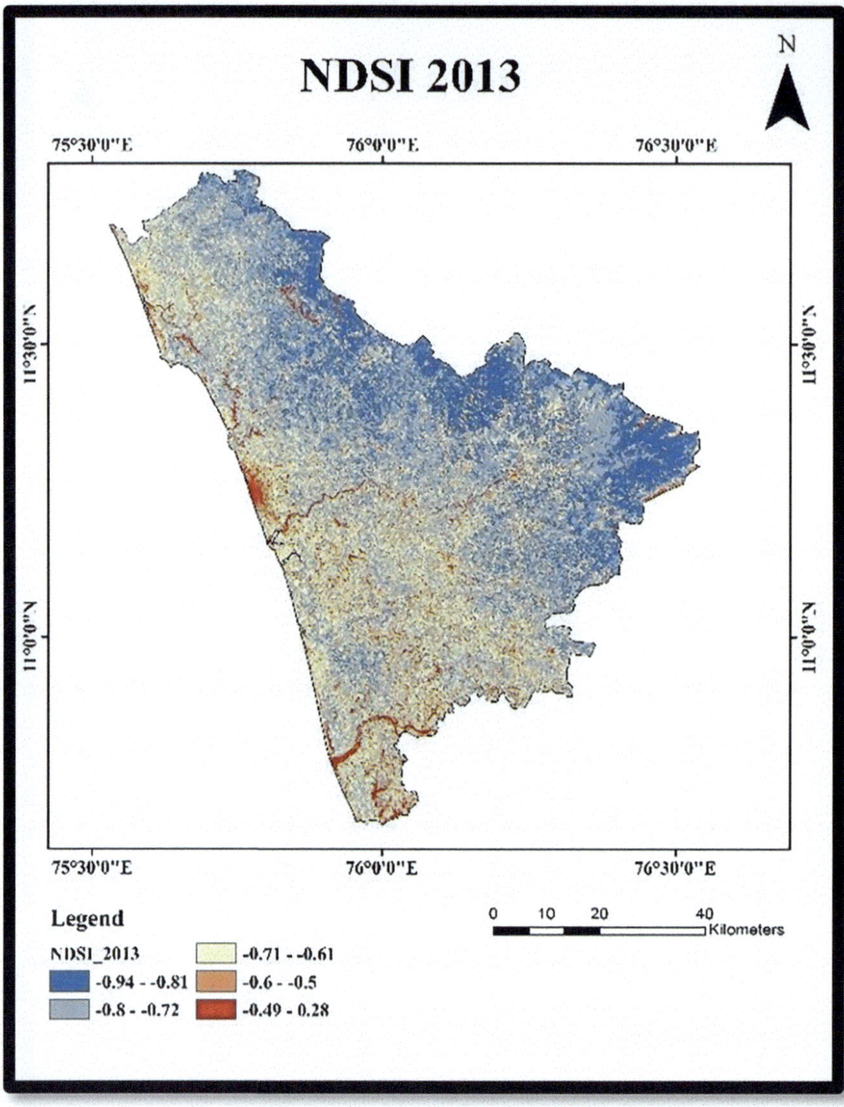

Fig. 17 NDSI 2013

distribution, potentially influenced by environmental factors, land use changes, and water management practices in the region.

The above given maps explain the spatiotemporal changes in salinity. We have classified the NDSI in five categories. It shows the ranges of salinity in different years. The primary category shows very low soil salinity level. The second category describes the low salinity level in the study area. The third category explains the moderate salinity level. The fourth one shows the high salinity level. The last classification shows very high soil salinity in the particular study area. Figure 10 shows that in 1993 the soil salinity in the study area was very less, while Fig. 11 shows NDSI map of 2003, when the salinity level in was more in the past ten years. Compared to 2003, the 2013 figures indicate that the salinity level is steadily increasing. Fig. 13 NDSI map shows that in 2023 there was high level of salinity in the study area. The rise in salinity levels within the estuaries situated in the mangrove region of Kadalundi–Vallikkunnu is commonly identified as an environmental stressor that has a detrimen-

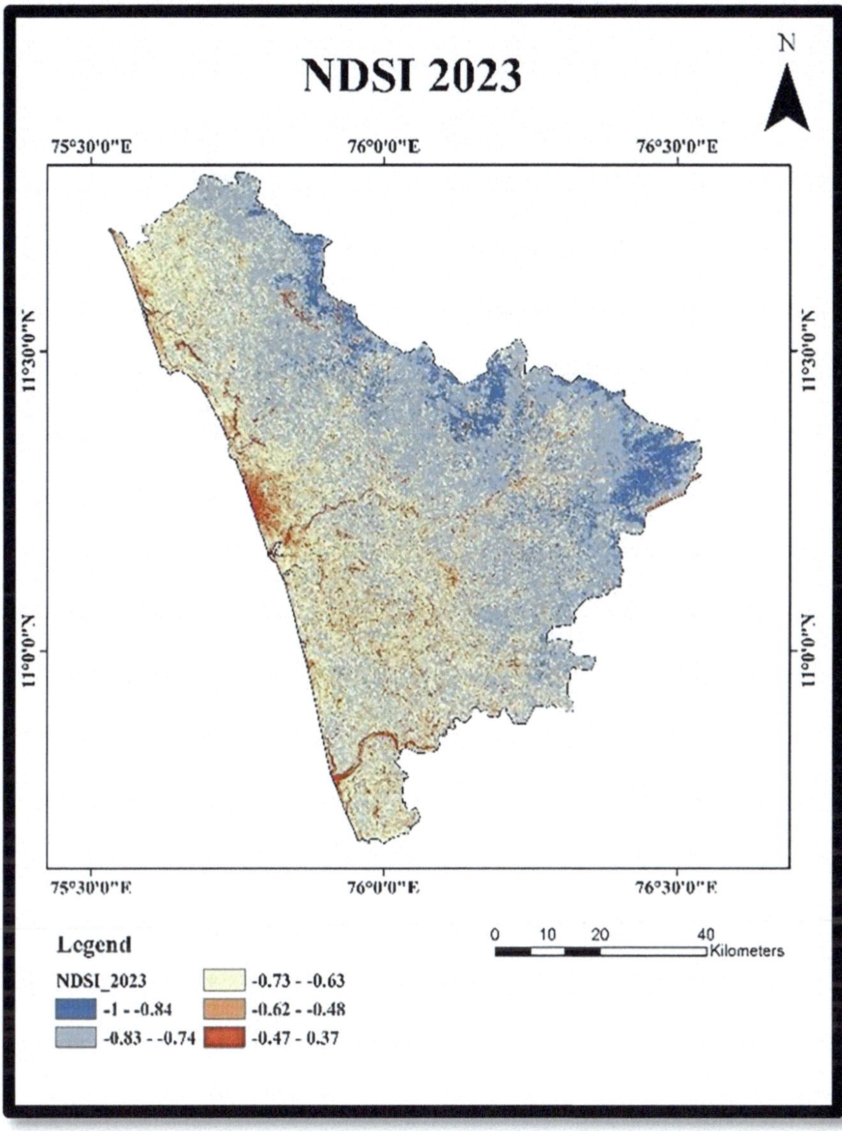

Fig. 18 NDSI 2023

tal impact on the well-being of the mangroves and the various ecosystem services they provide (Samanta et al., 2021).

22 Discussion

The study highlights critical transformations in land use and environmental conditions over the past three decades, driven largely by urbanization and anthropogenic activities. A substantial decline in vegetation and forest cover, amounting to 192.81 km^2, was observed alongside a notable increase of 284.75 km^2 in built-up areas, underscoring the rapid urban expansion at the expense of natural ecosystems. Waterbodies also experienced significant shrinkage, reducing from 225.03 km^2 in 1993 to 138.69 km^2 in 2023, signaling heightened pressure on critical water resources. These land transformations have adversely impacted vegetation health, as evidenced by NDVI analyses that reflect a considerable reduction in forest density and vitality, posing severe threats to biodiversity, ecological balance, and local climate regulation. The expansion of settlements into ecologically sensitive areas, particularly mangrove ecosystems, further exacerbates the loss of critical habitats that serve as buffers against coastal erosion and significant carbon sinks. In addition to these land-use changes, the study reveals substantial shifts in soil salinity patterns, with very high salinity zones decreasing from 1427.56 km^2 in 1993 to 501.36 km^2 in 2023, while medium salinity areas expanded. These variations indicate the interplay of changing agricultural practices, climatic factors, and land management policies influencing soil health and productivity. The mangrove ecosystems, vital for coastal resilience and biodiversity, have seen a marked decline due to settlement encroachments and environmental stressors. This degradation undermines their essential ecological functions, including shoreline stabilization and nutrient cycling. Collectively, these findings emphasize the urgent need for sustainable land management practices, enhanced conservation policies, and community-led initiatives to mitigate the adverse impacts of urbanization and environmental degradation while promoting ecological restoration and resource sustainability.

23 Conclusion

Mangroves represent diverse and remarkably productive ecological communities situated at the interface between land and sea Kadalundi–Vallikkunnu Community Reserve is the primary community reserve area of Kerala. They offer a wide range of vital ecosystem services, encompassing the supply of nourishment and hydration to numerous organisms; mitigation against severe consequences of natural calamities like cyclones and tsunamis; functioning as a substantial enduring reservoir for carbon; preservation of land-based sediments; and serving as a dwelling place for many species. Over the past 20 years, the Kadalundi–Vallikkunnu Mangrove ecosystem has experienced the influence of human activities and climatic impact, gradual climatic shifts, and severe weather occurrences. The anthropogenic actions within the populated regions of the Kozhikode and Malappuram exert a more pronounced cumulative effect on mangrove woodlands, increased salinity levels, relative average sea level elevation, and territorial erosion compared to earlier estimations. The primary waterbody in the region is the Kadalundi River along with its tributaries, complemented by minor reservoirs and small bodies of water, thus establishing the Kadalundi River as the predominant water source within the area of analysis. The impacts of anthropogenic degradation are frequently deemed to be irreversible. The depletion of coastal mangroves is an enduring global occurrence, with scholars currently contemplating the potential of a future devoid of mangroves due to shifting climatic conditions and human-induced disruptions. The rate of mangrove transformation exhibited temporal variability; the most substantial decline occurred between 2013 and 2023, resulting in forest reductions of 192.81 km^2, while 284.75 km^2 of area was obtained through urban development. In this research, we have categorized five primary land use/land cover classifications. Utilizing Landsat data, image processing methodologies, as well as software programs such as ArcGIS and ERDAS, we were able to delineate land use categories including developed lands, agricultural lands, forests, unused lands, and waterbodies. Within the specified research area, the predominant land use type is identified as forests, followed by urban areas, and subsequently cropland. There exist rural regions containing a combination of both developed and residential zones. The current investigation will be highly beneficial for analyzing alterations in land use and land cover within the Kadalundi River Basin. NDVI values from Landsat demonstrate a gradual decrease in the overall health status of the mangrove forest. The temporal variability in mangrove change did not exhibit a consistent pattern; the most substantial decline occurred between 2013 and 2023, resulting in a loss of

192.81 km^2 of forest cover, while a gain of 284.75 km^2 was attributed to settlement expansion.

References

Badola, R., & Hussain, S. A. (2005). Valuing ecosystem functions: An empirical study on the storm protection function of Bhitarkanika mangrove ecosystem, India. *Environmental Conservation, 32*(1), 85–92. https://doi.org/10.1017/S0376892905001967

Benton, M. J., & Twitchett, R. J. (2003). How to kill (almost) all life: The end-Permian extinction event. *Trends in Ecology and Evolution, 18*(7), 358–365). Elsevier Ltd. https://doi.org/10.1016/S0169-5347(03)00093-4

Bindu, K. B., & Suresh, Y. (2015). An analysis of land use/land cover in Kadalundi River basin using remote sensing and GIS. *International Journal of Geomatics and Geosciences, 6 (2), 1442-1449.*

Dahdouh-Guebas, F., Hettiarachchi, S., Lo Seen, D., Batelaan, O., Sooriyarachchi, S., Jayatissa, L. P., & Koedam, N. (2005). Transitions in ancient inland freshwater resource management in Sri Lanka affect biota and human populations in and around coastal lagoons. *Current Biology, 15*(6), 579–586. https://doi.org/10.1016/j.cub.2005.01.053

Ewel, K. C., Twilley, R. R., & Ong, J. E. (1998). Different kinds of mangrove forests provide different goods and services. *Source: Global Ecology and Biogeography Letters, 7*, 83-94.

FAO (2007) The State of Food and Agriculture: Paying Farmers for Environmental Services. Agricultural Development Economics Division (ESA), FAO, Rome.

Food and Agriculture Organization of the United Nations. (2004). *The state of world fisheries and aquaculture: 2004.* Food and Agriculture Organization of the United Nations.

Forest Survey of India. (2023). India state of forest report 2023 (Vol. 1). Forest Survey of India, Ministry of Environment, Forest and Climate Change.

Ghosh, A., Schmidt, S., Fickert, T., & Nüsser, M. (2015). The Indian Sundarban mangrove forests: History, utilization, conservation strategies and local perception. *Diversity, 7*(2), 149–169. https://doi.org/10.3390/d7020149

Giri, C., Ochieng, E., Tieszen, L. L., Zhu, Z., Singh, A., Loveland, T., Masek, J., & Duke, N. (2011). Status and distribution of mangrove forests of the world using earth observation satellite data. *Global Ecology and Biogeography, 20*(1), 154–159. https://doi.org/10.1111/j.1466-8238.2010.00584.x

Hema, M., & Devi, P. I. (2015). Economic valuation of mangrove ecosystems of Kerala, India. *Journal of Environmental Professionals Sri Lanka, 4*(1), 1-16.

Khaleel, K. M., & Jaleel, C. A. (2009). Environmental challenges to the mangrove wetlands of North Malabar (Kerala), India: Their sustainable development and influence on local people. *Knowledge and Management of Aquatic Ecosystems, 392 03.* https://doi.org/10.1051/kmae/2009009

Nagelkerken, I., Blaber, S. J. M., Bouillon, S., Green, P., Haywood, M., Kirton, L. G., Meynecke, J. O., Pawlik, J., Penrose, H. M., Sasekumar, A., & Somerfield, P. J. (2008). The habitat function of mangroves for terrestrial and marine fauna: A review. *Aquatic Botany, 89*(2), 155–185. https://doi.org/10.1016/j.aquabot.2007.12.007

Ragavan, P., Dubey, S. K., Dagar, J. C., Mohan, P. M., Ravichandran, K., Jayaraj, R. S. C., & Rana, T. S. (2019). Current understanding of the mangrove forests of India. In J. Dagar, R. Yadav, & P. Sharma (Eds.), Research developments in saline agriculture (pp. 257–304). Springer. https://doi.org/10.1007/978-981-13-5832-6_8

Samanta, S., Hazra, S., Mondal, P. P., Chanda, A., Giri, S., French, J. R., & Nicholls, R. J. (2021). Assessment and attribution of mangrove forest changes in the Indian sundarbans from 2000 to 2020. *Remote Sensing, 13*(24), 4957. https://doi.org/10.3390/rs13244957

Vinod, K.; Asokan, P.K.; Zacharia, P.U.; Ansar, C.P.; Vijayan, G.; Anasukoya, A.; Kunhi Koya, V.A., and Nikhiljith, M.K., 2019. Assessment of biomass and carbon stocks in mangroves of Thalassery estuarine wetland of Kerala, south-west coast of India. In: Jithendran, K.P.; Saraswathy, R.; Balasubramanian, C.P.; Kumaraguru Vasagam, K.P.; Jayasankar, V.; Raghavan, R.; Alavandi, S.V., and Vijayan, K.K. (eds.), BRAQCON 2019: World Brackishwater Aquaculture Conference. Journal of Coastal Research, Special Issue No. 86, pp. 209-217. Coconut Creek (Florida), ISSN 0749-0208

Walsh, G. E. (1974). MANGROVES: A REVIEW11Publication no. 154 from the Gulf breeze laboratory, Environmental Protection Agency, gulf breeze, Florida 32561 associate Laboratory of the National Environmental Research Center, Corvallis. In *Ecology of halophytes* (pp. 51–174). Elsevier. https://doi.org/10.1016/b978-0-12-586450-3.50008-1

Walton, M. E., Le Vay, L., Lebata, J. H., Binas, J., & Primavera, J. H. (2007). Assessment of the effectiveness of mangrove rehabilitation using exploited and non-exploited indicator species. *Biological Conservation, 138*(1–2), 180–188. https://doi.org/10.1016/j.biocon.2007.04.012

Zhang, C. G., Leung, K. K., Wong, Y. S., & Tam, N. F. Y. (2007). Germination, growth and physiological responses of mangrove plant (Bruguieragymnorrhiza) to lubricating oil pollution. *Environmental and Experimental Botany, 60*(1), 127–136. https://doi.org/10.1016/j.envexpbot.2006.09.002

India: An Emerging Role Model for Developing Tropical Cyclone Early Warning Systems

Sana Rafi, Mahendra Singh, Ashutosh Singh, Amir Ali Khan, and Azizur Rahman Siddiqui

1 Introduction

India, the world's second most populous nation, spans 3,287,590 square kilometers from the snow-covered Himalayan peaks to the tropical rainforests in the south. India is vulnerable to a variety of natural disasters such as floods, droughts, cyclones, tsunamis, earthquakes, landslides, avalanches, and forest fires because of its varied geoclimatic and topographic circumstances (Gupta et al., 2014; Kapur, 2009; Sati, 2011). Notably, regions such as the Himalayas, northeastern Gujarat, and the Andaman Nicobar Islands are particularly prone to seismic activity, covering about 59% of India's land area. According to the Ministry of Agriculture's Natural Disaster Management Division, 33% of India's land receives <750 mm of rainfall annually, making it drought-prone, while 35% receives (750–1125 mm). Approximately 7.5 million hectares of the country's 40 million flood-prone hectares are affected by floods each year, with states like Assam, Bihar, Gujarat, West Bengal, and Uttar Pradesh experiencing floods almost annually across various river basins. Floods and river erosion impact about 12% of the land, particularly in hilly regions where landslides and avalanches are also concerns. Cyclones threaten more than 71% (5300 km) of India's 7516 km coastline, affecting areas named Tamil Nadu, Puducherry, Andaman and Nicobar Islands, Andhra Pradesh, and Odisha on a regular basis. As per the Ministry of Home Affairs (MHA)'s annual report (2012–13), 27 out of India's 35 states and union territories face the risk of disasters, with many susceptible to multiple hazards. Coastal communities' livelihood security and the ecological health of India's coastal zones are already jeopardized by factors such as dense population, rapid urbanization, industrial growth, environmental degradation along the coast, and frequent natural calamities like cyclones and storms. The situation is worsened by rising sea levels due to global warming. Cyclones, floods, and tsunamis threaten India's eastern coastline. In the past 25 years, natural disasters have become more frequent globally, affecting developed and developing nations (Dilley et al., 2005; IFRC, 2015; Djalante, 2018; Chanda, 2018). Today, more people are vulnerable to these disasters than ever before, with those in less developed countries facing heightened risks.

S. Rafi · M. Singh · A. Singh (✉)
Department of Geography, School of Earth Science, Hemvati Nandan Bahuguna Garhwal University (A Central University), SRT Campus, Tehri, India

A. A. Khan
National Institute of Disaster Management, New Delhi, India

A. R. Siddiqui
Department of Geography, University of Allahabad, Prayagraj, India

Z. Khanam et al. (eds.), *Climate Change, Resilience and Ecosystem Services*, Sustainable Development Goals Series, https://doi.org/10.1007/978-3-031-96977-5_21

2 Early Warning System (EWS) for Multi-Hazards

As per the International Telecommunication Union (ITU), a multi-hazard early warning system (MHEWS) can manage various types of hazards and their consequences, whether occurring individually, simultaneously, cascading, or accumulating over time, considering potential interconnected impacts. For these systems to be effective, they require engagement from diverse stakeholders and active participation from at-risk communities. This ensures an enabling environment with suitable technology, regulatory frameworks, operational capacities, and clearly defined roles for all involved entities, including communities. Over the past 50 years, while economic losses have risen, fatalities from hydrometeorological hazards have declined. This progress is attributed to effective EWS, which leverages advancements in weather monitoring and forecasting, coupled with efficient communication and preparedness at local and national levels. India has been notably successful in implementing such systems.

Warning systems consist of four main operational elements that are evaluated collectively:

1. Identifying, tracking, forecasting hazards, and issuing alerts based on these assessments.
2. Assessing risks and integrating findings into warning messages.
3. Issuing timely warnings to relevant authorities and at-risk communities through authoritative sources.
4. Implementing community-based emergency preparations in response to warnings to mitigate damage to lives and livelihoods.

Several national, state, and local entities work together in India to handle these components. The system's functionality depends on these entities coordinating with one another seamlessly, as any malfunction in one area could compromise the system's efficacy. Early warning systems have received much attention worldwide in the last 10 years, particularly in the wake of the horrific tsunami that slammed into the Indian Ocean in December 2004. Soon after, to mark 10 years after the Yokohama Strategy for a Safer World was adopted, the United Nations International Strategy for Disaster Reduction (ISDR) hosted the Second World Conference on Disaster Reduction in Kobe, Hyogo, Japan. During this summit, 168 countries deliberated and adopted the "Hyogo Framework for Action 2005–2015: Building Disaster Resilience in Nations and Communities." The disaster risk management paradigm has changed due to these international initiatives; it is now more comprehensive and includes preparedness and preventative strategies in addition to post-disaster response. The Hyogo Framework lists "identifying, quantifying, and monitoring catastrophic risks and enhancing early warning systems"as a top priority. Early warning systems should be an essential component of any country's disaster risk management plan, the declaration emphasizes, to empower national governments and local communities to prepare for disasters proactively. Odisha was struck by a "super cyclone" on October 29–30, 1999, affecting all coastal districts with incredibly high wind speeds ranging from 170 to 185 miles per hour. Along with an unusual 16–23-foot storm surge and 48 hours of intense rain, the cyclone caused catastrophic flooding in key river basins. The severity of the cyclone resulted in the tragic loss of over 10,000 lives. The Orissa Relief Code was activated due to the extensive economic devastation. Odisha gained global prominence because the super cyclone occurred towards the conclusion of the United Nations International Decade for Natural Disaster Reduction (UN I.D.N.D.R). The significant number of fatalities in 1999 was due to deficiencies in disaster management, including inadequate coordination, communication failures, and a complacent approach. According to the Indian Meteorological Department (IMD), there was a lack of collaboration and coordination among government departments in Odisha before 1999. Technological capabilities were also limited, with weather forecasts being reliant on New Delhi and Kolkata, and communication of weather warnings being delayed. Just 2 days before the super cyclone struck, the IMD managed to issue early warnings. However, both the generation and dissemination of these warnings were hindered by underdeveloped communication systems. The lack of coordination was worsened by the absence of a central authority in Bhubaneswar to oversee relief and rescue operations. It was not until the mid-term review of the (UN, IDNDR) in Yokohama, Japan, in 1994, that the global community acknowledged the significant impact of disasters on developing countries (Enarson, 1998). Over the past two decades, Odisha has significantly enhanced its multi-hazard early warning systems following the devastating 1999 storm. Through deliberate efforts and incremental improvements, particularly at the grassroots level, the state has effectively reduced casualties and impacted populations. Efforts have included introducing community-level alerts, constructing multipurpose cyclone shelters, and establishing an Early Warning Dissemination System through the National Cyclone Risk Mitigation Project. These measures have significantly enhanced the community's ability to manage natural disasters. Compared to the nearly 10,000 fatalities during the 1999 storm, subsequent cyclones such as Phailin (2013), Hudhud (2014), Titli (2018), Fani (2019), Tauktae (2021), and Yaas (2021) have resulted in notably fewer deaths. Experts attribute this improvement to timely evacuations facilitated by advanced early warning systems. For instance, Cyclone Phailin in 2013 affected over 1.1 million people and caused losses estimated at approximately $4.24 billion. Cyclone Fani, which struck on May 3, 2019, with winds gusting up to 180–190 kilometers per hour, led to 64 fatalities and inflicted damages estimated at around Rs 24,176 crore. As Odisha prepares for Cyclone Yaas, expected to hit on May 26, 2021, experts have praised the state's disaster preparedness and resilience.

In contrast to the 1999 event, where warnings were issued just two days prior to impact, recent cyclones have been forecasted—five to six days ahead, allowing authorities ample time to take necessary precautions. In late 1999, the Odisha government was informed about the super cyclone's altered trajectory, but the state's communication networks failed abruptly, hindering effective monitoring of the situation. However, present-day cyclone warnings from the IMD are issued early and with high accuracy, closely matching observed and projected paths. This has instilled confidence not only among the public but also among officials, leading to a significant reduction in unnecessary evacuations. Moreover, widespread use of television and mobile phones has heightened public awareness of cyclones well ahead of their landfall. In contrast to 1999, when the chief secretary lacked a cell phone and backup communication lines, today's Odisha secretariat is equipped with internet connectivity.

From 1999 to 2013, referred to as the "inter-disaster period" by the district emergency officer of Jagatsinghpur, the Odisha government implemented a new disaster management system consisting of two distinct parts (Government of Odisha, 2012). Initially, there was heightened cooperation among national and state governments, the IMD, non-profit organizations, and vulnerable communities. Second, this revamped disaster management system incorporated advanced technology. This approach effectively addressed the coordination gaps, communication failures, and complacency observed during the 1999 super cyclone. After the disaster, the IMD utilized space technology, the Meteo France International synergy system, and advanced computing to improve their forecasting capabilities (Barik, 2014). Additionally, in 2007, the Indian government introduced the first Disaster Management Act, establishing a knowledge network that included key organizations such as the Indian Meteorological Department (IMD), Earth System Science Observation (ESSO), Indian Space Research Organization (ISRO), Central Water Commission (CWC), Geological Survey of India (GSI), and National Remote Sensing Centre (NRSC), among others (Srivastava, 2009).

During Cyclone Phailin, this network played a pivotal role in generating critical information that was effectively communicated to the vulnerable population. To disseminate essential information to those at risk, various communication platforms such as media, mobile text messaging, hotlines, and VSat were extensively utilized. Ahead of

Cyclone Phailin's impact, accurate foundational data played a crucial role in developing an effective response strategy. This data facilitated the efforts of frontline responders, leading to the evacuation of 1.2 million people from 18 districts (Mishra, 2013). This evacuation stands as one of India's largest emergency operations ever conducted. The successful execution of this large-scale operation was enabled by stakeholder coordination, access to vital information, detailed evacuation planning, flexibility in protocols, and the dedication of responders to safeguard lives. Leadership is essential for both developing an effective response system and combating complacent worldviews. In the aftermath of the superstorm, Odisha's Chief Minister, Mr. Naveen Patnaik, offered much-needed leadership. He declared October 29 as Odisha's Disaster Preparedness Day starting in 2000. As a result, a disaster-preparedness culture developed. Additionally, he worked hard to upgrade the state's infrastructure, essential to the operation of a disaster response system. Patnaik built concrete structures, bridges, roads, and multipurpose cyclone shelters with funding from the federal government and the World Bank. Well-maintained roads and proximity to storm shelters made evacuation easier during storm Phailin. During Cyclone Phailin, Patnaik exhibited strategic leadership (Jenkins, 1969) by making "saving lives" the top priority for all parties concerned in reducing the storm's effects. There was a notable decrease in mortality because of the local responders' successful communication of this command. With the implementation of an early warning system (EWS) for natural disasters, including cyclones and tsunamis, in April 2018, Odisha became the first state in India to protect inhabitants along its 480-kilometer coastline. With the goal of boosting the state's emergency operation centers and enhancing the distribution of catastrophe warnings, the EWDS is a powerful communication system. Currently, the coastal districts of Balasore, Bhadrak, Jagatsinghpur, Puri, and Ganjam are covered by the EWDS, which spans 1205 villages over 22 blocks. As part of the system, 122 watchtowers have been built within 1.5 km of the shore, allowing for the mass messaging and siren-based transmission of cyclone and tsunami warnings. The experience of Odisha provides valuable lessons for disaster preparedness and response strategies.

- First, when disaster management systems are skillfully coordinated, even poor nation-states may cut disaster casualties. Creating an effective response system hinged on having precise core data and ensuring seamless coordination and information sharing among key stakeholders. A contemporary disaster management system is envisioned as one that integrates human and technological elements. To enhance coordination and communication effectiveness, policymakers and UN bodies should prioritize both technological advancement and capacity building. Rather than operating independently, this system should integrate with early warning systems to optimize its impact.
- Second, the instance of Odisha shows how political leadership plays an increasingly important role before, during, and after a crisis. Effective disaster response systems are in line with the objective of saving lives in situations when political leadership is proactive. Political leadership can also help to foster a disaster-preparedness culture. The Chief Minister of Odisha during Phailin established an objective of "saving lives at whatever cost. As a result, all performers and responders banded together to attain this goal. In this sense, the UN and other international financial agencies might make a significant contribution by persuading political leaders to enact "priorities for action" for successful disaster management.
- Third, eliminating disaster-related mortality is critical and demonstrates the system's resilience. This ethos is now represented in the United Nations' Sendai Framework for Action's first global aim (2015–2030) (Preventionweb),[1] which calls for "global disaster mortality to be reduced by 2030." As

[1] http://www.preventionweb.net/files/43291_sendaiframeworkfordrren.pdf

demonstrated by the instance of Odisha, setting goals to lower mortality and creating a socio-technical disaster management system, in addition to encouraging a preparation culture, are essential measures in accomplishing the first global aim of the Sendai Framework. India donated to international organizations in addition to improving EWS in Odisha. RIMES, an intergovernmental body registered with the United Nations, is managed by 45 countries across the Asia Pacific and Africa. Established in 2009 following the 2004 Indian Ocean tsunami, RIMES operates from its regional center at the Asian Institute of Technology in Thailand. It was created through collaboration among India, Asian, and African partners to develop a regional early warning system that addresses multiple hazards, enhances information dissemination, and builds capacity for cross-border hazard preparedness and response.

India plays a significant role in RIMES, contributing both financially and in-kind resources. The National Centre for Medium-Range Weather Forecasting (NCMRWF) in Noida provides global outputs and high-performance computing capabilities to RIMES, enabling the creation of tailored products for member countries. RIMES facilitates the exchange of observational data among member states and validates outputs from NCMRWF. The Indian National Centre for Ocean Information Services (INCOIS) in Hyderabad supplies high-resolution ocean state prediction products through RIMES to meet the marine forecasting needs of member states (Reliefweb). Under the auspices of RIMES, the Ministry of Earth Sciences promotes specialized projects in research, education, communication, and the application of established science and technology. Disaster-related activities that are funded and implemented with a strategic approach can drastically cut down on the number of people killed and property destroyed by natural, man-made, and technological disasters. The Indian government supports RIMES financially to achieve the goals set by member states and collaborating countries. India remains dedicated to fostering regional engagement and capacity building through south-south cooperation, actively supporting RIMES initiatives. RIMES has collaborated effectively with the Odisha State Disaster Management Authority (OSDMA) to enhance the state's disaster management capabilities, focusing on improving quality, predictability, and response readiness. OSDMA benefits from RIMES' expertise in data analysis, validation of predictions, early warning systems, and preparedness for lightning, heatwaves, floods, droughts, and tsunamis, facilitated by autonomous weather stations across gram panchayats. RIMES also provides specialized training to enhance OSDMA's capacity to respond to warnings effectively. Additionally, RIMES assists in developing a comprehensive risk management system tailored to OSDMA's needs, integrating various data sources and servers. Despite facing major challenges such as Cyclone Phailin in 2013, Cyclone Hudhud in 2014, and numerous floods and droughts since 2000, Odisha's response to these disasters has received global acclaim. The United Nations has recognized Odisha as a model state for its exemplary disaster management during Cyclone Phailin.

Directors General of South Asian National Meteorological and Hydrological Services (NMHSs) and Permanent Representatives (PRs) to WMO convened in August 2009 at the Abdus Salam International Centre for Theoretical Physics (ICTP) in Trieste, Italy, and resolved unanimously to create the South Asian Climate Outlook Forum (SASCOF). SASCOF began operations in 2010, focusing on providing regular climate forecasts. Following the devastating Super Cyclone in 1999, IMD significantly upgraded its multi-hazard early warning system. It now provides forecasts for terrestrial, atmospheric, and oceanic conditions across various time scales: nowcasting (up to 6 h), short-range (1–2 days), medium-range (up to a few weeks), extended-range (up to 2 weeks), and seasonal weather and climate services (for the monsoon season from June to September).

3 Operation of MHEWS

Several effective early warning systems for meteorological and hydrological risks have demonstrated success. Examples include responses to Cyclones Phailin, Fani, and Hudhud, as well as the management of the Kerala floods in 2021 and the Assam Earthquake of 2021. UNDP has collaborated extensively with government officials, civil society groups, and volunteers at national and subnational levels, under government leadership, to enhance community preparedness for disaster risk management, particularly by strengthening early warning systems in tsunamis-affected states. India's experience serves as a global example, demonstrating that effective disaster alert systems combined with empowered communities can significantly reduce disaster-related fatalities. In response to the 2004 tsunami, India's government took significant measures to establish early warning systems. Under the auspices of the Ministry of Earth Sciences, the Indian National Centre for Ocean Information Services (INCOIS) in Hyderabad, Andhra Pradesh, developed the National Tsunami Early Warning System. The IMD implemented mechanisms to issue accurate alerts and provide real-time weather reports for disaster management agencies and emergency response teams. Following the tsunami, UNDP and other UN agencies offered technical assistance to develop protocols for damage assessment, monitor relief and recovery efforts, and address related issues. In Tamil Nadu, one of the states most impacted, Emergency Operation Centers were set up in all five coastal districts with public address systems and VHF wireless communication networks, supported by UNDP. Tamil Nadu is expanding this initiative to enhance early warning systems utilizing cutting-edge technologies, helping 36 million people living in highly populated coastal areas, with support from the World Bank. Similarly, the National Cyclone Risk Mitigation Project, which has implemented early warning and communication systems, involves the vulnerable states of Odisha and Andhra Pradesh (Relief Web).

Through a substantial collaboration with the Ministry of Home Affairs, UNDP has played a crucial role in strengthening disaster resilience in communities. This partnership, which now spans 150,000 high-risk communities, has grown to encompass over 300 million people and rank among the greatest community-based disaster risk management initiatives globally. Communities have been empowered with emergency response tools, knowledge of warning signals, and preparedness training materials distributed in both English and local languages. This ensures communities are well-informed about safety measures during disasters. During cyclones Hudhud and Phailin, the efficiency of this preparation was demonstrated. Prior to Cyclone Phailin, a variety of media outlets, including print, television, internet, fax, phone, text messaging, and public announcement systems, provided regular information on the cyclone's path and intensity. Officials in the 14 districts most at-risk were given satellite phones to guarantee the timely transmission of information and alarms during the hurricane.

4 Timely Action

During Cyclone Phailin, authorities in Odisha ensured that emergency control rooms operated 24/7 and deployed numerous rapid-response disaster field officers in advance for evacuations and rescue efforts. They also stocked food, water, and medical supplies, and mobilized earthmoving and road-clearing equipment to facilitate the distribution of aid to severely affected areas. Cyclone and flood shelters were inspected and readied to accommodate evacuees. Special attention was given to fishermen, who are often most affected during coastal disasters; they received training to recognize warning signs for their safety. INCOIS collaborated with the M.S. Swaminathan Research Foundation to establish the Fisher Friend Network, providing fishermen with ocean weather forecasts, early warnings of hazardous weather conditions, and guidance on optimal fishing zones, utilizing advanced communication technologies.

5 Summary

India's coastal governments are investing in early warning systems, which have been crucial in saving millions of lives, 10 years after the catastrophic tsunami that struck the Indian Ocean. With 1.2 billion people on Earth, more than a third of whom reside near storm-surge- prone coastlines that are prone to cyclones, tsunamis, and other natural disasters, these systems have repeatedly shown themselves to be life-saving. Two consecutive cyclones that have affected India's eastern coast in the previous 2 years are examples of recent occurrences.

Cyclone Hudhud in October 2014 caused widespread devastation and claimed 46 lives, whereas Cyclone Phailin, a year earlier, resulted in significantly fewer casualties due to proactive measures such as early evacuations, relocating 1.2 million people to safer areas. This contrasts sharply with earlier cyclones, such as the one in 1999 that claimed 10,000 lives, and the 2004 tsunami that killed over 10,000 people, highlighting substantial improvements in preparedness and response capabilities since then. Despite improvements in disaster preparedness, significant efforts are still needed across many parts of India to reduce the risk of catastrophes. In Jammu and Kashmir, a northern Himalayan region far from vulnerable coastal areas, irregular and intense rainfall events are increasingly common, leading to frequent floods (Basistha et al., 2009; Kulkarni et al., 2013; Shekhar et al., 2017) such as those seen in 2014, despite nearby areas experiencing near-drought conditions. Enhancing flood early warning systems is now a crucial priority to enhance the accuracy and timelines of flood forecasts. Efforts are focused on expanding these systems beyond coastal regions to inland areas. Climate change is introducing a new and complex vulnerability, posing a threat to the resilience of millions across India (IPCC, 2014; Oxfam International, 2008; Rafi, 2021). India's ability to adapt government structures, communities, and individuals to this evolving challenge will be crucial. The UNDP and the Ministry of Home Affairs are working together to improve state and district administrations' ability to reduce risks and deal with climate change. States and districts are already creating policies to address climate change and adapt to new realities, as well as giving priority to public investments and putting regulations in place to lessen the negative effects of climate-related disasters. It has been determined that several fundamental ideas are necessary for early warning systems to function well, irrespective of the institutional, political, social, or economic environment in any given nation.

- Political commitment is crucial: Early warning systems require robust backing from national and local governments, evident in unified disaster risk management policies, planning, laws, and budget allocations.
- Shared operational components: This entails coordinating efforts across federal, state, and local agencies to detect, monitor, and forecast hazards; incorporate risk assessments into emergency plans and alerts; disseminate prompt and reliable warnings; and ensure community preparedness by activating emergency protocols.
- Defined roles: All relevant parties, including technical organizations like meteorological services, must be properly recognized, their roles and duties delineated, and coordination mechanisms recorded in national and municipal plans, regulations, directives, and memorandums of understanding.
- Allocation of resources: Sufficient human, financial, and equipment resources are essential to support the capabilities of early warning systems at both national and local levels, ensuring their sustainability.
- Risk assessment: Data on exposure, vulnerability, and hazards are used to assess risk at various levels, directing the creation of warning messages and emergency preparations.
- Clear warning messages: Warnings need to be straightforward, coherent, and incorporate risk information, using universally recognizable formats (e.g., color codes, flags) issued by a credible and authoritative source.
- Timely distribution: Mechanisms for disseminating warnings must reliably reach

authorities, stakeholders, and vulnerable populations promptly.

- Integration into response strategies: Emergency response plans should account for hazard and risk levels, community characteristics, coordination mechanisms, and the roles of stakeholders.
- Education and training: Awareness of risks and training in emergency response are integrated into formal and informal educational initiatives, accompanied by regular drills to maintain operational readiness.
- Feedback and improvement: Effective feedback loops at all levels facilitate systematic evaluation and continuous improvement of the system over time.
- Countries facing multi-hazard risk management challenges can adapt these best practices, considering their unique historical, cultural, socioeconomic, institutional, capacity, and resource contexts to ensure sustainable system design and implementation.[2]

References

Barik, B. (2014). "Met Office goes digital," *The Telegraph*, February 18, 2014; and interview with the Director of Indian Meteorology Department in Bhubaneswar, July 22, 2014.

Basistha, A., Arya, D. S., & Goel, N. K. (2009). Analysis of historical changes in rainfall in the Indian Himalayas. *International Journal of Climatology: A Journal of the Royal Meteorological Society, 29*(4), 555–572.

Chanda, A. (2018). Impact of Natural Disasters on Socio-economic Condition of Population in Sikkim, Pawar Pub., New Delhi.

Dilley, M., Chen, R. S., Deichmann, U., Lerner-Lam, A., Arnold, M., Agwe, J., Buys, P., Kjekstad, O., Lyon, B., & Yetman, G. (2005). Natural disaster hotspots: A global risk analysis. In World Bank Disaster Risk Management Series (Issue 5).

Djalante, R. (2018). A systematic literature review of research trends and authorships on natural hazards, disasters, risk reduction and climate change in Indonesia. Natural Hazards and Earth System Sciences, 18(6), 1785-1810.

Enarson, E. (1998). Through women's eyes: A gender research agenda for disaster social science. *Disasters, 22*, 157–173.

Government of Odisha (Bhubaneswar: Office of the Special Relief Commissioner, 2012). *Procedures/guidelines for maintenance of records relating to the relief operations on account of natural calamities* (No. 768/SR), http://www.odisha.gov.in/disaster/src/Procedure_Guidelines/Maintenance_NC.pdf

Gupta, A. K., Nair, S. S., Wajih, S. A., Chopde, S., Gupta, G., & Aggarwal, G. (2014). *Mainstreaming climate change adaptation and disaster risk reduction into district level development plans: Training module*. National Institute of Disaster Management, Gorakhpur Environmental Action Group and Institute for Social and Environmental Transition.

IFRC. (2015). Focus on local actors , the key. http://reliefweb.int/sites/reliefweb.int/files/resources/1293600-World-DisastersReport-2015_en.pdf

IPCC. (2014). In Core Writing Team, R. K. Panchauri, & L. A. Meyer (Eds.), *Climate change 2014: SynthesisReport. Contribution of working groups I,II and III to the fifth assessment report of the intergovernmental panel on climate change* (p. 151). IPCC. https://unfccc.int/documents?f%5B0%5D=document_type%3A3499&search2=&search3=&page=0%2C0%2C56

Jenkins, G. M. (1969). The systems approach. *Journal of Systems Engineering, 1*, 3–49.

Kapur, A. (2009). *Vulnerable India, a geographical study of disasters*. SAGE pub. ISBN:978-81-321-0077-5(HB).

Kulkarni, A., Patwardhan, S., Kumar, K. K., Ashok, K., & Krishnan, R. (2013). Projected climate change in the Hindu Kush-Himalayan region by using the high-resolution regional climate model PRECIS. *Mountain Research and Development, 33*(2), 142–151. https://doi.org/10.1659/MRD-JOURNAL-D-11-00131.1

Mishra, B. N. (2013). Tryst with Phailin: The deadliest cyclone in 2013. *The Response, 13*, 5–7.

Oxfam International. (2008). *Rethinking disasters: Why death and destruction is not nature's fault but human failure*. South Asia Regional Centre.

Rafi, S. (2021) *Assessment of climate hazards in the Garhwal Himalayan Region using Geospatial Techniques*.

Sati, V. P. (2011). Climate disasters in the Himalaya : Risk and vulnerability. In *International conference on climate change and natural hazards in mountain area, September 2011* (pp. 1–6). https://doi.org/10.13140/2.1.3797.440

Shekhar, M., Bhardwaj, A., Singh, S., Ranhotra, P. S., Bhattacharyya, A., Pal, A. K., et al. (2017). Himalayan glaciers experienced significant mass loss during later phases of little ice age. *Scientific Reports, 7*(1), 1–14.

Srivastava, S. K. (2009). Making a technological choice for disaster management and poverty alleviation. *Disasters, 33*, 58–81.

[2] https://www.mei.edu/publications/learning-deaths-disasters-case-odisha-india

Climate Change, Geospatial Technologies and Sustainable Development: A Framework for Action

Tidal Flats Wetland System of the Estuarine Coast in the Lower Ganga Delta Under Climatic Stress

Ashis Kumar Paul, Anurupa Paul, Joydeb Sardar, and Sk Saharukh Ali

1 Introduction

The term tidal flat is very broad and covers a range of generally muddy low-gradient intertidal or supratidal surfaces (Amos, 1995; Dyer, 1998; Eisma et al., 1981; Daidu et al., 2013, Gao, 2019). According to the magnitude of exposures to waves and tidal currents, tidal flats can be classified into four major types in terms of their depositional environments: (i) Intertidal flat, (ii) Supratidal flat, (iii) Subtidal exposure flat, and (iv) Shore fringe tidal flat of the coastal marine environment, with increasing wave exposure from low to high (Paul, 2002; Paul et al., 2017, 2018, 2023; Mukherjee & Paul, 2023; Sardar et al., 2023). There are specific sedimentary features and facies associations of tidal flat formation in the Sundarbans.

The tidal flats wetland system occupies components of extensive mudflats, tidal shoaling flats, land cover of halophytic vegetations, and the delta plain estuaries with a network of tidal creeks and inlet channels at the lower part of the Ganga Delta. They are interacting and interplaying with each other to modify their forms and patterns by the energy drivers of tides, currents, surge waves, river floods, and storm winds in the low-lying coastal tracts. In the lower part of the Ganga Delta, the tidal river mouths of the Hugli, the Muriganga, the Saptamukhi, the Thakuran, the Matla, and the Ichhamati–Raimangal (Indian Part) were fluvial channels during the early Holocene epoch (6000 YBP), and gradually submerged by marine transgression with modification of fluvio-tidal deposits to transform into "ria" type of estuaries (Paul, 2002; Allison et al., 2003). Thus, the present-day tidal landforms are separated into deltaic islands by tidal river estuaries from west to east in the sedimentary depositional environment by fluvio-tidal and marine processes; under the halophytic vegetation cover tidal flats, sediments are transported into the system and nutrients are transported out from the system by the energy drivers.

Eight tidal flats wetland systems are selected from estuarine setting, tidal deltaic setting, and coastal marine setting of the lower part of the Ganga Delta for this study to show the climate change impacts and sea level rise threats using the satellite-derived data and field monitoring methods in the regional physiographic settings (Tables 1 and 2). The tidal flat wetland ecosystems processes are extensively explored based on the field-based ecological status monitored data and satellite-derived data to set the major research

A. K. Paul (✉) · S. S. Ali
Department of Geography, Vidyasagar University, Midnapore, West Bengal, India

A. Paul
Department of Remote Sensing and GIS, Vidyasagar University, Midnapore, West Bengal, India

J. Sardar
Centre for Environmental Studies, Vidyasagar University, Midnapore, West Bengal, India

Z. Khanam et al. (eds.), *Climate Change, Resilience and Ecosystem Services*, Sustainable Development Goals Series, https://doi.org/10.1007/978-3-031-96977-5_22

Table 1 Natural features of tidal flat wetland systems of the Sundarbans

Category of tidal flat	Major features	Processes involved
Supratidal flat	Levee surface with swamp forest	Rapid accretion
	Flats with swamp and tidal roots of mangroves	The slow rate of accretion
	Tidal creeks with marginal flats	Active drainage and soil piping
	Salt flats and marshes	Tidal drainage loss
	Saltpans and tidal pools	Tidal drainage loss and salt encrustation
Intertidal flat	Tidal mud flats with mangroves	Floccules of mud in the high tides
	Bare flat with break of slope	Erosion and accretion
	Low tide silt flat with ripples	Wash deposits
Subtidal flat	Emerged bars	Bidirectional tidal currents
	Shoaling flats of tidal river beds	Floccules of mud in the high tides and low tides
	Tidal sand ridges in a linear direction	Strong tidal currents of flood and ebb direction
Shore fringe tidal flat of coastal marine environment	Beach ridges	Tidal surges
	Linear swell basins	Inundations
	Sand spits (shore parallel features)	Longshore currents
	Backwater tidal flats	Submergence and emergence by tides
	Stepped mud banks on the sea face with gully channels	Storm surges, swell waves, tidal surges, and wave actions on the sea face

Table 2 Spatial variation of tidal inundation zones in the tidal flats wetland system of Indian Sundarbans

Types of tide	Occurrences	Areas affected
Daily high tides	Diurnal (including spring and neap)	Inter tidal region
Astronomical tides	Seasonal (equinoxes)	Part of the supratidal region
Storm tides	Infrequent (extreme storms)	Entire supratidal region

objectives of this study areas such as (i) classification of tidal flat wetland systems in the Indian Sundarban; (ii) the role of the natural features of tidal flat ecosystems into coastal risk reduction planning, and (iii) the assessment of the coastal stress on the tidal flats of Indian Sundarbans (Fig. 1).

2 Material and Methods of the Study Area

This study carries out several aspects of data including satellite remote sensing data, CARTOSAT DEM, sea level rise data PSMSL and AVISO, CPT tide data, INCOIS swell, wave and wind data, IMD cyclone frequency and climatic data, SOI toposheet, historical map, and other attribute data were used to explore and identify 14 stressors of the tidal flats wetland ecosystem in Sundarban deltaic tracts of the southwestern parts of the Bay of Bengal. The series of the Landsat satellite (https://earthexplorer.usgs.gov/) images were obtained from the Earth Explorer. The existing literature survey can help to identify the research gap and to find out some landscape alteration problems in tidal flat wetland systems of Indian Sundarbans. Therefore, this research aims to fill these knowledge gaps by adopting a multi-disciplinary approach to understanding the threats of tidal flat wetland ecosystems in the impact of climate change and sea level rise threats. Figure 2 shows the schematic workflow of the methodological flow which highlights the threats and suggests possible management techniques by the policy maker.

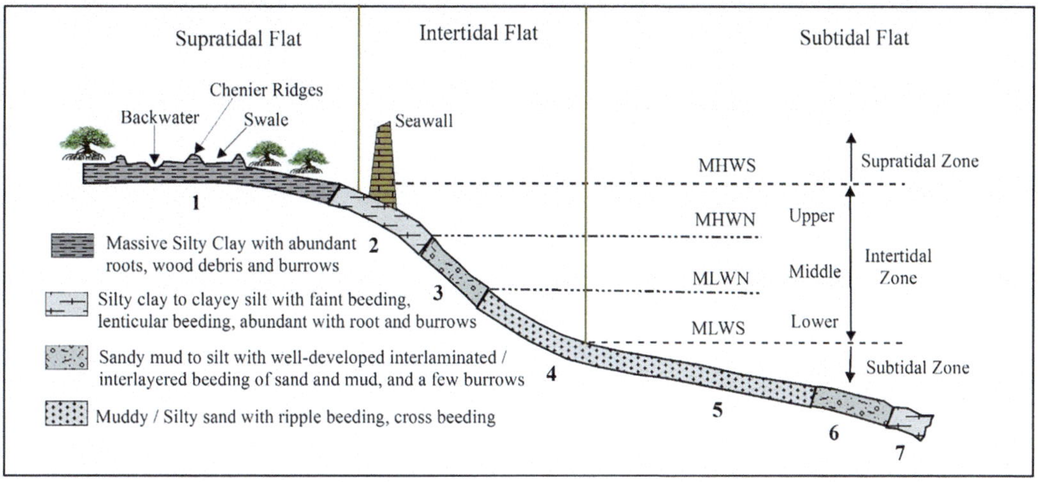

Fig. 1 Types of tidal flat wetland system (Indian Sundarban)

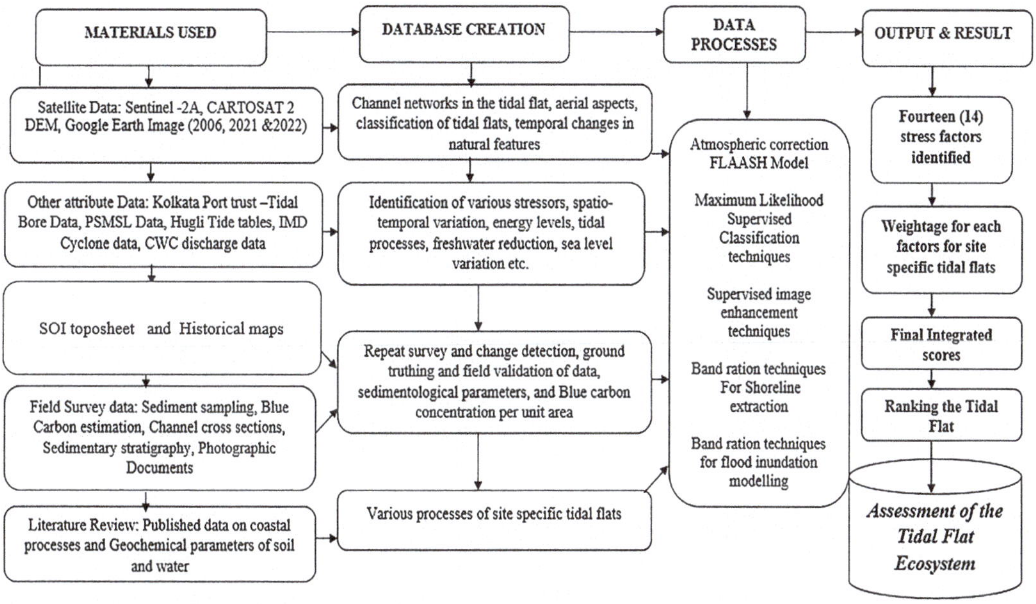

Fig. 2 The methodological flowchart showing the sequences of the study

3 Preprocessing of the Images

The multiple series of the Landsat datasets were used to extract the tidal flat wetland ecosystems in this study areas. The Landsat 8 OLI data comprises the multiple bands combinations (FCC for the bands 4, 3, and 2 for Landsat 5 TM and bands 5, 4, and 3 for Landsat 8 OLI/TIRS) were radiometrically and atmospherically normalized to reduce the scattering effects by using the ERDAS Imagine software. The radiometric correction procedure was used to convert the DN value to top-of-atmosphere radiance L_{TOA} by the applied algorithm of the FLAASH model. Equations 1 and 2 are applied by using the atmospheric correction tools to convert the radiance of VNIR bands into surface reflectance (Song et al., 2001; Chander et al., 2007)

$$L_{TOA} = \left(\frac{L_{\max\lambda} - L_{\min\lambda}}{QCAL_{\max} - QCAL_{\min\lambda}} \right) (DN - QCAL_{\min}) + L_{\min\lambda} \quad (1)$$

where $L_{max\lambda}$ and $L_{min\lambda}$ are the maximum and minimum spectral radiance of the sensor (W/m^{-2} sr^{-1}µm^{-1} and $QCAL_{max}$ and $QCAL_{min}$ are the maximum and minimum possible pixel values of Digital Number (DN) corresponding to $L_{max\lambda}$ and $L_{min\lambda}$

$$p = \frac{(L_{TOA} - L_p)\pi d^2}{ESUN_\lambda \cos\theta_z T_z} \quad (2)$$

where p refers to the surface reflectance, d refers to the Earth-sun distance ($ESUN_\lambda$ indicate the astronomical units refer to the mean exo-atmospheric solar irradiance [W/m^{-2} sr^{-1}µm^{-1}], T_z refers to the atmospheric transmission between ground and the top of the atmosphere, Lp is the radiance resulting from the interaction between aerosol and atmospheric particles, and θz is the solar zenith angle (degree) all the input parameters of the FLAASH (Fast Line-of-sight Atmospheric Analysis of Spectral Hypercubes) model (Sobrino et al., 2004).

The two spectral indices as Normalized Difference Vegetation Index (NDVI) and Normalized Difference Water Index (NDWI) were used to develop a new model of the Combined Mangrove Recognition Index to identify the healthier mangrove vegetation and classify the tidal flat wetland ecosystem in the tidal, estuarine, and sea face islands of this study area (McFeeters, 1996). The output indices range varies from m −1.0 to +1.0 (Gao, 1996). The higher accuracy Combined Mangrove Recognition Index (CMRI) spectral index identifies healthier mangrove-dominated areas using the satellite-derived algorithm techniques, whereas it is a more suitable index than the NDVI, Simple Ratio (SR), Leaf Area Index (LAI), Perpendicular Vegetation Index (PVI), Soil Adjusted Vegetation Index (SAVI), and Enhanced Vegetation Index (EVI). Equations 3, 4, and 5 were used to extract the tidal flat wetland ecosystem by using the spectral models and to validate the test and trained data set by the applied confusion matrix of the kappa coefficient (Eq. 6).

$$NDVI = \frac{\rho_{NIR} - \rho_{red}}{\rho_{NIR} + \rho_{red}} \quad (3)$$

$$NDWI = \frac{\rho_{green} - \rho_{NIR}}{\rho_{green} + \rho_{NIR}} \quad (4)$$

$$\text{CMRI} = (\text{NDVI} - \text{NDWI}) \quad (5)$$

$$k = \frac{N\sum_{i=1}^{r} X_{ll} - \sum_{i=1}^{r}(X_l + X + i)}{N^2 - \sum_{i=1}^{r}(X_l + X + i)} \quad (6)$$

The depth-wise intertidal soil samples (30–70 cm) were collected and dried in the oven at 105 °C for 24 h for a fortnight to estimate the soil bulk density (BD). The laboratory-based Loss-on-Ignition and Walkley-Black wet oxidation methods were used to estimate the soil SOC (soil organic carbon) amount (% C_{org}) and also finally calculate the blue carbon stock (BCS) using the estimated soil bulk density and soil organic carbon in the planted and natural forest areas of the southwestern Sundarbans (Eqs. 7, 8 and 9).

$$BD = \frac{W_a - W_f}{V} \quad (7)$$

where W_a and W_f are the initial collected soil sample and final dried soil sample weights (g), and V is the volume of the soil (cm^3).

$$SOC\,density(gcm^{-3}) = BD(gcm^{-3}) \times (\%C_{org}) \quad (8)$$

$$SOCorBCSincoredepth_{ij}(mgm^3) = SOCdensity_{ij} \times Coredepthinterval_{ij}(cm) \quad (9)$$

The discrete multivariate method of the kappa coefficient method was applied based on the three-confusion matrix such as producer accuracy, kappa coefficient, and user accuracy to calculate the accuracy assessment of the tidal flat classes during the 14 years from 2006 to 2022. The kappa statistic level of agreement ranges from <0.20 (poor) to > 0.80 (strong), whereas > 0.80 indicates the strong relation of the agreement with the limit of good accuracy. The temporal changes of the tidal flat morphology are assessed in the three island ecosystems of the Indian Sundarbans. The kappa coefficient results were 0.78 (2006) and 0.89 (2022) on Lothian Island, 0.75 (2006) and 0.82 on Patibunia Island, and 0.81 (2006) and (0.87) on Bhangaduni Island, respectively.

Fourteen stressor factors of the tidal flat morphology identified by considering the geomorphology settings of tidal and estuarine delta systems, morphodynamic and hydrodynamic behaviors, climate-related phenomena and anthropogenic pressure played a significant role in the modification of the tidal flat wetland ecosystem in Indian Sundarbans. The fourteen factors included: (1) the location of the tidal flat (LTF), (2) elevation ranges (EV), (3) tidal ranges (TR), (4) tidal channel density (TCD), (5) frequency of major and medium size channels (FMMSC), (6) the ratio between tidal flat and channel space (RBTFCS), (7) ratio between tidal flat and mangrove cover area (RBTFMCA), (8) reduction of freshwater flow (RFWF), (9) frequency of tidal bores (FTB), (10) frequency of cyclonic storms (FCS), (11) rate of sea level rise (RSLR), (12) concentration of saltpans (CS), (13) erosion and accretion rate (EAR), and (14) impacts of overwash sands and shifting sands (IOSSS) utilized to calculate the integrated scores of the stress factors of the tidal flat ecosystem in the deltaic ecosystem of the islands. The fuzzy logistic and AHP (analytic hierarchy process) and weighted sum methods were applied to integrate the final scores of the tidal flat ecosystem (Saaty, 2004). According to 14 criteria, weights are assigned on the basis of the importance of each stress factor in the tidal flat ecosystem. Finally, calculate the overall priority score and consistency ratio (CR or λ) to assess the integrated stress factors of the tidal flat wetland ecosystems on the basis of criterion importance.

4 Geographical Settings of the Study Area

This study area comprises the Chemagari tidal flats of Sagar Island; Mousuni Island tidal flats and Patibunia Island tidal flats in the Hugli estuarine delta; the sea-facing tidal flats of Lothian Island, Dhanchi Island, and Bhangaduni Island along the marine sections of Saptamukhi-Malta estuarine complex; the inner island tidal flats system of Herobhanga island; and Hamilton Island in the tidal delta (Fig. 3).

5 Results and Discussion

This study depicts the diversity of the tidal flats ecosystem in the Indian Sundarbans extending from west to east and from the sea face to the inland areas of the delta plain surface. A total of 14 stressors have been identified in the site-specific tidal flats using geospatial techniques, field records, monitoring geomorphological features, and other sources of aspect data. Finally, the integrated scores are estimated by assigning the weightage value for the site-specific tidal flats ecosystems to identify the final ranking of stressors in the region of the Lower Ganga Delta Plain.

6 Common Sub-Environment

Sediment size categorizes and bio-geomorphological characters of different sub-environments of tidal flats wetland system are explored at the Saptamukhi estuary fringe environment (Table 3).

The finely textured mangrove muds have high permeability and low hydraulic conductivity. However, the tidal flat sediment is characterized by a low content of organic matter and a high amount of swelling type of clay particles with low exchangeable sodium. The dominance of silts (74.86%) and lower amounts of clays (12.12%) and sands (13.02%) are also found in the texture of sediments in the Lothian Island tidal flat (Fig. 4).

7 Temporal Changes in Tidal Flats Wetland System

The physiographic settings of eight site-specific tidal flats wetland systems are categorized into (i) estuarine island tidal (Chemagari, Mousuni, and Patibunia Islands); (ii) sea facing island fringe tidal flats (Lothian Island, Dhanchi Island, and Bhangaduni Islands); and (iii) inland delta plain

Fig. 3 Tidal flats of the Indian Sundarban

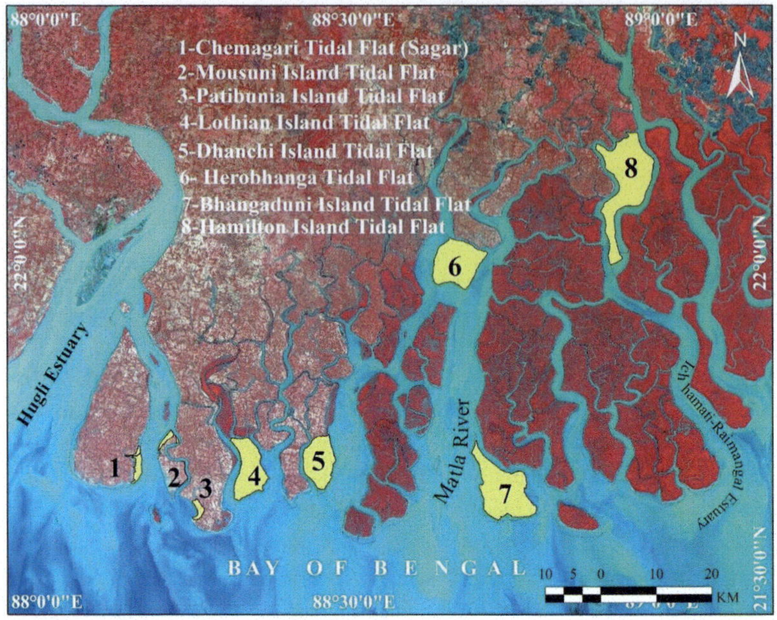

Table 3 Common sub-environments and sediment size categories of the tidal flats and mangrove swamps along the Saptamukhi estuary fringes (Haripur–Kalisthan point)

Sub-environment	Bio-geomorphological characters	Sediment size categories (in percent)		
		Sand	Silt	Clay
Subtidal region	Strong tidal current	48.4	24.0	28.0
Lower mudflat	Abundant bioturbation	46.7	27.3	24.0
Upper mudflat	Algal binding of sediment	44.4	24.4	31.2
Swamp floor	Mangroves and slow accumulation of sediment	40.0	30.1	29.9

tidal flats (Herobhanga and Hamilton Islands) of the Indian Sundarbans. Dynamic morphology of the tidal flats wetland system is assessed between the temporal span of 2006 and 2022 by using satellite-derived data. Patibunia Island is selected from the estuarine setting, Lothian and Bhangaduni Islands are considered from the sea-facing marine- dominated setting, and Hamilton Island is assessed from the inland tidal delta setting for temporal changes in tidal flat morphology in the study (Figs. 4, 5, 6, 7 and 8).

It is estimated that washover sands and shoaling flats increased in the expansion of decreased mangrove cover area and tidal creeks in Patibunia Island during the 17 years between 2006 and 2022 due to the storminess of the sea. Bhangaduni Island on the other hand registered significant losses of mangrove swamp cover areas, mudflats, tidal creeks, and sand spits in the sea-facing environment. The shore fringe tidal flats, mangrove cover areas, and tidal space have been decreased similarly in the sea-facing Lothian Island of Saptamukhi Estuary of the Indian Sundarbans between the timespan. Finally, Hamilton and Herobhanga Islands of inland tidal deltaic settings, the accretion rate of sediments, mangrove cover areas, and tidal creek space have been significantly increased by the study. Thus, marine influence, increased salinity, reduced discharges of freshwater supply, and sediments are major stressors that played significant roles in temporal changes of tidal flat wetland morphology in the region of the Lower Ganga Delta. Tidal siltation rates, freshwater supply, and active mangrove regeneration systems, however, recovered the inner tidal deltaic islands from the rapid change

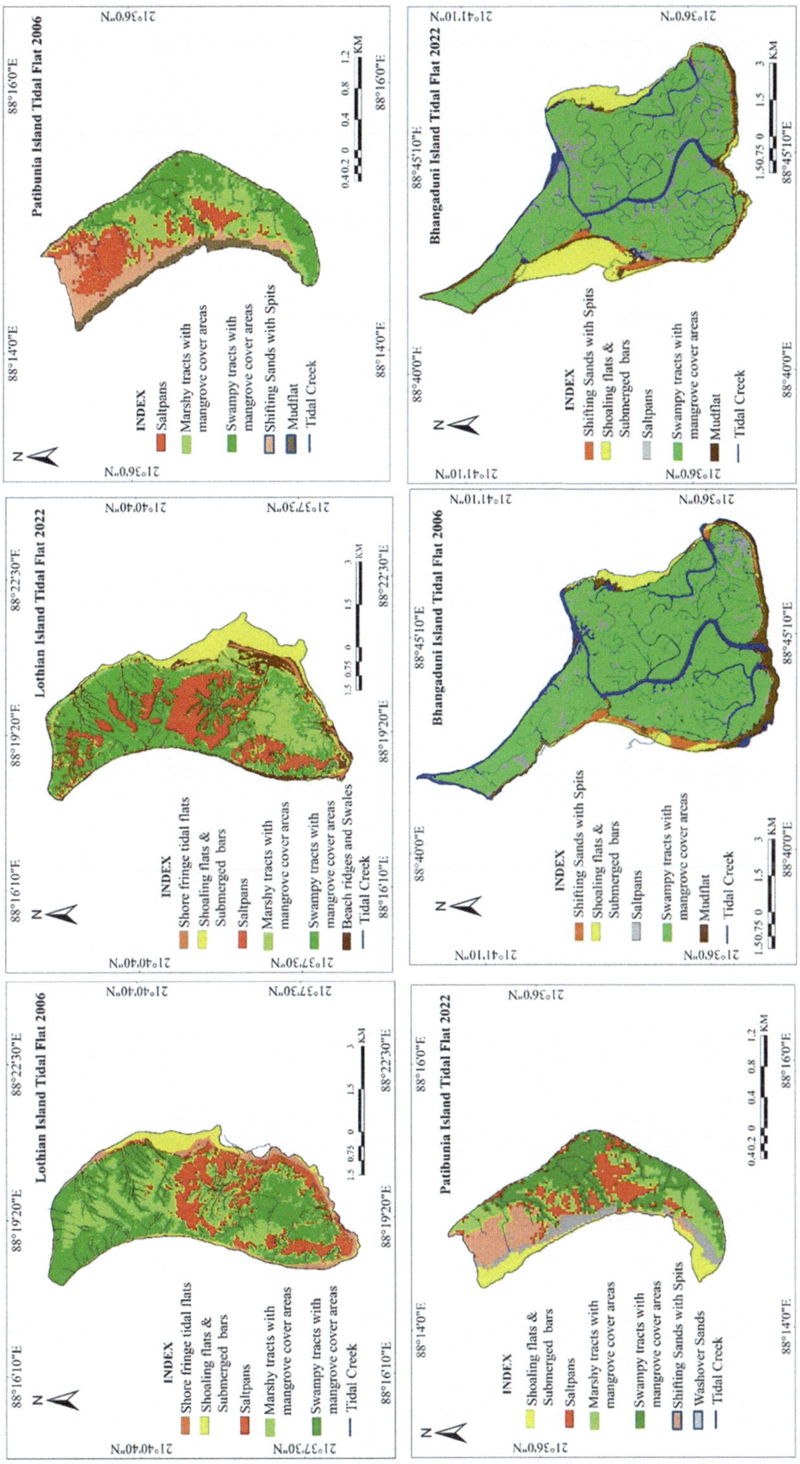

Fig. 4 Temporal changes in tidal flat morphology in Lothian Island, Patibunia Island, and Bhangaduni (2006–2022)

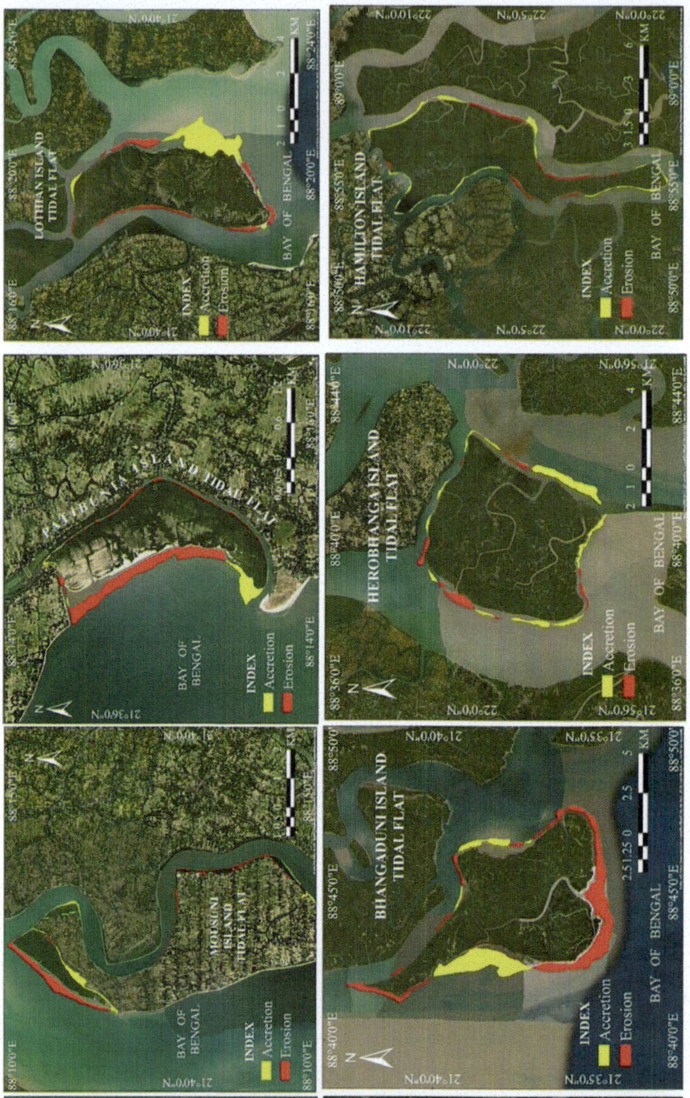

Fig. 5 Erosion and accretion rates in the tidal flat wetland systems of the study area

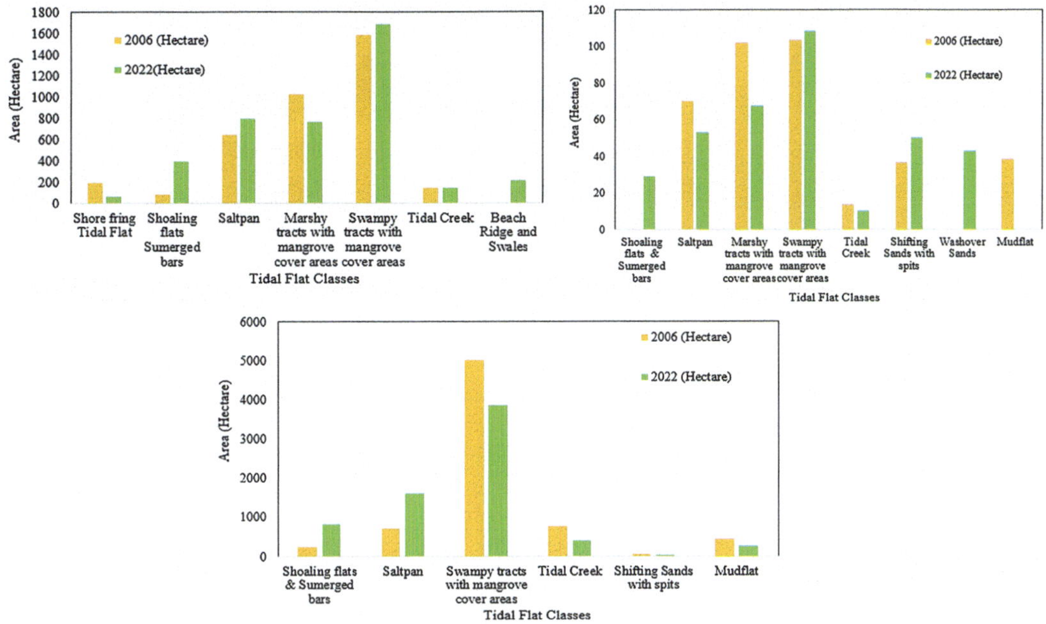

Fig. 6 Classification of tidal flat morphology on Lothian, Patibunia Island, and Bhangaduni Island

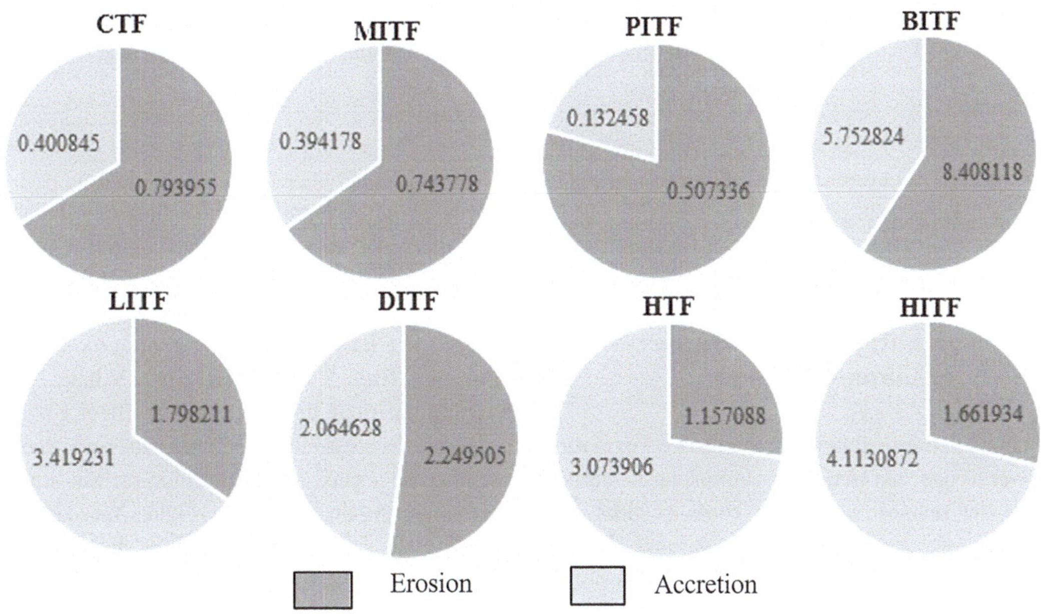

Fig. 7 Erosion and Accretion rates of the Tidal flat wetland system (CTF—Chemagari Tidal Flat, MITF—Mousuni Island Tidal Flat, PITF—Patibunia Island Tidal Flat, LITF—Lothian Island Tidal Flat, DITF—Dhanchi Island Tidal Flat, HTF—Herobhanga Tidal Flat, BITF—Bhangaduni Island Tidal Flat, HITF—Hamilton Island Tidal Flat)

Fig. 8 Tidal flats sub-environments with swampy forest of mangroves in Hamilton Island and mangrove plantation unit in Haripur tidal flats of Saptamukhi Estuary

Table 4 Blue carbon stock monitored from Chargheri and Haripur areas of the Sundarbans (mt or metric ton)

Location	AGB (mt/ha)	BGB (mt/ha)	Blue carbon stock (mg/m^3)	Remarks
Chargheri 22°06′06.08″N 88°55′24.05″E	750–800	150–280	98–120	Plantation forest (2010–2023)
Hamilton Island 22°05′57.31″N 88°55′58.61″E	1400–1800	400–600	170–235	Natural forest
Haripur 21°37′16.80″N 88°17′13.83″E	300–400	80–200	34–84	Plantation forest (2022–2023)
Lothian Island 21°42′12.82″N 88°18′16.17″E	1200–1400	300–400	150–210	Natural forest

of the wetland systems in the favorable environment conditions at present.

8 The Role of Wetlands in Mitigating the Impact of Climate Change

The tidal flats wetland of mangrove vegetation cover types can sequester atmospheric carbon into the marshy sediments, roots of tidal life, shoots, and canopy cover areas, and directly reduce the impact of warming conditions on the tropical deltaic lowland coasts. Using geospatial techniques, mangrove-dominated tidal marshes are classified into different subunits of dense forest, dwarfed forest, and bare flats for monitoring carbon stock. Depth- wise sediments are collected from the above sub-units of tidal marshes in the natural wetland system along with the planted mangrove wetland sites of the estuary fringe tidal flats to compare the rates of carbon sequestration between them. Thus, planting mangroves in the bare areas of tidal flats can reduce the impact of climate change through the sequestration of atmospheric carbon in the Sundarbans.

The study reveals that the planted mangrove forests of Chargheri Village at Satjelia Island (Gosaba Block) and Haripur Village at Saptamukhi tidal flats (Namkhana Block) have successfully restored the habitats and have sequestered a good amount of blue carbon in the region. During the events of cyclones Yass (2021) and Remal (2023), embankments behind the planted mangroves were well preserved in the low-lying deltaic wetlands due to the growth of such ecological buffers along the estuary fringe tidal flats with planted mangroves.

The monitored biomass (mt/ha) and blue carbon stock (mg/m^3) of the plantation sites and natural forests are compared to identify the status of mangrove plantation sites and their roles in minimizing the impacts of climate change in the Sundarbans (Table 4). Such assessment techniques will provide a good database for coastal managers and local administrators to understand the role of tidal flat wetland systems, as well as the necessity of restoration and conservation of mangrove wetlands in the tidal environment of the deltaic coast. Thus, more areas of tidal lands with increased cover of mangroves through plantation missions will mitigate the climate change impacts on the tropical coast of the Sundarbans (Fig. 8).

Table 5 Integrated score for the stress factors over the tidal flats of Indian Sundarbans

Sl. no.	Site-specific tidal flat	Integrated score	Category of stress
1	Lothian Island tidal flat	22.15	Very high (rank-I)
2	Bhangaduni Island tidal flat	18.9	High (rank-II)
3	Patibunia Island tidal flat	18.35	High (rank-II)
4	Chemagari tidal flat (Sagar)	17.85	Medium (rank-III)
5	Mousuni Island tidal flat	17.25	Medium (rank-III)
6	Hamilton Island tidal flat	15.95	Low (rank-IV)
7	Herobhanga tidal flat	14.9	Low (rank-IV)
8	Dhanchi Island tidal flat	14.65	Low (rank-IV)

9 Integrated Stressor Score and Ranking of the Island

The integrated score for the stress factors shows maximum value in favor of the Mousumi Island tidal flat wetland system by the above study. Patibunia Island and Chemagari tidal flats wetland systems of the Hugli downstream section on the other hand depict higher integrated scores for the estimated and calculated stressors of geomorphic, hydrologic, and biologic components (Table 5, Fig. 9, and Table 5).

Lothian Island of the Saptamukhi Estuary also shows moderate impacts of stressors particularly to the sea-facing areas and south-central parts of the island under marine influences, hypersalinity, and shortage of freshwater discharges. Climate change impacts and sea level rise threats will produce stresses to disturb the tidal flats wetland systems in the Indian Sundarbans.

10 Conclusion

There is significant diversity in the natural features of the tidal flat wetland ecosystem in the Sundarbans. Restoration of natural ecosystems of the tidal flats will ensure coastal resilience and provide opportunities for essential ecosystem

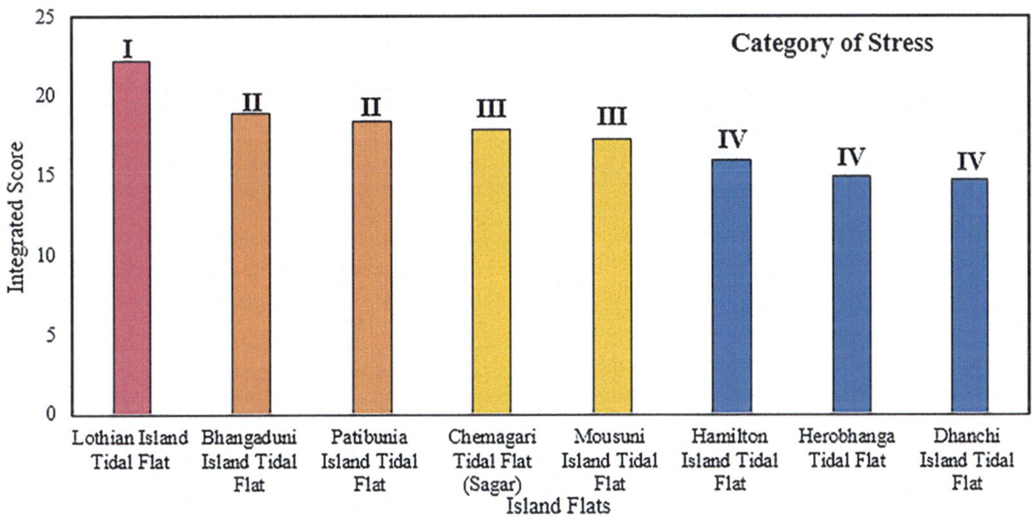

Fig. 9 Tidal flats under stress, Indian Sundarbans

services in the dynamic sedimentary depositional environment. The tidal flats covered with mangroves, salt marshes, and sedge plants sequester and store more carbon per unit area than terrestrial forests. Presently, they are affected by multiple stress factors in the estuarine, deltaic, and coastal marine environments. Lothian, Bhangaduni, and Patibunia Islands are estimated as very high to highly stressed tidal flat wetland systems in the Indian Sundarbans by the study. Mangrove cover tidal flat wetland systems are recognized for their role in mitigating climate change.

The wave attenuation process is enhanced by the presence of extensive tidal flat wetland systems during the occurrences of tidal surges and storm surges in the Lower Ganga Delta. The tidal flat wetland system ensures the coastal resiliencies in the sedimentary depositional coast. Mass-scale mangrove plantations can restore the tidal flat wetland systems in the dynamic coastal environment in risk reduction planning. Climate resilient embankment with an ecological buffer and physical buffer is possible if the riverside tidal flats are restored by mangrove plantations in the reclaimed islands of the Sundarbans.

References

Allison, M. A., Khan, S. R., Goodbred, S. L., Jr., & Kuehl, S. A. (2003). Stratigraphic evolution of the late Holocene Ganges–Brahmaputra lower delta plain. *Sedimentary Geology, 155*(3–4), 317–342.

Amos, C. (1995). Siliciclastic tidal flats. In G. M. E. Perillo (Ed.), *Geomorphology and sedimentology of estuaries, advances in sedimentology 53* (pp. 273–306). Elsevier.

Chander, G., Markham, B. L., & Barsi, J. A. (2007). Revised Landsat-5 thematic mapper radiometric calibration. IEEE Geoscience and remote sensing letters, 4(3), 490–494

Daidu, F., Yuan, W., & Min, L. (2013). Classifications, sedimentary features, and facies associations of tidal flats. *Journal of Palaeogeography, 2*(1), 66–80.

Dyer, K. R. (1998). The typology of intertidal mudflats. *Geological Society, London, Special Publications, 139*(1), 11–24.

Eisma, D. W. G. M., Mook, W. G., & Laban, C. (1981). An early Holocene tidal flat in the Southern Bight. In *Holocene marine sedimentation in the north Sea basin* (pp. 229–237). Wiley.

Gao, B. C. (1996). NDWI—A normalized difference water index for remote sensing of vegetation liquid water from space. *Remote Sensing of Environment, 58*(3), 257–266.

Gao, S. (2019). Geomorphology and sedimentology of tidal flats. In *Coastal wetlands* (pp. 359–381). Elsevier.

McFeeters, S. K. (1996). The use of the Normalized Difference Water Index (NDWI) in the delineation of open water features. *International Journal of Remote Sensing, 17*(7), 1425–1432.

Mukherjee, D., & Paul, A. K. (2023). Mangrove sensitivity to the ephemeral rise of sea waters in the Western Sundarban. In *Crisis on the coast and hinterland: assessing India's East Coast with geomorphological, environmental and remote sensing and GIS approaches* (pp. 143–154). Cham.

Paul, A. K. (2002). *Coastal geomorphology and environment* (pp. 1–342). ACB Publication.

Paul, A. K., Ray, R., Kamila, A., & Jana, S. (2017). *Mangrove degradation in the Sundarbans (in coastal wetlands: Alteration and remediation)* (pp. 357–392). Springer. https://doi.org/10.1007/978-3-319-56179-0

Paul, A. K., Kamila, A., & Ray, R. (2018). Natural threats and impacts to mangroves within the coastal fringing forests of India. In *Threats to mangrove forests: Hazards, vulnerability, and management* (Coastal research library) (Vol. 25, pp. 105–140). Springer.

Paul, A. K., Paul, A., Sardar, J., Ray, R., Khatun, K., Chand, S., et al. (2023). Mangroves in cyclone-battered Sundarbans, India: A geoinformatics-based multi-temporal study. In *Crisis on the coast and hinterland: Assessing India's East Coast with geomorphological, environmental and remote sensing and GIS approaches* (pp. 327–341). Cham.

Saaty, T. L. (2004). Decision making—The analytic hierarchy and network processes (AHP/ANP). *Journal of Systems Science and Systems Engineering, 13*, 1–35.

Sardar, J., Paul, A., Nayak, K., Naskar, S., Dey, S., Mallick, I., et al. (2023). Managing the coastal squeeze and wetland loss in Sagar Island in a sustainable framework using geospatial techniques. In *Crisis on the coast and hinterland: Assessing India's East Coast with geomorphological, environmental and remote sensing and GIS approaches* (pp. 353–365). Cham.

Sobrino, J. A., Jiménez-Muñoz, J. C., & Paolini, L. (2004). Land surface temperature retrieval from LANDSAT TM 5. *Remote Sensing of Environment, 90*(4), 434–440.

Song, C., Woodcock, C. E., Seto, K. C., Lenney, M. P., & Macomber, S. A. (2001). Classification and change detection using Landsat TM data: When and how to correct atmospheric effects? *Remote Sensing of Environment, 75*(2), 230–244.

Assessment of Annual Soil Loss using RUSLE: A Case Study of Upper Kangsabati River Basin

Subrata Bhunia, Sayanti Poddar, and Dipanwita Dutta

1 Introduction

Soil erosion is a pressing environmental issue across many regions in India, characterized by the displacement and deposition of soil particles from one location to another. This gradual and natural phenomenon involves the detachment, transportation, and accumulation of surface soil due to the forces of water or wind (Jain et al., 2001). Factors such as rainfall intensity, soil susceptibility to erosion, vegetation cover, and landform contribute to this process. While soil erosion is a natural occurrence, it becomes problematic when accelerated by human activities like deforestation, agricultural practices (such as ploughing and irrigation), overgrazing, land clearing, construction, mining, and urban expansion. Many nations consider soil erosion a critical issue, further aggravated by unsustainable farming techniques and poor land management practices (Lal, 2003; Pimentel, 2006). A thorough understanding of these factors is essential for crafting effective soil conservation measures to mitigate negative impacts on the environment and agriculture (Boardman & Poesen, 2006; Montgomery, 2007).

The consequences of soil erosion and land degradation are profound, leading to significant environmental, economic, and social challenges. The removal of the nutrient-rich topsoil results in reduced agricultural yields (Lal, 2001). The sedimentation of eroded soil in water bodies diminishes reservoir storage capacity and negatively affects aquatic ecosystems due to increased levels of suspended particles. Moreover, soil erosion often leads to the transport of pesticides and other harmful chemicals, contributing to water contamination and eutrophication (Zhou et al., 2008). Severe soil erosion can also indirectly heighten the risk of flooding and result in financial losses related to irrigation and hydroelectric infrastructure (Arya et al., 2001). In India, soil erosion affects around 329 million hectares, or 53% of the total land area, with an average annual loss of 16 t ha^{-1} $year^{-1}$ (Jain et al., 2001; Srinivas et al., 2002; Pandey et al., 2009). Approximately 5334 million tons of soil are eroded each year (Pandey et al., 2007). Despite efforts by the government to curb erosion and land degradation, the lack of precise information hampers the development of effective conservation strategies. Accurate quantification of soil loss and risk assessment is essential for identifying areas most affected by erosion and implementing appropriate land management practices. Field-based methods for measuring

S. Bhunia · D. Dutta
Department of Remote Sensing and GIS, Vidyasagar University, Midnapore, West Bengal, India

S. Poddar (✉)
Department of Geography, Asutosh College, University of Calcutta, Kolkata, West Bengal, India
e-mail: sayanti.poddar@asutoshcollege.in

Z. Khanam et al. (eds.), *Climate Change, Resilience and Ecosystem Services*, Sustainable Development Goals Series, https://doi.org/10.1007/978-3-031-96977-5_23

soil loss are often labor-intensive and restricted by the limited number of sampling sites, which may compromise the accuracy of estimating the spatial extent of soil erosion. As a result, monitoring and mapping soil loss over large areas remain challenging due to the substantial time and costs associated with traditional field-based approaches (Kumar & Kushwaha, 2013).

Zingg (1940) was the first one, who mathematically captured the effects of slope steepness and length on erosion. This was highlighted and popularized by Moore and Burch (1986). Smith (1976) later refined this equation by integrating additional factors related to cropping systems and support practices, which facilitated the creation of a graphical tool for selecting appropriate conservation practices tailored to specific soil conditions in the Midwest of the United States.

Browning (1979) and Renard et al. (1997), made further improvements to the equation by incorporating factors such as soil erodibility and management practices, resulting in comprehensive tables that provided relative values for various soils, crop rotations, and slope lengths. Their method focused on assessing slope-length limitations for different cropping systems on specific soils and varying degrees of slope, with or without the implementation of terracing, contouring, or strip cropping. Moore and Burch (1986) introduced a method to estimate soil losses from fields with claypan soils, offering soil-loss ratios for different slopes under contour farming, strip cropping, and terracing, and suggesting slope length limits for contour farming.

In recent decades, empirical models such as the Universal Soil Loss Equation (USLE), the Modified Universal Soil Loss Equation (MUSLE), and the Revised Universal Soil Loss Equation (RUSLE) have gained global recognition for their simplicity, low data requirements, and broad applicability (Perović et al., 2013; Borrelli et al., 2021). However, both USLE and MUSLE have faced criticism for their limitations in accurately predicting the spatial distribution of soil erosion. The RUSLE, an advanced version of USLE, improves soil loss estimation at the plot scale and enhances the spatial representation of soil erosion within a given area (Renard et al., 1997). The integration of geospatial techniques with the RUSLE model has proven to be effective due to its simplicity, accuracy, and cost-efficiency over large areas (Krishna Bahadur, 2009; Demirci & Karaburun, 2012; Dutta et al., 2015; Pradeep et al., 2015; Chandrika et al., 2025). This attempt aims to apply the RUSLE method in combination with remote sensing and geographic information systems (GIS) techniques to assess soil erosion in the Upper Kangsabati River Basin, located in the western part of West Bengal and the eastern part of Jharkhand. Additionally, the study will attempt to estimate the potential soil erosion in the Upper Kangsabati River Basin and analyze the causal factors, both physical and anthropogenic, contributing to soil erosion.

2 The Study Area

The Kangsabati River basin, known as the Kasai or Cossey River, is economically significant for West Bengal and India. The western plateau, part of the study area, receives low annual rainfall (100–200 cm) and frequently faces drought, particularly in Bankura and Purulia, where agriculture depends on irrigation (Chatterjee, 2016; Mukherjee et al., 2018). This research focuses on the Upper Kangsabati River basin, covering 3613.77 km^2 between latitudes 23°28′41.091″North to 22°42′48.16″North and longitudes 85°57′3.38″East to 86°48′53.8″East. The topography includes the Chota Nagpur Plateau's highlands, composed of Archaean and Carboniferous rocks, and the Rarh region's Deccan Plateau-derived soils (Sarkar & Sen, 2019; Mandal et al., 2020). The soil composition of the Kangsabati River basin is diverse, significantly impacting agriculture and land use. The western parts, including Bankura and Midnapore, feature laterite and lateritic soils, rich in iron and aluminum, formed through intense weathering under tropical conditions, and characterized by undulating terrain (Chatterjee, 2016; Mukherjee et al., 2018). Red soils, noted for their ferric oxide content, are also prevalent and support various crops with proper management. The fertile Vindhya alluvium soils, formed from river-borne alluvium, dominate the flat topography, essential for paddy and other crops (Chatterjee, 2016; Mukherjee et al., 2018). Colluvial and skeletal soils, rich in sand and gravel, are found in Bankura, posing agricultural challenges due to poor nutrient content and water retention. The region's geological formations, including Archaean igneous rocks and Carboniferous mudstone and quartzite in the Chota Nagpur Plateau, contribute to the soil diversity (Ghosh, 2018; Sarkar & Sen, 2019). This soil variety is crucial for land use, agricultural productivity, and soil conservation practices (Chatterjee, 2016; Mukherjee et al., 2018). The study area experiences a tropical monsoon climate, with annual rainfall between 100 and 200 cm, predominantly occurring from June to September, crucial for agriculture but also leading to soil erosion (Ghosh, 2018). Summer temperatures range from 26°C to 43°C, leading to higher evapotranspiration and the need for efficient water management. Winter temperatures, ranging from 10°C to 19°C, are ideal for crops such as wheat and mustard, although cooler temperatures can impact germination. These climatic factors play a critical role in agricultural cycles, soil health, and water management, necessitating strategic planning to maximize crop yields and support local farming communities (Ghosh, 2018). Land use includes water bodies, settlements, vegetation, agricultural land, fallow land, and sandy areas along the riverbanks. The economy of the Kangsabati River basin is primarily based on agriculture, with paddy as the staple crop, along with significant cultivation of potatoes, wheat, vegetables, and mustard (Mandal et al., 2020). The region's agricultural output is largely dependent on the monsoon climate and the variety of soil types present. Irrigation systems are vital, particularly in low-rainfall areas like Bankura and Purulia (Chatterjee, 2016; Mukherjee et al., 2018). Despite the challenges of soil erosion and climate variability, agriculture is the mainstay of the local economy, supporting a large portion of the population. Enhancing agricultural techniques and implementing soil conservation measures are essential for maintaining productivity and economic resilience in the basin (Ghosh, 2018) (Fig. 1).

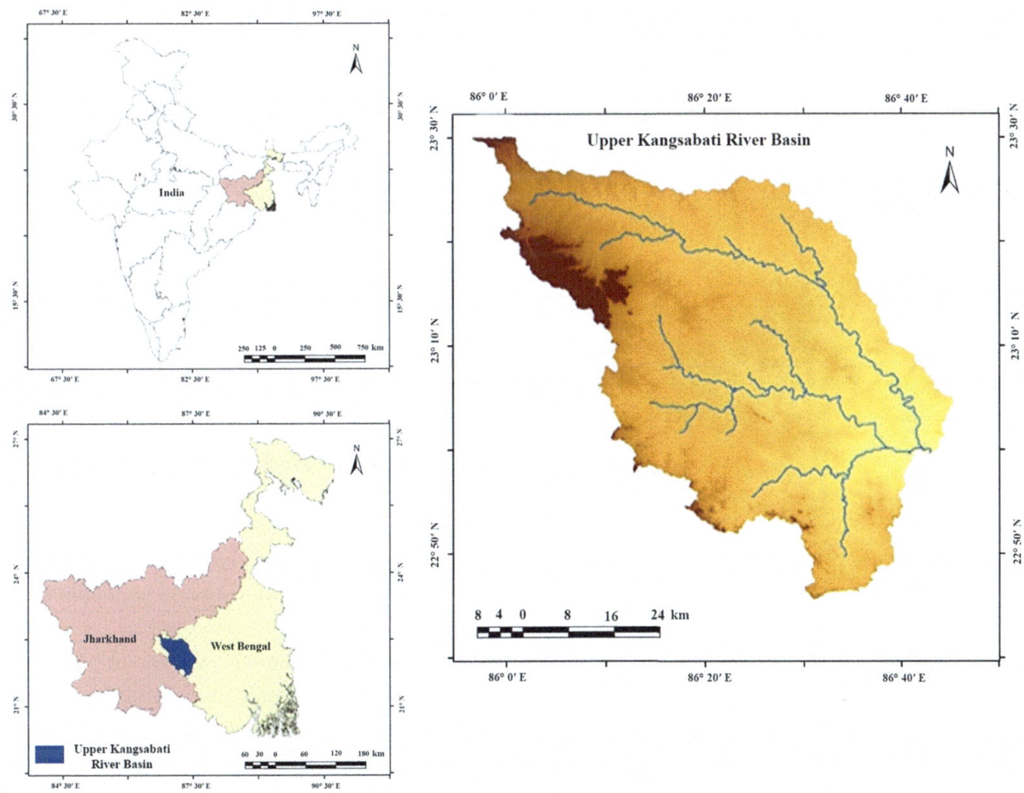

Fig. 1 Location map of the study area

3 Materials and Methods

3.1 Materials Used for Analysis

Sentinel-2A Data The Sentinel-2A data for 7th February, 2019 was used for preparing the land use and land cover map and also the NDVI (Normalized Difference Vegetation Index) map. It was obtained from the Earth Explorer, United States Geological Survey (USGS) site (https://earthexplorer.usgs.gov/) (Table 1).

The Indian Space Research Organization (ISRO) has made a significant contribution to digital mapping with the provision of Cartosat digital elevation models (DEMs). This development represents a substantial leap forward in the availability of high-quality elevation data, particularly for large regions within the tropics and other areas of the developing world. The Cartosat-1 DEMs, which feature a PAN sensor with a resolution of 2.5 m, offer a spatial resolution of 1 arc second (approximately 30.87 m). These DEMs are available in GeoTiff file format and each pixel is assigned a 16-bit value. The DEMs can be downloaded from ISRO's geoportal, Bhuvan (https://bhuvan.nrsc.gov.in).

Table 1 Bands of Sentinel 2A and their specification

Bands	Band name	Central wavelength(nm)	Spatial resolution(m)
Band 1	Coastal aerosol	442.2	60
Band 2	Blue	492.1	10
Band 3	Green	559.0	10
Band 4	Red	664.9	10
Band 5	Vegetation red edge	703.8	20
Band 6	Vegetation red edge	739.1	20
Band 7	Vegetation red edge	779.7	20
Band 8	NIR	832.9	10
Band 8A	Narrow NIR	864.0	20
Band 9	Water vapor	943.2	60
Band 10	SWIR-cirrus	1376.9	60
Band 11	SWIR	1610.4	20
Band 12	SWIR	2185.7	20

The meteorological data utilized in this study refers to rainfall data, developed by the Center for Hydrometeorology and Remote Sensing (CHRS) at the University of California, Irvine (UCI). This data is generated using neural network-based classification and approximation techniques to estimate rainfall rates from geostationary satellite data. The PERSIANN-Cloud Classification System (PERSIANN-CCS) is a real-time global satellite precipitation product with high resolution (0.04° × 0.04° or 4 km × 4 km) developed by CHRS at UCI. The PERSIANN-CCS system categorizes cloud-patch features based on cloud height, area, and texture variability derived from satellite imagery. Central to PERSIANN-CCS is a variable threshold cloud segmentation algorithm, which differs from the traditional constant threshold method by enabling the identification and separation of individual cloud patches. These patches are classified based on texture, geometric properties, dynamic evolution, and cloud top height. Rainfall values are then assigned to pixels within each cloud based on a curve that defines the relationship between rain rate and brightness temperature. The accumulated rainfall data used in this study covers a 12-year period (2007-2018) and is obtained from the CHRS Data Portal (https://chrsdata.eng.uci.edu/). Soil plays a crucial role in this research, and soil taxonomic classification is utilized in the calculations. The soil texture map employed in this study is sourced from the ICAR-National Bureau of Soil Survey and Land Use Planning (NBSS & LUP) office in Kolkata.

3.2 Methodology of RUSLE

The advancement of soil erosion models has greatly enhanced the understanding and management of soil erosion, offering a dependable level of accuracy crucial for conservation initiatives. Various models, such as the Universal Soil Loss Equation (USLE) and its revised version (RUSLE), the Water Erosion Prediction Project (WEPP), the Soil Erosion Model for

Mediterranean Regions (SEMMED), the Areal Non-point Source Watershed Environment Response Simulation (ANSWERS), the European Soil Erosion Model (EUROSEM), and the Soil and Water Assessment Tool (SWAT), have been applied in regional-scale soil erosion assessments (Boggs et al., 2001; Tian et al., 2009). Each of these models has unique features and specific applications. Among them, the USLE and RUSLE models are the most widely adopted globally due to their simplicity and compatibility with GIS technology (Millward and Mersey, 1999; Jain et al., 2001; Lu et al., 2004; Jasrotia Singh, 2006; Dabral et al., 2008; Kouli et al., 2009; Pandey et al., 2009; Bonilla & Vidal, 2010). The Revised Universal Soil Loss Equation (RUSLE), which Renard (1997) refined from the original USLE (Wischmeier & Smith, 1978), is commonly employed to estimate the potential for average annual soil erosion (Fig. 2).

$$A = R^{*} K^{*} LS^{*} P^{*} C$$

where

A = Estimated average soil loss in tons per acre per year
R = Rainfall erosivity factor
K = Soil erodibility factor
L = Slope length factor
S = Slope steepness factor
P = Support practice factor
C = Cover management factor

3.3 Pre-processing of the Datasets

Before integrating all the factors, the raw datasets were georeferenced using the Universal Transverse Mercator (UTM) projection system with the WGS 1984 datum. The boundary of the Upper Kangsabati watershed was extracted from the Cartosat DEM using the Spatial Analyst tool in ArcGIS 10.3. This extracted boundary was subsequently utilized to subset the Sentinel-2A satellite imagery, CHRS data, and the soil map.

3.4 Parameters of RUSLE

3.4.1 Rainfall Erosivity Factor (R)

Rainfall erosivity, an essential component of the RUSLE model, plays a significant role in soil erosion by indicating the potential of a storm event to cause soil displacement at a particular site. The R factor is determined by evaluating the total annual or seasonal rainfall erosivity of individual erosive storms, calculated as the product of the storm's total energy and its peak 30-minute intensity (Wischmeier & Smith, 1978). Given the direct correlation between rainfall erosivity and average annual precipitation, this relationship is frequently utilized to estimate the R value (Arnoldus, 1978; Singh et al., 1981; Kumar and Kushwaha, 2013). In this study, average annual rainfall data over a 12-year period (2007-2018)

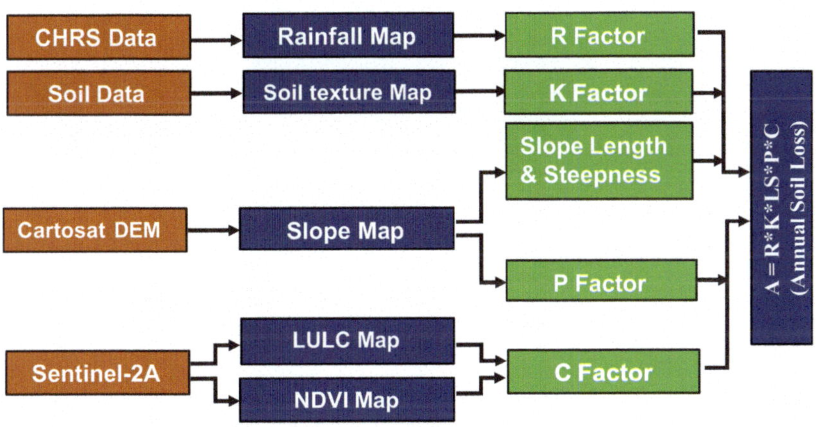

Fig. 2 Flowchart of methodology

was obtained from the Center for Hydrometeorology and Remote Sensing (CHRS). The R factor was subsequently estimated using the formula provided by Choudhury and Nayak (2003).

$$R = 79 + 0.363X\,a$$

where R is the rainfall erosivity, Xa is the average annual rainfall in mm over the study area.

3.4.2 Soil Erodibility Factor (K)

The soil erodibility factor (K) represents the combined impact of rainfall, runoff rate, and infiltration on soil loss, providing a quantitative measure of a soil type's susceptibility to erosion. While soil texture is the primary factor influencing soil erodibility, other factors such as soil structure, permeability, and organic matter content also play significant roles. The K factor is determined by analyzing the relationship between soil texture class and organic matter content.

3.4.3 Topographic Factor (LS)

In soil erosion modelling, the topographic factor (LS) combines the slope length factor (L) and the slope steepness factor (S). The slope length factor (L) represents the ratio of soil loss from a specific slope length compared to a standard slope length, while the slope steepness factor (S) denotes the ratio of soil loss from a given slope gradient relative to a standard slope gradient (Fig. 3).

These factors can be derived from a digital elevation model (DEM) using the unit stream power theory. In this study, slope steepness and flow accumulation were extracted from the Cartosat-1 DEM. The topographic factor (LS)

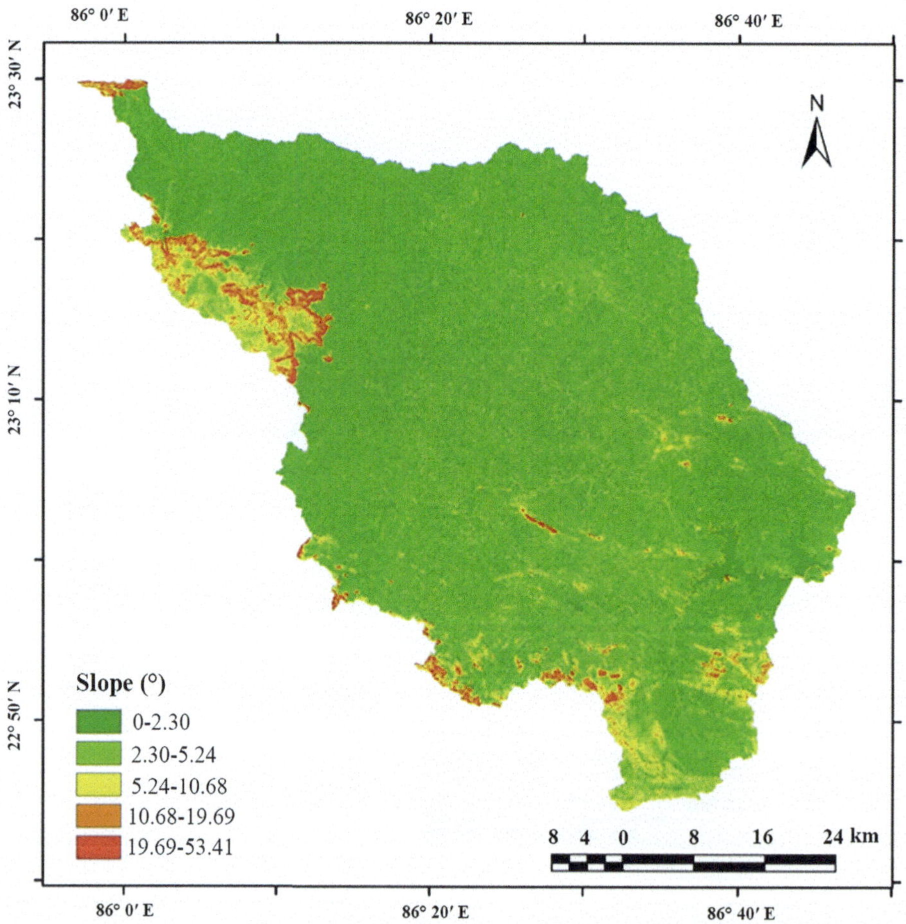

Fig. 3 Slope map

was subsequently calculated through raster computations within a GIS environment (Moore & Burch, 1986).

$$LS = ([Flow\,accumulation]^{*}$$
$$[cell\,size]/22.13)^{0.4*}(\sin slope/0.0896)^{1.3}$$

where LS is the slope length-slope steepness factor, cell size is the size of grid cell (for this study 30.87 m), and sin slope is the slope degree value in sin.

3.4.4 Support Practice Factor (P)

To reduce soil erosion in cultivated lands subjected to erosive rainfall, it is vital to enhance the natural protection provided by grass or dense crops with additional practices aimed at slowing down runoff. Key practices include contour farming, strip cropping, terracing, and constructing waterways to manage excess rainfall. In the Universal Soil Loss Equation (USLE), the conservation practice factor (P) is defined as the ratio of soil loss under a particular conservation practice compared to the soil loss resulting from conventional up-and- down slope farming (Wischmeier & Smith, 1978).

3.4.5 Cover Management Factor (C)

The C factor primarily reflects the percentage of vegetation cover and is defined as the ratio of soil loss from specific crops to that from bare, tilled test plots. The value of the C factor varies based on the type of vegetation, its growth stage, and the extent of ground cover. These elements are related to land use and act as a reduction factor for soil erosion susceptibility. Therefore, accurate land-use data is crucial for generating reliable C factor values. The crop and management factor (C) is determined from a land use/land cover map (Renard et al., 1997) (Fig. 4).

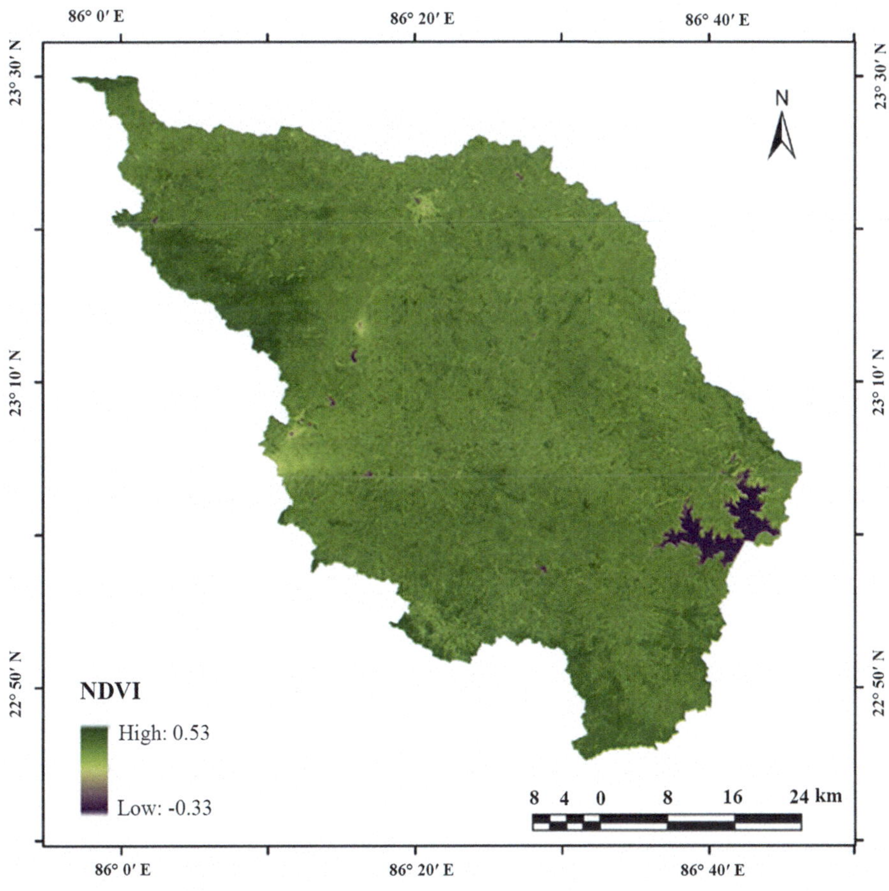

Fig. 4 NDVI map

4 Results and Discussions

RUSLE is an empirically-based, straightforward model used to predict the long-term average annual soil erosion rate on slopes. It integrates data on rainfall patterns, soil properties, topography, cropping systems, and management practices. In this study, an annual soil erosion rate map for the Upper Kangsabati River Basin was developed. Various data sources were utilized to derive the input factors for the RUSLE model, which were then stored as raster GIS layers in ArcGIS. The potential annual soil loss was determined by multiplying these factors (R, K, LS, C, and P), representing the geo-environmental conditions of the study area, within ArcGIS's spatial analyst extension (Renard et al., 1997). The study estimated an average soil erosion rate of 19.032 t ha^{-1} year^{-1} for the region.

4.1 RUSLE Parameters Estimation

The annual soil loss in the study area was assessed using the Revised Universal Soil Loss Equation (RUSLE) combined with geographic information system (GIS) techniques. The RUSLE parameters—including rainfall erosivity, soil erodibility, topographic factor, crop management, and support practice factors—were determined in either spatial or numerical formats. These parameters were subsequently used to evaluate the annual soil loss in the Upper Kangsabati River Basin.

4.1.1 Rainfall Erosivity Factor (R)

The rainfall erosivity factor serves as an index to quantify the erosive potential of a given rainfall event. It is calculated based on the rainfall's volume, intensity, and duration and can be derived from either a single storm or a series of storms to capture cumulative erosivity over time. On exposed soil surfaces, splash or raindrop erosion is the most common form of erosion. For this study, 12 years (2007–2018) of rainfall data obtained from the Center for Hydrometeorology and Remote Sensing (CHRS) were utilized to determine the R-factor. Spatial interpolation methods in ArcGIS were employed alongside the rainfall data to analyze the spatial variation in rainfall and rainfall erosivity within the study region. In this analysis of the Upper Kangsabati River Basin, the R-factor was derived using the available rainfall data. By applying spatial interpolation techniques in ArcGIS, combined with the rainfall data, the spatial distribution of both rainfall and its erosive potential was evaluated. The CHRS- derived rainfall estimates suggest that the area experiences an average annual rainfall erosivity ranging between 791.224 and 928.396 MJ ha^{-1} h^{-1} year^{-1} (Fig. 5).

4.1.2 Soil Erodibility Factor (K)

The soil erodibility factor (K) measures the likelihood of soil or surface material to erode its ease of transport, and the runoff volume and rate resulting from a specific rainfall input under standardized conditions. These conditions include a plot 22.6 m in length with a 9% slope, kept in continuous fallow and tilled along the slope (Kim, 2006). For this work, a generalized soil texture map was sourced from the ICAR-National Bureau of Soil Survey and Land Use Planning (NBSS & LUP) to develop the K factor map. The corresponding K values for the various soil types were calculated using the soil erodibility nomograph provided by the USDA in 1978 (Barnett et al., 1978), which accounts for factors such as particle size distribution, organic matter content, and soil permeability (Fig. 6).

K factor values were allocated to the corresponding soil types in the soil map to create the soil erodibility map. The K factor values range from 0 to 0.31. Lower K factor values are linked to soils with low permeability and low antecedent moisture content (Figs. 7, 8 and Table 2).

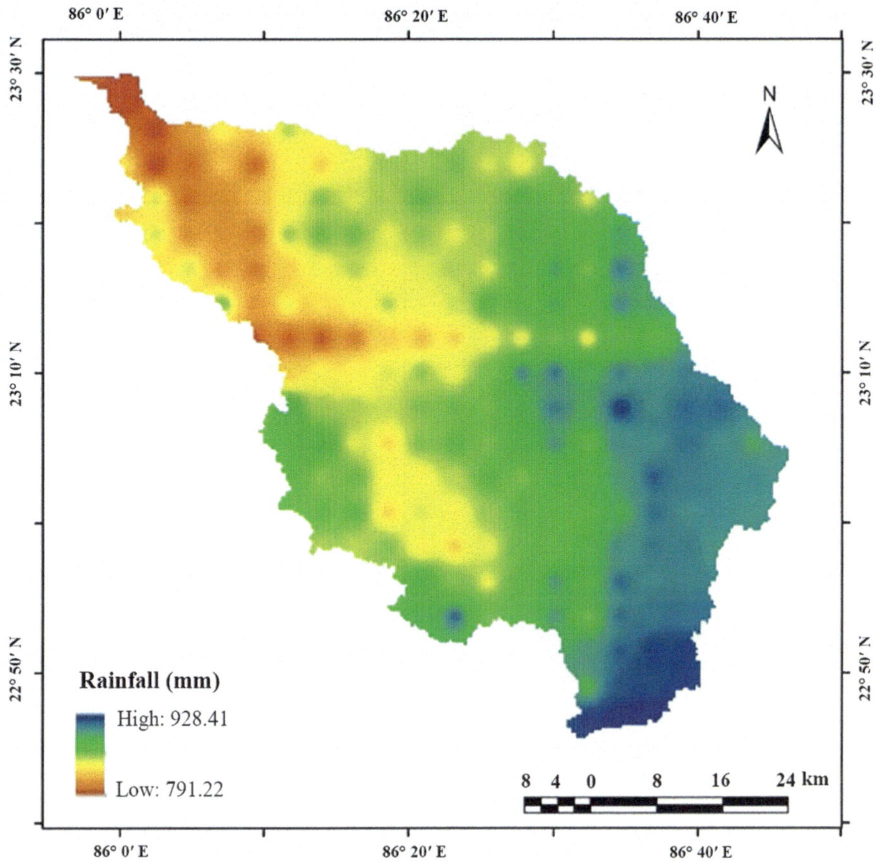

Fig. 5 Rainfall erosivity factor (R)

4.1.3 Topographic Factor (LS)

The topographic factor (LS) indicates the ratio of soil loss under specific conditions in comparison to that at a reference site with a slope gradient of 9% and a slope length of 22.6 m. This factor consists of two elements: slope length (L) and slope steepness (S). The slope length (L) component assesses the influence of slope length on soil erosion (Fig. 9).

Slope length (L) is defined as the distance from the point where overland flow begins to the point where the slope becomes gentle enough for deposition to occur, or where runoff water enters a well-defined channel. As a result, soil loss per unit area increases with an extension in slope length. The slope steepness factor (S) measures the impact of slope gradient on soil erosion, having a more significant effect on soil loss than slope length. As slope steepness rises, so does erosion, with the most intense erosion typically occurring on slopes with gradients between 10% and 25%. Unlike the slope length factor (L), which accounts for the effects of slope length, the Revised Universal Soil Loss Equation (RUSLE) does not differentiate between rill and inter-rill erosion when calculating the S factor's impact on soil loss (Renard et al., 1997; Lu et al., 2004; Krishna Bahadur, 2009). The combined LS factor

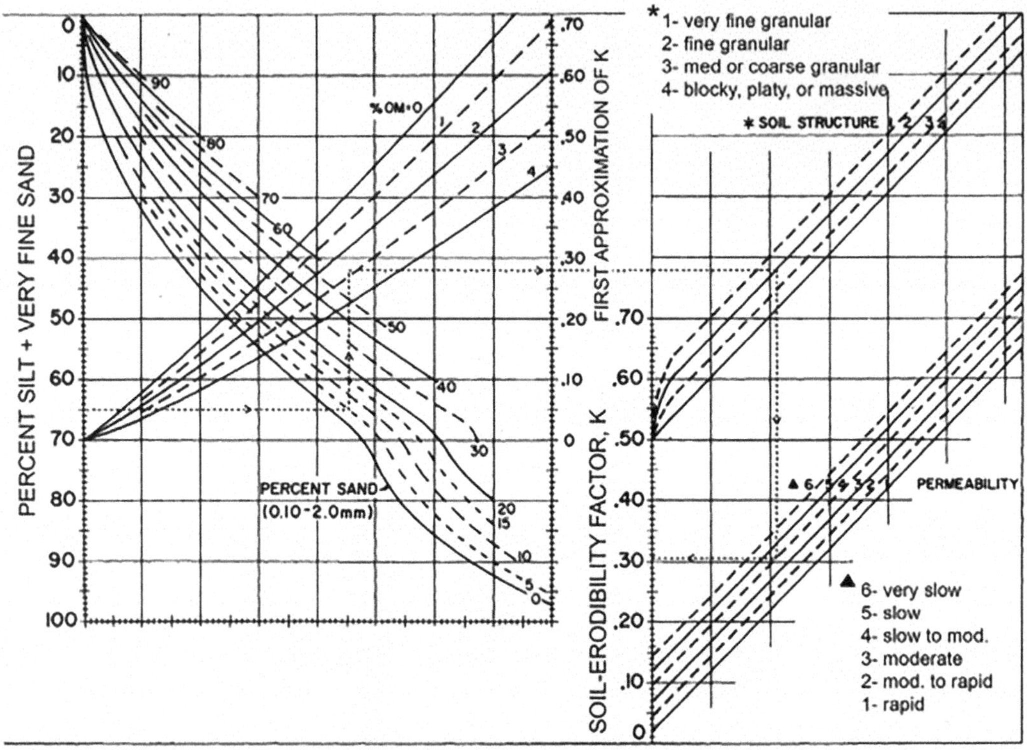

Fig. 6 Soil erodibility nomograph

for the watershed was computed using the ArcGIS Spatial Analyst extension with a Digital Elevation Model (DEM), based on the equation proposed by Moore and Burch (1986).

$$LS = (\left[Flow\,accumulation\right]^{*} \left[cell\,size\right] / 22.13)^{0.4*} \left(\sin slope / 0.0896\right)^{1.3}$$

where LS is the slope length-slope steepness factor, cell size is the size of grid cell (for this study 30.87 m), and sin slope is the slope degree value in sin.

The calculation of LS requires factors such as flow accumulation and slope steepness. The flow accumulation and slope steepness were computed from the DEM using ArcGIS Spatial analyst plus and Arc Hydro extension.

4.1.4 Support Practice Factor (P)

The support practice factor (P) is the ratio of soil loss from a specific conservation practice to that from conventional straight-row farming conducted up and down the slope. It is used to quantify the beneficial effects of these conservation practices. The P factor represents control measures that reduce erosion potential by affecting drainage patterns, runoff concentration, runoff speed, and the hydraulic forces exerted by runoff on the soil. P values range from 0 to 1, with values closer to 0 indicating more effective conservation practices and values nearer to 1 representing less effective practices. In this study, the P factor was determined based on the relationship between terracing and slope. The value of the support practice factor was calculated according to the cultivation method and

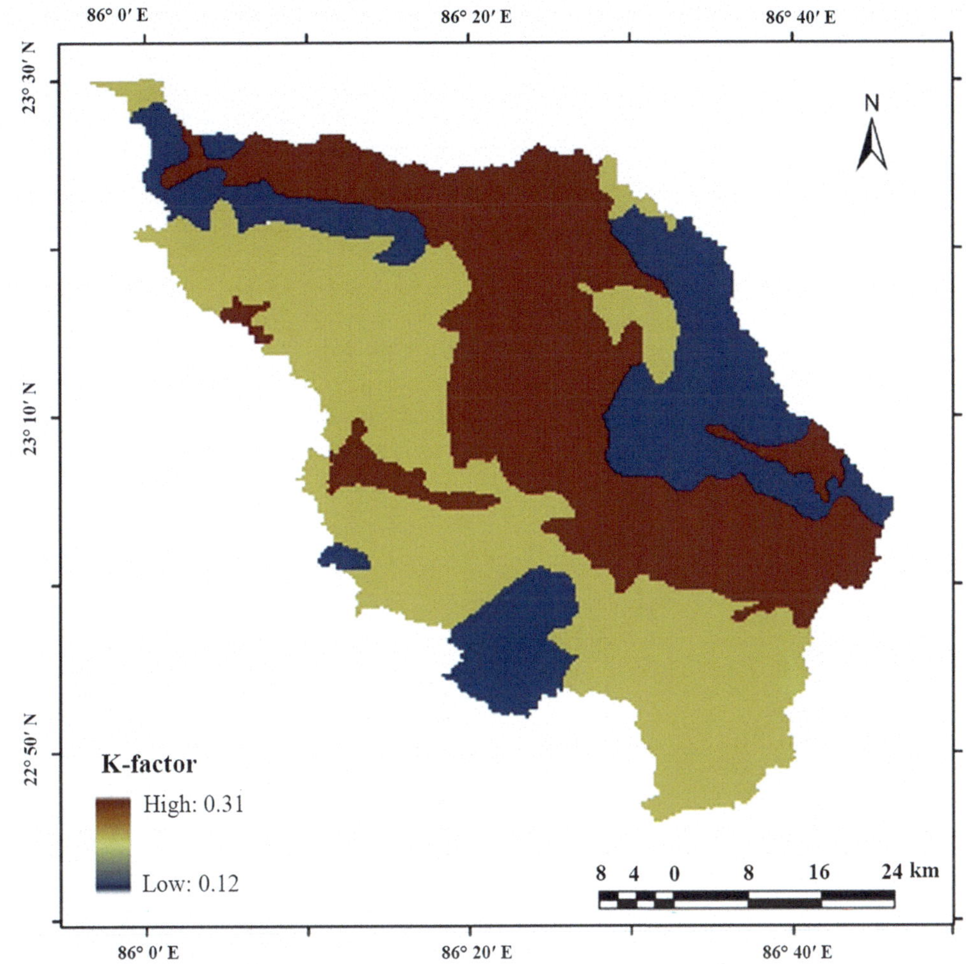

Fig. 7 Soil erodibility factor (K) map

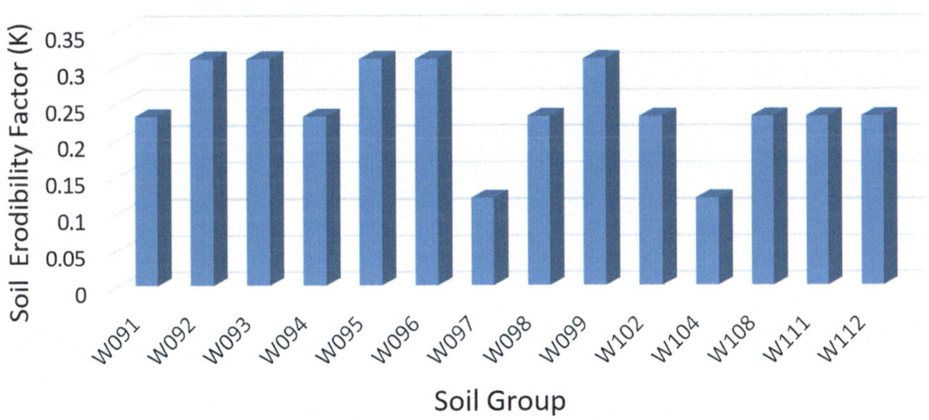

Fig. 8 Soil erodibility factor (K)

Table 2 K factor of different soil classes

Soil group	Textural classes	K-Factor
W091	Fine loamy	0.232
W092	Loamy	0.310
W093	Loamy	0.310
W094	Fine loamy	0.232
W095	Loamy	0.310
W096	Loamy	0.310
W097	Fine	0.120
W098	Fine loamy	0.232
W099	Loamy	0.310
W102	Fine loamy	0.232
W104	Fine	0.120
W108	Fine loamy	0.232
W111	Fine loamy	0.232
W112	Fine loamy	0.232

Source: ICAR-NBSS & LUP Handbook

slope (Park et al., 2011). To protect cultivated land from erosive rainfall, the natural shielding provided by grass or dense crops must be complemented with practices that slow runoff, thereby minimizing soil erosion. Key supporting practices include contour farming, strip cropping, terracing, and the construction of waterways to manage excess rainfall. In the Universal Soil Loss Equation (USLE), the conservation practice factor (P) is defined as the ratio of soil loss with a specific conservation practice to the corresponding soil loss from conventional up-and-down slope cultivation. In land use and land cover classification, P values are assigned differently, as outlined below (Fig. 10, Tables 3 and 4).

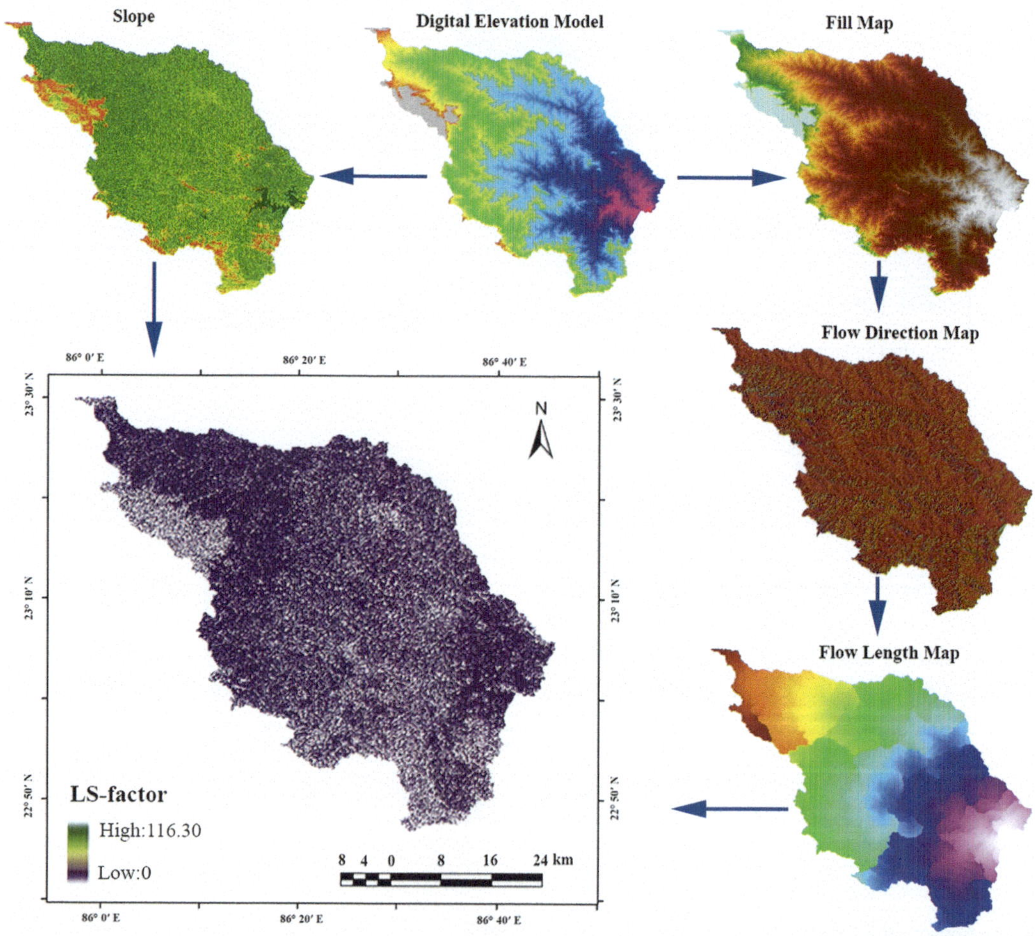

Fig. 9 Derivation of LS map

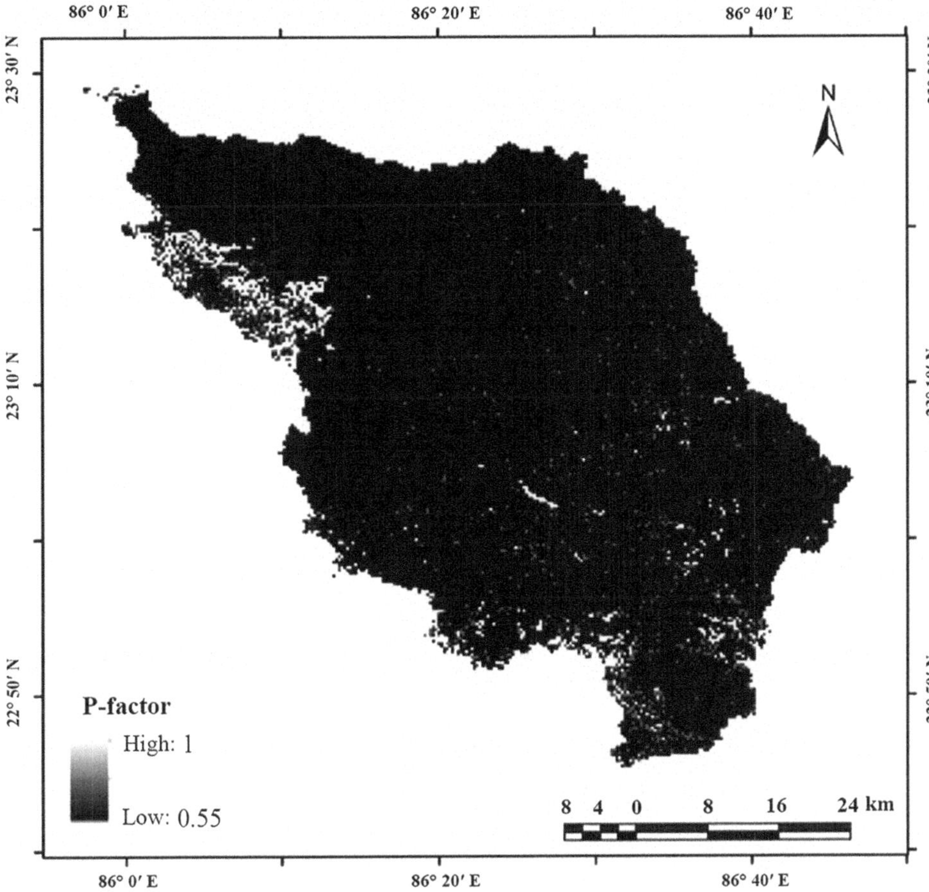

Fig. 10 Support practice factor (P) map

Table 3 P factor according to slope

Slope in degree	Contouring	Terracing
0–2.30	0.55	0.10
2.30–5.24	0.60	0.12
5.24–10.68	0.80	0.16
10.68–19.69	0.90	0.18
19.69–53.41	1.00	0.20

Table 4 P values assigned for different LULC classes

Sl. no.	Land use/land cover type	P values
1	Dense forest	1.0
2	Open forest	0.8
3	Waterbodies	1.0
4	Agricultural land	0.5
5	Settlement	1.0
6	Fallow land	0.9
7	Barren land	1.0

Source: USDA Handbook No. 282, 1981

4.1.5 Cover Management Factor (C)

Agricultural practices and management techniques are crucial in mitigating soil erosion. For example, as vegetation cover increases, soil loss rates decline exponentially (Panos Panagosa, et al., 2014). Toy et al. (1999) emphasized the significance of the C factor by stating,—The C- factor is perhaps the most important factor in RUSLE because it represents surface conditions that often are easily managed for erosion control and the values range from virtually 0 to slightly greater than 1, strongly influencing the soil loss rate.‖ As vegetative or artificial cover and soil biomass increase, the C factor values decrease. Well-protected soil has C factor values close to zero, while nearly bare soil has C factor values close to one. C factors are often derived using NDVI for soil loss assessment with RUSLE (Fig. 11).

These techniques utilize regression models to examine the relationship between C factor values, either directly measured in the field or obtained from reference tables, and NDVI values derived from remote sensing data. By employing linear regression analysis, it becomes feasible to estimate C factor values for land cover classes with unknown parameters. The main goal of this study is to improve erosion modeling in the Upper Kangsabati River Basin by estimating C factor values for various land cover classes using NDVI values through regression analysis.

The regression line that depicts the relationship between C and NDVI values, with R representing the correlation coefficient, demonstrates the strength of the regression analysis.

The regression equation was found as,

$$\left(C\ factor = 1.02 - 1.21^{*}\ NDVI\right)$$

The NDVI map for the study was generated from Sentinel-2A imagery dated 7th February 2019, using ERDAS IMAGINE 9.2 software.

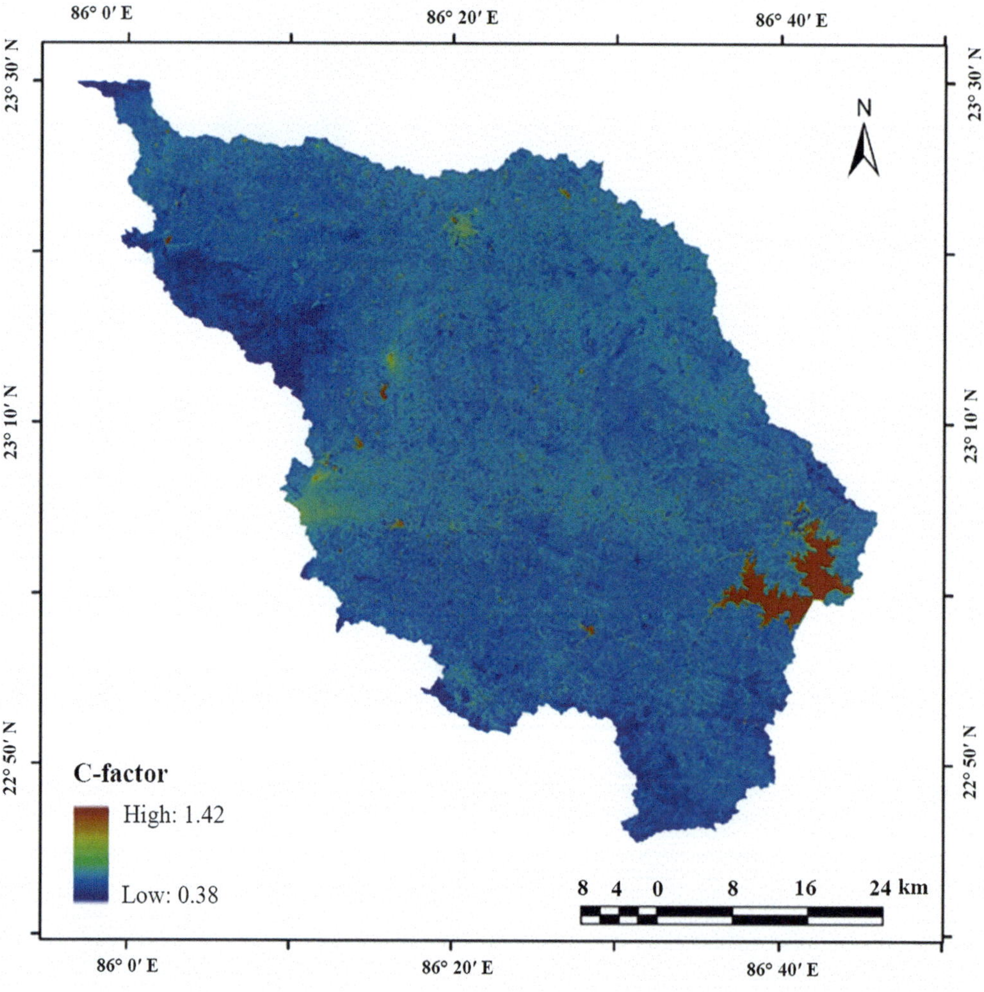

Fig. 11 Cover management (C) factor map

Subsequently, the final C factor map was produced by applying the regression equation to the NDVI map within the Spatial Analyst toolbox of ArcGIS 10.3 software.

4.2 Annual Soil Loss (A) Estimation:

The map of the Upper Kangsabati River Basin reveals an annual soil erosion rate that varies from 0 to 20.31 t ha^{-1} year^{-1}, with a standard deviation of 2.75 t ha^{-1} year^{-1}. High soil erosion rates are predominantly found in forested regions with steep slopes and in areas of fallow and barren land situated along the riverbanks. These high-risk zones are particularly susceptible due to the combined effects of steep gradients and insufficient vegetation cover, which exacerbate soil displacement. Conversely LULC categories, such as agricultural lands and settled areas, exhibit moderate-to-low erosion rates, indicating more stable soil conditions and effective land management practices. The spatial variation in erosion rates underscores the critical influence of topography and land cover on soil erosion dynamics. This highlights the necessity for targeted soil conservation strategies in the most vulnerable areas to mitigate erosion, preserve soil health, and ensure sustainable land productivity. Implementing measures such as reforestation, terracing, and maintaining vegetative cover in these high-risk areas can significantly reduce soil loss and enhance the overall resilience of the basin's landscape (Fig. 12).

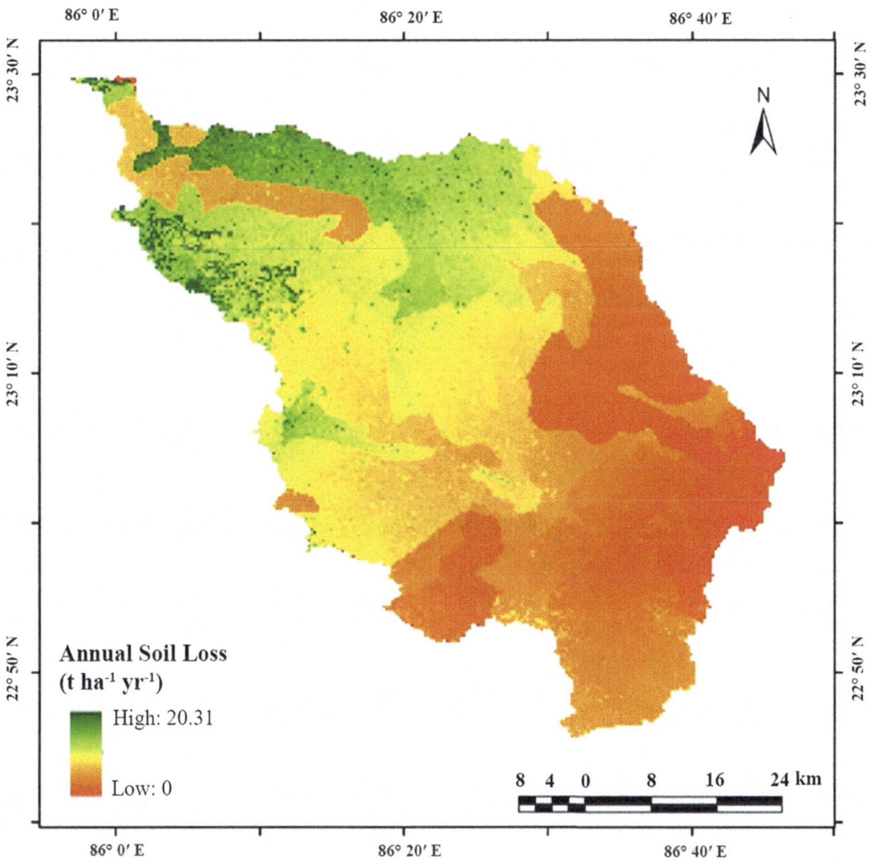

Fig. 12 Annual soil loss (A) factor map

5 Discussion

The future direction of this study involves several enhancements aimed at improving soil erosion management in the Upper Kangsabati River Basin. The integration of high-resolution remote sensing data and advanced geospatial techniques, such as LiDAR and UAV-based photogrammetry, will increase the spatial precision of erosion models like RUSLE, MUSLE, and USLE. Additionally, including climate change projections will enable the prediction of future erosion scenarios, which is essential for long-term watershed planning. The development of real-time monitoring systems utilizing Internet of Things (IoT) sensors and data analytics platforms will provide continuous data on soil moisture, rainfall, and sediment transport, aiding in early warning systems and adaptive management (Terranova et al., 2009, Benavldez et al., 2019). The use of machine learning algorithms will enhance predictive capabilities by identifying complex environmental relationships. Assessing the effectiveness of various soil conservation practices through field experiments and long-term monitoring will help develop best practice guidelines specific to the basin's conditions. Engaging local communities through participatory approaches and educational programs will ensure the sustainable implementation of soil conservation efforts. This comprehensive strategy, which combines advanced technology, field research, and community participation, aims to create resilient solutions for managing soil erosion and improving the health of the river basin ecosystem.

Today, soil erosion and degradation are major concerns for any river basin, necessitating the assessment and management of soil erosion and vulnerability zones (Dutta et al., 2015, Anache et al., 2021). Future research should focus on developing improved methods for deriving the conservation and management factor (P) to enhance soil erosion studies. This could involve the use of advanced remote sensing techniques for monitoring and evaluating conservation practices over time, employing machine learning algorithms to analyze large datasets, and identifying patterns for more accurate P factor calculations. Collaboration with local farmers and land managers to gather field data can also provide practical insights and validation for new methodologies (Vulevic and Dragovic 2017). The quality and source of GIS data are crucial; meticulous attention must be given to checking and preprocessing GIS data, including format conversion, geo-referencing, data interpolation, and registration. Implementing automated data quality assessment tools can streamline the preprocessing phase and reduce human error. Utilizing high-resolution satellite imagery and LiDAR data can significantly improve the accuracy of topographic and land-use datasets. Developing standardized protocols for data handling and preprocessing across different studies can enhance consistency and reliability. Validating soil erosion loss with locally available reference data is essential for improving result accuracy and reliability. Establishing long-term monitoring stations in various locations within the study area can provide continuous data for validation purposes. Involving local communities through citizen science initiatives can expand the dataset and increase engagement in soil conservation efforts. Applying cross-validation techniques with multiple datasets can help identify and correct discrepancies in erosion estimates.

Additionally, incorporating other soil erosion models, such as the Water Erosion Prediction Project (WEPP), with GIS can potentially increase the precision and applicability of erosion assessments. Combining the strengths of different models, such as using RUSLE for broad- scale assessments and WEPP for detailed field-level analyses, can provide a more comprehensive understanding of soil erosion dynamics. Exploring hybrid models that integrate empirical and process-based approaches can offer more robust predictions. Real-time data integration from weather stations and IoT sensors can enhance the responsiveness and accuracy of erosion models.

Investigating the impact of climate change on soil erosion patterns can provide valuable insights for future conservation planning. Developing climate-resilient erosion models that account for changing precipitation patterns and extreme weather events can help predict future erosion risks. Incorporating climate projection data into soil erosion models can guide long-term soil conservation strategies. Interdisciplinary research combining climatology, hydrology, and soil science can improve the understanding of climate change effects on soil erosion. Promoting the use of sustainable agricultural practices is crucial for reducing soil erosion. Encouraging conservation tillage, cover cropping, and agroforestry can significantly reduce erosion rates. Providing training and resources to farmers on sustainable land management practices can enhance the effectiveness of erosion control measures. Developing policies and incentives to support sustainable agriculture can promote widespread adoption of soil conservation practices. By addressing these points in greater detail and incorporating newer solutions, future studies can significantly enhance the accuracy, reliability, and applicability of soil erosion assessments.

6 Conclusion

A detailed evaluation of average annual soil loss in the Upper Kangsabati River Basin was carried out using the GIS-based RUSLE model, which integrated rainfall, soil, land use, and topographic data. The land use patterns within the basin reveal that areas with natural forest cover in the headwater regions experience minimal soil erosion, while regions impacted by human activities exhibit slightly higher erosion rates. Terrain alterations, coupled with elevated LS factors and significant rainfall, increase the susceptibility of these areas to soil erosion. The predicted soil loss and its spatial distribution provide a crucial basis for effective watershed management and sustainable land use planning. Areas identified with high and severe soil erosion levels require the immediate implementation of erosion control measures. While the current model effectively maps soil erosion, incorporating finer-scale data on rainfall intensity, soil texture, and field measurements could further enhance the accuracy and predictive power of the remote sensing and GIS-based analysis.

References

Anache, J. A. A., Wendland, E., Oliveira, P. T. S., Flanagan, D. C., & Nearing, M. A. (2021). Runoff and soil erosion modelled with RUSLE, remote sensing, and GIS in the Brazilian Cerrado. *Water, 13*(2), 118.

Arnoldus, H. M. J. (1978). An approximation of the rainfall factor in the universal soil loss equation. In M. de Boodts & D. Gabriels (Eds.), *Assessment of erosion*. Wiley.

Arya, S. L., Samra, J. S., & Aggarwal, R. K. (2001). Contribution of beneficiary participation in project effectiveness in watershed management projects: A case study in shivalik foothill region in Northern India. *Asia-Pacific Journal of Rural Development, 11*(1), 90–112.

Barnett, A. P., Dooley, A. E., & Smith, G. A. (1978). Soil erosion and sediment movement under sugarcane culture on the flat lands of southern Louisiana. *Transactions of the ASAE, 21*(6), 1144–1150.

Browning, G. M. (1979). Development for and of the universal soil loss equation. *Universal Soil Loss Equation: Past, Present, and Future, 8*, 1–5.

Benavldez, R., Jackson, B., Maxwell, D., & Norton, K. (2019). A review of the (revised) universal soil loss equation ((R)USLE): With a view to increasing its global applicability and improving soil loss estimates. *Earth Science Reviews, 198*, 102904.

Boardman, J., & Poesen, J. (2006). *Soil erosion in Europe*. Wiley.

Boggs, G., Evans, K. G., & Puig, P. (2001). GIS-based rapid assessment of erosion risk in a small catchment in the wet/dry tropics of Australia. *Land Degradation & Development, 12*(5), 417–434.

Bonilla, C. A., & Vidal, K. L. (2010). Estimation of soil erosion in five soil series in the Mediterranean climate of central Chile. *Chilean Journal of Agricultural Research, 70*(1), 95–101.

Borrelli, P., Robinson, D. A., Fleischer, L. R., Lugato, E., Ballabio, C., Alewell, C., et al. (2021). An assessment of the global impact of 21st century land use change on soil erosion. *Nature Communications, 12*(1), 3870.

Chatterjee, M. (2016). Drought and its impacts on agricultural productivity in Bankura and Purulia districts of West Bengal. *Indian Journal of Geography, 48*(2), 113–129.

Choudhury, P. R., & Nayak, D. R. (2003). Estimating the R factor from daily rainfall data for computation of soil loss. *Hydrological Sciences Journal, 48*(3), 493–503.

Chandrika, Kumar, M., Poddar, S., Kundu, A., Pal, A., Lakshmi, S., Kumar, R., & Denis, D. M. (2025). Estimating soil erosion in Surguja District Chhattisgarh (India) using the RUSLE model and geospatial techniques. In S. Das, P. C. Pandey, O. Mutanga, S. Tikle, M. Jiang, U. Chatterjee, & A. Anand (Eds.), *Application of geospatial technology and modelling on natural resources management: Current state challenges and sustainability. Advances in geographic information science* (pp. 361–390). Springer Nature. https://doi.org/10.1007/978-3-031-84698-4_16

Dabral, P. P., Baithuri, N., & Pandey, A. (2008). Soil erosion assessment in a hilly catchment of north eastern India using USLE, GIS and remote sensing. *Water Resources Management, 22*(12), 1783–1798.

Demirci, A., & Karaburun, A. (2012). Estimation of soil erosion using RUSLE in a GIS framework: A case study in the Buyukcekmece Lake watershed, northwest Turkey. *Environmental Earth Sciences, 66*(3), 903–913.

Dutta, D., Das, S., Kundu, A., & Taj, A. (2015). Soil erosion risk assessment using geo- informatics, RUSLE model and TRMM data in Sanjal watershed, Jharkhand. *Journal of Soil and Water Conservation, 14*(2), 123–131.

Ghosh, S. (2018). Hydrological impact of Kangsabati reservoir on the regional ecosystem. *Environmental Studies Review, 10*(4), 256–269.

Jain, M. K., Das, D., & Chaudhry, A. (2001). Soil erosion assessment and prediction using remote sensing and GIS. *Journal of the Indian Society of Remote Sensing, 29*(1), 25–35.

Jasrotia, A. S., & Singh, R. (2006). Modeling runoff and soil erosion in a catchment area using the GIS-based ANSWERS model. *Hydrological Sciences Journal, 51*(4), 599–611.

Kim, H. (2006). Assessment of soil erosion vulnerability using USLE and GIS: A case study in Korea. *Environmental Geology, 50*(5), 615–625.

Kouli, M., Soupios, P., & Vallianatos, F. (2009). Soil erosion prediction using the Revised Universal Soil Loss Equation (RUSLE) in a GIS framework, Chania, northwestern Crete, Greece. *Environmental Geology, 57*(3), 483–497.

Krishna Bahadur, K. C. (2009). Mapping soil erosion susceptibility using remote sensing and GIS: A case of the Upper Nam Wa watershed, Nan Province, Thailand. *Environmental Geology, 57*(3), 695–705.

Kumar, R., & Kushwaha, S. P. S. (2013). Modelling soil erosion risk based on RUSLE-3D using GIS in a Shivalik sub-watershed. *Journal of Earth System Science, 122*(2), 389–398.

Lal, R. (2001). Soil degradation by erosion. *Land Degradation & Development, 12*(6), 519–539.

Lal, R. (2003). Soil erosion and the global carbon budget. *Environment International, 29*(4), 437–450.

Lu, H., Moran, C. J., & Prosser, I. P. (2004). Modelling sediment delivery ratio over the Murray Darling basin. *Environmental Modelling & Software, 19*(7–8), 647–654.

Mandal, S., Roy, A., & Biswas, R. (2020). Soil formation and slope characteristics in the Rarh region. *Earth Sciences Journal, 8*(2), 78–89.

Millward, A. A., & Mersey, J. E. (1999). Adapting the RUSLE to model soil erosion potential in a mountainous tropical watershed. *Catena, 38*(2), 109–129.

Montgomery, D. R. (2007). Soil erosion and agricultural sustainability. *Proceedings of the National Academy of Sciences, 104*(33), 13268–13272.

Moore, I. D., & Burch, G. J. (1986). Physical basis of the length-slope factor in the Universal Soil Loss Equation. *Soil Science Society of America Journal, 50*(5), 1294–1298.

Mukherjee, S., Biswas, S., & Ghosh, A. (2018). Impact of climate change on rainfall and agriculture in the Western Plateau Region of West Bengal. *Journal of Climate Change, 4*(3), 145–157.

Pandey, A., Chowdary, V. M., & Mal, B. C. (2009). Sediment yield modeling of a forested watershed in the Himalayan region using RS and GIS. *Environmental Geology, 56*(7), 1335–1349.

Pandey, A., Chowdary, V. M., & Mal, B. C. (2007). Identification of critical erosion prone areas in the small agricultural watershed using USLE, GIS and remote sensing. *Water Resources Management, 21*(4), 729–746.

Park, S., Oh, C., Jeon, S., Jung, H., & Choi, C. (2011). Soil erosion risk in Korean watersheds, assessed using the revised universal soil loss equation. *Journal of Hydrology, 399*(3–4), 263–273.

Perović, V., Životić, L., Kadović, R., Đorđević, A., Jaramaz, D., Mrvić, V., & Todorović, M. (2013). Spatial modelling of soil erosion potential in a mountainous watershed of South-eastern Serbia. *Environmental Earth Sciences, 68*(1), 115–128.

Pradeep, G. S., Krishnan, M. N., & Vijith, H. (2015). Identification of critical soil erosion prone areas and annual average soil loss in an upland agricultural watershed of Western Ghats, using analytical hierarchy process (AHP) and RUSLE techniques. *Arabian Journal of Geosciences, 8*(6), 3697–3711.

Pimentel, D. (2006). Soil erosion: A food and environmental threat. *Environment, Development and Sustainability, 8*(1), 119–137.

Renard, K.G., Foster, G.R., Weesies, G.A., McCool, D.K., & Yoder, D.C. (1997). *Predicting soil erosion by water: A guide to conservation planning with the Revised Universal Soil Loss Equation (RUSLE).* USDA Handbook No. 703.

Sarkar, A., & Sen, S. (2019). Geomorphological features of the Chota Nagpur Plateau fringe. *Journal of Indian Geography, 55*(3), 201–217.

Singh, V., et al. (1981). Estimation of rainfall erosivity. *Journal of Hydrology, 49*(1–2), 61–77.

Smith, R. E. (1976). Approximations for vertical infiltration rate patterns. *Transactions of the ASAE, 19*(3), 505–0509.

Srinivas, C. V., Maji, A. K., Reddy, G. P. O., & Chary, G. R. (2002). Assessment of soil erosion using remote sensing and GIS in Nagpur district, Maharashtra for prioritisation and delineation of conservation units. *Journal of the Indian Society of Remote Sensing, 30*(4), 197–212.

Terranova, O., Antronico, L., Coscarelli, R., & Iaquinta, P. (2009). Soil erosion risk scenarios in the Mediterranean environment using RUSLE and GIS: An application model for Calabria (southern Italy). *Geoderma, 154*(3–4), 564–574.

Tian, Y., Gupta, H. V., & Sorooshian, S. (2009). A preliminary assessment of the impact of NASA's GRACE mission on hydrological cycle component estimation. *Journal of Hydrology, 370*(1–4), 1–4.

Toy, T. J., Foster, G. R., & Renard, K. G. (1999). RUSLE for mining, construction and reclamation lands. *Journal of Soil and Water Conservation, 54*(2), 462–467.

Vulevic, T., & Dragovic, N. (2017). Ranking subwatersheds using multi-criteria decision analysis in Serbia. *Journal of Environmental Management, 203*, 865–876.

Wischmeier, W.H., & Smith, D.D. (1978). *Predicting rainfall erosion losses: A guide to conservation planning*. USDA Handbook No. 537.

Zhou, P., Luukkanen, O., Tokola, T., & Nieminen, J. (2008). Effect of vegetation cover on soil erosion in a mountainous watershed. *Catena, 75*(3), 319–325.

Zingg, A. W. (1940). Degree and length of land slope as it affects soil loss in run-off. *Agricultural Engineering, 21*(2), 59–64.

Geospatial Technique in Forest Fragmentation Analysis Using Earth Observation Datasets

Pritam Bar, Prashant Kumar Srivastava, Shruti Kanga, Suraj Kumar Singh, and Bhartendu Sajan

1 Introduction

Forests are vital renewable natural resources that play an indispensable role in supporting ecosystems, economic growth, and environmental sustainability. They are home to a wide variety of plant and animal species, contributing to global biodiversity. Economically, forests provide essential goods such as wood, which are used as fuel and as raw material for industries including pulp, paper, furniture, and construction (Cakir et al., 2008). They also supply non-timber products such as gums, resins, dyes, tannins, lac, fibers, medicinal plants, aromatic herbs, and food resources like fruits, tubers, roots, and meat, supporting the livelihoods of many communities, including tribal populations. Additionally, forests promote small-scale and cottage industries, create employment opportunities, generate revenue, and boost tourism. From an environmental perspective, forests are crucial in maintaining ecological balance by regulating climate, enhancing rainfall, conserving soil, controlling floods, and reducing pollution. They improve the aesthetic value of landscapes and serve as carbon sinks, mitigating the effects of climate change. (Dewan et al., 2012) By sustaining life-supporting systems, forests contribute to the overall health of the planet, benefiting both current and future generations. Their role in providing raw materials, ecological services, and socioeconomic benefits makes them indispensable for sustainable development and the well-being of both nature and society (Giriraj et al., 2010).

Forests are not only a vital natural resource but also an essential part of the global ecosystem, offering a multitude of benefits that extend far beyond timber production. They play a crucial role in carbon sequestration, helping mitigate climate change by absorbing significant amounts of carbon dioxide from the atmosphere. Forests also regulate the water cycle, enhance groundwater recharge, and protect watersheds, which are critical for maintaining water quality and availability (Batar et al., 2017). By fostering biodiversity, forests support a wide range of flora and fauna,

P. Bar · B. Sajan
Centre for Climate Change and Water Research, Suresh Gyan Vihar University, Jaipur, India
e-mail: bhartendu.sajan@mygyanvihar.com

P. K. Srivastava
Institute of Environment and Sustainable Development, Banaras Hindu University, Varanasi, Uttar Pradesh, India

S. Kanga (✉)
Department of Geography, School of Environment and Earth Sciences, Central University of Punjab, Bathinda, India
e-mail: shruti.kanga@cup.edu.in

S. K. Singh
Centre for Climate Change and Water Research, Suresh Gyan Vihar University, Jaipur, India

Centre for Sustainable Development, Suresh Gyan Vihar University, Jaipur, India
e-mail: suraj.kumar@mygyanvihar.com

Z. Khanam et al. (eds.), *Climate Change, Resilience and Ecosystem Services*, Sustainable Development Goals Series, https://doi.org/10.1007/978-3-031-96977-5_24

many of which are yet to be discovered, contributing to the overall health of ecosystems. Furthermore, forests serve as habitats for indigenous and local communities who depend on them for sustenance and livelihood (Jha et al., 2005).

The economic value of forests cannot be overstated, as they are integral to various industries, including paper and pulp, furniture manufacturing, and pharmaceuticals. Forests provide raw materials for these industries, thereby creating employment opportunities, especially in rural areas, and supporting local economies (Sharma et al., 2017). The wood-based industry, for instance, is a significant contributor to the national economy, with forests offering not only timber but also fuelwood for cooking and heating in many parts of the world. Additionally, the collection of non-timber forest products (NTFPs) such as medicinal plants, honey, and mushrooms adds to the economic importance of forests. These products also support small-scale and cottage industries, further diversifying the economic benefits derived from forests (Foley et al., 2005).

Beyond their tangible economic benefits, forests offer intangible advantages such as aesthetic value, cultural significance, and recreational opportunities. Forested areas often attract tourists and nature enthusiasts, contributing to the growth of ecotourism and local tourism- based economies (Nagendra et al., 2009). Moreover, forests are integral to the cultural and spiritual practices of many communities, providing spaces for rituals, traditions, and a sense of identity. The protection of forests is not only an environmental necessity but also a key factor in achieving sustainable development goals, as forests contribute to social, economic, and environmental well-being. Therefore, the conservation and sustainable management of forests are critical for maintaining their role in supporting human life and biodiversity for future generations (Jha et al., 2005).

In addition to the numerous direct benefits forests provide, they also play a fundamental role in maintaining global ecological balance. Forests act as the Earth's lungs by absorbing carbon dioxide and releasing oxygen, making them essential in regulating atmospheric gases (Padalia et al., 2020). They are vital for climate regulation, acting as buffers against extreme weather events such as droughts, floods, and hurricanes. Forests help maintain the global temperature by stabilizing microclimates, preventing soil erosion, and reducing the intensity of natural disasters. Moreover, they play a crucial role in supporting the livelihoods of millions of people, especially in rural and remote areas, where forest resources are used for fuel, construction materials, and food. The economic activities surrounding forests, such as sustainable logging, forest-based tourism, and the harvesting of NTFPs, are vital in providing income and sustenance for local communities. Thus, preserving forests is not only necessary for environmental protection but also for ensuring the social and economic well-being of communities that depend on them.

2 Study Area

The study area for this research is Bageshwar District, located in the Eastern Kumaon region of Uttarakhand, India, covering an area of 2298 km^2. Geographically, it lies between latitudes 29° 40′50″ to 30° 19′10''North and longitudes 79° 28′5″ to 80° 9′40″ East. The district experiences a subtropical highland climate with an average annual temperature of 20.4 °C and receives around 1221 mm of precipitation annually. June is the warmest month with an average temperature of 27.3 °C. The region's diverse topography, ranging from hilly terrains to river valleys, along with its seasonal climate variations, significantly influences the local ecosystems, land use patterns, and agricultural practices, making it an ideal location to study environmental changes and the impacts of both natural and anthropogenic factors (Fig. 1).

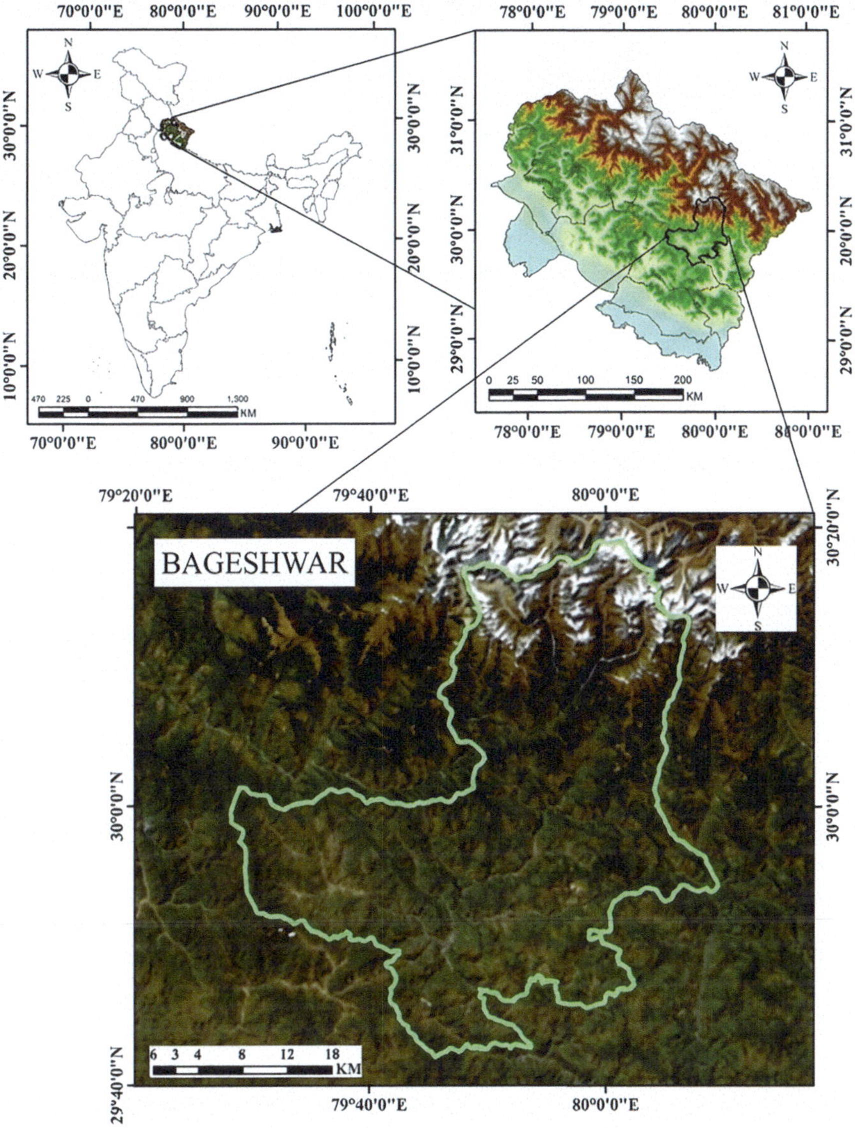

Fig. 1 Location map of study area

3 Materials and Methods

3.1 Land Cover Assessment Using Modis

The Terra and Aqua satellites are equipped with the MODIS (Moderate Resolution Imaging Spectroradiometer) instrument, which is designed for large-scale global monitoring of Earth's surface. With its remarkable swath width of 2330 km, MODIS captures comprehensive data of the entire planet every one to two days. This capability ensures frequent and detailed observations critical for monitoring dynamic changes on Earth's surface. MODIS collects data at three spatial resolutions—250, 500, and 1000 m—enabling flexibility in analyzing phenomena at varying scales. Its detectors are tuned to 36 spectral bands, spanning wavelengths from 0.405μm to 14.385μm. This wide spectral range makes MODIS highly versatile, allowing it to monitor a variety of surface, atmospheric, and oceanic conditions, from vegetation health and snow cover to sea surface temperature and atmospheric aerosols.

The data collected by MODIS and other sensors aboard the Terra and Aqua satellites are transmitted to Earth through the sophisticated Tracking and Data Relay Satellite System (TDRSS). This system ensures reliable communication with ground stations located in White Sands, New Mexico, where the initial reception occurs. Once received, the data is forwarded to the EOS Data and Operations System (EDOS) at NASA's Goddard Space Flight Center for further processing. The MODIS Adaptive Processing System (MODAPS) is responsible for generating critical data products, including Level 1A and Level 1B data, geolocation information, cloud masks, and higher-level land and atmosphere products. These processed products are then distributed among three Distributed Active Archive Centres (DAACs) for broader access. Additionally, the Ocean Colour Data Processing System (OCDPS) processes ocean-related data, ensuring that specific products like ocean color and sea surface properties are available for research and applications across various scientific domains.

The diverse data products derived from MODIS observations provide invaluable insights into the Earth's land, ocean, and atmospheric systems. These products are extensively used to study processes and detect trends across local, regional, and global scales. For instance, atmospheric data such as aerosol concentrations and cloud dynamics are essential for climate studies, while land-related products contribute to vegetation monitoring, deforestation analysis, and urbanization tracking. Ocean data, including sea surface temperature and chlorophyll concentrations, are critical for understanding marine ecosystems and supporting fisheries management. To ensure accessibility, MODIS products are hosted on various platforms. The LAADS website provides Level 1 and atmospheric products, while the Land Processes DAAC at the U.S. Geological Survey's EROS Data Center delivers land-related data. Cryosphere products, including snow and sea ice cover, are available from the National Snow and Ice Data Center (NSIDC) in Boulder, Colorado. Ocean-related data, such as ocean color and temperature, are distributed by the OCDPS at NASA's Goddard Space Flight Center. Additionally, users with x-band receiving equipment can receive real-time regional MODIS data directly via the Direct Broadcast Signal, providing a vital resource for immediate applications.

3.2 Modis Bands

MODIS (Moderate Resolution Imaging Spectroradiometer), a vital instrument aboard NASA's Terra and Aqua satellites, is specifically designed for detailed and extensive Earth observation. Its advanced capability to capture data across 36 spectral bands, spanning visible, near-infrared, and thermal infrared wavelengths, enables it to serve a wide range of scientific and practical applications. Among these, Bands 1 and 2, with a spatial resolution of 250 meters, are particularly significant for their ability to operate in the visible and near- infrared spectrum. These bands play a critical role in land cover classification, crop monitoring, and vegetation health assessment. By detecting subtle differences in light absorption and reflectance, they allow researchers to identify various types of vegetation, evaluate their growth conditions, and monitor changes over time, which are essential for agriculture, forestry, and environmental management.

3.3 Data Period

The data were derived as annual averages for three representative years spanning two decades—2001, 2011, and 2021. These years were selected due to the availability of MODIS data from 2001 to 2022. To enable a decadal comparison of land use and land cover (LULC) dynamics, data at ten-year intervals were utilized

3.4 Landscape Metrics Using Fragstats

PLAND is a proportion of the landscape that is covered by some specific patch type and it is computed by summing up all the areas (퐦2) of patches on that patch type, then dividing with the landscape's area in (퐦2) and finally multiplying by 100. It should be noted that the interior of any background belongs to total landscape area (A) (Lamine et al., 2017; Midha et al., 2010).

$$PLAND = Pi = \sum n\, j = i\, aijA$$

where

Pi = proportion of the landscape occupied by patch type (class) i.

aij = area (m^2) of patch in.

A = total landscape area (m^2).

UNIT: Percent.

$$\mathbf{0 < PLAND \leqq 100}$$

As patch class becomes rare in landscape, it moves closer to 0. If there is one type (patch) covering the whole image, then PLAND =100.

In ecology, PLAND is a measure that is used to quantify the proportion of each type of patch within a landscape. This measure is particularly useful because it tells us a relative amount instead of an absolute value such as total area, which might be misleading in some cases of comparing different-sized landscapes. Paraphrased version: It gives us a viewpoint as to how many types they are compared to other areas regardless if they are bigger or smaller than ours.

4 Number of Patches

The number of patches of the matching patch type (class) is denoted by NP. The total edge length of a specific patch type is calculated at the class level. This measure is less effective for comparisons between landscapes of different sizes than edge density (see below). However, when landscapes are of the same size, both total edge and edge density are unnecessary.

$$NP = ni$$

NP > 1, infinitely many.

When there is only one patch of the matching patch type in the landscape, or when the class only consists of one patch, NP = 1.

One easy way to gauge how much a patch type has been divided or fragmented is to count the number of patches that belong to that kind. While the quantity of patches within a class could hold significant importance for certain ecological processes, its interpretive value is frequently restricted as it does not provide information on the area, distribution, or density of patches. If we keep the following areas the same; area for the overall landscape and area for the various classes, the information that is shared by patch number equals that that is shared by mean patch size or patch density (PD). This makes the index potentially helpful to comprehend. The most useful indicator, though, is arguably the number of fixes when used as a foundation for other, easier to understand metrics. Be aware that this measure will change depending on whether the four-neighbor or eight-neighbor rule is used to delineate patches.

5 Patch Density

PD can be calculated by multiplying the number of patches of a given patch type by the total landscape area (m^2) and then multiplying the result by 10,000 and 100 to convert to 100 hectares. Note that any interior background present is included in the entire landscape area (A).

$$PD = ni \,/\, A^{*}\left(10000\right)^{*}\left(100\right)$$

where ni denotes the number of patches of a given patch type (class) in the landscape, and A represents the total area of the terrain (m^2). Patch density (PD) is expressed as the number of patches per 100 hectares and is constrained to values greater than zero (PD 0).

Since the highest PD occurs when each cell represents an individual patch, in the end PD must be limited by the grain size of the raster image. Thus, the largest number of patches per unit area would ultimately depend on cell sizes.

However, once every other cell in the focus class belongs to this class, we reach the peak density of patches of that class; this can be illustrated as a checkerboard pattern.

Patch density is one minor yet significant aspect of landscape design. With the exception of expressing the number of patches per unit area, patch density has the same basic function as the number of patches in an index, facilitating comparisons between landscapes of various sizes. Plain and simple, if the total land does not differ, both the density and the number of patches pass on the same information. Patch density alone, like patch count alone, often offers little information about the sizes and spatial distribution of patches, making it hard to understand. Be advised that the measurement will vary based on whether patches are drawn using the four-neighbor or eight-neighbor approach.

6 Largest Patch Index

LPI, or largest patch index, is calculated by taking the area (in square meters) of the largest patch of a given type and dividing it by the total landscape area (also in square meters). This ratio is then multiplied by 100 to convert it into a percentage. Essentially, LPI represents the percentage of the landscape (PLAND) that is made up by the largest patch. It's crucial to remember that any internal background seen is included in the overall landscape area (A).

$$CPI = (aij)\, ja = i / a^{*}(100)$$

where aim = area (m^2) of patch in. A = total landscape area (m^2).

$$\mathbf{0 < LPI \leqq 100}$$

When the largest patch of the appropriate patch type gets smaller and smaller, LPI gets closer to zero. When a single patch of the relevant patch type makes up the entire landscape, or when the largest patch makes up 100% of the landscape, LPI = 100.

At the class level, the largest patch index measures the portion of the overall landscape area that is made up of the largest patch. Thus, it is a straightforward indicator of supremacy.

7 Total Edge

The lengths (m) of all edge segments involving the matching patch type added together equal total edge (TE). If a landscape border is present, only the "real" edge (i.e., adjacent patches of different classes) is represented by the landscape boundary segments involving the corresponding patch type in TE. In the event that there is no landscape border, TE incorporates a user-specified percentage of segments of the border that involve the matching patch type. TE comprises a user-specified percentage of internal backdrop edge segments involving the matching patch type, regardless of the existence of a landscape border.

$$TE = \sum eik\, m\, k = I$$

In the landscape, eik represents the total edge length (meters) of patch type I, and is a sum of background segments in patch type I and the landscape border.

$$\mathbf{TE \geq 0, without\ limit}$$

When the user indicates that none of the landscape boundary and background edge should be handled as edges, and the entire landscape, including the landscape border, if it exists, is made up of the matching patch type, then TE = 0 occurs.

8 Related Circumscribing Circle

CIRCLE is equal to the product of the area (m^2) of the smallest circumscribing circle and the patch area (m^2).

$$SQU\ ARE = 1 - [aij / aij]$$

where

where a_{ij} denotes the area of patch i, measured in square meters (m^2).

aij = The size (m^2) of the smallest circle that can be used to surround patch in.

$$0 < \text{CIRCLE} < 1$$

CIRCLE gets closer to 0 for patches that are round and closer to 1 for patches that are long and straight, one cell wide.

Baker and Cain (1992) came up with the related circumscribing circle. Please note that patch size does not change this number. Patches may deviate from a perfectly circular shape due to the way grid data are stored; however, the area of the circle surrounding each patch is calculated geometrically using its radius.

9 Disjunct Core Area Density

The disjunct core area density (DCAD) is calculated by adding up the number of separate core areas in each patch of the right type and dividing that number by the total landscape area ($m2$), then multiplying that number by 10,000 and 100 to get 100 hectares. Keep in mind that the total scenery area (A) includes any background that is inside the image.

$$DCAD = \sum nj = Inije / A(10000)^{*}(100)$$

The number of core areas that don't connect in patch ij based on the given edge depths (m) is given by nijc. A = total ground area ($m2$). Number per 100 hectares.

$$DCAD \geq 0, without\ limit.$$

DCAD = 0 when TCA = 0 (i.e., every location within patches of the associated patch type are within the specified depth-of-edge distance(s) from the patch perimeters); in other words, when there are no core areas. Disjunct core area density, like its counterpart, patch density (PD), expresses number of disjunct core areas on a per unit area base that facilitates comparisons among landscapes of varying size. Of course, if total core area is held constant, then disjunct core area population and number of disjunct core areas give the same information.

10 Core Area

CORE refers to the area within a lot that extends beyond its boundaries to at least the designated border-distance depth. This area is divided by 10,000 to convert it to hectares. Core sectors adjacent to the landscape boundary are considered part of the background unless a landscape border exists. In such cases, boundary edge types are defined by the border.

$$CORE = aije^{*}(1/10000)$$

aije = core area (m^2) of patch *ij* based on given edge depths (m).

$$\text{CORE} \geq 0, \text{without limit.}$$

CORE = 0 when every location within the patch is within the given depth-of-edge distance from the patch perimeter. CORE approaches AREA as the stated depth-of-edge distance(s) decreases and as patch shape is simplified. Core area indicates the area in the patch greater than the specified depth-of-edge distance from the perimeter. Note that a single depth-of- edge distance can be used for all edges or the user can specify an edge depth file that offers unique distances for each pairwise combination of patch types.

11 Core Area Index

CAI equals the patch core area (m^2) divided by overall patch area (m^2), multiplied by 100 (to convert to a percentage); in other words, CAI equals the percentage of a patch that is core area.

$$CAI = aije / aij(100)$$

aije = core area (m^2) of patch ij based on given edge depths (m).

aij = the patch ij's size in square meters.0 $\leqq$ CAI < 100.

CAI = 0 when CORE = 0 (i.e., every location within the patch is within the given depth-of-edge distance(s) from the patch perimeter); that is, when the patch contains no core area. CAI approaches 100 when the patch, because of size, shape, and edge width, includes mostly core area. Core area index is a relative index that quantifies core area as a percentage of patch area (i.e., the percentage of the patch that is made of core area).

12 Euclidean Nearest-Neighbor Distance

Euclidean nearest-neighbor (ENN) means the distance (m) to the nearest neighboring patch of the same type, based on shortest edge-to-edge distance. Note that the edge-to-edge lengths are from cell center to cell center.

$$ENN = hij$$

where hij = distance (m) from patch ij to closest neighboring patch of the same type (class), based on patch edge-to-edge distance, computed from cell center to cell center. ENN > 0, without limit. ENN approaches 0 as the distance to the closest neighbor decreases. The minimum ENN is limited by the cell size, and is equal to twice the cell size when the eight-neighbor patch rule is used or the space between diagonal neighbors when the eight-neighbor rule is used. The upper limit is limited by the extent of the landscape. ENN is undefined and recorded as "N/A" in the "basename." Patch file if the patch has no neighbors (i.e., no other patches of the same class). Euclidean nearest-neighbor distance is perhaps the simplest measure of patch context and has been used widely to quantify patch isolation. Here, the nearest neighbor distance is defined using simple Euclidean geometry as the shortest straight-line distance between the focus patch and its nearest neighbor of the same class.

13 Effective Mesh Size

MESH means 1 divided by the total landscape area (m^2) multiplied by the sum of patch area (m^2) squared, summed across all patches in the landscape. Keep in mind that the total scenery area (A) includes any background that is inside the image.

$$MESH = \sum mi = I \sum nj = Iaij2 / A$$

where
aij = the patch ij's size in square meters.
A = total ground area (m^2).
cell size $\leq$ $MESH$ $\leq$ total landscape area (A).

The bottom limit of MESH is constrained by the cell size and is achieved when the landscape is maximally subdivided; that is, when every cell is a separate patch. MESH is maximum when the scenery consists of a single patch. Mesh is based on the cumulative patch area distribution and is understood as the size of the patches when the landscape is subdivided into S patches, where S is the value of the splitting index. Note, MESH is redundant with DIVISION above, i.e., they are perfectly, but inversely, correlated, but both measures are included because of differences in units and interpretation. DIVISION is understood as a probability, whereas MESH is given as an area. Conceptually and computationally, these two metrics are almost identical at the landscape level, and under most situations will return identical values. Specifically, AREA_AM gives the area-weight mean patch size, where the proportional area of each patch is based on total landscape area excluding any background (i.e., background is removed from the total landscape area). MESH also gives the area weighted mean patch size, but the proportional area of each patch is based on the total field area including any background.. Thus, if there is no internal background, these metrics will return identical numbers. If there is internal background, these measures will return different values, and the magnitude of the difference will depend on the proportional extent of background. In the latter case, the choice of metrics relies on how you want to consider background.

14 Results and Discussions

14.1 Land Use Map

The image analysis shows the regional spread of different land use groups in the study area between 2001 and 2021 (Fig. 2).

14.2 Percentage of Landscape

The land cover statistics for the years 2001, 2011, and 2021 provide a perceptive look at how the

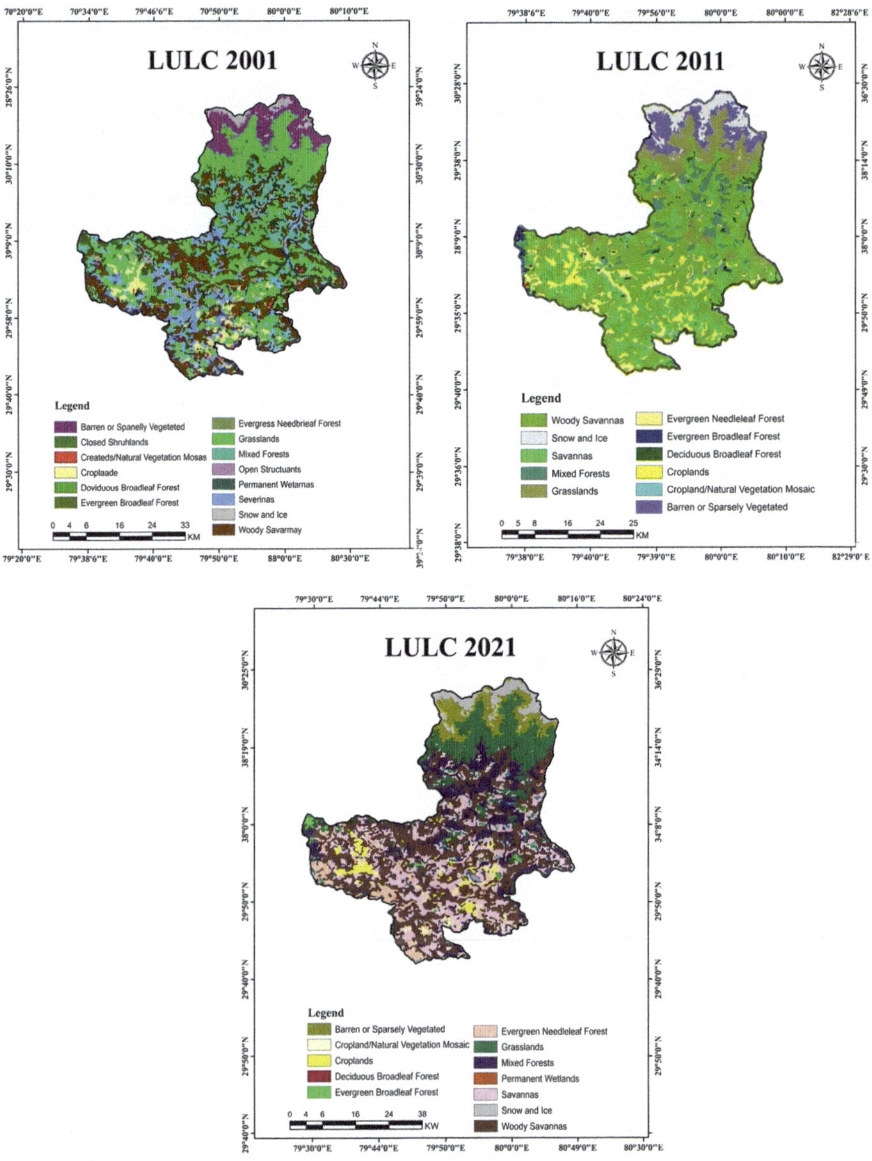

Fig. 2 Change detection maps of period of 2001–2021

environment is evolving. There have been notable changes in the extents of several kinds of forests, shrublands, savannas, grasslands, wetlands, croplands, and other land coverings. The dynamics of land use and cover during the last 20 years have been impacted by wider environmental, climatic, and human impacts, which are reflected in these shifts (Table 1).

14.3 Forest Types

The most noticeable change is seen in the mixed forests, which expanded from 7.84% in 2001 to 11.00% in 2021. This substantial increase suggests a trend toward more diverse forest ecosystems, potentially due to reforestation efforts or natural regeneration. Evergreen needleleaf for-

Table 1 Land use and land cover change

Type	2001	2011	2021
Evergreen needleleaf forest	2.87	4.9	6.37
Evergreen broadleaf forest	0.71	0.94	0.79
Deciduous broadleaf forest	0.25	0.28	0.13
Mixed forests	7.84	7.78	11
Closed shrublands	0.02	0	0
Open shrublands	0.03	0	0
Woody savannas	31.86	42.87	39.03
Savannas	15.36	11.44	17.78
Grasslands	27.84	16.82	13
Permanent wetlands	0.12	0	0.01
Croplands	4.01	4.89	2.18
Cropland/natural vegetation mosaic	0.61	0.18	0.02
Snow and ice	1.56	3.95	3.69
Barren or sparsely vegetated	6.92	5.94	6

ests also saw a considerable rise from 2.87% in 2001 to 6.37% in 2021. This could be attributed to afforestation projects or natural expansion in boreal and temperate regions where these forests are prevalent. Conversely, evergreen broadleaf forests exhibited a slight fluctuation, peaking at 0.94% in 2011 before declining to 0.79% in 2021. Deciduous broadleaf forests saw a decrease from 0.25% in 2001 to 0.13% in 2021, indicating a reduction in areas where these forests are found, possibly due to deforestation or land-use changes for agriculture or urban development.

14.4 Shrublands

Closed and open shrublands have diminished to zero by 2011, remaining nonexistent through 2021. This dramatic reduction might be due to their conversion to other land types, such as croplands or urban areas, or possibly due to climate change affecting the semi-arid regions where these shrublands are typically found.

14.5 Wetlands and Croplands

Permanent wetlands have nearly disappeared, from 0.12% in 2001 to a mere 0.01% in 2021. This alarming trend underscores the vulnerability of wetland ecosystems to drainage for agriculture, urban development, or climate change impacts. Croplands have shown a decline from 4.01% in 2001 to 2.18% in 2021, suggesting a reduction in agricultural land possibly due to urban expansion, land degradation, or shifts toward other land uses. The cropland/natural vegetation mosaic category has almost vanished, from 0.61% in 2001 to 0.02% in 2021, indicating a move toward more homogeneous land uses.

14.6 Snow and Ice

Snow and ice cover increased from 1.56% in 2001 to 3.95% in 2011 before slightly decreasing to 3.69% in 2021. These changes likely reflect the dynamics of glacial and ice cap extents influenced by climatic variations and global warming trends.

14.7 Barren or Sparsely Vegetated

The area classified as barren or sparsely vegetated has remained relatively stable, fluctuating slightly from 6.92% in 2001 to 6.00% in 2021. This stability suggests that regions with minimal vegetation cover have not experienced substantial changes in land use or climatic conditions (Table 2).

Between 2001 and 2021, the data shows how different types of land cover have changed. There are units on the graphs that show how much space each type takes up, and the changes that have happened over the last 20 years are due to climate, the environment, and people. More information about these changes can be found in Fig. 3.

14.8 Types of Forests

Evergreen needleleaf forests: From 93 units in 2001 to 88 units in 2011, there was a little decrease in this kind of forest, which increased to 103 units in 2021. The subsequent rise indicates effective reforestation initiatives or natural regeneration processes, whereas the earlier fall may have been caused by insect outbreaks or deforestation.

Evergreen broadleaf forests: In 2001, this category had 21 units; in 2011, it had 37 units; in 2021, it had 36 units, a minor decline. While the modest drop may imply some amount of forest degradation or changes in land use, the rise suggests a recovery phase or good conservation activities.

Deciduous broadleaf forests: From 12 units in 2001 to 8 units in 2021, the area covered by these forests has gradually shrunk. This decrease is indicative of continued deforestation or the transfer of these forest areas to urban or agricultural uses.

Mixed forests: From 99 units in 2001 to 128 units in 2021, these forests have consistently increased. This tendency may be explained by efforts to reforest, by the regeneration of natural forests, or by better forest management techniques that promote biodiversity.

Table 2 Number of patches

Type	2001	2011	2021
Evergreen needleleaf forest	93	88	103
Evergreen broadleaf forest	21	37	36
Deciduous broadleaf forest	12	10	8
Mixed forests	99	103	128
Closed shrublands	2	0	0
Open shrublands	2	0	0
Woody savannas	117	36	52
Savannas	280	245	153
Grasslands	156	145	87
Permanent wetlands	11	0	1
Croplands	72	81	47
Cropland/natural vegetation mosaic	35	11	2
Snow and ice	14	13	9
Barren or sparsely vegetated	18	12	10

14.9 Shrublands

Closed and open shrublands: By 2011, both types had completely vanished, and they remained at zero units until 2021. This loss points to major changes in land use maybe as a result of urbanization, agriculture, or the effects of climate change that are causing these ecosystems to disappear.

14.10 Grasslands and Savannas

Woody savannas: From 117 units in 2001 to 36 units in 2011, the area covered by woody savannas severely declined, and then it slightly increased to 52 units in 2021. The minor comeback might suggest some recovery or improvements in land management methods, whereas the rapid fall could be the consequence of clearing land for agricultural or urban growth.

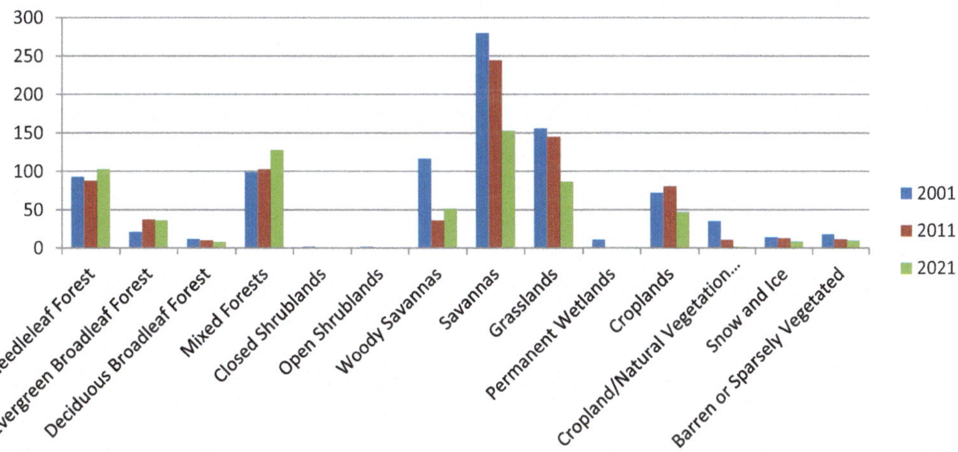

Fig. 3 Number of patches

Savannas: From 280 units in 2001 to 153 units in 2021, savannas had a major decline. This steady decrease points to either widespread land conversion for agriculture or other purposes, or adjustments to fire control techniques that have an impact on the ecosystems of the savanna.

Grasslands: From 156 units in 2001 to 87 units in 2021, the area of grasslands has continuously decreased. Large-scale agricultural development, urban sprawl, or other types of land-use change that diminish grassland ecosystems are probably the causes of this trend.

14.11 Wetlands and Croplands

Permanent wetlands: By 2011, they had all but disappeared, and by 2021, there was just one remaining unit. The susceptibility of wetlands to draining for agriculture, urbanization, or other activities is highlighted by this worrying trend.

Croplands: In 2001, there were 72 units; in 2011, there were 81 units; in 2021, there were 47 units. The early growth is consistent with the spread of agriculture, but the subsequent fall may be the result of land degradation, urbanization, or a change to other land uses. Cropland/natural vegetation mosaic: This category had a sharp drop from 35 units in 2001 to 2 units in 2021. This pattern suggests a shift towards more uniform land usage, which is probably the result of urbanization or more intensive agriculture methods.

14.12 Snow and Ice

Between 2001 and 2021, the area covered by snow and ice dropped from 14 units to 9 units. This decrease is probably the consequence of melting glaciers and ice caps brought on by climate change and global warming.

14.13 Barren or Sparsely Vegetated

The number of units in this category decreased from 18 in 2001 to 10 in 2021. The decline implies that certain regions could have undergone land use modifications or revegetation that has resulted in increased plant cover (Table 3).

Table 3 Patch density

Type	2001	2011	2021
Evergreen needleleaf forest	0.0417	0.0394	0.0461
Evergreen broadleaf forest	0.0094	0.0166	0.0161
Deciduous broadleaf forest	0.0054	0.0045	0.0036
Mixed forests	0.0443	0.0461	0.0573
Closed shrublands	0.0009	0	0
Open shrublands	0.0009	0	0
Woody savannas	0.0524	0.0161	0.0233
Savannas	0.1254	0.1097	0.0685
Grasslands	0.0699	0.0649	0.039
Permanent wetlands	0.0049	0	0.0004
Croplands	0.0322	0.0363	0.021
Cropland/natural vegetation mosaic	0.0157	0.0049	0.0009
Snow and ice	0.0063	0.0058	0.004
Barren or sparsely vegetated	0.0081	0.0054	0.0045

In 2001, 2011, and 2021, Fig. 4 show how much of each land cover type covered a certain area. Of the whole area that was examined, these partial numbers show what amount of each type of land cover there is. The environment and land use have changed a lot in the last 20 years, and these trends show how.

14.14 Types of Forests

Evergreen needleleaf forests: From 0.0417 in 2001 to 0.0394 in 2011, there was a minor decrease in the fractional coverage of evergreen needleleaf forests, which increased to 0.0461 in 2021. This pattern suggests that logging or land conversion caused the initial loss of forests, with reforestation initiatives or natural regrowth potentially driving the recovery in the following years.

Evergreen broadleaf forests: From 0.0094 in 2001 to 0.0166 in 2011, these forests saw a slight decline before finally reaching 0.0161 in 2021. While the minor recent reduction may be the result of localized deforestation or land use pres-

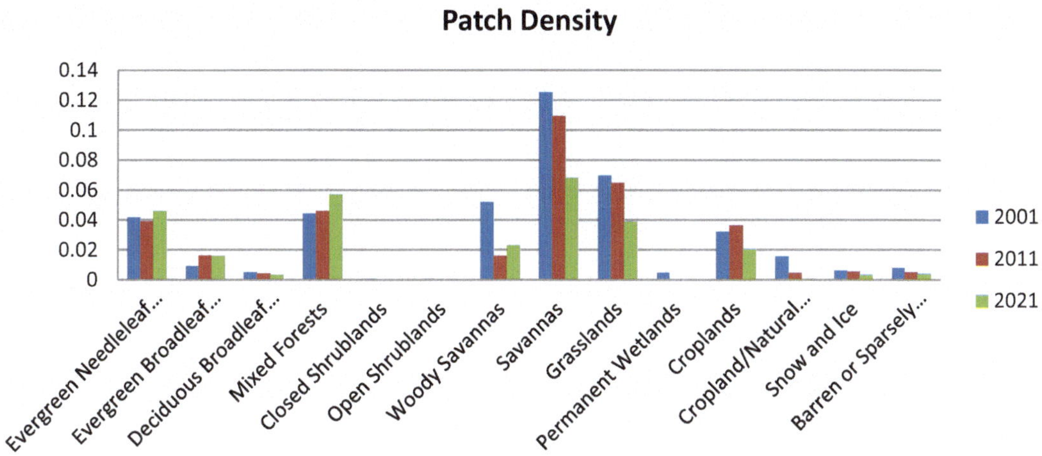

Fig. 4 Patch density

sures, the overall growth over the past 20 years implies that conservation and replanting activities have been successful.

Deciduous broadleaf forests: From 0.0054 in 2001 to 0.0036 in 2021, there has been a consistent decrease in this type of forest. The continuous decline suggests that deforestation is still occurring, most likely as a result of urbanization or agricultural growth that is shrinking the area of these woods.

Mixed forests: Between 2001 and 2021, the area covered by mixed forests increased significantly, from 0.0443 to 0.0573. The observed increase implies that mixed forest ecosystems are either spreading naturally or are undergoing intentional restoration, hence promoting ecological stability and biodiversity.

14.15 Shrublands

Closed and open shrublands: Both types were extant at 0.0009 in 2001, but by 2011, they had vanished, and as of 2021, they were nonexistent. The total disappearance of shrublands suggests that there have been major changes in land use, such as conversion to urban or agricultural regions, or that these ecosystems may have been diminished by climatic changes.

14.16 Grasslands and Savannas

Woody savannas: From 0.0524 in 2001 to 0.0161 in 2011, there was a sharp fall in woody savannas, followed by a modest recovery to 0.0233 in 2021. The minor comeback may reflect some degree of improved land management or natural regrowth, but the initial decline is probably the result of clearing land for development or agriculture.

Savannas: Between 2001 and 2021, savannas decreased dramatically, from 0.1254 to 0.0685. The steady decrease highlights how these ecosystems are altered by land conversion for grazing, agriculture, or urbanization.

Grasslands: Between 2001 and 2021, the coverage of grasslands dropped from 0.0699 to 0.0390. Similar reasons impacting savannas, like increased agricultural production and urbanization that reduces grassland ecosystems, are probably responsible for this negative trend.

14.17 Wetlands and Croplands

Permanent wetlands: By 2011, the area of permanent wetlands had all but vanished, and by 2021, it had only slightly recovered to 0.0004 of its former size. This decrease emphasizes how susceptible wetlands are to drainage for urban or agricultural uses, as well as the effects of climate change.

Croplands: Between 2001 and 2021, the amount of croplands increased from 0.0322 to 0.0363, then decreased to 0.0210. The first increase is consistent with the growth of agriculture, whereas the second decrease could be a sign of urbanization, deterioration of the land, or changes to other land uses.

This category decreased substantially from 0.0157 in 2001 to 0.0009 in 2021, indicating a near elimination of mixed land-use systems. The trend points toward increasing land-use homogenization, potentially resulting from intensified agriculture and urban expansion. The initial rise may correspond to agricultural development, whereas the later decline likely reflects urbanization or land-use conversion.

Cropland/natural vegetation mosaic: From 0.0157 in 2001 to 0.0009 in 2021, this category has all but disappeared. The decrease suggests a tendency towards more uniform land usage, most likely as a result of growing metropolitan areas or more intensive farming methods.

14.18 Snow and Ice

Between 2001 and 2021, the fractional coverage of snow and ice areas dropped from 0.0063 to 0.0040. This decrease is probably the result of melting glaciers and ice caps brought on by climate change and global warming.

14.19 Barren or Sparsely Vegetated

Between 2001 and 2021, the coverage of arid or poorly vegetated areas fell from 0.0081 to 0.0045. The decrease implies that certain regions have experienced land-use modifications or revegetation that promote a higher level of plant cover (Table 4 and Fig. 5).

Based on how much space they cover, the data shows how different types of land cover have changed over 20 years (2001, 2011, and 2021). By looking at these changes, we can learn a lot about how the world is changing and how people are affecting different communities. Here is a more in-depth look at the trends that have been seen.

Table 4 Largest patch index

Type	2001	2011	2021
Evergreen needleleaf forest	0.21	0.58	0.72
Evergreen broadleaf forest	0.23	0.25	0.21
Deciduous broadleaf forest	0.1	0.13	0.04
Mixed forests	1.37	0.82	1.07
Closed shrublands	0.01	0	0
Open shrublands	0.02	0	0
Woody savannas	19.97	41.01	37.36
Savannas	2.57	0.89	3.77
Grasslands	10.96	9.01	10.05
Permanent wetlands	0.02	0	0.01
Croplands	0.88	1.18	0.53
Cropland/natural vegetation mosaic	0.08	0.04	0.01
Snow and ice	0.64	2.69	1.99
Barren or sparsely vegetated	6.66	5.79	5.84

14.20 Forest Types

Evergreen needleleaf forests: There is a substantial increase in the coverage of evergreen needleleaf forests from 0.21 in 2001 to 0.58 in 2011 and further to 0.72 in 2021. This upward trend may be attributed to reforestation efforts, natural regrowth, or conservation policies aimed at protecting these forest types, which are prevalent in boreal and temperate regions.

Evergreen broadleaf forests: The coverage of evergreen broadleaf forests shows a slight increase from 0.23 in 2001 to 0.25 in 2011, followed by a decrease to 0.21 in 2021. The initial increase could be due to effective conservation measures or natural growth, while the subsequent decline might reflect deforestation or land conversion pressures.

Deciduous broadleaf forests: This category experiences a notable decline from 0.10 in 2001 to 0.04 in 2021. The continuous decrease indicates ongoing deforestation, possibly driven by agricultural expansion, urban development, or logging activities.

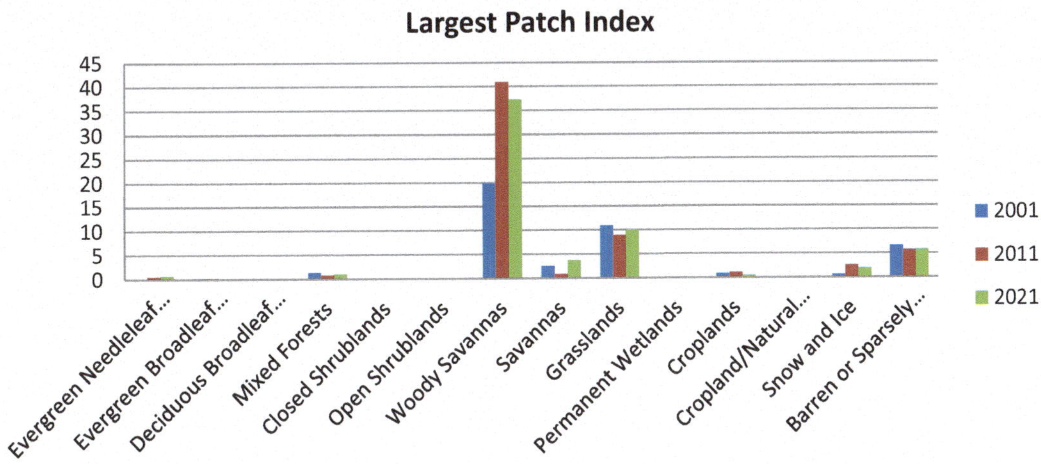

Fig. 5 Largest patch index

Mixed forests: Mixed forests saw an initial decline from 1.37 in 2001 to 0.82 in 2011, but then increased to 1.07 in 2021. The decline could be due to forest degradation or land conversion, while the subsequent rise suggests reforestation efforts or natural forest regeneration.

14.21 Shrublands

Closed and open shrublands: Both types of shrublands disappear completely by 2011 and remain absent in 2021, starting from 0.01 and 0.02, respectively, in 2001. This loss indicates significant land-use changes such as conversion to agriculture or urban areas, or climatic changes affecting these ecosystems.

14.22 Savannas and Grasslands

Woody savannas: The coverage of woody savannas shows a dramatic increase from 19.97 in 2001 to 41.01 in 2011, followed by a decrease to 37.36 in 2021. The initial surge might be due to changes in land management practices, such as reduced grazing pressure or fire suppression, while the slight decline indicates recent pressures on these landscapes. Savannas: There is a significant reduction from 2.57 in 2001 to 0.89 in 2011, followed by a substantial rebound to 3.77 in 2021. This fluctuation reflects dynamic land use changes, possibly driven by agricultural expansion and subsequent land restoration or conservation efforts.

Grasslands: The fractional coverage of grasslands declines from 10.96 in 2001 to 9.01 in 2011 but then increases slightly to 10.05 in 2021. The initial decline suggests conversion to agricultural lands or urbanization, while the slight recovery might indicate some successful restoration efforts or shifts in land use practices.

14.23 Wetlands and Croplands

Permanent wetlands: These ecosystems nearly disappear by 2011, dropping from 0.02 in 2001 to 0.00, with a minor recovery to 0.01 in 2021. The drastic decline underscores the vulnerability of wetlands to drainage for agriculture or development, while the minimal recovery indicates some efforts at wetland conservation.

Croplands: Croplands show an increase from 0.88 in 2001 to 1.18 in 2011, followed by a decrease to 0.53 in 2021. The initial rise corresponds to agricultural expansion, while the later decline could reflect urbanization, land degradation, or a shift toward other land uses.

Cropland/natural vegetation mosaic: This category declines significantly from 0.08 in 2001 to 0.01 in 2021. The steady decrease suggests a

Table 5 Total edge

Type	2001	2011	2021
Evergreen needleleaf forest	368,797	464,703	556,902
Evergreen broadleaf forest	66,254	105,172	99,612
Deciduous broadleaf forest	35,675	34,285	20,849
Mixed forests	674,120	612,036	816,358
Closed shrublands	3707	0	0
Open shrublands	3707	0	0
Woody savannas	2,337,877	2,257,724	2,412,007
Savannas	1,666,537	1,311,176	1,608,623
Grasslands	1,694,336	1,043,381	584,701
Permanent wetlands	22,239	0	1853
Croplands	395,669	478,602	223,317
Cropland/natural vegetation mosaic	94,516	28,262	3707
Snow and ice	102,855	166,793	148,260
Barren or sparsely vegetated	277,524	305,787	284,474

move towards more homogeneous land uses, likely due to intensified agricultural practices or urban development.

14.24 Snow and Ice

The coverage of snow and ice areas increases significantly from 0.64 in 2001 to 2.69 in 2011, then decreases to 1.99 in 2021. The initial increase could be due to variations in climatic conditions or better detection methods, while the subsequent decline is likely linked to global warming and melting of glaciers and ice caps.

14.25 Barren or Sparsely Vegetated

This category shows a decrease from 6.66 in 2001 to 5.79 in 2011, with a slight increase to 5.84 in 2021. The overall decline suggests that some barren areas have undergone revegetation or land-use changes that support more vegetation cover, while the recent slight increase could be due to land degradation or desertification.

14.26 Total Edge

Over a 20-year period from 2001 to 2021, the data shows significant changes in various land cover types. These shifts are influenced by numerous factors, including land use regulations, agricultural practices, natural biological processes, and climate change. Examining these changes sheds light on environmental trends and informs future conservation and land management strategies (Table 5 and Fig. 6).

14.27 Evergreen Needleleaf Forests

Evergreen needleleaf forests have shown a notable and steady increase over the past two decades. The area covered by these forests expanded from 368,797 hectares in 2001 to 556,902 hectares in 2021. This growth can be attributed to successful afforestation and reforestation initiatives, climate shifts favoring these forest types, and effective conservation measures protecting these regions from destruction. Natural forest regeneration in areas previously impacted by logging or agriculture likely also contributed to this rise.

14.28 Evergreen Broadleaf Forests

Evergreen broadleaf forests also saw significant expansion, increasing from 66,254 hectares in 2001 to 105,172 hectares in 2011, before slightly declining to 99,612 hectares in 2021. The initial growth is likely due to reforestation efforts and

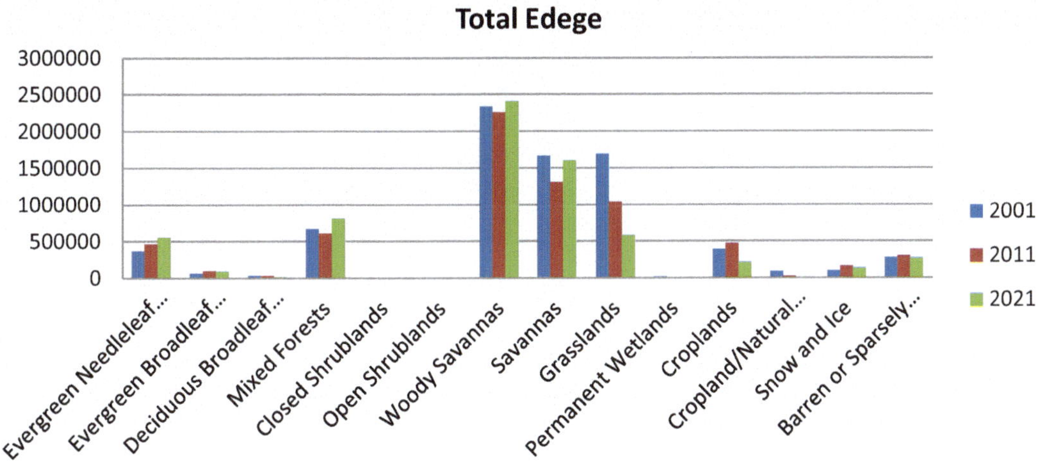

Fig. 6 Total edge

natural regeneration. However, the slight decrease between 2011 and 2021 may indicate pressures from urbanization, agricultural land conversion, and possibly the effects of climate change on forest health and growth patterns.

14.29 Deciduous Broadleaf Forests

The area of deciduous broadleaf forests has steadily decreased over the last 20 years, from 35,675 hectares in 2001 to 20,849 hectares in 2021. This decline suggests strong ecological pressures from urbanization, deforestation for agriculture, and unfavorable climatic conditions for these forest types. Considering their importance for biodiversity and ecosystem services, this decrease is concerning.

14.30 Mixed Forests

Mixed forests experienced a cyclical trend, with a decrease from 674,120 hectares in 2001 to 612,036 hectares in 2011, followed by a significant increase to 816,358 hectares in 2021. The initial decline may be due to land use changes and deforestation, while the subsequent rise could be attributed to successful reforestation efforts, natural forest succession, or environmental regulations favoring mixed forest ecosystems.

14.31 Shrublands

Both closed and open shrublands, covering 7414 hectares in 2001, had entirely vanished by 2011. This disappearance might result from conversion to agricultural land, urbanization, or biological succession where shrublands transition to forested areas. The loss of shrublands, which are crucial for many species, might indicate a broader ecological shift in the region.

14.32 Savannas and Woody Savannas

Woody savannas decreased from 2,337,877 hectares in 2001 to 2,257,724 hectares in 2011, then increased to 2,412,007 hectares in 2021. This pattern suggests resilience in these ecosystems, possibly due to natural regeneration or conservation efforts. In contrast, savannas saw a sharp decrease from 1,666,537 hectares in 2001 to 1,311,176 hectares in 2011, with a partial recovery to 1,608,623 hectares by 2021. The initial decline may be due to land use and climate changes, while the recovery could result from effective ecological management or the natural resilience of certain landscapes.

14.33 Grasslands

The area of grasslands saw a significant reduction from 1,694,336 hectares in 2001 to 584,701 hectares in 2021. This substantial decline points to widespread conversion of grasslands to agriculture, urban development, or overgrazing, further degrading these ecosystems. The loss of grasslands is especially concerning due to their role in biodiversity, carbon storage, and providing grazing space for livestock.

14.34 Permanent Wetlands

Permanent wetlands decreased from 22,239 hectares in 2001 to approximately 1853 hectares in 2021, almost disappearing. This substantial decrease likely results from large-scale land- use changes, such as urban growth or agricultural drainage, which adversely affect biodiversity, water quality, and flood management.

14.35 Croplands and Cropland/ Natural Vegetation Mosaic

Croplands saw a strong increase from 395,669 hectares in 2001 to 478,602 hectares in 2011, before declining to 223,317 hectares in 2021. This pattern may indicate changes in farming methods, such as shifts toward less land-intensive or more intensive agriculture. The sharp decline in the cropland/natural vegetation mosaic category, from 94,516 hectares in 2001 to just 3707 hectares in 2021, suggests a move toward more homogeneous land use and the disappearance of mixed-use landscapes.

14.36 Snow and Ice

The area covered by snow and ice increased significantly from 102,855 hectares in 2001 to 166,793 hectares in 2011, then slightly decreased to 148,260 hectares in 2021. These fluctuations could be due to periods of colder weather and increased snowfall, followed by a reduction in snowfall due to the impacts of global warming.

14.37 Barren or Sparsely Vegetated Areas

Sparsely vegetated or barren areas fluctuated slightly over time, increasing from 277,524 hectares in 2001 to 305,787 hectares in 2011, then decreasing to 284,474 hectares in 2021. These changes may result from climate variations, natural land degradation processes, and human activities like mining and deforestation (Table 6).

Over the two-decade span from 2001 to 2021, the data illustrates significant shifts in various land cover types, highlighting dynamic ecological and environmental transformations influenced by human activities, climatic changes, and natural processes. Each category of land cover reveals unique trends, reflecting broader environmental patterns and challenges (Fig. 7).

Table 6 Patch area

Type	2001	2011	2021
Evergreen needleleaf forest	69.01	124.4	138.17
Evergreen broadleaf forest	75.64	56.86	48.89
Deciduous broadleaf forest	46.51	62.25	37.57
Mixed forests	176.93	168.6	191.85
Closed shrublands	21.47	0	0
Open shrublands	32.2	0	0
Woody savannas	608.02	2658.79	1675.99
Savannas	122.51	104.26	259.41
Grasslands	398.5	259.07	333.58
Permanent wetlands	23.42	0	21.47
Croplands	124.32	134.89	103.68
Cropland/natural vegetation mosaic	38.64	37.08	21.47
Snow and ice	248.39	678.65	915.88
Barren or sparsely vegetated	858.64	1105.49	1339.47

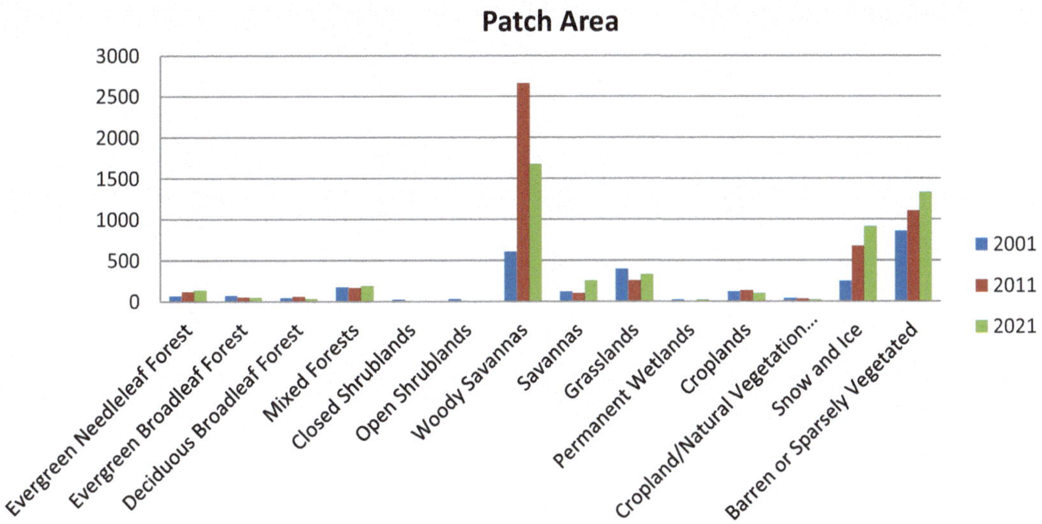

Fig. 7 Patch area graph

15 Discussions

The analysis of land cover assessment using MODIS data has provided significant insights into spatiotemporal changes over the study period, utilizing advanced tools and landscape metrics for an in-depth evaluation. The utility of MODIS, with its high revisit frequency and multi-resolution capabilities (250 m, 500 m, and 1000 m), proved instrumental in capturing decadal variations in land use and land cover (LULC) for the selected years of 2001, 2011, and 2021. The 36 spectral bands spanning visible to thermal infrared wavelengths enhanced the ability to monitor vegetation health, classify land cover types, and assess ecological changes. The integration of landscape metrics using Fragstats offered quantitative measures of landscape structure and fragmentation (Midha et al., 2010). For instance, the PLAND (Percentage of Landscape) metric highlighted the proportional dominance of specific land cover types, facilitating comparisons between years and identifying shifts in dominant land use classes. The increasing or decreasing trends in PLAND values provided a straightforward understanding of land cover transitions, such as urban expansion or deforestation. Furthermore, the Core Area Index (CAI) and Disjunct Core Area Density (DCAD) metrics highlighted changes in the integrity and connectivity of habitat patches. A decline in core areas over time, coupled with a rise in DCAD, suggested increasing landscape fragmentation, potentially isolating habitats and impacting ecosystem services (Reddy et al., 2013). This was especially evident in regions where natural vegetation gave way to urban or agricultural development. Lastly, the Euclidean Nearest Neighbor Distance (ENN) provided a spatial perspective on patch connectivity. An increase in ENN over time suggested that patches of similar land cover types were becoming more isolated, which could hinder ecological flows and processes such as species migration or nutrient cycling. In summary, the use of MODIS data in conjunction with landscape metrics provided a comprehensive understanding of land cover dynamics and structural changes. The observed trends reflect both natural processes and anthropogenic impacts, such as urban expansion, deforestation, and agricultural intensification. The study emphasizes the need for sustainable land management practices and conservation strategies to mitigate the adverse effects of landscape fragmentation and to promote ecological balance (Turner et al., 2015). Future research could incorporate higher-resolution data and multi-sensor approaches to further enhance the accuracy and applicability of LULC assessments.

16 Conclusions

This study successfully demonstrated the utility of MODIS satellite data in assessing long- term land cover changes using landscape metrics. By analyzing data spanning two decades (2001, 2011, and 2021), the research highlighted the increasing fragmentation of natural landscapes and the expansion of anthropogenic land uses. The findings emphasize the need for sustainable land management practices to mitigate the adverse ecological effects of land cover changes. The landscape metrics derived from Fragstats provided valuable insights into the spatial structure and composition of landscapes, offering a quantitative framework for monitoring and evaluating land use patterns. Metrics such as PLAND, NP, PD, and LPI served as effective indicators of landscape fragmentation and dominance, while CAI and TE highlighted the loss of core areas and edge effects, respectively. Future studies should focus on integrating high-resolution satellite data and additional landscape indices to further refine the understanding of LULC changes. Policymakers and planners can leverage these findings to prioritize conservation efforts, promote sustainable urban planning, and restore degraded ecosystems. This research underscores the importance of Earth observation data and landscape analysis tools in addressing global environmental challenges and supporting informed DECISION-making.

References

Batar, A. K., Watanabe, T., & Kumar, A. (2017). Assessment of Land-Use/Land-Cover Change and Forest Fragmentation in the Garhwal Himalayan Region of India. Environments, 4(2), 34. https://doi.org/10.3390/environments4020034

Cakir G, Sivrikaya F, Keleş S. (2008). Forest cover change and fragmentation using Landsat data in Maçka State Forest Enterprise in Turkey. Environ Monit Assess, 137(1–3):51–66. https://doi.org/10.1007/s10661-007-9728-9. Epub 2007 May 23. PMID: 17520337.

Dewan, A.M., Yamaguchi, Y. and Rahman, M.Z. (2012) Dynamics of Land Use/Cover Changes and the Analysis of Landscape Fragmentation in Dhaka Metropolitan. Bangladesh GeoJournal, 77, 315–330. https://doi.org/10.1007/s10708-010-9399-x

Foley, J. A., DeFries, R., Asner, G. P., Barford, C., Bonan, G., Carpenter, S. R., Chapin, F. S., Coe, M. T., Daily, G. C., Gibbs, H. K., Helkowski, J. H., Holloway, T., Howard, E. A., Kucharik, C. J., Monfreda, C., Patz, J. A., Prentice, I. C., Ramankutty, N., & Snyder, P. K. (2005). Global consequences of land use. Science, 309(5734), 570–574. https://doi.org/10.1126/science.1111772

Giriraj A, Murthy MS, Beierkuhnlein C. (2010). Evaluating forest fragmentation and its tree community composition in the tropical rain forest of Southern Western Ghats (India) from 1973 to 2004. Environ Monit Assess. 161(1–4):29–44. https://doi.org/10.1007/s10661-008-0724-5. Epub 2009 Jan 30. PMID: 19184490.

Jha, C. S., Goparaju, L., Tripathi, A., Gharai, B., Raghubanshi, A. S., & Singh, J. S. (2005). Forest fragmentation and its impact on species diversity: An analysis using remote sensing and GIS. Biodiversity and Conservation, 14, 1681–1698. https://doi.org/10.1007/s10531-004-0695-y

Lamine, S., Petropoulos, G. P., Singh, S. K., Szabó, S., Bachari, N. E. I., Srivastava, P. K., & Suman, S. (2017). Quantifying land use/land cover spatio-temporal landscape pattern dynamics from Hyperion using SVMs classifier and FRAGSTATS®. International Journal of Remote Sensing, 38(3), 862–878. https://doi.org/10.1080/10106049.2017.1307460

N. Lele, P.K. Joshi, S.P. Agrawal, Assessing forest fragmentation in northeastern region (NER) of India using landscape matrices, Ecological Indicators, Volume 8, Issue 5, 2008, Pages 657-663, ISSN 1470-160X, https://doi.org/10.1016/j.ecolind.2007.10.002

Liao J, Bearup D, Blasius B. (2017). Diverse responses of species to landscape fragmentation in a simple food chain. J Anim Ecol. 86(5):1169–1178. https://doi.org/10.1111/1365-2656.12702. Epub 2017 Jun 19. PMID: 28542896.

Midha, N., & Mathur, P. (2010). Assessment of forest fragmentation in the conservation priority Dudhwa landscape, India using FRAGSTATS computed class level metrics. *Journal of the Indian Society of Remote Sensing, 38*, 487–500.

Nagendra H, Paul S, Pareeth S, Dutt S. (2009). Landscapes of protection: forest change and fragmentation in Northern West Bengal, India. Environ Manage. 44(5):853–64. https://doi.org/10.1007/s00267-009-9374-9. Epub 2009 Sep 24. PMID: 19777293.

Padalia H, Ghosh S, Reddy CS, Nandy S, Singh S, Kumar AS. (2020). Assessment of historical forest cover loss and fragmentation in Asian elephant ranges in India. Environ Monit Assess. 191(Suppl 3):802. https://doi.org/10.1007/s10661-019-7696-5. PMID: 31989279.

Reddy, C. S., Sreelekshmi, S., Jha, C. S., & Dadhwal, V. K. (2013). National assessment of forest fragmentation in India: Landscape indices as measures of the effects of fragmentation and forest cover change. Ecological Engineering, 60, 453–464. https://doi.org/10.1016/j.ecoleng.2013.09.064

Sharma, M., Chakraborty, A., Garg, J. K., & Joshi, P. K. (2017). Assessing forest fragmentation in north-western Himalaya: A case study from Ranikhet forest range, Uttarakhand, India. Journal of Forestry Research, 28, 319–327. https://doi.org/10.1007/s11676-016-0311-5

Turner, M. G., & Gardner, R. H. (2015). Landscape ecology in theory and practice: Pattern and process (2nd ed.). Springer. https://doi.org/10.1007/978-1-4939-2794-4

Impact of Water Ecosystem on Soil Erosion: A Sub-Watershed Analysis in Shahzad River Watershed, Lalitpur, India, Using Geospatial Science

Tanzeel Khatoon, Furqan Ali Khan, Akram Javed, and Naushaba Naseem Ahmed

1 Introduction

Water ecosystems in the form of streams, rivers, lake estuaries, or oceans need to be conserved and managed to meet the demand and essentiality of sustainability. The key factor of watershed management is the sustainable utilization of natural resources without compromising the needs of future generations and negligibly affecting the environment (Bajirao et al., 2019). India where agriculture is a dominant activity; hence, it is important to understand the factors affecting land degradation (Kogo et al., 2020). Land degradation in the form of soil erosion is not only induced by anthropogenic activities rather; natural factors like elevation, slope, curvature, geology cropping pattern, and soil chemistry also play a crucial role in aggravating the erosional effect (Poesen, 2018). As reported by the UN (FAO & ITPS, 2015), nearly 20–30 Gt year^{-1} global soil erosion is caused by water.

Several concepts were developed, namely 'USLE: _Universal Soil Loss Equation', 'RUSLE: _Revised Universal Soil Loss Equation', 'MUSLE: _Modified Universal Soil Loss Equation', 'SYI: _Sediment Yield Index', and 'MA: _Morphometric Analysis' to quantify and predict soil erosion. The major drawback of the several models mentioned above requires parameters like vegetation, rainfall, and soil information as input to conduct the study, whereas the morphometric analysis approach requires a well-connected drainage pattern. Morphometry is a mathematical analysis to understand the dynamic nature of the watershed in terms of its surface expression, shape, and landform (Clark, 1966; Kaur et al., 2014; Khatoon & Javed, 2022). Moreover, it is an irreplaceable substitute for understanding the geo-hydrological properties of the water ecosystem even in the absence of a rain gauge (Romshoo et al., 2012; Sharma et al., 2022). Several approaches ranging from field observation to analytical modeling have been utilized to priori-

T. Khatoon (✉) · F. A. Khan
Interdisciplinary Department of Remote Sensing and GIS Applications, Aligarh Muslim University, Aligarh, Uttar Pradesh, India

A. Javed
Department of Geology, Aligarh Muslim University, Aligarh, Uttar Pradesh, India

N. N. Ahmed
Women's Collage, Geography Section, Aligarh Muslim University, Aligarh, Uttar Pradesh, India

Z. Khanam et al. (eds.), *Climate Change, Resilience and Ecosystem Services*, Sustainable Development Goals Series, https://doi.org/10.1007/978-3-031-96977-5_25

tize the watershed based on soil erosion. Several scholars using different approaches have conducted the study to prioritize the watershed for soil erosion and identify the area that possesses the maximum threat to the resource. Recent studies where morphometric parameters have been utilized to assess erosion are quoted as Altaf et al., 2014; Mushtaq et al., 2015; Sadhasivam et al., 2020; Bharath et al., 2021; Benzougagh et al., 2022; Rao et al., 2023; Ali et al., 2024.

In this study, the delineated area is latterly divided into 31 sub-watersheds using a geographic information system (GIS) platform to identify the most vulnerable zone for conservation and mitigation measures. The morphometric characteristics for each sub-watershed were estimated individually and subsequently prioritized for soil erosion by ranking the parameters using GIS techniques. Considering the entire watershed at once for conservation and mitigation is not helpful, rather identifying the hotspot of the problem (erosion in this case) is essential. The study's objective includes (1) calculation of the linear, arial, and relief parameters of each sub-watershed; (2) computation of compound parameters (Cp), priority ranking, and priority category; and (3) based on priority suggesting the mitigation plan to high priority area requires immediate attention.

2 Study Area

The research region is situated in the Lalitpur District of Uttar Pradesh, which includes the portion of Talebehat, Jakhaura, Bar, and Birdha blocks of the district. The watershed covers a total area of 1100 km^2 and stretches from 24° 25′ to 25° 03' North latitudes to 78° 17′ to 78° 39′ East longitude (Fig. 1). The slope of the area varies from 0° to 48°, while the highest and lowest elevations were 280 and 495 m amsl, respectively. Figure 2 displays a three-dimensional (3D) model of the watershed, derived using Global Mapper 13.

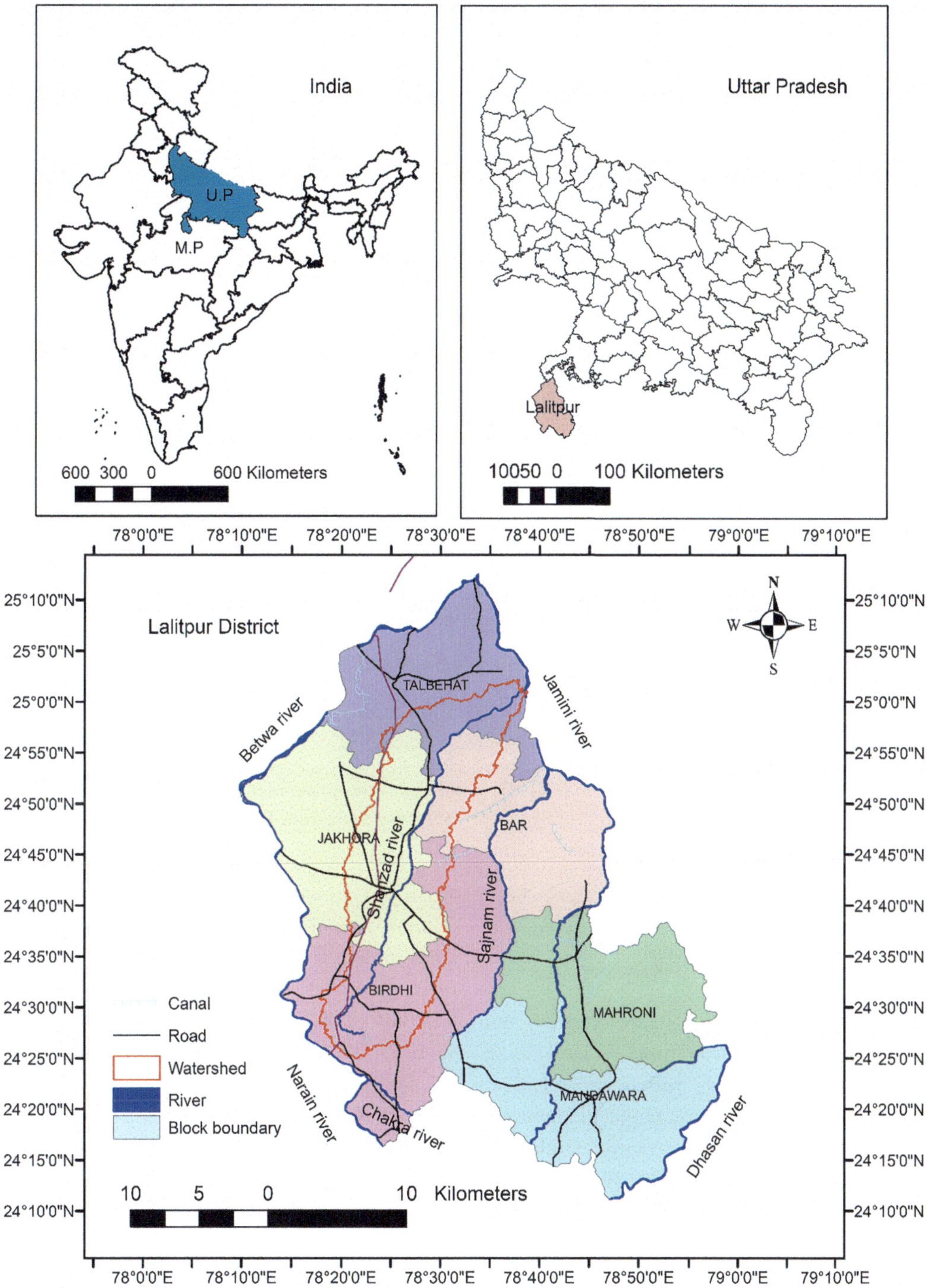

Fig. 1 Study map of the area

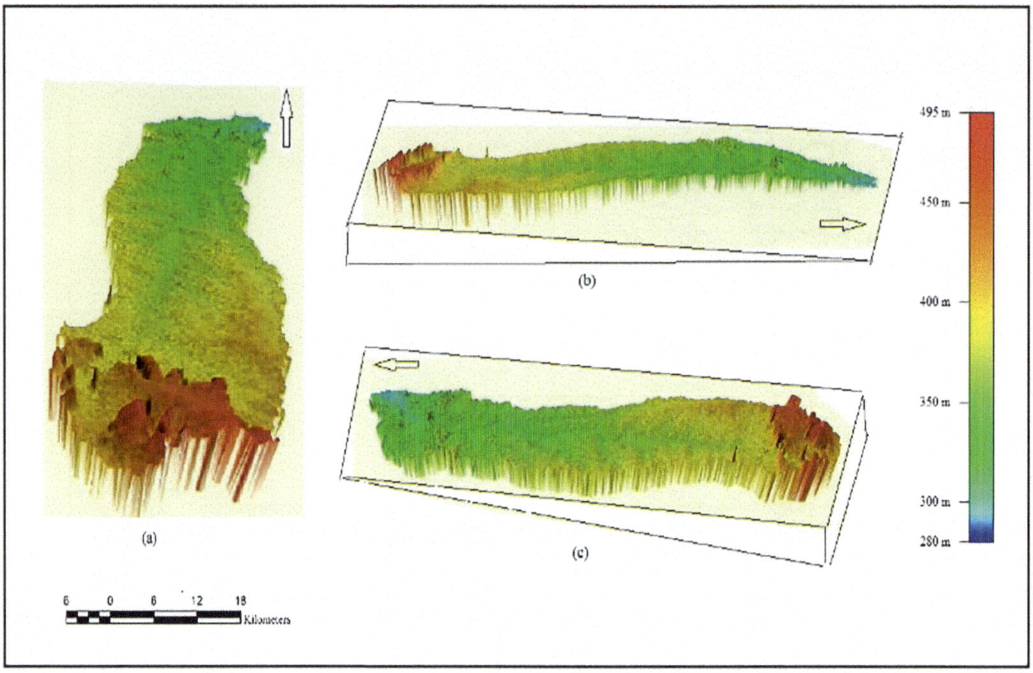

Fig. 2 (**a**) 3D map of the watershed with its (**b**) left and (**c**) right profile

Table 1 Data sources and its associated information

Sl. no.	Data	Source	Resolution/Scale
1	Shuttle radar topography Mission (SRTM)	https://earthexplorer.usgs.gov	30 m
2	Topographic map	Survey of India, New Delhi	1:50000

3 Data Sources and Methodology

The topographic maps were georeferenced, and projected in UTM considering WGS84 as a datum. The toposheets were clipped and mosaicked to prepare the base map and manually digitize the drainage network. Various module in the GIS environment was utilized to process the SRTM DEM tile to delineate the watershed and sub-watershed boundaries. Table 1 shows the utilized data and its associated information.

Strahler's method is adopted in stream order; moreover, characteristics like area (A), perimeter (P), and stream length (Lu) were computed in a GIS environment. The values of maximum watershed relief (Hmax) and minimum watershed relief (Hmin) were obtained from DEM after filling the sinks in the DEM. Aside from the basic morphometric parameters, each sub-watershed linear, areal, and relief morphometric parameter was estimated using the standard formulas listed in Table 2.

The initial ranking was allocated to different morphometric parameters in such a manner that all those factors having in linear relation with soil erosion susceptibility were given higher ranks e.g., the greatest value was denoted as one, the next value as two, and so forth. Whereas those parameters showing inverse relationships were assigned lower ranks for high parameter values i.e., the smallest value was ranked one, the next value as two, and so forth. Cp at the sub-watershed level was estimated by averaging the rank of individual parameters.

Table 2 Standard formulae utilized to calculate the morphometric parameters

Parameters and Aspects	Formulae/methods	Units	References
Area (A)	GIS software analysis	km^2	
Perimeter (P)	GIS software analysis	km	
Maximum elevation (H)	GIS software analysis	m	
Minimum elevation (h)	GIS software analysis	m	
Length (L_b)	maximum basin length	m	Nookaratnam et al. (2005)
Stream order (U)	Hierarchical rank	Dimensionless	Strahler (1964)
Stream number (N_u)	$N_{u1} + N_{u2} + \cdots + N_{un}$	Dimensionless	Horton (1945)
Stream length (L_u)	$L_{u1} + L_{u2} + \cdots + L_{un}$	km	Horton (1945)
Mean stream length (L_{sm})	Average stream length of all orders	km	Horton (1945)
Bifurcation ratio (R_b)	N_u/N_{u+1}	Dimensionless	Schumm (1956)
Stream length ratio (Rl)	L_u/L_{u-1}	Dimensionless	Horton (1945)
Mean bifurcation ratio (R_{bm})	Average of bifurcation ratios of all orders	Dimensionless	Schumm (1956)
Mean stream length ratio (R_{lm})	Average of stream length ratios of all orders	Dimensionless	
Stream frequency (F_s)	N_u/A	km^{-2}	Schumm (1956)
Drainage density (D_d)	L_u/A	km/km^2	Schumm (1956)
Drainage texture (D_t)	N_u/P	km^{-1}	Schumm (1956)
Length of overland flow (L_o)	$1/(2D_d)$	km	Horton (1945)
Drainage intensity (D_i)	F_s/D_d	km^{-1}	Faniran (1968)
Infiltration number (I_f)	$F_s \times D_d$	km^{-3}	Faniran (1968)
Circulatory ratio (Rc)	$12.57\ A/P^2$	Dimensionless	Miller (1953)
Elongation ratio (Re)	1.128 A^0.5	Dimensionless	Schumm (1956)
Form factor (Ef)	A/Lb^2	Dimensionless	Horton (1945)
Lemniscates ratio (K)	$Lb^2/4A$	Dimensionless	Chorley (1957)
Compactness coefficient (Cc)	$P/2(\pi A)\ 0.5$	Dimensionless	Horton (1945)
Constant of Channel maintenance (C)	1/Dd	km	Schumm(1956)
Relief (R)	H–h	km	Strahler (1952)
Relief ratio (Rh)	Bh/Lb	Dimensionless	Schumm (1956)
Relative relief (Rh)	$H \times 100/P$	Dimensionless	Melton (1957)
Ruggedness number (Rn)	$Bh \times Dd/1000$	Dimensionless	Strahler (1954)

Thus, based on the compound parameter, priority rank and priority category in the Shahzad watershed were executed.

4 Result and Discussion

The current investigation was conducted in the following steps:

- Delineate the watershed and sub-watershed boundary and update the stream pattern using SRTM DEM data having a resolution of 30 meters.
- Computation of morphometric parameters, namely linear, areal, and relief aspects.
- Categorization of sub-watersheds for identifying erosion-prone areas based on each sub-watershed's calculated compound values (Cp).

4.1 Morphometric Analysis Area (a)

A larger area intercepts high rainfall, leading to peak discharge and high runoff. The small area is occasionally associated with the highest levels of floods and sedimentation. The morphometric properties such as stream networks, shape, length, and relief parameters may contribute to this phenomenon. In the study, the smallest watershed (SW28) occupies an area of 4.06 km^2, whereas the largest (SW17) occupies 121.40 km^2.

4.2 Perimeter (P)

The peripheral length of the delineated watershed is termed as perimeter. It varies from 14.43 to 128.74 km, with SW28 being the shortest and SW17 being the longest.

4.3 Watershed Length (Lb)

The distance between the most remote point and the watershed outlet is termed watershed length. In this study, SW17 (41.82 km) is the longest sub-watershed, while SW3 (3.0 km) is the shortest.

4.4 Relief (R)

Relief in the area is described as the elevation difference between the highest and the lowest elevation value in the catchment‖. In this study, SW31 has a maximum relief of 129 m, and SW25 has a minimum relief of 26 m.

The above basic parameters of the watershed were calculated directly in the GIS environment (Table 3).

4.5 Stream Order (U) Stream Number (nu) and Stream Length (Lu)

According to Strahler (1964), the smallest finger type, as well as any unbranched tributaries, is termed first stream order. Two first-stream orders are combined to generate a second-stream order. Following that, the second-stream order combines the third, and so on‖. The Shahzad watershed is divided of 31 sub-watersheds, where 6th order is the highest order stream and lies in SW1, SW3, SW5, SW9, SW12, SW14, SW16, SW18, SW21, SW22, and SW26; 5th order for SW11, SW25, and SW26; 4th order for SW2, SW4, SW6, SW8, SW10, SW13, SW15, SW17, SW19, SW20, SW21, SW23, SW24, SW27, and SW31; 3rd order streams are present in all sub-watersheds except SW3, SW5, and SW25, while streams of 2nd and 1st order are present in all sub-watersheds.

The total number of streams of a particular order‖ is defined as the stream number (Nu) in the catchment area (Horton 1945). The number of streams is highest in SW17 (198) and the lowest in SW28 (11). Figure 3 shows the relationship between Nu and u of the sub-watersheds.

In general, first-order streams have the longest total length of stream segments, which reduces as stream order increases. However, in some sub-watersheds, the number of streams of different

Table 3 Basic morphometric characteristics of the watershed

SW	Area(A) km^2	Perimeter (P) km	Max elevation (H) m	Min elevation h m	Relief (R) m	Basin length (Lb) km
SW1	53.12	51.12	353	282	71	17.22
SW2	19.68	27.89	353	294	59	7.75
SW3	12.45	31.25	327	292	35	5.31
SW4	76.38	59.36	383	311	72	18.36
SW5	13.27	44.51	372	296	76	9.98
SW6	10.99	19.86	372	300	72	6.08
SW7	36.35	43.61	376	293	83	12.33
SW8	12.96	30.37	372	309	63	8.84
SW9	26.46	49.68	374	312	62	10.90
SW10	22.48	32.67	391	331	60	7.53
SW11	20.19	31.58	364	316	48	10.20
SW12	13.26	30.41	355	315	40	9.01
SW13	19.04	33.80	356	317	39	9.09
SW14	54.66	56.94	371	312	59	12.34
SW15	38.84	51.04	385	331	54	14.34
SW16	37.94	54.77	377	324	53	11.65
SW17	121.4	128.74	414	328	86	41.82
SW18	37.63	52.58	380	331	49	11.88
SW19	52.29	47.70	393	342	61	13.74
SW20	37.37	41.76	393	342	51	12.44
SW21	75.26	65.02	397	342	55	13.46
SW22	34.32	42.67	394	349	45	10.22
SW23	22.55	40.27	401	354	47	3.09
SW24	12.64	25.40	416	366	50	7.33
SW25	5.41	16.08	386	360	26	3.00
SW26	50.03	67.00	462	355	107	17.73
SW27	20.65	29.62	414	365	49	8.87
SW28	4.06	14.43	415	382	33	4.39
SW29	35.25	43.95	481	381	100	13.51
SW30	11.71	23.47	465	375	90	5.53
SW31	111.36	74.52	495	366	129	30.18

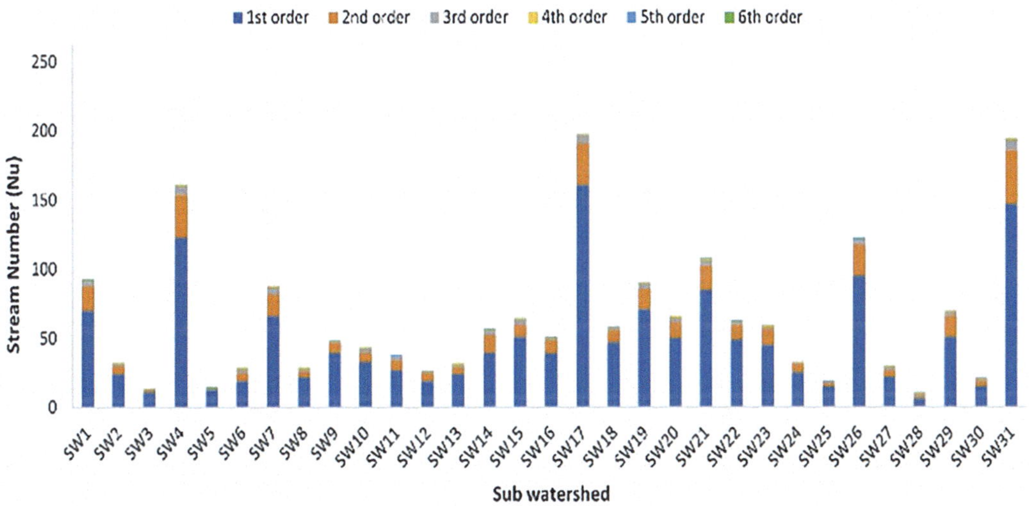

Fig. 3 Relationship between Nu and u of 31 sub-watersheds

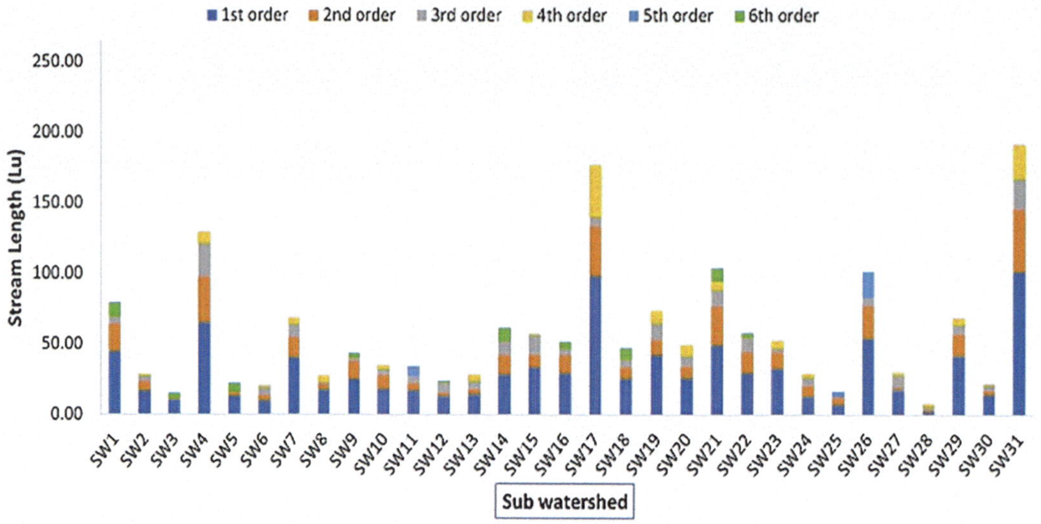

Fig. 4 Relationship between Lu and u of 31 sub-watersheds

orders deviates from what is obvious. In the current research, the lengths of the longest and smallest of the streams are SW31 (192.42 km) and SW27 (12.16 km), respectively. Figure 4 shows the relationship between Lu and u of the sub-watersheds.

4.6 Mean Bifurcation Ratio (Rbm)

Schumm (1956) defined Rbm as the average bifurcation ratio of all orders where the ratio of the number of the stream segments of a given order to the number of segments of the next higher orderl. Here, Rbm varies from 2.00 to 13.00. In this study, SW5 shows the highest value, and SW28 depicts the lowest Rb. Higher values of the bifurcation ratio imply that drainage patterns are controlled by the geological structure significantly, whereas lower values show less structural disturbance in sub-watersheds (Table 4).

4.7 Stream Frequency (Fs)

Horton (1932) introduced Fs as the total number of stream segments of all orders per unit areal. Here Fs exhibits a direct correlation with the drainage density except in SW3, SW5, SW11, SW14, SW16, SW25, SW28, and SW30 sub-watersheds. The study exhibits the highest Fs is at SW25 (3.512) and the lowest Fs is at SW14 (1.042).

4.8 Drainage Density (Dd)

Horton (1932) addressed the concept of Dd in American hydrologic literature, describing it as a measure of channel spacing. According to Nag (1998), areas with low Dd typically have resistant/permeable subsoil, thick green cover, and gentle relief. Conversely, high Dd is associated with weak/impermeable underlying materials, scanty vegetation, and hilly topography. Lower

Table 4 Sub-watershed analysis of linear aspect of Shahzad watershed

SW	Stream Order						Total no. of Streams (Nu)	Stream Length in Km (Lu)						Mean Stream Length(Lsm)						Stream Length Ratio(RL)					Bifurcation Ratio(Rb)					Mean Bifurcation Ratio (Rbm)	Total Stream Length (Lu)
	I	II	III	IV	V	VI		I	II	III	IV	V	VI	I	II	III	IV	V	VI	II/I	III/II	IV/III	V/IV	VI/V	I/II	II/III	III/IV	IV/V	V/VI		
SW1	70	18	4	0	0	1	93	44.71	19.8	4.99	0	0	10.3	0.63	1.1	1.24	0	0	10.3	0.44	0.25	0	0	0	3.88	4.5	0	0	0	4.190	79.81
SW2	24	6	2	1	0	0	33	16.81	6.67	3.85	1.56	0	0	0.7	1.11	1.92	1.56	0	0	0.39	0.57	0.4	0	0	4	3	2	0	0	3.000	28.9
SW3	11	2	0	0	0	1	14	10.7	0.53	0	0	0	4.23	0.97	0.26	0	0	0	4.23	0.04	0	0	0	0	5.5	0	0	0	0	5.500	15.48
SW4	123	31	7	1	0	0	162	65.25	32.74	23.38	8.11	0	0	0.53	1.05	3.34	8.11	0	0	0.5	0.71	0.34	0	0	3.96	4.42	7	0	0	5.127	129.49
SW5	13	1	0	0	0	1	15	13.29	2.6	0	0	0	6.72	1.02	2.6	0	0	0	6.72	0.19	0	0	0	0	13	0	0	0	0	13.000	22.63
SW6	19	6	3	1	0	0	29	9.96	3.98	5.55	1.38	0	0	0.52	0.66	1.85	1.38	0	0	0.39	1.39	0.24	0	0	3.16	2	3	0	0	2.720	20.88
SW7	66	16	5	1	0	0	88	40.29	15.28	8.69	4.47	0	0	0.61	0.95	1.73	4.47	0	0	0.37	0.56	0.51	0	0	4.12	3.2	5	0	0	4.107	68.75
SW8	22	4	2	1	0	0	29	17.64	4.2	1.46	4.57	0	0	0.8	1.05	0.73	4.54	0	0	0.23	0.34	3.1	0	0	5.5	2	2	0	0	3.167	27.84
SW9	40	7	1	0	0	1	49	25.21	12.82	1.88	0	0	3.52	0.63	1.83	1.88	0	0	3.52	0.5	0.14	0	0	0	5.71	7	0	0	0	6.355	43.45
SW10	33	7	3	1	0	0	44	18.39	9.61	4.27	2.93	0	0	0.55	1.37	1.42	2.93	0	0	0.52	0.44	0.68	0	0	4.71	2.33	3	0	0	3.347	35.23
SW11	27	7	3	0	1	0	38	17.17	5.37	4.64	0	7.35	0	0.63	0.76	1.54	0	7.35	0	0.31	0.86	0	0	0	3.85	2.33	0	0	0	3.090	34.55
SW12	19	6	1	0	0	1	27	13.19	2.66	6.6	0	0	1.49	0.69	0.44	6.6	0	0	1.49	0.2	2.48	0	0	0	3.16	6	0	0	0	4.580	23.95
SW13	24	5	2	1	0	0	32	14.33	3.44	6.12	4.63	0	0	0.59	0.68	3.06	4.63	0	0	0.24	1.77	0.75	0	0	4.8	2.5	2	0	0	3.100	28.54
SW14	40	13	3	0	0	1	57	28.47	14	9.87	0	0	9.5	0.71	1.07	3.29	0	0	9.5	0.49	0.7	0	0	0	3.07	4.33	0	0	0	3.700	61.85
SW15	51	9	4	1	0	0	65	33.52	9.38	14.07	0.89	0	0	0.65	1.04	3.51	0.89	0	0	0.27	1.5	0.06	0	0	5.66	2.25	4	0	0	3.970	57.87
SW16	39	10	1	0	0	1	51	29.17	13.21	4.32	0	0	5.16	0.74	1.32	4.32	0	0	5.16	0.45	0.32	0	0	0	3.9	10	0	0	0	6.950	51.87
SW17	161	30	6	1	0	0	198	98.39	35.28	6.53	37.19	0	0	0.61	1.17	1.08	37.19	0	0	0.35	0.18	5.69	0	0	5.36	5	6	0	0	5.453	177.4
SW18	47	9	2	0	0	1	59	25.81	8.27	5.1	0	0	8.13	0.54	0.91	2.55	0	0	8.13	0.32	0.61	0	0	0	5.22	4.5	0	0	0	4.860	47.32
SW19	71	15	4	1	0	0	91	42.16	11.65	11.57	9.14	0	0	0.59	0.77	2.89	9.14	0	0	0.27	0.99	0.78	0	0	4.73	3.75	4	0	0	4.160	74.53
SW20	50	12	3	1	0	0	66	25.99	8.69	7.09	8.06	0	0	0.51	0.72	2.36	8.06	0	0	0.33	0.81	1.13	0	0	4.16	4	3	0	0	3.720	49.84
SW21	85	18	3	1	0	1	108	49.55	27.71	11.63	5.49	0	9.78	0.58	1.53	3.87	5.49	0	9.78	0.55	0.41	0.47	0	0	4.72	6	3	0	0	4.573	104.18
SW22	49	11	2	0	0	1	63	29.51	15.26	10.25	0	0	3.31	0.6	1.38	5.12	0	0	3.31	0.51	0.67	0	0	0	4.45	5.5	0	0	0	4.975	58.34
SW23	45	12	2	1	0	0	60	32.76	11.97	2.65	5.49	0	0	0.72	0.99	1.32	5.49	0	0	0.36	0.22	2.07	0	0	3.75	6	2	0	0	3.917	52.88
SW24	25	6	1	1	0	0	33	12.96	7.92	5.94	2.3	0	0	0.51	1.32	5.94	2.3	0	0	0.61	0.75	0.41	0	0	4.16	6	1	0	0	3.720	29.13
SW25	15	3	0	0	1	0	19	7.24	5.6	0	0	3.71	0	0.48	1.86	0	0	3.71	0	0.77	0	0	0	0	5	0	0	0	0	5.000	34.47
SW26	95	23	3	0	1	1	123	54.45	23.55	5.56	0	17.9	0.35	0.57	1.02	1.85	0	17.9	0.35	0.43	0.23	0	0	0	4.13	7.66	0	0	0	5.895	101.83
SW27	22	5	2	1	0	0	30	17.56	2.15	8.59	1.75	0	0	0.79	0.43	4.29	1.75	0	0	0.12	3.99	0.2	0	0	4.4	2.5	2	0	0	2.967	12.16
SW28	6	2	2	1	0	0	11	2.34	1.91	2.12	1.65	0	0	0.39	0.95	1.06	1.65	0	0	0.81	1.1	0.77	0	0	3	1	2	0	0	2.000	25.94
SW29	51	15	3	1	0	0	70	41.9	15.6	6.78	5.12	0	0	0.82	1.04	2.26	5.12	0	0	0.37	0.43	0.75	0	0	3.4	5	3	0	0	3.800	69.41
SW30	15	4	2	1	0	0	22	14.4	3.04	3.9	1.06	0	0	0.96	0.76	1.95	1.06	0	0	0.21	1.28	0.27	0	0	3.75	2	2	0	0	2.583	22.41
SW31	147	39	8	1	0	0	195	101.62	44.54	21.89	24.35	0	0	0.69	1.14	2.73	24.35	0	0	0.43	0.49	1.11	0	0	3.76	4.87	8	0	0	5.543	192.42

drainage density out-turns into a coarse drainage texture, whereas fine drainage texture is a result of higher drainage density. SW25 (6.372) and SW28 (6.389) exhibit high drainage density, whereas all other areas display low drainage density, ranging from 0.589 to 2.345.

4.9 Drainage Texture (Dt)

Horton (1945), described Dt as the total number of stream segments of all orders per perimeter of that areal. The value of Dt <2 suggests very coarse, between 2 and 4 indicates coarse, 4 and 6 intermediate, 6 and8 fine, and > 8 indicates very fine drainage texture. In this scenario, the drainage density ranges from very coarse to fine texture.

4.10 Length of the Overland Flow (Lo)

Horton (1945) defines Lo as the length of water over the ground before it gets concentrated into definite stream channelsl. The Lo value in the watershed varies from 0.294 to 3.194.

4.11 Drainage Intensity (Di)

Faniran (1968), defined Di as the ratio of stream frequency to drainage densityl. In this study, SW27 has the highest drainage density (2.467), and SW28 shows the lowest value.

4.12 Infiltration Number (If)

Faniran (1968) demonstrated it as a product of drainage density and stream frequencyl. The If reflects the water transmission potential of a watershed. The If ranges from 0.855 (SW27) to 22.376 (SW25) in the watershed.

4.13 Circulatory Ratio (Rc)

The circularity ratio (Rc) is impacted by Lu and Fs, geology, LULC, climate, and basin geomorphology. The value of Rc varies from 0.084 to 0.35. The sub-watersheds having values <0.50 suggesting that they are elongated in nature.

4.14 Elongation Ratio (Re)

Schumm (1956) defined Re as the ratio between the diameter of the circle of the same area as the drainage basin and the maximum length of the basinl (Table 4). The watershed can be categorized into circular, oval, and less elongated on the basis of elongation ratio; the value of Re varies from 0.297 to 1.733 in the investigation region.

4.15 Form Factor (Ff)

Horton (1932), defined Ff as the ratio of basin area to square of the basin lengthl. It has been observed in the study that the Ff extends between 0.069 (SW17) and 2.361 (SW23). Higher values indicate a circular shape, while lower values indicate an elongated shape of the sub-watersheds.

4.16 Lemniscate Ratio (K)

Chorley (1957) utilized K to investigate the slope of the watershed. K values for the 31 sub- watersheds ranged from 0.105 to 3.601. The erosional risk is higher in the areas having higher K values.

4.17 Compactness Coefficient (Cc)

The Cc is defined as the proportion of the catchment's perimeter to the circumference of an equivalent circular areal. The Cc value in the region ranges from 1.689 to 3.446. In this study, SW5 has the highest compactness coefficient and SW6 has the lowest compactness coefficient value.

4.18 Relief Ratio (Rh) and Relative Relief (Rr)

The ratio of maximum relief to the horizontal distance along the longest dimension of the basin parallel to the principal drainage linel is termed as relief ratio (Schumm, 1956). Values of Rh range from 0.001 to 0.016 in the watershed. The formula quoted by Melton, 1957 is used to calculate relative relief (Rr) where the highest point and the perimeter of the watershed are used as input. It reflects the variation in height in the area. Rr varies from 0.322 to 2.876, SW28 shows the highest value, whereas SW17 shows the lowest relative relief in the watershed.

4.19 Ruggedness Number (Rn)

The surface undulation and roughness are determined by the ruggedness number (Strahler, 1954). It is one of the factors used to determine a basin's vulnerability to soil erosion. The investigation has determined that SW31 (0.222) is the higher value and SW27 (0.028) is the lower value.

5 Prioritization of Sub-Watershed

Morphometric parameters are an important input in mapping and estimating the erosion risk in the area. Parameters such as F_s, D_d, D_t, L_o, and R_{bm} are directly related to erodibility. In contrast, parameters like Rc, Re, Cc, Ff, and basin shape have a negative relationship with erosion (Nookaratnam et al., 2005). In this study, 16 different morphometric parameters were taken into account to calculate the compound parameter (Cp). Each of these parameters was assigned a rank. These ranks were then averaged separately to obtain the Cp value for individual sub-watersheds. Table 5 shows the value of each parameter and the number in the parenthesis indicate the assigned rank of each sub-watershed. Based on the rank, the compound parameters are calculated and the map (Fig. 5) is generated in the GIS platform to show the spatial distribution of the priority classes. The compound values range from 9.125 to 22.00 (Table 5) in the sub-watersheds which were later on classified into five intervals, namely 9.125–11.700, 11.701–14.275, 14.276–16.850, 16.851–19.425, and 19.426–22.000. Priority ranks of sub-watersheds were allocated on the basis of Cp values. Sub- watersheds having lower Cp values were given higher priority ranks, indicating a greater need for attention, whereas higher Cp values were given lower priority ranks. Consequently, the area was classed into five group: very high, high, medium, low, and very low (Fig. 5). This classification helps in identifying areas that require different levels of management and intervention.

Table 5 Priority ranks and priority category of sub-watersheds based on compound parameters

SW	Rbm	Fs	Dd	Dt	Lo	Di	If	R	Rh	Rr	Rn	Ff	Rc	Re	Cc	K	Compound Parameter(Cp)	Priority Ranking	Priority Category
SW1	4.19 [14]	1.75 [19]	1.502 [19]	1.819 [6]	0.751 [19]	1.165 [10]	2.63 [17]	0.071 [9]	0.004 [9]	0.691 [24]	0.106 [15]	0.179 [7]	0.255 [22]	0.477 [7]	1.978 [10]	1.395 [25]	14.500	12th	Medium
SW2	3 [26]	1.676 [22]	1.468 [22]	1.183 [18]	0.734 [22]	1.141 [11]	2.462 [21]	0.059 [14]	0.007 [6]	1.266 [7]	0.086 [18]	0.327 [23]	0.318 [30]	0.645 [23]	1.773 [2]	0.762 [9]	17.125	21st	Low
SW3	5.5 [6]	1.124 [30]	1.243 [29]	0.448 [30]	0.621 [29]	0.904 [27]	1.398 [28]	0.035 [26]	0.007 [6]	1.046 [14]	0.043 [27]	0.441 [29]	0.16 [6]	0.749 [29]	2.498 [26]	0.566 [3]	19.750	26th	Very Low
SW4	5.127 [8]	2.12 [9]	1.695 [16]	2.729 [1]	0.847 [16]	1.251 [5]	3.595 [11]	0.072 [8]	0.004 [9]	0.645 [29]	0.122 [12]	0.226 [13]	0.272 [27]	0.536 [13]	1.916 [5]	1.103 [19]	12.562	8th	High
SW5	13 [1]	1.13 [29]	1.705 [14]	0.337 [31]	0.852 [14]	0.666 [28]	1.927 [26]	0.076 [7]	0.008 [5]	0.836 [19]	0.129 [10]	0.133 [3]	0.084 [1]	0.411 [3]	3.446 [31]	1.876 [29]	15.687	17th	Medium
SW6	2.72 [28]	2.638 [4]	1.99 [7]	1.46 [13]	0.949 [9]	1.388 [2]	5.013 [5]	0.072 [8]	0.012 [3]	1.873 [4]	0.136 [8]	0.297 [22]	0.35 [31]	0.615 [22]	1.689 [1]	0.84 [10]	11.062	3rd	Very High
SW7	4.107 [16]	2.42 [7]	1.891 [10]	2.017 [3]	0.945 [10]	1.28 [4]	4.578 [8]	0.083 [6]	0.007 [6]	0.862 [18]	0.158 [7]	0.239 [16]	0.24 [17]	0.551 [16]	2.04 [15]	1.045 [16]	10.937	2nd	Very High
SW8	3.167 [23]	2.237 [8]	2.148 [5]	0.954 [24]	1.074 [5]	1.041 [21]	4.806 [7]	0.063 [10]	0.001 [11]	1.225 [8]	0.135 [9]	0.165 [6]	0.176 [9]	0.459 [6]	2.379 [23]	1.507 [26]	12.562	8th	High
SW9	6.355 [3]	1.851 [15]	1.642 [17]	0.986 [23]	0.821 [17]	1.127 [15]	3.04 [15]	0.062 [11]	0.006 [7]	0.753 [22]	0.101 [16]	0.222 [12]	0.134 [3]	0.532 [12]	2.724 [29]	1.122 [20]	14.812	13th	Medium
SW10	3.347 [22]	1.957 [12]	1.567 [18]	1.346 [14]	0.783 [18]	1.248 [6]	3.067 [14]	0.06 [13]	0.008 [5]	1.197 [9]	0.094 [17]	0.396 [27]	0.264 [24]	0.71 [27]	1.943 [8]	0.63 [5]	14.937	14th	Medium
SW11	3.09 [25]	1.931 [13]	1.711 [13]	1.234 [17]	0.855 [13]	1.128 [14]	3.305 [12]	0.048 [21]	0.005 [8]	1.153 [11]	0.082 [19]	0.194 [10]	0.254 [21]	0.496 [10]	1.982 [11]	1.288 [22]	15.000	15th	Medium
SW12	4.58 [12]	2.036 [10]	1.806 [11]	0.887 [28]	0.903 [11]	1.127 [15]	3.677 [10]	0.04 [24]	0.004 [9]	1.167 [10]	0.072 [22]	0.163 [5]	0.18 [10]	0.455 [5]	2.355 [22]	1.53 [27]	14.437	11th	Medium
SW13	3.1 [24]	1.68 [21]	1.499 [20]	0.946 [25]	0.749 [20]	1.121 [17]	2.519 [18]	0.039 [25]	0.004 [9]	1.053 [13]	0.058 [26]	0.23 [14]	0.209 [12]	0.541 [14]	2.185 [20]	1.084 [18]	18.500	22nd	Low
SW14	3.7 [21]	1.042 [31]	1.132 [30]	1 [22]	0.565 [30]	0.921 [26]	1.179 [29]	0.059 [14]	0.005 [8]	0.652 [28]	0.066 [24]	0.358 [25]	0.211 [13]	0.675 [25]	2.172 [19]	0.696 [7]	22.000	28th	Very Low
SW15	3.97 [17]	1.673 [23]	1.49 [21]	1.273 [16]	0.744 [21]	1.123 [16]	2.493 [19]	0.054 [16]	0.004 [9]	0.754 [21]	0.080 [20]	0.188 [8]	0.187 [11]	0.49 [8]	2.31 [21]	1.323 [24]	16.937	19th	Low
SW16	6.95 [2]	1.344 [28]	1.367 [26]	0.931 [27]	0.683 [26]	0.983 [24]	1.837 [27]	0.053 [17]	0.005 [8]	0.688 [26]	0.072 [22]	0.279 [21]	0.158 [5]	0.596 [21]	2.508 [27]	0.894 [11]	19.875	27th	Very Low

SW17	5.453 [7]	1.63 [24]	1.461 [23]	1.537 [10]	0.73 [23]	1.116 [18]	2.383 [22]	0.086 [5]	0.002 [10]	0.322 [31]	0.125 [11]	0.069 [1]	0.092 [2]	0.297 [1]	3.296 [30]	3.601 [31]	15.562	16th	Medium
SW18	4.86 [11]	1.567 [25]	1.258 [28]	1.122 [20]	0.628 [28]	1.246 [7]	1.971 [25]	0.049 [20]	0.004 [9]	0.723 [23]	0.061 [25]	0.266 [19]	0.171 [7]	0.582 [19]	2.418 [25]	0.937 [13]	19.000	23rd	Low
SW19	4.16 [15]	1.74 [20]	1.425 [24]	1.907 [4]	0.712 [24]	1.22 [8]	2.48 [20]	0.061 [12]	0.004 [9]	0.824 [20]	0.086 [18]	0.276 [20]	0.288 [28]	0.593 [20]	1.86 [4]	0.902 [12]	16.125	18th	Medium
SW20	3.72 [20]	1.766 [17]	1.334 [27]	1.58 [9]	0.666 [27]	1.324 [3]	2.355 [23]	0.051 [18]	0.004 [9]	0.941 [16]	0.068 [23]	0.241 [17]	0.269 [26]	0.554 [17]	1.927 [6]	1.035 [15]	17.062	20th	Low
SW21	4.573 [13]	1.435 [27]	1.384 [25]	1.661 [7]	0.692 [25]	1.063 [20]	1.986 [24]	0.055 [15]	0.004 [9]	0.611 [30]	0.076 [21]	0.415 [28]	0.223 [14]	0.727 [28]	2.114 [18]	0.601 [4]	19.250	24th	Low
SW22	4.975 [10]	1.835 [16]	1.7 [15]	1.476 [12]	0.849 [15]	1.079 [19]	3.12 [13]	0.045 [23]	0.004 [9]	0.923 [17]	0.076 [21]	0.328 [24]	0.236 [16]	0.646 [24]	2.054 [16]	0.76 [8]	16.125	18th	Medium
SW23	3.917 [18]	2.66 [3]	2.345 [3]	1.489 [11]	1.172 [3]	1.134 [12]	6.239 [3]	0.047 [22]	0.015 [2]	0.996 [15]	0.110 [14]	2.361 [31]	0.174 [8]	1.733 [31]	2.392 [24]	0.105 [1]	12.562	8th	High
SW24	3.72 [20]	2.61 [5]	2.305 [4]	1.299 [15]	1.152 [4]	1.132 [13]	6.016 [4]	0.05 [19]	0.007 [6]	1.638 [5]	0.115 [13]	0.235 [15]	0.246 [19]	0.547 [15]	2.015 [13]	1.062 [17]	11.687	5th	Very High
SW25	5 [9]	3.512 [1]	6.372 [2]	1.181 [19]	3.185 [2]	0.551 [29]	22.376 [1]	0.026 [28]	0.009 [4]	2.4 [2]	0.165 [6]	0.601 [30]	0.263 [23]	0.874 [30]	1.95 [9]	0.415 [2]	12.312	7th	High
SW26	5.895 [4]	2.458 [6]	2.035 [6]	1.835 [5]	1.017 [6]	1.207 [9]	5.004 [6]	0.107 [2]	0.006 [7]	0.69 [25]	0.217 [2]	0.159 [4]	0.14 [4]	0.45 [4]	2.672 [28]	1.57 [28]	9.125	1st	Very High
SW27	2.967 [27]	1.452 [26]	0.589 [31]	1.012 [21]	0.294 [31]	2.467 [1]	0.855 [30]	0.049 [20]	0.006 [7]	1.398 [6]	0.028 [28]	0.262 [18]	0.295 [29]	0.577 [18]	1.838 [3]	0.952 [14]	19.375	25th	Low
SW28	2 [30]	2.709 [2]	6.389 [1]	0.762 [29]	3.194 [1]	0.424 [30]	17.31 [2]	0.033 [27]	0.008 [5]	2.876 [1]	0.210 [3]	0.21 [11]	0.245 [18]	0.517 [11]	2.02 [14]	1.186 [21]	12.875	9th	High
SW29	3.8 [19]	1.985 [11]	1.969 [8]	1.592 [8]	0.984 [7]	1.008 [23]	3.91 [9]	0.1 [3]	0.007 [6]	1.094 [12]	0.196 [4]	0.193 [9]	0.229 [15]	0.495 [9]	2.088 [17]	1.294 [23]	11.437	4th	Very High
SW30	2.583 [29]	1.878 [14]	1.914 [9]	0.937 [26]	0.956 [8]	0.981 [25]	3.595 [11]	0.09 [4]	0.016 [1]	1.981 [3]	0.172 [5]	0.382 [26]	0.267 [25]	0.698 [26]	1.934 [7]	0.652 [6]	14.062	10th	High
SW31	5.543 [5]	1.751 [18]	1.728 [12]	2.616 [2]	0.863 [12]	1.013 [22]	3.025 [16]	0.129 [1]	0.004 [9]	0.664 [27]	0.222 [1]	0.122 [2]	0.252 [20]	0.394 [2]	1.992 [12]	2.044 [30]	11.937	6th	High

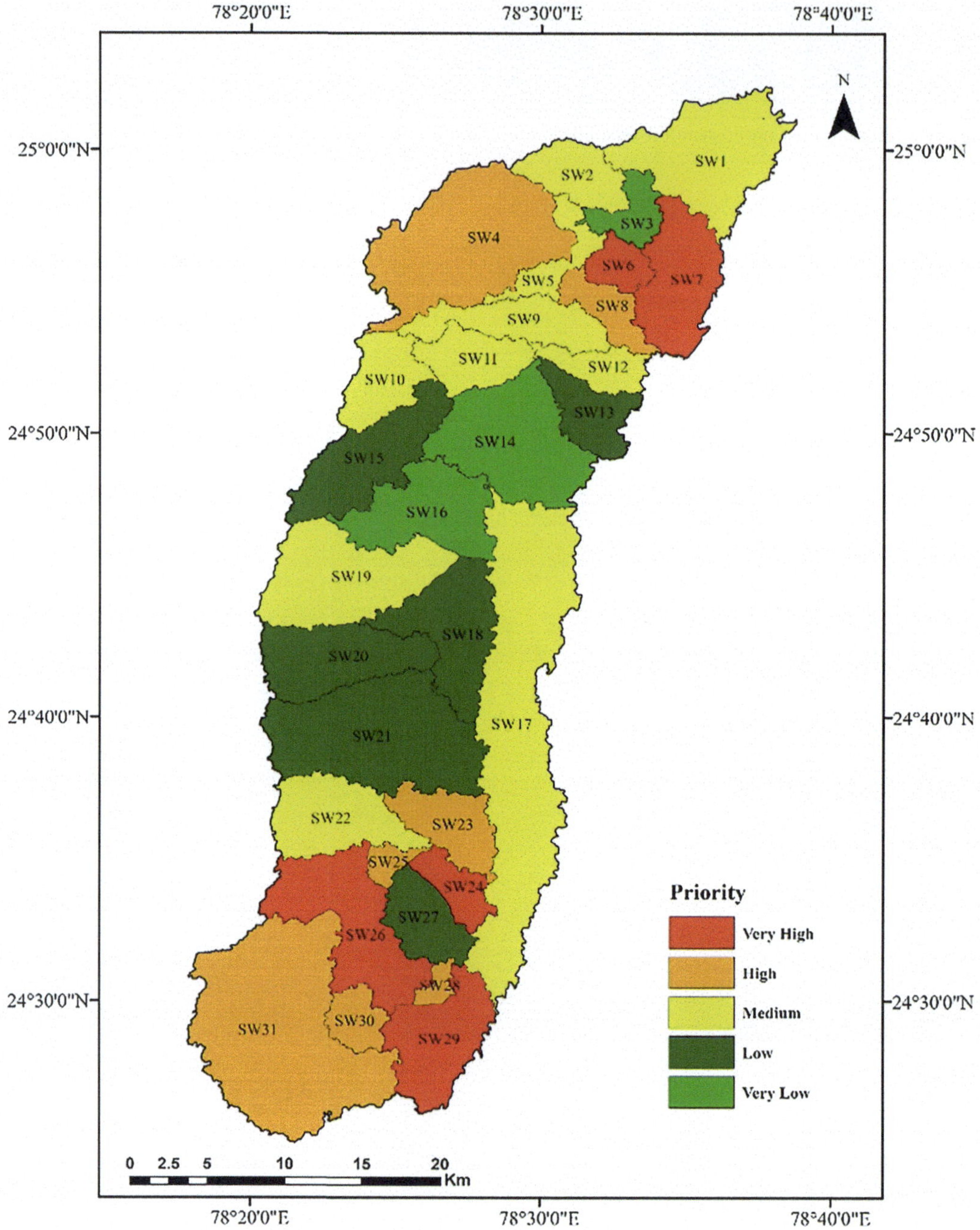

Fig. 5 Priority class of sub-watersheds based on morphometric analysis

6 Conclusions

Geospatial techniques were utilized for assessing and prioritizing the watershed for soil erosion by conducting morphometric analysis where linear, areal, and relief parameters were considered as input. The result reveals that SW6, SW7, SW24, SW26, and SW29 fall under a very high-priority zone concerning soil erosion susceptibility. The watershed namely SW4, SW8, SW23, SW25,

SW28, SW30 and SW31 falls under high priority. The value of Stream frequency is highest in SW25, followed by SW28 thereby indicating more runoff due to numerous streams in the area. The hilly region encounters high and very high priority areas in the southern part, where Vindhyan hills are exposed followed by the northern region of granitic/gneisses rock exposure. SW1, SW2, SW5, SW9, SW10, SW11, SW12, SW17, SW19, and SW22 were categorized into medium priority. SW13, SW15, SW18, SW20, SW21, and SW27 were assigned low. The very low priority was given to SW3, SW14, and SW16. In low- to very-low prioritized zones, factors like stream length, bifurcation ratio, drainage texture, stream frequency, and drainage intensity have shown insignificant effects on soil erosion. Keen observation of values indicates that the watershed SW26 is the most susceptible to soil erosion. Government and NGOs may initiate measures like afforestation or plantation drives in the hilly regions that come under the reserved forest area. Besides this in, vulnerable watersheds check dams may be constructed. The modus operandi employed in the study can be used further in different geographical or geological settings for investigating vulnerable zones, thereby suggesting the management and mitigation strategies for the conservation of the natural resources.

References

Ali, R., Sajjad, H., Masroor, M., Saha, T. K., Roshani, & Rahaman, M. H. (2024). Morphometric parameters based prioritization of watersheds for soil erosion risk in Upper Jhelum Sub-catchment, India. *Environmental Monitoring and Assessment, 196*, 82. https://doi.org/10.1007/s10661-023-12226-1

Altaf, S., Meraj, G., & Romshoo, S. A. (2014). Morphometry and land cover based multi-criteria analysis for assessing the soil erosion susceptibility of the western Himalayan watershed. *Environmental Monitoring and Assessment, 186*(12), 8391–8412. https://doi.org/10.1007/s10661-014-4012-2

Bajirao, T. S., Kumar, P. K., Kumar, A., Tarate, C., & Bajirao, S. (2019). Application of remote sensing and GIS for morphometric analysis of watershed: A review. *International Journal of Chemical Studies. IJCS, 7*(2), 709–713.

Benzougagh, B., Meshram, S. G., Dridri, A., Boudad, L., Baamar, B., Sadkaoui, D., & Khedher, K. M. (2022). Identification of critical watershed at risk of soil erosion using morphometric and geographic information system analysis. *Applied Water Science, 12*(1), 8. https://doi.org/10.1007/s13201-021-01532-z

Bharath, A., Kumar, K. K., Maddamsetty, R., Manjunatha, M., Tangadagi, R. B., & Preethi, S. (2021). Drainage morphometry based sub-watershed prioritization of Kalinadi basin using geospatial technology. *Environmental Challenges, 5*, 1–10. https://doi.org/10.1016/j.envc.2021.100277

Chorley, R. J. (1957). Climate and morphometry. *Journal of Geology, 65*, 628–638. https://doi.org/10.1086/626468

Clark, C. (1966). Morphometry from map. In G. H. Dury (Ed.), *Essay in geomorphology* (pp. 235–274). Elsevier.

Faniran, A. (1968). The index of drainage intensity—A provisional new drainage factor. *Australian Journal of Science, 31*, 328–330.

FAO, & ITPS (2015). *The status of the world's soil resources (Main report)*. ISBN 978-92-5-109004-6.

Horton, R. E. (1932). Drainage basin characteristics. *Transactions of the American Geophysical Union, 13*, 350–361.

Horton, R. E. (1945). Erosional development of streams and their drainage basins: Hydrophysical approach to quantitative morphology. *Geological Society of America Bulletin, 56*, 275–370.

Kaur, M., Singh, S., Verma, V. K., & Pateriy, B. (2014). Quantitative geomorphological analysis & landuse/landcover change detection of two sub- watersheds in Ne Region of Punjab, India. *The International Archives of the Photogrammetry, Remote Sensing and Spatial Information Sciences, XL-8*, 371–375. https://doi.org/10.5194/isprsarchives-XL-8-371-2014

Khatoon, T., & Javed, A. (2022). Morphometric behavior of Shahzad watershed, Lalitpur District, Uttar Pradesh, India: A geospatial approach. *Journal of Geographic Information System, 14*, 193–220. https://doi.org/10.4236/jgis.2022.143011

Kogo, B. K., Kumar, L., & Koech, R. (2020). Impact of land use/cover changes on soil erosion in Western Kenya. *Sustainability, 12*(22), 9740. https://doi.org/10.3390/su12229740

Melton, M. A. (1957). *An analysis of the relations among the elements of climate, surface properties and geomorphology*. Technical Report 11,. Department of Geology, Columbia University.

Miller, V. C. (1953). *A quantitative geomorphic study of drainage basin characteristics on the Clinch Mountain area, Virgina and Tennessee* (Proj. NR, Tech Rep 3) (pp. 389–402). Columbia University, Department of Geology, ONR.

Mushtaq, A., Dar, A. Q., Rather, N. A., & Din, S. A. U. (2015). Watershed management of Dal Lake Catchment (J&K) based on erosion intensity hazard using geospatial technique. *Elixir Civil Engineering, 80*, 31028–31034.

Nag, S. K. (1998). Morphometric analysis using remote sensing techniques in the Chaka Sub Basins, Purulia District, West Bengal. *Journal of Indian Society of Remote Sensing, 26*, 69–76. https://doi.org/10.1007/BF03007341

Nookaratnam, K., Srivastava, Y. K., Rao, V. V., Amminedu, E., & Murthy, K. S. (2005). Check dam positioning by prioritization of micro-watersheds using SYI model and morphometric analysis - remote sensing and GIS perspective. *Journal of the Indian Society of Remote Sensing, 33*(1), 25–38. https://doi.org/10.1007/BF02989988

Poesen, J. (2018). Soil erosion in the anthropocene: Research needs. *Earth Surface Processes and Landforms, 84*, 64–84. https://doi.org/10.1002/esp.4250

Rao, K. S., Barman, B. K., Neog, S., Barman, J., & Biswas, B. (2023). Morphometry- based prioritization of watersheds for soil and water conservation measures: a case study on Vanvate Lui basin, Mizoram, NorthEast India. *Arabian Journal of Geosciences, 16*(4), 279. https://doi.org/10.1007/s12517-023-11353-w

Romshoo, S. A., Bhat, S. A., & Rashid, I. (2012). Geoinformatics for assessing the morphometric control on hydrological response at watershed scale in the upper Indus basin. *Journal of Earth System Science, 121*(3), 659–686.

Sadhasivam, N., Bhardwaj, A., Pourghasemi, H. R., & Kamaraj, N. P. (2020). Morphometric attributes-based soil erosion susceptibility mapping in Dnyanganga watershed of India using individual and ensemble models. *Environmental Earth Sciences, 79*(14), 360. https://doi.org/10.1007/s12665-020-09102-3

Schumm, S. A. (1956). Evolution of drainage systems and slopes in badland, at Perth Amboy, New Jersey. *Geological Society of America Bulletin, 67*, 597–646.

Sharma, R., Rajwant, Singh, Y., Singh, N., & Sangra, R. (2022). Morphometric analysis of Baner, Neogal and Awa River Basins, Himachal Pradesh, India. *Journal of the Geological Society of India, 98*, 125–132. https://doi.org/10.1007/s12594-022-1938-9

Strahler, A. N. (1952). Dynamic basis of geomorphology. *Bulletin of the Geological Society of America, 63*, 923–938.

Strahler, A. N. (1954). Quantitative geomorphology of drainage basins and channel networks. In V. T. Chow (Ed.), *Handbook of applied hydrology* (Vol. 4, pp. 39–76). McGraw Hill.

Strahler, A. N. (1964). Quantitative geomorphology of drainage basins and channel networks. In V. T. Chow (Ed.), *Handbook of applied hydrology* (pp. 4–11). Mc-Graw Hill Book Company.

Spatiotemporal Assessment of Agricultural Drought in Punjab and Haryana: A Multi-Index Approach

Mohd Saqib, Aakib Yaqoob Mir, Md Ashif Ali, and Manal Ahmad

Agricultural drought is a term used to describe a period when there is no rain or moisture, in the soil, which has a negative impact on the growth and productivity of crops (Reddy, 2018). Unlike hydrological droughts that focus on deficits in rainfall or water supply, agricultural drought specifically looks at how it affects agriculture. This kind of drought can lead to the depletion of soil moisture and water for crops, ultimately resulting in lower agricultural yields (Bodner et al., 2015). They can have particularly devastating effects on farmers, rural communities, and the overall food supply chain (Malik et al., 2022). Agricultural droughts are influenced by factors such as irregular rainfall patterns, high temperatures, and increased evaporation rates. Human activities like land management, deforestation, and excessive extraction of groundwater can make agricultural systems more vulnerable to drought (Sheffield & Wood 2012). During droughts, farmers often face challenges including limited water availability for irrigation, increased competition for scarce water resources, crop failures, and financial losses (Haile et al., 2020). Mitigating the impacts of droughts requires implementing strategies for managing water resources efficiently along with using crop varieties that are resistant to drought. Early warning systems and policies that promote practices also play a crucial role (Van Ginkel & Biradar, 2021). With the population on the rise, it becomes increasingly important to comprehend and tackle the issues surrounding agricultural drought, especially in the countries that lie near the tropics (Miyan, 2015). This is vital for safeguarding food security, promoting farming practices, and ensuring that rural communities can withstand the impacts of changing climate patterns. To effectively address these challenges, it is crucial to foster cooperation, conduct research, and adopt technologies as part of our strategy to adapt to agricultural drought. Rain-fed agriculture often practiced in developing countries is more susceptible to the effects of climate change when contrasted with irrigation-based agricultural systems. Moreover, the repetitive occurrence of droughts and heavy reliance on rain-fed agriculture have led to food shortages and household food insecurity (Gebrehiwot et al., 2016; Gidey et al., 2018).

Throughout its historical existence, India has been susceptible to various environmental uncertainties, including droughts and heat waves (BMTPC, 1997). Between 1871 and 2002, the nation witnessed 22 notable drought years, characterized by precipitation levels falling below one standard deviation from the mean. The Centre for Research on the Epidemiology of Disasters (CRED) Emergency Database (EM-DAT) serves as a repository for documenting the impact of droughts in India. As per the

M. Saqib · A. Y. Mir (✉) · M. A. Ali · M. Ahmad
Department of Geography, Aligarh Muslim University, Aligarh, Uttar Pradesh, India

Z. Khanam et al. (eds.), *Climate Change, Resilience and Ecosystem Services*, Sustainable Development Goals Series, https://doi.org/10.1007/978-3-031-96977-5_26

database, the deleterious effects of droughts extended to approximately 1061 million individuals in India, resulting in the regrettable loss of 4.25 million lives during the period spanning 1900–2006 (Center for Research on Epidemiology of Disasters, 2006a, b). India receives rainfall of approximately 975 mm with more than 75% of it occurring during the crucial period from June to September. The agricultural sector in India heavily depends on these months for its performance (Department of Agriculture and Cooperation, 2004). The monsoon patterns bring less rainfall to northwest India while the southwest and northeast regions experience longer and intense rainy seasons. Around 68% of India's cultivated land, which receives rainfall ranging from 750 to 2000 mm, is highly susceptible to monsoon patterns. This vulnerability exposes them to droughts of varying durations and intensities (Shaw et al., 2005). While extensive research has been carried out in the past on drought-related topics, there is a notable lack of recent investigations into the evolving environmental conditions that may intensify drought patterns in India. Most studies predominantly focus on precipitation and temperature data, neglecting consideration for other crucial environmental factors (Mukherjee et al., 2018). A significant drawback of climate-based drought indicators is their inability to provide adequate spatial detail. Satellite by virtue of synoptic viewing ability and repeat coverage enables successful monitoring of drought-related vegetation conditions. A prominent limitation inherent in climate-based drought indicators resides in their inadequacy to provide spatial details. Conversely, satellites, leveraging their synoptic observational capabilities and repetitive coverage, prove instrumental in effectively monitoring the vegetation conditions associated with drought occurrences (Bayissa et al., 2019). Key indices such as the Vegetation Condition Index (VCI), Temperature Condition Index (TCI), and Vegetation Health Index (VHI) play a vital part in this process (Bento et al., 2018). The VCI is used to assess the greenness and overall health of vegetation, whereas the TCI gives vital information about temperature- induced stress on plants (Zambrano et al., 2016). The VHI combines VCI and TCI to provide a complete assessment of vegetation health. Additionally, Land Surface Temperature (LST) data are critical for detecting thermal anomalies associated with drought condition (Jamal et al., 2023). Collectively, these indicators provide spatially detailed and timely information on vegetation status, thereby permitting effective drought management and mitigation techniques.

1 Study Area

This research involves the two states of Punjab and Haryana, both of which are located in the northwestern part of India. These states heavily depend on rainfall for their agricultural practices. Despite the reality that India receives a significant amount of rainfall on average, these states face difficulties due to unpredictable precipitation patterns and occasional droughts. As reported by the Times of India (2015) this monsoon, 9 out of the 20 districts of Punjab have seen rainfall below 30% of normal. In Haryana, 10 out of 21 districts are in a similar situation and Haryana and Punjab faces 6th and 5th droughts in 11 years, respectively. In these states, the barren lands play an insignificant role in the nation's grain production, with crops like wheat, paddy, and cotton being neglected (Subramanya, 2005). However, the region experiences unpredictable precipitation, with varying amounts across its terrain, ranging from 900 mm in the foothills of the mountains to 400 mm in the plains (Paul & Singh, 2014). This inconsistency, coupled with high inter-annual variability, often leads to drought conditions, which adversely impact agricultural productivity and livelihoods. Similarly, Haryana, with its diverse agricultural landscape, suffers the impact of unevenly distributed rainfall, ranging from 300 to 800 mm (Ministry of Agriculture and Irrigation, 1976). The state heavily counts on crops like wheat, paddy, sugarcane, cotton, and oilseeds for its agricultural output. However, the vulnerability to drought is evident, as the fluctuating rainfall patterns pose challenges to crop

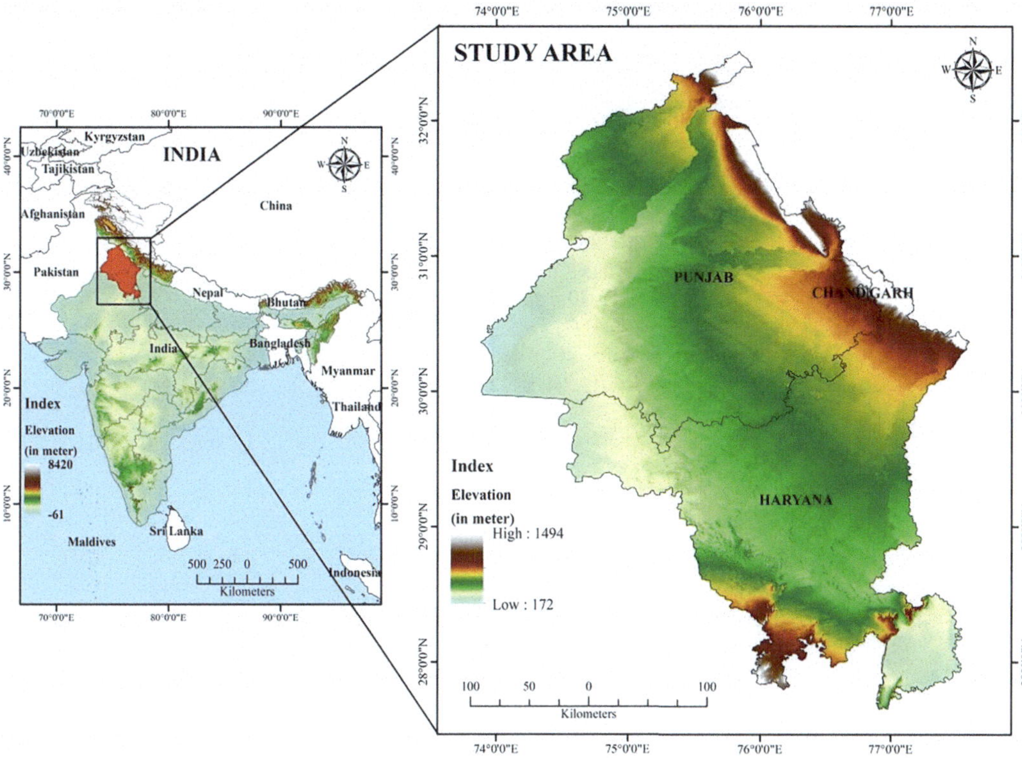

Fig. 1 Study area: Punjab and Haryana

cultivation and water availability. During periods of drought, both Punjab and Haryana experience water scarcity, reduced crop yields, and economic strain on farming communities (Dutta et al., 2015). The inadequate rainfall affects soil moisture levels, leading to crop failures and diminished agricultural output. Moreover, the socio-economic consequences of drought, such as food insecurity, financial hardship among farmers, and rural migration, aggravate the difficulties faced by these states. Attempts to alleviate the impact of drought in Punjab and Haryana often include strategies related to water management, such as collecting rainwater, implementing efficient irrigation practices, and growing drought-resistant crop varieties. The government also intervenes with relief measures, subsidies, and initiatives to conserve water, all aimed at supporting farmers and reducing the adverse effects of drought on agriculture and livelihoods (Fig. 1).

2 Database

The MODIS Terra Vegetation 16-Day NDVI (Normalized Difference Vegetation Index) dataset, providing 16-day composite NDVI values, and the MODIS Land Surface Temperature (LST) Daytime dataset, offering daily daytime surface temperature readings, were used to assess agricultural drought in Punjab and Haryana for the years 2003, 2013, and 2023. These indices help monitor vegetation health and surface temperatures, providing critical insights into drought conditions, crop performance, and the impacts of climate variability on agriculture in these regions over two decades (Table 1).

Table 1 Geo-spatial database

MODIS product	Product Id	Date of acquisition	File name	Spatial resolution	Scale factor
16-day NDVI	MOD13 Q1	30-09-2003	MOD13Q1.A2003273.h24v05.061. 2020109225005.hdf	250 m	0.0001
			MOD13Q1.A2003273.h24v06.061. 2020109224650.hdf	250 m	0.0001
		30-09-2013	MOD13Q1.A2013273.h24v05.061. 2021240225346.hdf	250 m	0.0001
			MOD13Q1.A2013273.h24v06.061. 2021240223122.hdf	250 m	0.0001
		30-09-2023	MOD13Q1.A2023273.h24v05.061. 2023290140742.hdf	250 m	0.0001
			MOD13Q1.A2023273.h24v06.061. 2023290140753.hdf	250 m	0.0001
Daytime LST	MOD11 A2	30-09-2003	MOD11A2.A2003273.h24v05.061. 2020145132039.hdf	1 km	0.02
			MOD11A2.A2003273.h24v06.061. 2020145132042.hdf	1 km	0.02
		30-09-2013	MOD11A2.A2013273.h24v05.061. 2021309152844.hdf	1 km	0.02
			MOD11A2.A2013273.h24v06.061. 2021309152843.hdf	1 km	0.02
		30-09-2023	MOD11A2.A2023273.h24v05.061. 2023282035952.hdf	1 km	0.02
			MOD11A2.A2023273.h24v06.061. 2023282040939.hdf	1 km	0.02

3 Methodology (Fig. 2)

3.1 Calculation of NDVI

The Normalized Difference Vegetation Index (NDVI) is an inherent characteristic of natural entities, representing a crucial surface parameter derived from the radiance emitted at a specific location. The NDVI values are utilized to evaluate the average emissivity of an element present on the Earth's surface, expressed mathematically by Eq. 1.

$$\mathrm{NDVI} = \frac{(NIR + \mathrm{RED})}{(NIR\,?\,\mathrm{RED})} \tag{1}$$

where

NIR stands for the near-infrared light reflected by vegetation.

RED stands for the red light reflected by vegetation.

The NDVI values range from −1 to +1, where higher values indicate healthier and denser vegetation, while lower values indicate sparse or unhealthy vegetation.

3.2 Calculation of VCI

The Vegetation Condition Index (VCI) is an important index for monitoring and assessing the health of the vegetation and the severity of the drought. It comes from the Normalized Difference Vegetation Index (NDVI), which compares the difference between red light (which vegetation absorbs) and near-infrared light (which vegetation strongly reflects) to determine the density and health of vegetation, as expressed in Eq. 2.

$$VCI = 100\,?\,\frac{(\mathrm{NDVI}\,?\,\mathrm{NDVI}_{\min})}{(\mathrm{NDVI}_{\max}\,?\,\mathrm{NDVI}_{\min})} \tag{2}$$

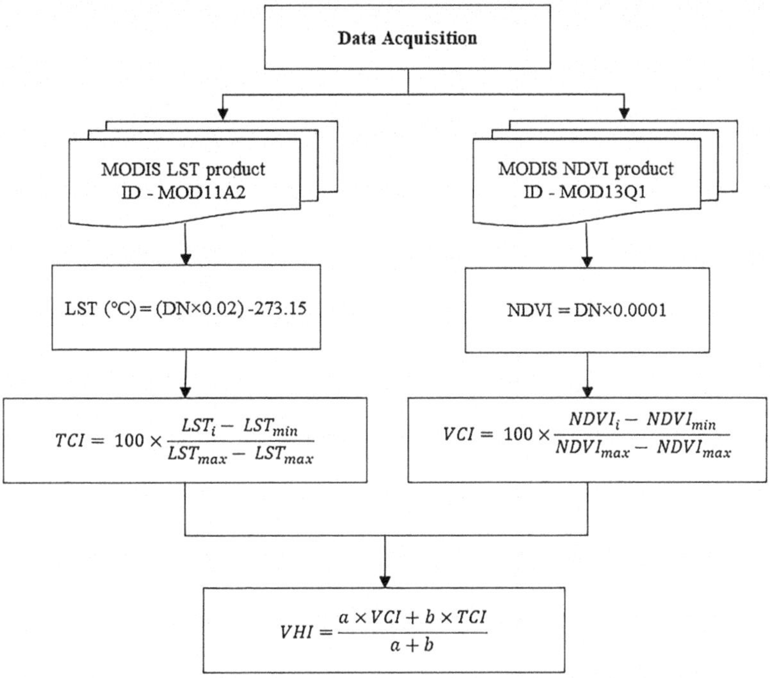

Fig. 2 Methodological flowchart of the study

where

NDVI is the current NDVI value.

$NDVI_{min}$ is the minimum NDVI value observed over a specific period.

$NDVI_{max}$ is the maximum NDVI value observed over the same period.

VCI values were divided into five classes where <10 values represent the extreme drought conditions and >40 represents no drought conditions.

3.3 Calculation of LST

To process MODIS LST Daytime data, the raw digital numbers (DN) must be converted to degrees Celsius using the scale factor provided in the dataset metadata. The formula is:

$$LST\,(?\mathrm{C}) = (DN\,?0.02)?273.15 \quad (3)$$

where 0.02 is the scale factor and 273.15 converts from Kelvin to Celsius. This conversion ensures accurate temperature readings essential for analyzing land surface temperature variations.

3.4 Calculation of TCI

The Temperature Condition Index (TCI) is a vital indicator for monitoring and evaluating the effects of temperature stress on vegetation and the severity of drought conditions (Jiao et al., 2016). It is based on Land Surface Temperature (LST) data, which measures Earth's surface temperature. TCI calculates the difference between the current temperature and the historical maximum and minimum temperature values, providing information about the thermal stress that vegetation experiences. It is expressed in Eq. 4:

$$TCI = \frac{(LST_{\mathrm{max}}\,?LST)}{(LST_{\mathrm{max}}\,?LST_{\mathrm{min}})}?100 \quad (4)$$

where

LST is the current Land Surface Temperature.

LST_{max} is the historical maximum LST.

LST_{min} is the historical minimum LST.

TCI values were divided into five categories from <10 to >40, with lower values indicating higher temperature stress and potentially severe drought conditions and higher values indicate favorable conditions for vegetation.

3.5 Calculation of VHI

The Vegetation Health Index (VHI) is a critical indicator for monitoring the overall health of vegetation, combining temperature and vegetation condition data. VHI combines the Temperature Condition Index (TCI) and the Vegetation Condition Index (VCI) to provide a comprehensive assessment of vegetation stress caused by temperature and moisture conditions. The VHI is stated in Eq. 5:

$$VHI = \frac{a?VCI + b?TCI}{a + b} \quad (5)$$

where
VHI is the Vegetation Health Index.
VCI is the Vegetation Condition Index.
TCI is the Temperature Condition Index.
a and b are weighting coefficients that determine the influence of VCI and TCI, respectively.

The VHI values were divided into five categories, from <10 to >40. Lower VHI values indicate higher stress on vegetation, potentially indicating extreme and severe drought or poor health conditions, while higher VHI values suggest favorable conditions for vegetation health and growth.

4 Results and Discussion

4.1 Normalized Difference Vegetation Index (NDVI)

Calculating NDVI is a pivotal tool in agricultural drought monitoring because it gives the reliable status of the vegetation cover and thus can help in the early detection of the effects of drought stress (Mullapudi et al., 2023). It enables spatial and temporal analysis of vegetation conditions and assists in resource and food productivity planning. The NDVI data enables decision-making to be made on time to reduce the effects of drought on agriculture. The NDVI maps of the years 2003, 2013, and 2023 offer a thorough temporal examination of vegetation health in Punjab and Haryana. In 2003, the NDVI values varied from −0.19 to 0.87, with the northern, eastern, and central areas mainly displaying green hues, indicating robust vegetation, whereas the southern and northwestern regions showed notable red patches, suggesting poorer vegetation health. By 2013, there was an enhancement in the NDVI values, ranging from −0.18 to 0.99, with an expansion of green areas and the previous red zones transitioning to yellow or light green shades, demonstrating an overall improvement in vegetation health throughout the decade, attributed to changes in rainfall patterns during that period. However, by 2023, a slight decline in the NDVI values was observed, ranging from −0.19 to 0.89, displaying a combination of green and yellow areas, with certain regions experiencing a slight deterioration, shifting toward yellow or red tones, indicating potential vegetation stress due to alterations in rainfall patterns or degradation compared to 2013. Geospatially, the northern, eastern, and central regions consistently upheld higher NDVI values, signifying stable vegetation health, while the southern regions exhibited fluctuations over the span of 20 years. These patterns imply that environmental elements, alterations in land usage, shifts in rainfall trends, and conservation endeavors could potentially be influencing factors. Alterations in precipitation patterns, temperature, agricultural expansion, deforestation, and urbanization likely had an impact on vegetation health, whereas enhancements could be credited to effective conservation initiatives (Fig. 3).

4.2 Vegetation Condition Index (VCI)

The Vegetation Condition Index (VCI) is a crucial tool for evaluating the health of vegetation, indicating the presence of drought stress, and providing guidance for agricultural and environmental management (Moussa Kourouma et al., 2021). Elevated VCI values are indicative of healthy vegetation, whereas diminished values signify the presence of severe drought stress. Analyzing VCI data and maps about Punjab and Haryana in 2003, 2013, and 2023 unveils noteworthy patterns. In 2003, a considerable 67.62% of the region encountered extreme and severe drought conditions, while 4.73% faced moderate and mild drought, leaving only 27.65% unaf-

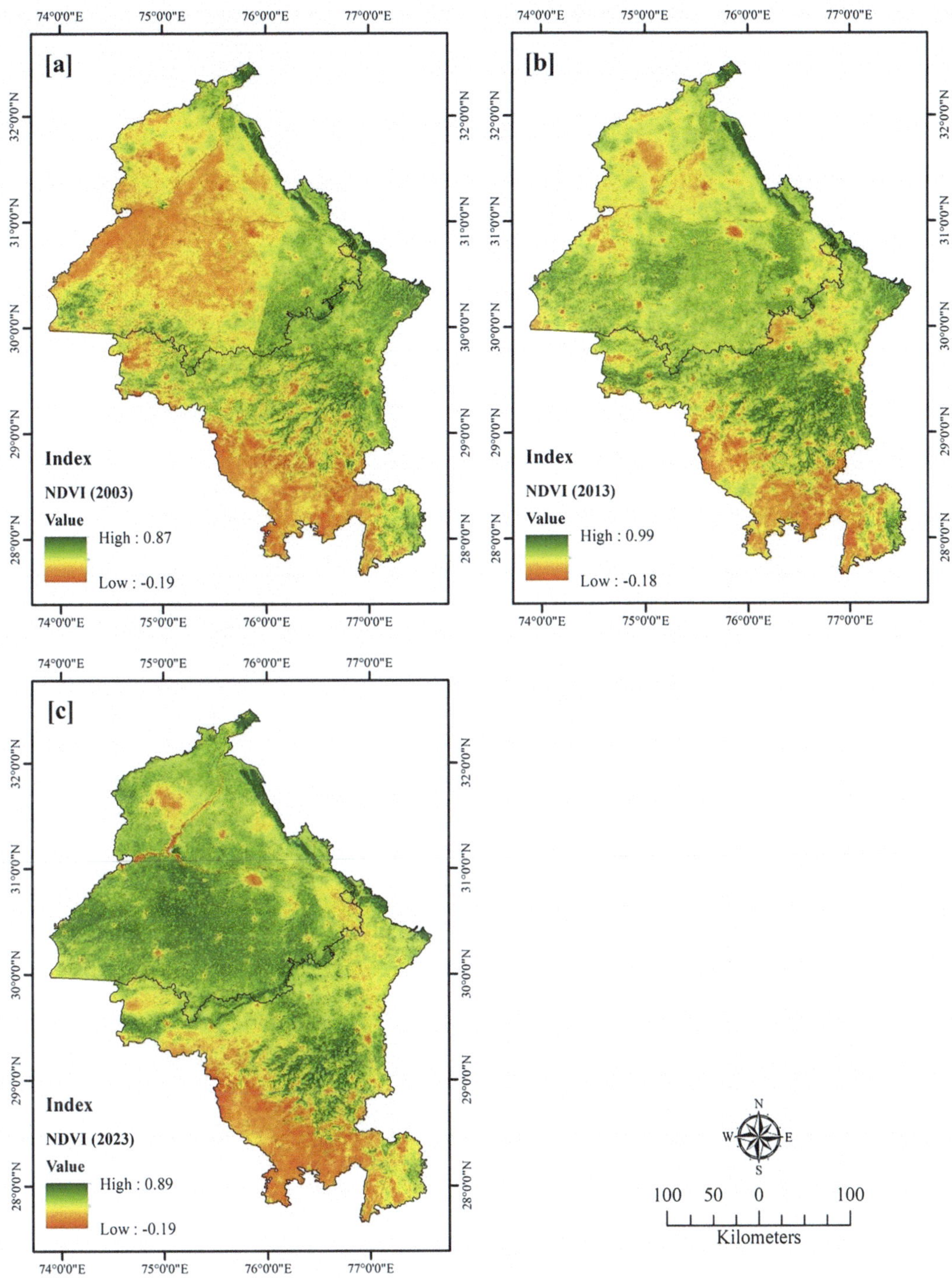

Fig. 3 Normalized Difference Vegetation Index (NDVI): (**a**) 2003, (**b**) 2013, and (**c**) 2023

fected by drought. However, significant improvement was observed by 2013 as the prevalence of extreme and severe drought decreased to 18.75%, moderate and mild drought areas expanded to 6.21%, and areas free from drought rose to 75.05% due to changes in rainfall patterns. Conversely, in 2023, the occurrence of extreme and severe drought again surged to

27.44%, moderate and mild drought areas decreased to 4.27%, and regions without drought experienced a slight reduction to 68.28% (Figs. 4, 5 and Table 2).

The maps of 2003 predominantly portray a red hue, signifying widespread drought, particularly in the southern and northwestern areas. Conversely, the 2013 VCI map evaluation reveals

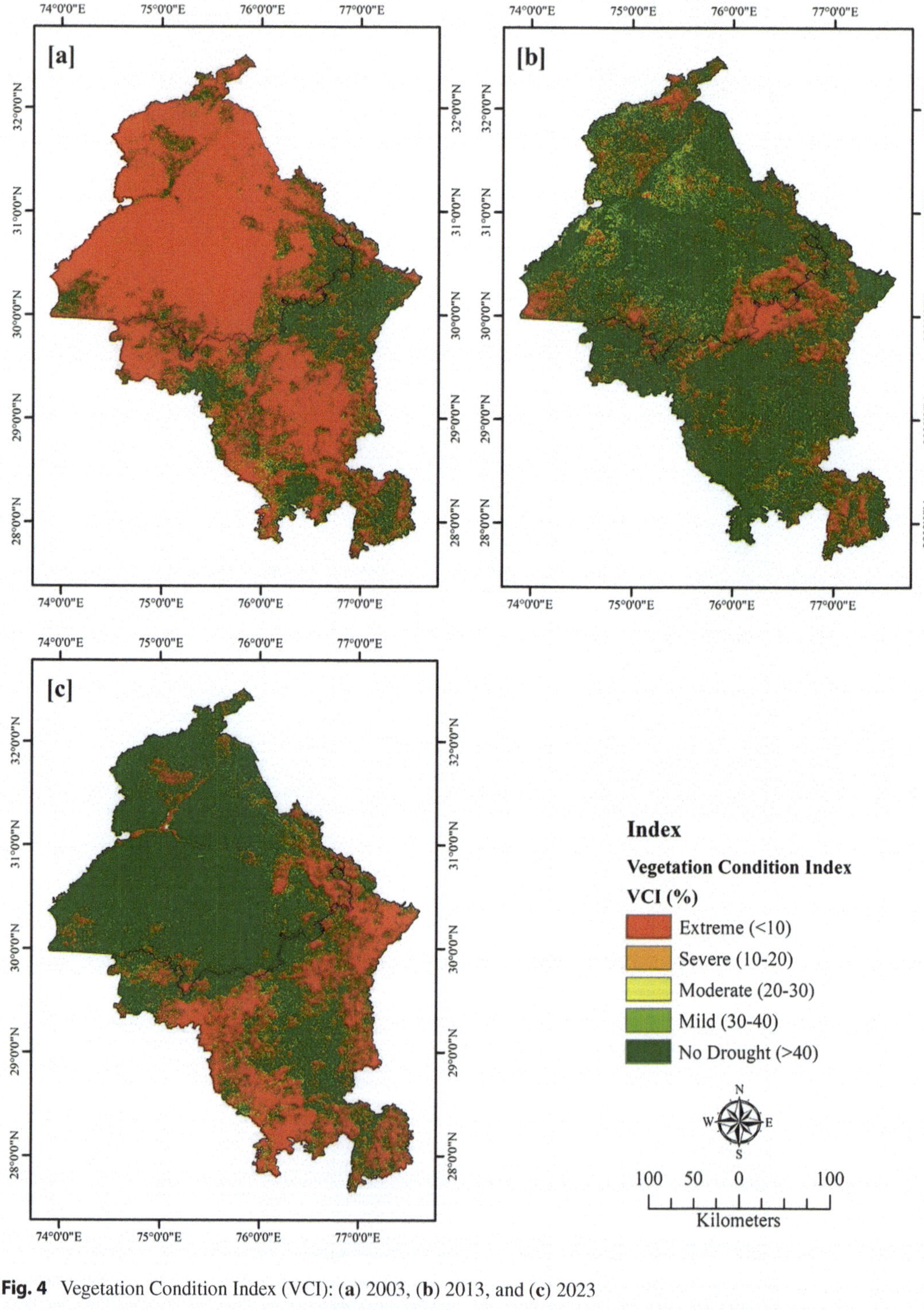

Fig. 4 Vegetation Condition Index (VCI): (**a**) 2003, (**b**) 2013, and (**c**) 2023

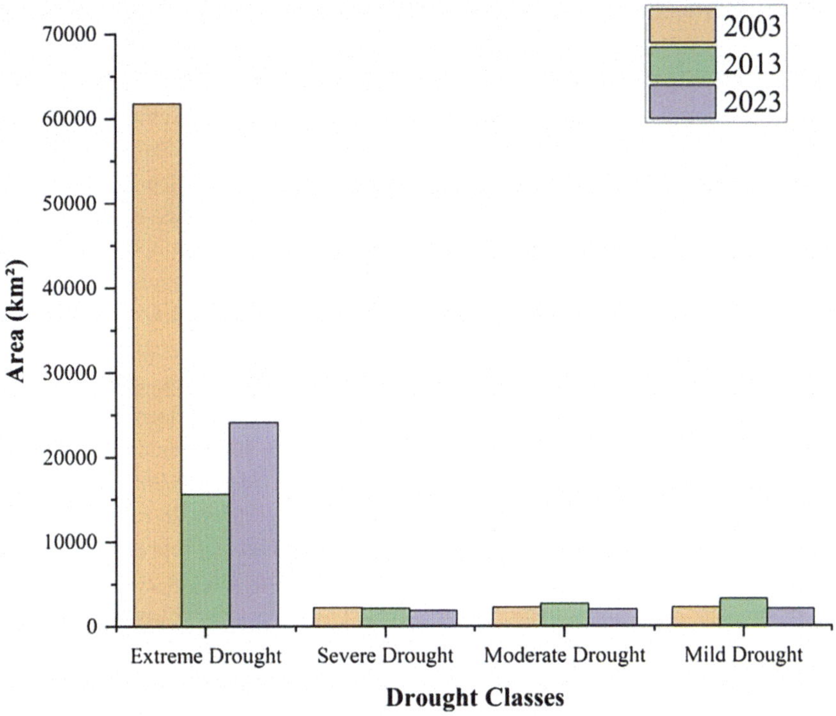

Fig. 5 VCI: Trend and area under drought classes

Table 2 VCI: area under different drought classes

Drought classes	VCI (range)	2003		2013		2023	
		Area (km^2)	Area (%)	Area (km^2)	Area (%)	Area (km^2)	Area (%)
Extreme drought	<10	61,744.17	65.25	15,591.38	16.48	24,084.85	25.45
Severe drought	10–20	2241.48	2.37	2143.81	2.27	1886.81	1.99
Moderate drought	20–30	2265.00	2.39	2648.92	2.80	1979.64	2.09
Mild drought	30–40	2216.55	2.34	3227.42	3.41	2063.87	2.18
No drought	>40	26,163.23	27.65	71,018.89	75.05	64,615.24	68.28
Total		**94,630.42**	**100.00**	**94,630.42**	**100.00**	**94,630.42**	**100.00**

vast expanses of green, symbolizing enhanced vegetation health, particularly in the northeastern and central zones. The 2023 analysis showcases a blend of green and red, with certain areas displaying progress while others confront renewed drought challenges. These trends underscore the impact of environmental elements, alterations in land use, and the necessity for continual monitoring and efficient management strategies to uphold vegetation well-being.

4.3 Land Surface Temperature

The Land Surface Temperature (LST) maps for Punjab and Haryana from 2003, 2013, and 2023 show a precise representation of temperature fluctuations across the region, providing important insights into the implications for vegetation health and drought. Temperatures in 2003 ranged from 24.99 °C to 44.13 °C, with the highest temperatures inclined to the south, and lower temperatures in the north and center. The greater temperatures in the southern regions can be linked to lesser vegetation, which lessens the cooling effect of evapotranspiration, as well as specific land use practices that lead to increased surface heating. By 2013, the temperature range has changed from 22.45 °C to 41.07 °C, with the southern region once again experiencing the greatest temperatures. Persistent high temperatures in these places can be attributed to continuous land use changes, decreasing vegetation cover, all of which contribute to higher land surface effect. Temperatures in 2023 ranged from 25.17 °C to 43.07 °C, following a regular pattern with the highest temperatures in the southern regions. The slight drop in maximum temperature over 2003 implies variability, possibly due to improved meteorological conditions or successful vegetation management measures. These trends highlight the dynamic nature of drought conditions, as well as the importance of continuous monitoring and adaptive management strategies. They emphasize the need for effective water management, sustainable agricultural practices, and conservation efforts to maintain and improve vegetation health in the face of climatic variability (Fig. 6).

4.4 Temperature Condition Index

Temperature Condition Index (TCI) is an important indicator to quantify the effects of temperature on vegetation growth and health, which is essential in agricultural drought monitoring. High TCI signifies favorable temperature conditions for vegetation while low TCI for vegetation stress occasioned by high temperatures. This index is very useful, especially in places such as Punjab and Haryana where temperature changes often impact on crop production and vegetation. With the help of TCI, one can identify the spatial and temporal changes of temperature stress to make better decisions in agriculture, resource management, or drought resistance (Dalezios et al., 2017) (Figs. 7, 8 and Table 3).

The findings indicate that in 2003, the majority of Punjab and Haryana had favorable temperatures, with 84.17% of the region not experiencing a drought, 13.10% of the area encountered extreme and severe drought, while a minor proportion (2.74%) fell into the categories of moderate and mild drought. Temperature-induced stress increased significantly by 2013, with 58.31% of the region suffering from extreme and severe drought. The percentage of the area without any drought conditions dropped sharply to 31.83%, while the area experiencing a moderate or mild drought increased to 9.86%. Between 2013 and 2023, there were some slight improvements, but severe temperature stress continued. Extreme and severe drought conditions still affected 52.16% of the region, with moderate and mild drought conditions covering 8.06%, while drought-free areas increased to 39.78%.

These trends are represented visually by the TCI maps. The map of 2003 shows broad green areas with little drought, with scattered red spots in the north, southwest, and east suggesting severe drought conditions. In contrast, the 2013 map shows widespread red and orange areas, indicating a considerable increase in extreme and severe drought conditions in the central and southern regions. The eastern and northwestern regions have a combination of yellow and green, signifying moderate to no drought, respectively. By 2023, the maps show a consistent expanse of red and orange areas, particularly in the central and northern parts, as well as the entire western side, which was drought-free in 2013, indicating that extreme and severe drought conditions will continue to prevail.

4.5 Vegetation Health Index (VHI)

The Vegetation Health Index (VHI) is critical in assessing vegetation status, determining drought impacts, and directing agricultural and environmental management methods (Bhuiyan et al., 2017). Increased VHI values indicate strong vegetation health, while decreased values suggest varying levels of drought-induced stress. VHI data from 2003, 2013, and 2023 show remarkable trends in vegetation health and drought conditions in Punjab and Haryana. In 2003, the region faced significant drought challenges, with a combined 6.65% of the area expe-

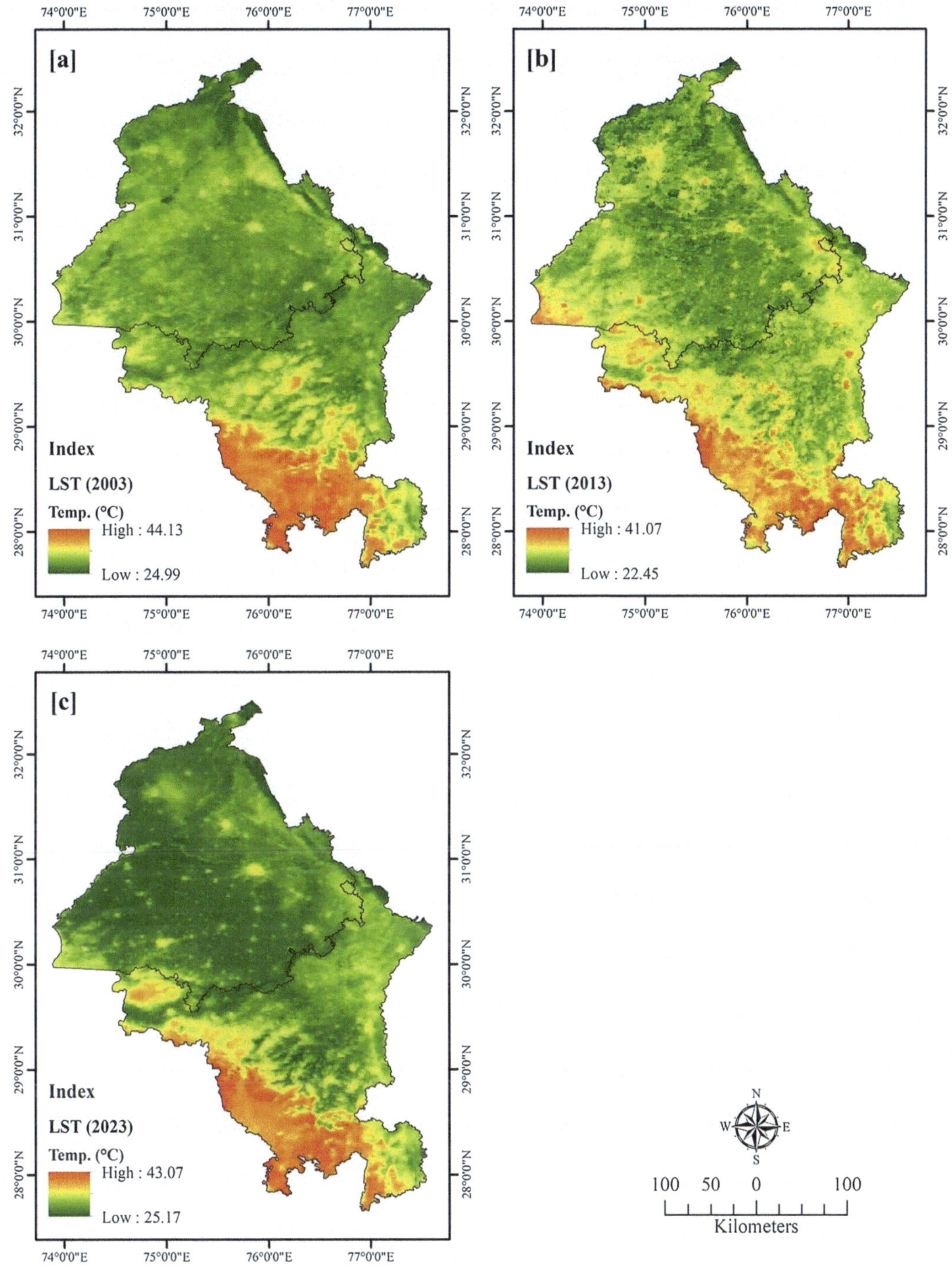

Fig. 6 Land Surface Temperature (LST): (**a**) 2003, (**b**) 2013, and (**c**) 2023

riencing extreme (4.86%) and severe (1.79%) drought conditions. Moderate (2.84%) and mild (3.06%) drought affected an additional 5.90% of the region. Areas free from drought constituted 87.44% of the area. By 2013, although extreme and severe drought increased slightly to 12.56% (8.12% extreme, 4.44% severe), moderate and mild drought expanded notably to

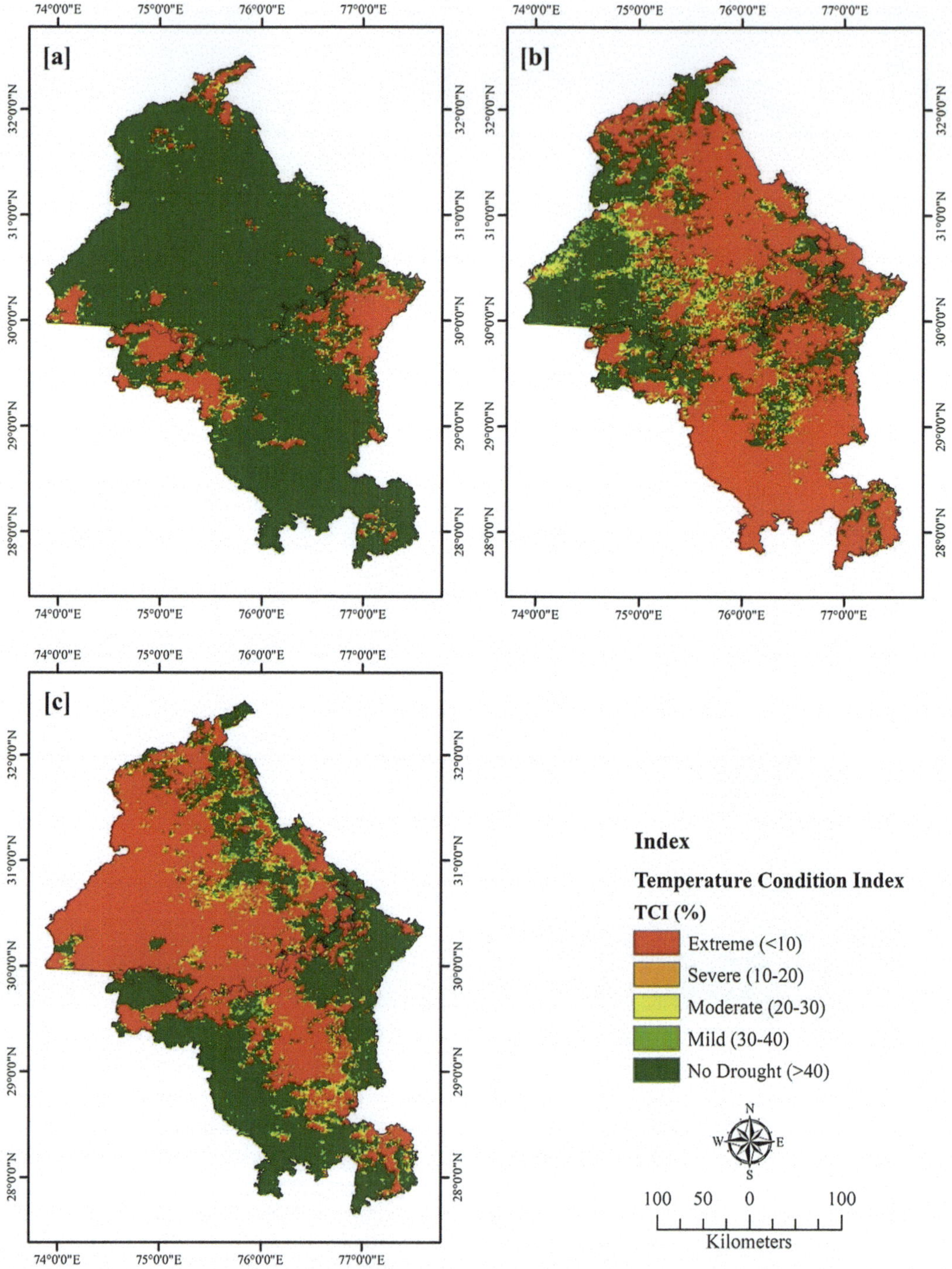

Fig. 7 Temperature Condition Index (TCI): (**a**) 2003, (**b**) 2013, and (**c**) 2023

15.46% (6.64% moderate, 8.82% mild). This period saw improvements in areas unaffected by drought, as area free from drought conditions decreased to 71.98%. In 2023, while extreme and severe drought decreased to 8.96% (6.95% extreme, 2.01% severe), moderate and mild drought reduced significantly to 5.16% (2.73% moderate, 2.43% mild). This reduction in

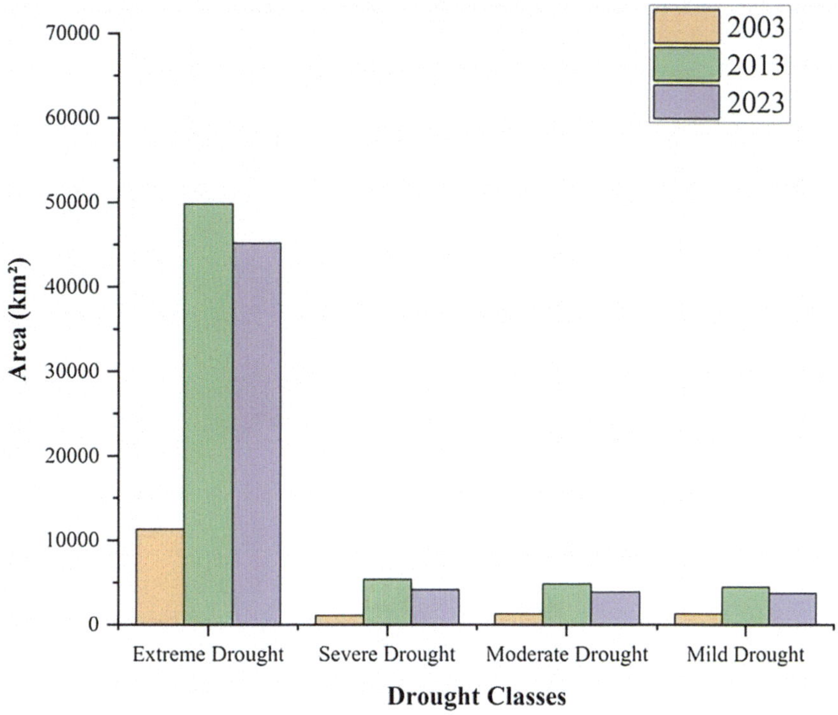

Fig. 8 TCI: trend and area under different classes

Table 3 TCI: area under different drought classes

	TCI (range)	2003		2013		2023	
Drought classes		Area (km^2)	Area (%)	Area (km^2)	Area (%)	Area (km^2)	Area (%)
Extreme drought	<10	11,298.23	11.94	49,798.33	52.62	45,182.77	47.75
Severe drought	10–20	1097.71	1.16	5380.83	5.69	4170.95	4.41
Moderate drought	20–30	1292.53	1.37	4859.72	5.14	3881.70	4.10
Mild drought	30–40	1293.52	1.37	4469.01	4.72	3748.57	3.96
No drought	>40	79,648.42	84.17	30,122.54	31.83	37,646.43	39.78
Total		**94,630.42**	**100.00**	**94,630.42**	**100.00**	**94,630.42**	**100.00**

drought-stricken areas in Punjab and Haryana occurred due to enhanced monsoonal precipitation and effective water management measures. As a result, these regions experienced a considerable increase in the percentage of land classified as drought-free, reaching 85% by the close of the year (Fig. 9 and Table 4).

The VHI maps for these years illustrate these trends vividly. The 2003 map depicted widespread drought conditions with notable red zones in the southern and northwestern areas. By 2013, improvements were visible with expanded green areas, particularly in central and northeastern regions, though areas of intensified drought stress were also evident. In 2023, the map continued to show a mixed pattern with green, yellow, and red areas, highlighting both improvements and ongoing drought challenges, particularly in the northern, western, and central parts of the region. These findings underscore the dynamic nature of drought resilience and emphasize the importance of adaptive management strategies in mitigating drought impacts effectively.

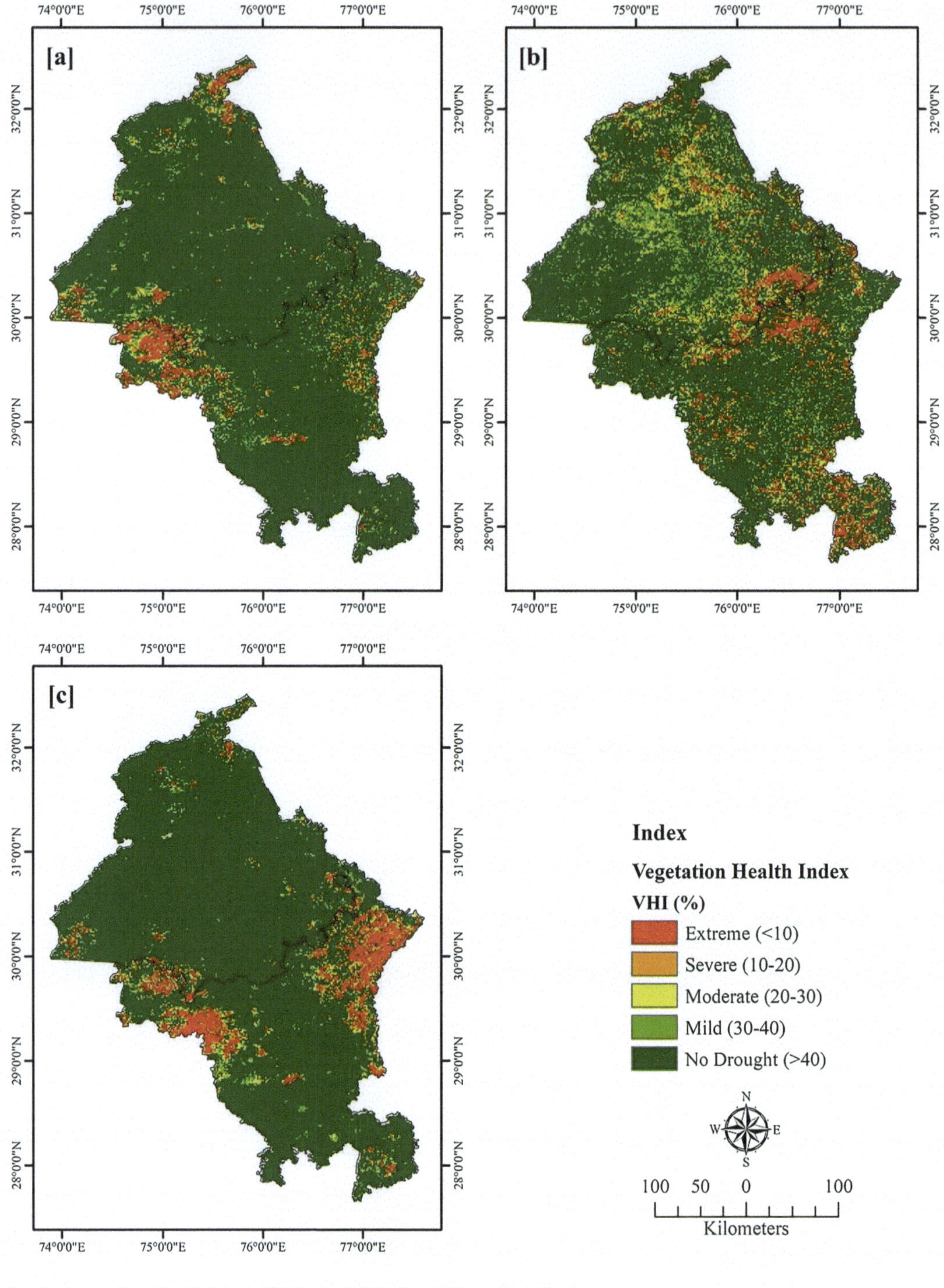

Fig. 9 Vegetation Health Index (VHI): (**a**) 2003, (**b**) 2013, and (**c**) 2023

Table 4 VHI: area under different drought classes

Drought classes	VHI (range)	2003		2013		2023	
		Area (km²)	Area (%)	Area (km²)	Area (%)	Area (km²)	Area (%)
Extreme drought	<10	4597.92	4.86	7684.61	8.12	6576.11	6.95
Severe drought	10–20	1693.74	1.79	4198.89	4.44	1899.93	2.01
Moderate drought	20–30	2691.15	2.84	6282.90	6.64	2582.55	2.73
Mild drought	30–40	2898.39	3.06	8346.44	8.82	2295.31	2.43
No drought	>40	82,749.23	87.44	68,117.57	71.98	81,276.51	85.89
Total		**94,630.42**	**100.00**	**94,630.42**	**100.00**	**94,630.42**	**100.00**

4.6 Correlation Between Different Indices

The correlation matrices for the years 2003, 2013, and 2023 demonstrate that the important agricultural drought indices VHI, TCI, LST, VCI, and NDVI are dynamically and constantly changing. Notable relationships in 2003 included a moderate negative correlation (−0.49) between NDVI and LST, reflecting the detrimental effects of elevated temperatures on vegetation productivity, and a moderate positive correlation (0.502) between VHI and TCI, indicating that temperature stress significantly influenced vegetation health. In 2013, the data showed a weak positive correlation (0.336) between VCI and NDVI, indicating that better vegetation condition scores corresponded with healthier vegetation, and a strong positive correlation (0.63) between VHI and TCI, suggesting that temperature anomalies had a significant impact on vegetation health. The relationships continued to change in 2023. There was a particularly strong positive correlation (0.991) between VCI and NDVI, likely attributable to the uniform responses of vegetation to prevailing climatic conditions, thereby reflecting consistent assessments of vegetation health across different regions. There were also strong negative correlations (−0.583 and −0.459) between VHI and TCI and VHI and LST, which highlighted the negative effects of temperature extremes on the health of vegetation. These results demonstrate the important impact that environmental stresses, technological developments, and climate change have on how these indices interact throughout time (Fig. 10).

4.7 Radar Chart of VHI

The radar chart depicts the intensity of droughts in 2003, 2013, and 2023, categorizing them as mild, moderate, severe, or extreme. Each axis represents a particular drought category, and the scales range from 0 to 9000. The green polygon for 2003 depicts moderate to severe drought conditions, with extreme drought levels approaching 5000. The blue polygon for 2013 depicts a year dominated by severe and extreme droughts, peaking at 9000 for extreme drought and almost 8000 for severe drought, with few mild and moderate droughts. The red polygon for 2023 shows a more equal distribution across all categories, with severe and intense drought levels peaking around 5000 and 7000, respectively, but mild and moderate droughts are also present. This visual comparison shows that 2013 experienced the most extreme and severe drought conditions, while 2003 had a more moderate range, and 2023, while still severe, had a greater spread over all drought categories (Fig. 11).

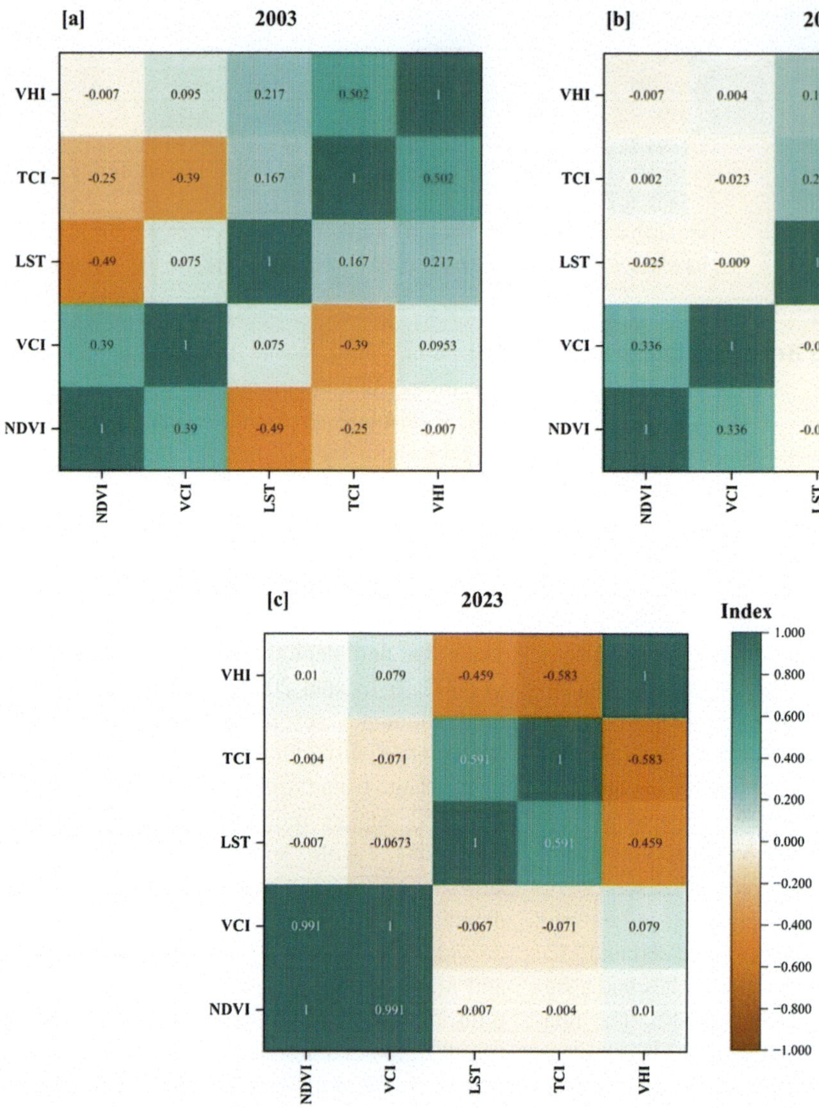

Fig. 10 Karl Pearson correlation among drought indices. Normalized: (**a**) 2003, (**b**) 2013, and (**c**) 2023

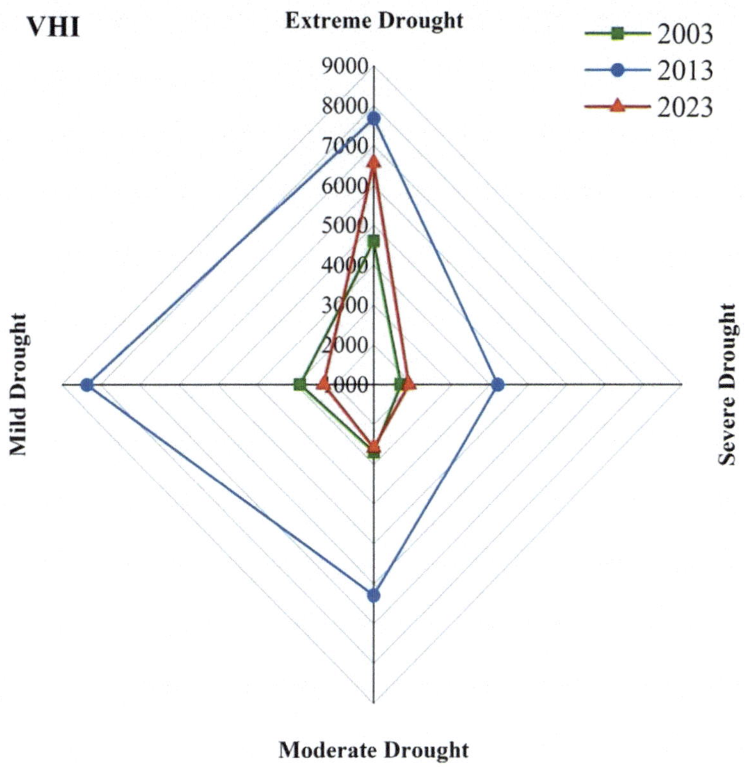

Fig. 11 Vegetation Health Index (VHI): (**a**) 2003, (**b**) 2013, and (**c**) 2023

5 Discussion

The spatiotemporal assessment of agricultural drought in Punjab and Haryana using a multi-index approach sheds light on the region's changing vegetation health and drought conditions over the past two decades. The analysis of NDVI (Normalized Difference Vegetation Index) data from 2003, 2013, and 2023 demonstrates considerable differences in vegetation health. In 2003, the NDVI values varied, with the northeastern and central regions having resilient vegetation and the southern and northwest displaying signs of poor vegetation health. By 2013, the NDVI readings had increased dramatically, with more green zones indicating healthier vegetation, most likely due to favorable rainfall patterns. However, by 2023, there has been a slight reduction in the NDVI values, with certain areas experiencing vegetative stress, potentially due to shifting rainfall patterns or environmental degradation.

The VCI (Vegetation Condition Index) data supports these conclusions. In 2003, a substantial portion of the region experienced exceptional to severe drought conditions. By 2013, there had been a dramatic improvement, with fewer places experiencing severe drought and more drought-free zones. However, by 2023, major and severe drought had returned, particularly in the southern and northwest regions. This reappearance emphasizes the impact of shifting rainfall patterns and maybe poor conservation measures.

Based on an LST (Land Surface Temperature) study, southern regions have persistently high temperatures throughout the years, which harms vegetation health and exacerbates drought conditions. Temperatures in 2003 ranged from 24.99 °C to 44.13 °C, with the southern regions having the highest temperatures. By 2013, the temperature range had shifted somewhat, from 22.45 °C to 41.07 °C, with comparable regional temperature patterns. Temperatures in 2023 ranged from 25.17 °C to 43.07 °C, indicating a continued tendency of high temperatures in the south, most likely due to lower plant cover and certain land use practices.

The correlation matrices depict the dynamic interactions between key drought indicators. In 2003, a moderate negative relationship between the NDVI and LST highlighted the detrimental effects of high temperatures on vegetation productivity. By 2013, there was a substantial positive relationship between the Vegetation Health Index (VHI) and the Temperature Condition Index (TCI), indicating that temperature anomalies have a significant impact on vegetation health. In 2023, a particularly strong positive relationship between the VCI and NDVI demonstrated that vegetation responds consistently to climatic conditions, whereas large negative correlations between VHI and TCI and VHI and LST highlighted the deleterious effects of temperature extremes on plant health.

These findings illustrate the complex interplay of environmental stressors, technological improvements, and climate change on vegetation health and conditions of drought in Punjab and Haryana. The study underlines the need for continuous monitoring, adaptive management strategies, and effective conservation efforts to mitigate the effects of climatic variability, improve vegetation health, and ensure sustainable agricultural practices in the region.

6 Conclusion

The spatiotemporal evaluation of agricultural drought in Punjab and Haryana, employing a multi-index approach, revealed useful insights into vegetation health and drought trends over the last two decades. By examining NDVI, VCI, LST, TCI, and VHI, this study has identified major trends and challenges to agricultural sustainability in the region. The study's major findings show that vegetation health has fluctuated significantly. In 2003, vegetative health was significantly harmed, particularly in the southern and northwestern regions, as indicated by reduced NDVI values and widespread dryness. Extreme temperatures ranging from 24.99 °C to 44.13 °C exacerbated vegetative stress during this time period. In 2013, however, vegetation health improved significantly, with higher NDVI values and less extreme drought conditions. During this time, favorable rainfall patterns and effective conservation techniques played critical roles. However, vegetation health declined in 2023, with the revival of extreme and severe drought conditions in the southern regions. This reduction highlights the variability of rainfall patterns and the potential for environmental deterioration. Temperature was consistently identified as a key factor influencing vegetation stress. High temperatures in the southern regions throughout the years exacerbated drought conditions. The LST analysis for 2003, 2013, and 2023 found that severe temperatures had a significant impact on vegetation quality. Correlation analysis revealed additional insights; a moderate negative correlation between the NDVI and LST in 2003 demonstrated the deleterious consequences of high temperatures, whereas a substantial positive correlation between VHI and TCI in 2013 suggested significant temperature stress. In 2023, a very significant positive connection between the VCI and NDVI indicated consistent vegetation responses to climatic circumstances, but large negative correlations between VHI and both TCI and LST demonstrated the deleterious effects of temperature extremes.

References

Bayissa, Y. A., Tadesse, T., Svoboda, M., Wardlow, B., Poulsen, C., Swigart, J., & Van Andel, S. J. (2019). Developing a satellite-based combined drought indicator to monitor agricultural drought: A case study for Ethiopia. *GIScience & Remote Sensing, 56*(5), 718–748.

Bento, V. A., Trigo, I. F., Gouveia, C. M., & DaCamara, C. C. (2018). Contribution of land surface temperature (TCI) to vegetation health index: A comparative study using clear sky and all-weather climate data records. *Remote Sensing, 10*(9), 1324.

Bhuiyan, C., Saha, A. K., Bandyopadhyay, N., & Kogan, F. N. (2017). Analyzing the impact of thermal stress on vegetation health and agricultural drought–a case study from Gujarat, India. *GIScience & Remote Sensing, 54*(5), 678–699.

BMTPC. (1997). *Vulnerability atlas of India, Parts I–III, building materials and technology promotion council*.

Bodner, G., Nakhforoosh, A., & Kaul, H. P. (2015). Management of crop water under drought: A review. *Agronomy for Sustainable Development, 35*, 401–442.

Center for Research on Epidemiology of Disasters. (2006a). *India country profile of natural disasters*. EM-DAT: The International Disaster Database. http://www.emdat.net/disasters/Visualisation/profiles/countryprofile.php. Cited 6 Mar 2006.

Center for Research on Epidemiology of Disasters. (2006b). https://www.cred.be/sites/default/files/CredCrunch07.pdf

Dalezios, N. R., Gobin, A., Alfonso, A. M. T., & Eslamian, S. (2017). Agricultural drought indices: Combining crop, climate, and soil factors. In *Handbook of drought and water scarcity* (pp. 73–89). CRC Press.

Department of Agriculture and Cooperation. (2004). *Drought 2002: A report* (p. 190). Department of Agriculture and Cooperation, Ministry of Agriculture.

Dutta, D., Kundu, A., Patel, N. R., Saha, S. K., & Siddiqui, A. R. (2015). Assessment of agricultural drought in Rajasthan (India) using remote sensing derived vegetation condition index (VCI) and standardized precipitation index (SPI). *Egyptian Journal of Remote Sensing and Space Science, 18*(1), 53–63.

Gebrehiwot, T., Veen, A., & Maathuis, B. (2016). Governing agricultural drought: Monitoring using the vegetation condition index. *Ethiopian Journal of Environmental Studies and Management, 9*(3), 354–371.

Gidey, E., Dikinya, O., Sebego, R., Segosebe, E., & Zenebe, A. (2018). Analysis of the long-term agricultural drought onset, cessation, duration, frequency severity and spatial extent using Vegetation Health Index (VHI) in Raya and its environs, Northern Ethiopia. *Environmental Systems Research, 7*, 13. https://doi.org/10.1186/s40068-018-0115-z

Haile, G. G., Tang, Q., Li, W., Liu, X., & Zhang, X. (2020). Drought: Progress in broadening its understanding. *WIREs Water, 7*(2), e1407.

Jamal, S., Saqib, M., Ahmad, W. S., Ahmad, M., Ali, M. A., & Ali, M. B. (2023). Unraveling the complexities of land transformation and its impact on urban sustainability through land surface temperature analysis. *Applied Geomatics, 15*(3), 719–741.

Jiao, W., Zhang, L., Chang, Q., Fu, D., Cen, Y., & Tong, Q. (2016). Evaluating an enhanced vegetation condition index (VCI) based on VIUPD for drought monitoring in the continental United States. *Remote Sensing, 8*(3), 224.

Malik, A., Li, M., Lenzen, M., Fry, J., Liyanapathirana, N., Beyer, K., et al. (2022). Impacts of climate change and extreme weather on food supply chains cascade across sectors and regions in Australia. *Nature Food, 3*(8), 631–643.

Ministry of Agriculture and Irrigation. (1976). *Report on progress and review part I*, National Commission on Irrigation, Government of India.

Miyan, M. A. (2015). Droughts in Asian least developed countries: Vulnerability and sustainability. *Weather and Climate Extremes, 7*, 8–23.

Moussa Kourouma, J., Eze, E., Negash, E., Phiri, D., Vinya, R., Girma, A., & Zenebe, A. (2021). Assessing the spatio-temporal variability of NDVI and VCI as indices of crops productivity in Ethiopia: A remote sensing approach. *Geomatics, Natural Hazards and Risk, 12*(1), 2880–2903.

Mukherjee, S., Mishra, A., & Trenberth, K. E. (2018). Climate change and drought: A perspective on drought indices. *Current Climate Change Reports, 4*, 145–163.

Mullapudi, A., Vibhute, A. D., Mali, S., & Patil, C. H. (2023). A review of agricultural drought assessment with remote sensing data: Methods, issues, challenges and opportunities. *Applied Geomatics, 15*(1), 1–13.

Paul, S., & Singh, O. P. (2014). A study on trends in meteorological parameters over Punjab and Haryana. *Mausam, 65*(4), 603–608.

Reddy, P. R. (2018). Dynamics of drought and present status of drought prediction-an overview. *The Journal of Indian Geophysical Union, 22*(2), 222–232.

Shaw, R., Prabhakar, S. V. R. K., & Fujieda, A. (2005). *Community level climate change adaptation and policy issues: A case study from Gujarat, India* (Graduate school of global environmental studies). Kyoto University.

Sheffield, J., & Wood, E. F. (2012). *Drought: Past problems and future scenarios*. Routledge.

Subramanya, K. (2005). *Engineering hydrology* (4th ed.). Tata McGraw-Hill Education.

Times of India. (2015). *Haryana faces 6th drought in 11 years Punjab 5th*. https://timesofindia.indiatimes.com/india/haryana-faces-6th-drought-in-11-years-punjab-5th/articleshow/48831261.cms#:~:text=Punjab%20currently%20has%20a%20monsoon,50%25%20of%20normal%20monsoon%20rainfall

Van Ginkel, M., & Biradar, C. (2021). Drought early warning in Agri-food systems. *Climate, 9*(9), 134.

Zambrano, F., Lillo-Saavedra, M., Verbist, K., & Lagos, O. (2016). Sixteen years of agricultural drought assessment of the BioBío region in Chile using a 250 m resolution Vegetation Condition Index (VCI). *Remote Sensing, 8*(6), 530.

A Study on the Dynamics of the East Kolkata Wetlands from 1767 to 2024 Using a Geospatial and Historical Approach

Bharghav Ghosh, Sunetra Ghosh, Arunima Chanda, Kiran Kumari Singh, Sudhakar Sharma, and Suraj Kumar Singh

The East Kolkata Wetlands (EKW), a Ramsar site and globally recognized model for sustainable wastewater reuse, underwent significant ecological and spatial transformations between 1767 and 2024. This study combines historical cartography, satellite imagery, and archival records to examine long-term changes in the wetlands' extent and function. Once sustained by the tidal Bidyadhari River, the wetland system began to decline following colonial-era hydraulic interventions and continued with rapid post-independence urbanization. Major losses occurred after the development of Salt Lake City and the Dhapa landfill, leading to over 40% reduction in wetland area in recent decades. Despite protective measures since 2002, EKW remains under threat from encroachment, pollution, and institutional neglect. Our findings emphasize the importance of integrating historical insight with modern geospatial tools to inform wetland conservation and urban planning. This interdisciplinary approach offers a replicable framework for monitoring and protecting urban wetlands under similar socio-environmental pressures globally.

B. Ghosh
Forest Asset Management & Urban Planning, Central University of Punjab, Bathinda, Punjab, India

S. Ghosh
University of Calcutta, Kolkata, West Bengal, India

A. Chanda
Jamia Millia Islamia University, Delhi, India

K. K. Singh
Department of Geography, University of Allahabad, Prayagraj, Uttar Pradesh, India

S. Sharma
Nirma University, Ahmedabad, India

S. K. Singh (✉)
Centre for Sustainable Development, Suresh Gyan Vihar University, Jaipur, Rajasthan, India
e-mail: suraj.kumar@mygyanvihar.com

1 Introduction

History has witnessed that every human civilization has developed along a source of water, may be along a river or a big lake since very ancient times. In Southern Eastern West Bengal a Civilization developed along the Bidyadhari River and the lower course of the Ganges River, which was called by the Greeks the Gangaridai or Vanga Civilization, where later modern urban centers developed, especially after the establishment of Calcutta city/Kolkata under the East India Company's rule in the middle of the eighteenth century. Before these scenarios, the Great

Z. Khanam et al. (eds.), *Climate Change, Resilience and Ecosystem Services*, Sustainable Development Goals Series, https://doi.org/10.1007/978-3-031-96977-5_27

Saltwater Wetland was the largest fishing ground of these areas and played an important role in environmental as well as human wellbeing. Its western part was surrounded by three major settlements, called Sutanuti in the North, Kalikata in the center along the Adi Ganga (the old course of the Ganges), Govindapur in the southern part, and a course of the Bidyadhari River passing through the center of the wetlands. Other courses of the Bidyadhari River bordered the eastern part of it. However, the development of Calcutta City under the Company's rule and ultimately the Capital of the Bengal Presidency, which attracted a huge number of people from other parts of the British company's Empire led to it speeding up the processes of urbanization. Thus, different canal systems developed especially to create the sewage outlet through the western course of the Bidyadhari River along the center of the wetlands, which was actually the beginning of the degradation process of the wetlands and adjoining drainage, including the Bidyadhari River. This degradation gained in momentum during post-Colonial India, especially since 1960s, when the City of Kolkata start expanding eastward. Today's wetlands and adjoining drainage are almost degraded, although since the 1980s steps have been taken to conserve them. This study mainly throws light on the spatial temporal analysis of wetland degradation and as a consequence the problems faced by the people living there.

Wetlands are the intermediate zone that lies between the aquatic and terrestrial systems, which is entirely or partially covered with water (MoEF, 2011), which may be replenished by rainwater or by adjoining natural drainage during high tidal seawater inflow into the wetland system. These wetlands, which play a very important role in climate mitigation, support both aquatic and terrestrial life, which has a very unique ecosystem, indirectly as well as directly helping the human habitat near it through economic and water supply, both drinking and as other uses (Ramsar_Convention, 2018). It is a suitable breeding and sprawling ground for fish. In Uganda these wetlands contribute a gross value of US$1,091,444 per year and income earned from wetland crop farming is US$ 417,536 to 25.09 million (Kakuru et al., 2013). The wetlands used to store carbon; thus, to reduce the emission of carbon (Ramsar_bn10, 2018) metropolitan cities such as Kolkata in West Bengal, India, which is responsible for emitting a huge amount of carbon/day, and the role of the East Calcutta Wetlands (EKW), prevent extra emission by storing it. Therefore, its soil has 200 times more carbon than its vegetation. The wetlands also plays a role in absorbing many non-point pollutants and helping with pollution abatement. Additionally, the wetlands control flooding by absorbing the flood water in its basin and reduce the speed of the flood (Bassi et al., 2014).

In the light of modernization, which is fueled by steady urbanization, associated with pollution, aquaculture (Meng et al., 2017) especially in wetland areas, led a threat to the existence of wetlands with shrinkage. In the twentieth century alone the wetland loss globally was between 64% and 71% (Ramsar_bn10, 2018). Since the eighteenth century, an 87% loss of natural wetlands has been surveyed, which reached at peak in the twentieth century, but the loss and its steadily converting wetland areas slowed down after the 1980s (Davidson, 2014). In Asia, in a 2014 survey by the NRSCC, in all the major 20 Ramsar sites there was about 1% wetland loss between 2001 and 2013. The reclamation of land in the Sanjiang Plain for agricultural and construction purposes cut the wetlands from the water supply, leading to a huge loss of wetland areas in the river plain of about 79% between 1954 and 2015 (Xu et al., 2019). The biggest wetland of Cambodia, i.e., Tonle Sap lake, supported a huge surrounding habitat, whose major water source came from Mekong river, but, owing to the construction of dams on the river, the source of water was disturbed and the forest areas around the lakes were converted into agricultural land. This, together with other human interference, led this largest wetland to lose its area (Sithirith, 2015). In Turkey, there are about 135 noted wetlands, but about 22 wetlands, such as Amik Lake of Hatay Province, Emen Lake of Kahramanmaraş Province, and the Yarma Marshes of Konya Province, are exposed to drainage for irrigation.

Armed with local political reasoning in the initial stages, the irrigation activities led to the expansion of agricultural lands and the later development of settlements by encroaching on the areas of wetlands, leading it to lose area, thus shrinking the wetland space in Turkey, which adversely affects the natural habitats and other contributions to the adjoining ecosystem. The remaining portion of wetlands is subjected to industrial pollution owing to contamination by industrial waste of those wetlands in later phases (Karadeniz et al., 2009). In the case of Valdivia in Chile, originally wetlands, but since the Spanish Colonial era and post-colonial era, these wetlands have been subjected to extreme degradation owing to continuous human interference, i.e., from the Spanish establishment of military installation followed by the activities of the free Chilian government in draining the wetlands for irrigation. Later, urbanization and its expansion led it to losing area, i.e., between 1992 and 2007, urban expansion encroached on 20 ha of wetlands (Dhawan et al., 2018).

The assessment of spatial changes in the wetland and its adjoining areas is carried out using contemporary government documents and maps in conjunction with GIS and remote sensing techniques. For example, in a study of Harike Wetland of Punjab State in India, which is a Ramsar site, has been subjected to loss of its area from 3114.6 ha in 1990 to 1330.4 ha in 2018. The reasons behind this wetland loss are weed infestation and human interference for agricultural activities. The spatio-temporal status of these wetlands is observed through the crucial visual interpretation of satellite images from the Landsat 5 (1990) dataset and the Landsat 8 (2018) database (Kumar & Singh, 2020). Dharmambudhi Lake of Bangaluru city in Southern India was once interlinked with adjoining lakes, but urbanization and sanitary development in the nineteenth to twentieth centuries led to the gradual degradation of these wetlands and now Dharmambudhi has been completely wiped out and converted into a bus station. These changes are analyzed with the application of GIS techniques in the mapping of spatio-temporal variations of Dharmambudhi lake to the bus station (Unnikrishnan et al., 2020). The degradation of iSimangaliso Wetland Park in South Africa, whose reasons behind the loss of wetlands are similar to those of the other wetlands discussed, i.e., anthropogenic interference, is observed in the given article from 1987–2017. The spatio-temporal analysis is used to understand the dynamics of the wetlands with satellite images of Landsat 5/Tm for 1987, 1997, 2007; Landsat 8 Oli_TIRS for 2017, and the use of the Normalized Difference Water Index (NDWI) with GIS, which provides results on the depletion of water around the wetlands (Orimoloye et al., 2018). In the Makurdi area in Nigeria, which is drained by the river Benue, the basin of Benue was once filled with numerous wetlands, but within 20 years (1996–2016) under the policies of the Nigerian Government, by expanding the agricultural land, a loss of wetlands of about 22.57% ensued, according to the spatio-temporal analysis through the use of satellite images from Landsat ETM (November 1996), ETM+ (November 2006), and OLI+ (December 2016), as well as ERDAS software, which was used to modify these images (Anule & Ujoh, 2017). From the above discussion, light is shed on some common points, i.e., almost all the wetlands are subject to loss of their area, mainly due to human interference such as rapid urbanization, expansion of agricultural activities and the associated encroachment of wetland areas, drainage of wetlands for irrigational purposes, and two discussions are on the blocking of the water supply to wetlands owing to the formation of dams on the upper course of the river, which leads its water to dry out, for example, in the case of Tonle Sap in Cambodia (Sithirith, 2015).

Let us focus on the study area, i.e., EKW and its adjoining drainage system, especially with reference to Bidyadhari River, which includes a discussion on the available information on the importance of the shrinkage of its areas and the associated impacts.

As mentioned in the introduction, this study area is a place where the Indus Valley civilization once flourished. Among all the sites excavated, Chandraketugarh city, which was a port city contemporary to Harappa city in Indus

Valley, flourished along the Bidyadhari River. These sites were excavated in 1905 by Taraknath Ghosh and many such contemporary sites were also excavated in the 1950s, 1960s, and 1970s (Srivastava & Das, 2019). This shows that these channels were used as a major navigational network with other parts of that ancient world (Bansal et al., 2015). This founding proved the importance of Bidyadhari River since ancient times.

In the words of Sir William Willcocks, who quoted during his 1930 lecture at the University of Calcutta from Ramayana and Mahabharat, that this region, i.e., the Lower Ganga Plain, which was an area of saltwater (EKW is also termed saltwater lakes), was inhabited by 50,000 people or King's subjects (Willcocks, 1930). These wetlands are a very important source of food as well as a good economic source. The 1987 report shows that about 14%, i.e., 3,27,009 of 24 Parganas' population, were dependent on riverine wetland fishing activities. The revenue derived from this wetland fishing is about GBP595/annum (Hunter, 1875). In the case of the EKW there were about 264 fish farms, which covered an area of 2858.65 ha in the form of bheris, bills, jheels, etc. (Ghosh & Chakravarti, 1999). Marshy land cultivation such as pat or jute cultivation was common along the periphery of the EKW recorded in Hunter's report in 1875 (Hunter, 1875). EKW are also a huge storage area for water; previously, it was well connected with the sea and was replenished by sea water during high tide (Ghosh & Sen, 1987). These wetlands also have same other utility and functions such as carbon sequestration, habitat for many birds, many types of fishes, etc. (Bassi et al., 2014).

The EKW of West Bengal was once replenished by the tidal water that subjected it to loss of its area since the development of different canals in Calcutta City, which connected with wetlands, especially during British Calcutta (Ghosh & Sen, 1987) and mainly when the Company's Government linked the City's main sewage outlet with the middle of Bidyadhari River, leading to steady sedimentation of its banks, which actually disconnected the wetlands and the Bay of Bengal; thus, the supply of plentiful tidal water to the wetlands stopped (Dutta, 2013; Dey & Bannerjee, 2013). Being an agriculturally rich region and given a high local demand for fish, most parts of the wetland areas are divided into bheris/bills for fishing ground as well as certain areas being converted into agricultural land, a major part of wetland loss started with the reclamation of land for first the Dhapa, which was for the city's waste dumping ground and later for the development of Salt Lake City since the 1950s–1960s in the northern part of the wetlands. Owing to real-estate interference (Ghosh & Sen, 1987; Dutta, 2013; Sengupta, 2018; Ghosh, 1999), the wetland loss between 1972 and 1991 of about 8.2 km^2 of wetlands was converted into built-up areas (Li et al., 2016). In 1991–2001 the wetland loss was 23.55%, which was followed by 7.34% between 2011 and 2017, mainly because of the eastward expansion of a new township adjoining the saltlake areas into the northern part of the wetlands and transformation of the land for aquaculture. These losses were observed by using geospatial techniques such as GIS, Landsat images (Ghosh et al., 2018), and Geographic Object-Based Image Analysis (Li et al., 2016). Besides the loss of these wetland areas, the long years of contamination of the City's sewage water and the later development of various industries, released a huge amount of waste water into the wetlands, about 250 million gallons (Bhattacharya et al., 2012), and the chemicals used in the associated agricultural farms, waste water from tanneries, and solid waste from dumping grounds degraded the quality of water in it, which directly and indirectly affected the whole habitat for those who depend on the wetlands, both economically and for survival (Ghosh, 1999; Hussan, 2016). Initially started by individuals, nongovernmental organizations, and later the Government, followed by Ramsar in 2002, this took various steps, formulated strategies, and ground planning with the perfect intention in the process of conservation of EKW by declaring it a Ramsar site in 2002, and setting up the EKW Management authority in 2006 with the available national wetland strategies (Ramsar Convention

Secretariate, 2016; Prasad et al., 2002; EKMA, 2010).

2 Objectives of the Study

- To map the evolution of ancient or traditional water bodies from 1767 to February 2024 using a geospatial approach.
- To discuss the history of the formation of the EKW.
- To shed a light on the conservational initiatives taken by individuals and authorities to protect the wetlands and drainage system.

3 Study Area

3.1 Locational Description

The study area is located between 22°48′00″ N to 22°24′00″N and 88°18′00E to 88°30′00″E latitude in the Presidency division of the Indian State of West Bengal. The entire study area lies at the conjuncture of three districts: Kolkata (KMC), North 24 Parganas (Bidhannagar), and South 24 Parganas (Bhangar 1) in the south eastern part of West Bengal state, i.e., the Ramsar site. This region experiences summer temperatures between April and June of about 29°C to 35°C, whereas winter temperatures are between about 17°C and 21°C. The average annual rainfall is approximately 1582 mm to 1612 mm and is mainly concentrated in the months of June, July, August, and September (Ghosh, 1999).

3.2 Drainage of the Study Area

The study area lies in the Lower Ganga Plain, a major part is located on the Bengal delta, and only the south eastern part of the study area is located on the active delta. The extension of the study area is from Bariti Bill to the EKW. This area is drained by the Bhagirathi or Hooghly river on the western side, which is the lower course of the Ganges river, and the eastern part of the study area is drained by the Bidyadhari River/Harua Gang/Nona Gang river. In the east the center of the study area lies in the EKW, which is situated on the eastern part of Kolkata City. The northern part of the EKW lies on 2 main streams, i.e., Saraswati Nadi, Sonai Nadi, (locally called Sonai Nadi) which is now a ruin, Labanyabati Nadi (Locally know as Nawai Nadi) which is found in the form of channels of ponds (Sharma, 1994), which originated from Nakuri Bill (now a ruin) near Jafarpur, which was the extended portion of the Bariti Bill. Several canals developed through Kolkata City and the major objective of these canal systems was use as a sewage outlet that passes through the wetlands and another part of these sewage projects of the British time extended up to Harua Gang/Nona Gang/Bidyadhari River through Bangur Kata Khal and the West East Canal. Bidyadhari River, which seems to have originated from the middle of the EKW joined the Matla River in the South. This river was once connected the wetlands to the sea (Bay of Bengal via the Matla River). Through a crucial study of the topographical maps, it may be said that this Bidyadhari River was the extension of the ruins of Sonai or Saraswati Nadi. With the passing of time and owing to the changing landcover for anthropogenic interference this channel was brought to ruin.

In recent time, most of the channels except for Bhagirathi/Hooghly rivers are tidally fed channels, owing to excessive sedimentation in the middle of its course (Dutta, 2013), either due to natural sedimentation or caused by anthropogenic activities. Here in this study, with the direction of the channel's location that is linked to the wetlands are delineated as:

- **Northern Drainage system**, which includes the Sonai/Saraswati river up to Bagjola Khal Chattopadhayah et al.,1049.
- **Southern Drainage system**, which includes Tolly Nala, a ruined channel of Bidyadhari River, and the remainder of Bidyadhari River (Hunter, 1876).
- **East Kolkata Wetlands and the City's canal system** (Fig. 1).

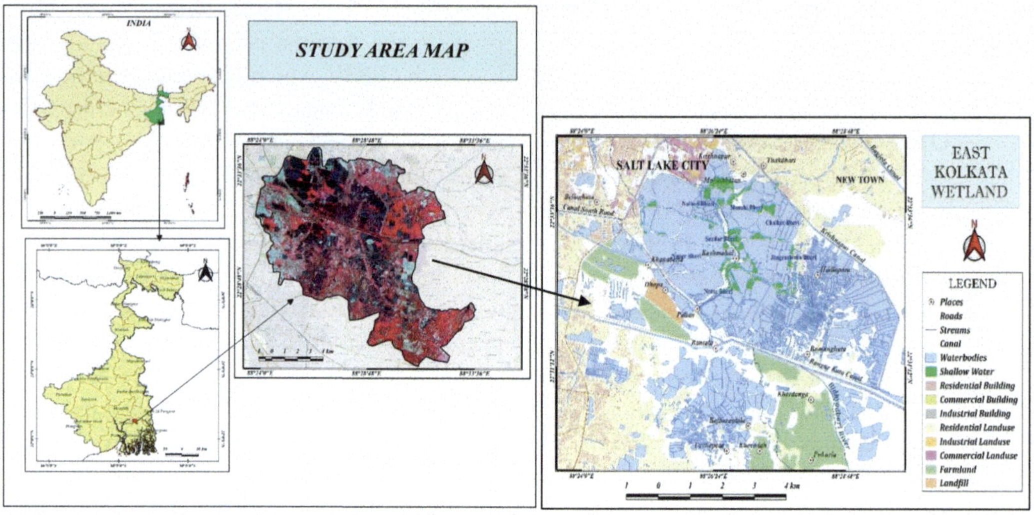

Fig. 1 Study area map

4 Methodology

4.1 Topographical Maps/Other Maps

In this study, the Rennell's maps show the situation of the drainage system and wetlands in the study area in the 1780s (Rennell, 1781). The approximate map of the 1880s is prepared on the basis of a description provided in W.W. Hunter's Statistical account of Bengal (Hunter, 1876). The map of the Survey of India (1:50,000) of covered the years1919–1970. The topographical maps are georeferenced and digitalized using Quantum GIS software, which is freely available open software with more than 50+ processing tools (Global Mapper, ArcGIS pro 2.8).

4.2 Field Survey and GPS Collection

A field survey has been conducted to collect a recent picture of the channels and wetlands, as well as the GPS coordinates by using the GPS Waypoint application. The survey was conducted on 25 February 2024 from EKW and Central Park. This survey focuses on EKW, especially the Bidyadhari River, Nalban Beel, Nuner Bheri, etc. The beels of the EKW are surveyed along the western and southern parts of the wetlands, i.e., the Nalban Beel, the beel near Nabadiganta, and Nuner Bheri, which are mainly used for fishing activities. The surveyed part along the old city sewage route where vegetable cultivation and fishing activities have recently been carried out, a north eastern portion called Dhapa is being used as a waste dumping ground, which is clearly visible from the Dhapa road.

From the Dhapa road to the Basanti highway, which is where the Bidyadhari River is seen to flow from north to south at Bantala, where the Bidyadhari crosses the East Canal (Fig. 2).

4.3 Satellite Image Datasets

After using the topographical maps up to 1955, the status of the study area is now being observed and analyzed by using the available satellite images of Landsat 1–5 MSS for 1972, Landsat 4–5 TM for 1988 and 2000, Sentinel 2A for 2019 (acquisition date: 31 December 2019) https://scihub.copernicus.eu/dhus, Landsat 9 for 2023 (acquisition date: 2 December 2020). https://earthexplorer.usgs.gov/, and Sentinel 2A was

Fig. 2 Field pictures

used for 2024 for ground validation along with a higher resolution. Cartosat 1 V2 and SRTM dem are used (Table 1 and Fig. 3).

A portion of this study was carried out by analyzing available contemporary texts such as statistical accounts by W.W. Hunter (Hunter, 1875), William Willcocks' lecture on Bengal irrigation from 1930 (Willcocks, 1930), Major James Rennell's description of the Bengal river from the Bengal Atlas, local articles such as from Lok Journal (Dutta, 2013), reports on the extensive renovation work on Saraswati/Sonai's remaining channel 1949 and 1996, other types of articles, reports, and papers by Dhrubajyoti Ghosh (Ghosh & Sen, 1987), who was the pioneer of Kolkata's wetland conservations, regarding the EKW and its adjoining drainage, especially Bidyadhari River, as the objective was to show the spatio-temporal changes. This spatial temporal analysis was carried out by using the interpretation of available maps after digitalization in QGIS software, especially for

Table 1 Details of datasets used in the study

Toposheets			Satellite datasets		
Map type/source	Date of acquisition	Usage	Map type/ source	Date of acquisition	Usage
Rennell's Map Thornton Maps	1767 and 1680	Analysis	Landsat 5 TM	1980, 1993, 1998, 2003, 2008, 2013	Feature extraction, i.e., indices
Gentleman Magazine 1757 Map	1757	Analysis	Landsat 9/OLI	2018, 2023, 2024	Feature extraction, i.e., indices
Topographic Map (Survey of India)	1922 and 1973	To show the status of river and wetlands in relation to urban expansion in digitalization by using ArcGIS Pro			
Esri Topographic Map	2024	To show the status of river and wetlands in relation to urban expansion in 2020, digitalization by using ArcGIS Pro			

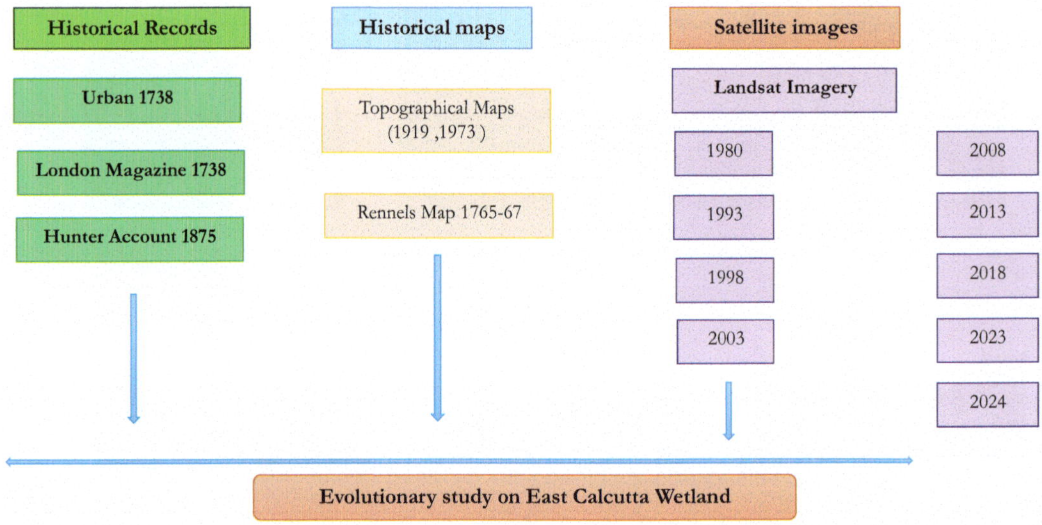

Fig. 3 Methodology of the study

the information before the 1970s conditions in the study area, which includes the map by Major J. Rennell of 1767, the topographical map of Calcutta 79 B of 1930, US Army Corps maps of 1955 Calcutta, and a 2020 map by using the osm and open sheet map in XYL tiler of Qgis. From 1972, the usage of satellite image information provided above, by using these images methods such as image classification of the spectral angular mapper Supervised Classification with accuracy assessments to understand the correlation among the three features of land use, i.e., drainage, wetlands, and settlements. This study is also descriptive in nature as field surveys have been carried out to observe the present condition of certain selected sites on the ground with GPS location using the GPS Waypoint application and the works of Dhrubajyoti Ghosh on the conservation of the EKW since the 1980s.

5 Discussion and Results

The shrinking of the EKW and the decline in channels of some important rivers are directly linked to the development of settlements such as Calcutta by the British East India Company, initially when the British trading company came to Bengal and chose this site for their official seat on the subcontinent (today Kolkata city), as this region is surrounded by the Hooghly river in the west, in the east vast wetlands and marshy land, and in the north a narrow tract left between the Hooghly river and the vast wetlands of saltwater. This provided a defensive position for their fortification and to protect them from the advancing Maratha Army and Nawab's raids in 24 Parganas. Soon they faced another problem: malaria. In 1742–1743, during the Maratha invasion of Bengal, the British company obtained special permission from Nawab and took it as an opportunity to fortify by cutting a ditch along the northern narrow tract of today's Kolkata. This canal extended from Bag Bazar of Suthanuti (North Calcutta) up to Entally in East Calcutta, and was called Maratha Ditch. This canal was later (1799) filled and a new road was constructed over it called Circular Road, which is today's APC and AJC roads (Boumik, 2006). In 1777, the Company's government excavated another canal called Tolly Nala, or Tolly canal, along the old course of the Bhagirathi (Hooghly) river or Adi Ganga (up to Garia) from the Hastings bank of today's Hooghly river south of Fort William to Bidyadhari River near Samukpota. As a result, Hooghly River was linked to Bidyadhari River. This the head water of Bidyadhari River was diverted, causing siltation in the lower reaches of the river. In 1803, the city sewage and other drainage systems were directly linked (artificially) to the Hooghly river against the natural slope. This proved problematic as the city got flooded during the monsoonal rainfall, causing the pollution of the river, which is a major source of drinking water in the city. Then, the Governor General of Bengal, Lord Wellesley, appointed a committee. The proposal of the committee was to construct several underground drainage systems for disposal sewage and flood water through the same conduit into saltwater lakes and finally into the Bay of Bengal through the Bidyadhari via the Matla River. This entire project was completed by 1884 under the British Crown through a series of canals and lock gates constructed across the wetlands and City. These canals are discussed below:

1. In 1810, Beliaghata canal was constructed from Entally to the EKW along the northern course of the Bidyadhari area; thus, the flood water also flowed into the wetlands directly.
2. In 1829, a circular canal was excavated from Bagbagar (North Calcutta) to Entally (East Calcutta), thus allowing the entry of water from the Hooghly river into the City and finally into the wetlands.
3. In 1830–1834, the excavation of several east and west canals such as Bagore Khal had as a consequence disturbances of the tidal equilibrium of the spill channel of the Bidyadhari River and the Harua and Kulti ranked rivers (Dutta, 2013).
4. In 1859, a canal was excavated that linked Ultadanga to the Beliaghata canal, which resulted in keeping the water level high in the subsoil and interfering with drainage.
5. In 1865, a plot of land (about 2.59 square miles) in the salt lake Dhapa Manpur region was handed over to the Calcutta municipal corporation, which was used as a dumping ground for the City's garbage, sewage, and fisheries. It was followed by the construction of railway lines along the north of the wetland area from Calcutta to Jessore and garbage cultivation along the sewage zone started in 1868. In 1883, the Dhapa lock gate was constructed, which locked the swift entry of tidal water into the Dhapa Entally region; thus, water in these regions was stagnant and turned into a breeding ground for dengue and malaria mosquitoes (Dutta, 2013; Ghosh & Sen, 1987).

The Committee's planning started again in 1897, and was directly responsible for the ruin of Bidyadhari River (which was used as a major source of the City's sewage outlet by the Company's Government and later the British Crown).

1. In 1897–1898 Bangar Khal was constructed, which linked the wetlands with Harua Gang and Kulti Gang drainage. This led to further

deterioration of the central lake channel, which had began before the 1890s. Around 1904, warning sign were submitted by environmentalist but no steps were taken from British Bengal Administration.

2. In 1910, initiatives were taken to construct Krishnapur Canal, which links the New Cut Canal with Bangur Kata Khal (canal), resulting in a spill area of more than 78 km^2 of the Bidyadhari River being cut off. In 1913, a second warning was given that Bidyadhari was in danger and ultimately in 1928, the Government of Bengal declared Bidyadhari a dead river. In the 1930s, the ruined part of this river was used for inland fisheries, which proved somewhat successful so that Bidyadhari River was in a position of ruin Ghosh, 1999 Chattopadhyah et al., 1949.

As an alternative for the Bidyadhari the Government of Bengal province took major initiatives toward the Harua Gang–Kulti Gang River which was connected to sea among the multiple channel of HaruaGang initiatives one of them is Arapanja Khal to the Matla River via the Piyali River. In 1945, under the chairpersonship of B.N. Dey, the Drainage Committee under the Government of Bengal Province presented a report on the current situation of Calcutta's sewage outlet through the Harua–Nona or Bidyadhari River which is well depicted in the land use/cover dynamics study of EKW by using the maps of nine different years covering 253 years of the period 1767–1955 years by using the digitalized version of the available 1767 map by Rennell and the 1930, 1955, and 2020 maps are used for identification of the recent wetlands and its adjoining drainage system. With departure of the British rulers from India in 1947, India along with Bengal Province was partitioned. This also partitioned a huge population from their own homeland and caused people to migrate from their place of origin to other parts of the country under other Indian Governments. Calcutta city as well as EKW also witnessed a huge influx of refugees from the eastern part of Bengal (East Bengal was under Pakistani occupation until 1971). Thus, they started living in the eastern part of Calcutta city and along the bheris. They chose fisheries and agriculture as their main source of income. Therefore, the number of settlements rose in East Calcutta leading to the disappearance of waterbodies seen in 2024.

5.1 Historical Record of the 1737 Earthquakes and Hypothetical Thought Process on the Formation of the East Kolkata Wetlands

According to the version of the EKW Authority, the wetlands of Kolkata were formed owing to the spillage of the Bidyadhari River's water into the submerged area, but these wetlands were not shown on the maps of 1680 (Fig. 4), etc., which may hypothetically be considered a land form of no importance for Europeans in the sense of defense and trading; thus, it was not mapped, or it can be hypothetically considered that these wetlands might have formed after the great Calcutta cyclone and earthquake of 1737, where the river water was raised up to 15 inches in 6 hours, destroying about 20,000 ships, boats, etc., with 300,000 recorded casualties, as there was huge destruction of villages and their houses when many ships were thrown into the villages. The water of the Ganges or Hooghly Rivers was raised up to 40 ft above the normal height of the water level of the Ganges River (Urban, 1738; London Magazine, 1738). Hypothetically, if the intensity of earthquakes had been as severe as recorded, then submergence might have occurred along the eastern banks of the Hooghly/locally called the Ganges River along the fortified city of Calcutta, possibly the cause of the formation of today's wetlands as the saline water through the Bidyadhari River reached this submerged area and was spilled. If we concentrate on the map published in 1757 after Nawab's men stormed the EIC's wooden fort of Calcutta (Fig. 5), where no wetlands were mapped, so the hypothetical history of the 1737 earthquakes for the formation of the EKW could be ignored, but more the point of natural defense and trading route importance for advancing Europeans can be considered that they might neglect to map this big waterbody owing to

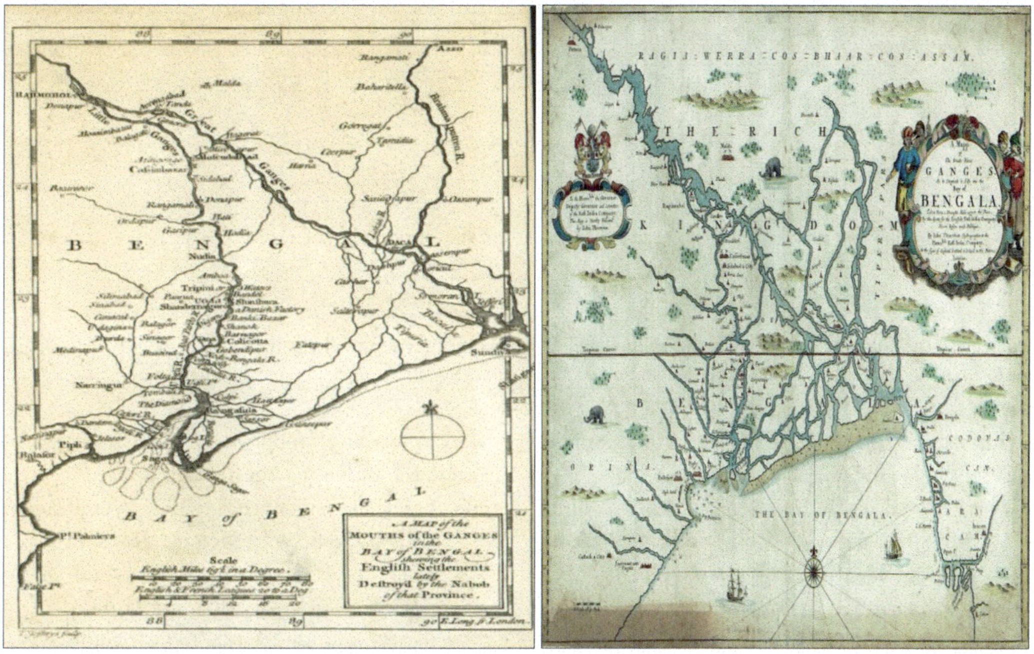

Fig. 4 Map of Bengal and surrounding areas by John Thornton (1680) with no wetlands mapped (https://mappingbengal.com/earlies-eic-map) and map of Bengal Subah in the aftermath of Nawab's seizure of Calcutta, with no wetlands mapped (by Gentleman's Magazine, 1757) https://mappingbengal.com/english-settlements

a lack of European importance because it might have some economic importance for natives. Therefore, the exact timing and precise reason behind the formation of these wetlands, then the Great Salt Lake (now EKW) are highly debatable (Fig. 4).

5.2 Dynamics of East Kolkata Wetlands and Study of Their Degradation Between 1765 and 2024

The EKW, recognized as a Ramsar site on 19 August 2002, represent a crucial ecological zone characterized by dynamic interactions between natural and anthropogenic processes. Analyzing this area's dynamics requires a multifaceted approach that incorporates contemporary reports, historical topographical maps, and modern remote-sensing techniques. The 1922 Survey of India map indicates that approximately 7600 ha of the 12,000 ha of the Ramsar area were covered by shallow and saline water, highlighting a predominantly aquatic ecosystem essential for local biodiversity, water management, and fisheries. Comparing this with the 1973 map reveals changes in land cover, likely driven by urban expansion, agricultural development, and evolving water management practices. The most recent 2024 map further illuminates current land-use patterns and the extent of urban encroachment. Over the decades, shifts in land use—marked by increased urbanization and agricultural activities—have likely diminished the area of shallow and saline waters, affecting the wetlands' ecological functions and habitat provision for diverse flora and fauna. Remote-sensing technologies, such as Landsat and Sentinel-2A imagery, facilitate detailed monitoring of land-cover changes over time, with Landsat providing a long historical archive for analyzing trends since the 1970s, and Sentinel-2A offering higher-resolution imagery for precise monitoring of wetland health, vegetation changes, and surface-water

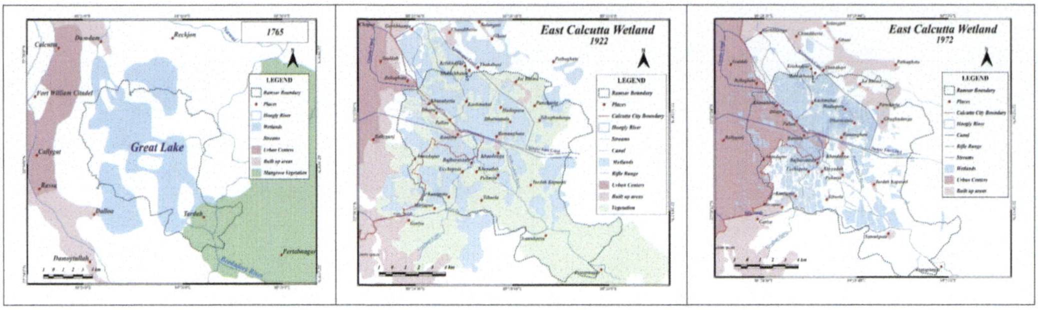

Fig. 5 Map of East Kolkata Wetlands and their surroundings in 1765 by Major Rennell, 1922, by the Survey of India, and 1972, by the Survey of India topographical map

dynamics (Dutta, 2013; Boumik, 2006; Ghosh & Sen, 1987) (Fig. 5).

The 1765 map was prepared by Major Rennell and is the first map to show the existence of EKW as great lakes, where about 118.2 km^2 out of 120.48 km^2 area is covered, although some areas such as today's Salt Lake City were a part of the wetlands. The map shows that there was a stream that flowed into the wetland area along Dum Dum and a river named Beedadory River at the southeastern part of the wetlands, which seems to source out from the Great Lake near Tardaha village. In 1765, the map shows the extent of mangrove vegetation along the southeastern part of the Great Lake. The map of 1765 also shows the "S"-shaped pattern of the wetlands, which seems to be a part of a river, as in its upper portion there was a flow with an unnamed stream (possibly Saraswati River) and the Bidyadhari River seems to be sourced from the wetlands and flow southeastern to Canning Port. On the 1922 map, which is digitalized from the 1922 Survey of India topographical map showing that the extensive spread of aquatic content is 109.58 km^2, which also includes some area of what is now Salt Lake City). In 1972 the Survey of India topographical map shows that only 58.68 km^2 of aquatic content was present, which is due to the huge influx of refugees owing to the Partition of Bengal Province and the rapidly developing settlement inside the wetland areas (Fig. 6).

Accuracy assessment performed based on Google Earth Pro dataset:

- Accuracy assessment for 1980: overall accuracy: **87%**; Kappa coefficient: **83%**
- Accuracy assessment for 1993: overall accuracy: **93%**; Kappa coefficient: **91.43%**
- Accuracy assessment for 1998: overall accuracy**: 93%**; Kappa Coefficient: **91.43%**
- Accuracy assessment for 2003: overall accuracy: **96%**; Kappa coefficient: **95.71%**
- Accuracy assessment for 2008: overall accuracy: **90%**; Kappa coefficient: **87.23%**
- Accuracy assessment for 2013: overall accuracy: **96%**; Kappa coefficient: **95.71%**
- Accuracy assessment for 2018: overall accuracy: **96.66%**; Kappa coefficient: **95.75%**
- Accuracy assessment for 2023: overall accuracy: **90%** and Kappa coefficient: **87.23%**
- Accuracy assessment for 2024: overall accuracy: **96.66%** and Kappa coefficient: **95.75%** (Table 2, Figs. 7, 8 and 9)

5.3 Early Cartography in the Eighteenth Century

The EKW, often referred to as the East Calcutta Wetlands, is a unique ecological zone situated along the eastern fringes of Kolkata, India. The earliest detailed depiction of this wetland area can be traced back to a 1765–1767 map prepared by Major Rennell. This map is instrumental in understanding the historical extent and ecological characteristics of the wetlands.

In the 1765 cartographic representation, the EKW is illustrated as an expansive water body

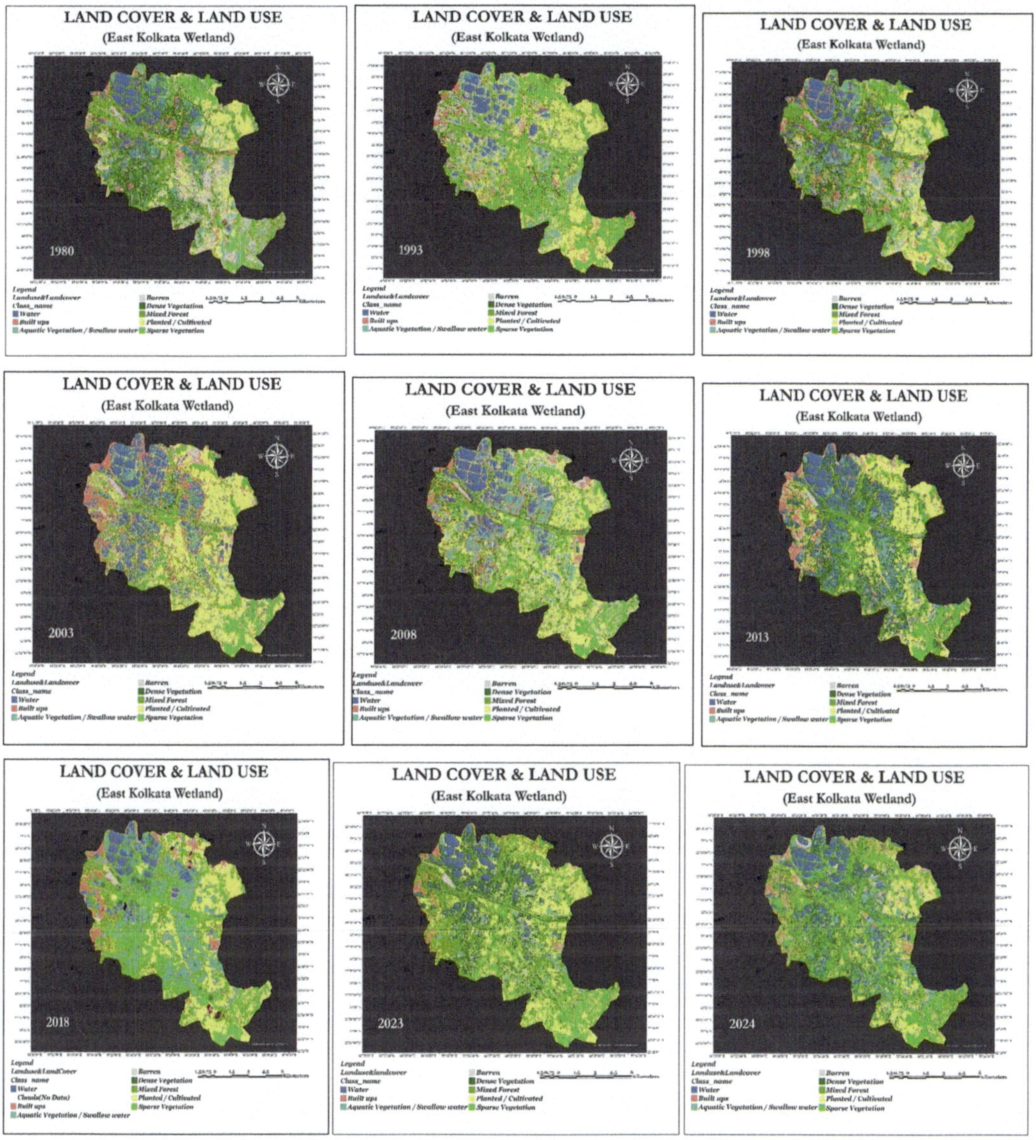

Fig. 6 Land use and landcover changes of the Ramsar Site of the East Kolkata Wetlands

Table 2 Land use and land cover dynamics based on satellite image classification from the 1980s to 2024

		Land Use & Land Cover								
Sl No	Year	Water	Built Up Areas	Aquatic Vegetation	Barren	Dense. Vegetation	Mixed. Vegetation	Cultivated	Sparse. Vegetation	Total Area
1	2024	9.83	4.37	8.01	2.12	18.98	29.00	35.80	12.37	**120.48**
2	2023	11.0658	4.3668	7.0173	1.7091	19.1072	28.003	34.9081	14.3055	**120.4828**
3	2018	12.1995	3.2319	10	0.7326	22.5314	19.628	34.597	17.5624	**120.4828**
4	2013	14.0994	6.77	3.6747	2.8337	29.3112	18.0171	36.8424	8.9343	**120.4828**
5	2008	15.165	7.85	4.25	2.97	29.44	16.28	36.55	7.9778	**120.4828**
6	2003	18.56	6.18	4.35	2.99	20.15	15.68	38.99	13.5828	**120.4828**
7	1998	23.7528	5.22	4.55	3.15	10.01	14.22	49.54	10.04	**120.4828**
8	1993	15.1416	4.81	14.8584	2.45	15.25	11.08	41.6528	15.24	**120.4828**
9	1980	19.3824	3.87	18.25	2.66	16.87	12.07	39.56	7.8204	**120.4828**

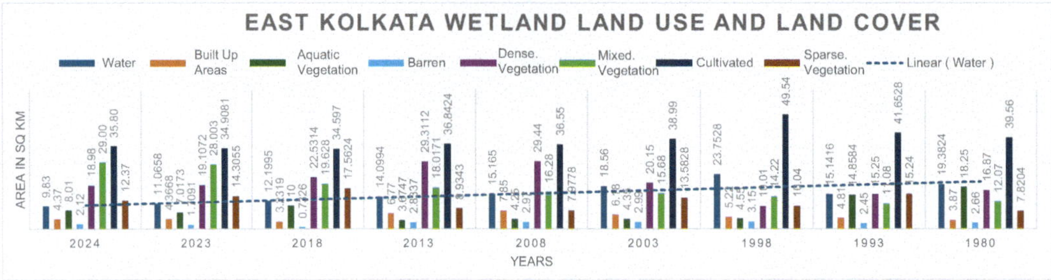

Fig. 7 Land use and land cover distribution for East Kolkata Wetland

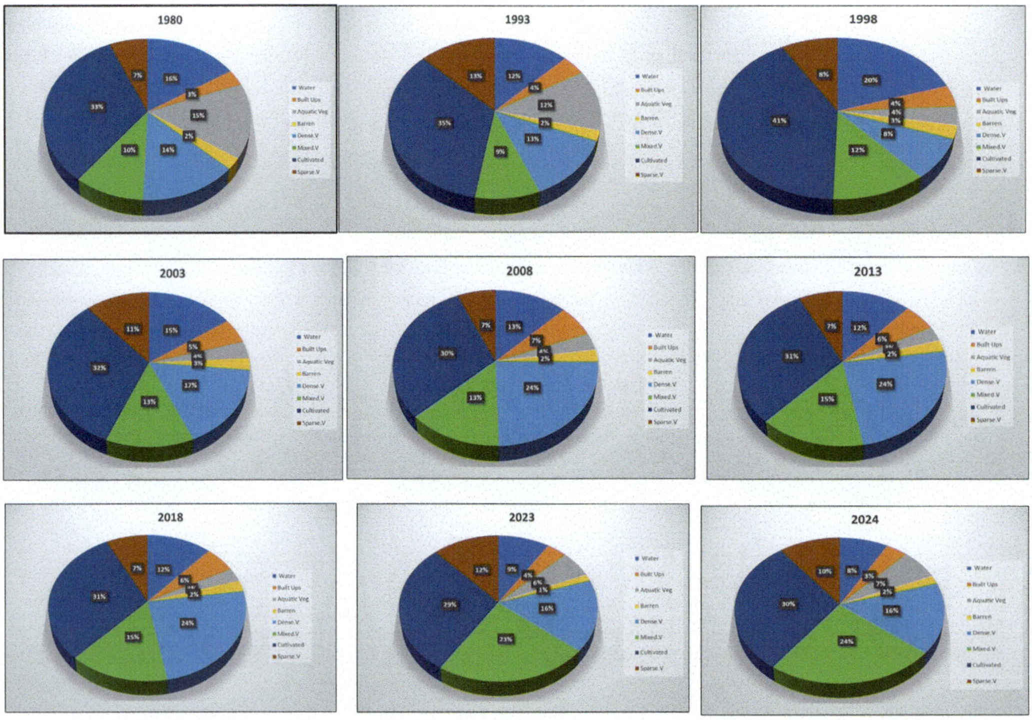

Fig. 8 Year-wise graphical representation of land cover from 1980 to 2024

that was mapped as a Great Salt Lake, roughly covering an area of 118.2 km^2 out of the total 120.48 km^2. The map delineates the wetlands' intricate network of watercourses and vegetation, capturing the extensive mangrove coverage along the southeastern part. Notably, this historical map also identifies a stream flowing into the wetlands from the Dum Dum region, and a river named the Beedadory, originating from near Tardaha village and flowing southeastward into the wetlands. The "S"-shaped pattern of the wetlands, as depicted on the map, suggests their function as part of a larger river system, potentially including the unnamed stream, which could be the Saraswati River and the Bidyadhari River flowing from the wetlands toward Canning Port.

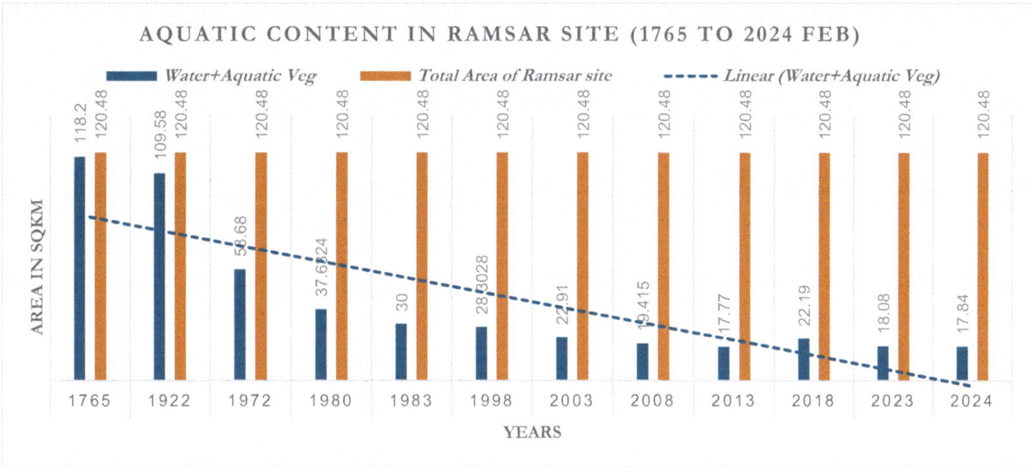

Fig. 9 Graphical representation showing the gradual degradation of aquatic content in the wetlands from 1765 to February 2024

5.4 Changes Throughout the Nineteenth and Early Twentieth Centuries

In 1922, a Survey of India topographical map offers a snapshot of the wetlands' condition at that time. This map indicates an extensive aquatic area of 109.58 km^2, which includes parts of what is now Salt Lake City. This period reflects a time when the wetlands were still relatively intact but beginning to show signs of encroachment and transformation. The aquatic vegetation and open water bodies were prominent features, reflecting the ecological health of the wetlands before the major urban development waves began to impact it.

By 1972, the Survey of India's topographical map presents a marked reduction in the aquatic content of the wetlands, down to 58.68 km^2. This significant decline can be attributed to several socio-political factors, notably the Partition of Bengal Province in 1947, which led to a large influx of refugees and accelerated urban development. The resultant pressure on the wetlands from rapid settlement and infrastructure development began to alter its natural state significantly.

The land cover of the wetland area is mostly composed of aquatic plants and open waterbodies inside the EKW Ramsar site, where the water content varied from 16% in 1980 to 8% in 2024. A crucial observation is that in 1998 there was a slight increase in the distribution of waterbodies because at the end of the 1980s the Government of West Bengal set up the Rajarhat township. The Government used the British colonial Land Acquisition Act (1894) to acquire 21 Mouzas of land. Later on, a few more Mouzas were acquired, e.g., Kalikapur, Patharghata, Muhammadpur, Natun Pukur, Chabna, and Hudarait. Seven thousand hectare of land from 21 Mouzas were acquired and the process was started in 1999 Haldar 1986 Ghosh & Sen, 1987. The total number of affected families amounted to nearly 2,50,000. Thus, the wetland areas along the northern portion came under direct threat from this recent development of cities and future pollution (Sengupta, 2018).

5.5 Conservation Efforts and Development Pressures

The latter half of the twentieth century and the early twenty-first century marked critical phases for the EKW in efforts at conservation. The 1980s saw a very important moment for the wetlands' history with **Dr. Dhrubajyoti Ghosh**, an eminent ecologist known for his advocacy for the preservation of wetlands, stepping in to address the encroachment issues. His intervention was crucial in halting illegal encroachments by real-estate developers in the 1980s and preventing the construction of a trade center inside the wetland area. During this time there was a notable increase from 3225.06 ha in 1988 to 3417.57 ha in 1990, demonstrating the positive impact of Dr. Ghosh's conservation efforts (Guardian, 2016).

5.6 RAMSAR Recognition and Role of the East Kolkata Wetlands Management Authority and Its Conservation Efforts

On 19 August 2002, the EKW were officially recognized as a Ramsar site. This led the Bengal Government to establish a management committee under the Chief Secretary of the state. The primary objective of this committee was to oversee the conservation and management of the EKW, employing high-resolution satellite data to delineate and monitor the area plot-wise. A subcommittee was subsequently formed to draft an outline management plan aimed at safeguarding the wetlands' hydro-ecological equilibrium (EKMA, 2010).

In 2006, the East Kolkata Wetlands Management Authority (EKWMA) was established under the East Kolkata Wetlands (Conservation and Management) Act, granting it statutory responsibilities for the wetlands' conservation and management. The primary role of the EKWMA involves maintaining and managing land use within the wetlands and preserving their unique recycling activities, which are integral to their inclusion on the Ramsar List. The functions of the EKWMA include demarcating wetland boundaries, detecting changes in ecological character and land use, enforcing land-use control, and preventing unauthorized developmental activities. The authority also focuses on preventing mining and similar operations, abating pollution, conserving biodiversity, promoting research, and raising awareness of the wetlands. To effectively manage the EKW, the EKWMA has implemented several key strategies. One of the core strategies is management zoning, which involves a multi-scalar and hierarchical inventory and assessment of wetlands on various scales to support planning and decision-making. The entire wetland area is designated a core zone, with stringent regulations under the East Kolkata Wetlands (Conservation and Management) Act guiding developmental activities. A buffer zone is established around the lake basin to ensure hydrological connectivity with freshwater and coastal processes, balancing ecosystem conservation with development. Water management is another critical focus area. The EKWMA has undertaken efforts to enhance hydrological connectivity by rejuvenating watercourses such as the Sunti, Nowaii, Nona Gang, Adi Ganga, Bidyadhari, and Piyali rivers. Rehabilitation of inflow regimes through selective dredging of the KMC network and primary and secondary channels within the wetlands has been prioritized. Upgrading the Palmer's Bazaar and Ballygunge Drainage and Pumping Stations is aimed at improving sewage delivery. Additionally, improving water quality involves managing the Dhapa landfill to reduce percolation into the wetlands, promoting waste segregation, and establishing ecosan units for comprehensive sanitation coverage. Biodiversity conservation efforts are also a key component of the EKWMA's strategy. This includes forming bird-protection committees to control poaching and enhancing fish biodiversity through the establishment of a center for the culture of indigenous fish species and units for breeding endangered species. Ecotourism development is encouraged through the creation of recreational facilities such as boardwalks, nature trails, and guided boat rides, which also serve as tools for community livelihood diversification.

A significant project undertaken by the EKWMA is the restoration of the Adi Ganga, supported by the National Lake Conservation Plan. This project involves constructing intercepting sewer lines, lifting stations, and sewage treatment plants, as well as de-silting and weed removal (EKMA, 2010).

5.7 Recent Developments and Current Status

The early 2000s saw continued challenges for the wetlands, including illegal occupation and construction activities. By 2001, the area of the water bodies had decreased to 23.75 km^2 from 4.55 km^2 in 1990, reflecting the ongoing pressures and encroachments. However, the Ramsar designation in August 2002 played a crucial role in stabilizing the situation. It facilitated the recovery of many areas that had been illegally built upon, leading to improved conservation and management efforts. By 2013, the built-up area had reduced to 6.77 km^2, reflecting a positive trend in conservation. This reduction continued into 2024, with built-up areas decreasing further to 4.37 km^2. This ongoing decrease in built-up areas highlights the effectiveness of conservation initiatives and the gradual restoration of the wetlands' natural environment.

To understand the situation of waterbodies and aquatic-vegetation/shallow-water areas in a precise way, the Normalized Difference Aquatic Vegetation Index and the Modified Normalized Difference Water Index for the EKW from 2013 to 2023 were applied. The selection was from 2013 because in 2011 the EKW were selected as a Ramsar site (Fig. 10 and Table 3).

This analyzes the changes in water content and aquatic vegetation within the EKW from 2013 to 2023, focusing on the impact of human activities that led to the change in wetland water content. In 2013, the water content of the EKW was 47.21014 km^2. The water content had decreased to 36.0751 km^2 in 2018, marking a loss of approximately 11.13504 km^2 over just 5 years. This reduction can be attributed to various factors, including land reclamation for urban development and the expansion of agricultural activities. These changes reflect broader trends in land use and urban expansion in the region, i.e., the eastward shift of Kolkata City's periphery. The decline in water content continued into 2023, with the area further reduced to 30.4767 km^2. This additional loss of 5.5984 km^2 suggests a persistent trend of encroachment and conversion of wetland areas for other uses, including residential and commercial developments, which have placed increasing pressure on the hydrology of the wetlands.

In contrast, the analysis of aquatic vegetation reveals a more complex dynamic. In 2013, aquatic vegetation covered 30.9287 km^2, reflecting a robust and healthy wetland ecosystem where plant species played critical roles in nutrient cycling, sediment stabilization, and habitat provision. By 2018, the coverage of aquatic vegetation had slightly decreased to 30.0194 km^2. Despite the reduction in water content, the minor decrease in vegetation indicates that the plant life of the wetlands managed to persist under changing conditions, although likely under stress from fluctuating water levels and quality. The situation had taken an intriguing turn by 2023, with the area of aquatic vegetation increasing to 39.076 km^2. This unexpected rise in vegetation coverage, despite the ongoing decrease in water content, suggests that certain aquatic plants have adapted to the reduced water conditions and may have proliferated owing to changes in water depth, nutrient availability, and reduced competition.

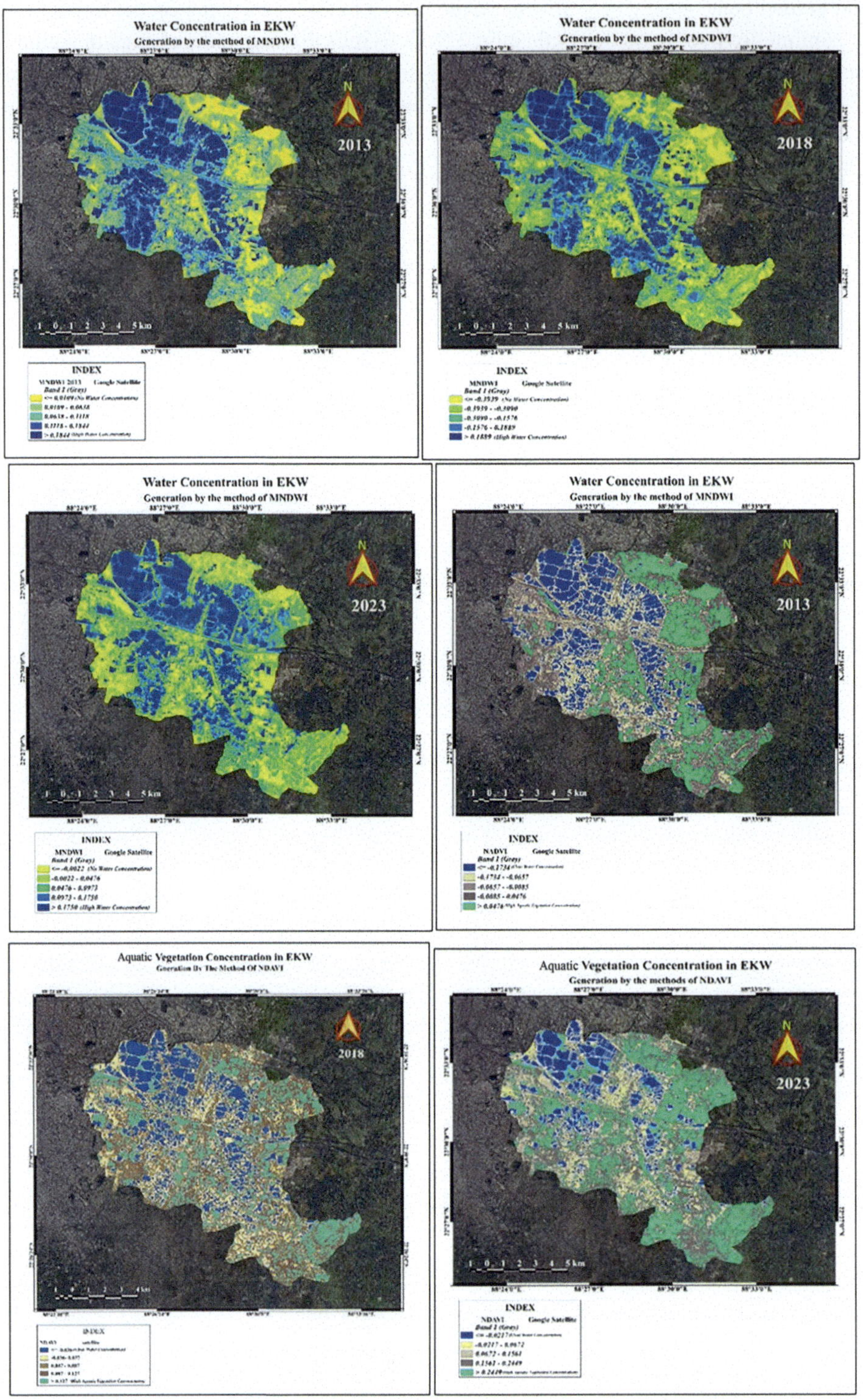

Fig. 10 Clean water and aquatic vegetation concentration in the Ramsar site from 2013 to 2023. EKW = East Kolkata Wetlands, MNDWI = Modified Normalized Difference Water Index, NDAVI = Normalized Difference Aquatic Vegetation Index

Table 3 Water and aquatic-vegetation concentration based on the Normalized Difference Aquatic Vegetation Index (NDAVI) and the Modified Normalized Difference Water Index (MNDWI)

Year	*Water Content (km2) MNDWI*	*Aquatic Vegetation(km2)NDAVI*
2013	**47.21014**	**30.9287**
2018	**36.0751**	**30.0194**
2023	**30.4767**	**39.076**

6 Conclusion

In conclusion, the alarming decline in water content within the EKW Ramsar site, which has dramatically fallen from 118.2 km^2 in the years 1765–1767 to just 17.84 km^2 by February 2024, underscores a critical environmental crisis that demands urgent attention. This reduction, revealed through comprehensive historical and geospatial analysis, highlights the effects of inadequate land management practices, unchecked expansion of debris fields, and uncontrolled settlement sprawl within the Ramsar site. These factors have severely compromised the hydrology and ecological health of this vital ecosystem, leading to cascading effects that threaten both the environment and local communities.

The EKWMA plays an important role in the efforts at preservation of the environment, community engagement, and the restoration of bheris within the Ramsar boundary. The authority can enforce land-use policies, ensure compliance with environmental regulations, and mobilize resources for conservation projects.

By prioritizing these actions and fostering collaboration among government agencies, local communities, and conservation organizations, we can work toward reversing the degradation of the EKW, most importantly, toward restoring the old channels such as the Bidyadhari and the Saraswati (Sonai), connecting Labanyabati (Nawai) with the wetlands, and letting the tidal water enter the wetland basin, which can be beneficial for both fishing and agrarian communities throughout the region.

References

Anule, P., & Ujoh, F. (2017). Geospatial analysis of wetlands degradation in Makurdi, Nigeria. *The International Archives of the Photogrammetry Remote Sensing and Spatial Information Sciences, XLII-2*, 1079–1080. https://doi.org/10.5194/isprs-archives-XLII-2-W7-1075-2017

Bansal, S., Pandey, V., Mandal, P., Paul, A., Saha, D., & Patnaik, P. (2015). *Chandraketugarh - Rediscovering a missing link in Indian history*. Indian Institute of Technology. http://www.iitkgpsandhi.org/Chandraketugarh_Report.pdf

Bassi, N., Kumar, M. D., Sharma, A., & Saradhi, P. P. (2014). Status of wetlands in India: Review of extent, ecosystem benefits, threats and management strategies. *Journal of Hydrology: Regional Studies*, 8–9. https://doi.org/10.1016/j.ejrh.2014.07.001

Bhattacharya, S., Ganguli, A., Bose, S., & Mukhopadhyay, A. (2012). Biodiversity, traditional practices and sustainability issues of East Kolkata Wetlands: A significant Ramsar site of West Bengal, (India). *Research & Review in Biosciences*.

Boumik, H. (2006). *Kolkata r pothghat* (p. 209). Lok, Sekeler path ghat.

Chattapahaya, B., Bannerjee, P., Dasgupta, K., Bagchi, N., Bhattercharjee, J., Basu Mallick, S., & Ghosh, K. L. (1949). *Memorandum (Reclamation of Bartti Beel and other smaller beels in the area for the purpose of irrigation, drainage, Anti Malaria and Grow more food measures) to Government of West Bengal*. Barrackpore Subdivisional Congress Committee.

Davidson, N. C. (2014). How much wetland has the world lost? Long-term and recent trends in global wetland area. *Marine and Freshwater Research*. https://doi.org/10.1071/MF14173

Dey, D., & Bannerjee, S. (2013). Change in Landuse pattern of East Kolkata Wetlands concern and consequences. *The Journal of National Institute of Urban Affairs, 33*, 155–156. https://doi.org/10.1177/0975425316652546

Dhawan, D., Lara, M., & Mezoughi, S. (2018). The growing importance of shrinking wetlands: a tale of three cities. *Taylor & Francis: Biodiversity*, 222–223. https://doi.org/10.1080/14888386.2018.1536562

Dutta, C. (2013). Kolkata r Jolabhumi: Poribesh Rokha i Bhumika. *Lok*, 51–52.

EKMA. (2010). *East Kolkata wetland*. East Kolkata Wetland Management Authority. http://www.ekwma.in/ek/wp-content/uploads/2015/09/NewsletterVolume1.pdf

Ghosh, D. (1999). Waste water utilization in East Calcutta Wetland. *Urban Waste Expertise Programme Occasional Paper, 5*(12), 13. https://www.ircwash.org/sites/default/files/351.1-99WA-15689.pdf

Ghosh, A., & Chakravarti, S. (1999). *Human interference and changing status*. Science and Culture.

Ghosh, D., & Sen, S. (1987). Ecological history of Calcutta's wetland conversion. *Environmental Conservation, 14*(3), 223–225. https://www.jstor.org/stable/44518089

Ghosh, S., Dinda, S., Das, C. N., & Das, K. (2018). *Analyzing risk factors for shrinkage and transformation of East Kolkata Wetland, India*. Spatial Information Research. https://doi.org/10.1007/s41324-018-0212-0

Guardian. (2016). *The miracle of Kolkata's wetland -and one man struggle to save them*. Guardian. https://www.theguardian.com/cities/2016/mar/09/kolkata-wetlands-india-miracle-environmentalist-flood-defence

Halder, D. (1986). *Ecology and fishery Management of Brackish Bheri water of West Bengal* (Bulletin No. 46). Central Inland fisheries Research Institute. http://www.cifri.res.in/Bulletins/Bulletin%20No.46.pdf

Hunter, W. W. (1875). *A statistical account of Bengal, 24 Parganas*. West Bengal District gazetteers, Department of Higher Education. Reprinted in 1998.

Hunter, W. W. (1876). *The statistical account of Bengal; Sundarbans*. West Bengal District Gazetteers, Department of Higher Education. 1998. Vol I, part II.

Hussan, A. (2016). Threats to fish diversity of East Kolkata Wetlands and conservation needs. *Scientific Article Aquaculture Times, 2*(6), 12–13. https://www.researchgate.net/publication/322910163

Kakuru, W., Turyahabwe, N., & Mugisha, J. (2013). Total economic value of wetlands products and services in Uganda. *The Scientific World Journal*. https://doi.org/10.1155/2013/192656

Karadeniz, N., Tırıl, A., & Baylan, E. (2009). Wetland management in Turkey: Problems, achievements and perspectives. *African Journal of Agricultural Research*, 1110. https://doi.org/10.5897/AJAR.9000763

Kumar, G., & Singh, K. K. (2020). Mapping and monitoring the selected wetlands of Punjab, India, using geospatial techniques. *Journal of the Indian Society of Remote Sensing, 10*. https://doi.org/10.1007/s12524-020-01104-9

Li, X., Mitra, C., Marzen, L., & Yang, Q. (2016). Spatial and temporal patterns of wetland cover changes in East Kolkata Wetlands, India from 1972 to 2011. *International Journal of Applied Geospatial Research, 7*, 9. https://doi.org/10.4018/ijagr.2016040101

London Magazine. (1738). *The London Magazine, foreign affairs*. Library of Congress Microfiche.

Meng, W., He, M., Hu, B., Mo, X., & Li, H. (2017). Status of wetlands in China: A review of extent, degradation, issues and recommendations for improvement. *Ocean & Coastal Management, Elsevier*, 53. https://doi.org/10.1016/j.ocecoaman.2017.06.0030964-5691

MoEF. (2011). *National wetland atlas*. Space Applications Centre (ISRO). https://www.indiawaterportal.org/sites/default/files/iwp/NWIA_National_atlas.pdf

Orimoloye, I. R., Kalumba, A. M., Mazinyo, S. P., & Nel, W. (2018). Geospatial analysis of wetland dynamics: Wetland depletion and biodiversity conservation of iSimangaliso Wetland, South Africa. *Journal of King Saud University*, 3–4. https://doi.org/10.1016/j.jksus.2018.03.004

Poff, N. L., & Zimmerman, J. K. (2010). Ecological responses to altered flow regimes: A literature review to inform the science and management of environmental flows. *Freshwater Biology, 55*(1), 194–205.

Prasad, S. N., Ramachandra, T. V., Ahalya, N., Sengupta, T., Kumar, A., Tiwari, A. K., Vijayan, V. S., & Vijauan, L. (2002). Conservation of wetlands of India—A review. *International Society for Tropical Ecology*, 178–179. 10.1.1.526.3559. https://citeseerx.ist.psu.edu/viewdoc/download;jsessionid=E28ED107CE628F7F920F1AF44EE0F4E7?doi=10.1.1.526.3559&rep=rep1&type=pdf

Ramsar Convention Secretariate. (2016). The Fourth Ramsar plan 2016–2024. In *Ramsar handbook for the wise use of wetlands* (Vol. 2, 5th ed.). Ramsar Convention Secretariate. https://www.ramsar.org/sites/default/files/hb2_5ed_strategic_plan_2016_24_e.pdf

Ramsar_bn10. (2018). *Wetland restoration for climate change resilience*. Ramsar Convention for Wetland, Ramsar briefing notes 10. https://www.ramsar.org/sites/default/files/documents/library/bn10_restoration_climate_change_e.pdf

Ramsar_Convention. (2018). *Scaling up wetland conservation, wise use and restoration to achieve the sustainable development goals*. Ramsar Convention on Wetlands.

Rennell, J. (1781). *A Bengal atlas*. Sahitya Samsad. Reprint Edition, 2016; Reprinted by Kalyan Rudra.

Sengupta, K. (2018). Shrinkage of East Kolkata wetlands and its effect. *International Journal of Academic Research and Development*, 57.

Sharma, B. K. (1994). *Water pollution*. Goel Publishing House. https://books.google.co.in/books?id=R8HtwoFsOogC&pg=PA57&redir_esc=y#v=onepage&q&f=false

Sithirith, M. (2015). The Governance of Wetlands in the Tonle Sap Lake, Cambodia. *Journal of Environmental Science and Engineering, 335*, 10.17265/2162-5263/2015.06.004.

Srivastava, T., & Das, S. (2019). *The heritage of chandraketugarh: A report*. St. Stephen's College. https://www.academia.edu/39881764

Unnikrishnan, H., Manjunathan, B., Narendra, H., & Broto, V. C. (2020). *Water governance and the colonial urban project: The Dharmambudhi lake in Bengaluru, India*. Taylor & Francis. https://doi.org/10.1080/02723638.2019.1709756

Urban. (1738). *The Gentleman's Magazine and Historical Chronicle* (Vol. VIII). Edw, Cave, jun.

Willcocks, W. (1930). *Lecture on Ancient system of Irrigation in Bengal and its application to modern problems*. University of Calcutta.

Xu, T., Weng, B., Yan, D., Kun, W., Li, X., & Bi, Y. (2019). Wetlands of international importance: Status, threats, and future protection. *International Journal for Environmental Research and Public Health*, 14. https://doi.org/10.3390/ijerph16101818

Simulation and Prediction of Land Use/Land Cover Changes Using CA-Markov Model: A Case Study of Bulandshahr District, India

Sania Farukh, Zeba Khanam, Sania Javed, and Tariq Usmani

1 Introduction

In recent decades, urbanization has expanded across the globe, impacting natural resources, environment, and urban landscape (Gharaibeh et al., 2020). It is a natural phenomenon influenced by a variety of important features that are crucial to the development and growth of a region (Lavalle et al., 2001). Natural vegetation and agricultural areas have been replaced as a result of rapid urbanization (Prabu & Dar, 2018). However, peri-urbanism has gained importance as an important topic for urban planners and policy makers all over the world, but especially in developing nations, where its impact on the physical and social environments is undeniable (Khanam et al., 2022; Patel et al., 2019). The local ecosystems are significantly impacted by the urban and peri-urban areas. Inadequate planning and haphazard management of productive agricultural land, forests, surface water bodies, and ground water resources caused loss of an ecosystem (Lin et al., 2018; Sannigrahi et al., 2018).

Nevertheless, people have migrated to peri-urban and rural suburbs for greater economic opportunities and well-connected transportation networks that altered the landscape of these areas as well as promoting the construction of built-up sites (Somvanshi et al., 2020). Furthermore, significant land use and land cover (LULC) changes brought about by the expansion of peri-urban areas in India have had an effect on the environment and society; for instance, unanticipated growth and changes in LULC have occurred in Uttar Pradesh, India, as a result of the extension of census towns close to statutory towns (Mishra et al., 2018). Several studies explored the distinctive qualities of peri-urban regions and how they are impacted by urbanization and economic development in order to address these concerns (Mondal & Banerjee, 2021). However, by careful consideration of land use changes, it is possible to manage natural resources sustainably while also meeting human needs (Şen et al., 2018). As a result, it became necessary to comprehend urban expansion patterns and trends in order to determine the best mitigation strategies for controlling and guiding this growth. Therefore, forecasting urban growth patterns has become essential for sustainable development and ecological protection (Aburas et al., 2016).

However, managing urban landscape offers several benefits for the environment. Planning an urban landscape requires decisions regarding urban land for how it will develop in the future. By predicting land use change over the time, it is possible to conduct effective and sustainable landscape planning studies (Cetin et al., 2008; Cetin &

S. Farukh · S. Javed · T. Usmani
Department of Geography, Aligarh Muslim University, Aligarh, Uttar Pradesh, India

Z. Khanam (✉)
Jawaharlal Nehru University, New Delhi, India

Z. Khanam et al. (eds.), *Climate Change, Resilience and Ecosystem Services*, Sustainable Development Goals Series, https://doi.org/10.1007/978-3-031-96977-5_28

Sevik, 2016; Yucedag et al., 2018). The Markov chain (MC) model's application makes it possible to comprehend and keep track of spatial land use patterns and activities for LULC-based development (Bagaeen & Hijazi, 2019; Mirkatouli et al., 2015). This model has shown to be essential given the dearth of historical data since it can predict the distribution of land cover in the future using just current land cover data (Lillesand et al., 2003). It can also be used to determine how much a city has developed and to put required control mechanisms in place. According to Dadhich and Hanaoka (2012) and Xu et al. (2016), this model consists of a succession of random values, the probabilities of which change over time depending on previous values. Additionally, this model is employed when describing changes in the landscape is challenging because it is a suitable tool for forecasting land use/cover changes (Kumar & Agrawal, 2023; Mubea et al., 2011).

The process of temporal and spatial land use changes has been studied in several studies employing remote sensing and satellite imagery techniques, and the cellular automata (CA) Markov model has been used to predict land use change (Mubea et al., 2011; Tripathy & Kumar, 2019). Few research works have attempted to combine Geographic Information System (GIS) tools with stochastic modeling approaches to examine peri-urban land use and land cover shifts, despite its simplicity and benefits. Markov chain methods are used in geospatial environmental modeling by Ghosh et al. (2017) and Du et al. (2018), whose works predicted changes in LULC using satellite data. Gong et al. (2015) projected the outcomes into the future by using the trajectories to mimic the geographical pattern of land use. Furthermore, Halmy et al. (2015) investigated land use changes in Egypt's desert region and used the Markov model to project future changes. Liu et al. (2015) also employed a CA-Markov model to examine the dynamics of the landscape pattern in an inland river delta in the city of Harbin, the tenth largest city in China.

Moreover, Khanam et al. (2022) and Naikoo et al. (2022) have worked on the LULC transition of Bulandshahr city and National Capital Region (NCR) of India using satellite data. Bulandshahr District is urbanizing rapidly and it is changing the ecosystem of the area. Over the past 20 years, the district has experienced rearrangement of land from adjacent districts, leading to a significant shift in the district's land use scenario (Bhatt & Sharma, 2020). Nonetheless, this study claims to be the only work that has been done on Bulandshahr District LULC prediction, because land use simulations in Bulandshahr District had never before utilized the combined CA-Markov model; its application to land use forecasts is unprecedented. In this light, this chapter attempts to analyze the LULC transitions and projections in peri-urban Bulandshahr District of Uttar Pradesh from 1995 to 2022 for 28 years. Furthermore, the land use maps have been predicted for the years 2025 and 2040 using spatial modeling (Markov Chain and Cellular Automata) in Land Change Modeler.

2 Study Area

Bulandshahr District is located in southern most part of Meerut division and extends between 28°4′–28°40′N latitudes and 77°18′–78°28′E longitudes. On the west it is separated from NCR of Delhi and Haryana, and by the newly constituted district of Gautam Budh Nagar. On the east it is separated from the districts of Moradabad and Badaun. The total geographical area of the district is 364,974 hectares (ha) (Fig. 1). The area is a flat plain with some varying uneven terrain along river banks. Bulandshahr District is a significant industrial hub. Potteries have been found in the Khurja region, as well as chemical, metal, and ceramics, which has resulted in significant increase in the district's population over the last few decades (Khanam et al., 2022). The district had a population of 2,662,347 in 1995, which increased to 3,499,171 in 2011, and in the next census the district is expected to have a population of more than a million (Census, 2011). In the last two decades the district encountered reshuffling of area from among the nearby districts; therefore, the scenario of land utilization has changed in the district vigorously (Bhatt & Sharma, 2020). This remarkable expansion has resulted in extensive modifications in LULC that have prompted a lot of research, subsequently

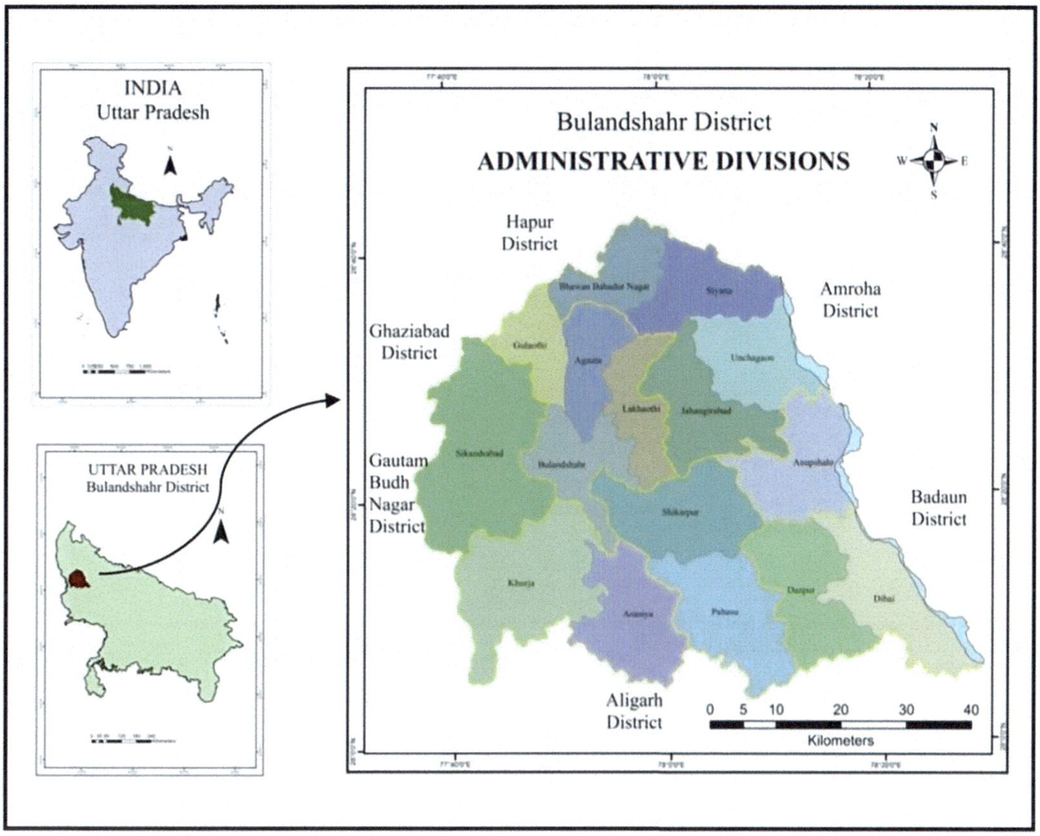

Fig. 1 Locational map of study area

(Naikoo et al., 2022; Lata & Lata, 2019; Suzanchi & Kaur, 2011).

3 Methodology

In this study, LULC maps of Bulandshahr District for the years 1995, 2005, 2015, and 2022 were generated with the help of satellite images; also, LULC maps for the years 2025 and 2040 were predicted with CA-Markov model, using GIS, ERDAS, TerrSet, and IDRISI software (Fig. 2).

3.1 Dataset and Preprocessing of Image

Remote sensing and GIS are a unique phenomenon that has played a major role in effective and well-organized mapping and monitoring enormous change of land use and land cover. In this study, Landsat datasets for 1995 Landsat 4–5 Thematic Mapper (TM), for 2005 Landsat 7 Enhanced Thematic Mapper Plus (ETM+), for 2015 Landsat 8 Operational Land Imager (OLI), and for 2022 Landsat 8 Operational Land Imager (OLI)/Thermal Infrared Sensor (TIRS) were utilized. The spatial resolution of all four Landsat datasets is 30 m, downloaded from the USGS website (https://earthexplorer.usgs.gov) (Table 1). Less than 10% of the sky was clouded over in the images that were collected from all the satellites. Image processing was done using remote sensing applications like ERDAS and ArcGIS. The radiometric correction was accomplished, and subsets from each of the Landsat sensors for several years were categorized using the Maximum Likelihood Classification methodology. Using supervised classification method, a multispectral composition-based class assign-

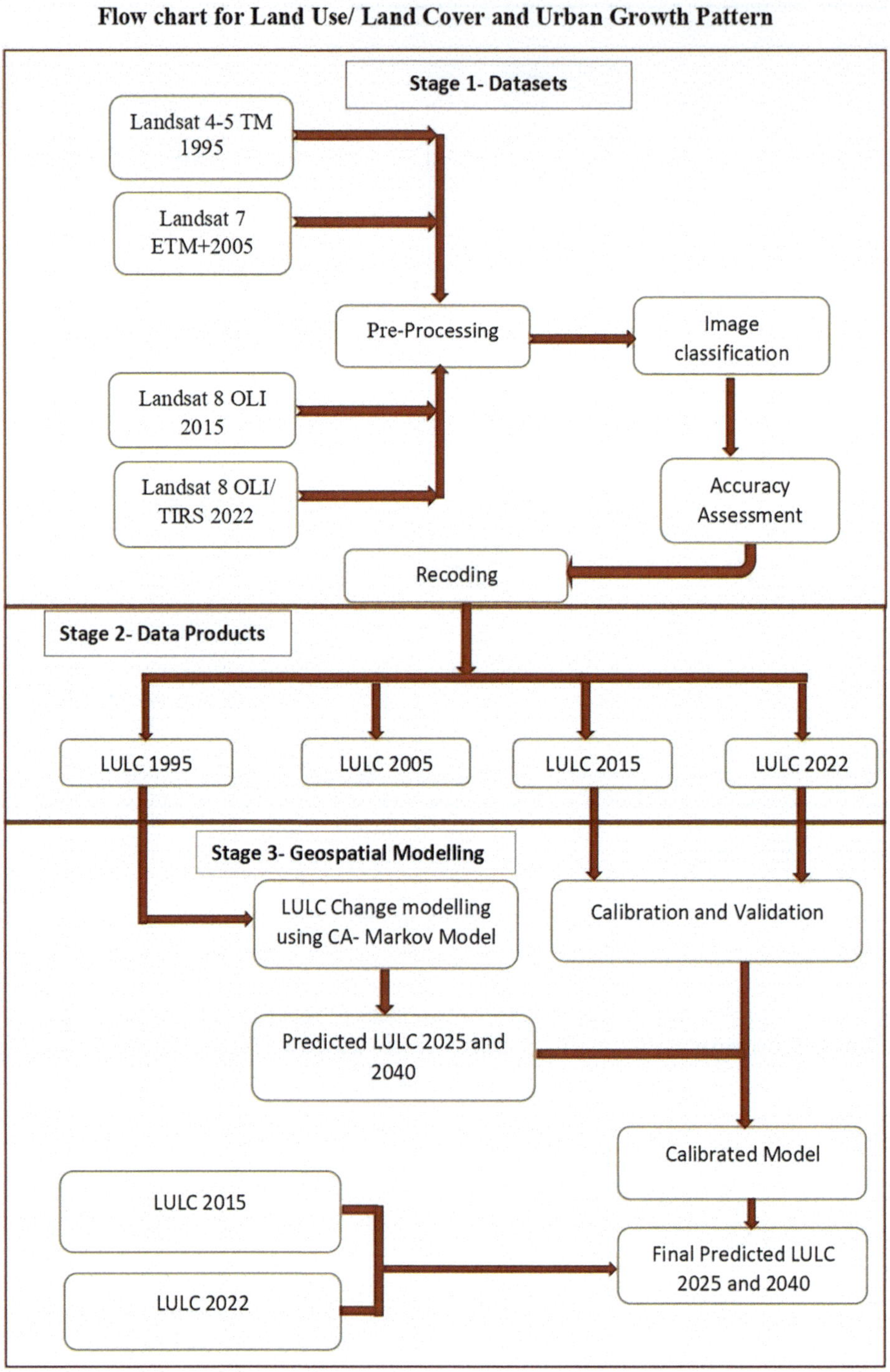

Fig. 2 Study flow chart

ment is made for each pixel on the map. In addition to these satellite photos, ground truthing was done with the help of Google Earth and Toposheet. Five types of LULC—water, built-up, agriculture, dense vegetation, and barren land—were identified over the 27 years between 1995 and

Table 1 Downloaded Landsat satellite images information from USGS

Year	Landsat satellite	Sensor	Resolution	Date
1995	Landsat 4–5	TM	30	19-10-1995
2005	Landsat 7	ETM+	30	22-10-2005
2015	Landsat 8	OLI	30	10-10-2015
2022	Landsat 8	OLI/TIRS	15	17-10-2022

2022 using the image processing signature file generating tool (Gounaridis et al., 2019; Brown et al., 2013).

3.2 Accuracy Assessment of LULC Maps

The final and most crucial step in the classification of the images is the accuracy assessment (Foody, 2002). It aims at quantifying the degree to which the pixels were accurately sampled into the relevant land cover classifications. The error matrix has shown to be the most effective way for determining how accurately categorizations performed using remotely sensed data were produced (Congalton, 1991). According to Lu and Weng (2007), this approach is often used to evaluate per-pixel categorization. For each classed map, the Kappa statistics/index was also created to assess the correctness of the findings. In this study, accuracy was performed for the years 1995, 2005, 2015, and 2022 using the standard Kappa coefficient, overall accuracy, producer accuracy, and user accuracy. Five categories were used to categorize the image: water, built-up, agriculture, dense vegetation, and barren land. The accuracy of the user is used to determine how frequently the class shown on the map is truly present. The accuracy of the producer is shown by the number of pixels that are accurately classified to a class and solely to that class. The connection between actual data and the classified information followed from using a confusion matrix (Pontius & Millones, 2011). An appropriate Kappa statistics for the stratified random sample was calculated using the equation below (Petropoulos et al., 2015).

$$\text{Kappa co-efficient} = \left(\left(T \times C\right) - G\right) / \left(T2 - G\right) \quad (1)$$

where G is the total sum of the multiplied total values, C is the observation of the properly recognized pixels, and T is the test pixels.

3.3 Cellular Automata

A popular spatial modeling method for simulating urban processes is called cellular automata (CA). The capacity of CA to predict proximity—a crucial spatial component that captures the rate at which land use patterns change—is what gives it its widespread appeal. According to CA, if the nearby regions likewise fall into that land use category, an area has a higher inclination to do so (Memarian et al., 2012). The core concept of CA is that past urban development patterns impact present development as a consequence of local interactions, which cumulatively form global urban growth patterns (Sante et al., 2010). A grid of cells divides the area in CA. On the area map, each cell is comparable to one pixel. One of the land use types is represented by the cellular state. According to a set of transition rules (the cell neighborhood), the state of the cell in the next stages is determined by its present state as well as the states of its neighbors. The two neighborhood types that are most frequently used in the state are Von Neumann and Moore (Memarian et al., 2012).

3.4 CA-Markov Chain Model

The CA-Markov model integration is useful for simulating and forecasting changes in land use (Singh et al., 2015). Cellular automata and a transition probability matrix produced by cross-tabulating two distinct images combine to form the CA-Markov model. A reliable method for modeling spatiotemporal dynamics is provided by the combination of the CA and Markov models (Wang & Zhang, 2001). Moreover, CA is combined with Markov to give the model a spatial quality. Stated differently, the CA-Markov chain has the ability to predict and simulate two-way transitions between any numbers of categories (Pontius & Malanson, 2005).

It is important to note that the land use cover change is demonstrated using a dynamic process called cellular automata. The literature on land use modeling appropriately uses a lot of these kinds of models. Due of their dynamic nature and ability to replicate, individual cells can represent both land parcels and self-growth interactions (Brown et al., 2004). Moreover, changes in neighboring cells and the current state can shed light on land use changes at any given site (cell), simulating two directions of growth. In spatial modeling, this approach is widely used to forecast future land use (Ye & Bai, 2008).

Markov chain model employs a transition probability matrix to depict potential changes between various land uses. The likelihood of switching a pixel's status from one land use to another is represented by each item in the matrix (Guan et al., 2011). To determine the likelihood of a transition between these two time periods, at least two separate maps of the same area with two different time stamps are required (Rimal et al., 2017). This Markov chain model, however, is unable to create model changes in spatial dimensions since it neglects to take nearby cells' influences into consideration. Because of this, the study of urbanization and land use change has seldom used MC (Arsanjani et al., 2013). Since the CA-MC technique combines the CA and MC strengths, it has been suggested to integrate both into a single framework (Guan et al., 2011; Arsanjani et al., 2013; Moghadam & Helbich, 2013; Aburas et al., 2016). Because it considers both spatial and temporal changes, the CA-MC technique is successful (Singh et al., 2015; Hamad et al., 2018). The transition rules of CA may be created in a variety of ways to take into account these components since CA-MC has an open structure that makes it simple to incorporate a range of external socio-economic and environmental aspects that have an influence on the urbanization processes (Santé et al., 2010). LULC maps of the Bulandshahr District from the years 2015 and 2022 were utilized in this investigation to predict 2025 and 2040 maps. TerrSet software was used to build the transition matrix.

3.5 Model Validation

The 2025 and 2040 land use maps were used to validate the model. The CA-MC model has their accuracy assessed, and the outcomes will be contrasted. Values above 80% (Eastman & Sangermano, 2006, Eastman, 2012) or 75% (Mitsova et al., 2011) are frequently associated with acceptable validity. The validation model identifies areas of agreement and disagreement between the reference (existing) and simulated maps. There is a quantifiable variance when cells in a category on the simulation map diverge from the same category on the reference map. A location disagreement occurs when a cell's placement in a category of the simulation map disagrees with the position of the same category on the reference map. This methodology's calculations were based on an idea that Pontius and Millones (2011) suggested. Additionally, the Kappa coefficient utilizing the confusion matrix was calculated by crossing and comparing with the real map approach. To determine Kappa indices, the TerrSet software's validation mechanism is used in conjunction with the actual known map for 2025 and 2040.

4 Results

4.1 Land Use and Land Cover Change in the Bulandshahr District from 1995 to 2022

The classification of Maximum Likelihood is used to examine changes in land use and land cover. The study area's classified LULC maps for 1995–2022 were classified into five categories of land use: aquatic bodies, built-up areas, agricultural, areas with thick vegetation, and barren surface. Over a period of the time, LULC changes were seen, ranging from 1995 to 2005, 2005 to 2015, and 2015 to 2022. Agriculture and built-up areas were discovered to be the primary land use category in the Bulandshahr District. The total area distributed among the various categories and the percentage of cover for LULC classes in the year 1995, 2005, 2015, and 2022 are shown in Table 2 (Fig. 3 a–c). Built-up areas grew quickly throughout the examined period. The built-up areas grew from 151,510.25 ha (4.33%) in 1995 to 24,159.29 ha (6.74%) in 2005 and built-up areas significantly increased as the overall area covered in 2015 increased to 28,311.51 ha (7.90%) and subsequently increased to 42,620.35 ha (11.89%) in 2022. Population growth contributed to the ongoing loss of agricultural land. In contrast, agriculture areas had a reduction in area overall as they went from 330,683.17 ha (92.28%) in 1995 to 323,039.95 ha (90.14%) in 2005. Agriculture areas' total area decreased between 2015 and 2022 from 318,547 ha (88.89% of all land area) in 2015 to 303,566.68 ha (84.71%) in 2022. The largest transition in agricultural class was into built-up area. On the other hand, the land use classes only underwent minor adjustment of water bodies, dense vegetation, and barren land. Dense vegetation in the study area increased from 4588.84 ha (1.28%) in 1995 to 4831.68 ha (1.35%) in 2005, followed by a rise in 2015 to 5445.86 ha (1.52%) and then to 6059.03 ha (1.69%) in 2022. The percentage cover of barren land showed marginal area change; it decreased from 0.79 to 0.70% within 1995–2005 and then minimal increase in 2015, i.e., 0.79 to 0.81% in 2022. Water bodies land declined throughout the time from 1.32% in 1995 to 1.07%, 0.90%, and 0.81% in the years 2005, 2015, and 2022, respectively (Table 3).

4.2 Accuracy Assessment of LULC Maps

The final maps of land use and land cover for the four research time periods were created by using an error/confusion matrix for 1995, 2005, 2015, and 2022 displayed in Fig. 3, having a map accuracy of 89% overall across all pictures. According to Lu and Weng (2007), this approach is often used to evaluate per-pixel categorization. For each classed map, the Kappa statistics/index was also created to assess the correctness of the findings. Each of the two created land use and land cover classification map has a Kappa value of 0.89. Due to this typically decent accuracy, this allowed for the subsequent analysis and change detection (Lea & Curtis, 2010).

For the four classed maps, an accuracy evaluation was done to verify if what was mapped matches what actually there was. The error matrix was used to measure the categorization accuracy. The Kappa coefficient, user and producer accuracy, and total accuracy were used to assess the four category maps' accuracy. For 1995, 2005, 2015, and 2022, the overall accuracies were 85%, 88%, 87%, and 89%. The four categorized maps satisfied the USGS's classification standard of an overall accuracy of 85% in identifying land use and land cover classes using remote sensor data. Additionally, the Kappa coefficients for all four classed maps were 0.836, 0.85, 0.837, and 0.875 in 1995, 2005, 2015, and 2022, respectively, thus higher than 0.75, indicating that the categorization is significantly better. The error matrix provided in Table 4 presents the land cover map's accuracy for each class and year as chosen by the user and the creator.

In the present analysis, mainly post-classification technique was employed. It is the most common method to detect change in LULC of an area. Table 4 shows the percentage of change in study area and change in each LULC class. To produce the LULC change map, pair of maps were overlaid between the dates. The cross-tabulation matrices show the major changes were observed from agriculture to built-up land. The outcome is the built-up area's expansion due to the conversion of agricultural and natural vegetation land during the period of 36 years from 1995 to 2022. Built-up area in Bulandshahr block has extended in north-west direction. The conversion of agriculture land into the built-up area shows a flourishing market, buildings, and other amenities during this period. As we know that population is increasing at very fast pace, it shows that the burden of population is directly on the agricultural land. For better facilities like education and health, people are migrating from nearby cit-

Table 2 LULC dynamics of Bulandshahr District from 1995 to 2022

Classes	1995		2005		2015		2022	
	Hectares	Percentage	Hectares	Percentage	Hectares	Percentage	Hectares	Percentage
Water bodies	4743.94	1.32	3835.67	1.07	3233.89	0.90	3194.6	0.89
Built-up	15,510.25	4.33	24,159.29	6.74	28,311.51	7.90	42,620.35	11.89
Agriculture	330,683.17	92.28	323,039.95	90.14	318,547.44	88.89	303,566.68	84.71
Dense vegetation	4588.84	1.28	4831.68	1.35	5445.86	1.52	6059.03	1.69
Barren land	2837.43	0.79	2495.91	0.70	2824.77	0.79	2919.72	0.81

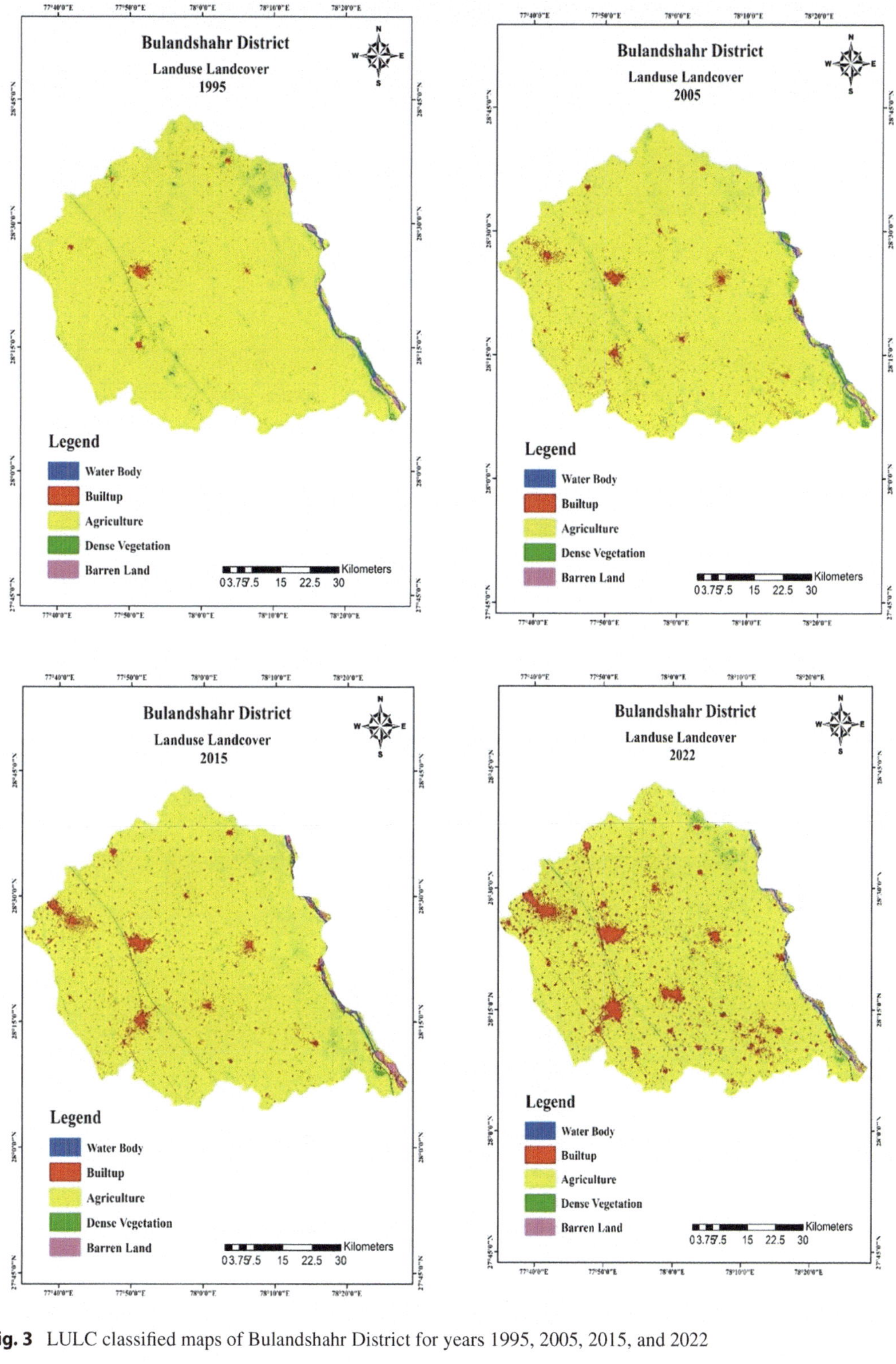

Fig. 3 LULC classified maps of Bulandshahr District for years 1995, 2005, 2015, and 2022

Table 3 Area changes for the various LULC classes between the years of 1995–2005, 2005–2015, and 2015–2022

	1995–2005		2005–2015		2015–2022	
Classes	Hectares	Percentage	Hectares	Percentage	Hectares	Percentage
Water bodies	−908.26	−0.25	−601.79	−0.17	−39.29	−0.01
Built-up	8649.04	2.41	4152.22	1.16	14,308.84	3.99
Agriculture	−7643.22	−2.13	−4492.51	−1.25	−14,980.76	−4.18
Dense vegetation	242.85	0.07	614.18	0.17	613.17	0.17
Barren land	−341.52	−0.10	328.86	0.09	94.95	0.03

Table 4 Accuracy assessment of LULC from 1995 to 2022

	1995		2005		2015		2022	
Classes	User's Accuracy	Producer Accuracy	User's Accuracy	Producer Accuracy	User's Accuracy	Producer Accuracy	User's Accuracy	Producer Accuracy
Water bodies	88.23 per cent	75 per cent	95.00 per cent	100 per cent	100.00 per cent	100 per cent	97.00 per cent	100.00 per cent
Built up	84.20 per cent	80 per cent	85 per cent	85 per cent	90.00 per cent	82 per cent	93.00 per cent	87.00 per cent
Agriculture	75 per cent	90 per cent	85 per cent	85 per cent	85 per cent	85 per cent	88.00 per cent	85 per cent
Dense Vegetation	95 per cent	95 per cent	95.00 per cent	90 per cent	90.00 per cent	81 per cent	93.00 per cent	89 per cent
Barren Land	85 per cent	85 per cent	80 per cent	80 per cent	70 per cent	73 per cent	85 per cent	83 per cent
Overall Accuracy	85 per cent		88 per cent		87 per cent		89 per cent	
Kappa Coefficient	0.8369		0.85		0.8375		0.875	

ies like Bulandshahr, Khurja, and Sikandrabad, which subsequently opens up the area for the better development. A decline in the rate of change of agriculture land on the peripheral parts of the district indicates that there is discontinuation in the agriculture activities. The substantial growth in built-up area in Bulandshahr municipality indicated that planned development is a necessity of the modern cities; otherwise, it would create problems in near future.

4.3 Spatiotemporal Changes of LULC (Change Matrix) from 1995 to 2002

Table 5 offers a depiction of areal change detection or alteration of LULC classes from 1995 to 2022. The figure effectively portrays the transition of area from one class to another, providing a clear representation of the changes in land cover composition over the specified time. Upon careful observation of Fig. 4, a discernible pattern emerges, particularly in the context of agricultural land. It becomes evident that the agricultural domain experienced a notable reduction in its expanse, yielding significant portions to other classes such as barren land and dense vegetation. Of notable significance is the substantial transformation witnessed from agriculture to built-up areas, signifying a pronounced shift toward urbanization. Table 6 presents a detailed account of the transitions in LULC classes within the study area over the span of 27 years, from 1995 to 2022. Notable transformations were observed across various categories. Specifically, a portion of agricultural land in 1995, amounting to 1383.77 ha, transitioned into barren land by 2022. This shift could be attributed to factors such as soil degradation or changes in agricultural practices. Additionally, a substantial area of agricultural land, total 24,779.95 ha, underwent a significant transformation into built-up areas over the same period, signaling extensive urbanization. Furthermore, areas previously designated for agriculture experienced a transition into regions characterized by dense vegetation, amounting to 5294.99 ha. This phenomenon may be attributed to natural regrowth or afforestation initiatives. Additionally, a minor portion of agricultural land, encompassing 1151.12 ha, transitioned into water bodies, indicating alterations in hydrological patterns. Table 6 also reveals that in 1995, a significant portion of barren land, total 1442.58 ha, underwent a transformation into agricultural land by 2022 and 648.13 ha of barren land transitioned into built-up areas during the same time frame. This transition indicates a shift in the characteristics or utilization of previously barren areas over the specified period. Furthermore, 2347.46 ha of land previously categorized as water bodies evolved into agricultural land, while 564.34 ha were converted into built-up areas. Notably, a smaller parcel of land, encompassing 151.07 ha, transformed into dense vegetation. Contrary to that, 3510.96 ha of dense vegetation transformed into agriculture and 430.86 ha transformed into built-up area since 1995 to 2022.

4.4 Prediction of LULC Change for 2025 by CA-ANN and CA-Markov Model

The predicted patterns of city expansion gradually displacing other key land types, such as forest and agricultural land, have been seen in prior urbanization studies in developing countries (Huang et al., 2008; Han & Song, 2015; Appiah et al., 2015; Rimal et al., 2017). The application of CA-Markov prediction model came up with

Table 5 Change matrix of Bulandshahr District from 1995 to 2022

LULC Classes		2022					
		Agriculture	**Barren Land**	**Built up**	**Dense Vegetation**	**Water bodies**	**Total**
1995	**Agriculture**	**288671.94**	1383.77	24779.95	5294.99	1151.12	321281.77
	Barren Land	1442.58	**648.13**	401.36	111.15	534.50	3137.72
	Built up	0.00	0.00	**15509.42**	0.00	0.00	15509.42
	Dense Vegetation	3510.96	189.19	430.86	**419.46**	247.53	4798.00
	Water bodies	2347.46	640.88	564.34	151.07	**1149.39**	4853.14
	Total	**292972.94**	**2862.14**	**41685.93**	**5976.67**	**382.54**	**343880.22**

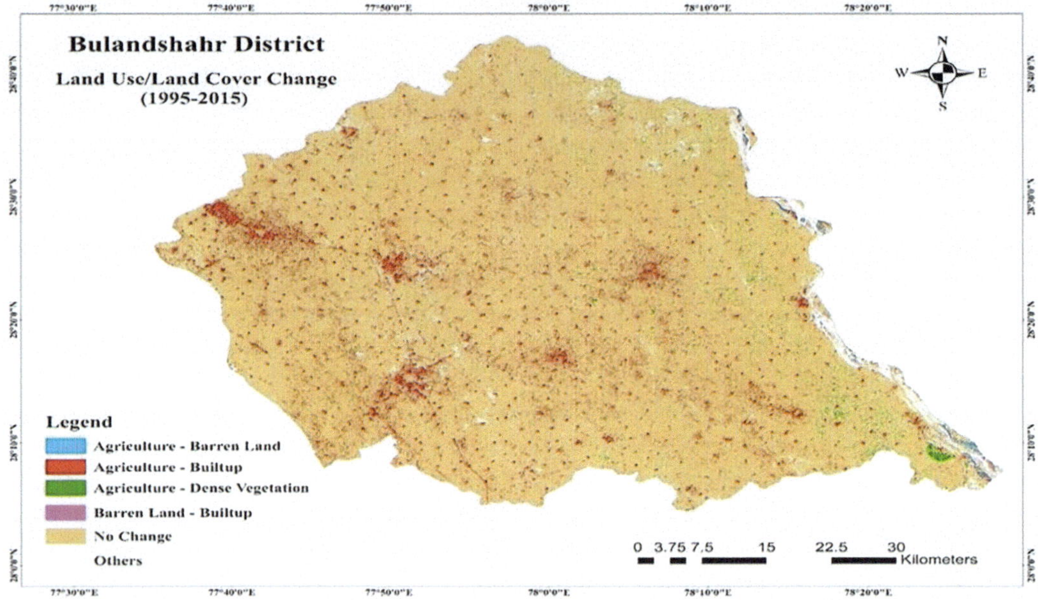

Fig. 4 LULC change matrix (1995–2022)

Table 6 LULC prediction maps of Bulandshahr District for 2025 and 2040

Classes	Stimulated 2025		Stimulated 2040	
	Hectares	Percentage	Hectares	Percentage
Water bodies	3195.41	0.89	3236.4	0.90
Built up	43126.96	12.03	52005.51	14.52
Agriculture	302565.48	84.43	294634.8	82.25
Dense Vegetation	6207.31	1.73	5514.84	1.54
Barren Land	3268.22	0.91	2828.79	0.79

the following result as shown in Table 6. The predicted LULC analyses for 2025 show that there will be further decrease in the water bodies (3195.41 ha; 0.89%) and agriculture (302,565.48 ha; 84.43%) type of land use in comparison to 2015 (Table 1). Built-up area 43,126.96 (12.03%) increased whereas dense vegetation 6207.31 ha (1.73%) and barren land 3268.22 ha (0.91%) has decresed in the total area of Bulandshahr city (Fig. 5).

The program IDRISI CA-Markov module is used to forecast LULC in 2040. To anticipate LULC change using a Markov chain, the transition area matrix from 2015 to 2022 is employed. This matrix obtained the data necessary to train the CA-Markov model. Weights are assigned to the driving elements as part of the Markov module analysis and a set of LULC transition conformance pictures acquired while running the MCE module, offering guidelines for the acceptability of LULC alterations. The built-up area rose by 14.52% from 2022 according to the LULC forecast findings for 2034. Additionally, the categories of dense vegetation, barren land, and agricultural land dropped between 2040 and 2025 by 0.19%, 0.12%, and 2.18%, respectively.

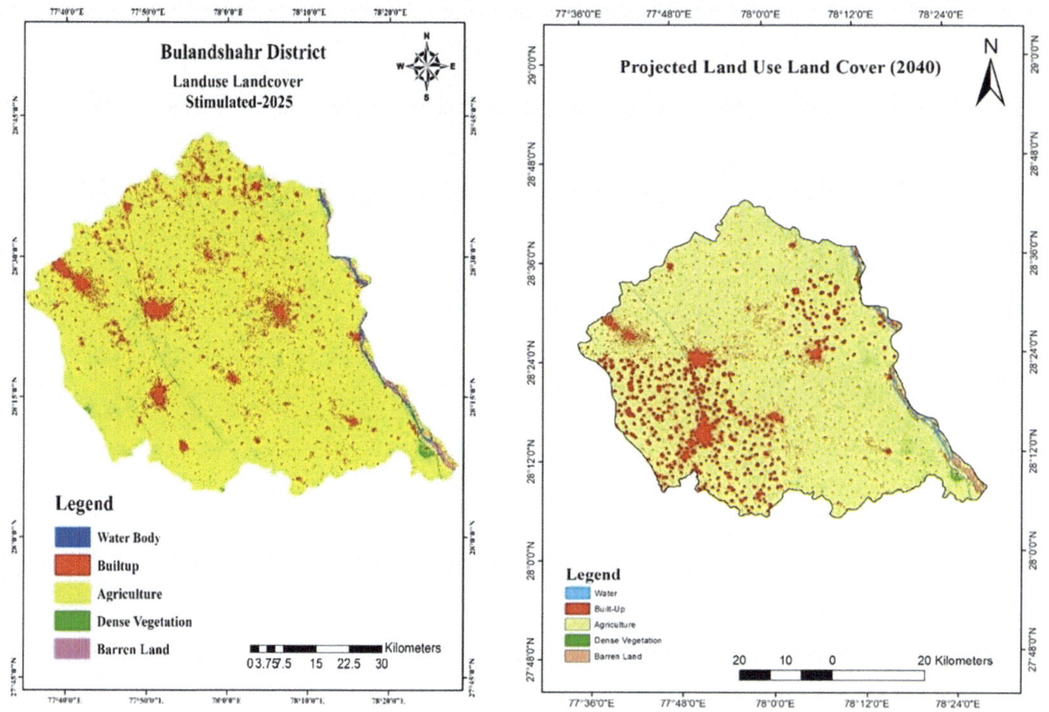

Fig. 5 Predicted maps of 2025 and 2040

The Bulandshahr District LULC shift was marked by a decrease in agricultural land and a rise in newly developed area, mostly driven by the settlement distance. Additionally, the characteristics of road systems and public facilities that are close to agricultural regions also aid in the growth of new settlement. This analysis demonstrates that the settlement class saw a growing tendency from 1995 to 2040. Settlement in the Bulandshahr District is expected to rise by 4.33% per year or by 14.22% over the course of 45 years (1995–2040), while agricultural land, dense vegetation, and barren land are expected to drop. A remarkable correlation has been noticed in population increase and land use changes (Figs. 6 and 7).

5 Policy Measures to Prevent Land Conversion

Overall, this study generally confirms that patterns of change in land use and land cover are crucial to examine if future changes in landscapes and their trajectory are to be forecast. Urban policy and urbanization policy are the two fundamental parts of urban policy. Urbanization policy addresses the location of urban places, both big and small, within the larger framework of human habitation. According to this viewpoint, urbanization policy addresses the spatial dimensions of development in both urban and rural areas (Gnaneshwar, 1995). Nevertheless, following the economic liberalization of the Indian economy in the 1990s, private real estate owners were granted the liberty to expand the number of residential complexes and collaborate with public enterprises to stimulate the development of infrastructure in areas beyond the city limits. Therefore, these private developers could afford more; they thereby drove the urban poor out of the urban perimeter. Furthermore, the rise of manufacturers and businesses in the Bulandshahr area over the past few decades has been tremendous, especially in the pottery industry (Sodhi, 2006). This has served as an attraction for migrants from other rural areas in Uttar Pradesh to Bulandshahr District. This area has turned into an economic growth hub as a result of the migration of the sur-

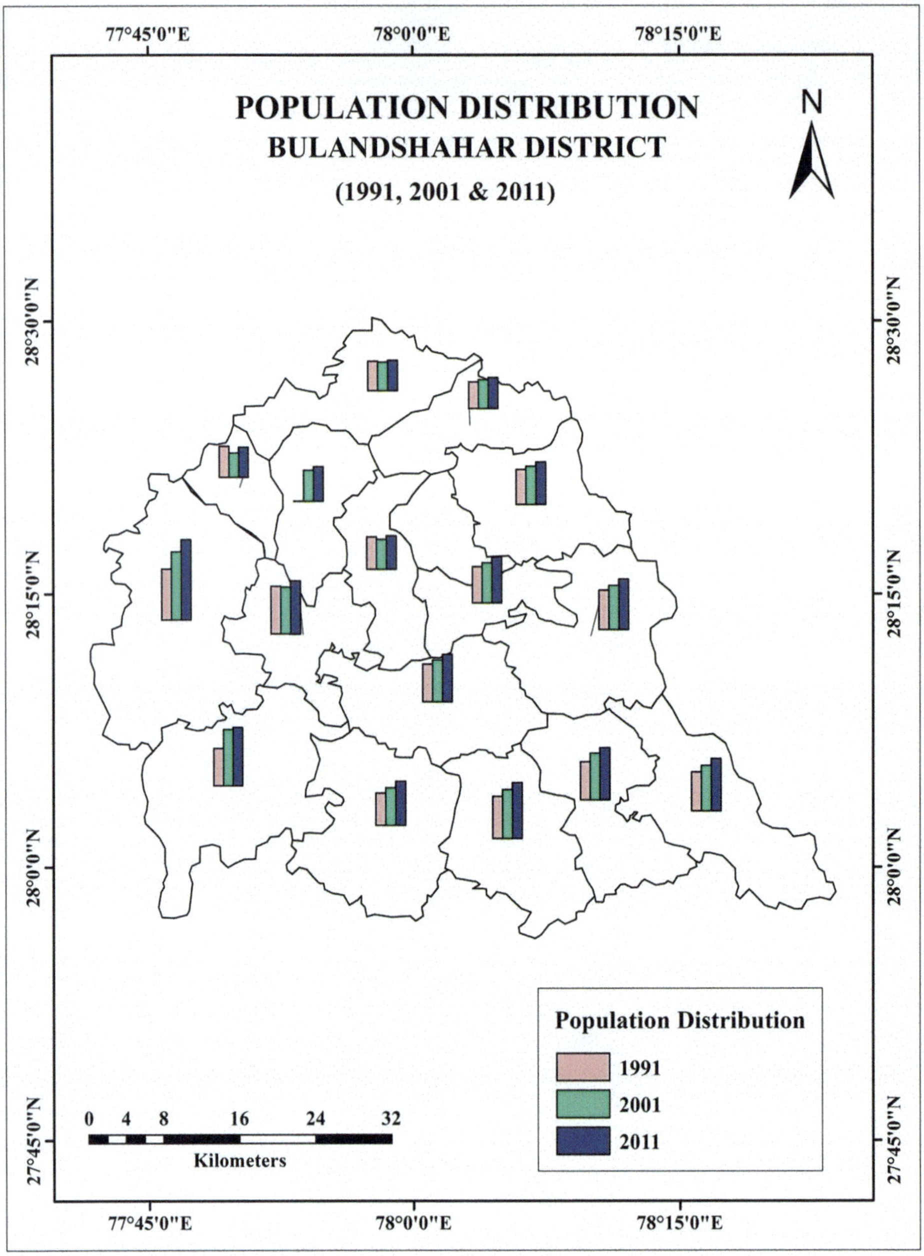

Fig. 6 Population distribution map of Bulandshahr District (1991, 2001, 2011)

rounding rural poor into urban area in search of jobs and industrial expansion, which led to the general development of the region's infrastructure (Aggarwal, 2010). The necessity to offer residential units for these in-migrants is urgent given the current increase in migrant populations and their needs (Mitra et al., 2013). Consequently, Indian cities are increasingly being surrounded by unplanned, haphazard suburban sprawl, even as they grow outward to accommodate increasing populations due to the growing migrant populations and their needs; it is imperative to supply residential units for these in-migrations (Pucher et al., 2007; Mitra et al., 2013). This has pushed

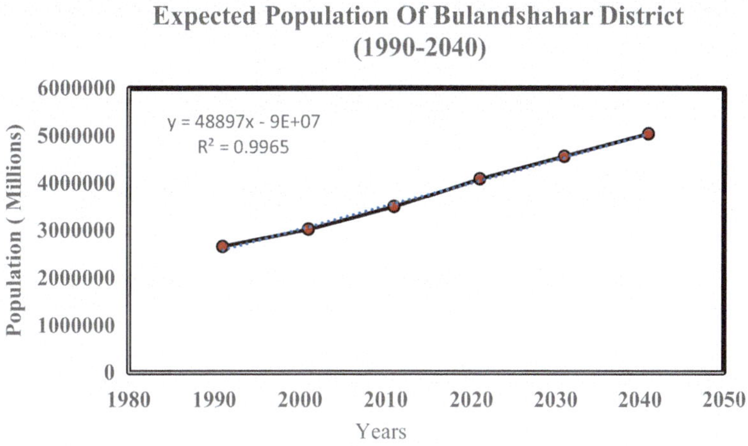

Fig. 7 Population change from 1991 to 2040 (predicted)

up urbanization, which would eventually cause built-up areas to spread outside of municipal control and around its periphery, resulting in unrestrained land conversion and an increase in land values (Naikoo et al., 2022; Khanam et al., 2022).

The fundamental natural resource that humans have access to is a land, which is used for a variety of purposes and in a variety of ways. Accordingly, the land and land use pattern of a region are directly impacted by any change in the human population (Nelson et al., 2010). Most population changes come in the form of increases brought either by migration or natural growth (Van Bavel, 2013). These gains mostly came at the price of reducing the utilization of agricultural land in the periphery. More residential land uses are being put to other purposes than unoccupied land. This study also discovered that whereas agricultural land fell by −4.18%, the built-up area grew by 3.99% during the same period. Natural systems become unsafe due to the growth of haphazard infrastructure, which affects living areas occupying the forefront. This exact point was demonstrated by a related study conducted in this context: more man-made living spaces lead to more "individual ecological footprints," or consumption habits, which in turn deteriorate environmental health (Banerji, 2021). In order to achieve this level of sustainable urban growth, planners must implement intricate government-mandated urban regulations that prohibit the conversion of wetlands into wastelands and wasteland into built-up regions. In order to assist planners in classifying areas to be chosen for urban growth and others to be left intact, such policies ought to set regulations (Biswas et al., 2020). Planners and administrators in the Bulandshahr District can make the most use of the study's results while trying to balance development in peri-urban areas.

6 Conclusion

For safe, sustainable development and efficient planning and use of natural resources, it is essential to comprehend the spatial pattern, size, and trends of land use and land cover changes. This work has shown that an integrated CA-Markov model—a combination of cellular automata and Markov models incorporated into IDRISI software—can be used to predict the spatial-temporal LULC pattern and trends of the Bulandshahr District for 2025 and 2040. This study, which uses a CA-Markov model to predict future land use and land cover changes of Bulandshahr District, is the first of its sort in terms of documented literature. To attain optimal and superior LULC outcomes in the future, the model was verified by the utilization of Kappa modifications, specifically K_{no}, $K_{location}$, and $K_{quantity}$. Based on the model validation results, it can be concluded that the CA-Markov model is a valuable tool for accurately predicting future spatial-temporal LULC dynamics in the terrain under study.

The population of Delhi NCR has increased approximately equally and constantly as a result of both immigration and natural growth. The fast changes in urban land uses brought on by Bulandshahr urban growth are more obvious in the city's periphery. The area covered by various built-up classifications, including residential, commercial, and industrial, as well as the amount of undeveloped land, have all increased significantly in the studied region. There are additional factors, such land speculation, that contribute to the loss of agricultural land besides urbanization. Poor agricultural yield, and salinity brought on by intensive irrigation, has also led to an increase in vacant land. Unplanned and unregulated urban growth in Bulandshahr District has led to haphazard building, mostly along main throughfares. In order to ensure that the anticipated uncontrolled expansion of bare land and built-up areas does not have a negative impact on the environment, planners, policy makers, and local communities in the studied landscape should prepare better strategies and land use policies in response to the undesired predicted LULC changes. Lastly, this chapter suggests conducting comparative research across India's various landscapes in order to adapt the CA-Markov model to other districts with comparable settings.

Declarations of Interests The authors declare that they have no known competing financial interest or personal relationships that could have appeared to influence the work reported in this chapter.

Ethics Approval Not applicable to this study.

Conflict of Interest The authors declare that they have no conflict of interest.

Consent of Participation Every author has equal participation.

Consent of Publication This research work is novel and nowhere has been submitted for publication.

Clinical Trial Number There is no clinical trial since it is a geographical case study.

Funding The authors declare that no funds, grants, or other support were received during the preparation of this manuscript. The corresponding author would like to thank UGC for providing SRF fellowship for PhD.

Data Availability All the data generated and analyzed during this study are available on request from the corresponding author.

References

Aburas, M. M., Ho, Y. M., Ramli, M. F., & Ash'aari, Z. H. (2016). The simulation and prediction of spatio-temporal urban growth trends using cellular automata models: A review. *International Journal of Applied Earth Observation and Geoinformation, 52*, 380–389.

Aggarwal, A. (2010). *Economic impacts of SEZs: Theoretical approaches and analysis of newly notified SEZs in India.*

Appiah, D. O., Hagan, D. O., Danso, J. A., & Nkansah-Boadu, F. (2015). Evaluating the effects of urban expansion on agricultural land use change in Kumasi, Ghana. *Journal of Land Use Science, 10*(3), 307–322.

Arsanjani, J. J., Helbich, M., Kainz, W., & Boloorani, A. D. (2013). Integration of logistic regression, Markov chain and cellular automata models to simulate urban expansion. *International Journal of Applied Earth Observation and Geoinformation, 21*, 265–275.

Bagaeen, S., & Hijazi, H. (2019). The role of land use change in developing city spatial models in Jordan: The case of the Irbid master plan (1970–2017). *Alexandria Engineering Journal, 58*(3), 861–875.

Banerji, S. (2021). Determinants of rural-urban differential in healthcare utilization among the elderly population in India. *BMC Public Health, 21*(1), 939.

Bhatt, G. D., & Sharma, J. K. (2020). Land use/cover change detection in the western part of Upper Indo-Gangetic Plains of Uttar Pradesh, India–a geospatial approach. *International Journal of Ecology and Environmental Sciences, 46*(1).

Biswas, M., Banerji, S., & Mitra, D. (2020). Land-use–land-cover change detection and application of

Markov model: A case study of Eastern part of Kolkata. *Environment, Development and Sustainability, 22*, 4341–4360.

Brown, D. G., Walker, R., Manson, S., & Seto, K. (2004). Modeling land use and land cover change. In *Land change science: observing, monitoring and understanding trajectories of change on the earth's surface* (pp. 395–409). Springer.

Brown, D. G., Verburg, P. H., Pontius, R. G., Jr., & Lange, M. D. (2013). Opportunities to improve impact, integration, and evaluation of land change models. *Current Opinion in Environmental Sustainability, 5*(5), 452–457.

Cetin, M., Musaoglu, N., & Tanik, A. (2008). Multitemporal assessment of land-use change in a rapidly urbanizing coastal region in Turkey using remote sensing. *Environmental Engineering Science, 25*(6), 917–928.

Cetin, M., & Sevik, H. (2016). Evaluating the recreation potential of Ilgaz mountain national park in Turkey. *Environmental Monitoring and Assessment, 188*(1), 52.

Congalton, R. G. (1991). A review of assessing the accuracy of classifications of remotely sensed data. *Remote Sensing of Environment, 37*(1), 35–46.

Du, G., Shin, K. J., Yuan, L., & Managi, S. (2018). A comparative approach to modelling multiple urban land use changes using tree-based methods and cellular automata: The case of Greater Tokyo Area. *International Journal of Geographical Information Science, 32*(4), 757–782.

Dadhich, P. N., & Hanaoka, S. (2012). Spatial investigation of the temporal urban form to assess impact on transit services and public transportation access. *Geo-spatial Information Science, 15*(3), 187–197.

District Census Handbook, 2011

Foody, G. M. (2002). Status of land cover classification accuracy assessment. *Remote Sensing of Environment, 80*(1), 185–201.

Ghosh, P., Mukhopadhyay, A., Chanda, A., Mondal, P., Akhand, A., Mukherjee, S., et al. (2017). Application of Cellular automata and Markovchain model in geospatial environmental modeling—A review. *Remote Sensing Applications: Society and Environment, 5*, 64–77.

Gong, W., Yuan, L., Fan, W., & Stott, P. (2015). Analysis and simulation of land use spatial pattern in Harbin prefecture based on trajectories and cellular automata—Markov modelling. *International Journal of Applied Earth Observation and Geoinformation, 34*, 207–216.

Gharaibeh, A., Shaamala, A., Obeidat, R., & Al-Kofahi, S. (2020). Improving land-use change modeling by integrating ANN with Cellular Automata-Markov Chain model. *Heliyon, 6*(9), e05092.

Gnaneshwar, V. (1995). Urban policies in India—Paradoxes and predicaments. *Habitat International, 19*(3), 293–316.

Gounaridis, D., Chorianopoulos, I., Symeonakis, E., & Koukoulas, S. (2019). A Random Forest-Cellular Automata modelling approach to explore future land use/cover change in Attica (Greece), under different socio-economic realities and scales. *Science of the Total Environment, 646*, 320–335.

Guan, D., Li, H., Inohae, T., Su, W., Nagaie, T., & Hokao, K. (2011). Modeling urban land use change by the integration of cellular automaton and Markov model. *Ecological Modelling, 222*(20–22), 3761–3772.

Hamad, R., Balzter, H., & Kolo, K. (2018). Predicting land-use/land cover changes using a CA-Markov model under two different scenarios. *Sustainability, 10*, 3421.

Han, L., & Song, C. (2015). Urban expansion dynamics and natural habitat loss in China: A multiscale landscape perspective. *Global Change Biology, 21*(1), 288–306.

Halmy, M. W. A., Gessler, P. E., Hicke, J. A., & Salem, B. B. (2015). Land use/land cover change detection and prediction in the north-western coastal desert of Egypt using Markov-CA. *Applied Geography, 63*, 101–112.

Huang, J., Lu, X., Sellers, J. M., & Jia, X. (2008). Spatiotemporal dynamics of urban expansion in the Beijing metropolitan area, China. *Landscape and Urban Planning, 85*(1–2), 85–96.

Khanam, Z., Siddiqui, M. A., Faridi, R. A., Ali, S. A., & Khan, Z. (2022). Analysing land use land cover changes and urban built-up growth using NDBI and Shannon's entropy model: A Spatio-temporal analysis of Bulandshahr City, India. *Asian Profile*.

Kumar, V., & Agrawal, S. (2023). A multi-layer perceptron–Markov chain based LULC change analysis and prediction using remote sensing data in Prayagraj district, India. *Environmental Monitoring and Assessment, 195*(5), 619.

Lata, S., & Lata, S. (2019). Agricultural land use patterns. In *Irrigation Water Management for Agricultural Development in Uttar Pradesh, India* (pp. 151–251). Springer.

Lavalle, C., Demicheli, L., Turchini, M., Casals-Carrasco, P., & Niederhuber, M. (2001). Monitoring megacities: The MURBANDY/MOLAND approach. *Development in Practice, 11*(2–3), 350–357.

Lea, C., & Curtis, A. C. (2010). *Thematic accuracy assessment procedures: National Park Service vegetation inventory, version 2.0.* Natural resource report NPS/2010/NRR—2010/204. National Park Service, Fort Collins, Colorado.

Liu, Y., Luo, T., Liu, Z., Kong, X., Li, J., & Tan, R. (2015). A comparative analysis of urban and rural construction land use change and driving forces: Implications for urban–rural coordination development in Wuhan, Central China. *Habitat International, 47*, 113–125.

Lu, D., & Weng, Q. (2007). A survey of image classification methods and techniques for improving classification performance. *International Journal of Remote Sensing, 28*(5), 823–870.

Lillesand, T. M., Kiefer, R. W., & Chipman, J. W. (2003). *Remote sensing and image interpretation* (Vol. 4, 5th ed., p. 763). Wiley. ISBN 0471451525.

Lin, X., Xu, M., Cao, C. P., Singh, R., Chen, W., & Ju, H. (2018). Land-use/land-cover changes and their influence on the ecosystem in Chengdu City, China during the period of 1992–2018. *Sustainability, 10*(10), 3580.

Memarian, H., Balasundram, S. K., Talib, J. B., Sung, C. T. B., Sood, A. M., & Abbaspour, K. (2012). *Validation of CA-Markov for simulation of land use and cover change in the Langat Basin, Malaysia.*

Mirkatouli, J., Hosseini, A., & Neshat, A. (2015). Analysis of land use and land cover spatial pattern based on Markov chains modelling. *City, Territory and Architecture, 2*(1), 1–9.

Mishra, V. N., Rai, P. K., Prasad, R., Punia, M., & Nistor, M. M. (2018). Prediction of spatio-temporal land use/land cover dynamics in rapidly developing Varanasi district of Uttar Pradesh, India, using geospatial approach: A comparison of hybrid models. *Applied Geomatics, 10*, 257–276.

Mitra, S., Mitra, T., & Bardhan, S. (2013). Conflicts in land and housing markets in Kolkata: Emergence of a divided city. In *49th Isocarp Congress.*

Mitsova, D., Shuster, W., & Wang, X. (2011). A cellular automata model of land cover change to integrate urban growth with open space conservation. *Landscape and Urban Planning, 99*(2), 141–153.

Moghadam, H. S., & Helbich, M. (2013). Spatiotemporal urbanization processes in the megacity of Mumbai, India: A Markov chains-cellular automata urban growth model. *Applied Geography, 40*, 140–149.

Mondal, D., & Banerjee, A. (2021). Exploring peri-urban dynamism in India: Evidence from Kolkata Metropolis. *Journal of Urban Management, 10*(4), 382–392.

Mubea, K. W., Ngigi, T. G., & Mundia, C. N. (2011). Assessing application of Markov chain analysis in predicting land cover change: A case study of Nakuru municipality. *Journal of Agriculture, Science and Technology, 12*(2).

Naikoo, M. W., Rihan, M., Shahfahad, Peer, A. H., Talukdar, S., Mallick, J., et al. (2022). Analysis of peri-urban land use/land cover change and its drivers using geospatial techniques and geographically weighted regression. *Environmental Science and Pollution Research*, 1–19.

Nelson, E., Sander, H., Hawthorne, P., Conte, M., Ennaanay, D., Wolny, S., et al. (2010). Projecting global land-use change and its effect on ecosystem service provision and biodiversity with simple models. *PLoS One, 5*(12), e14327.

Patel, S. K., Verma, P., & Shankar Singh, G. (2019). Agricultural growth and land use land cover change in peri-urban India. *Environmental Monitoring and Assessment, 191*, 1–17.

Petropoulos, G. P., Kalivas, D. P., Georgopoulou, I. A., & Srivastava, P. K. (2015). Urban vegetation cover extraction from hyperspectral imagery and geographic information system spatial analysis techniques: Case of Athens, Greece. *Journal of Applied Remote Sensing, 9*(1), 096088–096088.

Pontius, G. R., & Malanson, J. (2005). Comparison of the structure and accuracy of two land change models. *International Journal of Geographical Information Science, 19*(2), 243–265.

Pontius, R. G., Jr., & Millones, M. (2011). Death to Kappa: Birth of quantity disagreement and allocation disagreement for accuracy assessment. *International Journal of Remote Sensing, 32*(15), 4407–4429.

Prabu, P., & Dar, M. A. (2018). Land-use/cover change in Coimbatore urban area (Tamil Nadu, India)—A remote sensing and GIS-based study. *Environmental Monitoring and Assessment, 190*, 1–14.

Pucher, J., Peng, Z. R., Mittal, N., Zhu, Y., & Korattyswaroopam, N. (2007). Urban transport trends and policies in China and India: Impacts of rapid economic growth. *Transport Reviews, 27*(4), 379–410.

Rimal, B., Zhang, L., Keshtkar, H., Wang, N., & Lin, Y. (2017). Monitoring and modeling of spatiotemporal urban expansion and land-use/land-cover change using integrated Markov chain cellular automata model. *ISPRS International Journal of Geo-Information, 6*(9), 288.

Sannigrahi, S., Bhatt, S., Rahmat, S., Paul, S. K., & Sen, S. (2018). Estimating global ecosystem service values and its response to land surface dynamics during 1995–2015. *Journal of Environmental Management, 223*, 115–131.

Santé, I., García, A. M., Miranda, D., & Crecente, R. (2010). Cellular automata models for the simulation of real-world urban processes: A review and analysis. *Landscape and Urban Planning, 96*(2), 108–122.

Şen, G., Güngör, E., & Şevik, H. (2018). Defining the effects of urban expansion on land use/cover change: A case study in Kastamonu, Turkey. *Environmental Monitoring and Assessment, 190*, 1–13.

Singh, S. K., Mustak, S., Srivastava, P. K., Szabó, S., & Islam, T. (2015). Predicting spatial and decadal LULC changes through cellular automata Markov chain models using earth observation datasets and geo-information. *Environmental Processes, 2*, 61–78.

Sodhi, G. J. (2006). Traditional potters and technological change in a North Indian Town. *Sociological Bulletin, 55*(3), 367–382.

Somvanshi, S. S., Bhalla, O., Kunwar, P., Singh, M., & Singh, P. (2020). Monitoring spatial LULC changes and its growth prediction based on statistical models and earth observation datasets of Gautam Budh Nagar, Uttar Pradesh, India. *Environment, Development and Sustainability, 22*, 1073–1091.

Suzanchi, K., & Kaur, R. (2011). Land use land cover change in National Capital Region of India: A remote sensing & GIS based two decadal spatial-temporal analyses. *Procedia-Social and Behavioral Sciences, 21*, 212–221.

Tripathy, P., & Kumar, A. (2019). Monitoring and modelling spatio-temporal urban growth of Delhi using Cellular Automata and geoinformatics. *Cities, 90*, 52–63.

Van Bavel, J. (2013). The world population explosion: Causes, backgrounds and projections for the future. *Facts, Views & Vision in ObGyn, 5*(4), 281.

Wang, Y., & Zhang, X. (2001). A dynamic modeling approach to simulating socioeconomic effects on landscape changes. *Ecological Modelling, 140*(1–2), 141–162.

Xu, X., Du, Z., & Zhang, H. (2016). Integrating the system dynamic and cellular automata models to predict land use and land cover change. *International Journal of Applied Earth Observation and Geoinformation, 52*, 568–579.

Ye, B., & Bai, Z. (2008). Simulating land use/cover changes of Nenjiang County based on CA-Markov model. In *Computer and computing technologies in agriculture, Volume I: First IFIP TC 12 international conference on Computer and Computing Technologies in Agriculture (CCTA 2007), Wuyishan, China, August 18–20, 2007* (pp. 321–329). Springer US.

Yucedag, C., Kaya, L. G., & Cetin, M. (2018). Identifying and assessing environmental awareness of hotel and restaurant employees' attitudes in the Amasra District of Bartin. *Environmental Monitoring and Assessment, 190*(2), 60.

Effect of Spatiotemporal Dynamics of Urban Green Spaces (UGS) on Above and Below Ground Carbon Storage

L. Manawadu and V. P. I. S. Wijeratne

1 Introduction

Urban forests, grasslands, agricultural lands, water bodies, and other natural or semi-natural areas are collectively termed as green spaces (Danardono et al., 2021; Shi et al., 2024). These spaces contribute to the well-being of residents and provide numerous ecological benefits and services to cities. Among these benefits are the mitigation of the heat island effect (Danardono et al., 2021; Herath & Bai, 2024; Rahaman et al., 2021; Shi et al., 2024), the reduction of noise pollution, the enhancement of urban ecosystem circulation, and the fulfillment of recreational, aesthetic, and health-related needs (Shi et al., 2024; Wu et al., 2023). With the rapid urbanization and climate change, urban green spaces (UGS) have made significant contributions to mitigating the resulting urban heat island growth, particularly by reducing surface heat, extreme temperature events, and increased heat wave intensity through their cooling effect on surrounding areas (Rahaman et al., 2021).

Urbanization is an inevitable process of civilization and social development. It is estimated that by 2030, the global level of urbanization will exceed 60% (Wang et al., 2023). Since the Industrial Revolution, the rapid pace of urbanization and the increase in human activities have made global warming one of the most significant climate challenges worldwide. Urban areas, where human activities are most concentrated, are particularly impacted by this phenomenon (Wang et al., 2024). More than 90% of human economic activities occur in cities, which directly contribute at least 85% of global carbon emissions and are considered the main battleground for global carbon neutrality (Chao et al., 2023; Wang et al., 2023). The significant emissions of carbon dioxide, methane, and other greenhouse gases (GHGs) are leading to the exacerbation of the greenhouse effect (Luo et al., 2023). The greatest ecological risk and environmental crisis in history have been caused by the rise in carbon dioxide concentration, which has also led to extreme disaster events (Qu et al., 2023; Wang et al., 2023). Currently, reducing carbon emissions and increasing carbon sinks are the main ways to achieve urban carbon neutrality (Chao et al., 2023; Zhang et al., 2024).

Urban green spaces (UGS) are recognized as a "nature-based solution" that is vital for preserving ecosystem stability, enhancing human health, and fostering sustainable urban development (Shi et al., 2024). The conservation and expansion of urban green spaces are commonly employed as key strategies to mitigate the tension between urban environmental quality and rapid urbanization (Wu et al., 2023). Many countries and cities have formulated green policies and actively

L. Manawadu (✉) · V. P. I. S. Wijeratne
Department of Geography, Faculty of Arts, University of Colombo, Colombo, Sri Lanka
e-mail: lasan@geo.cmb.ac.lk

Z. Khanam et al. (eds.), *Climate Change, Resilience and Ecosystem Services*, Sustainable Development Goals Series, https://doi.org/10.1007/978-3-031-96977-5_29

planned urban green spaces to protect urban ecosystems and improve the quality of life for citizen. Urban green spaces are a crucial part of urban ecosystems, offering a range of environmental services such as noise reduction, air purification, rainwater management, and microclimate regulation, while also providing recreational opportunities, enhancing the sense of happiness among urban residents (Wu et al., 2023).

The urban green spaces in every city in the world are constantly depleting. The accompanying changes in UGS have had significant harmful impacts on sustainable environments as surface temperatures rise (Rahaman et al., 2021). Cities and urban areas are responsible for 37–85% of global anthropogenic greenhouse gas (GHG) emissions through their energy use, industry, transportation, waste management, and high levels of household consumption. In contrast, the direct emissions resulting from the conversion of forests, vegetation, and other carbon storage areas into urban development are relatively minor (Kukkonen et al., 2022). To address climate change, ensure healthy urban development, and optimize the carbon sequestration potential of urban green spaces, it is essential to understand their role in carbon sink functions and the factors influencing them (Shi et al., 2024). Urban green spaces can be pivotal in mitigating and adapting to climate change by reducing greenhouse gas emissions, enhancing urban resilience, and promoting sustainability (Pandey & Ghosh, 2023; Zhang et al., 2024). For instance, urban areas can act as both significant sources of greenhouse gases and potential carbon sinks. While cities contribute to greenhouse gas emissions through activities such as transportation and energy use, they also offer opportunities for carbon sequestration through urban infrastructure and green spaces (Pandey & Ghosh, 2023).

Carbon storage varies at different stages of urbanization and shows spatial differences across regions such as urban centers, suburbs, and rural areas. Studying the effects of urbanization on the spatiotemporal changes in carbon storage is a complex task (Fu et al., 2019). The level of understanding and planning of urban green spaces (UGS) differs globally, often reflecting the varying stages of development in different countries or regions. Developed nations that have undergone "development before governance" typically focus more on the strategic planning of UGS systems to support sustainable urban development (Wu et al., 2023). Carbon budgeting has become a crucial policy instrument for addressing climate change. As noted by Kinnunen et al. (2022), the Intergovernmental Panel on Climate Change (IPCC) aims to limit global warming to below 2 °C, which includes measures to enhance carbon capture and storage while reducing emissions. Likewise, many countries, cities, and companies are working toward carbon neutrality, where the residual greenhouse gas emissions from their activities are balanced by carbon sequestration efforts (Kinnunen et al., 2022).

Colombo, the commercial capital and largest city of Sri Lanka, features a landscape of lakes, canals, swamps, and both abandoned and active artificial rice fields (Dayathilake et al., 2020). Urbanization in Colombo continues to lead to significant changes in land use and land cover (LULC), transforming natural green environments into impermeable paved surfaces (Maheng et al., 2019). Thus, examining the effects of urban land use and land cover (LULC) changes on carbon storage is crucial in Sri Lanka. By analyzing spatial distribution maps of carbon storage in a region, the significance of different vegetation LULC types, such as urban green spaces (UGS), for carbon sequestration can be highlighted.

Considerable amounts of previous studies can be recognized to assess the carbon storage in UGS throughout the world. Different areas have been focused by researches (Bertrand Festus Nero, 2017; Danardono et al., 2021; Dayathilake et al., 2020; Fu et al., 2019; Kukkonen et al., 2022; Li et al., 2023, 2024; Liu et al., 2024; Luo et al., 2023; Qu et al., 2023; Rahaman et al., 2021; Shi et al., 2024; Tang et al., 2022; Wang et al., 2023, 2024, 2021; Zhang et al., 2024) and as pointed out by Rahaman et al. (2021), many previous studies have shown that urban green spaces provide various environmental benefits, such as mitigating climate change, reducing pollution, purifying air and water, regulating microclimate, protecting wildlife and biodiversity, flood control, and other environmental hazards. In addition, from the perspective of carbon sequestration and storage (CSS), there is almost no research that directly links individual carbon budgets with broader urban development models (Kinnunen et al., 2022).

As overviewed by Maheng et al. (2019), green space is a very subjective concept with many explanations and many studies have investigated the relationship between Urban Heat Island (UHI) and UGS strength in terms of composition and configuration. Composition refers to the density of vegetation and the diversity of land cover types, as well as the size and spatial distribution of green spaces. Some studies suggest that the composition of UGS may be more important for cooling effectiveness. Moreover, Luo et al. (2023) pointed out that few studies have been able to comprehensively estimate the carbon sequestration of vegetation and water, in order to analyze the current status and potential of carbon sequestration in cities with abundant vegetation and water resources. Herath and Bai (2024) noted that while numerous review studies have addressed the ecosystem services provided by Urban Green Infrastructure (UGI), there is a growing emphasis on exploring the benefits of UGI from the perspective of these ecosystem services.

Moreover, the literature has explored various methods for assessing urban green spaces (UGS) and carbon storage. According to Wu et al. (2023), there are currently few studies that integrate both quantitative and qualitative approaches to thoroughly examine the driving forces behind UGS changes, considering natural, urbanization, and political factors. Additionally, in statistical analyses, the use of quantitative evaluation methods remains prevalent. These methods include principal component analysis (PCA), partial least squares regression (PLSR), Gray relation analysis (GRA), spatial analysis techniques such as GeoDetector, and geographically weighted regression (GWR). To assess the carbon storage, some studies focus on primary data based on these methods (Dayathilake et al., 2020). It is a very common method for assessing carbon storage. However, satellite-based remote sensing data are now widely used for analyzing and evaluating environmental changes (Rahaman et al., 2021). In particular, the use of geographic information system (GIS) methods to estimate carbon sequestration has become a growing trend worldwide (Piyathilake et al., 2022).

This study has been conducted in Colombo city area and Colombo is the commercial capital city of Sri Lanka. It is highly urbanized area of the country and high rate of population growth can be recognized. Rapid urbanization and population growth of the city have contributed to diminishing UGS in the city area. Therefore, this study has been carried out on UGS and their impact on carbon storage of the urban area.

2 Objectives

This chapter is focusing on the impact of UGS dynamics on carbon storage. Under this main objective, the study has developed sub-objectives such as:

- To map and examine the spatiotemporal dynamics of UGS at Colombo urban area.
- To assess the carbon storage potential and changes of UGS in the Colombo urban area.

3 Methodology

3.1 Data Sources

Landsat 7 and 8 images were obtained for all estimations for 1995, 2005, 2015, and 2024. Carbon pool data for Sri Lanka were obtained from Piyathilake et al. (2022) to assess carbon storage. The study used UGS classification, verified using actual ground truths and Google Earth data.

3.2 Methods

The study is mainly based on GIS and remote sensing methods to assess the UGS changes and carbon storage in Colombo urban area. The Normalized Difference Vegetation Index (NDVI) was calculated using Google Earth Engine to evaluate urban green spaces (UGS), and the InVEST (Integrated Valuation of Ecosystem Services and Trade-offs) model was utilized to estimate the potential carbon storage within these spaces. Therefore, this chapter discussed the UGS dynamics in the Colombo urban area and the carbon storage contribution of the UGS throughout the period of 1995–2024.

3.2.1 InVEST Carbon Storage and Sequestration Model

The InVEST model (Integrated Valuation of Ecosystem Services and Trade-offs) is based on stand-alone software or on ArcGIS to transform script language into software that is commonly used for simplifying the storing of carbon cycle to calculate the ratio of dynamic to static energy for each unit in a given space. Raster data on land use and carbon pool stocks are used in the InVEST model for sequestering carbon and sequestration. Calculations may be made simpler by the sequestering carbon module of the InVEST model, but this benefit is transient. Soil carbon, soil organic carbon (found in both organic and mineral soils), dead organic carbon (found in litter, fallen or standing dead trees) and below-ground biomass (found in living root systems of plants) are the four main carbon pools that make up ecosystem carbon storage whose both the short-term and long-term sequestration of carbon in the environment can be measured. The land use data and the forest storing carbon pool, which are modeled to the land use and land cover rasters to determine each land type, are both essential to the accuracy of the investment model. The model uses the following equation to determine the total amount of carbon in the research area:

$C_{total} = C_{above} + C_{below} + C_{soil} + C_{dead}$ (i Qu et al., 2022; Xu et al., 2023)

$C_{total u} = (C_{above u} + C_{below u} + C_{soil u} + C_{dead u}) * Au$ (Xu et al., 2023)

In the equation, the terms Au and C_{total} refer to the total area of all land uses of type *u*; C_{below}, to the amount of carbon that is stored underground; C_{soil}, to the amount of organic carbon that is stored in soil; $_{below}$ and C_{dead}, to the amount of organic carbon that is stored in dead organisms.

4 Urban Green Spaces (UGS) and Its Dynamics in Colombo Urban Area from 1995 to 2024

Land use and land cover (LULC) in Colombo urban area underwent significant changes in between 1995 and 2024. With the purpose of analyzing LULC changes in the study area, it was categorized into five land-use types: water bodies and marshes, built-up areas, home gardens, paddy fields, and other natural areas and parks are the land use types. The analysis revealed the land use changes between 1995 and 2024, resulting in a decrease in water bodies and marsh lands by 1063 hectares (ha), while built-up areas and home gardens increased by 4621 ha and 9939 ha, respectively. Conversely, paddy fields had a decline of 4550 ha and the study found a significant decrease in the number of UGS, with a reduction of 8946 ha (Table 1).

The spatial analysis of data from 1990 to 2024 reveals a trend of an increasing number of built-up areas from the west to the east, decreasing green spaces. The built-up area has increased from 1410 ha in 1995–2005 to 3039 ha during 2005–2015 and 172 ha in 2015–2024. The study by Hoseini et al. (2024) examining land cover maps of Tehran, Iran, from 2003 to 2023 found that Districts 1 and 14 experienced a decrease in green spaces and barren lands, while built-up areas increased. This suggests that these specific urban areas lost natural areas in favor of development.

The most significant loss has been in water bodies and marshes in the study area, particularly in the northwest, which have changed into built-up areas (Fig. 1). The water bodies and mashes lost 89 ha in 1995–2005, 1045 ha in 2005–2015, and 107 ha in 2015–2024. This trend is mirrored in other cities around the world as well. A study by Liu et al. (2021) examining Shanghai, China, between 1990 and 2015 found that farmland and water bodies were the most affected green space

Table 1 Temporal dynamics of land use land cover (LULC) in Colombo urban area from 1995 to 2024 (in hectares)

LULC	1995	2005	2015	2024	Difference from 1995 to 2024
Water bodies and marshes	1811	1901	855	748	−1063
Built-up areas	3828	5238	8277	8449	4621
Home gardens	6953	9751	17,084	16,892	9939
Paddy fields	11,006	11,882	6835	6456	−4550
Other natural areas	9727	4555	275	781	−8946

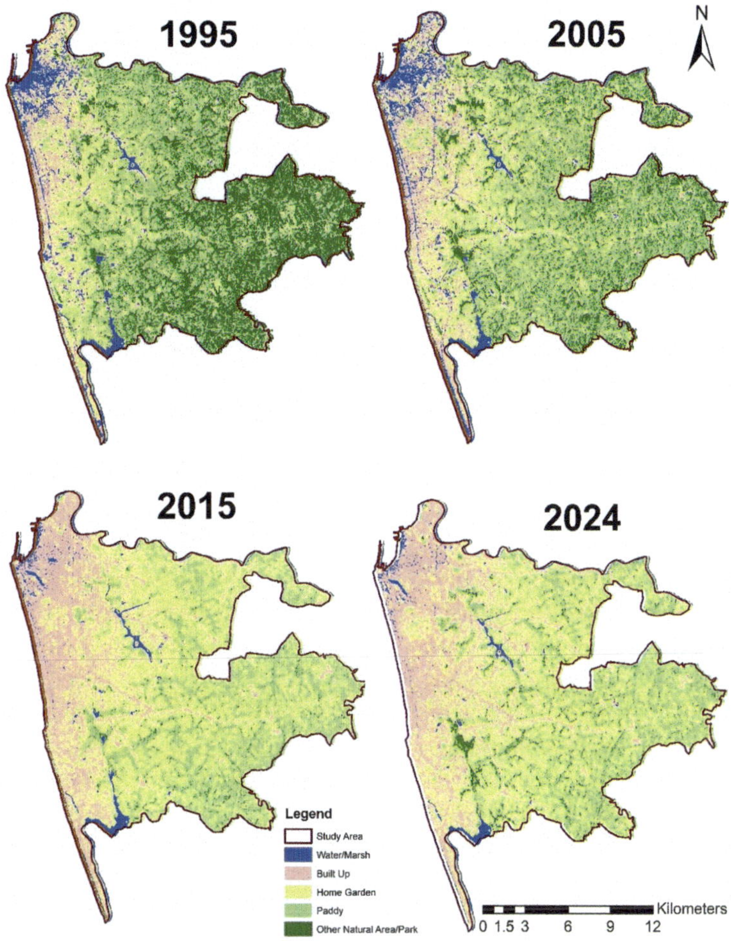

Fig. 1 Spatiotemporal changes of LULC in Colombo Urban Area from 1995 to 2024

types and water bodies and wetlands have shown sudden increase from 1995 to 2015 due to abandoned paddy lands in the area. A massive 1428.3 km^2 (95.3% of the total loss) was converted to built-up areas, with water bodies accounting for 25% of that loss. Forests and grasslands also declined, with forest area decreasing by 81.0% and grasslands losing 4.6 km^2. However, there were some positive signs, since the area transitioning from paddy fields to green space increased to 26.3 km^2 (Liu et al., 2021). Most of paddy lands have become abandoned in the urban area and these have grown with natural vegetation.

As revealed by the data, the paddy fields declined in area by 11,006 ha in 1995, 11,882 ha in 2005, 6835 ha in 2015, and 6456 ha in 2024. Overall, it could be identified a loss of 4550 ha in period of 2015–2024. It is worth to mention that paddy fields were heavily converted into built-up

Table 2 LULC change matrix for the period of 1995 to 2024 (in hectares)

LULC	Water bodies and mash	Built-up areas	Home gardens	Paddy fields	Other natural areas/ parks
Water bodies and marshes	579.60	1008.38	137.57	33.45	41.94
Built-up areas	89.58	2824.24	810.77	85.65	15.72
Home gardens	28.88	2949.09	3552.94	366.05	53.99
Paddy fields	29.34	1370.22	7613.44	1784.72	203.17
Other natural areas	15.81	290.504	4767.06	4177.98	464.88

areas, resulting in 1370.2 ha. Similar trends were observed in the Su-Xi-Chang region of China, where Fu et al. (2019) documented a significant conversion of cultivated land into urban areas between 1990 and 2000. This expansion occurred radially outward from the city center. In the same way, Liu et al. (2021) also noted a decrease in Shanghai's green space, particularly along an east–west axis, between 1990 and 2015. This suggests that suburban farmland is being absorbed by the growing city.

The LULC matrix shows that the highest land conversion from paddy fields to home gardens was 7613.44 ha, while the minimum land conversion was 15.72 ha from paddy areas to natural areas and parks. The higher conversion values were found in other natural areas and parks, covering 4767.06 ha for home gardens, 4177.98 ha for paddy fields, and 290.50 ha for built-up areas (Table 2). This decline was particularly noticeable in eastern regions compared to other areas. Further, it could be identified as a trend of decline from the west to the eastern wards (Fig. 1). As Li and Pussella (2017) observed in Colombo, Sri Lanka, land use is changing dramatically. Vacant land and forests are disappearing, replaced by houses, office buildings, and industrial zones. In another study by Li et al. (2024), researchers investigated the downtown area of Kaifeng, China. Their study area encompassed 26.41 km^2, with roughly half of that area consisting of residential and park green spaces. Specifically, the vegetation cover in Kumasi, Ghana, has also been experiencing a rapid decline. According to a study by Nero et al. (2016), the rate of decline has quadrupled in recent years (2009–2014) compared to the previous period (1986–2001). Currently, only 33% of the area remains covered in vegetation.

Rapid urbanization has a complex and uneven impact on green spaces. While some areas experience a loss of natural spaces, others might see a shift toward more managed parks and green areas. Further research and planning strategies that consider the full range of impacts and potential benefits are crucial for ensuring healthy and sustainable urban environments. Similarly, Wang et al. (2021) observed significant land use shifts in Hangzhou, China, where farmland was converted into settlements. This conversion led to a reduction in surface vegetation and carbon storage, negatively impacting the ecosystem in the area. These findings underscore how modifications in landscape structure can influence a city's capacity to reduce the urban heat island effect and enhance above-ground carbon storage. However, research also indicates the potential for positive outcomes. Sun et al. (2019) suggest that spatially explicit landscape assessment maps, which integrate both physical and social aspects of urban ecosystems, could be employed to evaluate the services provided by UGS. By understanding these services, urban green spaces could become a preferred strategy for mitigating climate change.

5 Spatial and Temporal Dynamics of Carbon Storage

5.1 Above-Ground Carbon

The above-ground carbon storage has been heavily concentrated in the eastern part of the study area. The area has been continuously reduced from 1995 to 2024. The study shows a decrease in areas with higher above-ground carbon storage from east to west, while the lowest above-average carbon concentration is increasing from west to east. This decrease is primarily in the east and gradually lessens toward the west, indicating a reduction in urban green spaces (Fig. 2). Remarkably, areas in the west that initially had low carbon storage have shown an increase in storage, with this increase moving eastward over time from 1995 to 2024. This eastward shift in carbon loss coincides with a decrease in urban green spaces in the east, suggesting that losing green spaces is linked to a decline in above-ground carbon storage. The study further analyzed the changes in carbon storage by classifying the area into four categories based on the amount of carbon stored as No carbon storage (0 tons), low carbon storage (0–0.5 tons), medium carbon storage (0.5–0.9 tons), and high carbon storage (0.9–2.7 tons).

The proportion of land falling into each class changed significantly between 1995 and 2024. The area with no carbon storage and the area with low carbon storage both increased significantly. The areas with 0 tons were at 17% in 1995 and have increased to 27.6% in 2024. Then the areas with 0–0.5 tons in carbon were 20.8% of the entire area and has increased till 50.6%, whereas areas with 0.5–0.9 carbon tons were 33% in 1995, which has significantly decreased to

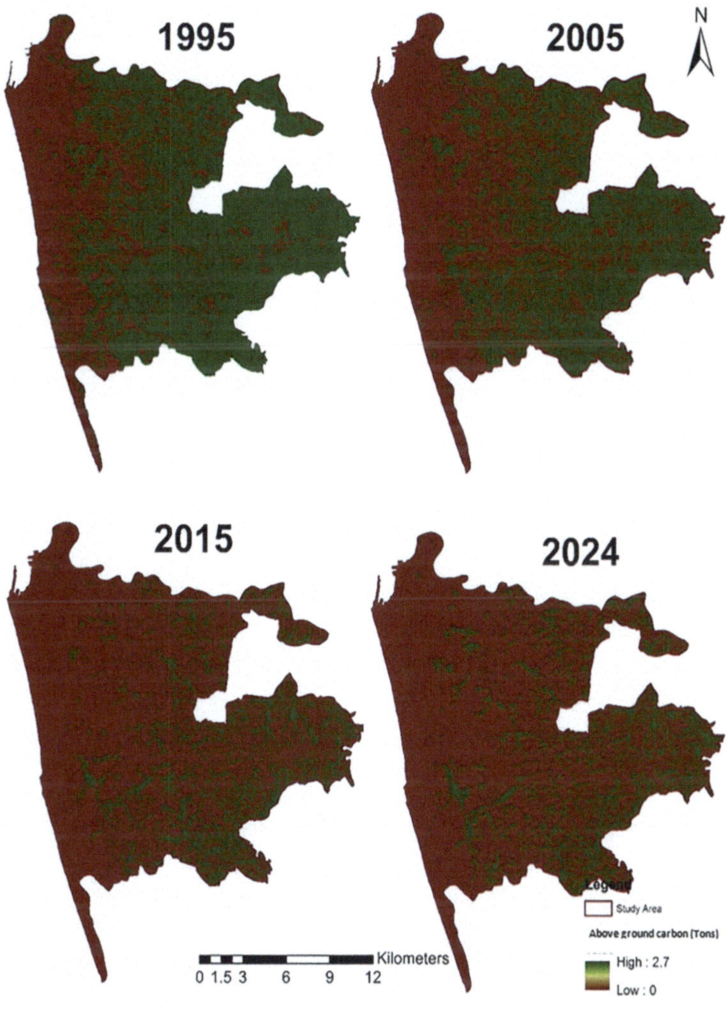

Fig. 2 Spatial changes in above ground carbon storage from 1995 to 2024. The green color legend shows the high carbon storage areas in the Colombo city

Table 3 Above ground carbon in 1995 and 2024

Above ground carbon (tons)	Area (ha)	Percentage	Area (ha)	Percentage
		1995		2024
0	5674	17.0	9214	27.6
0–0.5	6925	20.8	16,884	50.6
0.5–0.9	11,009	33.0	6457	19.4
0.9–2.7	9728	29.2	780	2.3

19.4%. Further, areas with heavy carbon content were 29.2% in 1995 and 2.3% in 2024. Conversely, the areas with medium and high carbon storage both decreased substantially (Table 3). This study highlights a disturbing trend of declining carbon storage in the eastern part of the area, likely caused by the loss of urban green spaces.

Land cover types significantly impacted above-ground carbon storage across the study period (1995–2024). The highest average above-ground carbon content, 1.77 tons/ha in 1995, was found in "other natural areas and parks." The values of other natural areas and parks steadily declined over the years: 1.22 tons/ha in 2005, 0.69 tons/ha in 2015, and 0.82 tons/ha in 2024. Conversely, home gardens exhibited the lowest average above-ground carbon content throughout the study period: 0.81 tons/ha in 1995, decreasing to 0.51 tons/ha in 2024. Notably, "water/marsh" and "built-up areas" were not included in the analysis due to the absence of significant above-ground carbon storage in these land cover types.

5.2 Below-Ground Carbon

From 1995 to 2024, the distribution of below-ground carbon concentrations changed significantly. The area with the highest concentration increased in the eastern part (Fig. 3). It could be seen that the area coverage with built-up areas was concentrated with the lowest below- ground carbon concentration since 1995–2024. The areas with 0 tons were at 16.77% in 1995 and have increased to 27.52% in 2024. Then the areas with 0–0.4 tons of carbon had a 54.33% increase from the entire area to 70.23%, and carbon tons with 0.45–2.7 tons had a 28.86% increase in 1995, which has significantly decreased to 2.2% (Table 4).

Land cover types significantly impacted above-ground carbon storage across the study period (1995–2024). The highest average above-ground carbon content, 1.61 tons/ha in 1995, was found in "other natural areas and parks." The values of other natural areas and parks steadily declined over the years: 1.01 tons/ha in 2005, 0.49 tons/ha in 2015, and 0.68 tons/ha in 2024. Conversely, home gardens exhibited the lowest average below-ground carbon content throughout the study period: 0.67 tons/ha in 1995, decreasing to 0.41 tons/ha in 2024. Notably, "water/marsh" and "built-up areas" were not included in the analysis due to the absence of significant below-ground carbon storage in these land cover types.

A significant reduction in both above- and below-ground carbon concentrations was observed over time. In 1995, total above-ground carbon reached 37,595.8 tons/ha, but this value declined to 1357 tons/ha by 2024. Similarly, below-ground carbon exhibited a substantial decrease, dropping from 32,937.1 tons/ha in 1995 to 10,853 tons/ha in 2024. On the other hand, the amounts of carbon stored in the above-ground and below-ground LULC types vary significantly. According to Paquit (2017), because it is difficult to quantify carbon data in deadwood in the field, only litter carbon data are typically taken into account when calculating carbon densities in dead matter.

This challenge arises from uncertainties regarding the transfer of carbon to the soil, deadwood, and litter, as well as emissions to the atmosphere (Piyathilake et al., 2022). Fu et al. (2019) found that the greatest losses in carbon stock occurred in developing urban areas due to rapid urbanization. The expansion of built-up areas often replaces other land uses, such as agricultural land, water bodies, and forests. This outward spread from developed urban centers leads to a significant reduction in carbon storage. In contrast, the study in Uva province revealed that 850,000 ha of land there store 9663.5 million

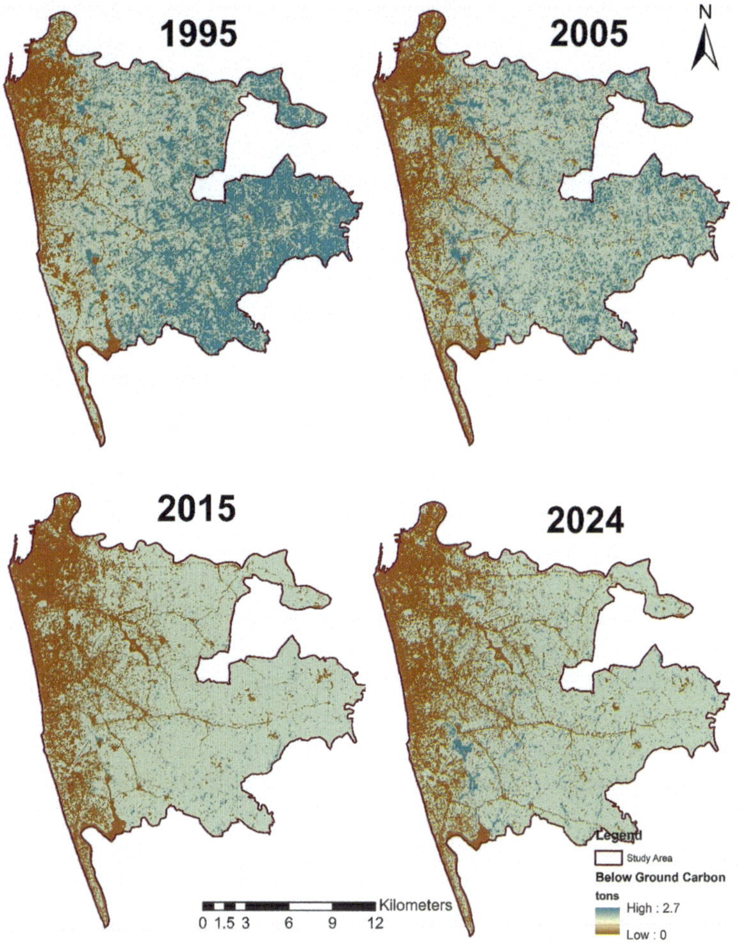

Fig. 3 Spatial changes in below carbon storage

Table 4 Below ground carbon from 1995 to 2024

Below-ground carbon (tons)	Area (ha)	Percentage	Area (ha)	Percentage
	1995		2024	
0	5593	16.77	9177	27.52
0–0.45	18,118	54.33	23,423	70.23
0.45–2.7	9626	28.86	741	2.22

tons of carbon, with natural forests having the highest carbon storage capacity. Land use and land cover (LULC) variations show that areas with rich vegetation hold more carbon.

Natural forests have the highest carbon storage capacity, accounting for 71% of the region's total carbon stock. This information can be used in developing country development plans, environmental policies, and environmental management plans (Piyathilake et al., 2022). The findings in Danardono et al. (2021) indicate that, in comparison to urban areas in temperate climates, carbon sequestration has a significant value in tropical urban areas. This situation arises from the fact that agricultural and garden areas still have green, open spaces. The monthly fluctuations in the value of carbon sequestration in this tropical urban area are attributed to variations in rainfall and human activities, including the conversion of agricultural land and plant types. One of the most relevant examples with a dynamic degree of 59.46%, we saw the grassland carbon sink in Xinzhou District's biggest shift range between 2000 and 2010. The carbon sinks of wetlands declined during this time, while those of both forests and water bodies increased. This was closely related to the ecological and natural preservation practices of the time, which involved converting farmland back to grassland and forest. Only the grassland carbon sink increased between 2010 and 2020, while the other carbon sink types decreased. During the specified period, the total carbon sinks in Xinzhou District exhibited an increasing trend, while all other districts displayed a decreasing trend. The carbon sink land observed less encroachment during this period of urban construction and land expansion because of its relative distance from the downtown area (Luo et al., 2023).

6 Carbon Storage Contribution of UGS

Urban green spaces play a crucial role in carbon storage, but their contribution has been declining over time. The contribution rate decreased from 83% in 1995 to 72% in 2024. This decline in carbon deposition within cities is likely to lead to rising temperatures and reduced urban resilience (Hoseini et al., 2024). Within these green spaces, some areas hold significantly more carbon than others. Home gardens, plantations, and natural forests are the top contributors, storing 23%, 15%, and 35% of the total carbon, respectively. Conversely, public parks, farmlands, grasslands, and cemeteries have a minimal impact, accounting for only 3.4%, 3.5%, 0.8%, and 0.4% of the total vegetation carbon, respectively. Furthermore, studies suggest that tropical freshwater wetlands hold immense potential for carbon storage (Dayathilake et al., 2020). This highlights the importance of incorporating such areas into carbon sequestration strategies in tropical regions.

From 1995 to 2024, land use has shifted, with increasing home lands and paddy fields, built-up areas, and decreasing urban green spaces, impacting both carbon concentration and contribution. Since 1995, urbanization has driven land use changes, and cities are expanding, leading to a decrease in urban green spaces (UGS). This decline in UGS impacts both the amount of carbon stored in cities and their contribution to mitigating climate change. Consequently, the Role of National Policy has grown. National climate change policies should aim for low-emission development. This can be achieved through sustainable energy, infrastructure, and carbon sink expansion. For instance, strategies include managing forests, agriculture, and aquatic resources, as exemplified by policies in Ghana (Nero et al., 2016). Land use and greening policies can mitigate the negative impacts of urbanization on UGS. Shanghai implemented successful policies since the 1980s, significantly increasing green space. Their "UGS System Plan" focused on protecting green spaces and planting grass, leading to positive outcomes (Liu et al., 2021). Then take the case of Colombo, a city in need of greening. Colombo, once a green city, faces challenges due to rapid urbanization and declining green space. A comprehensive green space plan for the city and suburbs is crucial for its transformation (Li & Pussella, 2017). Strategies for sustainable urban development and the studies suggest a multi-pronged approach such as: Expand cultivated land in rural areas, Implement vertical development in urban areas (e.g., high-rise buildings), and Plant trees within developed areas (e.g., urban forest parks). Carbon sequestration maps can guide government planning, environmental policies, and management practices (Fu et al., 2019).

7 Conclusion with Policy and Recommendation

LULC in the study area underwent significant changes between 1995 and 2024. Specifically, the built-up areas and home gardens have increased their distribution dramatically. In contrast, water bodies, marshes, paddy fields, and other natural areas and parks have declined. More importantly, the reduction of other natural areas and parks is the biggest decline. The decline could be identified in the eastern part of the study area. On the other hand, the built-up areas have been distributed from the western to the eastern wards by transforming green spaces. It is worth mentioning that the most land transformation could be identified from paddy fields to home gardens. The areas in the west where the built-up is highest, that initially had low carbon storage, have shown an increase in storage, with this increase moving eastward over time from 1995 to 2024. The highest average above-ground carbon content, 1.776574 tons/ha in 1995, was found in "other natural areas and parks," which has been reduced to 0.83 tons/ha in 2024, in particular land use. In the same way, the highest average below-ground carbon content, 1.62 tons/ha in 1995, was found in "other natural areas and parks," which has been reduced to 0.68 tons/ha in 2024, in particular land use. A significant reduction in both above- and below-ground carbon concentrations was observed over time. In 1995, total above-ground carbon reached 37,595.8 tons/ha, but this value declined to 13,577 tons/ha by 2024. Overall, it could be identified that the total carbon storage has been reduced with the land use changes in the area. In areas where green spaces are high, the carbon concentration is also high. Along with the changes in green spaces, low carbon storage has increased. It could be identified as urbanization is a significant process of LULC change, impacting land carbon pools. Cities are a dynamic and large part of the global carbon cycle due to their large and concentrated carbon fluxes. Maps of carbon sequestration and storage from different LULC types reveal patterns and carbon loss due to land use changes.

This approach offers individuals, communities, and the private sector improved ways to showcase their positive contributions to mitigating climate change in urban areas, as well as documented methods for assessing negative impacts. Additionally, the strengthening of green policies serves as a significant impetus for revising UGS, leading to their recovery, planning, and expansion. Analyzing changes in UGS provides valuable insights into how to enhance UGS planning, which can assist cities in improving their ecological environment and increasing urban value. These results can be useful in government development plans, environmental policy, and management planning. They also aid in prioritizing conservation lands and classifying reforestation sites for improved carbon storage and sequestration. Reforestation sites can be further improved through these databases. The study suggests focusing on urban–rural fringes and controlling urbanization development models to achieve carbon storage protection in rapidly urbanizing areas. Consequently, it should implement regional policies focusing on urban fringe areas, which serve as an ecological barrier and active area for urban expansion. These policies should clarify land expansion boundaries, promote smart growth in urban and rural areas, and promote compactness and multi-functionality.

References

Chao, M., Maimai, W., Hanzhang, L., Zhibo, C., & Xiaohui, C. (2023). A spatio-temporal neural network learning system for city-scale carbon storage capacity estimating. *IEEE Access, 11*, 31304–31322. https://doi.org/10.1109/ACCESS.2023.3260641

Danardono, S., Sunariya, M. I. T., Fikriyah, V. N., & Cholil, M. (2021). Spatiotemporal variation of terrestrial carbon sequestration in a tropical urban area: Case study in Surakarta District, Indonesia. *Quaestiones Geographicae, 40*(3), 5–20. https://doi.org/10.2478/quageo-2021-0020

Dayathilake, D. D. T. L., Lokupitiya, E., & Wijeratne, V. P. I. S. (2020). Estimation of aboveground and belowground carbon stocks in urban freshwater wetlands of Sri Lanka. *Carbon Balance and Management, 15*(1). https://doi.org/10.1186/s13021-020-00152-5

Fu, Q., Xu, L., Zheng, H., & Chen, J. (2019). Spatiotemporal dynamics of carbon storage in response to urbanization: A case study in the Su-xi-Chang region, China. *Processes, 7*(11). https://doi.org/10.3390/pr7110836

Herath, P., & Bai, X. (2024). *Benefits and co-benefits of urban green infrastructure for sustainable cities: Six current and emerging themes*. Springer. https://doi.org/10.1007/s11625-024-01475-9

Hoseini, S. H., Amiri, M. J., & Moarrab, Y. (2024). Analyzing spatiotemporal changes in urban green spaces' ecosystem service value and resilience. *International Journal of Human Capital in Urban Management, 9*(2), 217–234. https://doi.org/10.22034/IJHCUM.2024.02.03

Kinnunen, A., Talvitie, I., Ottelin, J., Heinonen, J., & Junnila, S. (2022). Carbon sequestration and storage potential of urban residential environment—A review. *Sustainable Cities and Society, 84*. https://doi.org/10.1016/j.scs.2022.104027

Kukkonen, M. O., Khamis, M., Muhammad, M. J., Käyhkö, N., & Luoto, M. (2022). Modeling direct above-ground carbon loss due to urban expansion in Zanzibar City Region, Tanzania. *Land Use Policy, 112*. https://doi.org/10.1016/j.landusepol.2021.105810

Li, L., & Pussella, P. G. R. N. I. (2017). Is Colombo city, Sri Lanka, secured for urban green space standards? *Applied Ecology and Environmental Research, 15*(3), 1789–1799. https://doi.org/10.15666/aeer/1503_17891799

Li, H., Hu, Y., Li, H., Ren, J., Shao, R., & Liu, Z. (2023). Assessing the impact of spatiotemporal evolution of urbanization on carbon storage in the mega-urban agglomeration area: Case study of Yangtze River Delta Urban Agglomeration, China. Sustainability (Switzerland), 15(19). https://doi.org/10.3390/su151914548.

Li, N., Deng, L., Yan, G., Cao, M., & Cui, Y. (2024). Estimation for refined carbon storage of urban green space and minimum spatial mapping scale in a plain city of China. *Remote Sensing, 16*(2). https://doi.org/10.3390/rs16020217

Liu, S., et al. (2021). Spatiotemporal dynamics of urban green space influenced by rapid urbanization and land use policies in Shanghai. *Forests, 12*(4). https://doi.org/10.3390/f12040476

Liu, Y., Fan, C., & Xue, D. (2024). A review of the effects of urban and green space forms on the carbon budget using a landscape sustainability framework. *Sustainability (Switzerland), 16*(5). https://doi.org/10.3390/su16051870

Luo, M., Liu, H., Gao, J., Tang, Y., Guo, L., Pi, J., & Yu, Y. (2023). Spatiotemporal variations and influencing factors of urban carbon sink: A case study of Wuhan, China. *Ecosystem Health and Sustainability, 9*. https://doi.org/10.34133/ehs.0133

Maheng, D., Ducton, I., Lauwaet, D., Zevenbergen, C., & Pathirana, A. (2019). The sensitivity of urban heat island to urban green space: A model-based study of City of Colombo, Sri Lanka. *Atmosphere, 10*(3). https://doi.org/10.3390/atmos10030151

Nero, B. F. (2017). *Urban green spaces enhance carbon sequestration and conserve biodiversity in cities of the Global South: Case of Kumasi, Ghana.*

Nero, B. F., Borgemeister, P. D. C., & Schmitt, P. D. D. C. B. (2016). *Urban green spaces enhance carbon sequestration and conserve biodiversity in cities of the Global South: Case of Kumasi*. Ghana.

Paquit, J. C., & Mindaña, F.-N. W. (2017). Modeling the spatial pattern of carbon stock in Central Mindanao University using InVEST tool. Journal of Biodiversity and Environmental Sciences, 10(4), 103-113.

Pandey, B., & Ghosh, A. (2023). Urban ecosystem services and climate change: A dynamic interplay. *Frontiers in Sustainable Cities, 5*. https://doi.org/10.3389/frsc.2023.1281430

Piyathilake, I. D. U. H., Udayakumara, E. P. N., Ranaweera, L. V., & Gunatilake, S. K. (2022). Modeling predictive assessment of carbon storage using InVEST model in Uva province, Sri Lanka. *Modeling Earth Systems and Environment, 8*(2), 2213–2223. https://doi.org/10.1007/s40808-021-01207-3

Qu, C., Li, W., Xu, J., & Shi, S. (2023). Blackland conservation and utilization, carbon storage, and ecological risk in green space: A case study from Heilongjiang Province in China. *International Journal of Environmental Research and Public Health, 20*(4). https://doi.org/10.3390/ijerph20043154

Qu, R.; He, L.; He, Z.; Wang, B.; Lyu, P.; Wang, J.; Kang, G.; Bai, W. A Study of Carbon Stock Changes in the Alpine Grassland Ecosystem of Zoigê, China, 2000–2020. Land 2022, 11, 1232. https://doi.org/10.3390/land11081232

Rahaman, S., Jahangir, S., Haque, M. S., Chen, R., & Kumar, P. (2021). Spatio-temporal changes of green spaces and their impact on urban environment of Mumbai, India. *Environment, Development and Sustainability, 23*(4), 6481–6501. https://doi.org/10.1007/s10668-020-00882-z

Shi, N., Yu, Y., Liang, S., Ren, Y., & Liu, M. (2024). Effects of urban green spaces landscape pattern on carbon sink among urban ecological function areas at the appropriate scale: A case study in Xi'an. *Ecological Indicators, 158*. https://doi.org/10.1016/j.ecolind.2023.111427

Sun, Y., Xie, S., & Zhao, S. (2019). Valuing urban green spaces in mitigating climate change: A city-wide estimate of aboveground carbon stored in urban green spaces of China's capital. *Global Change Biology, 25*(5), 1717–1732. https://doi.org/10.1111/gcb.14566

Tang, Y., Shao, Q., Shi, T., Lu, Z., & Wu, G. (2022). Spatiotemporal dynamics of forest ecosystem carbon budget in Guizhou: Customisation and application of the CBM- CFS3 model for China. *Carbon Balance and Management, 17*(1). https://doi.org/10.1186/s13021-022-00210-0

Wang, Z., Xu, L., Shi, Y., Ma, Q., Wu, Y., Lu, Z., Mao, L., Pang, E., & Zhang, Q. (2021). Impact of land use change on vegetation carbon storage during rapid urbanization: A case study of Hangzhou, China. *Chinese Geographical Science, 31*(2), 209–222. https://doi.org/10.1007/s11769-021-1183-y

Wang, H., Feng, Y., & Ai, L. (2023). Progress of carbon sequestration in urban green space based on bibliometric analysis. *Frontiers in Environmental Science, 11*. https://doi.org/10.3389/fenvs.2023.1196803

Wang, R. Y., Mo, X., Ji, H., Zhu, Z., Wang, Y. S., Bao, Z., & Li, T. (2024). Comparison of the CASA and InVEST models' effects for estimating spatiotemporal differences in carbon storage of green spaces in megacities. *Scientific Reports, 14*(1). https://doi.org/10.1038/s41598-024-55858-0

Wu, S., Wang, D., Yan, Z., Wang, X., & Han, J. (2023). Spatiotemporal dynamics of urban green space in Changchun: Changes, transformations, landscape patterns, and drivers. *Ecological Indicators, 147*. https://doi.org/10.1016/j.ecolind.2023.109958

Xu, X., Liu, J., Jiao, F., Zhang, K., Ye, X., Gong, H., et al. (2023). Ecological engineering induced carbon sinks shifting from decreasing to increasing during 1981–2019 in China. Sci. Total Environ. 864:161037. doi: 10.1016/j.scitotenv.2022.161037

Zhang, X., Huang, H., Tu, K., Li, R., Zhang, X., Wang, P., Li, Y., Yang, Q., Acerman, A. C., Guo, N., & Liu, Y. (2024). Effects of plant community structural characteristics on carbon sequestration in urban green spaces. *Scientific Reports, 14*(1). https://doi.org/10.1038/s41598-024-57789-2

Index

Z. Khanam et al. (eds.), *Climate Change, Resilience and Ecosystem Services*, Sustainable Development Goals Series, https://doi.org/10.1007/978-3-031-96977-5

The manufacturer's authorised representative in the EU is Springer Nature Customer Service Centre GmbH, Europaplatz 3, 69115 Heidelberg, Germany. If you have any concerns regarding our products, please contact ProductSafety@springernature.com

Printed and bound by CPI Group (UK) Ltd, Croydon, CR0 4YY
21/09/2026
02219964-0008